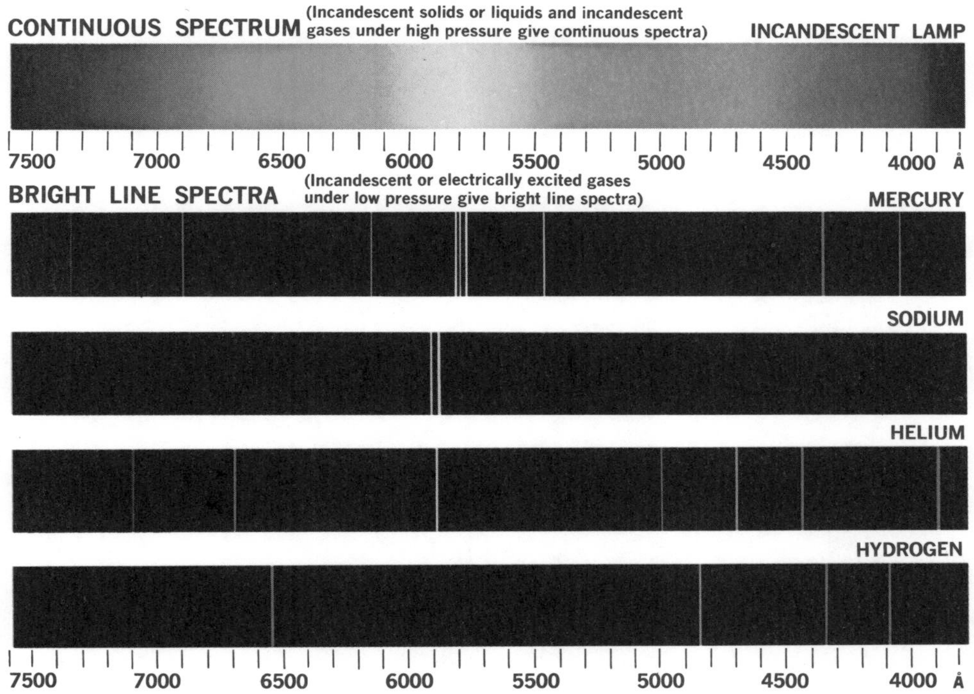

Useful Conversions and Combinations

$$1 \text{ eV} = 1.602 \times 10^{-19} \text{ J}$$
$$1 \text{ u} = 931.50 \text{ MeV}/c^2$$
$$1 \text{ Å} = 10^{-10} \text{ m} = 0.1 \text{ nm}$$
$$hc = 1.240 \times 10^3 \text{ eV} \cdot \text{nm} = 1.24 \times 10^4 \text{ eV} \cdot \text{Å}$$
$$\hbar c = 1.973 \times 10^2 \text{ eV} \cdot \text{nm} = 1.973 \times 10^3 \text{ eV} \cdot \text{Å}$$
$$kT = 0.02525 \text{ eV at } T = 300 \text{ K}$$
$$\frac{e^2}{4\pi\epsilon_0} = 1.440 \times 10^3 \text{ eV} \cdot \text{nm} = 1.440 \times 10^4 \text{ eV} \cdot \text{Å}$$

MODERN PHYSICS

SAUNDERS COLLEGE PUBLISHING
COMPLETE PACKAGE FOR TEACHING WITH
MODERN PHYSICS
by Raymond A. Serway, Clement J. Moses, and Curt A. Moyer

INSTRUCTOR'S MANUAL
with solutions and software instructions
R. A. Serway, C. J. Moses, and C. A. Moyer

SUPERCONDUCTIVITY SUPPLEMENT
R. A. Serway

OVERHEAD TRANSPARENCIES

MODERN PHYSICS

Raymond A. Serway
James Madison University

Clement J. Moses
Utica College of Syracuse University

Curt A. Moyer
Clarkson University

SAUNDERS GOLDEN SUNBURST SERIES
Saunders College Publishing
Philadelphia New York Chicago
San Francisco Montreal Toronto
London Sydney Tokyo

Text Typeface: Caledonia
Compositor: Progressive Typographers, Inc.
Acquisitions Editor: John Vondeling
Project Editor: Sally Kusch
Copy Editor: Jay Freedman
Art Director: Carol Bleistine
Art Assistant: Doris Bruey
Text Designer: Edward A. Butler
Cover Designer: Lawrence R. Didona
Text Artwork: Tom Mallon, Rolin Graphics
Layout Artist: Dorothy Chattin
Production Manager: Harry Dean

Cover Credit: Laser measurement. J Freis © THE IMAGE BANK.

Printed in the United States of America

MODERN PHYSICS

ISBN 0-03-004844-3

Library of Congress Catalog Card Number: 88-043248

PREFACE

This book is intended as a modern physics text for students majoring in science and engineering who have already completed an introductory calculus-based physics course. The contents of this text may be subdivided into two broad categories: an introduction to the theories of relativity and quantum physics (Chapters 1 through 8) and applications of elementary quantum theory to molecular, solid-state, nuclear, and particle physics (Chapters 9 through 15). Throughout the book our underlying objectives have been threefold:

1. To provide simple, clear, and mathematically uncomplicated explanations of physical concepts and theories of modern physics;
2. to clarify and show support for these theories through a broad range of current applications and examples. In this regard, we have attempted to answer questions such as: What holds molecules together? How do electrons tunnel through barriers? How do electrons move through solids? How can currents persist indefinitely in superconductors?
3. to enliven and humanize the text with brief sketches of the historical development of 20th century physics, including anecdotes and quotations from the key figures as well as interesting photographs of noted scientists and original apparatus.

Main Features of This Text

Topics The material covered in this book is concerned with fundamental topics in modern physics with extensive applications in science and engineering. Chapter 1 presents an introduction to the special theory of relativity, and includes an essay on general relativity. Chapters 2–4 present an historical and conceptual introduction to early developments in quantum theory, including important experimental work in support of the theory. Chapters 5–9 represent an introduction to the real "nuts and bolts" of quantum mechanics, covering the Schrödinger equation, tunneling phenomena, the hydrogen atom, multi-electron atoms, and a survey of statistical mechanics. The remainder of the book consists, for the most part, of applications of the theory set forth in earlier chapters to more specialized areas of modern physics. In particular, Chapter 10 treats the physics and properties of various types of lasers, Chapter 11 discusses the physics of molecules, Chapter 12 is an introduction to the physics of solids, Chapters 13 and 14 cover nuclear physics and methods of obtaining energy from nuclear reactions, and Chapter 15 treats the subject of particle physics.

Style We have attempted to write this book in a style that is clear, logical, and succinct, yet somewhat informal and relaxed, in the hope that students will find the text appealing and enjoyable to read. All new terms have been carefully defined, and we have tried to avoid jargon.

Worked Examples A large number of worked examples of varying difficulty are presented as an aid in understanding concepts. In many cases, these examples will serve as models for solving some end-of-chapter problems. The examples are set off with colored bars for ease of location, and most examples are given titles to describe their content.

Exercises Following Examples As an added feature, many of the worked examples are followed immediately by exercises with answers. These exercises are intended to make the textbook more interactive with the student and to test immediately the student's understanding of problem-solving techniques. The exercises represent extensions of the worked examples and are numbered in case the instructor wishes to assign them for homework.

Questions and Problems An extensive set of questions and problems is included at the end of each chapter with a total of 152 questions and 454 problems for the text. Answers to all odd-numbered problems are given at the end of the book. In general, the order of problems follows the order of topics presented in each chapter. Most of the questions serve to test the student's understanding of the concepts presented in that chapter, and many can be used to motivate classroom discussions.

Calculator/Computer Problems Two types of problems that require the use of a programmable calculator or a computer are given in many chapters. One type is a numerical problem that requires the student to write a simple program for its solution. The other type is an interactive computer exercise (such as Fourier synthesis of wave packets) which requires the use of special software that will be provided by the publisher upon request. The diskette to be provided contains 5 programs, which can be run on an IBM/PC (or IBM-compatible) microcomputer. These features will enable students to work with many realistic numerical problems and simulations and experience the power of the calculator and computer in scientific computations.

Units The international system of units (SI) is used throughout the text. Occasionally, where common usage dictates, other units are used (such as the angstrom, Å, and cm^{-1} so commonly used by spectroscopists) but all such units are carefully defined in terms of SI units.

Chapter Format Each chapter begins with a **preview,** which includes a brief discussion of chapter objectives and content. **Marginal notes** set in color are used to locate important concepts and equations in the text. Important statements are ***italicized*** or set in color, and important equations are set in a colored box for added emphasis and ease of review. Each chapter concludes with a **summary** which reviews the important concepts and equations discussed in that chapter.

In addition, many chapters contain **special topic sections,** which are clearly marked **optional** and set in a slightly smaller print. These sections expose the student to slightly more advanced material either in the form of current interesting discoveries or as fuller developments of concepts or calculations discussed in that chapter. Many of these special topic sections will be of particular interest to certain student groups such as chemistry majors, electrical engineers, or physics majors.

Guest Essays A unique feature of this text is the inclusion of extremely up-to-date and interesting material in the form of essays by guest authors. The essays cover a wide range of topics and are intended to convey an insider's view of exciting current developments in physics. Furthermore, the essay topics represent extensions and/or applications of the material covered in the chapter in which they are included. Some of the essay topics are recent developments in general relativity, the scanning tunneling microscope, and superconducting devices.

Mathematical Level Students using this text should have completed a comprehensive one-year calculus course as calculus is used throughout the text; however, we have made an attempt to keep physical ideas foremost so as not to obscure our presentations with overly elegant mathematics. Most steps are shown when basic equations are developed, but exceptionally long and detailed proofs that interrupt the flow of physical arguments have been placed in appendices or in optional sections.

Appendices and Endpapers The appendices in this text serve several purposes. Many consist of more lengthy derivations of key equations which are used in various portions of the text. In a few cases, the appendices are used to develop special topics such as the Big Bang theory of cosmology. In addition, the appendices contain physical constants, conversion factors, tables of atomic masses, and a list of Nobel Prize winners and their achievements. The endpapers inside the front cover of the book contain important physical constants and a table of conversions for quick reference, while a periodic table is included in the endpapers inside the back cover.

Ancillaries

The ancillaries available with this text include an Instructor's Manual containing solutions to all end-of-chapter problems and instructions for use of the software; modeling and simulation software which can be used in conjunction with many end-of-chapter problems in the text; a Superconductivity Supplement, and a set overhead transparencies. All of these ancillaries are available upon request from the publisher.

The software diskette contains five interactive programs which allow students to perform many simulations of problems in quantum mechanics as described in many of the end-of-chapter problems so designated. For example, students can compose matter-wave packets and study their evolution, or find the probability for particle transmission and reflection from various potential wells and barriers. The software will run on all IBM-compatible personal computers.

The Superconductivity Supplement was written because of the unprecedented surge of interest and excitement in this field since the discovery of high-temperature superconductors in 1986. The supplement introduces the student to the basic properties of superconductors and includes a survey of the important developments that have occurred in the last few years in high-temperature superconductivity. It contains all features found in a typical chapter of the text, including an extensive set of questions, problems, references, and suggested readings.

Teaching Options

As noted earlier, the text may be subdivided into two basic parts: Chapters 1–9, which contain an introduction to relativity, quantum physics, and statistical physics, and Chapters 10–15, which treat applications to lasers, molecules, the solid state, nuclear physics, and elementary particles. It is suggested that the first part of the book be covered sequentially. However, Chapter 1 on relativity may actually be covered at any time because $E^2 = p^2c^2 + m^2c^4$ is the only formula from this chapter that is essential for subsequent chapters. Chapters 10–15 are independent of each other and can be covered in any order with one exception: Chapter 14, Nuclear Physics Applications, should follow Chapter 13, Nuclear Structure.

A traditional sophomore or junior level modern physics course for science, mathematics, and engineering students should cover most of Chapters 1–9 and several chapters in the second part, depending on student major. For example, an audience consisting mainly of electrical engineering students might cover most of Chapters 1–9 with particular emphasis on tunneling and tunneling devices in Chapter 6, the Fermi–Dirac distribution in Chapter 9, lasers in Chapter 10, semiconductors, superconductivity, and superconducting devices in Chapter 12, and particle detectors in Chapter 14. Chemistry and chemical engineering majors could cover most of Chapters 1–9 with special emphasis on atoms in Chapter 8, classical and quantum statistics in Chapter 9, molecular bonding and spectroscopy in Chapter 11, and overlap integrals in Appendix D. Mathematics and physics majors should pay special attention to the unique development of operator methods and the concept of sharp and fuzzy observables introduced in Chapter 5. The deep connection of sharp observables with classically conserved quantities and the powerful role of sharp observables in shaping the form of system wavefunctions is developed more fully in Chapter 7 and Appendix C.

Our experience has shown that there is more material in this book than can be covered in a standard 3-credit-hour course. For this reason, one has to "pick-and-choose" from topics in the second part of the book as noted earlier. However, the text can be used in a two-semester sequence with some supplemental material such as one of many monographs on relativity, the Superconductivity Supplement available with this text, and selected readings from other sources in the areas of solid state, nuclear, and atomic physics. These suggested readings are listed at the end of each chapter.

Acknowledgements

We thank the following individuals for their careful reviews and suggestions for improvements of the manuscript:

Gordon Aubrecht, Ohio State University

Wendell A. Childs, U.S. Military Academy

David Cornell, Principia College

Sumner P. Davis, University of California at Berkeley

Frank Hereford, University of Virginia

Stanley Hirschi, Central Michigan University

James R. Huddle, U.S. Naval Academy

Drasko Jovanovic, Fermilab

Marcus Price, University of New Mexico

Jay Strieb, Villanova University

Martin Tiersten, City College of New York

T. A. Wiggins, Pennsylvania State University

George A. Williams, University of Utah

David M. Wolfe, University of New Mexico

We are grateful to the many students at Clarkson University and James Madison University who provided helpful comments and criticisms of the manuscript while it was tested in the classroom. We thank the professional staff at Saunders College Publishing Company for their fine work during the development and production of this text, especially Sally Kusch, project editor, Ellen Newman, Kate Pachuta, and John Vondeling. We thank Jay Freedman for his excellent copyediting of the entire manuscript, Mary Lou Glick for typing parts of the manuscript, and Suzon O. Kister for her unerring and valuable reference work. We also wish to acknowledge valuable discussions with many people including Don Chodrow, John R. Gordon, William Ingham, Christopher Moses, Joe Rudmin, and Dorn Peterson. We thank all of the authors of the essays included in this text for their valuable contributions and cooperation: Isaac Abella, Hans Courant, Roger A. Freedman, Clark A. Hamilton, Paul K. Hansma, Christopher T. Hill, and Clifford M. Will. Finally, we thank all of our families for their patience and continual support.

Raymond A. Serway
James Madison University
Harrisonburg, Virginia

Clement J. Moses
Utica College of Syracuse University
Utica, New York

Curt A. Moyer
Clarkson University
Potsdam, New York

CONTENTS

(Photograph courtesy of AIP Niels Bohr Library)

1. A. Piccard
2. E. Henriot
3. P. Ehrenfest
4. E. Herzen
5. Th. de Donder
6. E. Schroedinger
7. E. Verschaffelt
8. W. Pauli
9. W. Heisenberg
10. R.H. Fowler
11. L. Brillouin
12. P. Debye
13. M. Knudsen
14. W.L. Bragg
15. H.A. Kramers
16. P.A.M. Dirac
17. A.H. Compton
18. L.V. de Broglie
19. M. Born
20. N. Bohr
21. I. Langmuir
22. M. Planck
23. M. Curie
24. H.A. Lorentz
25. A. Einstein
26. P. Langevin
27. C.E. Guye
28. C.T.R. Wilson
29. O.W. Richardson

This unique photograph shows many eminent scientists who participated in the fifth international congress of physics held in 1927 by the Solvay Institute in Brussels, Belgium. At this and similar conferences held regularly from 1911 on, scientists were able to discuss and share the many dramatic developments in atomic and nuclear physics. This elite company of scientists includes fifteen Nobel Prize winners in physics and three in chemistry.

1

Relativity

At the end of the 19th century, scientists believed that they had learned most of what there was to know about physics. Newton's laws of motion and his universal theory of gravitation, Maxwell's theoretical work in unifying electricity and magnetism, and the laws of thermodynamics and kinetic theory were highly successful in explaining a wide variety of phenomena.

However, at the turn of the 20th century, a major revolution shook the world of physics. In 1900 Planck provided the basic ideas that led to the formulation of the quantum theory, and in 1905 Einstein formulated his brilliant special theory of relativity. The excitement of the times is captured in Einstein's own words: "It was a marvelous time to be alive." Both ideas were to have a profound effect on our understanding of nature. Within a few decades, these theories inspired new developments and theories in the fields of atomic physics, nuclear physics, and condensed matter physics.

Although modern physics has been developed during this century and has led to a multitude of important technological achievements, the story is still incomplete. Discoveries will continue to evolve during our lifetime, many of which will deepen or refine our understanding of nature and the world around us. It is still a "marvelous time to be alive."

1.1 INTRODUCTION

Light waves and other forms of electromagnetic radiation travel through free space at the speed $c = 3.00 \times 10^8$ m/s. As we shall see in this chapter, the speed of light is an upper limit for the speeds of particles and mechanical waves.

Most of our everyday experiences and observations deal with objects that move at speeds much less than the speed of light. Newtonian mechanics, and the early ideas on space and time, were formulated to describe the motion of such objects. This formalism is very successful for describing a wide range of phenomena. Although Newtonian mechanics works very well at low speeds, it fails when applied to particles whose speeds approach that of light. Experimentally, one can test the predictions of the theory at large speeds by accelerating an electron through a large electric potential difference. For example, it is possible to accelerate an electron to a speed of $0.99c$ by using a potential difference of several million volts. According to Newtonian mechanics, if the potential difference (as well as the corresponding energy) is increased by a factor of 4, then the speed of the electron should be doubled to $1.98c$. However, experiments show that the speed of the electron always remains *less* than the speed of light, regardless of the size of the accelerating voltage. Since Newtonian mechanics places no upper limit on the speed that a particle can attain, it is contrary to modern experimental results and is clearly a limited theory.

In 1905, at the age of only 26, Albert Einstein published his *special theory of relativity*.

> The relativity theory arose from necessity, from serious and deep contradictions in the old theory from which there seemed no escape. The strength of the new theory lies in the consistency and simplicity with which it solves all these difficulties, using only a few very convincing assumptions. . . .[1]

Although Einstein made many important contributions to science, the theory of relativity alone represents one of the greatest intellectual achievements of the 20th century. With this theory, one can correctly predict experimental observations over the range of speeds from $v = 0$ to velocities approaching the speed of light. Newtonian mechanics, which was accepted for over 200 years, is in fact a specialized case of Einstein's generalized theory. This chapter gives an introduction to the special theory of relativity, with emphasis on some of the consequences of the theory. A discussion of general relativity and some of its consequences and experimental tests is presented in a most interesting essay written by Clifford Will.

As we shall see, the special theory of relativity is based on two basic postulates:

The postulates of the special theory of relativity

1. The laws of physics are the same in all inertial reference systems. That is, basic laws such as $\Sigma F = ma$, have the same mathematical form for all observers moving at constant velocity with respect to each other.
2. The speed of light in vacuum is always measured to be 3×10^8 m/s, and the measured value is independent of the motion of the observer or of the motion of the source of light. That is, the speed of light is the same for *all* inertial observers.

Special relativity covers phenomena such as the slowing down of clocks and the contraction of lengths in moving reference frames as measured by a stationary observer. We shall also discuss the relativistic forms of momentum and energy and some consequences of the famous mass-energy equivalence formula, $E = mc^2$. You may want to consult a number of excellent books on relativity for more details on the subject.[2]

Although it is well known that relativity plays an essential role in contemporary theoretical physics, it also has many practical applications, including the design of accelerators and other devices that utilize high-speed particles. We shall have occasion to use relativity in some subsequent chapters of this text, but often only the outcome of relativistic effects will be presented.

1.2 THE PRINCIPLE OF RELATIVITY

In order to describe a physical event, it is necessary to establish a frame of reference, such as one that is fixed in the laboratory. You should recall from your studies in mechanics that Newton's laws are valid in *all* inertial frames of reference. Since an inertial frame of reference is defined as one in which Newton's first law is valid, one can say that *an inertial system is a system in*

Inertial frame of reference

[1] A. Einstein and L. Infeld, *The Evolution of Physics*, New York, Simon and Schuster, 1961.

[2] The following books are recommended for more details on the theory of relativity at the introductory level: E. F. Taylor and J. A. Wheeler, *Spacetime Physics*, San Francisco, W. H. Freeman, 1963; R. Resnick, *Introduction to Special Relativity*, New York, Wiley, 1968; A. P. French, *Special Relativity*, New York, Norton, 1968; other suggested readings are selected reprints on "Special Relativity Theory" published by the American Institute of Physics.

which a free body exhibits no acceleration. Furthermore, any system moving with constant velocity with respect to an inertial system is also an inertial system. There is no preferred frame. This means that the results of an experiment performed in a vehicle moving with uniform velocity will be identical to the results of the same experiment performed in the stationary laboratory.

> According to the **principle of Newtonian relativity**, the laws of mechanics are the same in all inertial frames of reference.

For example, if you perform an experiment while at rest in a laboratory, and an observer in a passing car moving with constant velocity also observes your experiment, the laboratory coordinate system and the coordinate system of the moving car are both inertial reference frames. Therefore, if you find the laws of mechanics to be true in the lab, the person in the moving car must agree with your observation. This also implies that no mechanical experiment can detect any difference between the two inertial frames. The only thing that can be detected is the relative motion of one frame with respect to the other. That is, the notion of *absolute* motion through space is meaningless.

Suppose that some physical phenomenon, which we call an *event,* occurs in an inertial system. The event's location and time of occurrence can be specified by the coordinates (x, y, z, t). We would like to be able to transform the space and time coordinates of the event from one inertial system to another moving with uniform relative velocity. This is accomplished by using a so-called *Galilean transformation.*

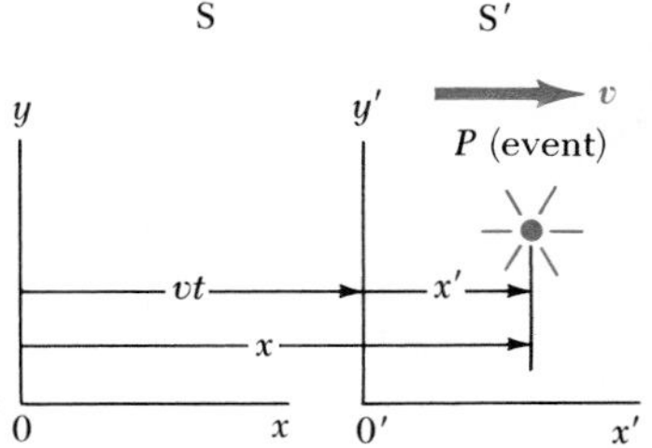

Figure 1.1 An event occurs at a point P. The event is observed by two observers in inertial frames S and S′, where S′ moves with a velocity v relative to S.

Consider two inertial systems S and S′, as in Figure 1.1. The system S′ moves with a constant velocity v along the xx' axes, where v is measured relative to the system S. We assume that an event occurs at the point P. The event might be the "explosion" of a flashbulb or a heartbeat. An observer in system S would describe the event with space-time coordinates (x, y, z, t), while an observer in system S′ would use (x', y', z', t') to describe the same event. As we can see from Figure 1.1, these coordinates are related by the equations

$$\boxed{\begin{aligned} x' &= x - vt \\ y' &= y \\ z' &= z \\ t' &= t \end{aligned}} \tag{1.1}$$

Galilean transformation of coordinates

These equations constitute what is known as a **Galilean transformation of coordinates.** Note that the fourth coordinate, time, is *assumed* to be the same in both inertial systems. That is, within the framework of classical mechanics, clocks are universal, so that the time of an event for an observer in S is the same as the time for the same event in S′. Consequently, the time interval between two successive events should be the same for both observers. Although this assumption may seem obvious, it turns out to be *incorrect* when treating situations in which v is comparable to the speed of light. In fact, this point represents one of the most profound differences between Newtonian concepts and the ideas contained in Einstein's theory of relativity.

Now suppose two events are separated by a distance Δx and a time interval Δt as measured by an observer in S. It follows from Equations 1.1 that the corresponding displacement $\Delta x'$ measured by an observer in S′ is given by $\Delta x' = \Delta x - v\,\Delta t$, where Δx is the displacement measured by an observer in S. Since $\Delta t = \Delta t'$, we find that

$$\frac{\Delta x'}{\Delta t} = \frac{\Delta x}{\Delta t} - v$$

If the two events correspond to the passage of a moving object (or wave pulse) past two "milestones" in frame S, then $\Delta x/\Delta t$ is the time-averaged velocity in S and $\Delta x'/\Delta t$ is the average velocity in S′. In the time limit $\Delta t \rightarrow 0$, this becomes

Galilean addition law for velocities

$$\boxed{u_x' = u_x - v} \tag{1.2}$$

where u_x and u_x' are the instantaneous velocities of the object relative to S and S′, respectively.

This result, which is called the **Galilean addition law for velocities** (or Galilean velocity transformation), is used in everyday observations and is consistent with our intuitive notion of time and space. However, as we shall soon see, these results lead to serious contradictions when applied to electromagnetic waves.

The Speed of Light

It is quite natural to ask whether the concept of Newtonian relativity in mechanics also applies to experiments in electricity, magnetism, optics, and other areas. For example, if one assumes that the laws of electricity and magnetism are the same in all inertial frames, a paradox concerning the velocity of light immediately arises. This can be understood by recalling that according to Maxwell's equations in electromagnetic theory, the velocity of light always has the fixed value of $(\mu_0\epsilon_0)^{-1/2} \approx 3.00 \times 10^8$ m/s. But this is in direct contradiction to what one would expect based on the Galilean addition law for velocities. According to this law, the velocity of light should *not* be the same in all inertial frames. For example, suppose a light pulse is sent out by an observer in a boxcar moving with a velocity $\boldsymbol{v}$ (Fig. 1.2). The light pulse has a velocity $\boldsymbol{c}$ relative to observer S' in the boxcar. According to the ideas of Newtonian relativity, the velocity of the pulse relative to the stationary observer S outside the boxcar should be $\boldsymbol{c} + \boldsymbol{v}$. This is in obvious contradiction to Einstein's theory, which postulates that the velocity of the light pulse is the same for all observers.

In order to resolve this paradox, one must conclude that either (1) the Galilean addition law for velocities is incorrect, or else (2) the laws of electricity and magnetism are not the same in all inertial frames. If the Galilean addition law for velocities were incorrect, we would be forced to abandon the seemingly "obvious" notions of absolute time and absolute length that form the basis for the Galilean transformations.

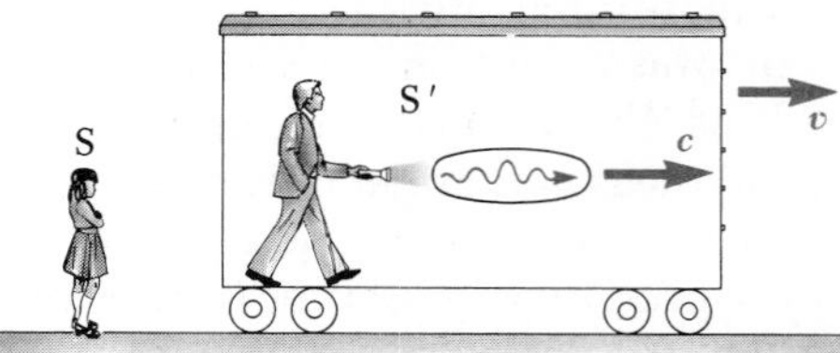

Figure 1.2 A pulse of light is sent out by a person in a moving boxcar. According to Newtonian relativity, the speed of the pulse should be $c + v$ relative to a stationary observer.

If we assume that the second alternative is true, then a preferred reference frame must exist in which the speed of light has the value c, and that it is greater or less than this value in any other reference frame in accordance with the Galilean addition law for velocities. It is useful to draw an analogy with sound waves, which propagate through a medium such as air. The speed of sound in air is about 330 m/s when measured in a reference frame in which the air is stationary. However, the measured speed of sound is greater or less than this value when measured from a reference frame that is moving with respect to the source of sound.

In the case of light signals (electromagnetic waves), recall that Maxwell's theory predicted that such waves must propagate through free space with a speed equal to the speed of light. However, Maxwell's theory does not require the presence of a medium for the wave propagation. This is in contrast to mechanical waves, such as water or sound waves, which do require a medium to support the disturbances. In the 19th century, physicists thought that electromagnetic waves also required a medium in order to propagate. They proposed that such a medium existed, and they gave it the name **luminiferous ether.** The ether was assumed to be present everywhere, even in free space, and light waves were viewed as ether oscillations. Furthermore, the ether had to have the unusual properties of being a massless but rigid medium, and would have no effect on the motion of planets or other objects. Indeed, this is a strange concept. Additionally, it was found that the troublesome laws of electricity and magnetism would take on their simplest form in a frame of reference at *rest* with respect to this ether. This frame was called the *absolute frame.* The laws of electricity and magnetism would be valid in this absolute frame, but they would have to be modified in any reference frame moving with respect to the ether frame.

As a result of the importance attached to this absolute frame, it became of considerable interest in physics to prove by experiment that it existed. A direct method for detecting the ether wind would be to measure its influence on the speed of light relative to a frame of reference on earth. If v is the velocity of the ether relative to the earth, then the speed of light should have its maximum value, $c + v$, when propagating downwind as shown in Figure 1.3a. Likewise, the speed of light should have its minimum value, $c - v$, when propagating upwind as in Figure 1.3b, and some intermediate value, $(c^2 - v^2)^{1/2}$, in the direction perpendicular to the ether wind as in Figure 1.3c. If the sun is assumed to be at rest in the ether, then the velocity of the ether wind would be equal to the orbital velocity of the earth around the sun, which has a magnitude of about 3×10^4 m/s. Since $c = 3 \times 10^8$ m/s, one should be able to detect a change in speed of about 1 part in 10^4 for measurements in the upwind or downwind directions. However, as we shall see in the next section, all attempts to detect such changes and establish the existence of the ether (and hence of the absolute frame) proved futile!

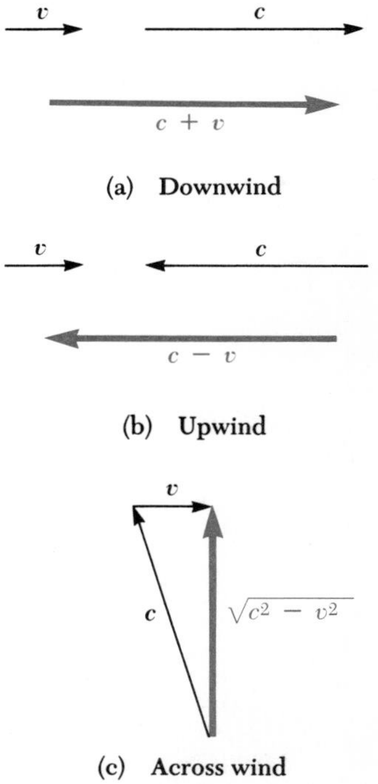

v c c + v

(a) Downwind

v c c − v

(b) Upwind

v c $\sqrt{c^2 - v^2}$

(c) Across wind

Figure 1.3 If the velocity of the ether wind relative to the earth is v, and c is the velocity of light relative to the ether, the speed of light relative to the earth is (a) $c + v$ in the downwind direction, (b) $c - v$ in the upwind direction, and (c) $(c^2 - v^2)^{1/2}$ in the direction perpendicular to the wind.

1.3 THE MICHELSON-MORLEY EXPERIMENT

The most famous experiment that was designed to detect small changes in the speed of light was performed in 1887 by A. A. Michelson and E. W. Morley (1838–1923).[3] We should state at the outset that the outcome of the experiment was *negative,* thus contradicting the ether hypothesis. The experiment

[3] A. A. Michelson and E. W. Morley, *Am. J. Sci.* **134:** 333 (1887).

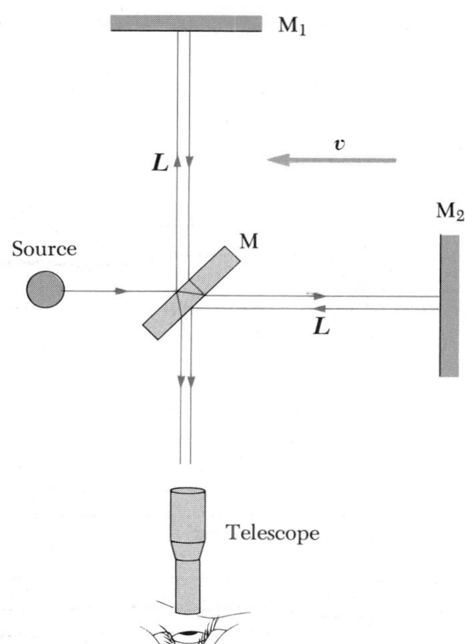

Figure 1.4 Diagram of the Michelson interferometer. A single beam is split into two rays by the partially silvered mirror M. According to the ether wind concept, the speed of light should be $c - v$ as the beam approaches mirror M_2 and $c + v$ after reflection.

was designed to determine the velocity of the earth with respect to the hypothetical ether. The experimental tool used was the Michelson interferometer shown in Figure 1.4. Suppose one of the arms of the interferometer is aligned along the direction of the motion of the earth through space. The motion of the earth through the ether would be equivalent to the ether flowing past the earth in the opposite direction. This ether wind blowing in the opposite direction should cause the speed of light as measured in the earth's frame of reference to be $c - v$ as it approaches the mirror M_2 in Figure 1.4 and $c + v$ after reflection. The speed v is the speed of the earth through space, and hence the speed of the ether wind, while c is the speed of light in the ether frame. The two beams of light reflected from M_1 and M_2 recombine, and an interference pattern consisting of alternating dark and bright bands or fringes is formed. During the experiment, the interference pattern was observed while the interferometer was rotated through an angle of 90 degrees. This rotation would change the speed of the ether wind along the direction of the arms of the interferometer. The effect of this rotation should have been to cause the fringe pattern to shift slightly but measurably. Measurements failed to show any change in the interference pattern! The Michelson-Morley experiment was repeated by other researchers under various conditions and at different locations, but the results were always the same: *no fringe shift of the magnitude required was ever observed.*

The negative results of the Michelson-Morley experiment meant that it was impossible to measure the absolute velocity of the earth (the orbital velocity) with respect to the ether frame. However, as we shall see in the next section, Einstein developed a postulate for his theory of relativity that places quite a different interpretation on these null results. In later years when more was known about the nature of light, the idea of an ether that permeates all of space was relegated to the ash heap of worn out concepts. Light is now understood to be *an electromagnetic wave, which requires no medium for its propaga-*

tion. As a result, the idea of an ether in which these waves could travel became unnecessary.

Details of the Michelson-Morley Experiment

In order to understand the outcome of the Michelson-Morley experiment, let us assume that the interferometer shown in Figure 1.4 has two arms of equal length L. First consider the beam traveling parallel to the direction of the ether wind, which is taken to be horizontal in Figure 1.4. As the beam moves to the right, its speed is reduced by the wind and its speed with respect to the earth is $c - v$. On its return journey, as the light beam moves to the left downwind, its speed with respect to the earth is $c + v$. Thus, the time of travel to the right is $L/(c - v)$ and the time of travel to the left is $L/(c + v)$. The total time of travel for the round trip along the horizontal path is

$$t_1 = \frac{L}{c+v} + \frac{L}{c-v} = \frac{2Lc}{c^2 - v^2} = \frac{2L}{c}\left(1 - \frac{v^2}{c^2}\right)^{-1}$$

Albert A. Michelson (1852–1931). (AIP Niels Bohr Library)

Now consider the light beam traveling perpendicular to the wind, which is the vertical direction in Figure 1.4. Since the speed of the beam relative to the earth is $(c^2 - v^2)^{1/2}$ in this case (see Fig. 1.3), then the time of travel for each half of this trip is $L/(c^2 - v^2)^{1/2}$, and the total time of travel for the round trip is

$$t_2 = \frac{2L}{(c^2 - v^2)^{1/2}} = \frac{2L}{c}\left(1 - \frac{v^2}{c^2}\right)^{-1/2}$$

Thus, the time difference between the light beam traveling horizontally and the beam traveling vertically is

$$\Delta t = t_1 - t_2 = \frac{2L}{c}\left[\left(1 - \frac{v^2}{c^2}\right)^{-1} - \left(1 - \frac{v^2}{c^2}\right)^{-1/2}\right]$$

Since $v^2/c^2 \ll 1$, this expression can be simplified by using the following binomial expansions after dropping all terms higher than second order:

$$(1 - x)^n \approx 1 - nx \qquad (\text{for } x \ll 1)$$

In our case, $x = v^2/c^2$, and we find

$$t_1 - t_2 = \Delta t = \frac{Lv^2}{c^3} \tag{1.3}$$

Because the speed of the earth in its orbital path is approximately 3×10^4 m/s, the speed of the wind should be at least this great. The two light beams start out in phase and return to form an interference pattern. Let us assume that the interferometer is adjusted for parallel fringes and that a telescope is focused on one of these fringes. The time difference between the two light beams gives rise to a phase difference between the beams, producing an interference pattern when they combine at the position of the telescope. A difference in the pattern should be detected by rotating the interferometer through 90° in a horizontal plane, such that the two beams exchange roles. This results in a net time difference of twice that given by Equation 1.3. Thus, the path difference which corresponds to this time difference is

$$\Delta d = c(2\ \Delta t) = \frac{2Lv^2}{c^2}$$

The corresponding phase difference $\Delta\phi$ is equal to $2\pi/\lambda$ times this path difference, where λ is the wavelength of light. That is,

$$\Delta\phi = \frac{4\pi L v^2}{\lambda c^2} \tag{1.4}$$

In the first experiments by Michelson and Morley, each light beam was reflected by mirrors many times to give an increased effective path length L of about 11 meters. Using this value, and taking v to be equal to 3×10^4 m/s, gives a path difference of

$$\Delta d = \frac{2(11 \text{ m})(3 \times 10^4 \text{ m/s})^2}{(3 \times 10^8 \text{ m/s})^2} = 2.2 \times 10^{-7} \text{ m}$$

This extra distance of travel should produce a noticeable shift in the fringe pattern. Specifically, calculations show that if one views the pattern while the interferometer is rotated through 90°, a shift of about 0.4 fringe should be observed. The instrument used by Michelson and Morley had the capability of detecting a shift in the fringe pattern as small as 0.01 fringe. However, *they detected no shift in the fringe pattern.* Since then, the experiment has been repeated many times by various scientists under various conditions, and no fringe shift has ever been detected. Thus, it was concluded that one cannot detect the motion of the earth with respect to the ether.

Many efforts were made to explain the null results of the Michelson-Morley experiment. For example, perhaps the earth drags the ether with it in its motion through space. To test this assumption, interferometer measurements were made at various altitudes, but again no fringe shift was detected. In the 1890s, G. F. Fitzgerald and H. A. Lorentz tried to explain the null results by making the following ad hoc assumption. They proposed that the length of an object moving along the direction of the ether wind would contract by a factor of $\sqrt{1 - v^2/c^2}$. The net result of this contraction would be a change in length of one of the arms of the interferometer such that no path difference would occur as it was rotated.

No experiment in the history of physics has received such valiant efforts to try to explain the absence of an expected result as did the Michelson-Morley experiment. The stage was set for the brilliant Albert Einstein, who solved the problem in 1905 with his special theory of relativity.

1.4 EINSTEIN'S PRINCIPLE OF RELATIVITY

In the previous section we noted the serious contradiction between the invariance of the speed of light and the Galilean addition law for velocities. In 1905, Albert Einstein (Fig. 1.5) proposed a theory that would resolve this contradiction but at the same time would completely alter our notion of space and time.[4] Einstein based his special theory of relativity on the following general hypothesis, which is called the **Principle of Relativity:**

All the laws of physics are the same in all inertial reference frames.

The postulates of special relativity

In Einstein's own words, ". . . the same laws of electrodynamics and optics will be valid for all frames of reference for which the equations of mechanics

[4] A. Einstein, "On the Electrodynamics of Moving Bodies," *Ann. Physik* **17:** 891 (1905). For an English translation of this article and other publications by Einstein, see the book by H. Lorentz, A. Einstein, H. Minkowski, and H. Weyl, *The Principle of Relativity*, Dover, 1958.

Figure 1.5 Albert Einstein (1879–1955), one of the greatest physicists of all times, was born in Ulm, Germany. As a child, Einstein was very unhappy with the discipline of German schools, and completed his early education in Switzerland at age 16. Because he was unable to obtain an academic position following graduation from the Swiss Federal Polytechnic School in 1901, he accepted a job at the Swiss Patent Office in Berne. During his spare time, he continued his studies in theoretical physics. In 1905, at the age of 26, he published four scientific papers that revolutionized physics. One of these papers, which won him the Nobel prize in 1921, dealt with the photoelectric effect. Another was concerned with Brownian motion, the irregular motion of small particles suspended in a liquid. The remaining two papers were concerned with what is now considered his most important contribution of all, the special theory of relativity. In 1915, Einstein published his work on the general theory of relativity, which relates gravity to the structure of space and time. One of the remarkable predictions of the theory is that strong gravitational forces in the vicinity of very massive objects cause light beams to deviate from a straight-line path. This and other predictions of the general theory of relativity have been experimentally verified (see the essay in this chapter by Clifford Will).

Einstein made many other important contributions to the development of modern physics, including the concept of the light quantum and the idea of stimulated emission of radiation, which led to the invention of the laser 40 years later. However, throughout his life he rejected the probabilistic interpretation of quantum mechanics when describing events on the atomic scale, in favor of a deterministic view. He is quoted as saying "God does not play dice with nature."

Einstein left Germany under the power of the Nazis in 1933, and spent his remaining years at the Institute for Advanced Study in Princeton, New Jersey. He devoted most of his later years to an unsuccessful search for a unified theory of gravity and electromagnetism.

hold good." This simple statement is a generalization of Newton's principle of relativity, and is the foundation of the special theory of relativity.

An immediate consequence of the Principle of Relativity is as follows:

> The speed of light in vacuum has the same value, $c = 3.00 \times 10^8$ m/s, in all inertial reference frames.

In other words, anyone who measures the speed of light will get the same value, c. This implies that the ether simply does not exist. The Principle of Relativity, together with the above statement, are often referred to as the **two postulates of special relativity.**

Although the Michelson-Morley experiment was performed before Einstein published his work on relativity, it is not clear that Einstein was aware of the details of the experiment. Nonetheless, the null result of the experiment can be readily understood within the framework of Einstein's theory. According to his Principle of Relativity, the premises of the Michelson-Morley experiment were incorrect. In the process of trying to explain the expected results, we stated that when light traveled against the ether wind its speed was $c - v$, in accordance with the Galilean addition law for velocities. However, if the state of motion of the observer or of the source has no influence on the value found for the speed of light, one will always measure the value to be c. Likewise, the light makes the return trip after reflection from the mirror at a speed of c, and not with the speed $c + v$. Thus, the motion of the earth should not influence the fringe pattern observed in the Michelson-Morley experiment, and a null result should be expected.

The Michelson-Morley experiment has been repeated many times, always giving the same null results. Modern versions of the experiment have compared the frequencies of resonant laser cavities of identical length oriented at right angles to each other. More recently, Doppler shift experiments using gamma rays emitted by a radioactive sample of ^{57}Fe placed an upper limit of about 5 cm/s on the ether wind velocity. These results have shown quite conclusively that the motion of the earth has no effect on the speed of light!

If we accept Einstein's theory of relativity, we must conclude that relative motion is unimportant when measuring the speed of light. At the same time, we must alter our common-sense notion of space and time and be prepared for some rather bizarre consequences.

1.5 CONSEQUENCES OF SPECIAL RELATIVITY

Almost everyone who has dabbled even superficially with science is aware of some of the startling predictions that arise because of Einstein's approach to relative motion. As we examine some of the consequences of relativity in this section, we shall find that they conflict with our basic notions of space and time. We shall restrict our discussion to the concepts of length, time, and simultaneity, which are quite different in relativistic mechanics than in Newtonian mechanics. For example, we shall see that *the distance between two points and the time interval between two events depend on the frame of reference in which they are measured.* That is, *there is no such thing as absolute length or absolute time in relativity.* Furthermore, *events at different locations that occur simultaneously in one frame are not simultaneous in another frame.*

Simultaneity and the Relativity of Time

A basic premise of Newtonian mechanics is that a universal time scale exists which is the same for all observers. In fact, Newton wrote that "Absolute, true, and mathematical time, of itself, and from its own nature, flows equably without relation to anything external." Thus, Newton and his followers simply took simultaneity for granted. In his special theory of relativity, Einstein abandoned this assumption. According to Einstein, *time interval measurements depend on the reference frame in which the measurement is made.*

Einstein devised the following thought experiment to illustrate this point. A boxcar moves with uniform velocity, and two lightning bolts strike the ends of the boxcar, as in Figure 1.6a, leaving marks on the boxcar and ground. The marks left on the boxcar are labeled A' and B', while those on the ground are labeled A and B. An observer at O' moving with the boxcar is midway between A' and B', while a ground observer at O is midway between A and B. The events recorded by the observers are the light signals from the lightning bolts.

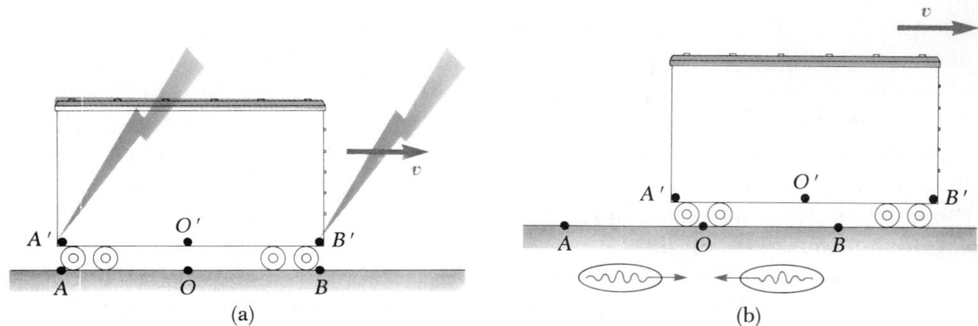

Figure 1.6 Two lightning bolts strike the ends of a moving boxcar. (a) The events appear to be simultaneous to the stationary observer at O, who is midway between A and B. (b) The events do not appear to be simultaneous to the observer at O', who claims that the front of the train is struck *before* the rear.

Let us assume that the two light signals reach the observer at O at the same time, as indicated in Figure 1.6b. This observer realizes that the light signals have traveled at the same speed over distances of equal length. Thus, the observer at O rightly concludes that the events at A and B occurred simultaneously. Now consider the same events as viewed by the observer on the boxcar at O'. By the time the light has reached the observer at O, the observer at O' has moved as indicated in Figure 1.6b. Thus, the light signal from B' has already swept past O', while the light from A' has not yet reached O'. According to Einstein, the observer at O' must find that light travels at the same speed as that measured by the observer at O. Therefore, the observer at O' concludes that the light reaches the front of the boxcar before it reaches the back. This thought experiment clearly demonstrates that the two events, which appear to be simultaneous to the observer at O, do not appear to be simultaneous to the observer at O'. In other words,

> two events that are simultaneous in one reference frame are in general not simultaneous in a second frame moving with respect to the first. That is, simultaneity is not an absolute concept.

At this point, you might wonder which observer is right concerning the two events. The answer is that *both are correct,* because the Principle of Relativity states that *there is no preferred inertial frame of reference.* Although the two observers reach different conclusions, both are correct in their own reference frame because the concept of simultaneity is not absolute. This, in fact, is the central point of relativity—any uniformly moving frame of reference can be used to describe events and do physics. There is nothing wrong with the clocks and meter sticks used to perform measurements. It is simply that time intervals and length measurements depend on the observer. Observers in different inertial frames of reference will always measure different time intervals with their clocks and different distances with their meter sticks. However, they will both agree on the laws of physics in their respective frames, since they must be the same for all observers in uniform motion. It is the alteration of time and space that allows the laws of physics (including Maxwell's equations) to be the same for all observers in uniform motion.

Time Dilation

In order to illustrate how time is altered, consider a vehicle moving to the right with a speed v as in Figure 1.7a. A mirror is fixed to the ceiling of the vehicle and an observer at O' at rest in this system holds a flash gun a distance d below the mirror. At some instant, the flash gun goes off and a pulse of light is released. Because the light pulse has a speed c, the time it takes the pulse to travel from the observer to the mirror and back again to the observer can be found from the definition of velocity:

$$\Delta t' = \frac{\text{distance traveled}}{\text{velocity}} = \frac{2d}{c} \tag{1.5}$$

where the prime notation indicates the time measured by the observer in the reference frame of the moving vehicle.

Now consider the same set of events as viewed by an observer at O in a stationary frame (Fig. 1.7b). According to this observer, the mirror and flash gun are moving to the right with a speed v. The sequence of events would

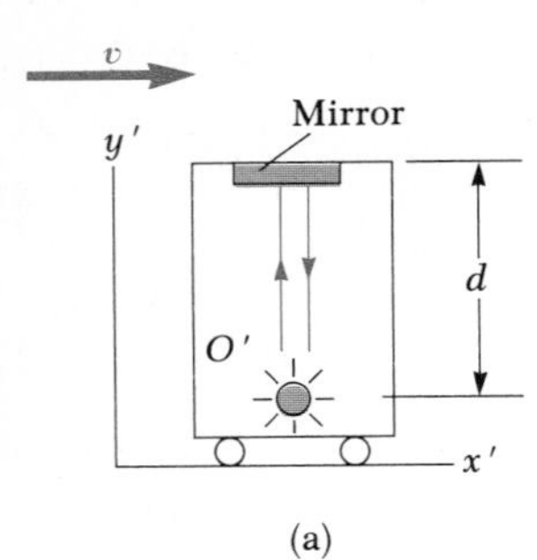

(a)

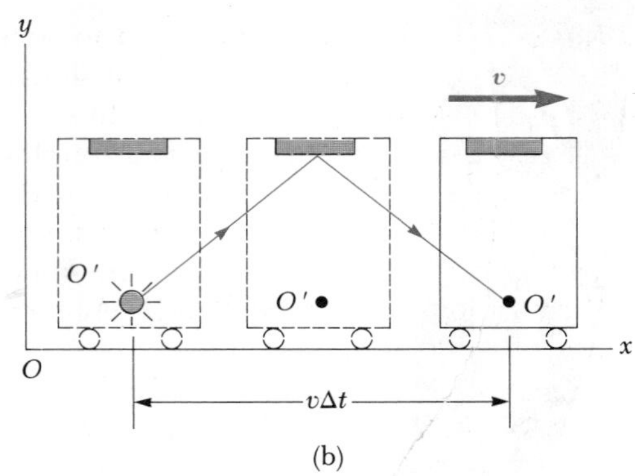

(b)

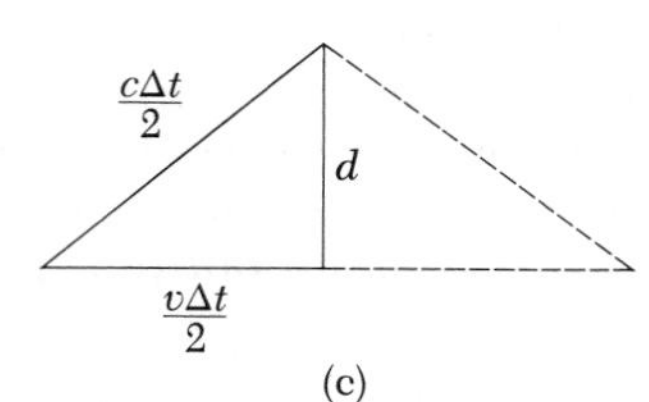

(c)

Figure 1.7 (a) A mirror is fixed to a moving vehicle, and a light pulse leaves O' at rest in the vehicle. (b) Relative to a stationary observer on earth, the mirror and O' move with a speed v. Note that the distance the pulse travels is greater than $2d$ as measured by the stationary observer. (c) The right triangle for calculating the relationship between Δt and $\Delta t'$.

appear entirely different as viewed by this observer. By the time the light from the flash gun reaches the mirror, the mirror will have moved a distance $v\ \Delta t/2$, where Δt is the time for the light pulse to travel from O' to the mirror and back, as measured by the stationary observer. In other words, the stationary observer concludes that, because of the motion of the system, if the light is to hit the mirror, it must leave the flash gun at an angle with respect to the vertical direction. Comparing Figures 1.7a and 1.7b, we see that the light must travel farther in the stationary frame than in the moving frame.

Now according to Einstein, the speed of light must be c as measured by both observers. Therefore, it follows that the time interval Δt measured by an observer in the stationary frame is longer than the time interval $\Delta t'$ measured by an observer in the moving frame. To obtain a relationship between Δt and $\Delta t'$, it is convenient to use the right triangle shown in Figure 1.7c. The Pythagorean theorem applied to this triangle gives

$$\left(\frac{c\ \Delta t}{2}\right)^2 = \left(\frac{v\ \Delta t}{2}\right)^2 + d^2$$

Solving for Δt gives

$$\Delta t = \frac{2d}{\sqrt{c^2 - v^2}} = \frac{2d}{c\sqrt{1 - \dfrac{v^2}{c^2}}} \tag{1.6}$$

Because $\Delta t' = 2d/c$, we can express Equation 1.6 as

Time dilation

$$\boxed{\Delta t = \frac{\Delta t'}{\sqrt{1 - \dfrac{v^2}{c^2}}} = \gamma\ \Delta t'} \tag{1.7}$$

where $\gamma = (1 - v^2/c^2)^{-1/2}$. This result says that the time interval measured by the observer in the stationary frame is *longer* than that measured by the observer in the moving frame (because γ is always greater than unity).

According to a stationary observer, a moving clock runs slower by the factor γ^{-1} than an identical stationary clock. This effect is known as **time dilation.**

The time interval $\Delta t'$ in Equation 1.7 is called the **proper time.** In general, proper time is defined as *the time interval between two events as measured by* *an observer who sees the events occur at the same place.* In our case, the observer at O' measures the proper time. That is, *proper time is always the time measured by an observer moving along with the clock.*

We have seen that moving clocks run slow by the factor γ^{-1}. This is true for ordinary mechanical clocks as well as for the light clock just described. In fact, we can generalize these results by stating that *all physical processes, including chemical reactions and biological processes, slow down relative to a stationary clock when they occur in a moving frame.* For example, the heartbeat of an astronaut moving through space would have to keep time with a clock inside the spaceship. Both the astronaut's clock and his heartbeat are slowed down relative to a stationary clock. The astronaut would not have any sensation of life slowing down in the spaceship.

Time dilation is a very real phenomenon that has been verified by various experiments. For example, muons are unstable elementary particles, which have a charge equal to that of an electron and a mass 207 times that of the electron. Muons can be produced by the absorption of cosmic radiation high in the atmosphere. These unstable particles have a lifetime of only 2.2 μs when measured in a reference frame at rest with respect to them. If we take 2.2 μs as the average lifetime of a muon and assume that their speed is close to the speed of light, we would find that these particles could travel a distance of only about 600 m before they decayed into something else (see Fig. 1.8a). Hence, they could not reach the earth from the upper atmosphere where they are produced. However, experiments show that a large number of muons *do* reach the earth. The phenomenon of time dilation explains this effect. Relative to an observer on earth, the muons have a lifetime equal to $\gamma\tau$, where $\tau = 2.2$ μs is the lifetime in a frame of reference traveling with the muons. For example, for $v = 0.99c$, $\gamma \approx 7.1$ and $\gamma\tau \approx 16$ μs. Hence, the average distance traveled as measured by an observer on earth is $\gamma v\tau \approx 4800$ m, as indicated in Figure 1.8b.

In 1976, experiments with muons were conducted at the laboratory of the European Council for Nuclear Research (CERN) in Geneva. Muons were injected into a large storage ring, reaching speeds of about $0.9994c$. Electrons produced by the decaying muons were detected by counters around the ring, enabling scientists to measure the decay rate, and hence the lifetime of the muons. The lifetime of the moving muons was measured to be about 30 times as long as that of the stationary muon (see Fig. 1.9), in agreement with the prediction of relativity to within two parts in a thousand.

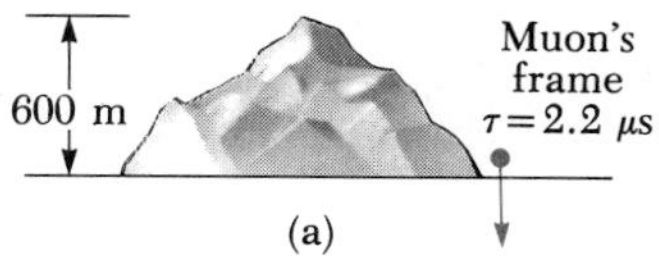

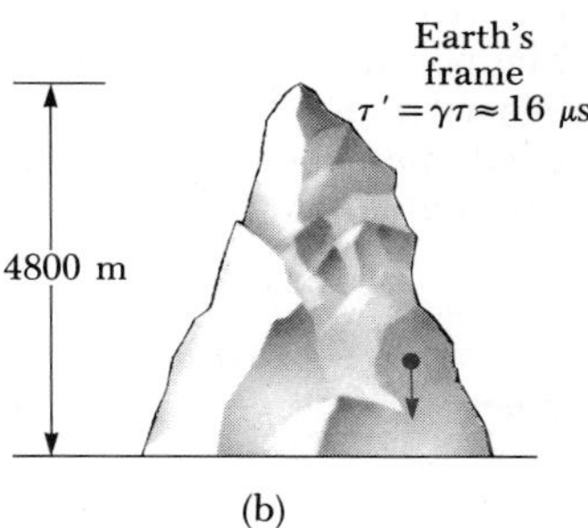

Figure 1.8 (a) Muons traveling with a speed of $0.99c$ travel only about 600 m as measured in the muons' reference frame, where their lifetime is about 2.2 μs. (b) The muons travel about 4800 m as measured by an observer on earth. Because of time dilation, the muons' lifetime is longer as measured by the earth observer.

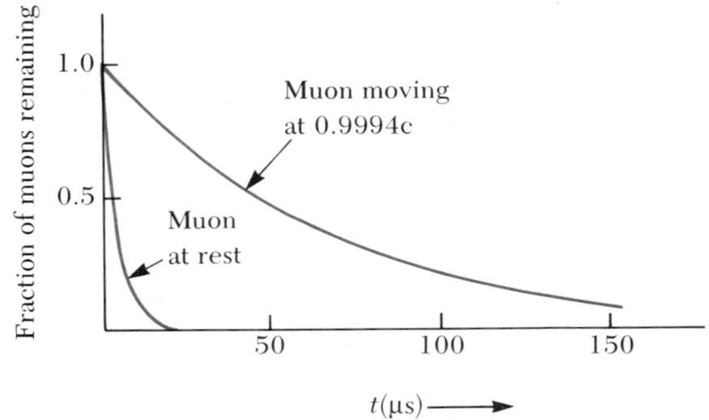

Figure 1.9 Decay curves for muons traveling at a speed of $0.9994c$ and for muons at rest.

The results of an experiment reported by Hafele and Keating provided direct evidence for the phenomenon of time dilation.[5] The experiment involved the use of very stable cesium beam atomic clocks. Time intervals measured with four such clocks in jet flight were compared with time intervals measured by reference atomic clocks located at the U.S. Naval Observatory. (Because of the earth's rotation about its axis, the ground-based clock is not in a true inertial frame.) In order to compare these results with the theory, many factors had to be considered, including periods of acceleration and deceleration relative to the earth, variations in direction of travel, and the weaker gravitational field experienced by the flying clocks compared to the earth-based clock. Their results were in good agreement with the predictions of the special theory of relativity. In their paper, Hafele and Keating report the following: "Relative to the atomic time scale of the U.S. Naval Observatory, the flying clocks lost 59 ± 10 ns during the eastward trip and gained 273 ± 7 ns during the westward trip. . . . These results provide an unambiguous empirical resolution of the famous clock paradox with macroscopic clocks."

EXAMPLE 1.1 What is the Period of the Pendulum? The period of a pendulum is measured to be 3 s in the inertial frame of the pendulum. What is the period of the pendulum when measured by an observer moving at a speed of $0.95c$ with respect to the pendulum?

Solution: In this case, the proper time is equal to 3 s. We can use Equation 1.7 to calculate the period measured by the moving observer. This gives

$$T = \gamma T' = \frac{1}{\sqrt{1 - \frac{(0.95c)^2}{c^2}}} T' = (3.2)(3 \text{ s}) = 9.6 \text{ s}$$

That is, the observer moving at a speed of $0.95c$ observes that the pendulum slows down.

The Twin Paradox

An interesting consequence of time dilation is the so-called twin paradox. Consider a controlled experiment involving 20-year-old twin brothers Speedo and Goslo. Speedo, the more adventuresome twin, sets out on a journey toward a star located 30 lightyears from earth. His spaceship is able to accelerate to a speed close to the speed of light. After reaching the star, Speedo becomes very homesick and immediately returns to earth at the same high speed. Upon his return, he is shocked to find that many things have changed. Old cities have expanded and new cities have appeared. Life styles, people's appearances, and transportation systems have changed dramatically. Speedo's twin brother Goslo has aged to about 80 years old and is now wiser, feeble, and somewhat hard of hearing. Speedo, on the other hand, has aged only about ten years. This is because his bodily processes had slowed down during his travels in space.

It is quite natural to raise the question, "Which twin actually traveled at a speed close to the speed of light, and therefore would be the one who did not age?" Herein lies the paradox: From Goslo's frame of reference, he was at rest while his brother Speedo traveled at a high velocity. On the other hand, according to the space traveler Speedo, it was he who was at rest while his brother zoomed away from him on earth and then returned. This leads to a contradiction as to which twin actually aged.

[5] J. C. Hafele and R. E. Keating, "Around the World Atomic Clocks: Relativistic Time Gains Observed," *Science*, July 14, 1972, p. 168.

In order to resolve this paradox, it should be pointed out that the trip is not as symmetrical as we may have led you to believe. Speedo, the space traveler, had to experience a series of accelerations and decelerations during his journey to the star and back home, and therefore is not always in uniform motion. This means that Speedo has been in a noninertial frame during a large part of his trip, so that predictions based on special relativity are not valid in his frame. On the other hand, the brother on earth has been in an inertial frame, and he can make reliable predictions based on the special theory. The situation is not symmetrical since Speedo experiences forces when his spaceship turns around, whereas Goslo is not subject to such forces. Therefore, the space traveler will indeed be younger upon returning to earth.

Length Contraction

We have seen that measured time intervals are not absolute, that is, the time interval between two events depends on the frame of reference in which it is measured. Likewise, the measured distance between two points depends on the frame of reference. The **proper length** of an object is defined as *the length of the object measured in the reference frame in which the object is at rest.* You should note that proper length is defined similarly to proper time, in that proper time is the time measured by a clock at rest relative to both events. However, proper length is *not* the distance between two points measured at the same time. The length of an object measured in a reference frame in which the object is moving is always less than the proper length. This effect is known as **length contraction.**

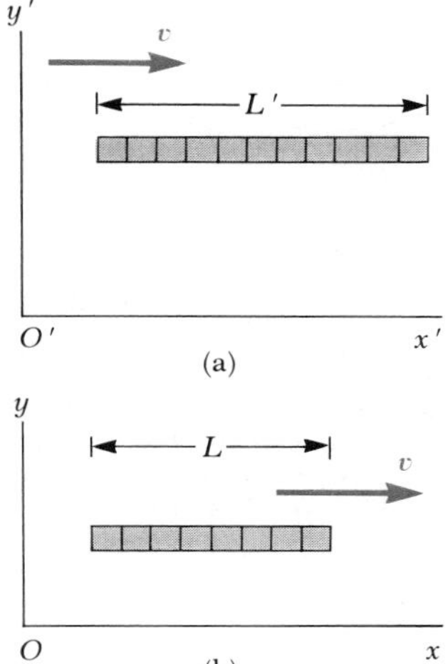

Figure 1.10 A stick moves to the right with a speed v. (a) The stick as viewed by a frame attached to it. (b) The stick as seen by an observer in a frame at rest relative to the stick. The length measured in the rest frame is *shorter* than the proper length, L', by a factor $(1 - v^2/c^2)^{1/2}$.

To understand length contraction quantitatively, let us consider a spaceship traveling with a speed v from one star to another, and two observers. An observer at rest on earth (and also assumed to be at rest with respect to the two stars) measures the distance between the stars to be L', where L' is the proper length. According to this observer, the time it takes the spaceship to complete the voyage is $\Delta t = L'/v$. What does an observer in the moving spaceship measure for the distance between the stars? Because of time dilation, the space traveler measures a smaller time of travel: $\Delta t' = \Delta t/\gamma$. The space traveler claims to be at rest and sees the destination star as moving toward the spaceship with speed v. Since the space traveler reaches the star in the time $\Delta t'$, he or she concludes that the distance, L, between the stars is shorter than L'. This distance measured by the space traveler is given by

$$L = v\,\Delta t' = v\,\Delta t/\gamma$$

Since $L' = v\,\Delta t$, we see that $L = L'/\gamma$ or

$$L = L'\left(1 - \frac{v^2}{c^2}\right)^{1/2} \tag{1.8}$$

Length contraction

According to this result,

> if an observer at rest with respect to an object measures its length to be L', an observer moving with a relative speed v with respect to the object will find it to be shorter than its rest length by the factor $(1 - v^2/c^2)^{1/2}$.

You should note that *the length contraction takes place only along the direction of motion.* For example, suppose a stick moves past a stationary earth observer with a speed v as in Figure 1.10. The length of the stick as measured

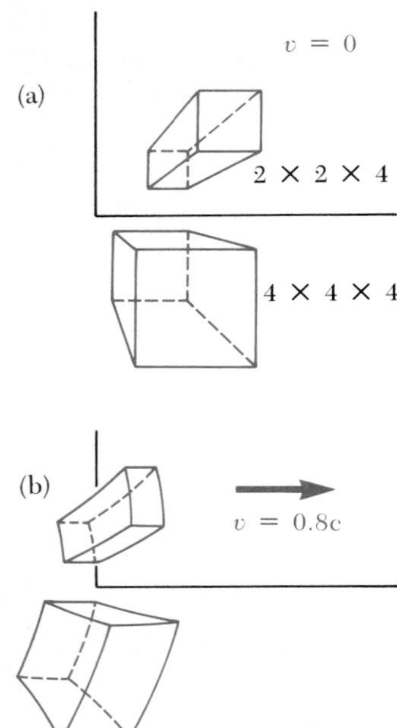

Figure 1.11 Computer-simulated photographs of a set of two rectangular boxes (a) at rest relative to the camera and (b) moving at a speed $v = 0.8c$ relative to the camera. [From G. D. Scott and M. R. Viner, *Am. J. Phys.* 33, 534 (1965).]

by an observer in the frame attached to it is the proper length L', as illustrated in Figure 1.10a. The length of the stick, L, as measured by the earth observer in the stationary frame is shorter than L' by the factor $(1 - v^2/c^2)^{1/2}$. Note that length contraction is a symmetrical effect: if the stick were at rest on earth, an observer in the moving frame would also measure its length to be shorter by the same factor $(1 - v^2/c^2)^{1/2}$.

It is important to emphasize that proper length and proper time are measured in *different* reference frames. As an example of this point, let us return to the example of the decaying muons moving at speeds close to the speed of light. An observer in the muon's reference frame would measure the proper lifetime, while an earth-based observer measures the proper height of the mountain in Figure 1.8. In the muon's reference frame, there is no time dilation, but the distance of travel is observed to be shorter when measured in this frame. Likewise, in the earth observer's reference frame, there is time dilation, but the distance of travel is measured to be the actual height of the mountain. Thus, when calculations on the muon are performed in both frames, one sees the effect of "offsetting penalties" and the outcome of the experiment is the same!

If an object in the shape of a box passing by could be photographed, its image would show length contraction, but its shape would also be distorted. This is illustrated in the computer-simulated drawings shown in Figure 1.11 for a box moving past an observer with a speed $v = 0.8c$. When the shutter of the camera is opened, it records the shape of the object at the instant light from it began its journey to the camera. Since light from different parts of the object must arrive at the camera at the same time (when the photograph is taken), light from more distant parts of the object must start its journey earlier than light from closer parts. Hence, the photograph records different parts of the object at different times. This results in a highly distorted image.

EXAMPLE 1.2 The Contraction of a Spaceship

A spaceship is measured to be 100 m long while it is at rest with respect to an observer. If this spaceship now flies by the observer with a speed of $0.99c$, what length will the observer find for the spaceship?

Solution: From Equation 1.8, the length measured by an observer in the spaceship is

$$L = L'\sqrt{1 - \frac{v^2}{c^2}} = (100\ \text{m})\sqrt{1 - \frac{(0.99c)^2}{c^2}} = 14\ \text{m}$$

Exercise 1 If the ship moves past the observer with a speed of $0.01c$, what length will the observer measure?
Answer: 99.99 m.

EXAMPLE 1.3 How High is the Spaceship?

An observer on earth sees a spaceship at an altitude of 435 m moving downward toward the earth with a speed of $0.97c$. What is the altitude of the spaceship as measured by an observer in the spaceship?

Solution: The moving observer in the ship finds the altitude to be

$$L = L'\sqrt{1 - \frac{v^2}{c^2}} = (435\ \text{m})\sqrt{1 - \frac{(0.97c)^2}{c^2}} = 106\ \text{m}$$

EXAMPLE 1.4 The Triangular Spaceship

A spaceship in the form of a triangle flies by an observer with a speed of $0.95c$. When the ship is at rest (Fig.

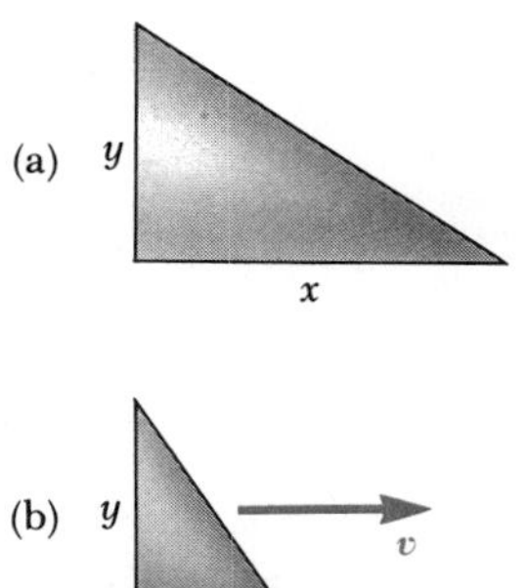

Figure 1.12 (Example 1.4) (a) When the spaceship is at rest, its shape is as shown. (b) The spaceship appears to look like this when it moves to the right with a speed v. Note that only its x dimension is contracted in this case.

1.12a), the distances x and y are found to be 50 m and 25 m, respectively. What is the shape of the ship as seen by an observer at rest when the ship is in motion along the direction shown in Figure 1.12b?

Solution: The observer sees the horizontal length of the ship to be contracted to a length of

$$L = L'\sqrt{1 - \frac{v^2}{c^2}} = (50 \text{ m})\sqrt{1 - \frac{(0.95c)^2}{c^2}} = 15.6 \text{ m}$$

The 25-m vertical height is unchanged because it is perpendicular to the direction of relative motion between the observer and the spaceship. Figure 1.12b represents the shape of the spaceship as seen by the observer at rest.

1.6 THE LORENTZ TRANSFORMATION

We have seen that the Galilean transformation is not valid when v approaches the speed of light. In this section, we shall derive the correct transformation equations that apply for all speeds in the range of $0 \leq v < c$. This transformation, known as the Lorentz transformation, was developed by H. A. Lorentz (1853–1928) in 1890. However, its real significance in a physical theory was first recognized by Einstein.

We shall derive the Lorentz transformation by considering a rocket moving with a speed v along the xx' axes as in Figure 1.13a. The frame of the rocket S′ is indicated with the coordinates (x', y', z', t'), while a stationary observer in S uses coordinates (x, y, z, t). A flashbulb mounted on the rocket emits a pulse of light at the instant that the origins of the two reference frames coincide.

At the instant the flashbulb goes off and the two origins coincide, we define $t = t' = 0$. The light signal travels as a spherical wave, where the origin of the wavefront is the fixed point O where the flash originated. At some later time, a point such as P on the spherical wavefront is at a distance r from O and a distance r' from O', as shown schematically in Figure 1.13b. According to Einstein's second postulate, the speed of light should be c for both observers. Hence, the distance to the point P on the wavefront as measured by an observer in S is given by $r = ct$, while the distance to the point P as measured by an observer in S′ is given by $r' = ct'$. That is,

$$r = ct \tag{1.8a}$$

$$r' = ct' \tag{1.8b}$$

If we accept Einstein's second postulate we must require that the times t and t' taken for the light to reach P be *different*. This is in contrast to the Galilean transformation, where $t = t'$.

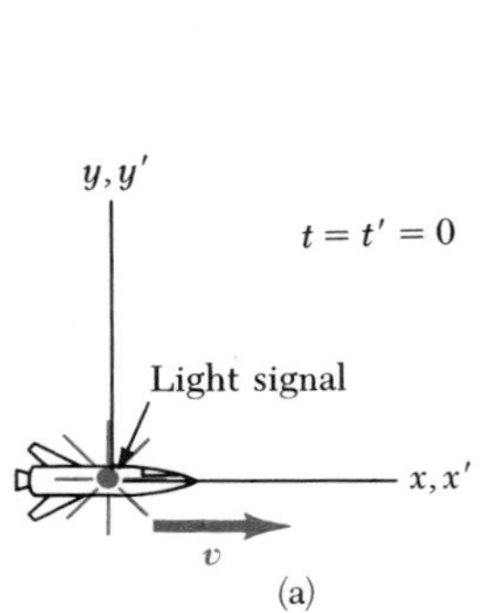

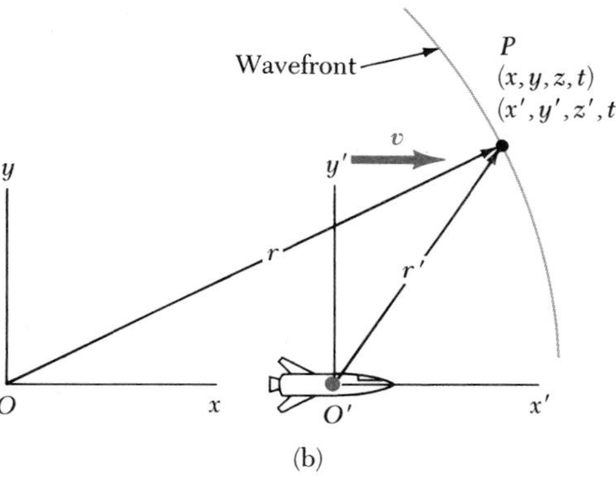

Figure 1.13 A rocket moves with a speed v along the xx' axes. (a) A pulse of light is sent out from the rocket at $t = t' = 0$ when the two systems coincide. (b) Coordinates of some point P on an expanding spherical wavefront as measured by observers in both inertial systems. (This figure is entirely schematic, and you should not be misled by the geometry.)

The radius of a sphere is given by the equation $r^2 = x^2 + y^2 + z^2$ as described by an observer in S. Likewise, the distance r' measured in S′ is given by $(r')^2 = (x')^2 + (y')^2 + (z')^2$. Hence, by squaring Equations 1.8a and 1.8b, we obtain the following expressions:

$$\text{Observer in S:} \qquad x^2 + y^2 + z^2 = c^2t^2 \tag{1.9}$$

$$\text{Observer in S}': \qquad (x')^2 + (y')^2 + (z')^2 = c^2(t')^2 \tag{1.10}$$

Since the motion of S′ is along the xx' axes, it follows that the y and z coordinates measured in the two frames are always equal. That is, they are unaffected by the motion along x, and therefore $y = y'$ and $z = z'$. Hence, by subtracting Equation 1.10 from Equation 1.9, we get

$$x^2 - (x')^2 = c^2t^2 - c^2(t')^2$$

$$x^2 - c^2t^2 = (x')^2 - c^2(t')^2 \tag{1.11}$$

Imposing the condition that $x = vt$ corresponds to $x' = 0$ and using Equation 1.11, we obtain for x' and t'

$$x' = \frac{x - vt}{\sqrt{1 - \dfrac{v^2}{c^2}}} \tag{1.12}$$

$$t' = \frac{t - (v/c^2)x}{\sqrt{1 - \dfrac{v^2}{c^2}}} \tag{1.13}$$

The algebra that gives Equations 1.12 and 1.13 is rather tedious and therefore is omitted. However, these equations can be verified by substituting Equations 1.12 and 1.13 into the right side of Equation 1.11.

We now summarize the **Lorentz transformation equations** for transforming from S to S′ as follows:

Lorentz transformation for $S \rightarrow S'$

$$\begin{aligned} x' &= \frac{x - vt}{\sqrt{1 - \dfrac{v^2}{c^2}}} = \gamma(x - vt) \\ y' &= y \\ z' &= z \\ t' &= \frac{t - \left(\dfrac{v}{c^2}\right)x}{\sqrt{1 - \dfrac{v^2}{c^2}}} = \gamma\left(t - \frac{v}{c^2}x\right) \end{aligned} \tag{1.14}$$

where the symbol γ (gamma) is defined as

$$\gamma \equiv \frac{1}{\sqrt{1 - \dfrac{v^2}{c^2}}} \tag{1.15}$$

If we wish to transform coordinates in the S′ frame to coordinates in the S frame, we simply replace v by $-v$ and interchange the primed and unprimed coordinates in Equation 1.14. The resulting transformation is given by

Inverse Lorentz transformation for S′ → S

$$\begin{aligned} x &= \frac{x' + vt'}{\sqrt{1 - \frac{v^2}{c^2}}} = \gamma(x' + vt') \\ y &= y' \\ z &= z' \\ t &= \frac{t' + \left(\frac{v}{c^2}\right) x'}{\sqrt{1 - \frac{v^2}{c^2}}} = \gamma\left(t' + \frac{v}{c^2} x'\right) \end{aligned} \tag{1.16}$$

Note that in the Lorentz transformation, t depends on both t' and x'. Likewise, t' depends on both t and x. This is unlike the case of the Galilean transformation, in which $t = t'$.

When $v \ll c$, the Lorentz transformation should reduce to the Galilean transformation. To check this, note that as $v \to 0$, $v/c^2 \ll 1$ and $v^2/c^2 \ll 1$, so that Equation 1.14 reduces in this limit to the Galilean coordinate transformation equations, given by

$$x' = x - vt \qquad y' = y \qquad z' = z \qquad t' = t$$

Lorentz Velocity Transformation

Let us now derive the Lorentz velocity transformation, which is the relativistic counterpart of the Galilean velocity transformation. Suppose that an unaccelerated object is observed in the S′ frame at x_1' at time t_1' and at x_2' at time t_2'. Its speed u_x' measured in S′ is given by

$$u_x' = \frac{x_2' - x_1'}{t_2' - t_1'} = \frac{dx'}{dt'} \tag{1.17}$$

Using Equation 1.14, we have

$$dx' = \frac{dx - v\,dt}{\sqrt{1 - \frac{v^2}{c^2}}}$$

$$dt' = \frac{dt - \frac{v}{c^2}\,dx}{\sqrt{1 - \frac{v^2}{c^2}}}$$

Substituting these into Equation 1.17 gives

$$u_x' = \frac{dx'}{dt'} = \frac{dx - v\,dt}{dt - \frac{v}{c^2}\,dx} = \frac{\frac{dx}{dt} - v}{1 - \frac{v}{c^2}\frac{dx}{dt}}$$

But dx/dt is just the velocity u_x of the object measured in S, and so this expression becomes

Lorentz velocity transformation for S → S′

$$u_x' = \frac{u_x - v}{1 - \frac{u_x v}{c^2}} \qquad (1.18)$$

Similarly, if the object has velocity components along y and z, the components in S′ are given by

$$u_y' = \frac{u_y}{\gamma\left(1 - \frac{u_x v}{c^2}\right)} \quad \text{and} \quad u_z' = \frac{u_z}{\gamma\left(1 - \frac{u_x v}{c^2}\right)} \qquad (1.19)$$

When u_x and v are both much smaller than c (the nonrelativistic case), we see that the denominator of Equation 1.18 approaches unity, and so $u_x' \approx u_x - v$. This corresponds to the Galilean velocity transformations. In the other extreme, when $u_x = c$, Equation 1.17 becomes

$$u_x' = \frac{c - v}{1 - \frac{cv}{c^2}} = \frac{c\left(1 - \frac{v}{c}\right)}{1 - \frac{v}{c}} = c$$

From this result, we see that an object moving with a speed c relative to an observer in S also has a speed c relative to an observer in S′—*independent* of the relative motion of S and S′. Note that this conclusion is consistent with Einstein's second postulate, namely, that the speed of light must be c with respect to all inertial frames of reference. Furthermore, the speed of an object can never exceed c. That is, the speed of light is the "ultimate" speed. We shall return to this point later when we consider the energy of a particle.

To obtain u_x in terms of u_x', we replace v by $-v$ in Equation 1.18 and interchange the roles of u_x and u_x'. This gives

Inverse Lorentz velocity transformation for S′ → S

$$u_x = \frac{u_x' + v}{1 + \frac{u_x' v}{c^2}} \qquad (1.20)$$

EXAMPLE 1.5 Relative Velocity of Spaceships
Two spaceships A and B are moving in *opposite* directions, as in Figure 1.14. An observer on the earth measures the speed of A to be $0.75c$ and the speed of B to be $0.85c$. Find the velocity of B with respect to A.

Solution: This problem can be solved by taking the S′ frame as being attached to spacecraft A, so that $v = 0.75c$ relative to an observer on the earth (the S frame). Spacecraft B can be considered as an object moving to the left with a velocity $u_x = -0.85c$ relative to the earth ob-

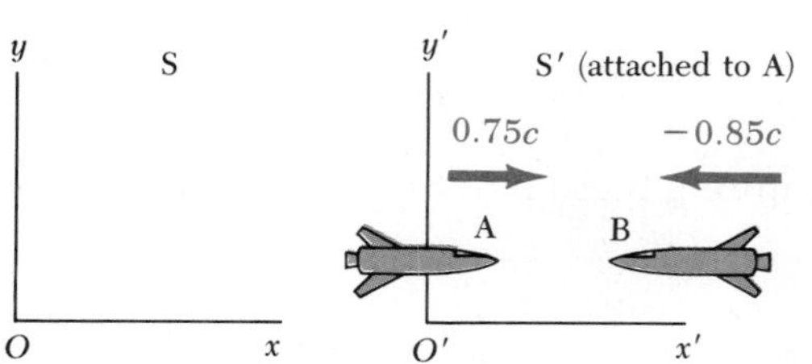

Figure 1.14 (Example 1.5) Two spaceships A and B move in *opposite* directions. The velocity of B relative to A is *less* than c and is obtained by using the relativistic velocity transformation.

server. Hence, the velocity of B with respect to A can be obtained using Equation 1.18:

$$u_x' = \frac{u_x - v}{1 - \frac{u_x v}{c^2}} = \frac{-0.85c - 0.75c}{1 - \frac{(-0.85c)(0.75c)}{c^2}}$$

$$= -0.9771c$$

The negative sign for u_x' indicates that spaceship B is moving in the negative x direction as observed by A. Note that the result is less than c. That is, a body whose speed is less than c in one frame of reference must have a speed less than c in *any other* frame. If the Galilean velocity transformation were used in this example, we would find that $u_x' = u_x - v = -0.85c - 0.75c = -1.6c$, which is greater than c.

EXAMPLE 1.6 The Speeding Motorcycle
Imagine a motorcycle rider moving with a speed of $0.8c$ past a stationary observer, as shown in Figure 1.15. If the rider tosses a ball in the forward direction with a speed of $0.7c$ relative to himself, what is the speed of the ball as seen by the stationary observer?

Solution: In this situation, the velocity of the motorcycle with respect to the stationary observer is $v = 0.8c$. The velocity of the ball in the frame of reference of the motorcyclist is $u_x' = 0.7c$. Therefore, the velocity, u_x, of the ball relative to the stationary observer is

$$u_x = \frac{u_x' + v}{1 + \frac{u_x' v}{c^2}} = \frac{0.7c + 0.8c}{1 + \frac{(0.7c)(0.8c)}{c^2}} = 0.962c$$

Exercise 2 Suppose that the motorcyclist moving with a speed $0.8c$ turns on a beam of light that moves away from him with a speed of c in the same direction as the moving motorcycle. What would the stationary observer measure for the speed of the beam of light?
Answer: c.

Figure 1.15 (Example 1.6) A motorcyclist moves past a stationary observer with a speed of $0.8c$ and throws a ball in the direction of motion with a speed of $0.7c$ relative to himself.

1.7 RELATIVISTIC MOMENTUM

We have seen that the principle of relativity is satisfied if the Galilean transformation is replaced by the more general Lorentz transformation. Therefore, in order to properly describe the motion of material particles within the framework of special relativity, we must generalize Newton's laws and the definition of momentum and energy. These generalized definitions of momentum and energy will reduce to the classical (nonrelativistic) definitions for $v \ll c$.

First, recall that the conservation of momentum states that when two bodies collide, the total momentum remains constant, assuming the bodies are isolated (that is, they interact only with each other). Suppose the collision is described in a reference frame S in which the momentum is conserved. If the velocities in a second reference frame S′ are calculated using the Lorentz transformation and the classical definition of momentum, $\mathbf{p} = m\mathbf{u}$, one finds that momentum *is not* conserved in the second reference frame. However, since the laws of physics are the same in all inertial frames, the momentum must be conserved in all systems. In view of this condition and assuming the Lorentz transformation is correct, we must modifiy the definition of momentum.

Our definition of relativistic momentum $\mathbf{p}$ must satisfy the following conditions:

1. The relativistic momentum must be conserved in all collisions.
2. The relativistic momentum must approach the classical value $m\mathbf{u}$ as $\mathbf{u} \rightarrow 0$.

The correct relativistic equation for the momentum that satisfies these conditions is given by the expression

Definition of relativistic momentum

$$p \equiv \frac{mu}{\sqrt{1-\dfrac{u^2}{c^2}}} = \gamma mu \qquad (1.21)$$

where u is the velocity of the particle. We use the symbol u for particle velocity rather than v, which is used for the relative velocity of two reference frames. The proof of this generalized expression for p is beyond the scope of this text. Note that when u is much less than c, the denominator of Equation 1.21 approaches unity, so that p approaches mu. Therefore, the relativistic equation for p reduces to the classical expression when u is small compared with c.

In many situations, it is useful to interpret Equation 1.21 as the product of a relativistic mass, γm, and the velocity of the object. Using this description, one can say that *the observed mass of an object increases with speed according to the relation*

$$\text{Relativistic mass} = \gamma m = \frac{m}{\sqrt{1-\dfrac{v^2}{c^2}}} \qquad (1.22)$$

where the relativistic mass is the mass of the object moving at a speed v with respect to the observer and m is the mass of the object as measured by an observer at rest with respect to the object. The quantity m is often referred to as the **rest mass** of the object. (In future chapters, we shall sometimes use the symbol m_0 to represent rest mass.) Note that Equation 1.22 indicates that *the mass of an object increases as its speed increases.* This prediction has been borne out through many observations on elementary particles, such as energetic electrons.

The relativistic force $\boldsymbol{F}$ on a particle whose momentum is p is defined by the expression

$$\boldsymbol{F} \equiv \frac{dp}{dt} \qquad (1.23)$$

where p is given by Equation 1.21. This expression is identical to the classical statement of Newton's second law, which says that force equals the time rate of change of momentum.

EXAMPLE 1.7 Momentum of an Electron

An electron, which has a rest mass of 9.11×10^{-31} kg, moves with a speed of $0.75c$. Find its relativistic momentum and compare this with the momentum calculated from the classical expression.

Solution: Using Equation 1.21 with $u = 0.75c$, we have

$$p = \frac{mu}{\sqrt{1-\dfrac{u^2}{c^2}}}$$

$$= \frac{(9.11 \times 10^{-31}\ \text{kg})(0.75 \times 3 \times 10^8\ \text{m/s})}{\sqrt{1-\dfrac{(0.75c)^2}{c^2}}}$$

$$= 3.10 \times 10^{-22}\ \text{kg}\cdot\text{m/s}$$

The incorrect classical expression would give

$$\text{Momentum} = mu = 2.05 \times 10^{-22}\ \text{kg}\cdot\text{m/s}$$

Hence, the correct relativistic result is 50% greater than the classical result!

It is left as a problem (Problem 41) to show that the acceleration a of a particle decreases under the action of a constant force, in which case $a \propto (1 - u^2/c^2)^{3/2}$. Furthermore, as the velocity approaches c, the acceleration caused by any finite force approaches zero. Hence, it is impossible to accelerate a particle from rest to a speed equal to or greater than c.

1.8 RELATIVISTIC ENERGY

We have seen that the definition of momentum and the laws of motion required generalization to make them compatible with the principle of relativity. This implies that the relation between work and energy must also be modified.

In order to derive the relativistic form of the work-energy theorem, let us start with the definition of work done by a force F and make use of the definition of relativistic force, Equation 1.23. That is,

$$W = \int_{x_1}^{x_2} F\,dx = \int_{x_1}^{x_2} \frac{dp}{dt}\,dx \tag{1.24}$$

where we have assumed that the force and motion are along the x axis. In order to perform this integration, we make repeated use of the chain rule for derivatives in the following manner:

$$\left(\frac{dp}{dt}\right) dx = \left(\frac{dp}{du}\frac{du}{dt}\right) dx = \frac{dp}{du}\left(\frac{du}{dx}\frac{dx}{dt}\right) dx$$

$$= \frac{dp}{du}\,u\,\frac{du}{dx}\,dx = \frac{dp}{du}\,u\,du$$

Since p depends on u according to Equation 1.21, we have

$$\frac{dp}{du} = \frac{d}{du}\frac{mu}{\sqrt{1-\frac{u^2}{c^2}}} = \frac{m}{\left(1-\frac{u^2}{c^2}\right)^{3/2}} \tag{1.25}$$

Using these results, we can express the work as

$$W = \int_0^u \frac{dp}{du}\,u\,du = \int_0^u \frac{mu}{\left(1-\frac{u^2}{c^2}\right)^{3/2}}\,du$$

where we have assumed that the particle is accelerated from rest to some final velocity u. Evaluating the integral, we find that

$$W = \frac{mc^2}{\sqrt{1-\frac{u^2}{c^2}}} - mc^2 \tag{1.26}$$

Recall that the work-energy theorem states that the work done by a force acting on a particle equals the change in kinetic energy of the particle. Since the initial kinetic energy is zero, we conclude that the work W is equivalent to the relativistic kinetic energy K, that is,

$$\boxed{K = \frac{mc^2}{\sqrt{1-\frac{u^2}{c^2}}} - mc^2} \tag{1.27}$$

Relativistic kinetic energy

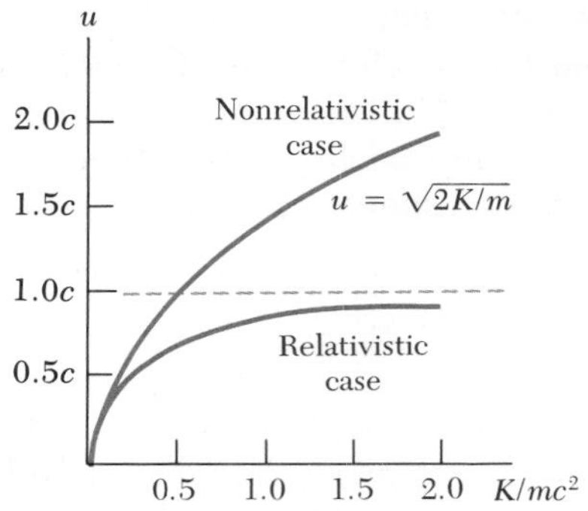

Figure 1.16 A graph comparing relativistic and nonrelativistic kinetic energy. The energies are plotted versus speed. In the relativistic case, u is always *less* than c.

This equation has been confirmed by experiments using high-energy particle accelerators. At low speeds, where $u/c \ll 1$, Equation 1.27 should reduce to the classical expression $K = \frac{1}{2}mu^2$. We can check this by using the binomial expansion $(1 - x^2)^{-1/2} \approx 1 + \frac{1}{2}x^2 + \cdots$ for $x \ll 1$, where the higher-order powers of x are neglected in the expansion. In our case, $x = u/c$, so that

$$\frac{1}{\sqrt{1 - \frac{u^2}{c^2}}} = \left(1 - \frac{u^2}{c^2}\right)^{-1/2} \approx 1 + \frac{1}{2}\frac{u^2}{c^2} + \cdots$$

Substituting this into Equation 1.27 gives

$$K \approx mc^2\left(1 + \frac{1}{2}\frac{u^2}{c^2} + \cdots\right) - mc^2 = \frac{1}{2}mu^2$$

which agrees with the classical result. A graph comparing the relativistic and nonrelativistic expressions is given in Figure 1.16. Note that in the relativistic case, the particle speed never exceeds c, regardless of the kinetic energy. The two curves are in good agreement when $u \ll c$.

It is useful to write the relativistic kinetic energy in the form

$$K = \gamma mc^2 - mc^2 \tag{1.28}$$

where

$$\gamma = \frac{1}{\sqrt{1 - \frac{u^2}{c^2}}}$$

The constant term mc^2, which is independent of the speed, is called the **rest energy** of the particle. The term γmc^2, which depends on the particle speed, is therefore the sum of the kinetic and rest energies. We define γmc^2 to be the **total energy** E, that is,

Definition of total energy

$$\boxed{E = \gamma mc^2 = K + mc^2} \tag{1.29}$$

or

Energy-mass equivalence

$$\boxed{E = \frac{mc^2}{\sqrt{1 - \frac{u^2}{c^2}}}} \tag{1.30}$$

This, of course, is Einstein's famous mass-energy equivalence equation. The relation $E = \gamma mc^2$ shows that *mass is a form of energy*. Furthermore, this result shows that a small mass corresponds to an enormous amount of energy. This concept has revolutionized the field of nuclear physics.

In many situations, the momentum or energy of a particle is known rather than its speed. It is therefore useful to have an expression relating the total energy E to the relativistic momentum p. This is accomplished by using the expressions $E = \gamma mc^2$ and $p = \gamma mu$. By squaring these equations and subtracting, we can eliminate u (Problem 18). The result, after some algebra, is

Energy-momentum relation

$$\boxed{E^2 = p^2c^2 + (mc^2)^2} \tag{1.31}$$

When the particle is at rest, $p = 0$, and so we see that $E = mc^2$. That is, the total energy equals the rest energy. As we shall discuss in the next chapter, it is well established that there are particles that have zero mass, such as photons (quanta of electromagnetic radiation). If we set $m = 0$ in Equation 1.31, we see that

$$\boxed{E = pc} \tag{1.32}$$

This equation is an *exact* expression relating energy and momentum for photons and neutrinos, which always travel at the speed of light.

Finally, note that since the mass m of a particle is independent of its motion, m must have the same value in all reference frames. On the other hand, the total energy and momentum of a particle depend on the reference frame in which they are measured, since they both depend on velocity. Since m is a constant, then according to Equation 1.31 the quantity $E^2 - p^2c^2$ must have the same value in all reference frames. That is, $E^2 - p^2c^2$ is *invariant* under a Lorentz transformation.

When dealing with electrons or other subatomic particles, it is convenient to express their energy in electron volts (eV), since the particles are usually given this energy by acceleration through a potential difference. The conversion factor is

$$1 \text{ eV} = 1.60 \times 10^{-19} \text{ J}$$

For example, the mass of an electron is 9.11×10^{-31} kg. Hence, the rest energy of the electron is

$$mc^2 = (9.11 \times 10^{-31} \text{ kg})(3.00 \times 10^8 \text{ m/s})^2 = 8.20 \times 10^{-14} \text{ J}$$

Converting this to eV, we have

$$mc^2 = (8.20 \times 10^{-14} \text{ J})(1 \text{ eV}/1.60 \times 10^{-19} \text{ J}) = 0.511 \text{ MeV}$$

where 1 MeV $= 10^6$ eV.

EXAMPLE 1.8 The Energy of a Speedy Electron
An electron moves with a speed $u = 0.85c$. Find its total energy and kinetic energy in eV.

Solution: Using the fact that the rest energy of the electron is 0.511 MeV together with Equation 1.30 gives

$$E = \frac{mc^2}{\sqrt{1 - \frac{u^2}{c^2}}} = \frac{0.511 \text{ MeV}}{\sqrt{1 - \frac{(0.85c)^2}{c^2}}}$$

$$= 1.90(0.511 \text{ MeV}) = 0.970 \text{ MeV}$$

The kinetic energy is obtained by subtracting the rest energy from the total energy:

$$K = E - mc^2 = 0.970 \text{ MeV} - 0.511 \text{ MeV} = 0.459 \text{ MeV}$$

EXAMPLE 1.9 The Energy of a Speedy Proton
The total energy of a proton is three times its rest energy. (a) Find the proton's rest energy in eV.

$$\begin{aligned}\text{Rest energy} = mc^2 &= (1.67 \times 10^{-27}\text{ kg})(3 \times 10^8 \text{ m/s})^2 \\ &= (1.50 \times 10^{-10}\text{ J})(1 \text{ eV}/1.60 \times 10^{-19}\text{ J}) \\ &= 938 \text{ MeV}\end{aligned}$$

(b) With what speed is the proton moving?

Since the total energy E is three times the rest energy, Equation 1.30 gives

$$E = 3mc^2 = \frac{mc^2}{\sqrt{1 - \dfrac{u^2}{c^2}}}$$

$$3 = \frac{1}{\sqrt{1 - \dfrac{u^2}{c^2}}}$$

Solving for u gives

$$\left(1 - \frac{u^2}{c^2}\right) = \frac{1}{9} \quad \text{or} \quad \frac{u^2}{c^2} = \frac{8}{9}$$

$$u = \frac{\sqrt{8}}{3}c = 2.83 \times 10^8 \text{ m/s}$$

(c) Determine the kinetic energy of the proton in eV.

$$K = E - mc^2 = 3mc^2 - mc^2 = 2mc^2$$

Since $mc^2 = 938$ MeV, $K = 1876$ MeV.

(d) What is the proton's momentum?

We can use Equation 1.31 to calculate the momentum with $E = 3mc^2$:

$$E^2 = p^2c^2 + (mc^2)^2 = (3mc^2)^2$$

$$p^2c^2 = 9(mc^2)^2 - (mc^2)^2 = 8(mc^2)^2$$

$$p = \sqrt{8}\,\frac{mc^2}{c} = \sqrt{8}\,\frac{(938 \text{ MeV})}{c} = 2653\,\frac{\text{MeV}}{c}$$

Note that the unit of momentum is written MeV/c for convenience.

1.9 CONFIRMATIONS AND CONSEQUENCES OF RELATIVITY THEORY

The special theory of relativity has been confirmed by a number of experiments. One important experiment concerned with the decay of muons, and time dilation in the muon's reference frame, was discussed in Section 1.5. This section describes further evidence in favor of Einstein's special theory of relativity.

One of the first predictions of relativity that was experimentally confirmed is the variation of momentum with velocity. Experiments were performed as early as 1909 on electrons, which can easily be accelerated to speeds close to c through the use of electric fields. When an energetic electron enters a magnetic field $\mathbf{B}$ with its velocity vector perpendicular to $\mathbf{B}$, a magnetic force is exerted on the electron, causing it to move in a circle of radius r. In this situation, the relativistic momentum of the electron is given by $p = eB/r$. From the relativistic equivalent of Newton's second law, $\mathbf{F} = d\mathbf{p}/dt$, the variation in momentum with kinetic energy can be checked experimentally. The results of such experiments on electrons and other charged particles support the relativistic expressions.

The release of enormous quantities of energy in nuclear fission and fusion processes is a manifestation of the equivalence of mass and energy. The conversion of mass into energy is, of course, the basis of atomic and hydrogen bombs, the most powerful and destructive weapons ever constructed. In fact, all reactions that release energy do so at the expense of mass (including chemical reactions). In a conventional nuclear reactor, the uranium nucleus undergoes fission, a reaction that results in several lighter fragments having considerable kinetic energy. In the case of ^{235}U (the parent nucleus), which undergoes spontaneous fission, the fragments are two lighter nuclei and two neutrons. The total mass of the fragments is *less* than that of the parent nucleus by some amount Δm. The corresponding energy Δmc^2 associated with this mass difference is *exactly* equal to the total kinetic energy of the fragments. This kinetic energy is then used to produce heat and steam for the generation of electrical power.

Next, consider the basic fusion reaction in which two deuterium atoms combine to form one helium atom. This reaction is of major importance in current research and development of controlled-fusion reactors. The decrease in mass which results from the creation of one helium atom from two deuterium atoms is calculated to be $\Delta m = 4.25 \times 10^{-29}$ kg. Hence, the corresponding excess energy which results from one fusion reaction is given by the expression $\Delta mc^2 = 3.83 \times 10^{-12}$ J $= 23.9$ MeV. To appreciate the magnitude of this result, if 1 g of deuterium is converted into helium, the energy released is about 10^{12} J! At the 1988 cost of electrical energy, this would be worth about \$50 000.

EXAMPLE 1.10 Binding Energy of the Deuteron
The mass of the deuteron, which is the nucleus of "heavy hydrogen," is not equal to the sum of the masses of its constituents, which are the proton and neutron. Calculate this mass difference and determine its energy equivalence.

Solution: Using atomic mass units (u), we have

$$m_p = \text{mass of proton} = 1.007276 \text{ u}$$

$$m_n = \text{mass of neutron} = 1.008665 \text{ u}$$

$$m_p + m_n = 2.015941 \text{ u}$$

Since the mass of the deuteron is 2.013553 u, we see that the mass difference Δm is 0.002388 u. By definition, $1 \text{ u} = 1.66 \times 10^{-27}$ kg, and therefore

$$\Delta m = 0.002388 \text{ u} = 3.96 \times 10^{-30} \text{ kg}$$

Using $E = \Delta mc^2$, we find that

$$E = \Delta mc^2 = (3.96 \times 10^{-30} \text{ kg})(3 \times 10^8 \text{ m/s})^2$$

$$= 3.56 \times 10^{-13} \text{ J}$$

$$= 2.23 \text{ MeV}$$

Therefore, the minimum energy required to separate the proton from the neutron of the deuterium nucleus (the binding energy) is 2.23 MeV.

1.10 SUMMARY

The two basic postulates of the ***special theory of relativity*** are as follows:

1. All the laws of physics must be the same for all observers moving at constant velocity with respect to each other.
2. In particular, the speed of light must be the same for all inertial observers, independent of their relative motion.

In order to satisfy these postulates, the Galilean transformations must be replaced by the **Lorentz transformations** given by

$$\begin{aligned} x' &= \gamma(x - vt) \\ t' &= \gamma\left(t - \frac{v}{c^2}x\right) \\ y' &= y \\ z' &= z \end{aligned} \tag{1.14}$$

where

$$\gamma = \frac{1}{\sqrt{1 - \dfrac{v^2}{c^2}}} \qquad (1.15)$$

In these equations, it is assumed that the primed system moves with a speed v along the xx' axes.

The relativistic form of the **velocity transformation** is

$$u_x' = \frac{u_x - v}{1 - \dfrac{u_x v}{c^2}} \qquad (1.18)$$

where u_x is the speed of an object as measured in the S frame and u_x' is its speed measured in the S′ frame.

Some of the consequences of the special theory of relativity are as follows:

1. Clocks in motion relative to an observer appear to be slowed down by a factor γ. This is known as **time dilation.**
2. Lengths of objects in motion appear to be contracted in the direction of motion.
3. Events that are simultaneous for one observer are not simultaneous for another observer in motion relative to the first.

These three statements can be summarized by saying that duration, length, and simultaneity are not absolute concepts in relativity.

The relativistic expression for the **momentum** of a particle moving with a velocity $\boldsymbol{u}$ is

$$p \equiv \frac{mu}{\sqrt{1 - \dfrac{u^2}{c^2}}} = \gamma mu \qquad (1.21)$$

where

$$\gamma = \frac{1}{\sqrt{1 - \dfrac{u^2}{c^2}}}$$

The relativistic expression for the **kinetic energy** of a particle is

$$K = \gamma mc^2 - mc^2 \qquad (1.28)$$

where mc^2 is called the **rest energy** of the particle.

The total energy E of a particle is related to the mass through the famous **energy-mass equivalence** expression:

$$E = \gamma mc^2 = \frac{mc^2}{\sqrt{1 - \dfrac{u^2}{c^2}}} \qquad (1.30)$$

Finally, the relativistic momentum is related to the total energy through the equation

$$E^2 = p^2c^2 + (mc^2)^2 \qquad (1.31)$$

SUGGESTIONS FOR FURTHER READING

1. See the references contained in footnotes 1 through 5 of this chapter.
2. H. Bondi, *Relativity and Common Sense*, Science Study Series, Garden City, N.Y., Doubleday, 1964.
3. J. Bronowski, "The Clock Paradox," *Sci. American*, February, 1963, p. 134.
4. A. Einstein, *Out of My Later Years*, New York, World Publishing, 1971.
5. A. Einstein, *Ideas and Opinions*, New York, Crown, 1954.
6. G. Gamow, *Mr. Tompkins in Wonderland*, New York, Cambridge University Press, 1939.
7. L. Infeld, *Albert Einstein*, New York, Scribner's, 1950.
8. J. Schwinger, *Einstein's Legacy*, Scientific American Library, New York, W. H. Freeman, 1985.
9. R. S. Shankland, "The Michelson-Morley Experiment," *Sci. American*, November 1964, p. 107.
10. R. Skinner, *Relativity for Scientists and Engineers*, New York, Dover Publications, 1982.

QUESTIONS

1. What two measurements will two observers in relative motion *always* agree upon?
2. A spaceship in the shape of a sphere moves past an observer on earth with a speed $0.5c$. What shape will the observer see as the spaceship moves past?
3. An astronaut moves away from the earth at a speed close to the speed of light. If an observer on earth could make measurements of the astronaut's size and pulse rate, what changes (if any) would he or she measure? Would the astronaut measure any changes?
4. Two identically constructed clocks are synchronized. One is put in orbit around the earth while the other remains on earth. Which clock runs slower? When the moving clock returns to earth, will the two clocks still be synchronized?
5. Two lasers situated on a moving spacecraft are triggered simultaneously. An observer on the spacecraft claims to see the pulses of light simultaneously. What condition is necessary in order that a stationary observer agree that the two pulses are emitted simultaneously?
6. When we say that a moving clock runs slower than a stationary one, does this imply that there is something physically unusual about the moving clock?
7. When we speak of time dilation, do we mean that time passes more slowly in moving systems or that it simply appears to do so?
8. List some ways our day-to-day lives would change if the speed of light were only 50 m/s.
9. Give a physical argument which shows that it is impossible to accelerate an object of mass m to the speed of light, even with a continuous force acting on it.
10. It is said that Einstein, in his teenage years, asked the question, "What would I see in a mirror if I carried it in my hands and ran at the speed of light?" How would you answer this question?
11. Since mass is a form of energy, can we conclude that a compressed spring has more mass than the same spring when it is not compressed? On the basis of your answer, which has more mass, a spinning planet or an otherwise identical but nonspinning planet?
12. Suppose astronauts were paid according to the time spent traveling in space. After a long voyage at a speed near that of light, a crew of astronauts return and open their pay envelopes. What will their reaction be?
13. What happens to the density of an object as its speed increases?
14. Consider the incorrect statement, "Matter can neither be created nor destroyed." How would you correct this statement in view of the special theory of relativity?

PROBLEMS

1. In a laboratory frame of reference, an observer notes that Newton's second law is valid. Show that it is also valid for an observer moving at a constant speed relative to the laboratory frame.
2. Show that Newton's second law is not valid in a reference frame moving past the laboratory frame of Problem 1 with a constant acceleration.
3. A 2000-kg car moving with a speed of 20 m/s collides with and sticks to a 1500-kg car at rest at a stop sign. Show that momentum is conserved in a reference frame moving with a speed of 10 m/s in the direction of the moving car.
4. A billiard ball of mass 0.3 kg moves with a speed of 5 m/s and collides elastically with a ball of mass 0.2 kg moving in the opposite direction with a speed of 3 m/s. Show that momentum is conserved in a frame of reference moving with a speed of 2 m/s in the direction of the second ball.
5. With what speed will a clock have to be moving in order to run at a rate that is one half the rate of a clock at rest?
6. How fast must a meter stick be moving if its length is observed to shrink to 0.5 m?
7. A clock on a moving spacecraft runs 1 s slower per day

relative to an identical clock on earth. What is the speed of the spacecraft? (*Hint:* For $v/c \ll 1$, note that $\gamma \approx 1 + v^2/2c^2$.)

8. A meter stick moving in a direction parallel to its length appears to be only 75 cm long to an observer. What is the speed of the meter stick relative to the observer?
9. A spacecraft moves at a speed of $0.9c$. If its length is L_0 when measured from in the spacecraft, what is its length measured by a ground observer?
10. An atomic clock is placed in a jet airplane. The clock measures a time interval of 3600 s when the jet moves with a speed of 400 m/s. What corresponding time interval does an identical clock held by an observer on the ground measure? (See hint in Problem 7.)
11. The average lifetime of a pi meson in its own frame of reference is 2.6×10^{-8} s. (This is the proper lifetime.) If the meson moves with a speed of $0.95c$, what is (a) its mean lifetime as measured by an observer on earth and (b) the average distance it travels before decaying, as measured by an observer on earth?
12. Two spaceships approach each other, each moving with the *same* speed as measured by an observer on the earth. If their *relative* speed is $0.7c$, what is the speed of each spaceship?
13. Show that the Lorentz transformation equations given by Equations 1.12 and 1.13 satisfy Equation 1.11.
14. An electron moves to the right with a speed of $0.90c$ relative to the laboratory frame. A proton moves to the right with a speed of $0.70c$ relative to the electron. Find the speed of the proton relative to the laboratory frame.
15. An observer on earth observes two spacecraft moving in the *same* direction toward the earth. Spacecraft A appears to have a speed of $0.5c$, and spacecraft B appears to have a speed of $0.8c$. What is the speed of spacecraft A measured by an observer in spacecraft B?
16. Calculate the momentum of a proton moving with a speed of (a) $0.01c$, (b) $0.5c$, (c) $0.9c$.
17. An electron has a momentum that is 90% larger than its classical momentum. (a) Find the speed of the electron. (b) How would your result change if the particle were a proton?
18. Show that the energy-momentum relationship given by $E^2 = p^2c^2 + (mc^2)^2$ follows from the expressions $E = \gamma mc^2$ and $p = \gamma mu$.
19. A proton moves with the speed of $0.95c$. Calculate its (a) rest energy, (b) total energy, and (c) kinetic energy.
20. An electron has a kinetic energy five times greater than its rest energy. Find (a) its total energy and (b) its speed.
21. Find the speed of a particle whose total energy is 50% greater than its rest energy.
22. A proton in a high-energy accelerator is given a kinetic energy of 50 GeV. Determine the (a) momentum and (b) speed of the proton.
23. Determine the energy required to accelerate an electron from (a) $0.50c$ to $0.75c$ and (b) $0.90c$ to $0.99c$.
24. A radium isotope decays by emitting an α particle to a radon isotope $^{222}_{86}Rn$ according to the decay scheme $^{226}_{88}Ra \rightarrow {}^{222}_{86}Rn + {}^{4}_{2}He$. The masses of the atoms are 226.0254 (Ra), 222.0175 (Rn), and 4.0026 (He). How much energy is released as the result of this decay?
25. Consider the decay $^{55}_{24}Cr \rightarrow {}^{55}_{25}Mn + e^-$, where e^- is an electron. The ^{55}Cr nucleus has a mass of 54.9279 u, and the ^{55}Mn nucleus has a mass of 54.9244 u. (a) Calculate the mass difference in MeV. (b) What is the maximum kinetic energy of the emitted electron?
26. Calculate the binding energy in MeV per nucleon (proton or neutron) in the isotope $^{12}_{6}C$. Note that the mass of this isotope is exactly 12 u, and the masses of the proton and neutron are 1.007276 u and 1.008665 u, respectively.
27. An astronaut at rest on earth has a heartbeat rate of 70 beats/min. What will this rate be when the astronaut is traveling in a spaceship at $0.9c$ as measured (a) by an observer also in the ship and (b) by an observer at rest on the earth?
28. The muon is an unstable particle that spontaneously decays into an electron and two neutrinos. If the number of muons at $t = 0$ is N_0, the number at time t is given by $N = N_0 e^{-t/\tau}$ where τ is the mean lifetime, equal to 2.2 μs. Suppose the muons move at a speed of $0.95c$ and there are 5×10^4 muons at $t = 0$. (a) What is the observed lifetime of the muons? (b) How many muons remain after traveling a distance of 3 km?
29. An electron has a velocity of $0.75c$. Find the velocity of a proton which has (a) the same kinetic energy as the electron and (b) the same momentum as the electron.
30. An atomic clock is placed in a jet airplane. The clock measures a time interval of 3600 s when the jet moves with a speed of 400 m/s. What corresponding time interval does an identical clock held by an observer on the ground measure? *Hint:* for $v/c \ll 1$,

$$\frac{1}{\gamma} \approx 1 - \frac{v^2}{c^2}$$

31. A rod of length L_0 moves with a speed v along the horizontal direction. The rod makes an angle of θ_0 with respect to the x' axis. (a) Show that the length of the rod as measured by a stationary observer is given by $L = L_0[1 - (v^2/c^2)\cos^2\theta_0]^{1/2}$. (b) Show that the angle that the rod makes with the x axis is given by the expression $\tan\theta = \gamma\tan\theta_0$. These results show that the rod is both contracted and rotated. (Take the lower end of the rod to be at the origin of the primed coordinate system.)
32. A cube of density ρ and edge L moves in a direction parallel to one of its edges with a speed v, comparable to c. What is (a) the volume of the cube and (b) the density of the cube as measured by a laboratory observer?

33. *Speed of light in a moving medium.* The motion of a medium such as water influences the speed of light. This effect was first observed by Fizeau in 1851. Consider a light beam passing through a horizontal column of water moving with a speed v. (a) Show that if the beam travels in the same direction as the flow of water, the speed of light measured in the laboratory frame is given by

$$u = \frac{c}{n}\left(\frac{1 + nv/c}{1 + v/nc}\right)$$

where n is the index of refraction of the water. (*Hint:* Use the velocity transformation relation, Equation 1.20, and note that the velocity of the water with respect to the *moving* frame is given by c/n.) (b) Show that for $v \ll c$, the expression above is, to a good approximation,

$$u \approx \frac{c}{n} + v - \frac{v}{n^2}$$

34. *Time dilation in an atom.* The atoms in a gas move in random directions with thermal velocities. Because of their motion with respect to a laboratory observer, the frequency at which they radiate is shifted by time dilation. (a) If an atom radiates at a frequency f_0 in its rest frame, show that the observed frequency in the laboratory frame is given by $f = (1 - v^2/c^2)^{1/2} f_0$. (b) If $v \ll c$, show that the fractional change in frequency is given by $\Delta f/f = -v^2/2c^2$. (c) Evaluate the fractional change in frequency for a hydrogen atom at a temperature of 300 K. (*Hint:* From the kinetic theory of gases, $M\overline{v}^2/2 = 3kT/2$.)

35. *Doppler effect for light.* If a light source moves with a speed v relative to an observer, there is a shift in the observed frequency analogous to the Doppler effect for sound waves. Show that the observed frequency f_0 is related to the true frequency f through the expression

$$f_0 = \sqrt{\frac{c \pm v_s}{c \mp v_s}}\, f$$

where the upper signs correspond to the source approaching the observer and the lower signs correspond to the source receding from the observer. [*Hint:* In the moving frame S′, the period is the proper time interval and is given by $T = 1/f$. Furthermore, the wavelength measured by the observer is given by $\lambda_0 = (c - v_s)T_0$, where T_0 is the period measured in s.]

36. *The red shift.* A light source recedes from an observer with a speed v_s, which is small compared with c. (a) Show that the fractional shift in the measured wavelength is given by the approximate expression

$$\frac{\Delta\lambda}{\lambda} \approx \frac{v_s}{c}$$

This result is known as the *red shift,* since the visible light is shifted toward the red. (Note that the proper period and measured period are approximately equal in this case.) (b) Spectroscopic measurements of light at $\lambda = 397$ nm coming from a galaxy in Ursa Major reveal a red shift of 20 nm. What is the recessional speed of the galaxy?

37. The free neutron is known to decay into a proton, an electron, and an antineutrino $\overline{\nu}$ (of zero rest mass) according to

$$\text{n} \longrightarrow \text{p} + \text{e}^- + \overline{\nu}$$

This is called *beta decay,* and will be discussed further in Chapter 15. The decay products are measured to have a total kinetic energy of 0.781 MeV. Show that this observation is consistent with the excess energy predicted by the Einstein mass-energy relationship.

38. Protons in an accelerator at the Fermi National Laboratory, near Chicago, IL, are accelerated to an energy of 400 times their rest energy. (a) What is the speed of these protons? (b) What is their kinetic energy in MeV?

39. Consider a stick that is at rest in the S′ reference frame as in Figure 1.10. Assume that the ends of the stick have coordinates x_1' and x_2' as measured in S′, so that its proper length is $L' = x_2' - x_2'$. Using the Lorentz transformation equations, show that an observer in the S frame measures the stick's length to be $L = L'/\gamma = (1 - v^2/c^2)^{1/2} L'$. (That is, show that the stick is contracted along its direction of motion.)

40. Show that Newton's second law in relativistic form can be expressed as

$$F = m\left(1 - \frac{v^2}{c^2}\right)^{-3/2} \frac{dv}{dt}$$

where m is the rest mass of an object and v is its speed.

41. A charged particle moves along a straight line in a uniform electric field E with a speed v. If the motion and the electric field are both in the x direction, (a) show that the acceleration of the charge q is given by

$$a = \frac{dv}{dt} = \frac{qE}{m}\left(1 - \frac{v^2}{c^2}\right)^{3/2}$$

(b) Discuss the significance of the dependence of the acceleration on the speed. (c) If the particle starts from rest at $x = 0$ at $t = 0$, how would you proceed to find the speed of the particle and its position after a time t has elapsed?

42. Recall that the magnetic force on a charge q moving with velocity $\boldsymbol{v}$ in a magnetic field $\boldsymbol{B}$ is equal to $q\boldsymbol{v} \times \boldsymbol{B}$. If a charged particle moves in a circular orbit with a fixed speed v in the presence of a constant magnetic field, use Newton's second law to show that the frequency of its orbital motion is given by

$$f = \frac{qB}{2\pi m}\left(1 - \frac{v^2}{c^2}\right)^{1/2}$$

Essay

THE RENAISSANCE OF GENERAL RELATIVITY

Clifford M. Will
McDonnell Center for the Space Sciences, and Washington University

During the two decades from 1960 to 1980, the subject of general relativity experienced a rebirth. Despite its enormous influence on scientific thought in its early years, by the late 1950s general relativity had become a sterile, formalistic subject, cut off from the mainstream of physics. It was thought to have very little observational contact, and was believed to be an extremely difficult subject to learn and comprehend.

Yet by 1970, general relativity had become one of the most active and exciting branches of physics. It took on new roles both as a theoretical tool of the astrophysicist and as a playground for the elementary-particle physicist. New experimental tests verified its predictions in unheard-of ways, and to remarkable levels of precision. Fields of study were created, such as "black-hole physics" and "gravitational-wave astronomy," that brought together the efforts of theorists and experimentalists. One of the most remarkable and important aspects of this renaissance of relativity was the degree to which experiment and observation motivated and complemented theoretical advances.

This was not always the case. In deriving general relativity during the final months of 1915, Einstein himself was not particularly motivated by a desire to account for observational results. Instead, he was driven by purely theoretical ideas of elegance and simplicity. His goal was to produce a theory of gravitation that incorporated in a natural way both the special theory of relativity that dealt with physics in inertial frames, and the principle of equivalence, the proposal that physics in a frame falling freely in a gravitational field was in some sense equivalent to physics in an inertial frame.

Once the theory was formulated, however, he did try to confront it with experiment by proposing three tests. One of these tests was an immediate success—the explanation of the anomalous advance in the perihelion of Mercury of 43 arcseconds per century, a problem that had bedeviled celestial mechanicians of the latter part of the 19th century. The next test, the deflection of light by the Sun, was such a success that it produced what today would be called a "media event." The measurements of the deflection, amounting to 1.75 arcseconds for a ray that grazes the Sun, by two teams of British astronomers in 1919, made Einstein an instant international celebrity. However, these measurements were not all that accurate, and subsequent measurements weren't much better. The third test, actually proposed by Einstein in 1907, was the gravitational redshift of light, but it was a test that remained unfulfilled until 1960, by which time it was no longer viewed as a true test of general relativity.

Cosmology was the other area where general relativity was believed to have observational relevance. Although the general relativistic picture of the expansion of the universe from a "big bang" was compatible with observations, there were problems. As late as the middle 1950s the measured values of the universal expansion rate implied that the universe was younger than the Earth! Even though this "age" problem had been resolved by 1960, cosmological observations were still in their infancy, and could not distinguish between various alternative models.

The turning point for general relativity came in the early 1960s, when discoveries of unusual astronomical objects such as quasars demonstrated that the theory would have important applications in astrophysical situations. Theorists found new ways to understand the theory and its observable consequences. Finally, the technological revolution of the last quarter century, together with the development of the interplanetary space program, provided new high-precision tools to carry out experimental tests of general relativity.

After 1960, the pace of research in general relativity and in an emerging field called "relativistic astrophysics" began to accelerate. New advances, both theoretical and observational, came at an ever increasing rate. They included the discovery of the cosmic background radiation; the analysis of the synthesis of helium from hydrogen in the big bang; observations of pulars and of black-hole candidates; the development of the theory of relativistic stars and black holes; the theoretical study of gravitational radiation and the beginning of an experimental program to detect it; improved versions of old tests of general relativity, and brand new tests, discovered after 1960; the discovery of the binary pulsar, which provided evidence for gravity waves; the analysis of quantum effects outside black holes and of black hole evaporation; the discovery of a gravitational lens; and the beginnings of a unification of gravitation theory with the other interactions and with quantum mechanics.

During the two decades following 1960, general relativity rejoined the world of physics and astronomy. Research in relativity took on an increasingly interdisciplinary flavor, spanning such subjects as celestial mechanics, pure mathematics, experimental physics, quantum mechanics, observational astronomy, particle physics, and theoretical astrophysics.

EINSTEIN'S EQUIVALENCE PRINCIPLE

The foundations of general relativity are actually quite old, dating back to the equivalence principle of Galileo and Newton. In Newton's view, the principle of equivalence stated that all objects accelerate at the same rate in a gravitational field regardless of their mass or composition. This equality has been verified abundantly over the years, including classic experiments by the Hungarian physicist Baron Lorand von Eötvös around 100 years ago and recent experiments at Princeton and Moscow State Universities. The accuracy of these tests is better than one part in 10^{11}. Einstein's insight was the recognition that, to an observer inside a freely falling laboratory, not only should objects float as if gravity were absent as a consequence of this equality, but also *all* laws of nongravitational physics, such as electromagnetism and quantum mechanics, should behave as if gravity were truly absent.

This idea is now known as the Einstein equivalence principle, and it was a key step, because it implied the converse: that in a reference frame where gravity is felt, such as in a laboratory on the Earth's surface, the effects of gravitation on physical laws can be obtained simply by mathematically transforming the laws from a freely falling frame to the laboratory frame. According to the branch of mathematics known as differential geometry, this is the same as saying that space-time is curved; in other words, that the effects of gravity are *indistinguishable* from the effects of being in curved space-time.

GRAVITATIONAL REDSHIFT

An immediate consequence of Einstein's principle of equivalence is the gravitational redshift effect. It is not a consequence of general relativity itself, although Einstein believed that it was. One version of a gravitational redshift experiment measures the frequency shift between two identical clocks (meaning any device that produces a signal at a well-defined, steady frequency), placed at rest at different heights in a gravitational field. To derive the frequency shift from the equivalence principle, one imagines a freely falling frame that is momentarily at rest with respect to one clock at the moment the clock emits its signal. Because special relativity is valid in that frame, the frequency of the signal is unaffected by gravity as it travels from one clock to the other. However, by the time the signal reaches the second clock, the frame has picked up a downward velocity because it is in free fall in the gravitational field, and therefore, as seen by the falling frame, the second clock is moving upward. Thus the frequency seen by the second clock will appear to be shifted from the standard value by the Doppler effect. For small differences in height h between clocks, the shift in the frequency Δf is given by

$$\frac{\Delta f}{f} = \frac{v_{\text{fall}}}{c} = \frac{gh}{c^2}$$

where g is the local gravitational acceleration and c is the speed of light. If the receiver is at a lower height than the emitter, the received signal is shifted to higher frequencies ("blueshift"), while if the receiver is higher, the signal is shifted to lower frequencies ("redshift"). The generic name for the effect is "gravitational redshift."

The first and most famous high-precision redshift measurement was the Pound-Rebka experiment of 1960, which measured the frequency shift of gamma-ray photons from the decay of iron-57 (^{57}Fe) as they ascended or descended the Jefferson Physical Laboratory tower at Harvard University.

The most precise gravitational redshift experiment performed to date was a rocket experiment carried out in June 1976. A "hydrogen maser" atomic clock was flown on a Scout D rocket to an altitude of 10 000 km, and its frequency was compared to that of a similar clock on the ground by radio signals. After the effects of the rocket's motion were taken into account, the observations confirmed the gravitational redshift to 0.02 percent.

Because of these kinds of experiments, physicists are now convinced that spacetime *is* curved, and that the correct theory of gravity must be based on curved spacetime. That does not automatically imply general relativity, however, since it is not the only such theory. To test the specific predictions of general relativity for the amount and nature of space-time curvature, other experiments are needed.

DEFLECTION OF LIGHT

The first of these is a test that made Einstein's name a household word: the deflection of light. According to general relativity, a light ray that passes the Sun at a distance d is deflected by an angle $\Delta\theta = 1\rlap{.}''75/d$, where d is measured in units of the solar radius, and the notation $''$ denotes seconds of arc. Half of the deflection can be calculated by appealing only to the principle of equivalence. The idea is to determine what observers in a sequence of free-falling frames all along the trajectory of the photon would find for the deflection from one frame to the next. It turns out that the net deflection over all the frames is $0\rlap{.}''875/d$. The same result can be obtained by calculating the deflection of a particle by a massive body using ordinary Newtonian theory, and then taking the limit in which the particle's velocity becomes that of light. But the result of both versions is the deflection of light only relative to local straight lines, as defined for example by rigid rods laid end-to-end. However, because of the curvature of space, a local straight line defined by rods passing close to the Sun is itself bent relative to straight lines far from the Sun by an amount that yields the remaining half of the deflection.

The prediction of the bending of light by the Sun was one of the great successes of general relativity. Confirmation by the British astronomers Eddington and Crommelin of the bending of optical starlight observed during a total solar eclipse in the first months following World War I helped make Einstein a celebrity. However, those measurements had only 30 percent accuracy, and succeeding eclipse experiments weren't much better; the results were scattered between one half and twice the Einstein value, and the accuracies were low.

However, the development of long-baseline radio interferometry produced a method for greatly improved determinations of the deflection of light. Radio interferometry is a technique of combining widely separated radio telescopes in such a way that the direction of a source of radio waves can be measured by determining the difference in phase of the signal received at the different telescopes. Modern interferometry has the ability to measure angular separations and changes in angles as small as 10^{-4} arcseconds. Coupled with this technological advance is a series of heavenly coincidences: each year groups of strong quasars pass near the Sun (as seen from the Earth). The idea is to measure the differential deflection of radio waves from one quasar relative to those from another as they pass near the Sun. A number of measurements of this kind were made almost annually over the period 1969–1975, yielding a confirmation of the predicted deflection to 1.5 percent.

The 1979 discovery of the "double" quasar Q0957 + 561 converted the deflection of light from a test of relativity to an important effect in astronomy and cosmology. The "double" quasar was found to be a multiple image of a single quasar caused by the gravitational lensing effect of a galaxy or a cluster of galaxies along the line of sight between us and the quasar. Several other such lenses have been found.

Closely related to light deflection is the "Shapiro time delay," a retardation of light signals that pass near the Sun. For instance, for a signal that grazes the Sun on a round trip from Earth to Mars at superior conjunction (when Mars is on the far side of the Sun), the round trip travel time is increased over what Newtonian theory would give by about 250 μs. The effect decreases with increasing distance of the signal from the Sun.

In the two decades following radio-astronomer Irwin Shapiro's 1964 discovery of this effect, several high-precision measurements have been made using the technique of radar-ranging to planets and spacecraft. Three types of targets were employed: planets such as Mercury and Venus, free-flying spacecraft such as Mariners 6 and 7, and combinations of planets and spacecraft, known as "anchored spacecraft," such as the Mariner 9 Mars orbiter and the 1976 Viking Mars landers and orbiters. The Viking experiments produced dramatic results, agreeing with the general relativistic prediction to one part in a thousand. This corresponded to a measurement accuracy in the Earth-Mars distance of 30 meters.

PERIHELION SHIFT

The explanation of the anomalous perihelion shift of Mercury's orbit was another of the triumphs of general relativity. This had been an unsolved problem in celestial mechanics for over half a century, since the announcement by Le Verrier in 1859 that, after the perturbing effects of the planets on Mercury's orbit had been accounted for, there remained in the data an unexplained advance in the perihelion of Mercury. The modern value for this discrepancy is 42.98 arcseconds per century. Many ad hoc proposals were made in an attempt to account for this excess, including the existence of a new planet Vulcan near the Sun, or a deviation from the inverse square law of gravitation. But each was doomed to failure. General relativity accounted for the anomalous shift in a natural way. Radar measurements of the orbit of Mercury since 1966 have led to improved accuracy, so that the relativistic advance is known to about 0.5 percent.

Other measurements carried out since 1960 have given further support to general relativity. Observations of the motion of the Moon using laser ranging to a collection of specially designed reflectors, deposited on the lunar surface during the Apollo and Luna missions, have shown that the Moon and the Earth accelerate toward the Sun equally to a part in 10^{11}, an important confirmation of the equivalence principle applied to planetary bodies. Other measurements of planetary and lunar orbits have shown that the gravitational constant is a true constant of nature; it does not vary with time as the universe ages. Ambitious experimenters are currently designing an experiment using supercooled quartz gyroscopes to be placed in Earth orbit, to try to detect an important general relativistic effect called the "dragging of inertial frames," caused by the rotation of the Earth.

General relativity has passed every experimental test to which it has been put, and many alternative theories have fallen by the wayside. Most physicists now take the theory for granted, and look to see how it can be used as a practical tool in physics and astronomy.

GRAVITATIONAL RADIATION

One of these new tools is gravitational radiation, a subject that is almost as old as general relativity itself. By 1916, Einstein had succeeded in showing that the field equations of general relativity admitted wavelike solutions analogous to those of electromagnetic theory. For example, a dumbell rotating about an axis passing at right angles through its handle will emit gravitational waves that travel at the speed of light. They also carry energy away from the rotating dumbell, just as electromagnetic waves carry energy away from a light source. That was about all that was heard on the subject for over 40 years, primarily because the effects associated with gravitational waves were extremely tiny, unlikely (it was thought) ever to be of experimental or observational interest.

But in 1968, the idea of gravitational radiation was resurrected by the stunning announcement by Joseph Weber that he had detected gravitational radiation of extraterrestrial origin by using massive aluminum bars as detectors. A passing gravitational wave acts as an oscillating gravitational force field that alternately compresses and extends the bar in the lengthwise direction. However, subsequent observations by other researchers using bars with sensitivity that was claimed to be better than Weber's failed to confirm Weber's results. His results are now generally regarded as a false alarm, although there is still no good explanation for the "events" that he recorded in his bars.

Nevertheless, Weber's experiments did initiate the program of gravitational-wave detection, and inspired other groups to build better detectors. Currently a dozen laboratories around the world are engaged in building and improving upon the basic "Weber bar" detector, some using bigger bars, some using smaller bars, some working at room temperature, and some working near absolute zero. The goal in all cases is to reduce noise from thermal, electrical, and environmental sources as much as possible to have a hope of detecting the very weak oscillations produced by a gravitational wave. For a bar of one meter length, the challenge is to detect a variation in length smaller than 10^{-20} meters, or 10^{-5} of the radius of a proton.

Another important type of detector is the laser interferometer, currently under development at several laboratories. This device operates on the same principle as the interferometer used by Michelson and Morley, but uses a laser as the light source. A passing gravitational wave will change the length of one arm of the apparatus relative to the other, and cause the interference pattern to vary. One ambitious proposal is to build two independent interferometers, each with arm lengths of 4 km, and separated by 1000 km.

It is hoped that these detectors will eventually be sensitive enough to detect gravitational waves from many sources, both in our galaxy and in distant galaxies, such as collapsing stars, double star systems, colliding black holes, and possibly even gravitational waves left over from the big bang. When this happens, a new field of "gravitational-wave astronomy" will be born.

Although gravitational radiation has not been detected directly, we know that it exists, through a remarkable system known as the binary pulsar. Discovered in 1974 by radio astronomers Russell Hulse and Joseph Taylor, it consists of a pulsar (which is a rapidly spinning neutron star) and a companion star in orbit around each other. Although the companion has not been seen directly, it is also believed to be a neutron star. The two bodies are circling so closely—their orbit is about the size of the Sun, and the period of the orbit is eight hours—that the effects of general relativity are very large. For example, the periastron shift (the analogue of the perihelion shift for Mercury) is more than 4 degrees per *year*.

Furthermore, the pulsar acts as an extremely stable clock, its pulse period of approximately 59 milliseconds drifting by only a quarter of a nanosecond per year. By measuring the arrival times of radio pulses at Earth, observers were able to determine the motion of the pulsar about its companion with amazing accuracy. For example, the accurate value for the orbital period is 27906.98163 seconds, and the orbital eccentricity is 0.617127.

Like a rotating dumbell, an orbiting binary system should emit gravitational radiation, and in the process lose some of its orbital energy. This energy loss will cause the pulsar and its companion to spiral in toward each other, and the orbital period to shorten. According to general relativity, the predicted decrease in the orbital period is 75 microseconds per year. The observed decrease rate is in agreement with the prediction to about 4 percent. This confirms the existence of gravitational radiation and the general relativistic equations that describe it.

BLACK HOLES

One of the most important and exciting aspects of the relativity renaissance is the study of and search for black holes. The subject began in the early 1960s as astrophysicists looked for explanations of the quasars, warmed up in the early 1970s following the possible detection of a black hole in an X-ray source, became white hot in the middle 1970s when black holes were found theoretically to evaporate, and continues today as one of the most active branches of general relativity.

However, the first glimmerings of the black hole idea date back to the 18th century, in the writings of a British amateur astronomer, the Reverend John Michell. Reasoning on the basis of the corpuscular theory that light would be attracted by gravity in the same way that ordinary matter is attracted, he noted that light emitted from the surface of a body such as the Earth or the Sun would be reduced in velocity by the time it reached great distances (Michell of course did not know special relativity). How large would a body of the same density as the Sun have to be in order that light emitted from it would be stopped and pulled back before reaching infinity? The answer he obtained was 500 times the diameter of the Sun. Light could therefore never escape from such a body. In today's language, such an object would be a supermassive black hole of about 100 million solar masses.

Although the general relativistic solution for a nonrotating black hole was discovered by Karl Schwarzschild in 1916, and a calculation of gravitational collapse to a black hole state was performed by J. Robert Oppenheimer and Hartland Snyder in 1939, black hole physics didn't really begin until the middle 1960s, when astronomers confronted the problem of the energy output of the quasars, and the mathematician Roy Kerr discovered the rotating black-hole solution.

A black hole is formed when a star has exhausted the thermonuclear fuel necessary to produce the heat and pressure that support it against gravity. The star begins to collapse, and if it is massive enough, it continues to collapse until the radius of the star approaches a value called the gravitational radius or Schwarzschild radius. In the nonrotating spherical case, this radius has a value given by $2GM/c^2$, where M is the mass of the star. For a body of one solar mass, the gravitational radius is about 3 km; for a body of the mass of the Earth, it is about 9 mm. An observer sitting on the surface of the star sees the collapse continue to smaller and smaller radii, until both star and observer reach the origin $r = 0$, with consequences too horrible to describe in detail. On the other hand, an observer at great distances observes the collapse to slow down as the radius approaches the gravitional radius, a result of the gravitational redshift of the light signals sent outward. However, in this case, the redshifting or slowing down becomes so extreme that the star appears almost to stop just outside the gravitational radius. The distant observer never sees any signals emitted by the falling observer once the latter is inside the gravitational radius. Any signal emitted inside can never escape the sphere bounded by the gravitational radius, called the "event horizon."

Since the middle 1960s the laws of "black hole physics" have been established and codified. For example, it was discovered that the Kerr (rotating) and Schwarzschild (nonrotating) solutions are unique; there are no other black-hole solutions in general relativity. Inside every black hole resides a "singularity," a pathological region of spacetime where the gravitational forces become infinite, where time comes to an end for any observer unfortunate enough to hit the singularity, indeed where all the laws of physics break down. Fortunately the event horizon of the black hole prevents any bizarre phenomena that such singularities might cause from reaching the outside world (this notion has been dubbed "cosmic censorship"). In 1974, Stephen Hawking discovered that the laws of quantum mechanics applied to the physics outside a black hole required it to evaporate by the creation of particles with a thermal energy spectrum, and to have an associated temperature and entropy. The temperature of a Schwarzschild black hole is $T = hc^3/8\pi kGM$, where h is Planck's constant and k is Boltzmann's constant. This discovery demonstrated a remarkable connection between gravity, thermodynamics, and quantum mechanics, which helped renew the theoretical quest for a grand synthesis of all the fundamental interactions. For black holes of astronomical masses, however, the evaporation is completely negligible, since for a solar-mass black hole, $T \approx 10^{-6}$ K.

Although a great deal is known about black holes in theory, rather less is known about them observationally. There are several instances in which the evidence for the existence of black holes is impressive, but in all cases it is indirect. For instance, in the X-ray source Cygnus X1, the source of the X-rays is believed to be a collapsed object with a mass larger than about 6 solar masses in orbit around a giant star. The object cannot be a neutron star, because general relativity predicts that neutron stars must be lighter than about 3 solar masses. Thus the object must be a black hole. The X-rays are emitted by matter pulled from the surface of the companion star and sent into a spiralling orbit around the black hole. Similarly, there is evidence in the centers of certain galaxies, such as M87 and possibly even our own, of collapsed objects of between 10^2 and 10^8 solar masses. A black hole model is consistent with the observations, although it is not necessarily required. Accretion of matter onto supermassive rotating black holes may produce the jets of outflowing matter that are observed in many quasars and active galactic nuclei. These and other astrophysical processes that might aid in the detection of black holes are being studied by relativists and astrophysicists.

COSMOLOGY

The other area in which there has been a renaissance for general relativity is cosmology. Although Einstein in 1917 first used general relativity to calculate a model for the universe as a whole, the subject was not considered a serious branch of physics until the 1960s, when astronomical observations lent firm credence to the idea that the universe was expanding from a "big bang." For instance, observations of the rates at which galaxies are receding from us, coupled with improved determinations of their distances, implied that the universe was at least 10 billion years old, a value that was consistent with other observations such as the age of the Earth and of old star clusters.

In 1965 came the discovery of the cosmic background radiation by Arno Penzias and Robert Wilson. This radiation is the remains of the hot electromagnetic blackbody radiation that dominated the universe in its earlier phase, now cooled to 3 kelvins by the subsequent expansion of the universe. Next came calculations of the amount of helium that would be synthesized from hydrogen by thermonuclear fusion in the very early universe, around 1000 seconds after the big bang. The amount, approximately 25 percent by weight, was in agreement with the abundances of helium observed in stars and in interstellar space. This was an important confirmation of the hot big bang picture, because the amount of helium believed to be produced by fusion in the interiors of stars is woefully inadequate to explain the observed abundances.

Today, the general relativistic hot big-bang model of the universe has broad acceptance, and cosmologists now focus their attention on more detailed issues, such as how galaxies and other large-scale structures formed out of the hot primordial soup, and on what the universe might have been like earlier than 1000 seconds, all the way back to 10^{-36} s (and some brave cosmologists are going back even further) when the laws of elementary-particle physics may have played a major role in the evolution of the universe.

ACCEPTANCE OF GENERAL RELATIVITY

One of the outgrowths of the renaissance of general relativity that has occurred since 1960 has been a change in attitude about the importance and use of the theory. Its importance as a fundamental theory of the nature of spacetime and gravitation has not been diminished in the least; if anything, it has been enhanced by the flowering of research in the subject that has taken place. Its importance as a foundation for other theories of physics has been strengthened by current searches for unified quantum theories of nature that incorporate gravity along with the other interactions.

But the real change in attitude about general relativity has been in its use as a tool in the real world. In astrophysical situations, general relativity plays a central role in the study of neutron stars, black holes, gravitational lenses, relativistic binary star systems, and the universe as a whole. Gravitational radiation may one day provide a completely new observational tool for exploring and examining the cosmos.

Relativity even plays a role in everyday life. For example, the gravitational redshift effect on clocks *must* be taken into account in satellite-based navigation systems, such as the US Air force's Global Positioning System, in order to achieve the required positional accuracy of a few meters or time transfer accuracy of a few nanoseconds.

To general relativists, always eager to find practical consequences of their subject, these have been very welcome developments!

Suggestions for Further Reading

Davies, P. C. W. *The Search for Gravity Waves*, Cambridge, England, Cambridge University Press, 1980.

Greenstein, G. *Frozen Star: Of Pulsars, Black Holes and the Fate of Stars*, New York, Freundlich Books, 1984.

Weinberg, S. *The First Three Minutes: A Modern View of the Origin of the Universe*, New York, Basic Books, 1977.

Will, C. M. *Was Einstein Right? Putting General Relativity to the Test*, New York, Basic Books, 1986.

Essay Questions

1. Two stars are a certain distance from each other as seen in the night sky. Half a year later, the stars are now overhead during the day, and the Sun is now located midway between the two stars. Because of the deflection of light by the Sun, do the stars now appear closer together or farther apart?
2. When gravitational waves are emitted by the binary pulsar, the double-star system loses energy. As a consequence, its orbital velocity increases, and thus its kinetic energy increases. How is this possible?
3. The temperature of a black hole is inversely proportional to its mass, so that when it gains energy (which is equivalent to mass), its temperature goes down, and when it loses energy, its temperature goes up. What do you conclude about the sign of its heat capacity? Can a black hole ever be in thermal equilibrium with an infinite reservoir at a fixed temperature?

Essay Problems

1. What is the gravitational frequency shift between two atomic clocks separated by 1000 km, both at sea level?
2. A pair of identical twin scientists set out to test the gravitational redshift effect in the following way. One climbs to the top of Mount Baldy and lives there for a year, while the other remains behind in the laboratory at the base of the mountain. When they are reunited after one year, one of the twins is slightly older than the other. Which one is older and why? If Mount Baldy is 5000 m high, what is the difference in age between the twins after one year?
3. Calculate the deflection of light for light rays that pass the Sun at distances of 2 solar radii, 10 solar radii, and 100 solar radii.
4. Gravitational waves travel with the same speed as light, 3×10^{10} cm/s. Calculate the wavelength of gravitational waves emitted by the following sources: (a) a star collapsing to form a black hole, emitting waves with a frequency around 1000 Hz, and (b) the binary pulsar, emitting waves with a period of four hours (one half the orbital period).
5. What is the fractional decrease in the orbital period of the binary pulsar per year? Estimate how long the binary system will continue before gravitational radiation will reduce the orbital period to zero (in other words will bring the stars so close together that they will merge into one object).
6. Calculate the mass of a black hole (in solar mass units) whose Schwarzschild radius is equal to the radius of the Earth's orbit around the Sun (1.5×10^8 km). One solar mass is about 2×10^{33} g.
7. Show that a billion-ton black hole has a Schwarzschild radius comparable to the radius of a proton (1 ton $\approx 10^6$ g, proton radius $\approx 10^{-13}$ cm). What kind of object on Earth weighs a billion tons?
8. Determine the mass and radius of a black hole whose apparent density is equal to that of water, 1 g/cm^3. Assume that the volume of a black hole is given by $4\pi R^3/3$, and express the mass in solar mass units.
9. Calculate the mass of a black hole whose temperature is room temperature.

2

The Quantum Theory of Light

Physicists at the beginning of the 20th century, following the lead of Newton and Maxwell, might have rewritten the biblical story of creation as follows:

In the beginning He created the heavens and the earth—

$$F = G\frac{mm'}{r^2} = ma$$

and He said "Let there be light"—

$$\oint \mathbf{E}\cdot d\mathbf{A} = \frac{Q}{\epsilon_0} \qquad \oint \mathbf{E}\cdot d\mathbf{s} = -\frac{d\Phi_B}{dt}$$

$$\oint \mathbf{B}\cdot d\mathbf{A} = 0 \qquad \oint \mathbf{B}\cdot d\mathbf{s} = \mu_0 I + \epsilon_0\mu_0\frac{d\Phi_E}{dt}$$

Actually, in addition to the twin pillars of mechanics and electromagnetism erected by the giants Newton and Maxwell, there was a third sturdy support for physics in 1900—thermodynamics and statistical mechanics. Classical thermodynamics was the work of many men (Carnot, Mayer, Helmholtz, Clausius, Lord Kelvin). It is especially notable because it starts with two simple propositions and gives solid and conclusive results independent of detailed physical mechanisms. Statistical mechanics, founded by Maxwell, Clausius, Boltzmann,[1] and Gibbs, uses the methods of probability theory to calculate averages and fluctuations from the average for systems containing many particles or modes of vibration. It is interesting that quantum physics started not with a breakdown of Maxwell's or Newton's laws applied to the atom but with a problem of classical thermodynamics—that of calculating the intensity of radiation at a given wavelength from a heated cavity. We also note that the desperate solution to this radiation problem was found by a thoroughly classical thermodynamicist, Max Planck, in 1900. Indeed, it is significant that both Planck and Einstein returned again and again to the simple and general foundation of thermodynamics and statistical mechanics as the only certain bases for the new quantum theory. Although we shall not follow the original thermodynamic arguments completely, we shall see in this chapter how Planck arrived at the correct spectral distribution for cavity radiation by allowing only certain energies for the oscillators in the cavity walls. We shall also see how Einstein extended this quantization of energy to light itself, thereby brilliantly explaining the photoelectric effect. We shall conclude our brief history of the quantum of light with a discussion of the scattering of light by electrons

[1] Upon whose tombstone is written $S = k \log W$, a basic formula of statistical mechanics, attributed to Boltzmann.

(Compton effect), which showed conclusively that the light quantum carried both momentum and energy. Finally, we describe some interesting applications of the blackbody distribution and of the gravitational mass of the photon to astronomy and cosmology in Section 2.8 and Appendix B.

2.1 HERTZ'S EXPERIMENTS—LIGHT AS AN ELECTROMAGNETIC WAVE

It is ironic that the same experimentalist who so brilliantly confirmed that the "newfangled" waves of Maxwell actually existed and possessed the same properties as light also sowed the seeds of destruction of the electromagnetic wave theory of light. To understand this irony let us briefly review the theory of electromagnetism developed primarily by the great Scottish physicist James Clerk Maxwell between 1865 and 1873.

Maxwell was primarily interested in the effects of electric current oscillations in wires. According to his theory, an alternating flow of current would set up fluctuating electric and magnetic fields in the region surrounding the original disturbance. Moreover, these waves were predicted to have a frequency equal to the frequency of the current oscillations. *In addition, and most importantly, Maxwell's theory predicted that the radiated waves would behave in every way like light:* electromagnetic waves would be reflected by metal mirrors, would be refracted by dielectrics like glass, would exhibit polarization and interference, and would travel through vacuum with a speed of 3.0×10^8 m/s. Naturally this led to the unifying and simplifying postulate that light was also a type of Maxwell wave or electromagnetic disturbance, created by extremely high frequency electric oscillators in matter. At the end of the 19th century the precise nature of these charged submicroscopic oscillators was unknown (Planck called them resonators), but physicists assumed that somehow they were able to emit light waves whose frequency was equal to the oscillator's frequency of motion.

Even at this time, however, it was apparent that this model of light emission was incapable of direct experimental verification, since the highest electrical frequencies then attainable were about 10^9 Hz and visible light was known to possess a frequency a million times higher. But Hertz did the next best thing. (See Figure 2.1.) In a series of brilliant and exhaustive experiments, he showed that Maxwell's theory was correct and that an oscillating electric current does indeed send out electromagnetic waves that possess every characteristic of light except the same wavelength. Using a simple spark gap oscillator consisting of two short stubs terminated in small metal spheres separated by an air gap of about half an inch, he applied pulses of high voltage, which caused a spark to jump the gap and produce a high frequency electric oscillation of about 5×10^8 Hz. This oscillation or ringing occurs while the air gap remains conducting and charge surges back and forth between the spheres until electrical equilibrium is established. Using a simple loop antenna with a small spark gap as the receiver, Hertz very quickly succeeded in detecting the radiation from his spark gap oscillator, even at distances of several hundred meters. Moreover, he found the detected radiation to have a wavelength of about 60 cm, corresponding to the oscillator frequency of 5×10^8 Hz. (Recall that $c = \lambda f$, where λ is the wavelength and f is the frequency.)

Figure 2.1 Heinrich Hertz (1857–1984), an extraordinarily gifted German experimentalist.

In an exhaustive tour de force, Hertz next proceeded to show that these electromagnetic waves could be reflected, refracted, focused, polarized, and made to interfere—in short, he convinced almost every leading physicist of the period that Hertzian waves and light waves were equivalent. The classical

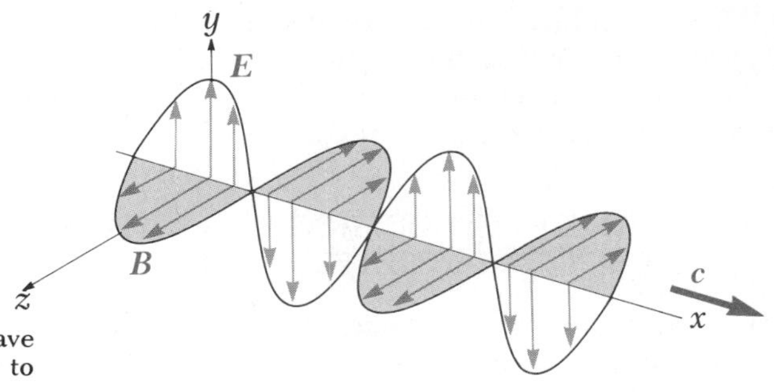

Figure 2.2 A light or radio wave far from the source according to Maxwell and Hertz.

model for light emission was an idea whose time had come. It spread like wildfire. The idea that light was an electromagnetic wave radiated by submicroscopic electric oscillators was applied in rapid succession to the transmission of light through solids, to reflection from metal surfaces, and perhaps most impressively to the newly discovered Zeeman effect. In 1896 Pieter Zeeman discovered that a strong magnetic field changes the frequency of the light emitted by a glowing gas. In an impressive victory for Maxwell, it was found that Maxwell's equations correctly predicted (in most cases) the change of vibration of the electric oscillators and hence the change in frequency of the light emitted. Maxwell, with Hertz behind the throne, reigned supreme, for he had united the independent kingdoms of electricity, magnetism, and light! (See Figure 2.2.)

A terse remark made by Hertz ends our discussion of his brilliant confirmation of the electromagnetic wave nature of light. In describing his oscillator he notes that "it is essential that the pole surfaces of the spark gap should be frequently repolished" to assure reliable operation of the spark.[2] Apparently this result was initially very mysterious to Hertz. In an effort to resolve this mystery, he later carefully investigated this side effect and concluded that it was the ultraviolet light from the initial spark acting on a clean metal surface that caused current to flow more freely between the poles of the spark gap. In the process of verifying the electromagnetic wave theory of light, Hertz had discovered the photoelectric effect, a phenomenon which would ultimately be the undoing of the classical wave theory of light!

2.2 BLACKBODY RADIATION

The tremendous success of Maxwell's theory of light emission immediately led to attempts to apply it to a long-standing puzzle about radiation—the so-called "blackbody" problem. Basically the problem is to predict the radiation intensity at a given wavelength emitted by a glowing solid at a specific temperature. Instead of launching immediately into Planck's solution of this problem, let us develop a feeling for its importance to classical physics by a quick review of its history.

Thomas Wedgwood, Darwin's relative and a renowned maker of china, seems to have been the first to note the universal character of all heated objects. In 1792 he observed that all the objects in his ovens, regardless of

[2] H. Hertz, *Ann. Physik* (Leipzig), **33:** 983, 1887.

their chemical nature, size, or shape, became red at the same temperature. This crude observation was sharpened considerably by the advancing state of spectroscopy, so that by the mid-1800s it was known that glowing solids emit continuous spectra rather than the bands or lines emitted by heated gases. (See Figure 2.3.) In 1859 Gustav Kirchhoff proved a theorem as important as his circuit loop theorem when he showed by thermodynamics that any body in thermal equilibrium with radiation[3] has a ratio of emissivity to absorption power that depends only on f, the frequency of light emitted, and T, the absolute temperature of the body. Thus we have

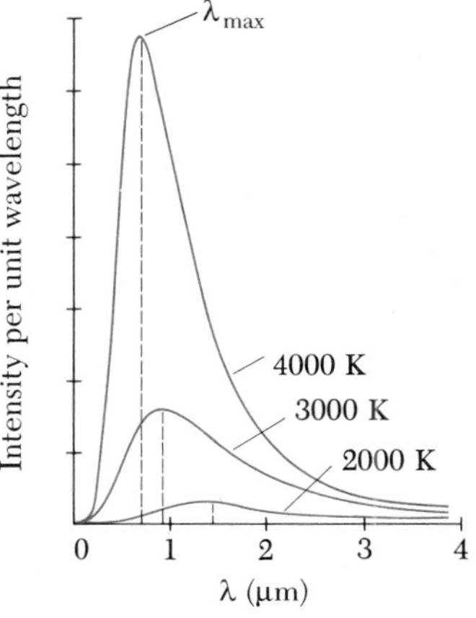

Figure 2.3 Emission from a glowing solid. Note that the amount of radiation emitted (the area under the curve) increases with increasing temperature.

$$\frac{E_f}{A_f} = J(f, T) \tag{2.1}$$

where E_f is the emissivity or intensity emitted by the body per unit frequency, and A_f is the absorption power or fraction of the incident intensity absorbed by the body at a particular frequency. A *black body* is defined as an object with $A_f = 1$ for all frequencies. Equation 2.1 shows that the emissivity of a black body depends *only* on the temperature and frequency and not at all on the physical and chemical nature of the black body. Thus for a black body Kirchhoff's theorem becomes

$$E_f = J(f, T) \tag{2.2}$$

The term black body arises from the fact that such an object absorbs all the electromagnetic radiation falling on it and consequently appears black. Since absorption and emission are connected by Kirchhoff's theorem, we see that a black body is also an ideal radiator. In practice a small opening in any heated cavity, such as a port in an oven, behaves like a black body since such an opening traps radiation (Fig. 2.4.) Note also that if the direction of the radiation is reversed in Figure 2.4, the light emitted by a small opening is in thermal equilibrium with the walls, since it has been absorbed and reemitted many times.

[3]An example of a body in equilibrium with radiation would be closed oven walls at a fixed temperature and the radiation within the oven cavity. To say that radiation is in thermal equilibrium with the oven walls means that the radiation has exchanged energy with the walls many times and is homogeneous, isotropic, and unpolarized. In fact, thermal equilibrium of radiation within a cavity can be considered to be quite similar to the thermal equilibrium of a fluid within a constant temperature container—both will cause a thermometer in the center of the cavity to achieve a final stationary temperature equal to that of the container.

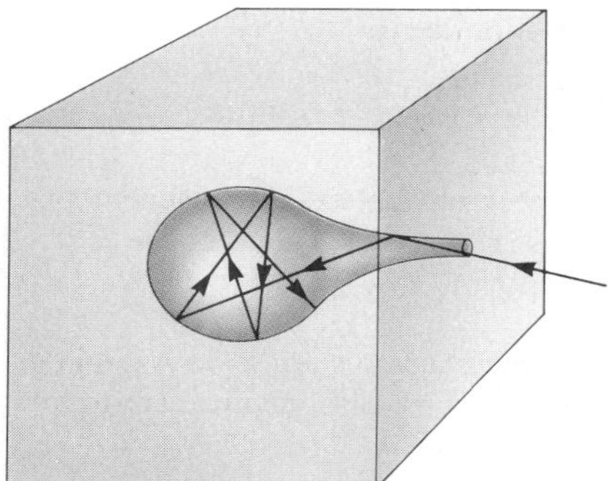

Figure 2.4 Trapping of radiation by a black body.

The next important development in the quest to understand the universal character of the radiation emitted by glowing solids came from the Austrian physicist J. Stefan in 1879. He found experimentally that the *total emissivity* of a hot solid was proportional to the fourth power of its absolute temperature. Since the total emissivity is the intensity radiated at all frequencies, the Stefan-Boltzmann law may be written as

Stefan-Boltzmann law

$$E = \int_0^\infty E_f \, df = \sigma T^4 \tag{2.3}$$

where E is the power per unit area emitted by the black body, E_f is the power per unit area per unit frequency emitted by the black body, T is the absolute temperature of the body, and σ is the Stefan-Boltzmann constant, given by $\sigma = 5.67 \times 10^{-8}\,\mathrm{W \cdot m^{-2} \cdot K^{-4}}$. A body that is not an ideal radiator will obey the same general law but with a coefficient, a, less than 1:

$$E = a\sigma T^4 \tag{2.4}$$

Only five years later another impressive confirmation of Maxwell's electromagnetic theory of light occurred when Boltzmann derived Stefan's law from a combination of thermodynamics and Maxwell's equations.

EXAMPLE 2.1 Stefan's Law Applied to the Sun
Estimate the surface temperature of the sun from the following information. The sun's radius is given by $R_s = 7.0 \times 10^8$ m. The average earth-sun distance is $R = 1.5 \times 10^{11}$ m. The power per unit area (total emissivity) from the sun is measured at the earth to be 1400 W/m². Assume that the sun is a black body.

Solution: For a black body, we take $a = 1$, so Equation 2.4 gives

$$E(R_s) = \sigma T^4 \tag{2.5}$$

Since we have the total emissivity at the earth, $E(R)$, we need the connection between $E(R)$ and $E(R_s)$. This comes from the conservation of energy:

$$E(R_s) \cdot 4\pi R_s^2 = E(R) \cdot 4\pi R^2$$

or

$$E(R_s) = E(R) \cdot R^2/R_s^2$$

Using Equation 2.5 we have

$$T = \left[\frac{E(R) \cdot R^2}{\sigma R_s^2}\right]^{1/4}$$

or

$$T = \left[\frac{(1400\ \mathrm{W/m^2})(1.5 \times 10^{11}\ \mathrm{m})^2}{(5.67 \times 10^{-8}\ \mathrm{W/m^2 \cdot K^4})(7.0 \times 10^8\ \mathrm{m})^2}\right]^{1/4}$$

$$= 5800\ \mathrm{K}$$

EXAMPLE 2.2 The Surface Temperature of the Sun Revisited!
As can be seen in Figure 2.3, *the wavelength marking the maximum power emission of a black body, λ_{max},* shifts toward shorter wavelengths as the black body gets hotter. Even this simple effect of $\lambda_{max} \propto T^{-1}$ was not definitely established experimentally until about 20 years after Kirchhoff's seminal paper had started the search to find $J(f, T)$. In 1893 Wilhelm Wien proposed a general form for the black body distribution law that gave the correct experimental behavior of λ_{max} with temperature. This law is called *Wien's displacement law* and may be written:

$$\lambda_{max} T = 2.898 \times 10^{-3}\ \mathrm{m \cdot K} \tag{2.6}$$

where λ_{max} is the wavelength in meters corresponding to maximum intensity and T is the absolute temperature. Assuming that the peak sensitivity of the human eye (which occurs at about 500 nm—blue-green light) coincides with λ_{max} for the sun (a black body), we can calculate the sun's surface temperature:

$$T = \frac{2.898 \times 10^{-3}\ \mathrm{m \cdot K}}{500 \times 10^{-9}\ \mathrm{m}} = 5800\ \mathrm{K}$$

Thus we have good agreement between measurements made at all wavelengths (Example 2.1) and at the maximum intensity wavelength.

Exercise 1 How nice that the sun's emission is peaked at the same wavelength as our eyes! Can you account for this?

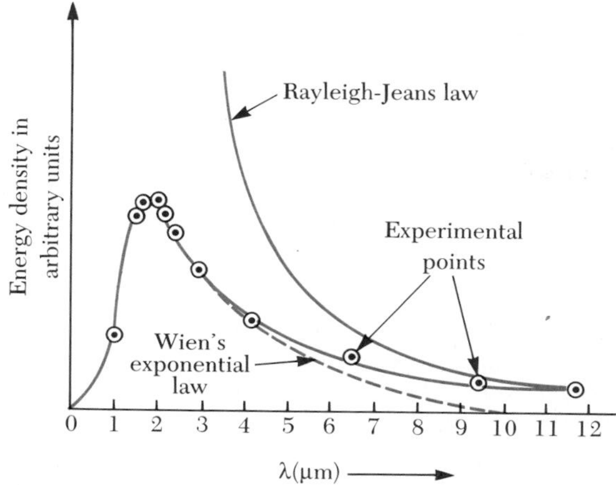

Figure 2.5 Discrepancy between Wien's law and experimental data for a black body at 1500 K.

Spectral energy density of a black body

We have been discussing the power radiated per unit area per unit frequency by the black body, $J(f, T)$, but *it is more convenient to consider the spectral energy density or energy per unit volume per unit frequency within the black body cavity,* $u(f, T)$. For light in equilibrium with the walls, the power emitted per square centimeter of opening is simply proportional to the energy density of the light in the cavity. Since the cavity radiation is isotropic and unpolarized, one can average over direction to show that the constant of proportionality between $J(f, T)$ and $u(f, T)$ is $c/4$, where c is the speed of light. Thus

$$J(f, T) = u(f, T)c/4 \tag{2.7}$$

The famous displacement law of W. Wien, which was as far as one could get using only thermodynamics and Maxwell's laws, was originally derived by using an energy density $u(f, T)$ of the form

$$u(f, T) = Af^3\, e^{-\beta f/T} \tag{2.8}$$

where A and β are constants. This important result was known as *Wien's exponential law;* it resembled and was loosely based on Maxwell's velocity distribution for gas molecules. Within a year the great German spectroscopist Friedrich Paschen (also the discoverer of the Paschen Series in the spectrum of hydrogen) had confirmed Wien's guess by working in the then difficult infrared range of 1 to 8 microns and at temperatures of 400 to 1600 K.[4]

As can be seen in Figure 2.5, Paschen had made most of his measurements in the maximum energy region of a body heated to 1500 K and had found good agreement with Wien's exponential law. In 1900, however, Lummer and Pringsheim extended the measurements to 18 microns and Rubens and Kurlbaum went even farther—to 60 microns. Both teams concluded that Wien's law failed in this region (Fig. 2.5). The experimental setup used by Rubens and

[4] We should point out the great difficulty in making blackbody radiation measurements and the singular advances made by German spectroscopists in the crucial areas of blackbody sources, sensitive detectors, and techniques for operating far into the infrared region. In fact, it is dubious whether Planck would have found the correct blackbody law as quickly without his close association with the brilliant experimentalists at the Physikalisch Technische Reichsanstalt of Berlin (a sort of German National Bureau of Standards)—Otto Lummer, Ernst Pringsheim, Heinrich Rubens, and Ferdinand Kurlbaum.

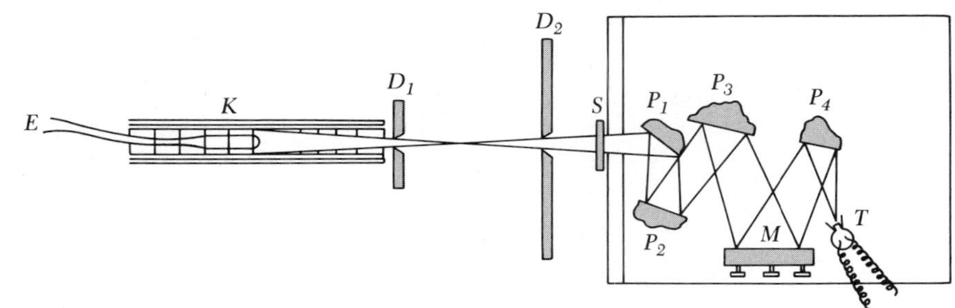

Figure 2.6 Apparatus for measuring blackbody radiation at a single wavelength in the far infrared region. The experimental technique that disproved Wien's law and was so crucial to the discovery of the quantum theory was the method of residual rays (Restrahlen). In this technique, one isolates a narrow band of far infrared radiation by causing white light to undergo multiple reflections from alkali halide crystals ($P_1 - P_4$). Since each alkali halide has a maximum reflection at a characteristic wavelength, quite pure bands of far infrared radiation may be obtained with repeated reflections. These pure bands can then be directed onto a thermopile (T) to measure intensity.

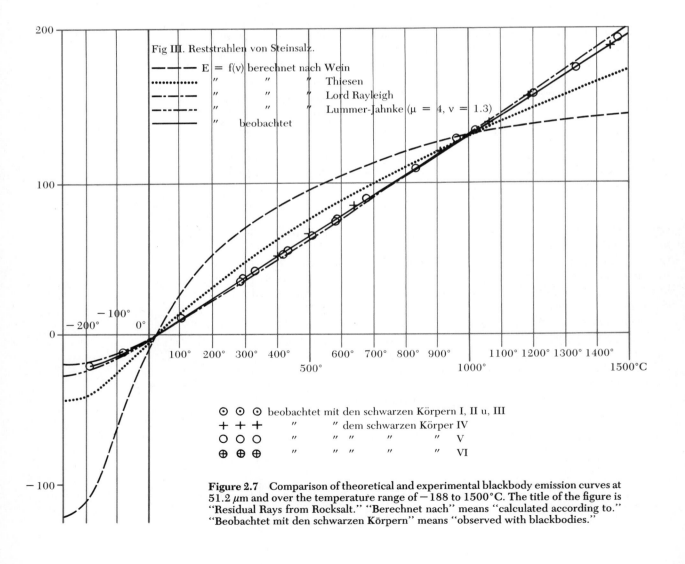

Figure 2.7 Comparison of theoretical and experimental blackbody emission curves at 51.2 μm and over the temperature range of −188 to 1500°C. The title of the figure is "Residual Rays from Rocksalt." "Berechnet nach" means "calculated according to." "Beobachtet mit den schwarzen Körpern" means "observed with blackbodies."

Kurlbaum is shown in Figure 2.6. It is interesting to note that these historic experiments involved the measurement of blackbody intensity at a fixed wavelength and variable temperature. Typical results measured at $\lambda = 51.2\ \mu$m and over the temperature range of -200 to $+1500$°C are shown in Figure 2.7 from the classic paper by Rubens and Kurlbaum.

Enter Planck

On a Sunday evening early in October of 1900, Max Planck discovered the famous blackbody formula, which truly ushered in the quantum theory (Fig. 2.8). Planck's proximity to the Reichsanstalt experimentalists was extremely important for his discovery—earlier in the day he had heard from Rubens that his latest measurements showed that $u(f, T)$, the spectral energy density, was proportional to T for long wavelengths or low frequency. Planck knew that Wien's law agreed well with the data at high frequency and indeed had been working hard for several years to derive Wien's exponential law from the principles of statistical mechanics and Maxwell's laws. Interpolating between the two limiting forms (Wien's exponential law and an energy density proportional to temperature), he immediately found a general formula which he sent to Rubens, on a postcard, the same evening. His formula was[5]

$$\boxed{u(f, T) = \frac{8\pi hf^3}{c^3}\left(\frac{1}{e^{hf/kT} - 1}\right)} \tag{2.9}$$

where h is Planck's constant $= 6.626 \times 10^{-34}$ J·s and k is Boltzmann's constant $= 1.380 \times 10^{-23}$ J/K. Note that at high frequencies, where $hf/kT \gg 1$,

$$\frac{1}{e^{hf/kT} - 1} \approx e^{-hf/kT}$$

so that

$$u(f, T) = \frac{8\pi hf^3}{c^3}\left(\frac{1}{e^{hf/kT} - 1}\right) \approx \frac{8\pi hf^3}{c^3}\, e^{-hf/kT}$$

[5] Planck originally published his formula as

$$u(\lambda, T) = \frac{C_1}{\lambda^5}\left(\frac{1}{e^{C_2/\lambda T} - 1}\right)$$

where $C_1 = 8\pi ch$ and $C_2 = hc/k$. He then found best fit values to the experimental data for C_1 and C_2 and evaluated $h = 6.55 \times 10^{-34}$ J·s and $k = N_A/R = 1.345 \times 10^{-23}$ J/s. As R, the universal gas constant, was fairly well known at the time, this technique also resulted in another method for finding N_A, Avogadro's number.

Figure 2.8 Max Planck (1858–1947). The work leading to the "lucky" blackbody radiation formula was described by Planck in his Nobel Prize acceptance speech (1920): "But even if the radiation formula proved to be perfectly correct, it would after all have been only an interpolation formula found by lucky guess-work and thus, would have left us rather unsatisfied. I therefore strived from the day of its discovery, to give it a real physical interpretation and this led me to consider the relations between entropy and probability according to Boltzmann's ideas. After some weeks of the most intense work of my life, light began to appear to me and unexpected views revealed themselves in the distance." (AIP Niels Bohr Library, W. F. Meggers Collection)

and we recover Wien's exponential law, Equation 2.8. At low frequencies, where $hf/kT \ll 1$,

$$\frac{1}{e^{hf/kT}-1} = \frac{1}{1+\dfrac{hf}{kT}+\cdots-1} \approx \frac{kT}{hf}$$

and

$$u(f, T) = \frac{8\pi hf^3}{c^3}\left(\frac{1}{e^{hf/kT}-1}\right) \approx \frac{8\pi f^2}{c^3}\,kT$$

This result shows that the spectral energy density is proportional to T in the low-frequency or so-called classical region.

We should emphasize that Planck's work entailed much more than clever mathematical manipulation. For more than six years Planck labored to find a rigorous derivation of the blackbody distribution curve. He was driven, in his own words, by the fact that the emission problem "represents something absolute, and since I had always regarded the search for the absolute as the loftiest goal of all scientific activity, I eagerly set to work." This work was to occupy most of his life as he strove to give his formula an ever deeper physical interpretation and to reconcile the elementary quantum with classical theory.

The Quantum of Energy

Planck's original theoretical justification of Equation 2.9 is rather abstract since it involves arguments based on entropy, statistical machanics, and several theorems proved earlier by Planck concerning matter and radiation in equilibrium.[6] We shall give arguments that are easier to visualize physically yet attempt to convey the spirit and revolutionary impact of Planck's original work.

Planck was convinced that blackbody radiation was produced by submicroscopic electric oscillators, which he called resonators. He assumed that the walls of a glowing cavity were composed of literally billions of these resonators (whose exact nature was unknown), all vibrating at different frequencies. Hence, according to Maxwell, each oscillator should emit radiation with a frequency corresponding to its vibration frequency. According to classical theory, an oscillator of frequency f can have any value of energy and can change its amplitude continuously as it radiates any fraction of its energy. This is where Planck made his revolutionary proposal. In order to secure agreement with experiment, Planck had to assume that the total energy of a resonator with mechanical frequency f could only be an integral multiple of hf or

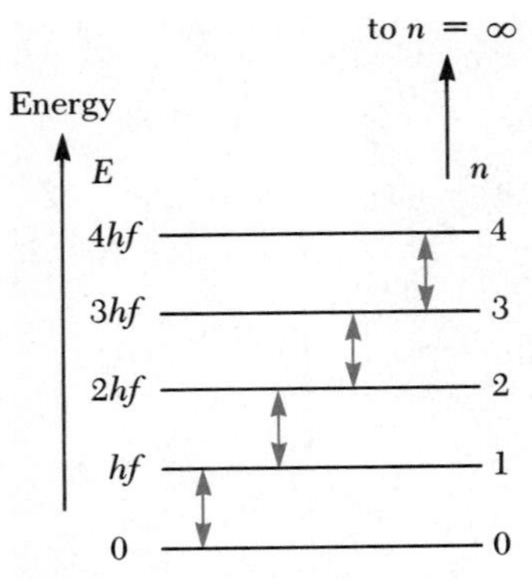

Figure 2.9 Allowed energy levels for an oscillator of natural frequency f. Allowed transitions are indicated by vertical arrows.

$$\boxed{E_{\text{resonator}} = nhf} \qquad n = 1,2,3,\ \ldots \tag{2.10}$$

where h is Planck's constant, $h = 6.626 \times 10^{-34}$ J·s. In addition, he concluded that emission of radiation of frequency f occurred when a resonator dropped to the next lowest energy state. Thus the resonator can change its energy only by the difference ΔE according to

$$\Delta E = hf \tag{2.11}$$

That is, it cannot lose just any fraction of its total energy. Figure 2.9 shows the quantized energy levels and allowed transitions proposed by Planck.

[6] M. Planck, *Ann. Physik*, **4:** 553, 1901.

EXAMPLE 2.3 A Quantum Oscillator versus a Classical Oscillator

Let us illustrate the implications of Planck's conjecture that *all* oscillating systems of natural frequency f have discrete energies $E = nhf$ and that the smallest change in energy of the system is given by $\Delta E = hf$.

(a) First compare an atomic oscillator sending out 540 nm light (green) to one sending out 700 nm light (red) by calculating the minimum energy change of each. For the green quantum,

$$\Delta E_{\text{green}} = hf = \frac{hc}{\lambda} = \frac{(6.63 \times 10^{-34}\ \text{J}\cdot\text{s})(3.00 \times 10^{8}\ \text{m/s})}{540 \times 10^{-9}\ \text{m}}$$
$$= 3.68 \times 10^{-19}\ \text{J}$$

Actually, the joule is much too large a unit of energy for describing atomic processes; a more appropriate unit of energy is the electron volt (eV). The electron-volt takes the charge on the electron as its unit of charge. By definition, an electron accelerated through a potential difference of one volt has an energy of 1 eV. An electron volt may be converted to joules by noting that

$$E = V \cdot q = 1\ \text{eV} = 1.602 \times 10^{-19}\ \text{C} \cdot 1\ \text{J/C}$$
$$= 1.602 \times 10^{-19}\ \text{J}$$

It is also most useful to have expressions for h and hc in terms of electron volts. These are:

$$h = 4.14 \times 10^{-15}\ \text{eV}\cdot\text{s}$$
$$hc = 1.240 \times 10^{-6}\ \text{eV}\cdot\text{m} = 1240\ \text{eV}\cdot\text{nm}$$

Returning to our example, we see that the minimum energy change of an atomic oscillator sending out green light is

$$\Delta E_{\text{green}} = \frac{3.68 \times 10^{-19}\ \text{J}}{1.602 \times 10^{-19}\ \text{J/eV}} = 2.30\ \text{eV}$$

For the red quantum we have

$$\Delta E_{\text{red}} = \frac{hc}{\lambda} = \frac{(6.63 \times 10^{-34}\ \text{J}\cdot\text{s})(3.00 \times 10^{8}\ \text{m/s})}{700 \times 10^{-9}\ \text{m}}$$
$$= 2.84 \times 10^{-19}\ \text{J} = 1.77\ \text{eV}$$

Note that the quantum of energy is not uniform under all conditions like the quantum of charge—the quantum of energy is proportional to the natural frequency of the oscillator. Note, too, that the high frequency of atomic oscillators produces a measurable quantum of energy of several electron-volts.

(b) Now consider a pendulum undergoing small oscillations with length $\ell = 1$ meter. According to classical theory, if air friction is present, the amplitude and consequently the energy decrease *continuously* with time as shown in Figure 2.10a. Actually, *all* vibrating systems are quantized (according to Equation 2.10) and lose energy in discrete packets or quanta, hf. This would lead to a decrease of the pendulum's energy in a stepwise manner as shown in Figure 2.10b. We shall show that there is no contradiction between quantum theory and the observed behavior of laboratory pendulums and springs.

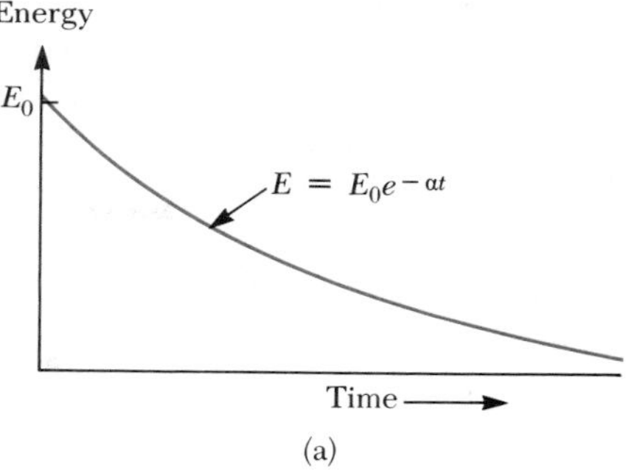

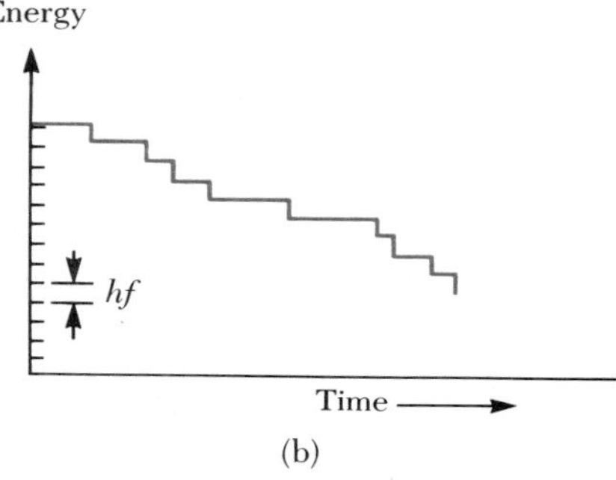

Figure 2.10 (a) Observed classical behavior of a pendulum. (b) Predicted quantum behavior of a pendulum.

An energy change of one quantum corresponds to

$$\Delta E = hf$$

where the pendulum frequency f is

$$f = \frac{1}{2\pi}\sqrt{\frac{g}{\ell}} = 0.50\ \text{Hz}$$

Thus

$$\Delta E = (6.63 \times 10^{-34}\ \text{J})(0.50\ \text{s}^{-1})$$
$$= 3.3 \times 10^{-34}\ \text{J}$$
$$= 2.1 \times 10^{-15}\ \text{eV}$$

Since the total energy of a pendulum of mass m and length ℓ displaced through an angle θ is

$$E = mg\ell(1 - \cos\theta)$$

we have for a typical pendulum with $m = 100$ g, $\ell = 1$ m, and $\theta = 10°$,

$$E = (0.1\ \text{kg})(9.8\ \text{m/s}^2)(1\ \text{m})(1 - \cos 10°) = 0.015\ \text{J}$$

Thus the fractional change in energy, $\Delta E/E$, is unobservably small:

$$\frac{\Delta E}{E} = \frac{3.3 \times 10^{-34}\ \text{J}}{1.5 \times 10^{-2}\ \text{J}} = 2.2 \times 10^{-32}$$

Note that the energy quantization of large vibrating systems is unobservable because of their low frequencies. Hence there is no contradiction between Planck's quantum postulate and the behavior of macroscopic oscillators.

Exercise 2 Calculate the quantum number, n, for this pendulum with $E = 1.5 \times 10^{-2}$ J.
Answer: 4.55×10^{31}

Exercise 3 An object of mass m on a spring of stiffness k oscillates with an amplitude A about its equilibrium position. Suppose that $m = 300$ g, $k = 10$ N/m, and $A = 10$ cm. (a) Find the total energy. (b) Find the mechanical frequency of vibration of the mass. (c) Calculate the change in amplitude when the system loses one quantum of energy.
Answers: (a) $E_{\text{Total}} = 0.050$ J (b) $f = 0.92$ Hz
(c) $\Delta E_{\text{quantum}} = 6.1 \times 10^{-34}$ J, so $\Delta A \approx -\frac{\Delta E}{\sqrt{2Ek}} = -6.1 \times 10^{-34}$ m

We have to this point been concentrating on the remarkable quantum properties of single oscillators of frequency f. Planck explained the continuous spectrum of the black body by assuming that the heated walls contained resonators vibrating at many different frequencies, each emitting light at the same frequency as its vibration frequency. By considering the conditions leading to equilibrium between the wall resonators and the radiation in the blackbody cavity, he was able to show that the spectral energy density could be expressed as the product of the number of oscillators having frequency f denoted by $N(f)$, and the average energy emitted per oscillator, $\overline{E}$. Thus we have

$$u(f, T) = N(f)\, \overline{E} \tag{2.12}$$

Furthermore, Planck showed that (see Appendix A for details)

$$N(f) = \frac{8\pi f^2}{c^3} \tag{2.13}$$

Substituting Equation 2.13 into Equation 2.12 gives

$$u(f, T) = \frac{8\pi f^2}{c^3}\, \overline{E} \tag{2.14}$$

This result shows that the spectral energy density at the frequency f is proportional to the average oscillator energy. Furthermore, since $u(f, T)$ approaches zero at high frequencies (see Fig. 2.5), $\overline{E}$ must tend to zero faster than $1/f^2$. The fact that the mean oscillator energy must become extremely small when the frequency becomes high guided Planck in the development of his theory. We shall also shortly see that the failure of the classical Rayleigh-Jeans theory to follow such a law led to the "ultraviolet catastrophe" — the prediction of an infinite spectral energy density at high frequencies in the ultraviolet region.

2.3 RAYLEIGH-JEANS LAW

Planck's law and the Rayleigh-Jeans law (the classical theory of blackbody radiation) are so important and offer so many examples of important theoretical techniques that it is instructive to examine them in more detail. Although Planck's law somewhat superceded the Rayleigh-Jeans law, we shall discuss Lord Rayleigh's derivation first since it is a more direct classical calculation.

While Planck concentrated on the thermal equilibrium of cavity radiation with oscillating electric charges in the cavity walls, Rayleigh concentrated directly on the electromagnetic waves in the cavity. Rayleigh reasoned that the

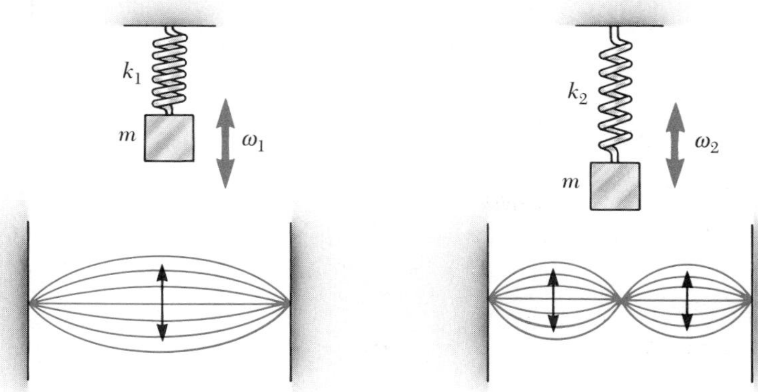

Figure 2.11 A one-dimensional harmonic oscillator is equivalent to a plane polarized electromagnetic standing wave.

standing electromagnetic waves in the cavity could be considered to have a temperature T, since they constantly exchange energy with the walls and could cause a thermometer to reach the same temperature as the walls. Further, he considered a standing polarized electromagnetic wave to be equivalent to a one-dimensional oscillator (Fig. 2.11). Using this idea, he expressed the energy density as a product of the number of standing waves (oscillators) and the average energy per oscillator. Taking $N(f)\,df$ to be the number of standing waves or normal modes per unit volume with frequency between f and $f + df$, and $\overline{E}$ to be the average energy per mode, we have

$$u(f, T)\,df = N(f)\,\overline{E}\,df$$

Note that $u(f, T)\,df$ is the energy per unit volume in the frequency range of f to $f + df$. We can show that $\overline{E} = kT$ from Boltzmann's distribution law. According to this fundamental law of statistical mechanics, the probability P of finding an individual system (such as a molecule or an oscillator) with energy E above some minimum energy, E_0, in an ensemble of systems at temperature T is

$$P(E) = P_0 e^{-(E-E_0)/kT} \tag{2.15}$$

where P_0 is the probability that a system has the minimum energy. In the case of a *discrete* set of allowed energies, the average energy, $\overline{E}$, is given by

$$\overline{E} = \frac{\Sigma\, E \cdot P(E)}{\Sigma\, P(E)} \tag{2.16}$$

where the denominator (a weighting factor) has been included in case probabilities have been used that do not sum to one. *In the classical case considered by Rayleigh, an oscillator could have any energy in a continuous range from 0 to* ∞. Thus the sums in Equation 2.16 must be replaced with integrals, and the expression for E becomes

$$\overline{E}_{\text{classical}} = \frac{\displaystyle\int_0^\infty E e^{-E/kT}\,dE}{\displaystyle\int_0^\infty e^{-E/kT}\,dE} = kT$$

The calculation of $N(f)$ is a bit more complicated but is of importance here as well as in the free electron model of metals. Appendix A gives the derivation of the density of modes, $N(f)$. One finds

$$N(f)\,df = \frac{8\pi f^2}{c^3}\,df \tag{A.6}$$

Density of standing waves in a cavity

or in terms of wavelength

$$N(\lambda)\, d\lambda = \frac{8\pi}{\lambda^4}\, d\lambda \tag{A.7}$$

The spectral energy density is simply the density of modes multiplied by kT, or

$$u(f, T)\, df = \frac{8\pi f^2}{c^3}\, kT\, df \tag{2.17}$$

In terms of wavelength,

Rayleigh-Jeans blackbody law

$$u(\lambda, T)\, d\lambda = \frac{8\pi}{\lambda^4}\, kT\, d\lambda \tag{2.18}$$

However, as one can see from Figure 2.5, this classical expression does not agree with the experimental data. Furthermore, Equation 2.18 diverges as $\lambda \rightarrow 0$, predicting unlimited energy emission in the ultraviolet region. One is forced to conclude that rigorous classical theory fails miserably to explain blackbody radiation.

2.4 THE DERIVATION OF THE PLANCK BLACKBODY FORMULA

As mentioned earlier, Planck concentrated on the energy states of resonators in the cavity walls and used the condition that the resonators and cavity radiation were in equilibrium to determine the spectral quality of the radiation. By thermodynamic reasoning (and apparently unaware of Rayleigh's derivation), he arrived at the same expression for $N(f)$ as Rayleigh. However, Planck arrived at a different form for $\overline{E}$ by postulating discrete values of energy for his resonators. Because it is simpler and more direct, we shall continue with Rayleigh's point of view and *quantize the energy of the standing waves or modes in the blackbody cavity to derive Planck's law. It is important to realize, however, that historically Planck quantized only the material resonators in the cavity walls and maintained the Maxwellian wave nature of light in the cavity.*

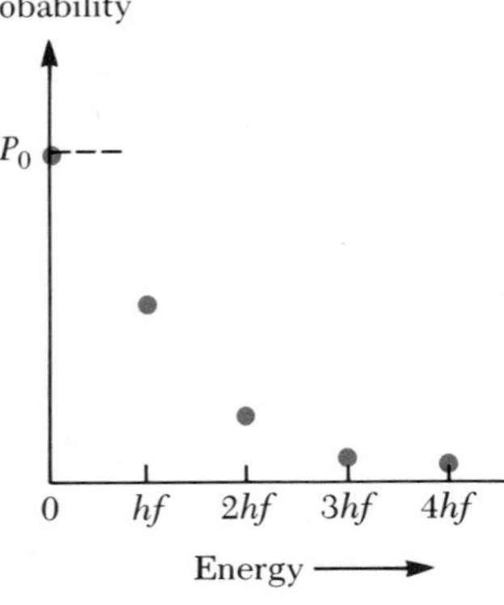

Figure 2.12 Probability that an oscillator of frequency f at temperature T is in one of its quantized energy states, nhf.

We shall now derive Planck's blackbody formula by quantizing the standing electromagnetic waves in the cavity. Instead of Rayleigh's assumption that a standing wave can have any energy from 0 to ∞, we postulate that a standing wave of frequency f can have only *discrete energies* $E = nhf$. We again apply the Boltzmann formula to find the probability that an oscillator is in its nth state, where now the probability of occupancy is $P(n) = P_0 e^{-nhf/kT}$. Note that in this case the probability is defined at discrete points as shown in Figure 2.12. The expression for the average energy of a standing wave, Equation 2.16, now becomes

$$\overline{E} = \frac{\sum_{n=0}^{\infty} (nhf)\, P_0 e^{-nhf/kT}}{\sum_{n=0}^{\infty} P_0 e^{-nhf/kT}} \tag{2.19}$$

After some manipulations (shown below), Equation 2.19 reduces to

$$\overline{E} = \frac{hf}{e^{hf/kT} - 1} \tag{2.20}$$

Using

$$\frac{1}{1-r} = \sum_{n=0}^{\infty} r^n$$

gives

$$\sum_{n=0}^{\infty} e^{-nhf/kT} = \frac{1}{(1-e^{-hf/kT})}$$

Substituting this into Equation 2.19 yields

$$\overline{E} = hf(1-e^{-hf/kT}) \sum_{n=0}^{\infty} ne^{-nhf/kT} \tag{2.21}$$

To evaluate the sum, we let $\alpha = hf/kT$, and note that

$$\sum ne^{-n\alpha} = -\frac{d}{d\alpha}\sum_{n=0}^{\infty} e^{-n\alpha}$$

$$= -\frac{d}{d\alpha}\left(\frac{1}{1-e^{-\alpha}}\right) = \frac{e^{-\alpha}}{(1-e^{-\alpha})^2}$$

Therefore,

$$\sum_{n=0}^{\infty} ne^{-nhf/kT} = \frac{e^{-\alpha}}{(1-e^{-\alpha})^2}$$

and the expression for $\overline{E}$, Equation 2.21, becomes

$$\overline{E} = \frac{hfe^{-hf/kT}}{1-e^{-hf/kT}} = \frac{hf}{e^{hf/kT}-1}$$

which is the desired result, Equation 2.20.

Multiplying $\overline{E}$ by $N(f)$ gives the Planck distribution formula:

Planck blackbody law

$$u(f,T) = \frac{8\pi f^2}{c^3}\left(\frac{hf}{e^{hf/kT}-1}\right) \tag{2.9}$$

This result shows that the ultraviolet catastrophe is avoided because the $\overline{E}$ term dominates the f^2 term at high frequencies. We can qualitatively understand why $\overline{E}$ tends to zero at high frequencies by noting that the first allowed oscillator level (hf) is so large compared to the average thermal energy available (kT) that there is almost zero probability that the first state is occupied.

In summary, Planck arrived at his blackbody formula by making two startling assumptions: (1) the energy of a charged oscillator of frequency f is limited to discrete values nhf and (2) during emission or absorption of light, the change in energy of an oscillator is hf. But Planck was every bit the "unwilling revolutionary," as he was known. From most of Planck's early correspondence one gains the impression that his concept of quantization was really a desperate calculational device, and moreover a device that applied only in the *one case of blackbody radiation.* It remained for the incarnation of physics in the 20th century, the great Albert Einstein, to elevate quantization to the level of a universal phenomenon by showing that both light itself and the vibrating atoms in a crystal are quantized.

2.5 LIGHT QUANTIZATION AND THE PHOTOELECTRIC EFFECT

We now turn to the year 1905, in which the next important development in quantum theory took place. The year 1905 was an incredible one for the "willing revolutionary" Albert Einstein (Fig. 2.13). In this year Einstein produced three immortal papers on three different topics, each revolutionary, and each worthy of a Nobel prize in itself. All three papers contained balanced, symmetric, and unifying new results achieved by spare and clean logic and simple mathematics. The first work, entitled "A Heuristic[7] Point of View About the Generation and Transformation of Light," formulated the theory of light quanta, and explained the photoelectric effect.[8] The second paper was entitled "On the Motion of Particles Suspended in Liquids as Required by the Molecular-Kinetic Theory of Heat." It explained Brownian motion and provided strong proof of the reality of atoms.[9] The third paper, which is perhaps his most famous, contains the invention of the theory of special relativity[10] and is entitled "On the Electrodynamics of Moving Bodies." It is interesting to note that when Einstein was awarded the Nobel Prize in 1922, the Swedish Academy judged his greatest contribution to physics to have been the theory of the photoelectric effect. No mention was made at all of his then arcane and controversial theory of relativity!

Figure 2.13 Albert Einstein just prior to his immensely creative year of 1905. (By permission of the Hebrew University of Jeruslem, Einstein Archives. AIP Niels Bohr Library)

Let us turn now to the paper concerning the light quantum,[8] in which Einstein crossed swords with Maxwell and challenged the unqualified successes of the wave theory of light. Einstein recognized an inconsistency between Planck's quantization of oscillators in the walls of the black body and his insistence that the cavity radiation consisted of classical electromagnetic waves. By showing that the change in entropy of blackbody radiation was like the change in entropy of an ideal gas consisting of independent particles, Einstein reached the conclusion that light itself is composed of "grains" or discontinuous quanta of energy.[11] Furthermore, he asserted that light interacting with matter also consists of quanta, and he brilliantly worked out the implications for photoelectric and photochemical processes. His explanation of the photoelectric effect offered such convincing proof that light consists of energy quanta that we should describe it in more detail. First, however, we need to consider the main experimental features of the photoelectric effect.

As noted earlier, Hertz first established that clean metal surfaces emit charges when exposed to ultraviolet light. In 1888 Hallwachs discovered that the emitted charges are negative, and in 1899 J. J. Thomson showed that the emitted charges are electrons. He did this by measuring the charge to mass ratio of the particles produced by ultraviolet light, and even succeeded in measuring e separately by a cloud chamber technique (see Chapter 3).

The last crucial discovery before Einstein's explanation was made in 1902 by Philip Lenard, who was studying the photoelectric effect with intense

[7] A heuristic argument is one that is plausible and enlightening but not rigorously justified.

[8] A. Einstein, *Ann. Physik,* **17:** 132, 1905 (March).

[9] A. Einstein, *Ann. Physik,* **17:** 549, 1905 (May).

[10] A. Einstein, *Ann. Physik,* **17:** 891, 1905 (June).

[11] Einstein, as Planck before him, fell back on the unquestionable solidity of thermodynamics and statistical mechanics to derive his revolutionary results. At the time it was well known that the probability, W, for n independent gas atoms to be in a partial volume V of a larger volume V_0 is $(V/V_0)^n$. Einstein showed that light of frequency f and total energy E enclosed in a cavity obeys an identical law, where in this case W is the probability that all the radiation is in the partial volume and $n = E/hf$.

carbon arc light sources. He found that the maximum kinetic energy of photoelectrons, K_{max}, *does not* depend on the *intensity* of the exciting light.[12] Although he was unable to establish the precise relationship, Lenard also indicated that K_{max} increases with light *frequency.* A typical apparatus used to measure the maximum kinetic energy of photoelectrons is shown in Figure 2.14. K_{max} is easily measured by applying a retarding voltage and gradually increasing it until the most energetic electrons are stopped and the photocurrent becomes zero. At this point,

$$K_{max} = \tfrac{1}{2}m_e v_{max}^2 = eV_s \qquad (2.22)$$

where m_e is the mass of the electron, v_{max} is the maximum electron velocity, e is the electronic charge, and V_s is the stopping voltage. A plot of the type found by Lenard is shown in Figure 2.15a; it illustrates that K_{max} is independent of light intensity I. The increase in current (or number of electrons per second) with increasing light intensity shown in Figure 2.15 was expected and could be explained classically. However, the result that K_{max} *does not* depend on the intensity was completely unexpected. It implied that the emitted electron could absorb only up to a fixed amount of the incident light energy, rejecting the remaining energy in light of high intensity. Conversely, for feeble light sources, the electron would remain in the solid until it had accumulated the same fixed amount, and would then be ejected from the metal with this energy.

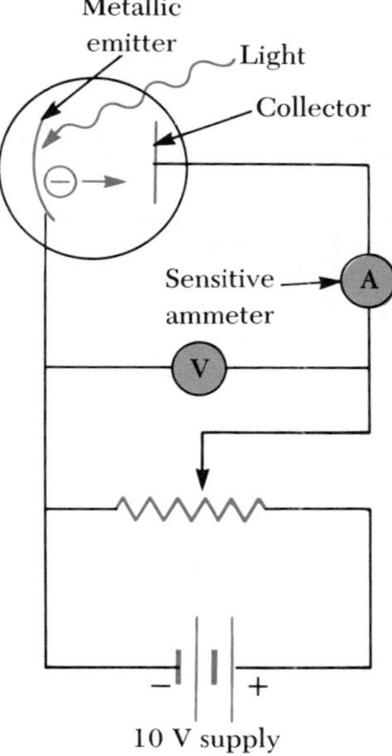

Figure 2.14 Photoelectric effect appartus.

Although it was not quantitatively established by Lenard in 1902, the dependence of K_{max} on frequency also shows unexpected results. A *threshold frequency,* f_0, is found for each metal, below which no photoelectrons are emitted no matter how intense the exciting radiation. Above the threshold frequency, K_{max} or V_s is found to be simply proportional to the light frequency, as shown in Figure 2.15b.

Einstein's explanation of the photoelectric effect was as brilliant for what it focused on as for what it omitted. For example, he stressed that Maxwell's classical theory had been immensely successful in describing the progress of light through space *over long time intervals,* but that a different theory might be needed to describe *momentary interactions* of light and matter, as in light emission by oscillators or the transformation of light energy to kinetic energy of the electron in the photoelectric effect. He also focused only on the energy aspect of the light, and avoided models or mechanisms concerning the conversion of the quantum of light energy to kinetic energy of the electron. In short,

[12] Electrons are emitted from the metal with a range of velocities.

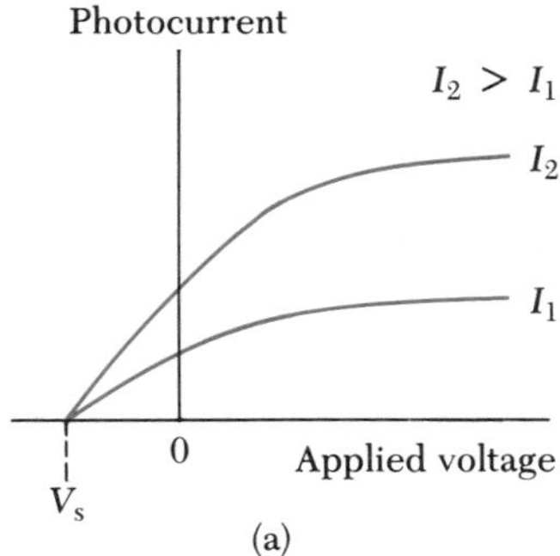

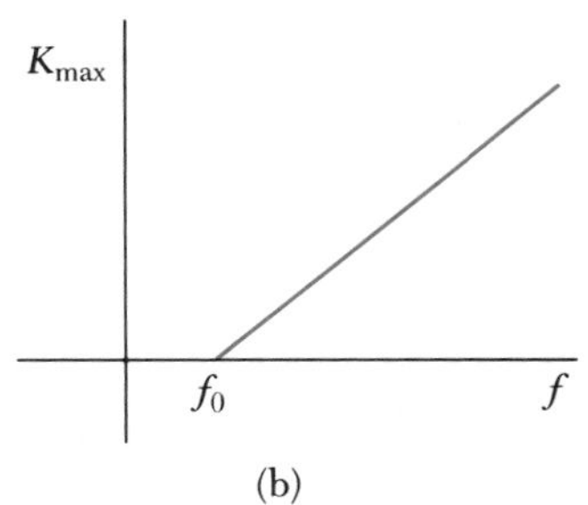

Figure 2.15 (a) A plot of photocurrent vs. applied voltage. The graph shows that K_{max} is independent of light intensity I for light of fixed frequency. (b) A graph showing the dependence of K_{max} on light frequency. Note that the results are independent of I.

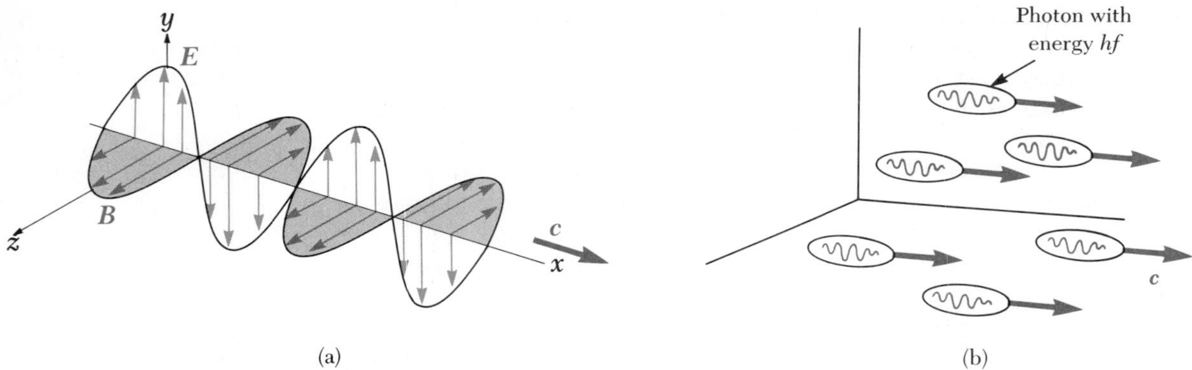

Figure 2.16 (a) A classical view of a traveling light wave. (b) Einstein's photon picture of "a traveling light wave."

he introduced only those ideas necessary to explain the photoelectric effect. He maintained that the energy of light is not distributed evenly over the classical wave front, but is concentrated in discrete regions (or in "bundles") called quanta. A suggestive image, not to be taken too literally, is shown in Figure 2.16b. Einstein's simple picture of the photoelectric effect was that a light quantum gives *all* its energy, hf, to a single electron in the metal. Electrons emitted from the surface of the metal possess the maximum kinetic energy, K_{max}. Thus, according to Einstein, the maximum kinetic energy for these emitted electrons is

Einstein's theory of the photoelectric effect

$$K_{max} = hf - \phi \tag{2.23}$$

where ϕ is the work function of the metal, which corresponds to the minimum energy with which an electron is bound in the metal. Table 2.1 lists values of work functions measured for different metals.

Equation 2.23 beautifully explained the puzzling independence of K_{max} and intensity found by Lenard. For a fixed light frequency f, an increase in light intensity means more photons and more photoelectrons per second, while K_{max} remains unchanged according to Equation 2.23. In addition, Equation 2.23 explained the phenomenon of threshold frequency. Light of threshold frequency f_0, which has just enough energy to knock an electron out of the metal surface, causes the electron to be released with zero kinetic energy. Setting $K_{max} = 0$ in Equation 2.23 gives

$$f_0 = \phi/h \tag{2.24}$$

Thus the variation in threshold frequency for different metals is produced by the variation in work function. We also note that light with $f < f_0$ has insufficient energy to free an electron. Consequently, the photocurrent is zero for $f < f_0$.

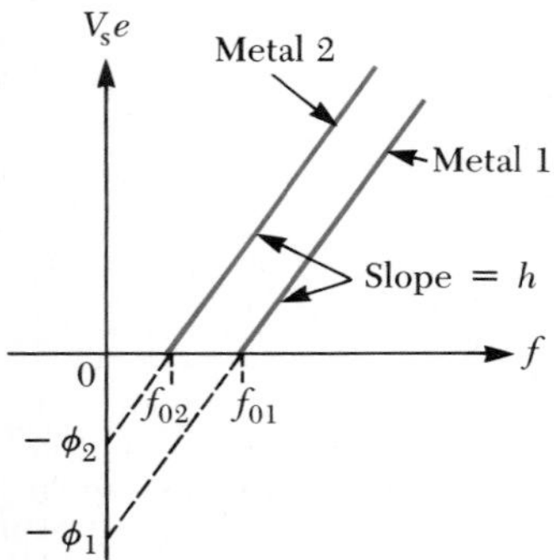

Figure 2.17 Universal characteristics of all metals undergoing the photoelectric effect.

TABLE 2.1 Work Functions of Selected Metals

Metal	Work Function, ϕ, in eV
Na	2.28
Al	4.08
Cu	4.70
Zn	4.31
Ag	4.73
Pt	6.35
Pb	4.14
Fe	4.50

With any theory, one looks not only for explanations of previously observed results, but also for new predictions. This was indeed the case here, as Equation 2.23 predicted the result (new in 1905) that K_{max} should vary linearly with f for any material and that the slope of the K_{max} versus f plot should yield the *universal constant h* (see Fig. 2.17). In 1916 the American physicist Robert Millikan reported photoelectric measurement data, from which he substantiated the linear relation between K_{max} and f and determined h with a precision of about 0.5%.[13]

[13] R. A. Millikan, *Phys. Rev.*, **7**: 355, 1916. Some of the experimental difficulties in the photoelectric effect were the lack of strong monochromatic uv sources, small photocurrents, and large effects of rough and impure metal surfaces on f_0 and E_{max}. Millikan brilliantly circumvented these difficulties by using alkali metal cathodes, which are sensitive in the visible to about 600 nm (thus making it possible to use the strong visible lines of the mercury arc), and machining fresh alkali surfaces while the metal sample was held under high vacuum.

EXAMPLE 2.4 The Photoelectric Effect in Zinc Philip Lenard determined that photoelectrons released from zinc by ultraviolet light were stopped by a voltage of 4.3 V. Find K_{max} and v_{max} for these electrons.

Solution:

$$K_{max} = eV_s = (1.6 \times 10^{-19}\ \text{C})(4.3\ \text{V}) = 6.88 \times 10^{-19}\ \text{J}$$

To find v_{max} we set the work done by the electric field equal to the change in the electron's kinetic energy, to obtain

$$\tfrac{1}{2} m_e v_{max}^2 = eV_s$$

or

$$v_{max} = \sqrt{\frac{2eV_s}{m_e}} = \sqrt{\frac{2(6.88 \times 10^{-19}\ \text{J})}{9.11 \times 10^{-31}\ \text{kg}}}$$

$$= 1.23 \times 10^6\ \text{m/s}$$

Thus a 4.3 eV electron is rather energetic and moves with a speed of about a million meters per second. Note, however, that this is still only about 0.4% of the speed of light, so relativistic effects are negligible.

EXAMPLE 2.5 The Photoelectric Effect for Iron Suppose that light of total intensity 1 μW/cm² falls on a clean iron sample 1 cm² in area. Assume that the iron sample reflects 96% of the light and that only 3% of the absorbed energy lies in the violet region of the spectrum above the threshold frequency.

(a) What intensity is actually available for the photoelectric effect?

Since only 4% of the incident energy is absorbed, and only 3% of this energy is able to produce photoelectrons, the intensity available is

$$I = (0.03)(0.04)I_0 = (0.03)(0.04)(1\ \mu\text{W/cm}^2)$$

$$= 1.2\ \text{nW/cm}^2$$

(b) Assuming that all the photons in the violet region have an effective wavelength of 250 nm, how many electrons will be emitted per second?

For an efficiency of 100%, one photon of energy hf will produce one electron, so

$$\text{number of electrons/s} = \frac{1.2 \times 10^{-9}\ \text{W}}{hf}$$

$$= \frac{\lambda(1.2 \times 10^{-9})}{hc}$$

$$= \frac{(250 \times 10^{-9}\ \text{m})(1.2 \times 10^{-9}\ \text{J/s})}{(6.6 \times 10^{-34}\ \text{J}\cdot\text{s})(3.0 \times 10^8\ \text{m/s})}$$

$$= 1.5 \times 10^9$$

(c) Calculate the current in the phototube in amperes.

$$i = (1.6 \times 10^{-19}\ \text{C})(1.5 \times 10^9\ \text{electrons/s})$$

$$= 2.4 \times 10^{-10}\ \text{A}$$

A sensitive electrometer is needed to detect this small current.

(d) If the cutoff frequency is $f_0 = 1.1 \times 10^{15}$ Hz, find the work function, ϕ, for iron.

From Equation 2.24 we have

$$\phi = hf_0 = (4.14 \times 10^{-15}\ \text{eV}\cdot\text{s})(1.1 \times 10^{15}\ \text{s}^{-1})$$

$$= 4.5\ \text{eV}$$

(e) Find the stopping voltage for iron if photoelectrons are produced by light with $\lambda = 250$ nm.

From the photoelectric equation, we have

$$eV_s = hf - \phi$$

$$= \frac{hc}{\lambda} - \phi$$

$$= \frac{(4.14 \times 10^{-15}\ \text{eV}\cdot\text{s})(3.0 \times 10^8\ \text{m/s})}{250 \times 10^{-9}\ \text{m}} - 4.5\ \text{eV}$$

$$= 0.46\ \text{eV}$$

Thus the stopping voltage is 0.46 V.

Exercise 4 The following data are found for photoemission from calcium:

λ (in nm)	253.6	313.2	365.0	404.7
V_s (in V)	1.95	0.98	0.50	0.14

Plot V_s versus f, and from the graph obtain Planck's constant, the threshold frequency, and the work function for calcium.

2.6 THE COMPTON EFFECT AND X-RAYS

Although Einstein introduced the concept that light consists of pointlike quanta of *energy* in 1905, he did not directly treat the *momentum* carried by light until 1919. In that year, in a paper treating a molecular gas in thermal equilibrium with electromagnetic radiation (statistical mechanics again!), Einstein concluded that a light quantum of energy E travels in a single direction (unlike a spherical wave) and carries a momentum directed along its line of motion of E/c or hf/c. In his own words, "if a bundle of radiation causes a molecule to emit or absorb an energy packet hf, then momentum of quantity hf/c is transferred to the molecule, directed along the line of the bundle for absorption and opposite the bundle for emission."

After developing the first theoretical justification for photon momentum, and treating the photoelectric effect much earlier, it is curious that Einstein bowed out of the picture and carried the treatment of photon momentum no farther. The theoretical treatment of photon-particle collisions had to await the insight of Peter Debye and Arthur Compton. In 1923 both men independently realized that the scattering of x-ray photons from electrons could be explained by treating photons as pointlike particles with energy hf and momentum hf/c, and by conserving relativistic energy and momentum of the photon-electron pair in a collision.[14,15] This remarkable development completed the particle picture of light by showing that photons, in addition to carrying energy hf, carry momentum and scatter like particles. Before treating this in detail, we will first give a brief introduction to the important topic of x-rays.

X-Rays

X-rays were discovered in 1895 by the German physicist Wilhelm Roentgen. He found that a beam of high-speed electrons allowed to fall on a target, in this case the glass walls of a cathode ray tube (Fig. 2.18), produced a new and extremely penetrating type of radiation. Within months of Roentgen's discovery the first medical x-ray pictures were taken, and within several years it became evident that x-rays are electromagnetic vibrations similar to light but with extremely short wavelengths and great penetrating power (see Fig. 2.19). Rough estimates obtained from the diffraction of x-rays by a narrow slit showed x-ray wavelengths of about 10^{-10} m, which is of the same order of magnitude as the atomic spacing in crystals. Since the best artificially ruled gratings of the time had spacings of 10^{-7} m, Max von Laue in Germany and W. H. Bragg and W. L. Bragg (a father and son team) in England suggested using single crystals such as calcite as natural three-dimensional gratings, the periodic atomic arrangement in the crystals constituting the grating rulings.

[14] P. Debye, *Phys. Zeitschr.*, **24:** 161, 1923. In this paper, Debye acknowledges Einstein's pioneering work on the quantum nature of light.
[15] A. H. Compton, *Phys. Rev.*, **21:** 484, 1923.

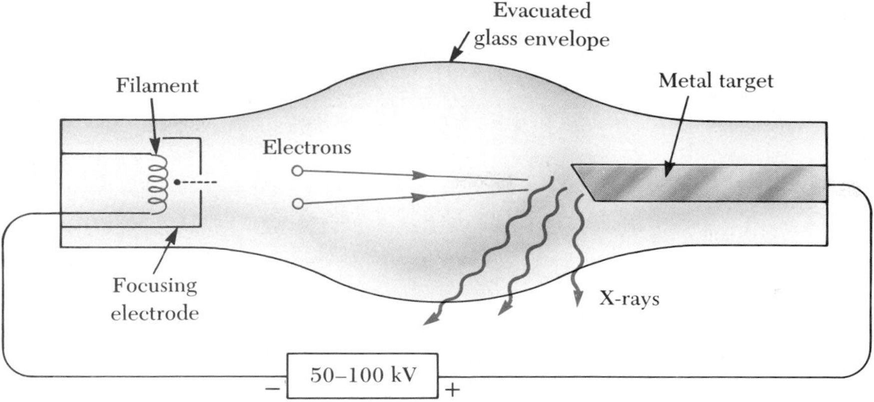

Figure 2.18 X-rays are produced by bombarding a metal target (copper, tungsten, and molybdenum are common) with energetic electrons having energies of 50 to 100 keV.

A particularly simple method of analyzing the scattering of x-rays from parallel crystal planes was proposed by William Lawrence Bragg in 1912. Consider two successive planes of atoms as shown in Figure 2.20. Note that adjacent atoms *in a single plane* A will scatter constructively if the angle of incidence, θ_i, equals the angle of reflection, θ_r. Atoms in *successive planes* (A and B) will scatter constructively at an angle θ if the path length difference for rays (1) and (2) is a whole number of wavelengths, $n\lambda$. From the diagram we have for constructive interference

$$AB + BC = n\lambda \qquad n = 1, 2, 3, \ldots$$

and because $AB = BC = d \sin\theta$, it follows that

$$\boxed{n\lambda = 2d\sin\theta} \qquad n = 1, 2, 3, \ldots \qquad (2.26)$$

Bragg equation

where n is the order of the intensity maximum, λ is the x-ray wavelength, d is the spacing between planes, and θ is the angle of the intensity maximum measured from plane A. Note that there are several maxima for a fixed d and λ

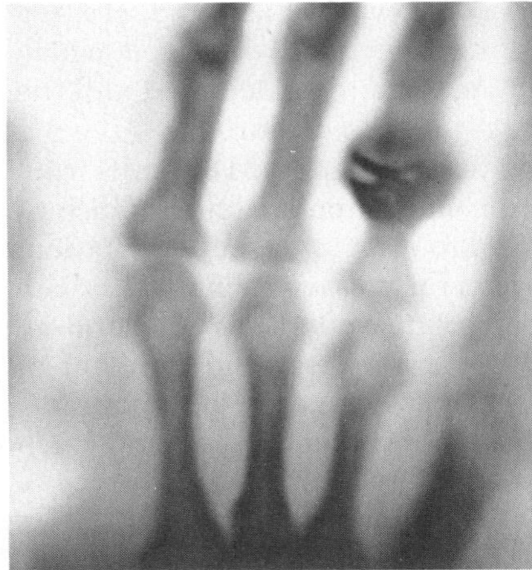

Figure 2.19 One of the first images made by Roentgen using x-rays (December 22, 1895).

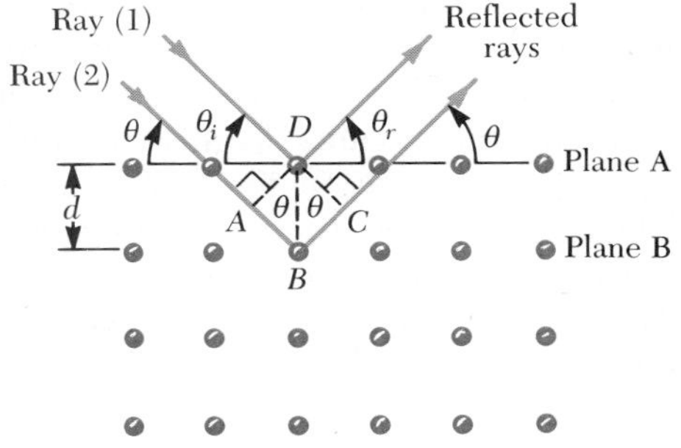

Figure 2.20 Bragg scattering of x-rays from successive planes of atoms. Constructive interference occurs for *ABC* equal to an integral number of wavelengths.

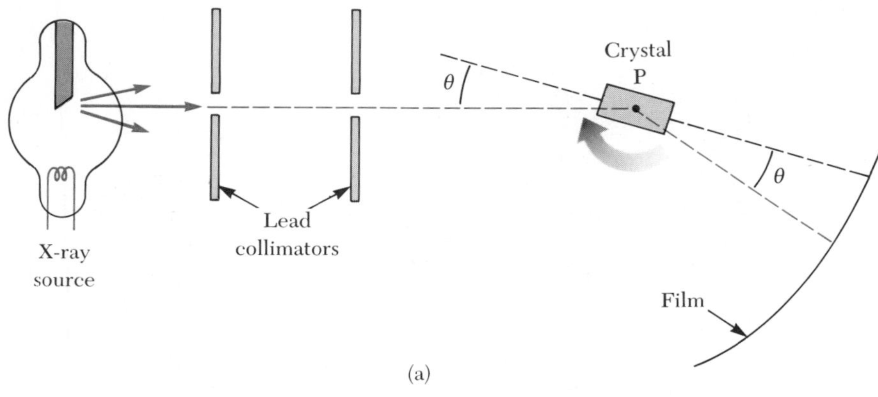

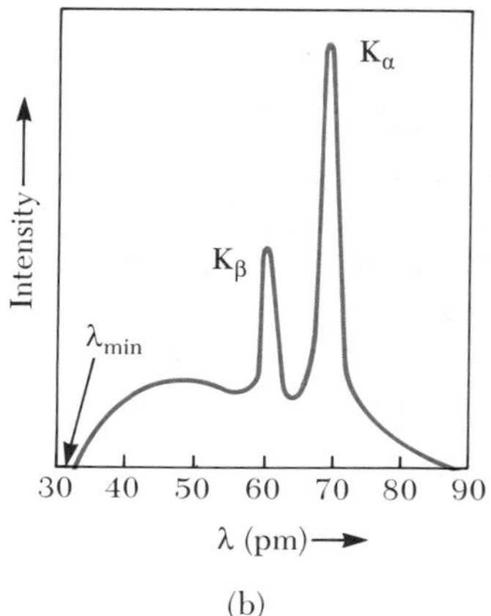

Figure 2.21 (a) A Bragg crystal x-ray spectrometer. The crystal is rotated about an axis through *P*. (b) The x-ray spectrum of a metal target consists of a broad continuous spectrum plus a number of sharp lines, which are due to the characteristic x-rays. Those shown were obtained when 35-keV electrons bombarded a molybdenum target. Note that 1 pm = 10^{-12} m = 10^{-3} nm.

corresponding to $n = 1, 2, 3, \ldots$. Equation 2.26 is known as the Bragg equation; it was used by the Braggs to determine the spectrum of wavelengths emitted by a particular metal target. A diagram of a Bragg x-ray spectrometer and a typical spectrum (intensity versus λ) are shown in Figures 2.21a and 2.21b. The crystal is slowly rotated until a strong reflection is observed, which means that Equation 2.26 holds. If d is known, λ can be calculated and the entire x-ray spectrum emitted by a given target can be explored.

The broad continuous x-ray spectrum shown in Figure 2.21b results from glancing or indirect scattering of electrons from metal atoms. In such collisions only part of the electron's energy is converted to electromagnetic radiation. This radiation is called bremsstrahlung (German for braking radiation), which refers to the radiation given off by any charged particle when it is decelerated. The minimum continuous x-ray wavelength is found to be independent of target composition and depends only on the tube voltage, V. It may be explained by attributing it to the case of a direct electron-atom collision in which all of the electron's kinetic energy is converted to electromagnetic energy in the form of a single x-ray photon. For this case we have

$$eV = hf = \frac{hc}{\lambda_{\min}}$$

or

$$\lambda_{\text{min}} = \frac{hc}{eV} \tag{2.27}$$

where V is the x-ray tube voltage.

The sharp lines in the x-ray spectrum vary with target composition and provide evidence for discrete energy levels of inner (tightly bound) atomic electrons in the range of thousands of electron volts. In simple terms an x-ray line (characteristic x-ray) may be pictured to arise when an energetic incident electron removes an electron from a target atom, leaving an inner shell vacancy. An outer shell electron immediately fills this vacancy, and the outer shell electron's excess energy is given off as an x-ray photon.

EXAMPLE 2.6 X-Rays from Molybdenum

Molybdenum's innermost (K-shell) electrons have an energy of about −20 000 eV. The outermost (M-shell) electrons have an energy of about −200 eV. Find the wavelength of the x-ray line emitted when one of the K-shell electrons has been removed from the atom by electron bombardment and its vacancy is filled by an M-shell electron.

Solution:

$$hf = \frac{hc}{\lambda} = E_{\text{upper}} - E_{\text{lower}}$$

Since $f = \lambda/c$, we have

$$\lambda = \frac{hc}{E_{\text{upper}} - E_{\text{lower}}} = \frac{12\,400\ \text{eV}\cdot\text{Å}}{[-200 + 20\,000]\ \text{eV}} = 0.626\ \text{Å}$$

[Note that one angstrom (Å) is a unit of length, equal to 0.1 nm, frequently used in atomic physics.]

Exercise 5 (a) Why is the total energy of K- and M-shell electrons negative? (b) What is the minimum tube voltage that will produce x-rays of wavelength 0.63 Å?
Answer: (b) 20 kV

The Compton Effect

Let us now turn to the year 1922 and the spectacular experimental confirmation by the American Arthur Holly Compton that x-ray photons behave like particles with momentum hf/c. For some time prior to 1922, Compton and his coworkers had been accumulating evidence that showed that classical wave theory failed to explain the scattering of x-rays from free electrons. In particular, classical theory predicted that incident radiation of frequency f_0 should accelerate an electron in the direction of propagation of the incident radiation, and that it should cause forced oscillations of the electron and reradiation at frequency f, where $f \leq f_0$.[16] (See Fig. 2.22a.) Also according to classical theory, the frequency or wavelength of the scattered radiation should depend on the length of time the electron was exposed to the incident radiation as well as on the intensity of the incident radiation.

Imagine the surprise when Compton showed, experimentally, that the wavelength shift of x-rays scattered at a given angle is absolutely independent of the intensity of radiation and the length of exposure, and depends only on the scattering angle. Figure 2.22b shows the quantum picture of the transfer of momentum and energy between an individual x-ray photon and an electron.

[16] This decrease in frequency of the reradiated wave is caused by a double Doppler shift, first because the electron is receding from the incident radiation, and second because the electron is a moving radiator as viewed from the fixed lab frame. See *Quantum Theory,* D. Bohm, Prentice-Hall, 1961, p. 35.

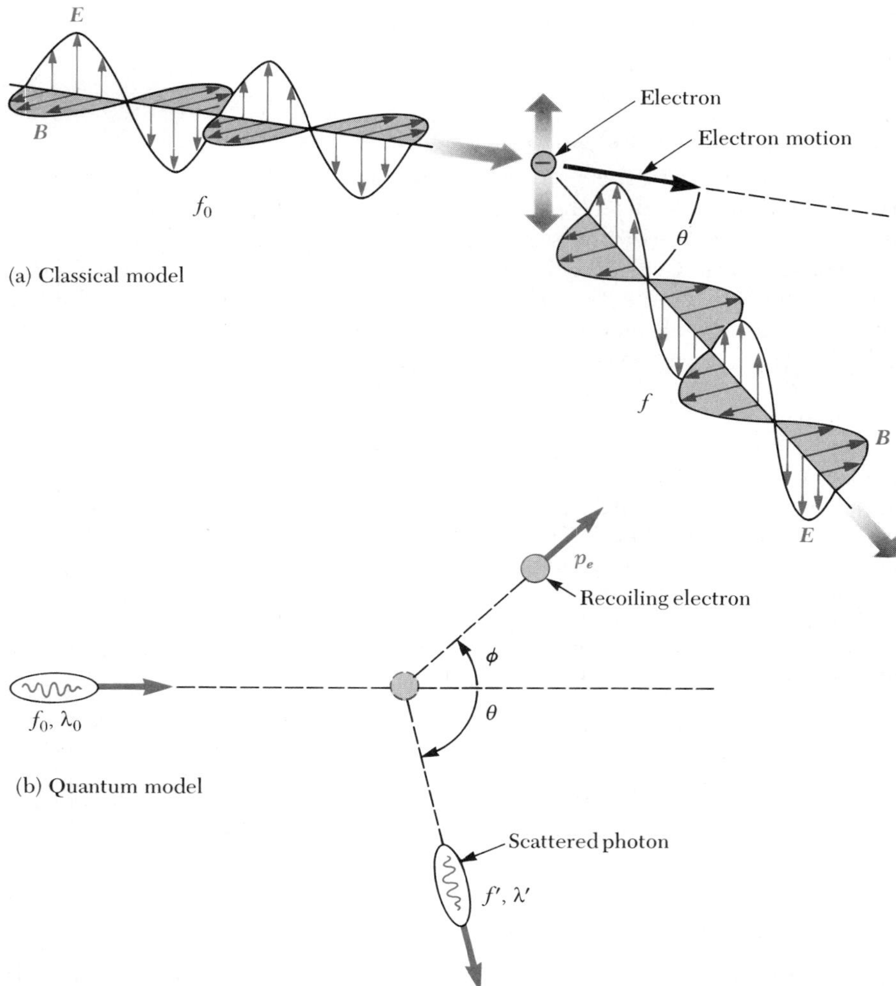

Figure 2.22 X-ray scattering from an electron: (a) the classical model, (b) the quantum model.

A schematic diagram of the apparatus used by Compton is shown in Figure 2.23a. In his original experiment Compton measured the dependence of scattered x-ray intensity on wavelength at three different scattering angles of 45°, 90° (shown in Fig. 2.23a), and 135°. The wavelength was measured with a rotating crystal spectrometer, and the intensity was determined by an ionization chamber that generated a current proportional to the x-ray intensity. Monochromatic x-rays of wavelength $\lambda_0 = 0.71$ Å constituted the incident beam. A carbon target with a low atomic number, Z, was used since atoms with small Z have more loosely bound (free) electrons. The experimental intensity versus wavelength plots observed by Compton for scattering angles of 0°, 45°, 90°, and 135° are shown in Figure 2.23b. They show two peaks, one at λ_0 and a shifted peak at λ'. The peak at λ_0 is caused by x-ray scattering from tightly bound electrons, which have an effective mass equal to that of the entire carbon atom. The shifted peak at λ' is caused by the scattering of x-rays from free electrons, and it was predicted by Compton to depend on scattering angle as

Compton effect

$$\lambda' - \lambda_0 = \frac{h}{m_e c}(1 - \cos\theta) \tag{2.28}$$

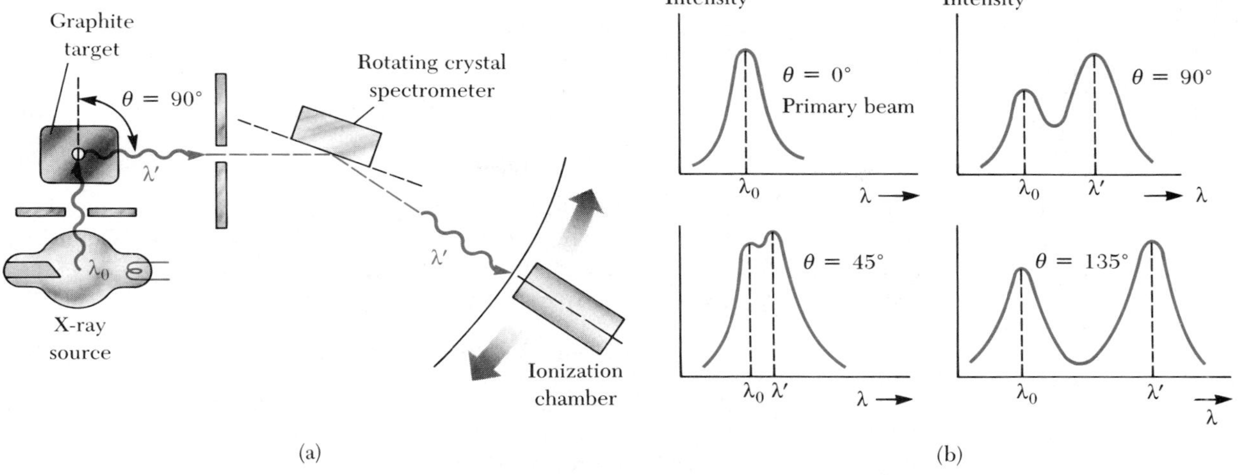

Figure 2.23 (a) Schematic diagram of Compton's apparatus. (b) Scattered intensity versus wavelength for Compton scattering at $\theta = 0°$, $45°$, $90°$, and $135°$.

where m_e = electron mass; h/m_ec is called the *Compton wavelength* of the electron and has a currently accepted value of

$$\frac{h}{m_ec} = 0.0243 \text{ Å} = 0.00243 \text{ nm}$$

Compton's careful measurements confirmed Equation 2.28 completely and resulted in a Compton wavelength of 0.0242 Å, in excellent agreement with the currently accepted value. It is fair to say that these sterling results were the first to convince most American physicists of the fundamental validity of the quantum theory!

Let us now turn to the derivation of Equation 2.28 by a treatment which assumes that the photon exhibits particle-like behavior and collides elastically with a free electron like a billiard ball. Figure 2.24 shows the photon-electron collision. Since the electron recoils at speeds close to the speed of light, we must conserve both relativistic energy and momentum. The expression for conservation of energy gives

$$E + m_ec^2 = E' + E_e \tag{2.29}$$

Energy conservation

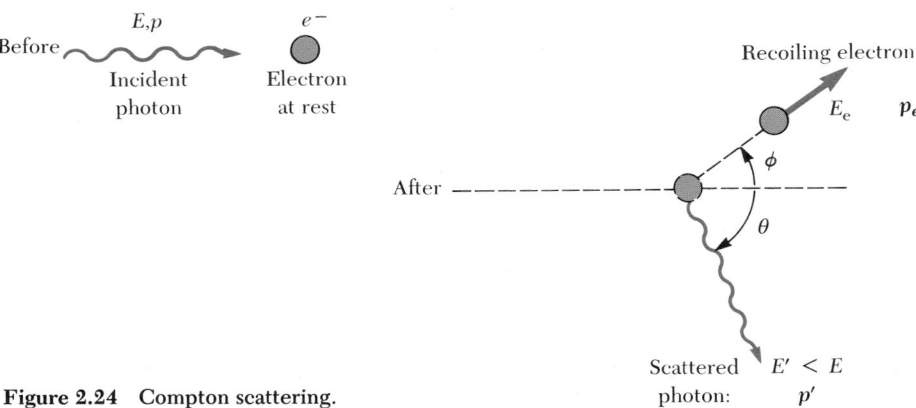

Figure 2.24 Compton scattering.

where E is the energy of the incident photon, E' is the energy of the scattered photon, m_ec^2 is the rest energy of the electron, and E_e is the recoil energy of the electron. Likewise, from momentum conservation we have

Momentum conservation

$$p = p' \cos\theta + p_e \cos\phi \tag{2.30}$$

$$p' \sin\theta = p_e \sin\phi \tag{2.31}$$

where p is the momentum of the incident photon, p' is the momentum of the scattered photon, and p_e is the recoil momentum of the electron. Equations 2.30 and 2.31 may be solved simultaneously to eliminate ϕ, the electron scattering angle, which gives the following expression for p_e^2:

$$p_e^2 = (p')^2 + p^2 - 2pp' \cos\theta \tag{2.32}$$

At this point it is necessary, paradoxically, to invoke the wave nature of light in order to explain the particle-like behavior of photons. We have already seen that the energy of a photon and the frequency of the associated light wave are related by $E = hf$. If we assume that, in addition, a photon obeys the relativistic expression $E^2 = p^2c^2 + m_0^2c^4$, and that *a photon has a rest mass of zero,* we have

$$p_{\text{photon}} = \frac{E}{c} = \frac{hf}{c} = \frac{h}{\lambda} \tag{2.33}$$

Here again we have a paradoxical situation; a particle property, the photon momentum, is given in terms of a wave property, λ, of an associated light wave. If the relations $E = hf$ and $p = hf/c$ are substituted into Equations 2.29 and 2.32, these become respectively

$$E_e = hf - hf' + m_ec^2 \tag{2.34}$$

and

$$p_e^2 = \left(\frac{hf'}{c}\right)^2 + \left(\frac{hf}{c}\right)^2 - \frac{2h^2ff'}{c^2}\cos\theta \tag{2.35}$$

Since we are not interested in the total energy and momentum of the electron, E_e and p_e may be eliminated by substituting Equations 2.34 and 2.35 into the expression for the electron's relativistic energy,

$$E_e^2 = p_e^2c^2 + m_e^2c^4$$

After some algebra (see Problem 41), this gives the Compton shift formula:

$$\lambda' - \lambda_0 = \frac{h}{m_ec}(1 - \cos\theta) \tag{2.28}$$

EXAMPLE 2.7 The Compton Shift for Carbon
X-rays of wavelength $\lambda = 0.20$ nm are aimed at a block of carbon. The scattered x-rays are observed at an angle of 45° to the incident beam. Calculate the wavelength of the scattered x-rays at this angle.

Solution: The shift in wavelength of the scattered x-rays is given by Equation 2.28. Taking $\theta = 45°$, we find that

$$\Delta\lambda = \frac{h}{m_ec}(1 - \cos\theta)$$

$$= \frac{6.63 \times 10^{-34}\ \text{J}\cdot\text{s}}{(9.11 \times 10^{-31}\ \text{kg})(3.00 \times 10^8\ \text{m/s})}(1 - \cos 45°)$$

$$= 7.11 \times 10^{-13}\ \text{m} = 0.000711\ \text{nm}$$

Hence, the wavelength of the scattered x-ray at this angle is

$$\lambda = \Delta\lambda + \lambda_0 = 0.200711 \text{ nm}$$

Exercise 6 Find the fraction of energy lost by the photon in this collision.
Answer: Fraction $= \Delta E/E = 0.00355$.

EXAMPLE 2.8 X-ray Photons versus Visible Photons
In this example we will deal with the following question: Why are x-ray photons used in the Compton experiment, rather than visible light photons? To answer this question, we shall first calculate the Compton shift for scattering at 90° from graphite for the following cases: (1) very high energy γ-rays from cobalt, $\lambda = 0.0106$ Å; (2) x-rays from molybdenum, $\lambda = 0.712$ Å; and (3) green light from a mercury lamp, $\lambda = 5461$ Å.

Solution: (a) In all cases, the Compton shift formula gives $\Delta\lambda = \lambda' - \lambda_0 = (0.0243 \text{ Å})(1 - \cos 90°) = 0.0243 \text{ Å} = 0.00243$ nm. That is, regardless of the incident wavelength, the same small shift is observed. However, the fractional change in wavelength, $\Delta\lambda/\lambda_0$, is quite different in each case:

γ-rays from cobalt:

$$\frac{\Delta\lambda}{\lambda_0} = \frac{0.0243 \text{ Å}}{0.0106 \text{ Å}} = 2.29$$

X-rays from molybdenum:

$$\frac{\Delta\lambda}{\lambda_0} = \frac{0.0243 \text{ Å}}{0.712 \text{ Å}} = 0.0341$$

Visible light from mercury:

$$\frac{\Delta\lambda}{\lambda_0} = \frac{0.0243 \text{ Å}}{5461 \text{ Å}} = 4.45 \times 10^{-6}$$

Since both incident and scattered wavelengths are simultaneously present in the beam, they can be easily resolved only if $\Delta\lambda/\lambda_0$ is a few percent or if $\lambda_0 \leq 1$ Å.

(b) The so-called free electrons in carbon are actually conduction electrons with a binding energy of about 4 eV. Why may this binding energy be neglected for x-rays with $\lambda_0 = 0.712$ Å?

The energy of a photon with this wavelength is

$$E = hf = \frac{hc}{\lambda} = \frac{12\,400 \text{ eV}\cdot\text{Å}}{0.712 \text{ Å}} = 17\,400 \text{ eV}$$

Thus the electron binding energy of 4 eV is negligible in comparison to the incident x-ray energy.

2.7 PARTICLE-WAVE COMPLEMENTARITY

As we have seen, the Compton effect offers ironclad evidence that when light interacts with matter it behaves as if it were composed of particles with energy hf and momentum h/λ. Yet the very success of Compton's theory raises many questions. If the photon is a particle, what can be the meaning of the "frequency" and "wavelength" of the particle, which determine its energy and momentum? Is light in some sense simultaneously a wave and a particle? Although photons possess no rest mass, is there a simple expression for a "moving" photon mass? Are the inertial and gravitational masses equal for the photon (as they are for material particles), and if so, do photons experience gravitational attraction? What is the spatial extent of a photon, and how does an electron absorb or scatter a photon?

While answers to some of these questions are possible, it is well to be aware that some demand a view of atomic processes that is too pictorial and literal. Many of these questions issue from the viewpoint of classical mechanics, where all matter and energy are seen in the context of colliding billiard balls or water waves breaking on a shore. Quantum mechanics gives light a more fluid and flexible nature by implying that different experimental conditions elicit either the wave properties or the particle properties of light. In fact, *both views are necessary and complementary.* Neither model can be used exclusively to describe radiation adequately. A complete understanding is obtained only if the two models are combined in a complementary manner.

The physicist Max Born, an important contributor to the foundations of quantum mechanics, had this to say about the particle-wave dilemma:

> The ultimate origin of the difficulty lies in the fact (or philosophical principle) that we are compelled to use the words of common language when we wish to describe a phenomenon, not by logical or mathematical analysis, but by a picture appealing to the imagination. Common language has grown by everyday experi-

ence and can never surpass these limits. Classical physics has restricted itself to the use of concepts of this kind; by analyzing visible motions it has developed two ways of representing them by elementary processes: moving particles and waves. There is no other way of giving a pictorial description of motions—we have to apply it even in the region of atomic processes, where classical physics breaks down.

Every process can be interpreted either in terms of corpuscles or in terms of waves, but on the other hand it is beyond our power to produce proof that it is actually corpuscles or waves with which we are dealing, for we cannot simultaneously determine all the other properties which are distinctive of a corpuscle or of a wave, as the case may be. We can therefore say that the wave and corpuscular descriptions are only to be regarded as complementary ways of viewing one and the same objective process, a process which only in definite limiting cases admits of complete pictorial interpretation . . . [17]

Thus we are left with an uneasy compromise between wave and particle concepts and must accept, at this point, that both are necessary to explain the observed behavior of light. Further considerations of the dual nature of light and indeed of all matter will be taken up again in Chapters 3 and 4.

2.8 INERTIAL AND GRAVITATIONAL MASS OF THE PHOTON (Optional)

It is interesting to speculate on how far the particle model of light may be carried. Encouraged by the successful particle explanation of the photoelectric and Compton effects, one may ask whether the photon possesses gravitational mass, and whether photons will experience an observable deflection when attracted by large masses such as those of the sun or earth.

In order to investigate these questions, let us recall that the Compton effect shows that photons carry momentum $p = mv = hf/c$. Since photons travel at $v = c$, the inertial mass of the photon, m_i, is given by

$$m_i = hf/c^2 \tag{2.36}$$

Note that the inertial mass determines how the photon responds to an applied force such as that exerted on it during a collision with an electron. The gravitational mass of an object determines the force of attraction of that object to another, such as the earth. It is a *remarkable* fact that the inertial mass of all material bodies is equal to the gravitational mass. (Recent measurements show that these two quantities are equal to within 1 part in 10^{10}.) The classic result which shows this equality is that all bodies fall with the same acceleration near the earth's surface, provided that air resistance is negligible.[18]

[17] M. Born, *Atomic Physics*, Hafner Publishing Co., New York, fourth edition, 1946, p. 92.

[18] Consider a falling platinum ball of inertial mass $m_i(\text{Pt})$ and gravitational mass $m_g(\text{Pt})$. Applying Newton's second law gives

$$m_i(\text{Pt})a(\text{Pt}) = \frac{GM_E m_g(\text{Pt})}{R_E^2}$$

where $a(\text{Pt})$ is the acceleration of the platinum ball, and M_E and R_E are the gravitational mass and the radius of the earth, respectively. A second ball of different mass composed of wood, and denoted by $m_i(\text{w})$ and $m_g(\text{w})$, falls according to a similar expression:

$$m_i(\text{w})a(\text{w}) = \frac{GM_E m_g(\text{w})}{R_E^2}$$

Dividing the two expressions, and noting that $a(\text{Pt}) = a(\text{w}) = g$, we have

$$\frac{m_i(\text{Pt})}{m_g(\text{Pt})} = \frac{m_i(\text{w})}{m_g(\text{w})}$$

Although this result only asserts that the inertial and gravitational masses of different materials have the same ratio, the ratio is made equal to one by a suitable choice of G.

Let us assume that the photon, like other material objects, also has a gravitational mass equal to its inertial mass. In this case a photon falling from a height H should increase in energy by mgH and thus increase in frequency. In fact, experiments have been carried out[19] that show this increase and confirm that the photon indeed has a gravitational mass of hf/c^2. Figure 2.25 shows a schematic representation of the experiment. An expression for f' in terms of f may be derived by applying conservation of energy to the photon at points A and B.

$$KE_B + PE_B = KE_A + PE_A$$

Since the photon's kinetic energy is $E = pc = hf$ and its potential energy is mgH where $m = hf/c^2$, we have

$$hf' + 0 = hf + \left(\frac{hf}{c^2}\right) gH$$

or

$$f' = f\left(1 + \frac{gH}{c^2}\right) \tag{2.37}$$

The fractional change in frequency, $\Delta f/f$, is given by

$$\frac{\Delta f}{f} = \frac{f' - f}{f} = \frac{gH}{c^2} \tag{2.38}$$

For $H = 50$ m, we find

$$\frac{\Delta f}{f} = \frac{(9.8 \text{ m/s}^2)(50 \text{ m})}{(3.0 \times 10^8 \text{ m/s})^2} = 5.4 \times 10^{-15}$$

This incredibly small shift has actually been measured (with difficulty)! The shift amounts to only about 1/250 of the linewidth for monochromatic γ-ray photons.

The increase in frequency for a photon falling inward suggests a decrease in frequency for a photon that escapes outward to infinity against the gravitational pull of a star (see Fig. 2.26). This effect, known as a *gravitational red shift* (not to be confused with a Doppler shift), would cause a photon emitted in the blue region to be shifted in frequency toward the red end of the spectrum. An expression for the red shift may be derived by once again conserving photon energy:

$$[KE + PE]_{R=\infty} = [KE + PE]_{R=R_s}$$

Using hf for the photon's kinetic energy and $-GMm/R$ for its potential energy, where m is the photon's inertial mass, yields

$$hf' - 0 = hf - \frac{GM}{R_s}\left(\frac{hf}{c^2}\right) \tag{2.39}$$

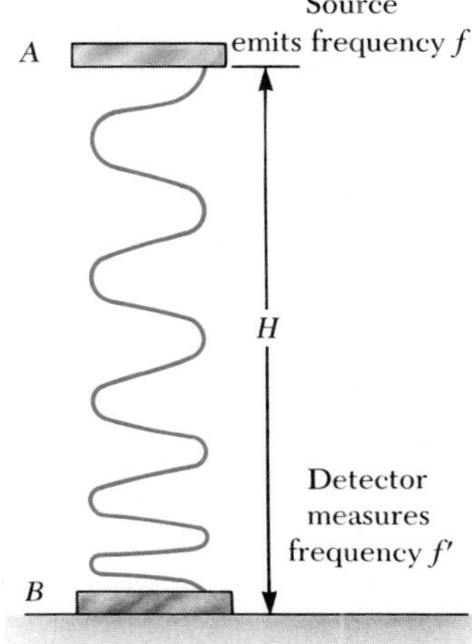

Figure 2.25 Schematic diagram of the falling-photon experiment.

[19] R. V. Pound and G. A. Rebka, Jr., *Phys. Rev. Lett.*, **4:** 337, 1960.

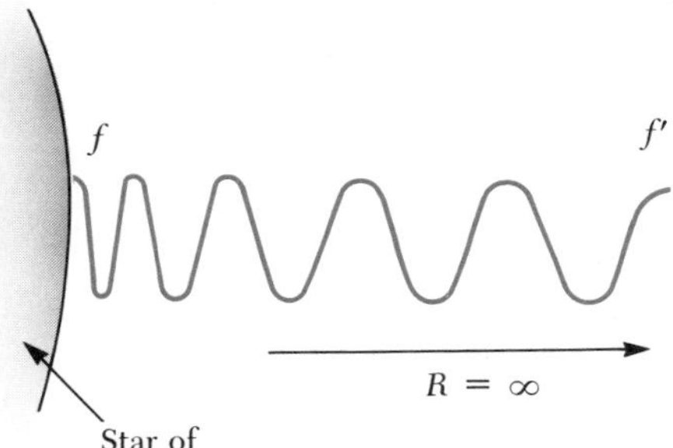

Figure 2.26 Gravitational red shift from a high-density star.

or

$$f' = f\left(1 - \frac{GM}{R_s c^2}\right) \tag{2.40}$$

EXAMPLE 2.8 The Red Shift for a White Dwarf
White dwarf stars are extremely massive, compact stars that have mass on the order of our sun's mass concentrated in a volume similar to that of the earth. Calculate the gravitational red shift for 300 nm light emitted from such a star.

Solution: We can write Equation 2.40 in the alternate form

$$\frac{f' - f}{f} = \frac{\Delta f}{f} = \frac{GM}{R_s c^2}$$

Using the values

$$M = \text{mass of sun} = 1.99 \times 10^{30}\ \text{kg}$$

$$R_s = \text{radius of earth} = 6.37 \times 10^{6}\ \text{m}$$

$$G = 6.67 \times 10^{-11}\ \text{N}\cdot\text{m}^2/\text{kg}^2$$

we find

$$\frac{\Delta f}{f} = \frac{(6.67 \times 10^{-11}\ \text{N}\cdot\text{m}^2/\text{kg}^2)(1.99 \times 10^{30}\ \text{kg})}{(6.37 \times 10^{6}\ \text{m})(3.00 \times 10^{8}\ \text{m/s})^2}$$

$$= 2.31 \times 10^{-4}$$

Since $\frac{\Delta f}{f} \approx \frac{df}{f}$, and $df = -\frac{c}{\lambda^2}\,d\lambda \left(\text{from } f = \frac{c}{\lambda}\right)$, we find $\left|\frac{df}{f}\right| = \frac{d\lambda}{\lambda}$. Therefore, the shift in wavelength is

$$\Delta\lambda = (300\ \text{nm})(2.31 \times 10^{-4}) = 0.0695\ \text{nm} \approx 0.7\ \text{Å}$$

Note that this is a red shift, so the observed wavelength would be 300.07 nm.

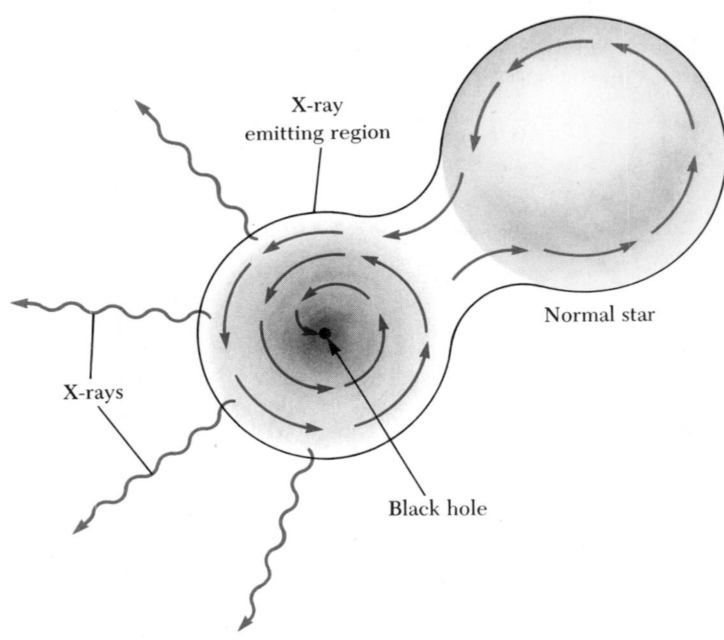

Figure 2.27 Schematic diagram of x-rays generated by matter from a normal star falling into a black hole.

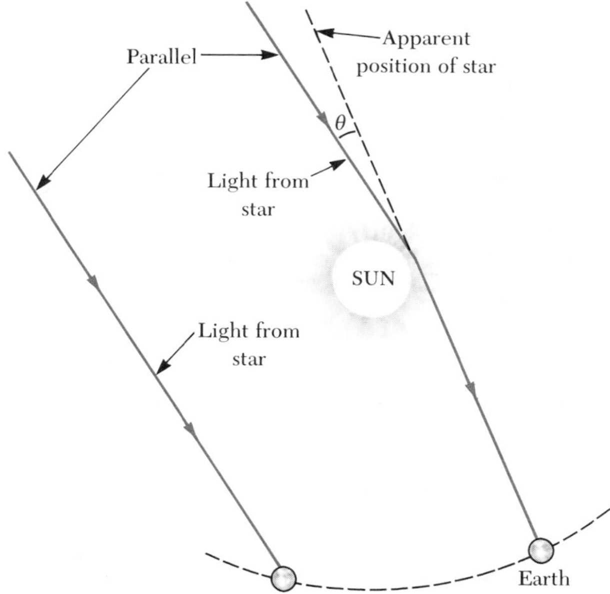

Figure 2.28 Bending of starlight passing close to the sun.

One more observation about Equation 2.40 is irresistible. Is it possible for a very massive star in the course of its life cycle to become so dense that the term $GM/R_s c^2$ becomes greater than 1? In that case Equation 2.39 shows that the photon cannot escape from the star, since escape requires more energy than the photon initially possesses. Such a star is called a *black hole* because it emits no light and acts like a celestial vacuum cleaner for all nearby matter and radiation. Even though the black hole itself is not luminous, it may be possible to observe it indirectly in two ways. One way is through the gravitational attraction the black hole would exert on a normal luminous star if the two constituted a binary star system. In this case the normal star would orbit the center of mass of the black hole–normal star pair, and the orbital motion might be detectable. A second indirect technique for "viewing" a black hole would be to search for x-rays produced by inrushing matter attracted to the black hole. Although the black hole itself would not emit x-rays, an x-ray emitting region of roughly stellar diameter should be observable, as shown in Figure 2.27. X-rays are produced by heating of the infalling matter as it circulates, is compressed, and eventually falls into the black hole. Such an intense nonluminous point source of x-rays has been detected in the constellation of the Swan. This source, designated Cygnus X-1, is believed by most astronomers to be a black hole; it possesses a luminosity or power output of 10^{30} W in the 2–10 keV x-ray range.

We shall conclude this section by discussing an extremely direct proof that the photon possesses gravitational mass. The phenomenon is the deflection of light passing close to the sun, an effect that can be predicted theoretically and has been observed. A proper treatment of this effect requires the theory of relativity,[20] but we shall obtain an order-of-magnitude answer by assuming that the photon's deflection is small. The basic idea is to measure the position of a star when its light does not pass close to the sun, and compare this position to the star's apparent position when its light passes close to the sun. This latter measurement is possible only when the sun's disk is obscured by the moon during a solar eclipse as in Figure 2.28.

To find an approximate numerical value for the deflection, it is necessary to assume only that the photon has a gravitational mass m and that its trajectory can be approximated by straight line segments as shown in Figure 2.29. Note that a photon at P' has velocity components v_x and v_y related by

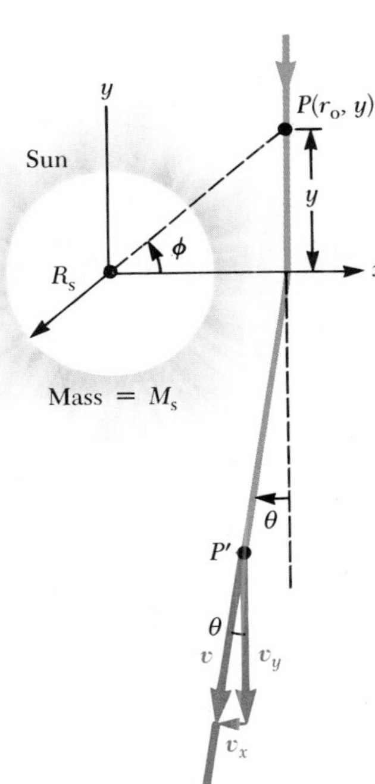

Figure 2.29 Deflection of light by the sun.

[20] L. I. Schiff, *Am. J. Phys.*, **28:** 340, 1961.

$$\tan\theta = \frac{v_x}{v_y} \tag{2.41}$$

Since θ is assumed to be small, $\tan\theta \approx \theta$. Furthermore, $v_y \approx c$, so Equation 2.41 becomes

$$\theta \approx \frac{|v_x|}{c} \tag{2.42}$$

In order to find a value for θ, we need to calculate the x-component of the photon's velocity produced by the sun's gravitational attraction. To find v_x, note that the change in momentum in the x-direction is

$$\Delta p_x = mv_x = \int F_x\, dt$$

The x-component of the gravitational force on a photon with mass m located at point P is

$$F_x = \frac{GM_s m}{(r_0{}^2 + y^2)} \cos\phi = \frac{GM_s m r_0}{(r_0{}^2 + y^2)^{3/2}} \tag{2.43}$$

Equation 2.43 strictly holds as long as the photon is undeflected. However, since θ is small it is reasonable to assume that Equation 2.43 is valid both above and below the x-axis. Thus,

$$v_x \approx GM_s \int_{y=-\infty}^{\infty} \frac{r_0\, dt}{(r_0{}^2 + y^2)^{3/2}}$$

To convert dt to dy, we use the definition $v_y = dy/dt$, which gives $dt = dy/v_y \approx dy/c$. Therefore v_x can be expressed as

$$v_x \approx \frac{2GM_s r_0}{c} \int_{y=\infty}^{0} \frac{dy}{(r_0{}^2 + y^2)^{3/2}}$$

Since

$$\int_{\infty}^{0} \frac{dy}{(r_0{}^2 + y^2)^{3/2}} = \frac{y}{r_0{}^2(y^2 + r_0{}^2)^{1/2}}\Bigg]_{\infty}^{0} = -\lim_{y\to\infty} \frac{y}{r_0{}^2(y^2 + r_0{}^2)^{1/2}} = -\frac{1}{r_0{}^2}$$

we obtain

$$v_x \approx -\frac{2GM_s}{cr_0}$$

Substituting this result into Equation 2.42, we find the angle of deflection to be

$$\theta \approx \frac{2GM_s}{c^2 r_0} \tag{2.44}$$

For light just grazing the sun's surface, we take $r_0 = R_s = 6.96 \times 10^8$ m. Substituting this value and the values for G, M_s, and c into Equation 2.44 gives

$$\begin{aligned} \theta &= \frac{(2)(6.67 \times 10^{-11}\ \mathrm{N \cdot m^2/kg^2})(1.99 \times 10^{30}\ \mathrm{kg})}{(3.00 \times 10^8\ \mathrm{m/s})^2(6.96 \times 10^8\ \mathrm{m})} \\ &= 4.24 \times 10^{-6}\ \mathrm{rad} = 2.43 \times 10^{-4}\ \mathrm{deg} \\ &= 0.874\ \text{seconds of arc} \end{aligned}$$

Einstein was the first (in recent times) to realize that the bending of light by the sun might well be detectable.[21] In 1911 he published the result given by Equation

[21] J. G. von Soldner, following up on a question raised by Newton on the gravitational deflection of light, found in 1801 that light corpuscles grazing the sun should have $\theta = 0.84''$.

2.44 and urged astronomers to look for the effect.[22] However, in 1915 he realized that Equation 2.44 was incorrect because space is non-Euclidean, and that the expression should be increased by a factor of 2, or $\theta = 1.74$ seconds of arc. When an eclipse expedition to the island of Principe in 1919 found a value of $\theta = 1.61 \pm 0.30$ seconds of arc, a tremendous outpouring of praise was showered on Einstein and he became, almost overnight, a legendary scientific hero. Perhaps dearer to Einstein than all the publicity in the press were the comments of J. J. Thomson who, presiding at the Royal Society meeting at which Einstein's predictions were confirmed, said: "This is the most important result obtained in connection with the theory of gravitation since Newton's day. . . . The result is one of the highest achievements of human thought."

2.9 SUMMARY

The work of Maxwell and Hertz in the late 1800s conclusively showed that light, heat radiation, and radio waves were all electromagnetic waves differing only in frequency and wavelength. Thus it astonished scientists to find that the spectral distribution of radiation from a heated cavity could not be explained in terms of classical wave theory. Planck was forced to introduce the concept of the quantum of energy in order to derive the correct blackbody formula. According to Planck, the atomic oscillators responsible for blackbody radiation can have only *discrete* or quantized *energies* given by

$$E = nhf \tag{2.10}$$

where n is an integer, h is Planck's constant, and f is the oscillator's natural frequency. Using general thermodynamic arguments Planck was able to show that $u(f, T)$, the blackbody radiation energy per unit volume with frequency between f and $f + df$, could be expressed as the product of the number of oscillators per unit volume in this frequency range, $N(f)\,df$, and the average energy emitted per oscillator, $\overline{E}$. That is,

$$u(f, T)\,df = N(f)\overline{E}\,df$$

If $\overline{E}$ is calculated by allowing a *continuous* distribution of oscillator energies, the incorrect Rayleigh-Jeans law is obtained. If $\overline{E}$ is calculated from a *discrete* set of oscillator energies (following Planck), the correct blackbody radiation formula is obtained:

$$u(f, T) = \frac{8\pi f^2}{c^3}\left(\frac{hf}{e^{hf/kT} - 1}\right) \tag{2.9}$$

While Planck quantized the energy of atomic oscillators, Einstein extended the concept of quantization to light itself. In Einstein's view, light of frequency f consists of a stream of particles, called *photons*, each with energy $E = hf$. The photoelectric effect, a process in which electrons are ejected from a metallic surface when light of sufficiently high frequency is incident on the surface, can be simply explained with the photon theory. According to this theory, the maximum kinetic energy of the ejected photoelectron, K_{max}, is given by

$$K_{max} = hf - \phi \tag{2.23}$$

where ϕ is the work function of the metal.

[22] A. Einstein, *Ann. Physik.*, **35:** 398, 1911.

Although the idea that light consists of a stream of photons with energy hf was put forward in 1905, the idea that these photons also carry momentum was not seriously considered until 1923. In this year it was found that x-rays scattered from free electrons suffer a shift in wavelength known as the Compton shift. When an x-ray of frequency f is viewed as a *particle* with *energy hf* and *momentum hf/c*, x-ray–electron scattering can be correctly analyzed to yield the Compton shift formula:

$$\Delta\lambda = \frac{h}{m_e c}(1 - \cos\theta) \tag{2.28}$$

Here m_e is the mass of the electron and θ is the x-ray scattering angle.

The striking success of the photon theory in explaining interactions between light and electrons contrasts sharply with the success of classical wave theory in explaining the polarization, reflection, and interference of light. This leaves us with the dilemma of whether light is a wave or a particle. The currently accepted view suggests that light has both wave and particle characteristics, and that these characteristics together constitute a complementary view of light.

Although the photon has no rest mass, it does possess a moving mass of hf/c^2. In fact, one finds that the photon is attracted gravitationally by other masses as if it were a material particle with mass hf/c^2. This gravitational interaction of photons with matter has actually been observed in cases such as the deflection of starlight passing close to the sun.

QUESTIONS

1. What assumptions were made by Planck in dealing with the problem of blackbody radiation? Discuss the consequences of these assumptions.
2. If the photoelectric effect is observed for one metal, can you conclude that the effect will also be observed for another metal under the same conditions? Explain.
3. Suppose the photoelectric effect occurs in a gaseous target rather than a solid. Will photoelectrons be produced at *all* frequencies of the incident photon? Explain.
4. How does the Compton effect differ from the photoelectric effect?
5. What assumptions were made by Compton in dealing with the scattering of a photon from an electron?
6. An x-ray photon is scattered by an electron. What happens to the frequency of the scattered photon relative to that of the incident photon?
7. Why does the existence of a cutoff frequency in the photoelectric effect favor a particle theory for light rather than a wave theory?
8. All objects radiate energy. Why, then, are we not able to see all objects in a dark room?
9. Which has more energy, a photon of ultraviolet radiation or a photon of yellow light?
10. What effect, if any, would you expect the temperature of a material to have on the ease with which electrons can be ejected from it in the photoelectric effect?
11. Some stars are observed to be reddish, and some are blue. Which stars have the higher surface temperature? Explain.
12. When wood is stacked on a special elevated grate (which is commercially available) in a fireplace, there is formed beneath the grate a pocket of burning wood whose temperature is higher than that of the burning wood at the top of the stack. Explain how this device provides more heat to the room than a conventional fire does and thus increases the efficiency of the fireplace.

PROBLEMS

1. Calculate the energy of a photon whose frequency is (a) 5×10^{14} Hz, (b) 10 GHz, (c) 30 MHz. Express your answers in eV.
2. Determine the corresponding wavelengths for the photons described in Problem 1.
3. An FM radio transmitter has a power output of 100 kW and operates at a frequency of 94 MHz. How many photons per second does the transmitter emit?
4. The average power generated by the sun has the value 3.74×10^{26} W. Assuming the average wavelength of the sun's radiation to be 500 nm, find the number of photons emitted by the sun in 1 s.

5. The temperature of your skin is approximately 35°C. What is the wavelength at which the peak occurs in the radiation emitted from your skin?
6. A sodium-vapor lamp has a power output of 10 W. Using 589.3 nm as the average wavelength of the source, calculate the number of photons emitted per second.
7. The photocurrent of a photocell is cut off by a retarding potential of 9.92 V for radiation of wavelength 250 nm. Find the work function for the material.
8. The work function for potassium is 2.24 eV. If potassium metal is illuminated with light of wavelength 350 nm, find (a) the maximum kinetic energy of the photoelectrons and (b) the cutoff wavelength.
9. Molybdenum has a work function of 4.2 eV. (a) Find the cutoff wavelength and threshold frequency for the photoelectric effect. (b) Calculate the stopping potential if the incident light has a wavelength of 200 nm.
10. When cesium metal is illuminated with light of wavelength 300 nm, the photoelectrons emitted have a maximum kinetic energy of 2.23 eV. Find (a) the work function of cesium and (b) the stopping potential if the incident light has a wavelength of 400 nm.
11. Consider the metals lithium, beryllium, and mercury, which have work functions of 2.3 eV, 3.9 eV, and 4.5 eV, respectively. If light of wavelength 300 nm is incident on each of these metals, determine (a) which metals exhibit the photoelectric effect and (b) the maximum kinetic energy for the photoelectron in each case.
12. Light of wavelength 500 nm is incident on a metallic surface. If the stopping potential for the photoelectric effect is 0.45 V, find (a) the maximum energy of the emitted electrons, (b) the work function, and (c) the cutoff wavelength.
13. The active material in a photocell has a work function of 2 eV. Under reverse-bias conditions (where the polarity of the battery in Figure 2.14 is reversed), the cutoff wavelength is found to be 350 nm. What is the value of the bias voltage?
14. A light source of wavelength λ illuminates a metal and ejects photoelectrons with a maximum kinetic energy of 1 eV. A second light source with half the wavelength of the first ejects photoelectrons with a maximum kinetic energy of 4 eV. What is the work function of the metal?
15. Calculate the energy and momentum of a photon of wavelength 500 nm.
16. X-rays of wavelength 0.200 nm are scattered from a block of carbon. If the scattered radiation is detected at 90° to the incident beam, find (a) the Compton shift $\Delta\lambda$ and (b) the kinetic energy imparted to the recoiling electron.
17. X-rays with an energy of 300 keV undergo Compton scattering from a target. If the scattered rays are detected at 30° relative to the incident rays, find (a) the Compton shift at this angle, (b) the energy of the scattered x-ray, and (c) the energy of the recoiling electron.
18. X-rays with a wavelength of 0.040 nm undergo Compton scattering. (a) Find the wavelength of photons scattered at angles of 30°, 60°, 90°, 120°, 150°, and 180°. (b) Find the energy of the scattered electrons corresponding to these scattered x-rays. (c) Which one of the given scattering angles provides the electron with the greatest energy?
19. Figure 2.30 shows the stopping potential versus incident photon frequency for the photoelectric effect for sodium. Use these data points to find (a) the work function, (b) the ratio h/e, and (c) the cutoff wavelength. (Data taken from R. A. Millikan, Phys. Rev. **7:** 362 [1916].)

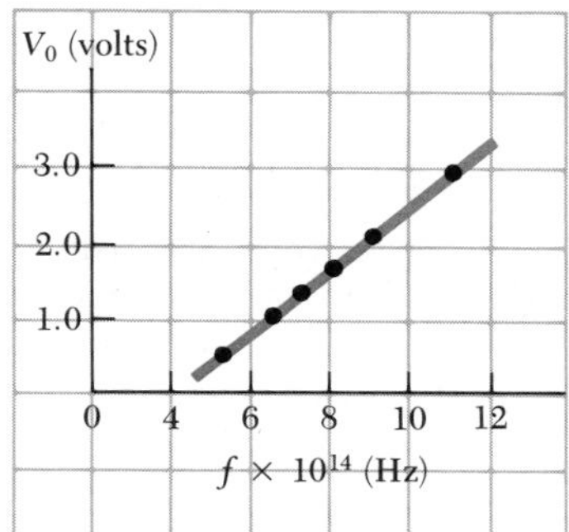

Figure 2.30 (Problem 19) Some of Millikan's original data for sodium.

20. A 2-kg mass is attached to a massless spring of force constant $k = 25$ N/m. The spring is stretched 0.4 m from its equilibrium position and released. (a) Find the total energy and frequency of oscillation according to classical calculations. (b) Assume that the energy is quantized and find the quantum number, n, for the system. (c) How much energy would be carried away in a one-quantum change?
21. Photons of wavelength 450 nm are incident on a metal. The most energetic electrons ejected from the metal are bent into a circular arc of radius 20 cm by a magnetic field whose strength is equal to 2×10^{-5} T. What is the work function of the metal?
22. Gamma rays (high-energy photons) of energy 1.02 MeV are scattered from electrons that are initially at rest. If the scattering is *symmetric,* that is, if $\theta = \phi$, find (a) the scattering angle θ and (b) the energy of the scattered photons.
23. If the maximum energy given to an electron during Compton scattering is 30 keV, what is the wavelength of the incident photon?
24. A photon of initial energy 0.1 MeV undergoes Compton scattering at an angle of 60°. Find (a) the energy of the scattered photon, (b) the recoil energy of the electron, and (c) the recoil angle of the electron.

25. Show that a photon cannot transfer all of its energy to a free electron. (*Hint:* Note that energy and momentum must be conserved.)

26. Show that the ratio of the Compton wavelength to the de Broglie wavelength for a *relativistic* electron is given by

$$\frac{\lambda_c}{\lambda} = \left[\left(\frac{E}{m_0c^2}\right)^2 - 1\right]^{1/2}$$

where E is the total energy of the electron and m_0 is its rest mass.

27. The total power per unit area radiated by a black body at a temperature T is given by the expression $E = \sigma T^4$. (a) Show that this power per unit area is given by

$$\frac{c}{4}\int_0^\infty u(\lambda, T)\, d\lambda = \sigma T^4$$

where $u(\lambda, T)$ is given by Planck's radiation law

$$u(\lambda, T) = \frac{8\pi hc}{\lambda^5(e^{hc/\lambda kT} - 1)}$$

This result is known as the *Stefan-Boltzmann law.* To carry out the integration, you should make the change of variable $x = hc/\lambda kT$ and use the fact that

$$\int_0^\infty \frac{x^3\, dx}{e^x - 1} = \frac{\pi^4}{15}$$

(b) Show that the Stefan-Boltzmann constant σ has the value

$$\sigma = \frac{2\pi^5k^4}{15c^2h^3} = 5.67 \times 10^{-8}\ \mathrm{W/m^2 \cdot K^4}$$

28. (a) Use the results of Problem 27 to calculate the total power radiated per unit area by a tungsten filament at a temperature of 3000 K. (Assume that the filament is an ideal radiator.) (b) If the tungsten filament of a light bulb is rated at 75 W, what is the surface area of the filament? (Assume that the main energy loss is due to radiation.)

29. Consider the problem of the distribution of blackbody radiation described in Figure 2.3. Note that as T *increases,* the wavelength λ_{max} at which $u(\lambda, T)$ reaches a maximum shifts toward shorter wavelengths. (a) Show that there is a general relationship between temperature and λ_{max} which states that $T\lambda_{max} =$ constant (*Wien's displacement law*). (b) Obtain a numerical value for this constant. [*Hint:* Start with Planck's radiation law and note that the slope of $u(\lambda, T)$ is zero when $\lambda = \lambda_{max}$.]

30. In the Compton scattering event illustrated in Figure 2.22b, the scattered photon has an energy of 120 keV and the recoiling electron has an energy of 40 keV. Find (a) the wavelength of the incident photon, (b) the angle θ at which the photon is scattered, and (c) the recoil angle ϕ of the electron.

31. An excited iron (Fe) nucleus (mass 57 u) decays to its ground state with the emission of a photon. The energy available from this transition is 14.4 keV. (a) By how much is the photon energy reduced from the full 14.4 keV as a result of having to share energy with the recoiling atom? (b) What is the wavelength of the emitted photon?

32. Find the energy of an x-ray photon that can impart a maximum energy of 50 keV to an electron by Compton collision.

33. Compton used photons of wavelength 0.0711 nm. (a) What is the energy of these photons? (b) What is the wavelength of the photons scattered at an angle of 180° (backscattering case)? (c) What is the energy of the backscattered photons? (d) What is the recoil energy of the electrons in this case?

34. A photon undergoing Compton scattering has an energy of 80 keV, and the electron recoils with an energy of 25 keV. (a) Find the wavelength of the incident photon. (b) Find the angle at which the photon is scattered. (c) Find the angle at which the electron recoils.

35. In a Compton collision with an electron, a photon of violet light ($\lambda = 4000$ Å) is backscattered through an angle of 180°. How much energy (eV) is transferred to the electron in this collision? Compare your result with the energy this electron would acquire in a photoelectric process with the same photon. Could violet light eject electrons from a metal by Compton collision? Explain.

36. The essentials of calculating the number of modes of vibration of waves confined to a cavity may be exposed by considering a one-dimensional example. (a) Calculate the number of modes (standing waves of different wavelength) with wavelengths between 2.0 cm and 2.1 cm that can exist on a string with fixed ends which is 2 m long. (*Hint:* use $n(\lambda/2) = L$, where $n = 1, 2, 3, 4, 5.\ .\ .\ .$ Note that a specific value of n defines a specific mode or standing wave with different wavelength.) (b) Calculate, in analogy to our three-dimensional calculation, the number of modes per unit wavelength per unit length, $\dfrac{\Delta n}{L\,\Delta\lambda}$. (c) Show that in general the number of modes per unit wavelength per unit length for a string of length L is given by

$$\frac{1}{L}\left|\frac{dn}{d\lambda}\right| = \frac{2}{\lambda^2}$$

Does this expression yield the same numerical answer as found in (a)? (d) Under what conditions is it justified to replace $\left|\dfrac{\Delta n}{L\,\Delta\lambda}\right|$ with $\left|\dfrac{dn}{L\,d\lambda}\right|$? Is the expression $n = 2L/\lambda$ even a continuous function?

37. *Classical Zeeman Effect or the Triumph of Maxwell's Equations!* As pointed out in Section 2.1, Maxwell's

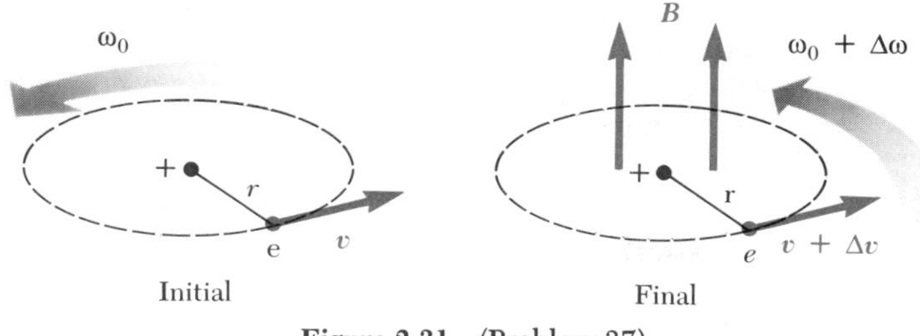

Figure 2.31 (Problem 37)

equations may be used to predict the change in emission frequency when gas atoms are placed in a magnetic field. Consider the situation shown in Figure 2.31. Note that the application of a magnetic field perpendicular to the orbital plane of the electron induces an electric field, which speeds up the electron as shown. (a) Using $\oint \boldsymbol{E}\cdot d\boldsymbol{s} = -\frac{d\Phi_B}{dt}$ show that the electric field which speeds up the electron is given by

$$E = \frac{r}{2}\frac{dB}{dt}$$

(b) Using $F\,dt = m\,dv$, calculate the change in velocity Δv of the electron. Show that if r remains constant,

$$\Delta v = \frac{erB}{2m_e}$$

(c) Find the change in angular frequency, $\Delta\omega$, of the electron and calculate the numerical value of $\Delta\omega$ for B equal to 1 T. Note that this is also the change in frequency of the light emitted according to Maxwell's equations. Find the fractional change in frequency $\Delta\omega/\omega$ for an ordinary emission line of 500 nm. (d) Actually, the original emission line at ω_0 is split into three components at $\omega_0 - \Delta\omega$, ω_0, and $\omega_0 + \Delta\omega$. The line at $\omega_0 + \Delta\omega$ is produced by atoms with electrons rotating as shown in Figure 2.31, while the line at $\omega_0 - \Delta\omega$ is produced by atoms with electrons rotating in the opposite sense. The line at ω_0 is produced by atoms with electronic planes of rotation oriented parallel to $\boldsymbol{B}$. Explain.

38. X-radiation from a K-shell transition in molybdenum (0.626 Å) is incident on a crystal with adjacent atomic planes spaced 4.0×10^{-10} m apart. Find the three smallest angles at which intensity maxima occur in the diffracted beam.

39. As a single crystal is rotated in an x-ray spectrometer (Fig. 2.21a), many parallel planes of atoms besides AA and BB produce strong diffracted beams. Two such planes are shown in Figure 2.32. (a) Determine geometrically the interplanar spacings d_1 and d_2 in terms of d_0. (b) Find the angles (with respect to the surface plane AA) of the $n = 1$, 2, and 3 intensity maxima from planes with spacing d_1. Let $\lambda = 0.626$ Å and $d_0 =$

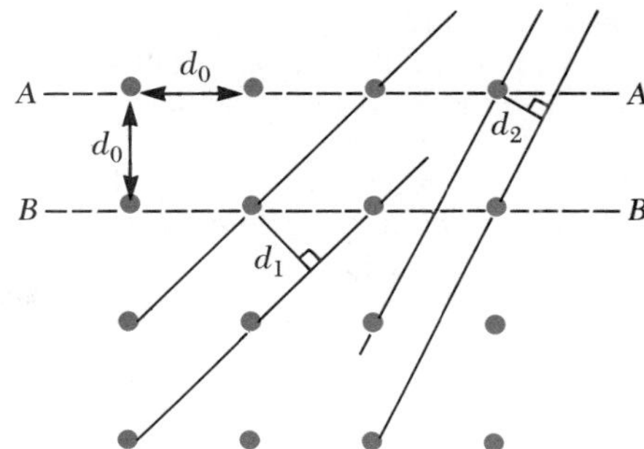

Figure 2.32 (Problem 39) Atomic planes in a cubic lattice.

4.00 Å. Note that a given crystal structure (e.g., cubic) has interplanar spacings with characteristic ratios, which produce characteristic diffraction patterns. Thus measurement of the angular position of diffracted x-rays may be used to infer the crystal structure.

40. *The determination of Avogadro's number with x-rays.* X-rays from a molybdenum target (0.626 Å) are incident on an NaCl crystal, which has the atomic arrangement shown in Figure 2.33. If NaCl has a density of 2.17 g/cm³ and the $n = 1$ diffraction maximum from planes separated by d is found at $\theta = 6.41°$, compute Avogadro's number. (*Hint:* First determine d. Using Figure 2.33, determine the number of NaCl molecules per primitive cell and set the mass per unit volume of the primitive cell equal to the density.)

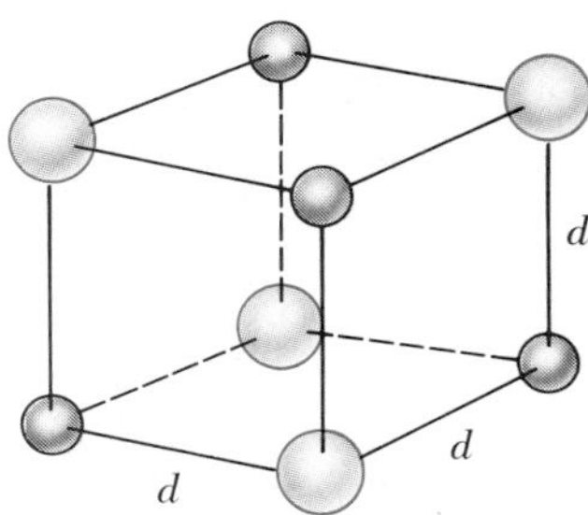

Figure 2.33 (Problem 40) The primitive cell of NaCl.

41. Show that the Compton formula

$$\lambda' - \lambda_0 = \frac{h}{m_e c}(1 - \cos\theta)$$

results when expressions for the electron energy (Equation 2.34) and momentum (Equation 2.35) are substituted into the relativistic energy expression,

$$E_e^2 = p_e^2c^2 + m_e^2c^4$$

*42. In deriving expressions for the change in frequency of a photon falling or rising in a gravitational field, we have assumed a small change in frequency and a constant photon mass of hf/c^2. Suppose that a star is so dense that Δf is not small. (a) Show that f', the photon frequency at ∞, is related to f, the photon frequency at the star's surface, by

$$f' = fe^{-GM_s/R_sc^2}$$

(b) Show that this expression reduces to Equation 2.40 for small M_s/R_s. (*Hint:* The decrease in photon energy, $h\,df$, as the photon moves dr away from the star is equal to the work done against gravity, $F_G\,dr$.)

*43. If our sun were to contract and become a black hole, (a) what would its radius be and (b) by what factor would its density increase?

*Problems marked with an asterisk relate to an optional section.

3

The Particle Nature of Matter

3.1 INTRODUCTION

In Chapter 2 we reviewed the evidence for the wave nature of electromagnetic radiation and dealt with the major experimental puzzles of the first quarter of the 20th century, which required a particle-like behavior of radiation for their solution. In particular, we discussed Planck's revolutionary idea of energy quantization of oscillators in the walls of a perfect radiator, Einstein's extension of energy quantization to light in his explanation of the photoelectric effect, and Compton's further confirmation of the existence of the photon as a particle carrying momentum in x-ray scattering experiments.

In this chapter we shall examine the evidence for the particle nature of matter. We shall pay only passing tribute to the early atomists and concentrate instead on the developments from 1800 onward that dealt with the composition of atoms. In particular, we shall review the ingenious and fascinating experiments that led to the discoveries of the electron, the proton, the nucleus, and the important Rutherford-Bohr planetary model of the atom.

3.2 ATOMIC NATURE OF MATTER

To say that the world is made up of atoms is, today, commonplace. Because students often accept the atomic picture of reality without question, they miss out on the rich and fascinating story of how atoms were shown to be real. The discovery and proof of the graininess of the world seem especially fascinating to us for two reasons. First, because of the size of individual atoms, measurements of atomic properties are usually indirect and necessarily involve clever manipulations of large scale measurements to infer properties of microscopic particles. Second, the historical evolution of ideas about atomicity shows clearly the real way in which science progresses. This progression is often nonlinear, and it involves an interdependence of physics, chemistry, and mathematics, and the convergence of many different lines of investigation.

There is also an exalted romance in honoring the great atomists who were able to pick out organizing principles from the confusing barrage of marketplace ideas of their time: Democritus and Leucippus, who speculated that the unchanging substratum of the world was atoms in motion; the debonair Lavoisier, who established the conservation of matter in so many careful chemical experiments; Dalton, who perceived the atomicity of nature in the law of multiple proportions of compounds; Avogadro, who in a most obscure and little appreciated paper, postulated that all pure gases at the same temperature and pressure have the same number of molecules per unit volume; and Maxwell,[1] who showed with his molecular-kinetic theory of gases how macro-

[1] Maxwell was a genius twice over. Either his theory of electricity and magnetism or his kinetic theory of gases would qualify him for that rank.

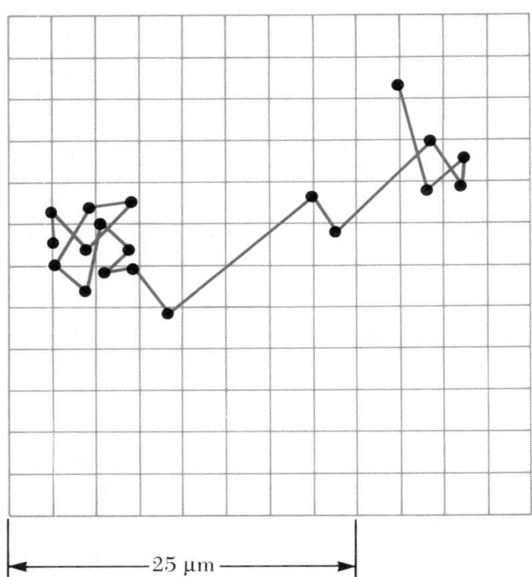

Figure 3.1 Brownian motion of a mastic grain. Successive horizontal positions of a mastic grain observed at 30 s intervals. The colored lines show the *net* displacements of the mastic grain between observations, not its actual path in space. The radius of the mastic grain was 0.5×10^{-6} m. From Jean Perrin, *Atoms*, translated by D. L. Hammick, New York, D. Van Nostrand Co., 1923.

scopic quantities such as pressure and temperature could be derived from averages over distributions of molecular properties. The list could run on and on. We abbreviate it by naming Jean Perrin and the ubiquitous Albert Einstein,[2] who carried on very important theoretical and experimental work concerning Brownian motion, the zig-zag motion of small suspended particles caused by molecular impacts. (See Figure 3.1 for an illustration.) Their work resulted in additional confirmation of the atomic-molecular hypothesis and resulted in improved values of Avogadro's number as late as the early 1900s.

3.3 THE COMPOSITION OF ATOMS

We now turn our attention to answering the dangerous leading question, "If matter is primarily composed of atoms, what are atoms composed of?" Again, we can point to some primary discoveries that showed that atoms are composed of negatively charged particles orbiting a dense positively charged nucleus. These were:

1. *The discovery of the law of electrolysis in 1833 by Michael Faraday.* Through careful experimental work on electrolysis, Faraday showed that the mass of an element liberated at an electrode is directly proportional to the charge transferred and to the atomic weight of the liberated material, but is inversely proportional to the valence of the freed material.
2. *The identification of cathode rays as electrons and the measurement of the charge to mass ratio (e/m) of these particles by J. J. Thomson in 1897.* Thomson measured the properties of negative particles emitted from different metals in different gases and found a constant value of e/m. He thus came to the revolutionary conclusion that the electron is a constituent of all matter!
3. *The precise measurement of the electronic charge by Robert Millikan in 1909.* By combining his result with Thomson's e/m value (about half of the cur-

[2] Much of Einstein's earliest work was concerned with the molecular analysis of solutions and determinations of molecular radii and Avogadro's number. See A. Pais, *'Subtle is the Lord . . .' The Science and the Life of Albert Einstein,* New York, Oxford University Press, 1982, Chapter 5.

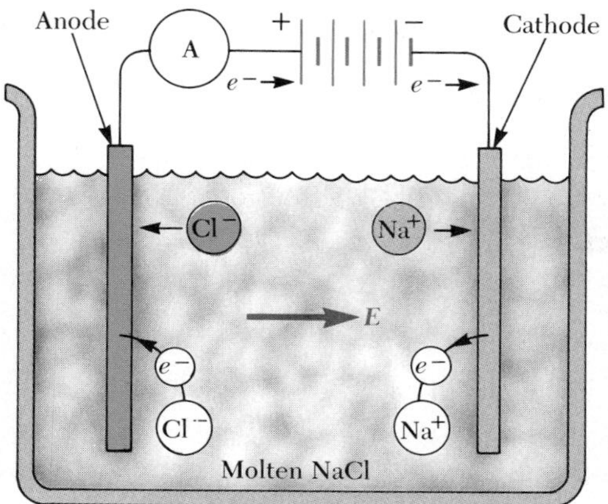

Figure 3.2 Electroylsis of molten NaCl.

rently accepted value), Millikan showed unequivocally that particles about 1000 times less massive than the hydrogen atom exist.

4. *The establishment of the nuclear model of the atom by Ernest Rutherford and co-workers Geiger and Marsden in 1910.* By scattering fast moving α-particles from metal foil targets, Rutherford established that atoms consist of a compact positively charged nucleus (diameter $\sim 10^{-14}$ m) surrounded by a swarm of orbiting electrons (electron cloud diameter $\sim 10^{-10}$ m).

Let us describe these developments in more detail. We start with a brief example of Faraday's experiments, in particular the electrolysis of molten common salt (NaCl). Faraday found that if 96 500 coulombs of charge (1 Faraday) are passed through such a molten solution, 23 grams of Na would deposit on the cathode and 35.5 grams of chlorine gas would bubble off the anode (Fig. 3.2). In this case exactly one gram atomic weight or mole of each element is released, since both are monovalent. For divalent and trivalent elements, exactly $\frac{1}{2}$ and $\frac{1}{3}$ of a mole would be released. As expected, doubling the quantity of charge passed doubles the mass of the neutral element liberated. Faraday's results may be given in equation form as

$$m = \frac{(q)(\text{molar weight})}{(96\ 500\ \text{coulombs})(\text{valence})} \tag{3.1}$$

Faraday's law of electrolysis

where m is the mass of the liberated substance in grams, q is the total charge passed in coulombs, the molar weight is in grams, and the valence is dimensionless.

EXAMPLE 3.1 The Electrolysis of $BaCl_2$

How many grams of barium and chlorine (cough) will you get if you pass a current of 10 A through molten $BaCl_2$ for 1 hour? Barium has a molar weight of 137 g and a valence of 2. Chlorine has a molar weight of 35.5 g and a valence of 1.

Solution: Using Equation 3.1 and $q = it$, we have

$$m_{\text{Ba}} = \frac{(q)(\text{molar weight})}{(96{,}500\ \text{C})(\text{valence})}$$

$$= \frac{(10\ \text{C/s})(3600\ \text{s})(137\ \text{g})}{(96{,}500\ \text{C})(2)} = 25.6\ \text{g}$$

$$m_{\text{Cl}} = \frac{(10\ \text{C/s})(3600\ \text{s})(35.5\ \text{g})}{(96{,}500\ \text{C})(1)} = 13.2\ \text{g}$$

Figure 3.3 The original *e/m* tube used by J. J. Thomson (after Figure 1.3, p. 7, R. L. Sproull and W. A. Phillips, *Modern Physics,* 3rd ed., New York, John Wiley & Sons, 1980).

Faraday's law of electrolysis is explained in terms of an atomic picture shown in Figure 3.2. Charge passes through the molten solution in the form of ions, which possess an excess or deficiency of one or more electrons. Under the influence of the electric field produced by the battery, these ions move to the anode or cathode, where they respectively lose or gain electrons and are liberated as neutral atoms.

Although it was far from clear in 1833, Faraday's law of electrolysis confirmed three important parts of the atomic picture. First, it offered proof that matter consists of molecules, and that molecules consist of atoms. Second, it showed that charge is quantized, since only integral numbers of charges are transferred at the electrodes. Third, it showed that the subatomic parts of atoms are positive and negative charges, although the mass and the size of the charge of these subatomic particles remained unknown.

The next major step elucidating the composition of atoms was taken by Joseph John (J. J.) Thomson. His discovery in 1897[3] that the "rays" seen in low pressure gas discharges were actually caused by negative particles (electrons) ended a debate dating back nearly 30 years: were cathode rays material particles or waves of some type? Contrary to our rather blasé present acceptance of the electron, many of Thomson's distinguished contemporaries responded initially with utter disbelief to the idea that electrons are a constituent of all matter. Much of the opposition to Thomson's discovery stemmed from the fact that it required the abandonment of the recently established concept of the atom as an indivisible entity. Thomson's discovery of the electron disturbed this newly established order in atomic theory and provoked startling new developments—Rutherford's nuclear model (1909), and within four years the first satisfactory theory of the emission of light by atomic systems, the Bohr model of the atom.

Figure 3.3 shows the original vacuum tube used by Thomson in his *e/m* experiments. Figure 3.4 shows the various parts of the Thomson apparatus for easy reference. Electrons are accelerated from the cathode to the anode, collimated by slits in the anodes, and then allowed to drift into a region of crossed (perpendicular) electric and magnetic fields. One can actually see a huge electromagnet used by Thomson to produce the ***B*** field in Figure 3.5. The simultaneously applied ***E*** and ***B*** fields are first adjusted to produce an undeflected beam. If the ***B*** field is then turned off, the ***E*** field alone produces a measurable beam deflection on the phosphorescent screen. From the size of

[3] J. J. Thomson, *Phil. Mag.* **44:** 269, 1897.

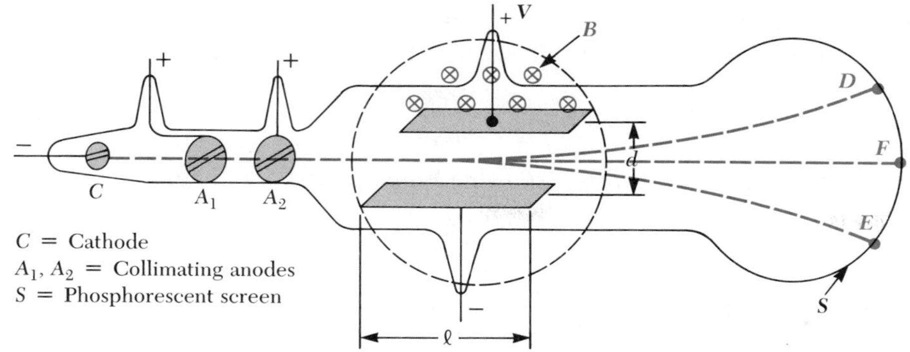

Figure 3.4 A diagram of Thomson's e/m tube (patterned after J. J. Thomson, *Philosophical Magazine* (5) **44:** 293, 1897. Electrons subjected to an electric field alone land at D, while those subjected to a magnetic field alone land at E. When both electric and magnetic fields are present and properly adjusted, the electrons experience no net deflection and land at F.

the deflection and the measured values of $\boldsymbol{E}$ and $\boldsymbol{B}$, the charge to mass ratio, e/m, may be determined.[4] The truly ingenious feature of this experiment is the manner in which Thomson measured v_x, the horizontal velocity component of the beam. He did this by balancing the magnetic and electric forces. In effect he created a velocity selector, which could select out of the beam those particles having a velocity within a narrow range of values. This device was extensively used in the first quarter of the 20th century in q/m measurements on many particles and in early mass spectrometers.

To gain a clearer picture of the Thomson experiment, let us analyze the electron's motion in his apparatus. Figure 3.6 shows the trajectory of a beam of negative particles entering the $\boldsymbol{E}$ and $\boldsymbol{B}$ field regions with horizontal velocity v_x. Consider first an $\boldsymbol{E}$ field alone between the plates. For this case v_x remains constant throughout the motion because there is no force acting in the x-direction. The y-component of velocity, v_y, is constant everywhere except between the plates, where the electron experiences a constant upward acceleration due to the electric force and consequently moves in a parabolic path. In order to solve for the deflection angle, θ, we must solve for v_x and v_y. Because v_y initially is zero, the electron leaves the plates with a y-component of velocity given by

$$v_y = a_y t \tag{3.2}$$

Figure 3.5 J. J. Thomson in the Cavendish Laboratory, University of Cambridge (picture is courtesy of Cavendish Laboratory).

[4] Thomson actually measured horizontal velocity of electrons in his apparatus at $\sim 0.1c$. Consequently there is little relativistic mass increase and we are technically referring to the ratio charge/rest mass or e/m_0.

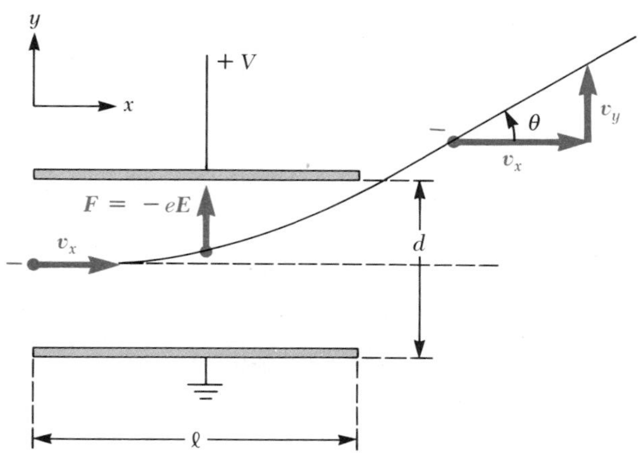

Figure 3.6 Deflection of negative particles by an electric field.

Since $a_y = F/m = Ee/m = Ve/md$, and $t = \ell/v_x$, where d and ℓ are the dimensions of the region between the plates and V is the applied potential, we obtain

$$v_y = \frac{V\ell e}{mv_x d} \tag{3.3}$$

From Figure 3.3, $\tan\theta = v_y/v_x$, so using Equation 3.3 we obtain

$$\tan\theta = \frac{V\ell}{v_x^2 d}\left(\frac{e}{m}\right) \tag{3.4}$$

Assuming small deflections, $\tan\theta \approx \theta$, so we have

$$\theta \approx \frac{V\ell}{v_x^2 d}\left(\frac{e}{m}\right) \tag{3.5}$$

Note that θ, the beam deflection, V, the voltage applied to the horizontal deflecting plates, and d and ℓ, the spacing and length, respectively, of the horizontal deflecting plates can all be measured. Hence, one needs only to measure v_x to determine e/m. Thomson determined v_x by applying a $\boldsymbol{B}$ field and adjusting its magnitude to just balance the deflection of the still present $\boldsymbol{E}$ field. Equating the magnitudes of the electric and magnetic forces gives

$$qE = qv_x B$$

or

$$v_x = \frac{E}{B} = \frac{V}{Bd} \tag{3.6}$$

Substituting Equation 3.6 into Equation 3.5 immediately yields a formula for e/m entirely in terms of measurable quantities:

$$\frac{e}{m} = \frac{V\theta}{B^2\ell d} \tag{3.7}$$

The currently accepted value of e/m is 1.758803×10^{11} C/kg. Although Thomson's original value was only about 1.0×10^{11} C/kg, prior experiments on the electrolysis of hydrogen ions had given q/m values for hydrogen of about 10^8 C/kg. It was clear almost immediately that Thomson had discovered a particle with a mass about 1000 times smaller than the smallest atom! In his observations, Thomson noted that the e/m ratio was independent of the discharge gas and the cathode metal. Furthermore, the particles emitted in a low pressure discharge gas were found to be the same as those observed in the photoelectric effect. Based on these observations, Thomson concluded that these particles must be a universal constituent of all matter. Man had achieved his first glimpse into the subatomic world!

EXAMPLE 3.2 Deflection of an Electron Beam by *E* and *B* Fields

Using the accepted e/m value, calculate the magnetic field required to produce a deflection of 0.20 radians in Thomson's experiment, assuming the values $V = 200$ V, $\ell = 5$ cm, and $d = 1.5$ cm (the approximate values used by Thomson). Compare this value of B to the earth's magnetic field.

Solution: Since $e/m = V\theta/B^2\ell d$, solving for B gives

$$B = \sqrt{\frac{V\theta}{\ell d(e/m)}}$$

so

$$B = \left[\frac{(200\text{ V})(0.20\text{ rad})}{(0.05\text{ m})(0.015\text{ m})(1.76 \times 10^{11}\text{ C/kg})}\right]^{1/2}$$

$$= [3.03 \times 10^{-7}\ \text{V}\cdot\text{kg/m}^2\cdot\text{C}]^{1/2}$$

$$= [3.03 \times 10^{-7}\ \text{N}^2/(\text{m/s})^2\cdot\text{C}^2]^{1/2}$$

$$= 5.5 \times 10^{-4}\ \text{N/(m/s)}\cdot\text{C}$$

$$= 5.5 \times 10^{-4}\ \text{T} \quad \text{or} \quad 5.5\ \text{gauss}$$

As the earth's magnetic field has a magnitude of about 0.5 gauss, we require a field 11 times as strong as the earth's field.

Exercise 1 Find the horizontal velocity v_x for this case.
Answer: 2.4×10^7 m/s $= 0.08c$, where c is the velocity of light.

Millikan's Value of the Elementary Charge

In 1897 Thomson had been unable to determine e or m separately. However, about two years later this great British experimentalist had bracketed the accepted value of e (1.602×10^{-19} C) with values of 2.3×10^{-19} C for charges emitted from zinc illuminated by ultraviolet light and 1.1×10^{-19} C for charges produced by ionizing x-rays and radium emissions. He was also able to conclude that "*e* is the same in magnitude as the charge carried by the hydrogen atom in the electrolysis of solutions." The technique used by Thomson and his students to measure e is especially interesting because it represents the first use of the cloud chamber technique in physics, and also formed the starting point for the famous Millikan oil-drop experiment. Charles Wilson, one of Thomson's students, had discovered that ions act as condensation centers for water droplets when damp air is cooled by a rapid expansion. Thomson used this idea to form charged clouds by using the apparatus shown in Figure 3.7a. Here Q is the measured total charge of the cloud, W is the measured weight of the cloud, and v is the rate of fall or terminal velocity. Thomson assumed that the cloud was composed of spherical droplets having a constant mass (no evaporation), and that the drag force D on a single droplet is given by Stokes's law,

$$D = 6\pi a\xi v \tag{3.8}$$

where a is the droplet radius, ξ is the viscosity of air, and v is the terminal velocity of the droplet. The following procedure was used to find a and w, the weight of a single drop. Since v is constant, the droplet is in equilibrium under

Robert Millikan (1868–1953). Although Millikan studied Greek as an undergraduate at Oberlin College, he fell in love with physics during his graduate training after teaching school for a few years. He was the first student to receive a PhD in physics from Columbia University in 1895. Following postdoctoral work in Germany under Planck and Nernst, Millikan obtained an academic appointment at the University of Chicago in 1910 where he worked with Michelson. He received the Nobel Prize in 1923 for his famous experimental determination of the electronic charge. He is also remembered for his careful experimental work to verify the theory of the photoelectric effect deduced by Einstein. Following World War I, he transferred to the California Institute of Technology, where he worked in atmospheric physics and remained until his retirement.

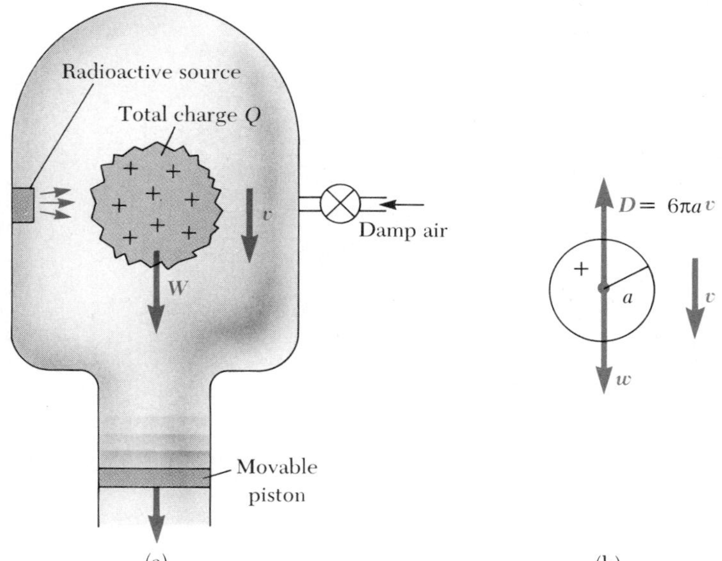

Figure 3.7 (a) A diagram of Thomson's apparatus for determining e. (b) A single droplet in the cloud.

the combined action of its weight, w, and the drag force, D, as shown in Figure 3.7b. Hence, we require that $w = D$, or

$$w = \tfrac{4}{3}\pi a^3 \rho g = D = 6\pi a \xi v$$

so

$$a = \sqrt{\frac{9\xi v}{2\rho g}} \tag{3.9}$$

where ρ is the mass density of the droplet and g is the acceleration of gravity. Once w is obtained, the number of drops n (or number of ions) is given by W/w and the electronic charge e is equal to Q/n, assuming that each droplet carries only one ion. Although ingenious, this method is inaccurate because (1) the theory applies only to a single particle, (2) the particles are all assumed to be identical, and (3) the theory is then compared to experiments performed on a cloud.

The tremendous advance of Millikan was made possible by his clever idea of making the experiment "fit" the theory. By observing single droplets he eliminated the problems of assuming all particles to be identical and of making uncertain measurements on a cloud. Millikan's basic idea was to measure the rate of fall of a single drop acted on by the force of gravity alone, then to measure its upward velocity in an opposing electric field, apply Stoke's law to determine the radius of the drop and its mass, and hence determine the electronic charge.[5]

A schematic of the Millikan apparatus is shown in Figure 3.8. Oil droplets charged by friction in an atomizer are allowed to pass through a small hole in the upper plate of a parallel-plate capacitor. If these droplets are illuminated

[5] Actually the idea of allowing charges to "fall" under a combined gravitational and electric field was first applied to charged clouds by H. A. Wilson in 1903.

Figure 3.8 A schematic view of the Millikan oil-drop apparatus.

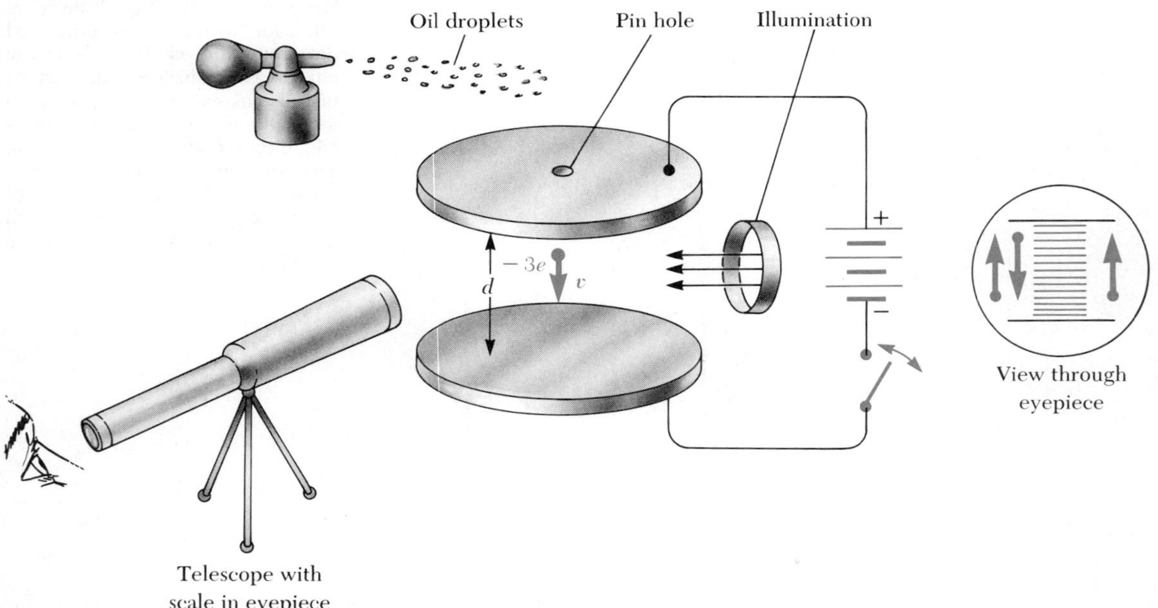

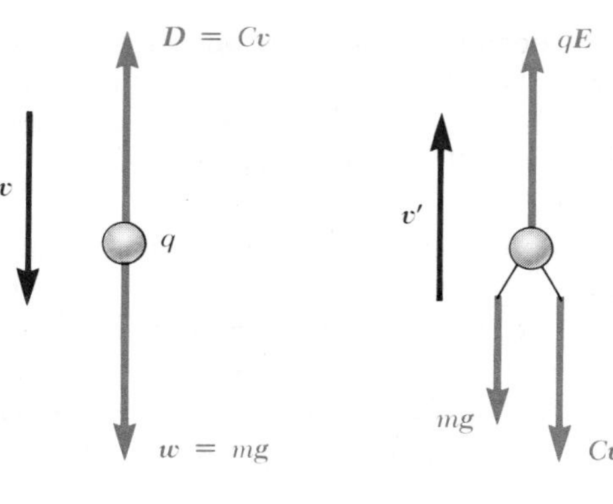

Figure 3.9 The forces on a charged oil droplet in the Millikan experiment.

from the side, they appear as brilliant stars against a dark background, and the rate of fall of individual drops may be determined.[6] If an electrostatic field of several thousand volts is applied to the capacitor plates the drop may move slowly upward, typically at rates of *hundredths* of a centimeter per second. Since the rate of fall is comparable, a single droplet with constant mass and radius may be followed for hours, alternately rising and falling, by simply turning the electric field on and off. The atomicity of charge is shown directly by the observation that after a long series of measurements of constant upward velocities one observes a discontinuous change or jump to a different *upward* velocity (higher or lower). This discontinuous change is caused by the attraction of an ion to the charged droplet. Such changes become more frequent when a source of ionizing radiation is placed between the plates.

Millikan's determination of the electronic charge

The quantitative analysis of the Millikan experiment starts with Newton's second law applied to the oil drop, $\Sigma F_y = ma_y$. Since the drag force is large, a constant velocity of fall is quickly achieved and all measurements are made for the case $a_y = 0$, or $\Sigma F_y = 0$. If we assume that the drag force is proportional to the velocity ($D = Cv$), and refer to Figure 3.9, we find

$$Cv - mg = 0 \qquad \text{(field off)}$$

$$qE - mg - Cv' = 0 \qquad \text{(field on)}$$

Eliminating C from these expressions gives

$$q = \frac{mg}{E}\left(\frac{v + v'}{v}\right) \tag{3.10}$$

When the droplet undergoes a discontinuous change in its upward velocity from v' to v_1' (m, g, E, and v remaining constant), its new charge q_1 is given by

$$q_1 = \frac{mg}{E}\left(\frac{v + v_1'}{v}\right) \tag{3.11}$$

Dividing Equation 3.10 by Equation 3.11 gives

$$\frac{q}{q_1} = \frac{v + v'}{v + v_1'} \tag{3.12}$$

[6] Perhaps the reason for the failure of "Millikan's Shining Stars" as a poetic and romantic image has something to do with the generations of physics students who have experienced hallucinations, near blindness, migraine attacks, etc. while repeating his experiment!

Equation 3.12 constitutes a remarkably direct and powerful proof of the atomicity of charge, because if successive velocity ratios are ratios of whole numbers, successive charges on the drop must be multiples of the same elementary charge! Millikan's experimental results beautifully confirmed this atomicity of charge in 1911 to within about 1% accuracy.[7]

Up to this point our arguments have been quite general and have assumed only that the drag force on the droplet is proportional to its velocity. To determine the actual value of the electronic charge, e, the mass of the drop must be determined, as can be seen from Equation 3.10. As noted earlier, the droplet radius a may be determined from the application of Stokes's law. This value of a, in turn, can be used to find m from the oil density, ρ. In this procedure, a is given by

$$a = \sqrt{\frac{9\xi v}{2\rho g}} \tag{3.9}$$

and the mass of the droplet can be expressed as

$$m = \rho \cdot \text{Volume} = \rho \tfrac{4}{3}\pi a^3 \tag{3.13}$$

An example of this technique for determining e using Stokes's law is given in Example 3.3.

[7] R. A. Millikan, *Phys. Rev.* 1911, p. 349.

EXAMPLE 3.3 Experimental Determination of e

In a Millikan experiment the distance of rise or fall of a droplet is 0.60 cm and the average time of fall (field off) is 21.0 s. The observed successive rise times are 46.0, 15.5, 28.1, 12.9, 45.3, and 20.0 s.

(a) Prove that charge is quantized.

Solution: Charge is quantized if q/q_1, q/q_2, q/q_3, and so on are ratios of small whole numbers, since

$$\frac{q}{q_1} = \frac{v+v'}{v+v_1'}, \qquad \frac{q_1}{q_2} = \frac{v+v_1'}{v+v_2'}, \qquad \text{etc.}$$

Thus

$$v = \frac{\Delta y}{\Delta t} = \frac{0.60\text{ cm}}{21.0\text{ s}} = 0.0286\text{ cm/s}$$

$$v' = \frac{\Delta y}{\Delta t} = \frac{0.60\text{ cm}}{46.0\text{ s}} = 0.0130\text{ cm/s}$$

$$v_1' = 0.60/15.5 = 0.0387\text{ cm/s}$$

$$v_2' = 0.60/28.1 = 0.0214\text{ cm/s}$$

$$v_3' = 0.60/12.9 = 0.0465\text{ cm/s}$$

$$v_4' = 0.60/45.3 = 0.0132\text{ cm/s}$$

$$v_5' = 0.60/20.0 = 0.0300\text{ cm/s}$$

so

$$\frac{q}{q_1} = \frac{v+v'}{v+v_1'} = \frac{0.0286+0.0130}{0.0286+0.0387} = 0.618 \approx 3/5$$

$$\frac{q_1}{q_2} = \frac{v+v_1'}{v+v_2'} = \frac{0.0286+0.0387}{0.0286+0.0214} = 1.35 \approx 4/3$$

$$\frac{q_2}{q_3} = \frac{v+v_2'}{v+v_3'} = \frac{0.0286+0.0214}{0.0286+0.0465} = 0.666 \approx 2/3$$

$$\frac{q_3}{q_4} = \frac{v+v_3'}{v+v_4'} = \frac{0.0286+0.0465}{0.0286+0.0132} = 1.80 \approx 9/5$$

$$\frac{q_4}{q_5} = \frac{v+v_4'}{v+v_5'} = \frac{0.0286+0.0132}{0.0286+0.0300}$$

$$= 0.713 \approx 8/11 \text{ or } 7/10$$

(b) If the oil density is 858 kg/m³ and the viscosity of air is 1.83×10^{-5} kg/m·s, find the radius, volume, and mass of the drop used in this experiment.

Solution: The radius of the drop is given by

$$a = \sqrt{\frac{9\xi v}{2\rho g}}$$

$$= \left[\frac{9(1.83\times10^{-5}\text{ kg/m}\cdot\text{s})(0.0286\times10^{-2}\text{ m/s})}{2(858\text{ kg/m}^3)(9.81\text{ m/s}^2)}\right]^{1/2}$$

$$= [2.80\times10^{-12}\text{ m}^2]^{1/2}$$

$$= 1.67\times10^{-6}\text{ m or } 1.67\text{ microns}$$

The volume is given by

$$V = \tfrac{4}{3}\pi a^3 = 1.95\times10^{-17}\text{ m}^3$$

The mass is

$$m = \rho V = (858 \text{ kg/m}^3)(1.95 \times 10^{-17} \text{ m}^3)$$
$$= 1.67 \times 10^{-14} \text{ kg}$$

(c) Calculate the charge on each drop, and from these results determine the electronic charge. Assume a plate separation of 1.60 cm and a potential difference of 4550 V for the parallel plate capacitor.

Solution: In order to calculate the charge on each drop using Equation 3.10, we first need to calculate the electric field, E. Thus,

$$E = \frac{V}{d} = \frac{4550 \text{ V}}{0.0160 \text{ m}} = 2.84 \times 10^5 \text{ V/m}$$

Now we can find the charge on each drop.

$$q = \left(\frac{mg}{E}\right)\left(\frac{v + v'}{v}\right)$$
$$= \frac{(1.67 \times 10^{-14} \text{ kg})(9.81 \text{ m/s}^2)}{(2.84 \times 10^5 \text{ V/m})}\left(\frac{0.0286 + 0.0130}{0.0286}\right)$$
$$= 8.39 \times 10^{-19} \text{ C}$$

Likewise, $q_1 = 13.6 \times 10^{-19}$ C, $q_2 = 10.1 \times 10^{-19}$ C, $q_3 = 15.2 \times 10^{-19}$ C, $q_4 = 8.43 \times 10^{-19}$ C, and $q_5 = 11.8 \times 10^{-19}$ C.

In order to find the average value of e, we shall use the fact that at the time of Millikan's work e was known to be between 1.5×10^{-19} C and 2.0×10^{-19} C. Dividing q through q_5 by these values gives the range of integral charges on each drop. Thus, the range of q is $8.39/1.5 = 5.6$ electronic charges to $8.39/2.0 = 4.2$ electronic charges. Similarly, q_1 has 9.1 to 5.8 charges, q_2 has 6.7 to 5.1 charges, q_3 has 10.1 to 7.6 charges, q_4 has 5.6 to 4.2 charges, and q_5 has 7.9 to 5.9 charges. Since there must be an integral number of charges on each drop, we pick an integer in the middle of the allowed range. Therefore, in terms of the electronic charge e, we conclude that $q = 5e$, $q_1 = 8e$, $q_2 = 6e$, $q_3 = 9e$, $q_4 = 5e$, and $q_5 = 7e$. Using the values found above, we find

$$e = q/5 = 1.68 \times 10^{-19} \text{ C}$$
$$e_1 = q_1/8 = 1.70 \times 10^{-19} \text{ C}$$
$$e_2 = q_2/6 = 1.68 \times 10^{-19} \text{ C}$$
$$e_3 = q_3/9 = 1.69 \times 10^{-19} \text{ C}$$
$$e_4 = q_4/5 = 1.69 \times 10^{-19} \text{ C}$$
$$e_5 = q_5/7 = 1.69 \times 10^{-19} \text{ C}$$

Taking the average of these values, we find the value of the electronic charge to be $e = 1.688 \times 10^{-19}$ C.

Stokes's law, as Millikan was aware, is only approximately correct for tiny spheres moving through a gas. The expression $D = 6\pi a\xi v$ holds quite accurately for a 0.1 cm radius ball bearing moving through a liquid or for any case where the moving object radius, a, is large compared to the mean free path, L, of the surrounding molecules. In the Millikan experiment, however, a is of the same order of magnitude as the mean free path of air at STP. Consequently Stokes's law overestimates the drag force, since the droplet actually moves for appreciable times through a frictionless "vacuum." Millikan corrected Stokes's law by using a drag force given by

$$D = \frac{6\pi a\xi v}{1 + \alpha \dfrac{L}{a}} \tag{3.14}$$

and found that $\alpha = 0.81$ gave the most consistent values of e for drops of different radii. Further corrections to Stokes's law were made by Perrin and Roux, and corrections to Stokes's law and the correct value of e remained a controversial issue for more than 20 years. The currently accepted value of the electronic charge is

$$|e| = 1.602189 \times 10^{-19} \text{ C}$$

Electronic charge

The Rutherford Model of the Atom

The early years of the 20th century were generally a period of incredible ferment, change, and activity in physics, including the advent of relativity, quantum theory, and atomic and subatomic physics. In fact, hardly had the reality of pristine, indivisible atoms been established ("They are the only material things which still remain in the precise condition in which they first began to exist" wrote Maxwell in 1872), when Thomson announced their divisibility in 1899: "Electrification essentially involves the splitting of the atom, a part of the mass of the atom getting free and becoming detached from the original atom." Further daring assaults on the indivisibility of the chemical atom came from the experimental work on radioactivity by Marie Curie and by Ernest Rutherford and Frederick Soddy, who explained radioactive transformations of elements in terms of the emission of subatomic particles. The porosity of atoms was also known before 1910 from Lenard's experiments, which showed that electrons are easily transmitted through thin metal and mica foils. All these discoveries, plus the suspicion that the intricate spectral lines emitted by atoms must be produced by something rattling around inside the atom, led to various proposals concerning the internal structure of atoms. The most famous of these early atomic models was the Thomson "plum-pudding" model (1898). This proposal viewed the atom as a homogeneous sphere of uniformly distributed mass and positive charge in which were embedded, like raisins in a plum pudding, negatively charged electrons. Although such models did possess electrical stability against collapse or explosion of the atom, they failed to explain the rich line spectra of even the simplest atom, hydrogen.

The key to understanding the mysterious line spectra and the correct model of the nucleus were both furnished by Ernest Rutherford and his students Hans Geiger and Ernst Marsden through a series of experiments conducted from 1909 to 1914. Noticing that a beam of finely collimated α-particles[8] was broadened upon passing through a metal foil, yet easily penetrated the thin film of metal, they embarked upon a series of experiments to probe the distribution of mass within the atom by observing in detail the scattering of α-particles from foils. These experiments ultimately led Rutherford to the discovery that most of the atomic mass and all of the positive charge lie in a minute central nucleus of the atom. The accidental chain of events and the clever capitalization upon the accidental discoveries leading up to Rutherford's monumental nuclear theory of the atom are nowhere better described than in Rutherford's own essay summarizing the development of the theory of atomic structure[9]:

Rutherford's nuclear model

> ". . . I would like to use this example to show how you often stumble upon facts by accident. In the early days I had observed the scattering of α-particles, and Dr. Geiger in my laboratory had examined it in detail. He found, in thin pieces of heavy metal, that the scattering was usually small, of the order of one degree. One day Geiger came to me and said, 'Don't you think that young Marsden, whom I am training in radioactive methods, ought to begin a small research?' Now I had thought that, too, so I said, 'Why not let him see if any α-particles can be scattered through a large angle?' I may tell you in confidence that I did not believe that they would be, since we knew that the α-particle was a very fast, massive particle, with a great deal of energy, and you could show that if the scattering was due to the accumulated effect of a number of small scatterings the chance of an α-particle's being scattered backwards was very small. Then I remember two or three days later Geiger coming to me in great excitement and saying, 'We have been able to get some of the α-particles coming backwards . . . ' It was quite the most incredible event that has ever happened to me in my life. It was almost as incredible as if you fired a 15-inch shell at a piece of

[8] α-Particles are the nuclei of helium atoms and possess a charge of $+2e$.

[9] An essay on "The Development of the Theory of Atomic Structure," 1936, Lord Rutherford, published in *Background to Modern Science*, New York; Macmillan Company, 1940.

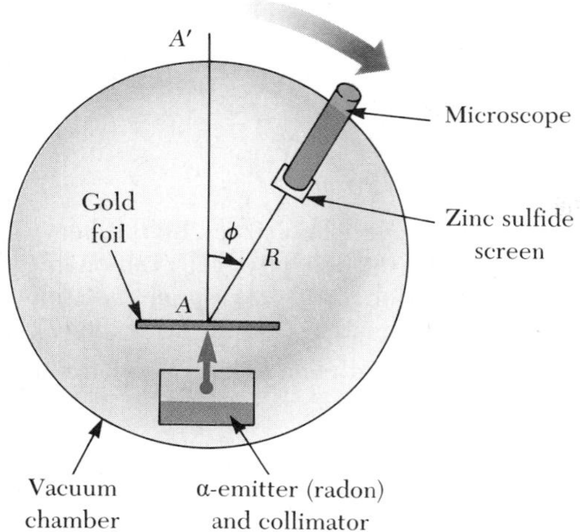

Figure 3.10 A schematic view of Rutherford's α-scattering apparatus.

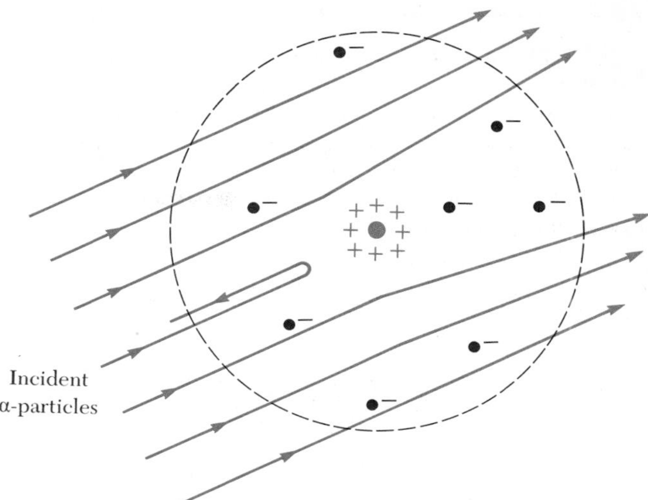

Figure 3.11 Scattering of α-particles by a dense, positively charged nucleus.

tissue paper and it came back and hit you. On consideration, I realized that this scattering backwards must be the result of a single collision, and when I made calculations I saw that it was impossible to get anything of that order of magnitude unless you took a system in which the greater part of the mass of the atom was concentrated in a minute nucleus. It was then that I had the idea of an atom with a minute massive center carrying a charge. I worked out mathematically what laws the scattering should obey, and I found that the number of particles scattered through a given angle should be proportional to the thickness of the scattering foil, the square of the nuclear charge, and inversely proportional to the fourth power of the velocity. These deductions were later verified by Geiger and Marsden in a series of beautiful experiments."

The essential experimental features of Rutherford's apparatus are shown in Figure 3.10. Alpha-particles emitted with speeds of about 2×10^9 cm/s struck a thin gold foil several thousand atomic layers thick. Most of the α's passed straight through the foil (AA') (again showing the porosity of the atom), but some were scattered at an angle ϕ. The number of scattered α's at each angle per unit area and per unit time was measured by counting the scintillations produced by scattered α's on the ZnS screen. These scintillations were counted with the aid of the microscope.

Rutherford's basic insight was that since the mass (four times the proton mass) and kinetic energy of the α's are large, even a nearly head-on collision with a single proton would deflect the α-particle only slightly and knock the proton straight ahead. Multiple scattering of the α-particles in the foil accounted for the small broadening (about 1°) originally observed by Rutherford, but it could not account for the occasional large-scale deflections. On the other hand,[10] if all of the protons in an atom are assumed to be concentrated at a single central point and not spread out throughout the atom, the electric repulsion experienced by an incident α-particle in a head-on collision becomes much greater. In addition, the mass of the heavy gold atom is concentrated at the nucleus and large deflections of the α-particle could be experienced in a *single* collision with the massive nucleus. This situation is shown in Figure 3.11.

[10] A dangerous expression that never fails to bring to mind Truman's infamous request: "If you know of any one-handed economists, bring them to me."

EXAMPLE 3.4 Collision of an α-Particle with a Proton

(a) An α-particle of mass m_α and speed v_α strikes a stationary proton with mass m_p. If the collision is elastic and head-on, show that the speed of the proton after the collision, v_p, and the speed of the α-particle after the collision, v'_α, are given by

$$v_p = \left(\frac{2m_\alpha}{m_\alpha + m_p}\right) v_\alpha$$

and

$$v'_\alpha = \left(\frac{m_\alpha - m_p}{m_\alpha + m_p}\right) v_\alpha$$

Solution: Since the collision is elastic, the total kinetic energy is conserved; therefore

$$\tfrac{1}{2}m_\alpha v_\alpha^2 = \tfrac{1}{2}m_\alpha v'^2_\alpha + \tfrac{1}{2}m_p v_p^2 \quad (1)$$

Conservation of momentum for this one-dimensional collision yields

$$m_\alpha v_\alpha = m_p v_p + m_\alpha v'_\alpha \quad (2)$$

Solving Equation (1) for $(m_\alpha v'_\alpha)^2$ yields

$$(m_\alpha v'_\alpha)^2 = m_\alpha(m_\alpha v_\alpha^2 - m_p v_p^2) \quad (3)$$

Solving Equation (2) for $(m_\alpha v'_\alpha)^2$ and equating this to Equation (3) gives

$$(m_\alpha v_\alpha)^2 + (m_p v_p)^2 - 2m_\alpha m_p v_\alpha v_p = (m_\alpha v_\alpha)^2 - m_\alpha m_p v_p^2$$

or

$$(m_p v_p)(m_p v_p - 2m_\alpha v_\alpha + m_\alpha v_p) = 0$$

The solutions to this equation are

$$v_p = 0$$

and

$$v_p = \left(\frac{2m_\alpha}{m_\alpha + m_p}\right) v_\alpha \quad (4)$$

Since the proton must move when struck by the heavy α-particle, Equation (4) is the only physically reasonable solution for v_p. The solution for v'_α follows immediately from the substitution of Equation (4) into Equation (2).

(b) Calculate the percent change in velocity for an α-particle colliding with a proton.

Solution: Since an α-particle consists of two protons and two neutrons, $m_\alpha = 4m_p$. Thus,

$$v_p = \left(\frac{2m_\alpha}{m_\alpha + m_p}\right) v_\alpha = \left(\frac{8m_p}{5m_p}\right) v_\alpha = 1.60\ v_\alpha$$

$$v'_\alpha = \left(\frac{m_\alpha - m_p}{m_\alpha + m_p}\right) v_\alpha = \left(\frac{3m_p}{5m_p}\right) v_\alpha = 0.60\ v_\alpha$$

The percent change in velocity of the α-particle is

$$\%\text{ change in } v_\alpha = \left(\frac{v'_\alpha - v_\alpha}{v_\alpha}\right) \times 100\% = -40\%$$

Exercise 2 An α-particle with initial velocity v_α undergoes an elastic, head-on collision with an electron initially at rest. Using the fact that an electron's mass is about 1/2000 of the proton mass, calculate the final velocities of the electron and α-particle, and the percent change in velocity of the α-particle.

Answers: $v_e = 1.998 v_\alpha$, $v'_\alpha = 0.9998 v_\alpha$, and the percent change in $v_\alpha = -0.02\%$.

In his analysis, Rutherford assumed that large angle scattering is produced by a single nuclear collision and that the repulsive force between an α-particle and a nucleus separated by a distance r is given by Coulomb's law,

$$F = k\frac{(2e)(Ze)}{r^2} \quad (3.15)$$

where $+2e$ is the charge on the α, $+Ze$ is the nuclear charge, and k is the Coulomb constant. With this assumption, Rutherford was able to show that the number of α-particles entering the detector per unit time, Δn, at an angle ϕ is given by

$$\Delta n = \frac{k^2 Z^2 e^4 N n A}{4R^2(\frac{1}{2}m_\alpha v_\alpha^2)^2 \sin^4(\phi/2)} \quad (3.16)$$

Here R and ϕ are defined in Figure 3.10, N is the number of nuclei per unit area of the foil (and is thus proportional to the foil thickness), n is the total number of α-particles incident on the target per unit time, and A is the area of the detector. The dependence of scattering on foil thickness, α-particle veloc-

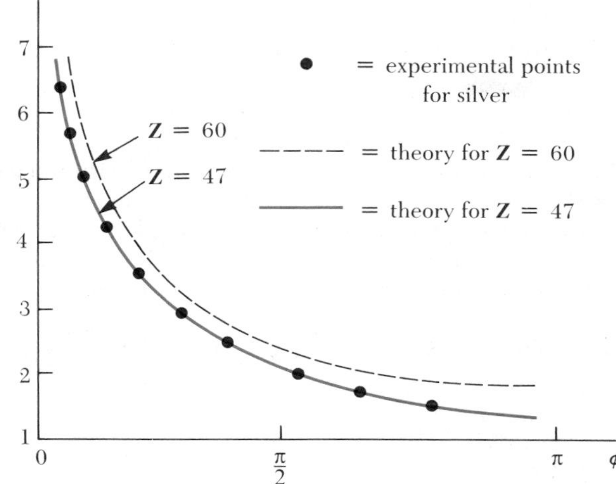

Figure 3.12 Comparison of theory and experiment for α-particle scattering from a silver foil. (From E. Rutherford, J. Chadwick, and J. Ellis, *Radiations from Radioactive Substances.* Cambridge University Press, Cambridge, 1951.)

ity, and scattering angle was almost immediately confirmed experimentally by Geiger and Marsden.[11] Since the values of atomic number (Z) were unknown at the time, the scattering dependence on Z could not be directly checked. Turning the argument around, however, and assuming the correctness of Equation 3.16, one could find the value of Z that gave the best fit of this equation to the experimental data points. An illustration of this sensitive technique for determining Z for a silver foil is shown in Figure 3.12. Note that changing Z produces only a vertical shift in the graph and not a change in shape.

Much of the remarkable experimental work of the ingenious Lord Rutherford can be credited to an ability to use his current discoveries to probe even deeper into nature's mysteries. For example, he turned his studies of the transmission of radioactive particles through matter into a sensitive and delicate technique for probing the atom. Another example was his clever technique for measuring the size of the nucleus. Realizing that Equation 3.15 would hold only if the α-particle did not deform or penetrate the scattering nucleus, he systematically looked for departures from his scattering equation for high energy α's, with the idea of calculating the distance of closest approach of the α to the nucleus. This distance of closest approach is roughly equal to the nuclear radius and may be found from energy conservation. Thus we set the kinetic energy of the α at infinity equal to the potential energy of the system (α + target nucleus) at closest approach to the nucleus, or

$$\tfrac{1}{2}m_\alpha v_\alpha{}^2 = k\,\frac{(Ze)(2e)}{r} \tag{3.17}$$

Equation 3.17 may then be solved for r to determine the distance of closest approach.

Rutherford was confronted with the experimental dilemma that no failures of Equations 3.15 or 3.16 were found for heavy metal foils with the most energetic naturally occurring α-particles available to him (~8 MeV). Showing

[11] "Deflection of α-Particles through Large Angles," H. Geiger and E. Marsden, *Phil. Mag.* (6) **25:** 605, 1913.

characteristic economy, instead of embarking on a particle accelerator program,[12] he made use of metals like aluminum with lower Z's and hence lower Coulomb barriers to α-penetration.

EXAMPLE 3.5 Estimate of the Radius of the Aluminum Nucleus

In 1919 Rutherford was able to show a breakdown in Equation 3.16 for 7.7 MeV α-particles scattered at large angles from aluminum nuclei ($Z = 13$). Estimate the radius of the aluminum nucleus from these facts.

Solution: Rutherford's scattering formula will no longer hold when α-particles begin to penetrate or touch the nucleus. Let us assume that when the α-particle is very far from the aluminum nucleus, its kinetic energy is 7.7 MeV. This is also the total energy of the system (α-particle plus aluminum nucleus), since the aluminum nucleus is at rest and the potential energy is zero for an infinite separation of particles. When the α-particle is at the point of closest approach to the aluminum nucleus, its kinetic energy is zero, and for a head-on collision we take their separation to be the radius of the aluminum nucleus. At this point, the kinetic energy of the system is zero, and the total energy is just the potential energy of the system. Applying conservation of energy gives

$$K_\alpha = \text{Potential energy at closest approach} = \frac{k(Ze)(2e)}{r}$$

or

$$r = k\frac{2Ze^2}{K_\alpha}$$

$$= \frac{2(13)(1.60 \times 10^{-19}\ \text{C})^2(8.99 \times 10^9\ \text{N}\cdot\text{m}^2/C^2)}{(7.7 \times 10^6\ \text{eV})(1.60 \times 10^{-19}\ \text{J/eV})}$$

$$= 4.9 \times 10^{-15}\ \text{m}$$

The overall success of the Rutherford nuclear model was truly striking. Rutherford and his students had shown that the model of the atom with a massive nucleus of diameter 10^{-14} m and atomic number equal to approximately half of the atomic weight agreed in all particulars with experiment. As with all great discoveries, however, the idea of the nuclear atom raised a swarm of questions at the next deeper level: (1) If there are only Z positive charges in the nucleus, what composes the other half of the atomic weight? (2) What provides the cohesive force to keep many protons confined in the incredibly small distance of 10^{-14} m? (3) How do the electrons move around the nucleus, and how does their motion account for the observed spectral lines?

Rutherford had no precise answer to the first question. He speculated that the difference between Z and the atomic weight could be accounted for by groupings of α-particles and by neutral particles each consisting of a bound electron-proton pair. The second conjecture seemed especially satisfying since it built the atom out of the most fundamental particles then known to exist. It also incorporated the existence of intra-nuclear electrons, which was known because they appear in beta decay.

In answer to the second question, Rutherford cautiously held that electrical forces provided the cement to hold the nucleus together. He wrote, "The nucleus, though of minute dimensions, is in itself a very complex system consisting of positively and negatively charged bodies bound closely together by intense electrical forces." In fact, it was not until 1921 that it was clearly recognized that Coulomb's law fails at very short distances and that a completely new and very strong type of force binds protons together. Interestingly it was James Chadwick, the discoverer of the neutron, who first recognized that a new force field of much more than electric intensity was at work in the nucleus.[13] Perhaps Rutherford's magnificient achievement of explaining α-

[12] Rutherford was famous for the remark to his young graduate students, "There is no money for apparatus—we shall have to use our heads" (A. Keller, *Infancy of Atomic Physics:* Hercules in His Cradle, Oxford, Clarendon Press, 1983, p. 215).

[13] J. Chadwick and E. S. Biele, *Phil. Mag.* **42:** 923, 1921.

Figure 3.13 Bohr (on the right) and Rutherford (on the left) were literal as well as intellectual supports for each other. This photograph of Bohr and Rutherford sitting back to back was taken at a rowing regatta in June, 1923. (AIP Niels Bohr Library, and *Physics Today*, October 1985, an issue devoted to Bohr.)

scattering with the Coulomb law blinded him to the possibility that this law failed within the nucleus.

The answer to the third question was not to be given by Rutherford. That was to be the masterwork of Niels Bohr (Fig. 3.13). Even so, with characteristic insight, Rutherford speculated on a planetary model of the atom, or more precisely that negative charges revolved around the dense positive core as the planets revolved around the sun.[14] Now the stage was truly set for Bohr!

3.4 THE BOHR ATOM

> "Bohr's original quantum theory of spectra was one of the most revolutionary, I suppose, that was ever given to science, and I do not know of any theory that has been more successful . . . I consider the work of Bohr one of the greatest triumphs of the human mind."
>
> Lord Rutherford
>
> "Then it is one of the greatest discoveries."
>
> Albert Einstein, upon hearing of Bohr's theoretical calculation of the Rydberg constants for hydrogen and singly ionized helium

Revolution followed revolution. In April of 1913, only four years after Rutherford's initial work on the nuclear atom, a young Danish physicist, Niels Bohr (who had recently been working with both Thomson and Rutherford), published a three-part paper that shook the world of physics to its foundations.[15] Not only did this young rebel give the first successful theory of atomic line spectra, but in the process he overthrew some of the most cherished principles of the reigning king of electromagnetism, James Clerk Maxwell.

From our point of view, Bohr's model may seem only a reasonable next step, but it appeared astounding, confounding, and incredibly bold to his contemporaries. As mentioned earlier, both Thomson and Rutherford realized that the electrons must revolve about the nucleus in order to avoid falling directly into it. They, along with Bohr, also realized that according to Max-

[14] Thomson and Nagaoka had worked even earlier with planetary atomic models in 1904.

[15] N. Bohr, "On the Constitution of Atoms and Molecules," *Phil. Mag.* **26:** 1, 1913. Also, N. Bohr, *Nature* **92:** 231, 1913.

well's theory, centripetally accelerated charges revolving with frequency f should radiate light waves of frequency f, as confirmed by Hertz in 1888. Unfortunately, pushed to its logical conclusion, this classical model leads to disaster.

As the electron radiates energy, its orbit radius steadily decreases and its frequency of revolution increases. This leads to an ever increasing frequency of emitted radiation and an ultimate catastrophic collapse of the atom as the electron plunges into the nucleus (Fig. 3.14).

These deductions of electrons falling into the nucleus and a continuous emission spectrum from elements were boldly circumvented by Bohr. He simply postulated that classical radiation theory, which had been confirmed by Hertz's detection of radio waves using large circuits, did not hold for atomic sized systems. Moreover, he drew on the work of Planck and Einstein as sources of the correct theory of atomic systems. He overcame the problem of a classical electron that continually lost energy by applying Planck's ideas of quantized energy levels to orbiting atomic electrons. Thus he postulated that electrons in atoms are generally confined to certain stable, nonradiating energy levels and orbits known as ***stationary states.*** He applied Einstein's concept of the photon to arrive at an expression for the frequency of light emitted when the electron jumps from one stationary state to another. Thus if ΔE is the separation of two possible electronic stationary states, then $\Delta E = hf$, where h is Planck's constant and f is the frequency of the emitted light regardless of the frequency of the electron's orbital motion.

In addition, where another might have left a wild and lawless gap between the revolutionary laws that apply to atomic systems and those that hold for classical systems, Bohr provided a gentle and refined continuum in the form of the **correspondence principle.** This philosophical principle states that the predictions for macroscopic quantum systems must agree with or correspond to the predictions of classical physics. We may state the correspondence principle another way as

Correspondence principle

$$\lim_{n\to\infty} [\text{Quantum Physics}] = [\text{Classical Physics}]$$

where n is a characteristic quantum number of the system, such as the principal quantum number for hydrogen. In the hands of Bohr, the correspondence principle became a masterful tool to test revolutionary quantum results, and it also became a source of fundamental postulates about atomic systems. As we shall see, Bohr used reasoning of this type to arrive at the concept of the quantization of electronic orbital angular momentum in his 1913 paper.

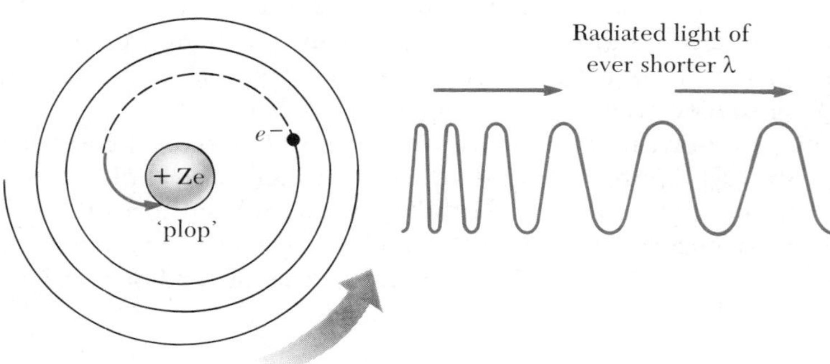

Figure 3.14 The classical model of the nuclear atom.

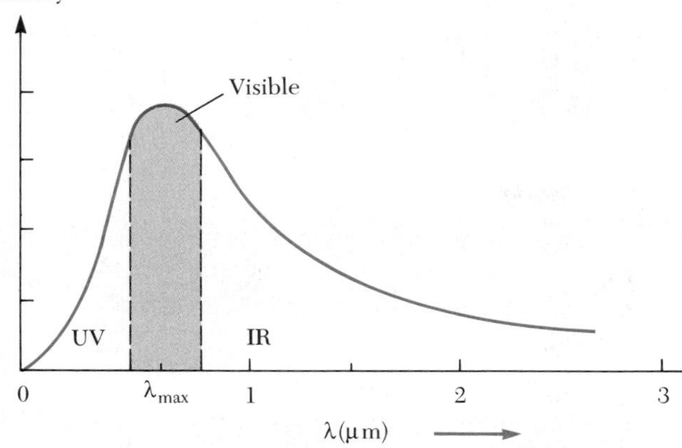

Figure 3.15 Intensity versus wavelength for a body heated to 6000 K.

Spectral Series

Before looking in detail at the first successful theory of atomic dynamics, we shall review the experimental work on line spectra that served as the impetus for and the clear confirmation of the first quantum theory of the atom. As already pointed out in Chapter 2, glowing solids and liquids (and even gases at the high densities found in stars) emit a continuous distribution of wavelengths. This distribution exhibits a common shape for the intensity versus wavelength curve, and the peak in this curve shifts toward shorter wavelengths with increasing temperature. This universal "blackbody" curve is shown in Figure 3.15.

In sharp contrast to this continuous spectrum is the **discrete line spectrum** emitted by a low-pressure gas subject to an electric discharge. When the light from such a low-pressure gas discharge is examined with a spectroscope, it is found to consist of a few bright lines of pure color on a generally dark background. This contrasts sharply with the continuous rainbow of colors seen when a glowing solid is viewed through a spectroscope. Furthermore, as can be seen from Figure 3.16, the wavelengths contained in a given line spectrum are characteristic of the particular element emitting the light. The simplest

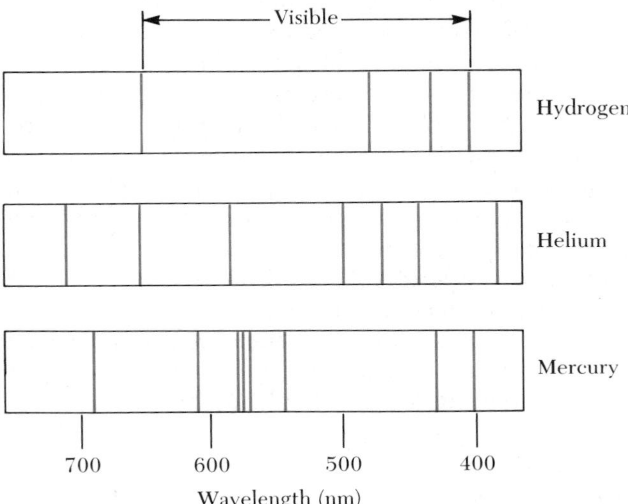

Figure 3.16 Emission line spectra of a few representative elements.

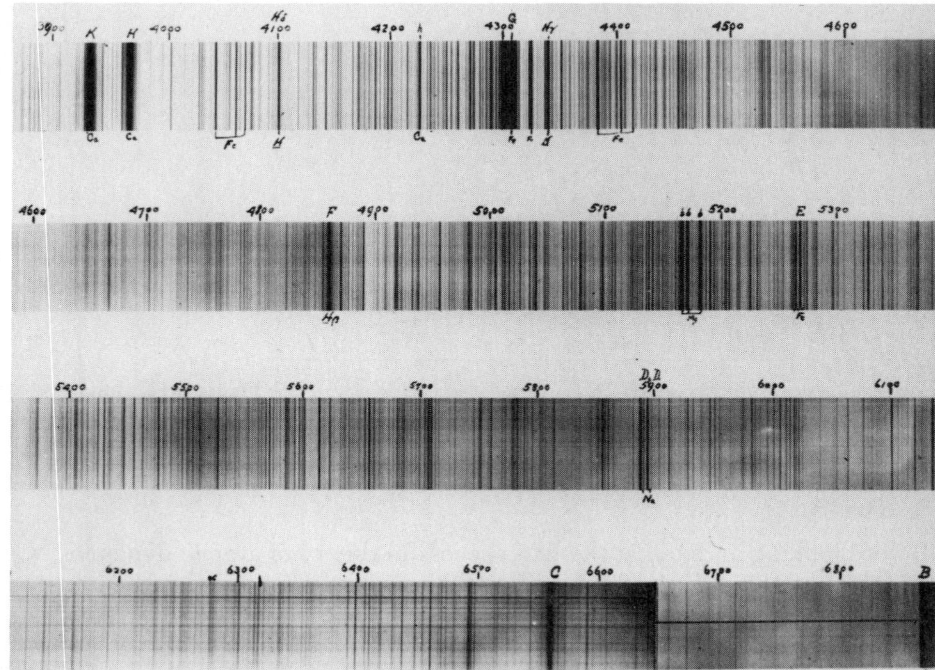

Figure 3.17 The spectrum of the sun as viewed with a high resolution spectrograph, showing many dark Fraunhofer lines. Wavelengths are indicated above the spectrum in Ångstrom units (10^{-8} cm). (Palomar Observatory photograph.)

line spectrum is observed for atomic hydrogen, and we shall describe this spectrum in detail. Other atoms such as mercury, helium, and neon give completely different line spectra. Since no two elements emit the same line spectrum, this pheonmenon represents a practical and sensitive technique for identifying the elements present in unknown samples. In fact, by 1860 spectroscopy had advanced so far in the hands of G. R. Kirchhoff (yes . . . the same fellow who brought you Kirchhoff's loop theorem) and R. W. Bunsen (of Bunsen burner fame) at the University of Heidelberg that they were able to discover two new elements, rubidium and cesium, by observing new sequences of spectral lines in mineral samples. Improvements in instruments and techniques resulted in an enormous growth in spectroscopy in Europe from 1860 to 1900. Even the European public imagination was captured by spectroscopy when spectroscopic techniques were used to show that "celestial" meteorites consisted only of known earth elements after all.

Kirchhoff's immense contribution to spectroscopy is also shown by another advance he made in 1859—the foundation of **absorption spectroscopy** and the explanation of Fraunhofer's dark D-lines in the solar spectrum.[16] In 1814 Joseph Fraunhofer had passed the continuous spectrum from the sun through a narrow slit and then through a prism. He observed the surprising result of nearly 1000 fine dark lines or gaps in the continuous rainbow spectrum of the sun, and he assigned the letters A, B, C, D . . . to the most prominent dark lines. These lines and many more are shown in Figure 3.17. Kirchhoff correctly deduced that the mysterious dark lines are produced by a cloud of vaporized atoms in the sun's outer, cooler layers, which absorb at discrete frequencies the intense continuous radiation from the center of the

[16] G. Kirchhoff, Monatsber. Berlin, 1859, p. 662.

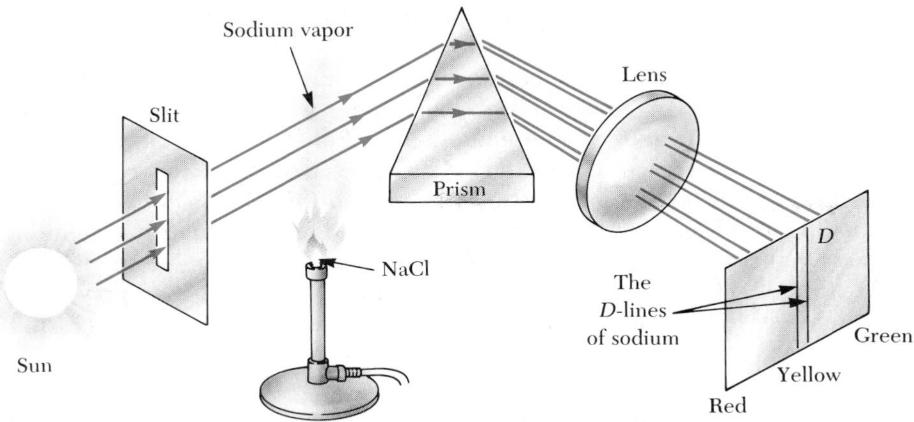

Figure 3.18 Kirchhoff's experiment explaining the Fraunhofer D-lines. The D lines darken noticeably when sodium vapor is introduced between the slit and the prism.

sun. Further he showed that the Fraunhofer D-lines were produced by vaporized sodium, and that they had the same wavelengths as the strong yellow lines in the emission spectrum of sodium. Kirchhoff also correctly deduced that all of Fraunhofer's dark lines should be attributable to absorption by different elements present in the sun. In a single stroke he opened the way to determining the elemental composition of stars trillions of miles from the earth. His ingenious and simple method for demonstrating the presence of sodium vapor in the solar atmosphere is shown in schematic form in Figure 3.18.

Today absorption spectroscopy is certainly as important as emission spectroscopy for qualitative and quantitative analyses of elements and molecular groups. In general one obtains an absorption spectrum by passing light from a continuous source [whether in the ultraviolet (uv), visible (vis), or infrared (IR) regions] through a gas or dilute solution of the element being analyzed. The absorption spectrum consists of a series of dark lines superimposed on the otherwise continuous spectrum emitted by the source. It is found that each line in the absorption spectrum of an element coincides with a line in the emission spectrum of that same element; however, not all of the emission lines are present in an absorption spectrum. The differences between emission and absorption spectra are complicated in general and depend on the temperature of the absorbing vapor, which in turn determines the numbers of atoms in different energy states in the absorbing vapor.

An interesting use of the coincidence of absorption and emission lines is made in the atomic absorption spectrometer, a device routinely used to measure parts per million (ppm) of metals in unknowns. For example, if sodium is to be measured, a sodium lamp emitting a line spectrum is chosen as the light source. The unknown is heated in a hot flame (usually oxy-acetylene) to vaporize the sample, to break the chemical bonds of sodium to other elements, and to produce a gas of elemental sodium. The spectrometer is then tuned to a wavelength for which both absorption and emission lines exist (say one of the D-lines at 588.99 or 589.59 nm), and the amount of darkening or decrease in intensity is measured with a sensitive photomultiplier. The decrease in intensity is a measure of the sodium concentration. With proper calibration, concentrations of 0.1 ppm can be measured with this extremely selective technique. Atomic absorption spectroscopy has been a useful technique in analyzing heavy metal contamination of the food chain. For example, the first determinations of high levels of mercury in tuna fish were made with atomic absorption.

From 1860 to 1885 spectroscopic measurements accumulated voluminously, burying frenzied theoreticians under a mountain of data. Accurate measurements of four visible emission lines of hydrogen had recently been made by Anders Ångström, when in 1885 a Swiss school teacher, Johann Jakob Balmer, published a paper with the unpretentious title "Notice Concerning the Spectral Lines of Hydrogen." By trial and error Balmer had found a formula that correctly predicted the wavelengths of Ångström's four visible lines: H_α (red), H_β (green), H_γ (blue), and H_δ (violet). Figure 3.19 shows these and other lines in the emission spectrum of hydrogen. Balmer gave his formula in the form

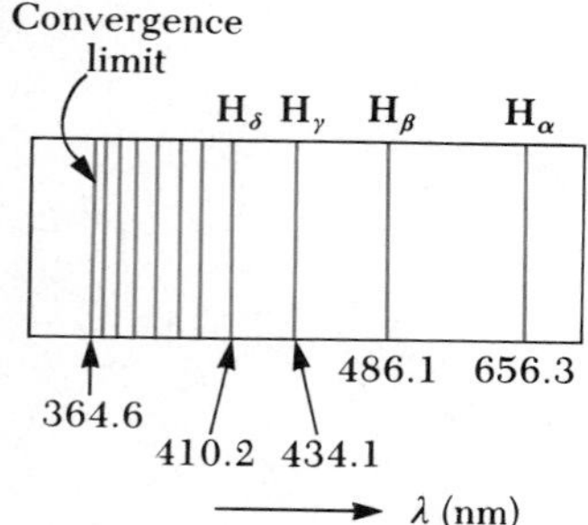

Figure 3.19 The Balmer series of spectral lines for hydrogen (emission spectrum).

$$\lambda\ (\text{cm}) = C_2\left(\frac{n^2}{n^2 - 2^2}\right) \qquad n = 3, 4, 5\ \ldots \tag{3.18}$$

where $C_2 = 3645.6 \times 10^{-8}$ cm, a constant called the convergence limit because it gave the wavelength of the line with largest n value ($n = \infty$). Also, note that $n = 3, 4, 5, \ldots$ where H_α has $n = 3$, H_β has $n = 4$, and so forth. Although only four lines were known to Balmer when he started his paper, by the time he had finished 10 more lines in the violet and ultraviolet had been measured. Much to his delight and satisfaction, these lines agreed with his empirical formula to within 0.1%! Encouraged by his success and because he was a bit of a numerologist, Balmer suggested that other hydrogen series might exist of the form

$$\lambda = C_3\left(\frac{n^2}{n^2 - 3^2}\right) \qquad n = 4, 5, 6\ \ldots \tag{3.19}$$

$$\lambda = C_4\left(\frac{n^2}{n^2 - 4^2}\right) \qquad n = 5, 6, 7\ \ldots \tag{3.20}$$

As we now know, his speculations were correct and these series do indeed exist. In today's notation all of these series are given by a single formula

$$\boxed{\frac{1}{\lambda} = R\left(\frac{1}{n_f^2} - \frac{1}{n_i^2}\right)} \tag{3.21}$$

where n_f and n_i are integers corresponding to the final and initial quantum states. The Rydberg constant R is the same constant for all series and has the value

$$R = 1.0973732 \times 10^7\ \text{m}^{-1} \tag{3.22}$$

Note that for a given series, n_f has the same value. Furthermore, for a given series $n_i = n_f + 1, n_f + 2, \ldots$. Table 3.1 lists the name of each series (named after their discoverers) and the integers or quantum numbers that define the series.

TABLE 3.1 Some Spectral Series for the Hydrogen Atom

Lyman Series (uv)	$n_f = 1$	$n_i = 2, 3, 4\ \ldots$
Balmer Series (vis-uv)	$n_f = 2$	$n_i = 3, 4, 5\ \ldots$
Paschen Series (IR)	$n_f = 3$	$n_i = 4, 5, 6\ \ldots$
Brackett Series (IR)	$n_f = 4$	$n_i = 5, 6, 7\ \ldots$
Pfund Series (IR)	$n_f = 5$	$n_i = 6, 7, 8\ \ldots$

Now that we have looked at the general principles of Bohr's model of hydrogen and at the detailed experimental spectra already discovered by 1913, let us examine in detail Bohr's quantum theory. The basic ideas of the Bohr theory as it applies to an atom of hydrogen are as follows:

Assumptions of the Bohr theory

1. The electron moves in circular orbits about the proton under the influence of the Coulomb force of attraction, as in Figure 3.20. So far nothing new!
2. Only certain orbits are stable. These stable orbits are ones in which the electron does not radiate. Hence the energy is fixed or stationary, and ordinary classical mechanics may be used to describe the electron's motion.
3. Radiation is emitted by the atom when the electron "jumps" from a more energetic initial stationary state to a lower state. This "jump" cannot be visualized or treated classically. In particular, the frequency f of the photon emitted in the jump *is independent of the frequency of the electron's orbital motion.* The frequency of the light emitted is related to the change in the atom's energy and is given by the Planck-Einstein formula

$$\boxed{E_\mathrm{i} - E_\mathrm{f} = hf} \tag{3.23}$$

where E_i is the energy of the initial state, E_f is the energy of the final state, and $E_\mathrm{i} > E_\mathrm{f}$.
4. The size of the allowed electron orbits is determined by an additional quantum condition imposed on the electron's orbital angular momentum. Namely, the allowed orbits are those for which the electron's orbital angular momentum about the nucleus is an integral multiple of $\hbar = h/2\pi$,

$$\boxed{mvr = n\hbar} \qquad n = 1, 2, 3, \ldots \tag{3.24}$$

Using these four assumptions, we can now calculate the allowed energy levels and emission wavelengths of the hydrogen atom. Recall that the electrical potential energy of the system shown in Figure 3.20 is given by $U = qV = -ke^2/r$, where k (the Coulomb constant) has the value $1/4\pi\epsilon_0$. Thus, the total energy of the atom, which contains both kinetic and potential energy terms, is given by

$$E = K + U = \tfrac{1}{2}mv^2 - k\frac{e^2}{r} \tag{3.25}$$

Applying Newton's second law to this system, we see that the Coulomb attractive force on the electron, ke^2/r^2, must equal the mass times the centripetal acceleration of the electron, or

$$\frac{ke^2}{r^2} = \frac{mv^2}{r}$$

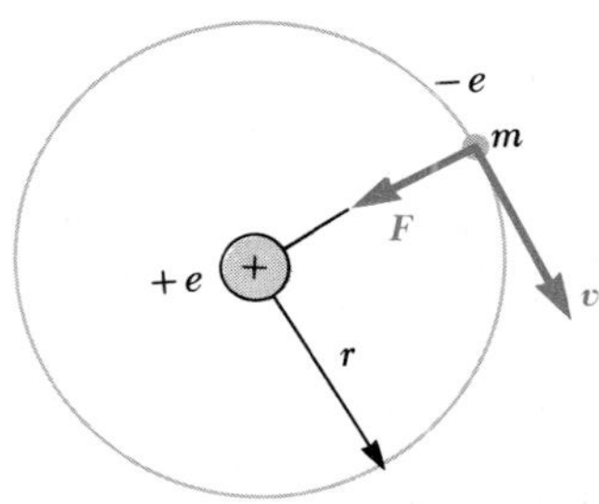

Figure 3.20 Diagram representing Bohr's model of the hydrogen atom.

From this expression, we immediately find the kinetic energy to be

$$K = \frac{mv^2}{2} = \frac{ke^2}{2r} \tag{3.26}$$

Substituting this value of K into Equation 3.25 gives the total energy of the atom as

$$E = -\frac{ke^2}{2r} \tag{3.27}$$

Note that the total energy is negative, indicating a bound electron-proton system. This means that energy in the amount of $ke^2/2r$ must be added to the atom to just remove the electron and make the total energy zero. An expression for r, the radius of the stationary orbits, may be obtained by solving Equations 3.24 and 3.26 for v and equating the results:

Radii of Bohr orbits in hydrogen

$$r_n = \frac{n^2\hbar^2}{mke^2} \qquad n = 1, 2, 3, \ldots \tag{3.28}$$

The orbit for which $n = 1$ has the smallest radius, and is called the **Bohr radius,** a_0, has the value

$$a_0 = \frac{\hbar^2}{mke^2} = 0.529 \text{ Å} = 0.529 \times 10^{-10} \text{ m} \tag{3.29}$$

The fact that Bohr's theory gave an accurate value for the radius of hydrogen from first principles without any empirical calibration of orbit size was considered a striking triumph for this theory. The first three Bohr orbits are shown to scale in Figure 3.21.

The quantization of the orbit radii immediately leads to energy quantization. This can be seen by substituting $r_n = n^2a_0$ into Equation 3.27, giving for the allowed energy levels

Energy levels of hydrogen

$$E_n = -\frac{ke^2}{2a_0}\left(\frac{1}{n^2}\right) \qquad n = 1, 2, 3, \ldots \tag{3.30}$$

Inserting numerical values into Equation 3.30 gives

$$E_n = -\frac{13.6}{n^2} \text{ eV} \qquad n = 1, 2, 3, \ldots \tag{3.31}$$

The lowest stationary or non-radiating state is called the **ground state,** has $n = 1$, and has an energy $E_1 = -13.6$ eV. The next state or **first excited state** has $n = 2$, and an energy $E_2 = E_1/2^2 = -3.4$ eV. An energy level diagram showing the energies of these discrete energy states and the corresponding quantum numbers is shown in Figure 3.22. The uppermost level, corresponding to $n = \infty$ (or $r = \infty$) and $E = 0$, represents the state for which the electron is removed from the atom. The minimum energy required to ionize the atom (that is, to completely remove an electron in the ground state from the proton's influence) is called the **ionization energy.** As can be seen from Figure 3.22, the ionization energy for hydrogen based on Bohr's calculation is 13.6 eV. This constituted another major achievement for the Bohr theory, since the ionization energy for hydrogen had already been measured to be precisely 13.6 eV.

Equation 3.30 together with Bohr's third postulate can be used to calculate the frequency of the photon emitted when the electron jumps from an outer orbit to an inner orbit:

$$f = \frac{E_i - E_f}{h} = \frac{ke^2}{2a_0h}\left(\frac{1}{n_f^2} - \frac{1}{n_i^2}\right) \tag{3.32}$$

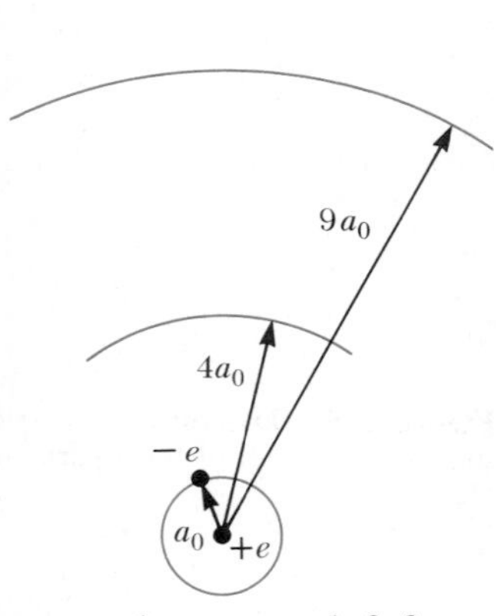

Figure 3.21 The first three Bohr orbits for hydrogen.

Since the quantity actually measured is wavelength, it is convenient to convert frequency to wavelength using $c = f\lambda$ to get

$$\boxed{\frac{1}{\lambda} = \frac{f}{c} = \frac{ke^2}{2a_0hc}\left(\frac{1}{n_f^2} - \frac{1}{n_i^2}\right)} \qquad (3.33)$$

Emission wavelengths of hydrogen

The remarkable fact is that the *theoretical* expression, Equation 3.33, is identical to Balmer's *empirical* relation

$$\frac{1}{\lambda} = R\left(\frac{1}{n_f^2} - \frac{1}{n_i^2}\right) \qquad (3.34)$$

provided that the combination of constants $ke^2/2a_0hc$ is equal to the experimentally determined Rydberg constant, $R = 1.0973732 \times 10^7\ m^{-1}$. When Bohr demonstrated the agreement of these two quantities to a precision of about 1% late in 1913, it was recognized as the crowning achievement of his new theory of quantum mechanics. Furthermore, Bohr showed that all of the spectral series for hydrogen have a natural interpretation in his theory. These spectral series are shown as transitions between energy levels in Figure 3.22.

Bohr immediately extended his model for hydrogen to other elements in which all but one electron had been removed. Ionized elements such as He^+, Li^{++}, and Be^{+++} were suspected to exist in hot stellar atmospheres, where frequent atomic collisions occurred with enough energy to completely remove one or more atomic electrons. Bohr showed that many mysterious lines observed in the sun and several stars could not be due to hydrogen, but were correctly predicted by his theory if attributed to singly ionized helium. In general, to describe a single electron orbiting a fixed nucleus of charge $+Ze$, Bohr's theory gives

$$r_n = (n^2)\frac{a_0}{Z} \qquad (3.35)$$

and

$$E_n = -\frac{ke^2}{2a_0}\left(\frac{Z^2}{n^2}\right) \qquad n = 1, 2, 3, \ldots \qquad (3.36)$$

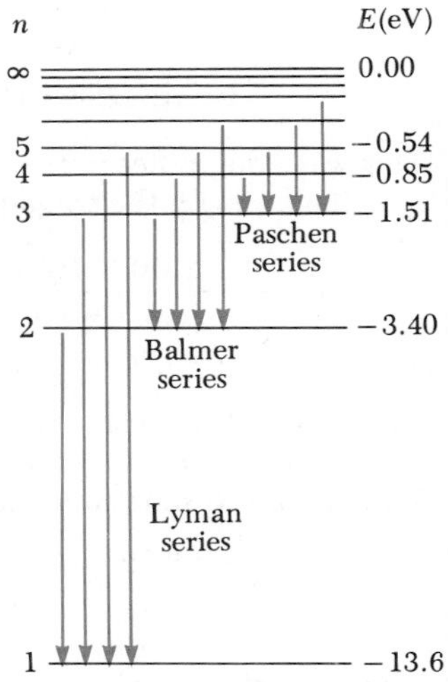

Figure 3.22 An energy level diagram for hydrogen. In such diagrams the discrete allowed energies are plotted on the vertical axis. Nothing is plotted on the horizontal axis, but the horizontal extent of the diagram is made large enough to show allowed transitions. Note that the quantum numbers are given on the left.

EXAMPLE 3.6 Spectral Lines from the Star ζ-Puppis
The mysterious lines observed by Pickering in 1896 in the spectrum of the star ζ-Puppis fit the empirical formula

$$\frac{1}{\lambda} = R\left(\frac{1}{(n_f/2)^2} - \frac{1}{(n_i/2)^2}\right)$$

Show that these lines can be explained by the Bohr theory as originating from He^+.

Solution: He^+ has $Z = 2$. Thus the allowed energy levels are given by Equation 3.36 as

$$E_n = \frac{ke^2}{2a_0}\left(\frac{4}{n^2}\right)$$

Using $hf = E_i - E_f$ we find

$$f = \frac{E_i - E_f}{h} = \frac{ke^2}{2a_0h}\left(\frac{4}{n_f^2} - \frac{4}{n_i^2}\right)$$

$$= \frac{ke^2}{2a_0h}\left(\frac{1}{(n_f/2)^2} - \frac{1}{(n_i/2)^2}\right)$$

or

$$\frac{1}{\lambda} = \frac{f}{c} = \frac{ke^2}{2a_0hc}\left(\frac{1}{(n_f/2)^2} - \frac{1}{(n_i/2)^2}\right)$$

This is the desired solution, since $R \equiv ke^2/2a_0hc$.

EXAMPLE 3.7 An Electronic Transition in Hydrogen
The electron in the hydrogen atom makes a transition from the $n = 2$ energy state to the ground state (corresponding to $n = 1$). Find the wavelength and frequency of the emitted photon.

Solution: We can use Equation 3.34 directly to obtain λ, with $n_i = 2$ and $n_f = 1$:

$$\frac{1}{\lambda} = R\left(\frac{1}{n_f^2} - \frac{1}{n_i^2}\right)$$

$$\frac{1}{\lambda} = R\left(\frac{1}{1^2} - \frac{1}{2^2}\right) = \frac{3R}{4}$$

$$\lambda = \frac{4}{3R} = \frac{4}{3(1.097 \times 10^7\ \text{m}^{-1})}$$

$$= 1.215 \times 10^{-7}\ \text{m} = 121.5\ \text{nm}$$

This wavelength lies in the ultraviolet region.

Since $c = f\lambda$, the frequency of the photon is

$$f = \frac{c}{\lambda} = \frac{3.00 \times 10^8\ \text{m/s}}{1.215 \times 10^{-7}\ \text{m}} = 2.47 \times 10^{15}\ \text{Hz}$$

Exercise 3 What is the wavelength of the photon emitted by hydrogen when the electron makes a transition from the $n = 3$ state to the $n = 1$ state?

Answer: $\frac{9}{8R} = 102.6$ nm.

EXAMPLE 3.8 The Balmer Series for Hydrogen

The Balmer series for the hydrogen atom corresponds to electronic transitions that terminate in the state of quantum number $n = 2$, as shown in Figure 3.22. (a) Find the longest-wavelength photon emitted and determine its energy.

Solution: The longest-wavelength photon in the Balmer series results from the transition from $n = 3$ to $n = 2$. Using Equation 3.34 gives

$$\frac{1}{\lambda} = R\left(\frac{1}{n_f^2} - \frac{1}{n_i^2}\right)$$

$$\frac{1}{\lambda_{max}} = R\left(\frac{1}{2^2} - \frac{1}{3^2}\right) = \frac{5}{36}R$$

$$\lambda_{max} = \frac{36}{5R} = \frac{36}{5(1.097 \times 10^7\ \text{m}^{-1})} = 656.3\ \text{nm}$$

This wavelength is in the red region of the visible spectrum.

The energy of this photon is

$$E_{photon} = hf = \frac{hc}{\lambda_{max}}$$

$$= \frac{(6.626 \times 10^{-34}\ \text{J}\cdot\text{s})(3.00 \times 10^8\ \text{m/s})}{656.3 \times 10^{-9}\ \text{m}}$$

$$= 3.03 \times 10^{-19}\ \text{J} = 1.89\ \text{eV}$$

We could also obtain the energy of the photon by using the expression $hf = E_3 - E_2$, where E_2 and E_3 are the energy levels of the hydrogen atom, which can be calculated from Equation 3.31. Note that this is the lowest-energy photon in this series since it involves the smallest energy change.

(b) Find the shortest-wavelength photon emitted in the Balmer series.

Solution: The shortest-wavelength photon in the Balmer series is emitted when the electron makes a transition from $n = \infty$ to $n = 2$. Therefore

$$\frac{1}{\lambda_{min}} = R\left(\frac{1}{2^2} - \frac{1}{\infty}\right) = \frac{R}{4}$$

$$\lambda_{min} = \frac{4}{R} = \frac{4}{1.097 \times 10^7\ \text{m}^{-1}} = 364.6\ \text{nm}$$

This wavelength is in the ultraviolet region and corresponds to the series limit.

Exercise 4 Find the energy of the shortest-wavelength photon emitted in the Balmer series for hydrogen.

Answer: 3.40 eV.

Although the theoretical derivation of the line spectrum was a remarkable feat in itself, the scope, breadth, and impact of Bohr's monumental achievement is truly seen only when it is realized what else he treated in his three-part paper of 1913:

(a) He explained the limited number of lines seen in the absorption spectrum of hydrogen compared to the emission spectrum.

(b) He explained the emission of x-rays from atoms.

(c) He explained the nuclear origin of β-particles.

(d) He explained the chemical properties of atoms in terms of the electron shell model.

(e) He explained how atoms associate to form molecules.

Two of these topics, the comparison of absorption and emission by hydrogen and the shell structure of atoms, are of such general importance that they deserve a bit more explanation.

We have already pointed out that a gas will absorb at wavelengths that correspond exactly to some emission lines, but that all lines present in emission are not seen as dark absorption lines. Bohr explained absorption as the reverse of emission; that is, an electron in a given energy state can absorb only a photon of the exact frequency required to produce a "jump" from a lower energy state to a higher energy state. Ordinarily hydrogen atoms are in the ground state ($n = 1$) and so only the high-energy Lyman series corresponding to transitions from the ground state to higher energy states is seen in absorption. The longer wavelength Balmer series corresponding to transitions originating in the first excited state ($n = 2$) is not seen because the average thermal energy of each atom is insufficient to raise the electron to the first excited state. That is, the number of electrons in the first excited state is insufficient at ordinary temperatures to produce measurable absorption.

EXAMPLE 3.9 Hydrogen in Its First Excited State
Calculate the temperature at which many atoms of monatomic hydrogen will be in the first excited state ($n = 2$). What series should be prominent in absorption at this temperature? (Calculate both from $N_2/N_1 = \exp(-\Delta E/kT)$ and from $\frac{3}{2}kT =$ average thermal energy.)

Solution: At room temperature almost all hydrogen atoms are in the ground state with an energy of -13.6 eV. The first excited state ($n = 2$) has an energy equal to $E_2 = -3.4$ eV. Therefore, the hydrogen atoms must gain an energy of 10.2 eV to reach the first excited state. If the atoms are to obtain this energy from heat we must have

$$\tfrac{3}{2}kT = \text{average thermal energy/atom} \approx 10.2 \text{ eV}$$

or

$$T = \frac{10.2 \text{ eV}}{(3/2)k} = \frac{10.2 \text{ eV}}{(1.5)(8.62 \times 10^{-5} \text{ eV/K})}$$

$$= 79\ 000 \text{ K}$$

Let us check this result by using the Boltzmann distribution. In Chapter 9 we shall see that the probability of finding an atom with energy E at temperature T is given by

$$P(E) = P_0 e^{-(E-E_0)/kT}$$

where P_0 is the probability of finding the atom in the ground state of energy E_0. From this expression, it follows that the ratio of the number of atoms in two different energy levels in thermal equilibrium at temperature T is given by

$$\frac{N_2}{N_1} = \frac{P(E_2)}{P(E_1)} = e^{-(E_2-E_1)/kT}$$

where N_2 is the number in the upper level, N_1 is the number in the lower level, and ΔE is the energy separation of the two levels. Let us use this equation to determine the temperature at which approximately 10% of the hydrogen atoms are in the $n = 2$ state.

$$\frac{N_2}{N_1} = 0.10 = e^{-(10.2 \text{ eV})/kT}$$

or

$$\ln(0.10) = -\frac{10.2 \text{ eV}}{kT}$$

Solving for T gives

$$T = -\frac{10.2 \text{ eV}}{k \ln(0.10)} = -\frac{10.2 \text{ eV}}{(8.62 \times 10^{-5} \text{ eV/K}) \ln(0.10)}$$

$$T = 51\ 000 \text{ K}$$

Thus the two estimates agree in order of magnitude and show that the Balmer series will only be seen in absorption if the absorbing gas is quite hot as in a stellar atmosphere.

The final comment on Bohr's work concerns his development of electronic shell theory to treat multi-electron atoms. In part II of his paper he attempted to find stable electronic arrangements subject to the conditions that the total angular momentum of all the electrons is quantized and, simultaneously, that the total energy is a minimum. This is a difficult problem, and one that becomes more difficult as more electrons are introduced into a system. Nevertheless Bohr had considerable success, as he was able to show that neutral hydrogen could easily add another electron to become H^-, and that neutral helium was particularly stable with a closed innermost shell of two electrons and a high ionization potential. He also proposed that lithium ($Z = 3$) had an electronic arrangement consisting of two electrons in one orbit near the nucleus and the third in a large, loosely bound outer orbit. This explains the tendency of lithium atoms to lose an electron and "take a positive charge in chemical combinations with other elements." Although we cannot afford the luxury of looking in detail at all of Bohr's predictions about multi-electron atoms, his basic ideas of shell structure are as follows:

(a) Electrons of elements with higher atomic number form stable concentric rings, with definite numbers of electrons allowed for each ring or shell.

(b) The number of electrons in the outermost ring determines the valency.[17]

With almost magical insight, Bohr ended part II of his classic paper with the explanation of the similar chemical properties of the iron group (Fe, Co, Ni) and the rare earths, which have atomic numbers that progressively increase by one and would not normally be expected to be chemically alike. The answer, according to Bohr, is that the configuration of electrons in the outer-

[17] G. N. Lewis, an American chemist, contributed much to our understanding of shell structure in 1916, building on Bohr's remarkable foundation.

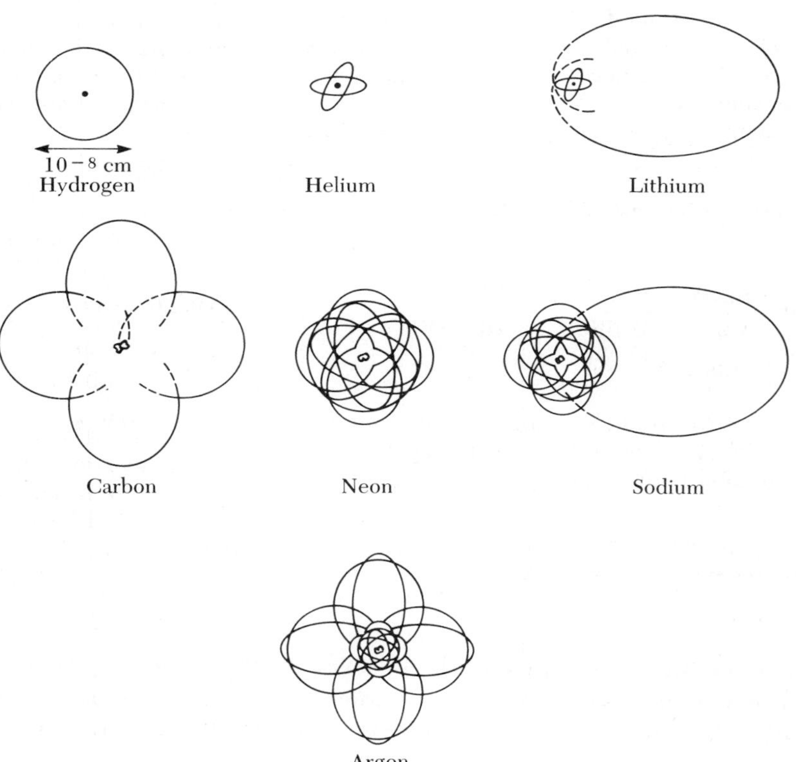

Figure 3.23 Bohr's sketches of electronic orbits.

most ring of these elements is identical and that it is energetically more favorable to add electrons to inner rings. At the risk of encouraging some to take the idea of electronic orbits too seriously, Figure 3.23 shows some sketches of electronic orbits as drawn by Bohr in the early 1900s.

Postscript: Why is Angular Momentum Quantized?

Both Bohr's idea of discrete, nonradiating energy states and the emission postulate for atoms were foreshadowed by Planck's quantization of the energy of blackbody oscillators and by Einstein's treatment of the photoelectric effect. However, the concept of angular momentum quantization seems to have sprung full blown from Bohr's Gedankenkuche (thought kitchen), as so aptly expressed by Einstein. Indeed, in some of his later writings Bohr emphasized the point of view that the quantization of angular momentum was a postulate, underivable from any deeper law, and that its validity depended simply on the agreement of his model with experimental spectra.

What is most interesting is that in his 1913 paper Bohr ingeniously showed that the quantization of angular momentum is a consequence of the smooth and gradual emergence of classical results from quantum theory in the limit of large systems. In particular Bohr argued, according to his correspondence principle, that Maxwell's classical laws of electromagnetic radiation and the quantum conditions for emission and energy quantization ($\Delta E = hf$) *must simultaneously hold for the case of extremely large electronic orbits.* This case is shown in Figure 3.24. In this figure, r_1 and r_2 are the radii of two large adjacent orbits that are separated in energy by an amount $\Delta E = E_1 - E_2$, and f is the orbital frequency of the electron, which is assumed to be approximately constant in a transition between large orbits. Now consider the emission of a photon of energy hf' when the electron makes a transition from r_1 to r_2. Since we have assumed that $r_1 \approx r_2 = r$ and $f_1 \approx f_2 = f$, the kinetic energy of the electron does not change during the transition. Thus if energy is to be conserved in the emission process, the energy carried off by the photon must come from the work done on the electron by the centripetal force:

$$hf' = F_c(\Delta r) = \frac{mv^2}{r}(\Delta r) = m(4\pi^2 f^2 r)(\Delta r) \tag{3.37}$$

If we now assume that for such large orbits Maxwell's law must hold and the frequency of the light emitted, f', must equal the orbital frequency of the circulating electron, f, we have

$$hf = m4\pi^2 f^2 r\ \Delta r \tag{3.38}$$

or

$$\hbar = m2\pi fr\ \Delta r = mv\ \Delta r \tag{3.39}$$

Since the change in angular momentum for this case is

$$\Delta(mvr) = mv\ \Delta r \qquad (\Delta m, \Delta v = 0)$$

we have

$$\Delta(mvr) = \hbar \tag{3.40}$$

for the change in electronic angular momentum for a transition between adjacent electronic orbits. In general, if the least change in angular momentum is $\hbar$, we may say that the total angular momentum has a value equal to whole multiples of $\hbar$ or

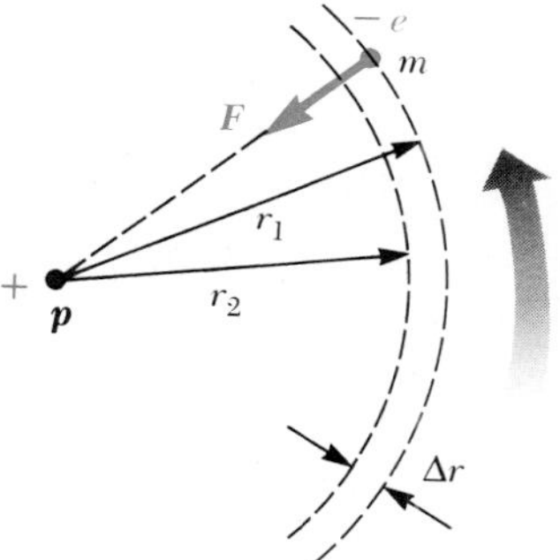

Figure 3.24 The classical limit of the Bohr atom. Note that r_1 and r_2 are the radii of two large adjacent quantum orbits, in which the electron has orbital frequencies of f_1 and f_2. We assume $r_1 \approx r_2$ and $f_1 \approx f_2$.

$$mvr = n\hbar \qquad n = 1, 2, 3, \ldots \tag{3.41}$$

Bohr's truly great insight was to realize that although Equation 3.41 was derived for the case of large electron orbits, it was a universal quantum principle applicable to all systems and of wider applicability than Maxwell's law of radiation. Such a bold and farseeing vision was characteristic of the man so aptly described by Einstein in the following quote: "That this insecure and contradictory foundation [physics from 1910 to 1920] was sufficient to enable a man of Bohr's unique instinct and tact to discover the major laws of the spectral lines and of the electron shells of the atoms together with their significance for chemistry appeared to me like a miracle—and appears to me as a miracle even today. This is the highest form of musicality in the sphere of thought."

3.5 SUMMARY

The determination of the composition of atoms relies heavily on four classic experiments:

1. **Faraday's law of electrolysis,** which may be stated as

$$m = \frac{(q)(\text{molar weight})}{(96\ 500\ C)(\text{valence})} \tag{3.1}$$

where m is the mass liberated at an electrode and q is the total charge passed through the solution. Faraday's law shows that atoms are composed of positive and negative charges and that atomic charges always consist of multiples of some unit charge.

2. **J. J. Thomson's determination of e/m** and that the electron, with e/m about 2000 times that of the proton, is a part of all atoms. Thomson measured e/m of electrons from a variety of elements by measuring the deflection of an electron beam by an electric field. He then applied a magnetic field to just cancel the electric deflection in order to determine the electron velocity. The charge to mass ratio in the Thomson experiment is given by

$$\frac{e}{m} = \frac{V\theta}{B^2 \ell d} \tag{3.7}$$

where V/d is the applied electric field, ℓ is the length of the vertical deflecting plates, θ is the deflection produced by the electric field, and B is the applied magnetic field.

3. **Millikan's determination of the fundamental charge, e.** By balancing the electric and gravitational force on individual oil drops, Millikan was able to determine the fundamental electric charge and to show that charges always occur in multiples of e. The quantum of charge may be determined from the relation

$$ne = \left(\frac{mg}{E}\right)\left(\frac{v + v'}{v}\right) \tag{3.11}$$

where n is an integer, m is the mass of the drop, E is the electric field, v is the terminal velocity of the drop with field off (falling), and v' is the terminal velocity of the drop with field on (rising).

4. **Rutherford's scattering of α-particles from gold atoms,** which established the nuclear model of the atom. By measuring the rate of scattering of α-particles into an angle ϕ, Rutherford was able to establish that most of the

mass and all of the positive charge of an atom, $+Ze$, are concentrated in a minute volume of the atom with a radius of about 10^{-14} m.

The explanation of the motion of electrons within the atom and of the rich and elaborate series of spectral lines emitted by the atom was given by Bohr. Bohr's theory was based partly on classical mechanics and partly on some startling new quantum ideas. **Bohr's postulates** were:

1. Electrons move about the nucleus in circular orbits determined by Coulomb's and Newton's laws.
2. Only certain orbits are stable. The electron does not radiate energy in these special orbits, which are specified by the quantization of angular momentum:

$$mvr = n\hbar \qquad n = 1, 2, 3, \ldots \tag{3.24}$$

where $\hbar$ is Planck's constant divided by 2π.

3. A spectral line of frequency f is emitted when an electron jumps from an initial orbit of energy E_i to a final orbit of energy E_f, where

$$hf = E_i - E_f \tag{3.23}$$

These postulates lead to quantized orbits and quantized energies for a single electron orbiting a nucleus with charge $+Ze$, given by

$$r_n = \frac{n^2 a_0}{Z} \tag{3.35}$$

and

$$E_n = -\frac{13.6Z^2}{n^2}\ \text{eV} \tag{3.36}$$

where n is an integer and $a_0 = \hbar^2/mke^2 = 0.529$ Å is the Bohr radius.

SUGGESTIONS FOR FURTHER READING

1. G. Holton, *Introduction to Concepts and Theories in Physical Science,* Reading, MA, Addison-Wesley, 1952. This is an accurate, humane, and eminently readable overview of the history and progression of scientific concepts from the Greeks to quantum theory.
2. J. Perrin, *Atoms,* translated by D. L. Hammick, New York, D. Van Nostrand Co., 1923. This is a superb account at first hand of the evidence for atoms at the beginning of the twentieth century.
3. G. Thomson, *J. J. Thomson and the Cavendish Laboratory in His Day,* New York, Doubleday and Co., 1965. This book offers a fascinating account of the e/m experiment and other works by Thomson's son.
4. R. A. Millikan, *Electrons (+ and −), Protons, Neutrons, Mesotrons, and Cosmic Rays,* Chicago, University of Chicago Press, 1947. This book gives detailed accounts of the determination of e.
5. *Physics Today,* October 1985, Special Issue: Niels Bohr Centennial. This magazine contains three articles dealing with Bohr's scientific, social, and cultural contributions to the physics community. Many interesting photos of the key players are included.

QUESTIONS

1. The Bohr theory of the hydrogen atom is based upon several assumptions. Discuss these assumptions and their significance. Do any of these assumptions contradict classical physics?
2. Suppose that the electron in the hydrogen atom obeyed classical mechanics rather than quantum mechanics. Why should such a *hypothetical* atom emit a continuous spectrum rather than the observed line spectrum?
3. Can the electron in the ground state of the hydrogen atom absorb a photon of energy (a) *less* than 13.6 eV and (b) *greater* than 13.6 eV?

PROBLEMS

1. Calculate the wavelengths of the first three lines in the Balmer series for hydrogen.
2. Calculate the wavelengths of the first three lines in the Lyman series for hydrogen.
3. (a) What value of n is associated with the Lyman series line in hydrogen whose wavelength is 102.6 nm? (b) Could this wavelength be associated with the Paschen or Brackett series?
4. Use Equation 3.35 to calculate the radii of the first, second, and third Bohr orbits of hydrogen.
5. (a) Construct an energy level diagram for the He^+ ion, for which $Z = 2$. (b) What is the ionization energy for He^+?
6. Construct an energy level diagram for the Li^{2+} ion, for which $Z = 3$.
7. What is the radius of the first Bohr orbit in (a) He^+, (b) Li^{2+}, and (c) Be^{3+}?
8. A hydrogen atom initially in its ground state ($n = 1$) absorbs a photon and ends up in the state for which $n = 3$. (a) What is the energy of the absorbed photon? (b) If the atom returns to the ground state, what photon energies could the atom emit?
9. A photon is emitted from a hydrogen atom that undergoes an electronic transition from the state $n = 3$ to the state $n = 2$. Calculate (a) the energy, (b) the wavelength, and (c) the frequency of the emitted photon.
10. What is the energy of the photon that could cause (a) an electronic transition from the $n = 4$ state to the $n = 5$ state of hydrogen and (b) an electronic transition from the $n = 5$ state to the $n = 6$ state?
11. (a) Calculate the longest and shortest wavelengths for the Paschen series. (b) Determine the photon energies corresponding to these wavelengths.
12. Find the potential energy and kinetic energy of an electron in the ground state of the hydrogen atom.
13. A hydrogen atom is in its ground state ($n = 1$). Using the Bohr theory of the atom, calculate (a) the radius of the orbit, (b) the linear momentum of the electron, (c) the angular momentum of the electron, (d) the kinetic energy, (e) the potential energy, and (f) the total energy.
14. *Positronium* is a hydrogen-like atom consisting of a positron (a positively charged electron) and an electron revolving around each other. Using the Bohr model, find the allowed radii (relative to the center of mass of the two particles) and the allowed energies of the system.
15. Use Bohr's model of the hydrogen atom to show that when the atom makes a transition from the state n to the state $n - 1$, the frequency of the emitted light is given by

$$f = \frac{2\pi^2 mk^2e^4}{h^3}\left[\frac{2n-1}{(n-1)^2n^2}\right]$$

Show that as $n \rightarrow \infty$, the expression above varies as $1/n^3$ and reduces to the classical frequency one would expect the atom to emit. (*Hint:* To calculate the classical frequency, note that the frequency of revolution is $v/2\pi r$, where r is given by Equation 3.28.) This is an example of the correspondence principle, which requires that the classical and quantum models agree for large values of n.

16. *The Auger process.* An electron in chromium makes a transition from the $n = 2$ state to the $n = 1$ state without emitting a photon. Instead, the excess energy is transferred to an outer electron (in the $n = 4$ state), which is ejected by the atom. (This is called an *Auger process,* and the ejected electron is referred to as an *Auger electron.*) Use the Bohr theory to find the kinetic energy of the Auger electron.
17. A *muon* is a particle with a charge of $-e$ and a mass equal to 207 times the mass of an electron. Muonic lead is formed when ^{208}Pb captures a muon. According to the Bohr theory, what are the radius and energy of the ground state of muonic lead?
18. A muon (Problem 17) is captured by a deuteron (an ^2H nucleus) to form a muonic atom. (a) Find the energy of the ground state and the first excited state. (b) What is the wavelength of the photon emitted when the atom makes a transition from the first excited state to the ground state?
19. Wavelengths of spectral lines depend, to some extent, on the nuclear mass. Determine the shift in the first line of the Balmer series (relative to hydrogen) for (a) deuterium (^2H) and (b) tritium (^3H).
20. An electron initially in the $n = 3$ state of a one-electron atom of mass M undergoes a transition to the $n = 1$ ground state. Show that the recoil velocity of the atom from emission of a photon is given approximately by

$$v = \frac{8hR}{9M}$$

21. In a Millikan oil drop experiment, the condenser plates are spaced 2.00 cm apart, the potential across the plates is 4000 V, the rise or fall distance is 4.0 mm, the density of the oil droplets is 0.80 g/cm^3, and the viscosity of the air is 1.81×10^{-5} kg·m^{-1}s^{-1}. The average time of fall in the absence of an electric field is 15.9 s. The following different rise times in seconds are observed when the field is turned on: 36.0, 17.3, 24.0, 11.4, 7.54. (a) Find the radius and mass of the drop used in this experiment. (b) Calculate the charge on each drop, and show that charge is quantized by considering both the size of each charge and the amount of charge gained (lost) when the rise time changes. (c) Determine the electronic charge from this data. You may assume that e lies between 1.5 and 2.0×10^{-19} C.
22. Actual data from one of Millikan's early experiments are as follows:
$a = 0.000276$ cm
$\rho = 0.9561$ g/cm^3
average time of fall = 11.894 s

rise or fall distance = 10.21 mm
plate separation = 16.00 mm
average potential difference between plates = 5085 V
sequential rise times in seconds: 80.708, 22.366, 22.390, 22.368, 140.566, 79.600, 34.748, 34.762, 29.286, 29.236
Find the average value of e by requiring that the difference in charge for drops with different rise times be equal to an *integral* number of elementary charges.

23. Using the Faraday (96 500 C) and Avogadro's number, determine the electronic charge. Explain your reasoning.

24. *Weighing a copper atom in an electrolysis experiment.* A standard experiment involves passing a current of several amperes through a copper sulfate solution ($CuSO_4$) for a period of time and determining the mass of copper plated onto the cathode. If it is found that a current of 1.00 A flowing for 3600 s deposits 1.185 g of copper, find (a) the number of copper atoms deposited (use $e = 1.60 \times 10^{-19}$ C), (b) the weight of a copper atom, and (c) the molar weight of copper.

25. Show that Balmer's formula, $\lambda = C_2\left(\frac{n^2}{n^2 - 2^2}\right)$, reduces to the Rydberg formula, $\lambda = R\left(\frac{1}{2^2} - \frac{1}{n^2}\right)$, provided that $\frac{2^2}{C_2} = R$. Check that $\frac{2^2}{C_2}$ has the same numerical value as R.

26. A mystery particle enters the region between the plates of a Thomson apparatus as shown in Figure 3.6. The deflection angle θ is measured to be 0.20 radians (downwards) for this particle when $V = 2000$ V, $\ell = 10.0$ cm, and $d = 2.00$ cm. If a perpendicular magnetic field of magnitude 4.57×10^{-2} T is applied simultaneously with the electric field, the particle passes through the plates without deflection. (a) Find q/m for this particle. (b) Identify the particle. (c) Find the horizontal velocity with which the particle entered the plates. (d) Must we use relativistic mechanics for this particle?

27. Figure 3.25 shows a cathode ray tube for determining e/m without applying a magnetic field. In this case v_x may be found by measuring the rise in temperature when a known amount of charge is stopped in a target. If V, ℓ, d, D, and y are measured, e/m may be found. Show that

$$\frac{e}{m} = \frac{yv_x^2d}{V\ell\left(\frac{\ell}{2} + D\right)}$$

28. A parallel beam of α-particles with fixed kinetic energy is normally incident on a piece of gold foil. (a) If 100 α-particles per minute are detected at 20°, how many will be counted at 40, 60, 80, and 100°? (b) If the kinetic energy of the incident α-particles is doubled, how many scattered α-particles will be observed at 20°? (c) If the original α-particles were incident on a copper foil of the same thickness, how many scattered α-particles would be detected at 20°? Note that $\rho_{Cu} = 8.9$ g/cm^3 and $\rho_{Au} = 19.3$ g/cm^3.

29. It is observed that α-particles with kinetic energies of 13.9 MeV and higher, incident on Cu foils, do not obey Rutherford's $(\sin \phi/2)^{-4}$ law. Estimate the nuclear size of copper from this observation, assuming that the Cu nucleus remains fixed in a head-on collision with an α-particle.

30. A hydrogen atom initially in the $n = 3$ state decays to the ground state with the emission of a photon. (a) Calculate the wavelength of the emitted photon. (b) Estimate the recoil momentum of the atom, and the kinetic energy of the recoiling atom. Where does this energy come from?

31. Calculate the frequency of the photon emitted by a hydrogen atom making a transition from the $n = 4$ to the $n = 3$ state. Compare your result with the frequency of revolution for the electron in these two Bohr orbits.

32. Calculate the longest and shortest wavelengths in the Lyman series for hydrogen, indicating the underlying electronic transition that gives rise to each. Are any of the Lyman spectral lines in the visible spectrum? Explain.

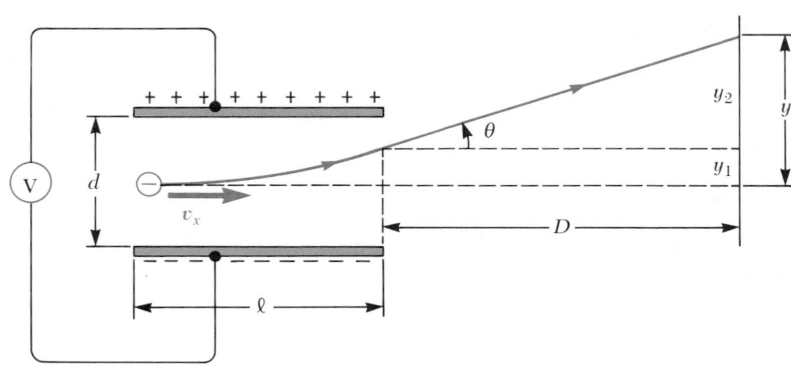

Figure 3.25 (Problem 27)

4

Matter Waves

In the previous chapter we discussed many important discoveries and theoretical concepts concerning the *particle nature of matter.* We shall now point out some of the shortcomings of these theories and introduce the fascinating and bizarre *wave properties of particles.* Especially notable are Count Louis de Broglie's remarkable ideas about how to represent an electron (and other particles) as a wave, and the experimental confirmation of de Broglie's hypotheses by the electron diffraction experiments of Davisson and Germer. We shall also see how the notion of representing a particle as a localized wave or wave group leads naturally to limitations on simultaneously measuring position and momentum of the particle. Finally, we shall discuss the passage of electrons through a double slit as a way of understanding the wave-particle duality of matter.

4.1 THE PILOT WAVES OF DE BROGLIE

By the early 1920s scientists recognized that the Bohr theory contained many inadequacies:

1. It failed to predict the observed intensities of spectral lines.
2. It had only limited success in predicting emission and absorption wavelengths for multi-electron atoms.
3. It failed to provide an equation of motion governing the time development of atomic systems starting from some initial state.
4. It overemphasized the particle nature of matter and could not explain the newly discovered wave-particle duality of light.
5. It did not supply a general scheme for "quantizing" other systems, especially those without periodic motion.

Figure 4.1 Louis de Broglie was the scion of an aristocratic French family that produced marshals, ambassadors, foreign ministers, and at least one Duke, his older brother Maurice de Broglie. Louis de Broglie came rather late to theoretical physics as he first studied history. Only after serving as a radio operator in World War I did he follow the lead of his older brother and begin his studies of physics. Maurice de Broglie was an outstanding experimental physicist in his own right and conducted experiments in the palatial family mansion in Paris.

The first bold step toward a new mechanics of atomic systems was taken by Louis Victor de Broglie in 1923 (Fig. 4.1). In his doctoral dissertation he postulated that *because photons have wave and particle characteristics, perhaps all forms of matter have wave as well as particle properties.* This was a highly revolutionary idea with no experimental confirmation at that time. According to de Broglie, electrons had a dual particle/wave nature. Accompanying every electron was a wave (not an electromagnetic wave!), which guided or "piloted" the electrons through space. He explained the source of this assertion in his 1929 Nobel Prize acceptance speech:

> On the one hand the quantum theory of light cannot be considered satisfactory since it defines the energy of a light corpuscle by the equation $E = hf$ containing the frequency f. Now a purely corpuscular theory contains nothing that enables us to define a frequency; for this reason alone, therefore, we are compelled, in the case of light, to introduce the idea of a corpuscle and that of periodicity simultaneously. On the other hand, determination of the stable motion of electrons in the atom introduces integers, and up to this point the only phenomena involving integers in physics were those of interference and of normal modes of

vibration. This fact suggested to me the idea that electrons too could not be considered simply as corpuscles, but that periodicity must be assigned to them also.

Let us follow de Broglie's reasoning in more detail. He proposed that the wavelength and frequency of a *matter wave* associated with any moving object be given by

$$\boxed{\lambda = \frac{h}{p}} \tag{4.1}$$

de Broglie wavelength

and

$$f = E/h \tag{4.2}$$

where h is Planck's constant, p is the relativistic momentum, and E is the total relativistic energy of the object. Recall from Chapter 1 that p and E can be written as

$$p = \gamma m_0 v = mv \tag{4.3}$$

and

$$E^2 = p^2c^2 + m_0{}^2c^4 = m^2c^4 \tag{4.4}$$

where $\gamma = (1 + v^2/c^2)^{-1/2}$, m_0 is the rest mass of the object, and $m = \gamma m_0$. Note that Equations 4.1 and 4.2 were originally applied to photons, but in that case the rest mass $m_0 = 0$, so $E_{\text{photon}} = pc$. Solving for the velocity (actually the phase velocity or velocity of a wave-crest) of these matter waves gives

$$v_p = f\lambda = \frac{E}{h}\cdot\frac{h}{p} = \frac{mc^2}{mv} = \frac{c^2}{v} \tag{4.5}$$

At first sight this relationship seems to be in conflict with relativity theory since v, the speed of a material particle, is necessarily *less* than c, making the phase velocity greater than c. The problem can be resolved by recognizing that a single sinusoidal matter wave of infinite extent traveling at a constant velocity cannot properly represent a small particle localized in space. What is needed is a superposition or sum of many sinusoidal waves with different wavelengths, which interfere to form a wave group. This wave group with limited extent can then be shown to have a velocity (technically the group velocity, v_g) less than c and identical to the particle's velocity. This argument is shown schematically in Figure 4.2 and will be treated in detail after a brief

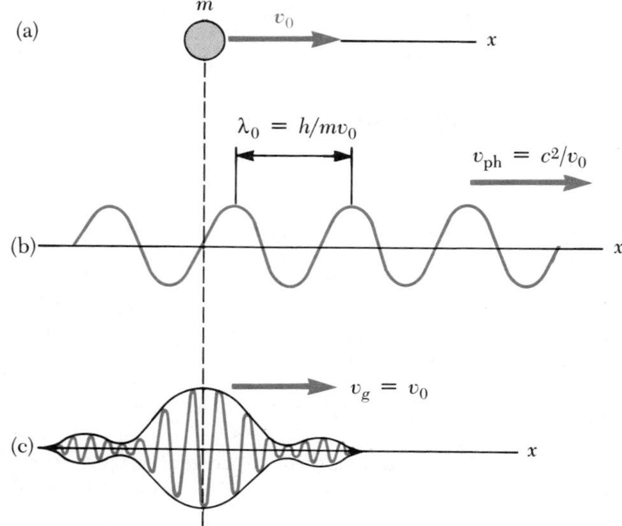

Figure 4.2 Representing a particle with matter waves: (a) particle of mass m and velocity v_0. (b) Single matter wave of infinite extent. (c) Superposition of many matter waves with a spread of wavelengths centered on $\lambda_0 = h/mv_0$.

discussion of the application of de Broglie's ideas to atomic systems. In this way we will develop some physical understanding for de Broglie's postulates before launching into more mathematical arguments.

Quantization of Angular Momentum in the Bohr Model

Bohr's model of the atom has many shortcomings and problems. For example, as the electrons revolve around the nucleus, how does one visualize the fact that only certain electronic energies are allowed? Why do all atoms of a given element have precisely the same physical properties regardless of the infinite variety of starting velocities and positions of the electrons in each atom?

De Broglie's great insight was to recognize that wave theories of matter handle these problems neatly by means of interference. For example, a plucked guitar string, while initially subjected to a wide range of wavelengths, will support only standing wave patterns that have nodes at each end. Thus only a discrete set of wavelengths is allowed for standing waves, while other wavelengths not included in this discrete set rapidly vanish by destructive interference. This same reasoning can be applied to electron matter waves bent into a circle around the nucleus. Although initially a continuous distribution of wavelengths may be present corresponding to a distribution of initial electron velocities, most wavelengths and velocities rapidly die off. The residual standing wave patterns thus account for the identical nature of all atoms of a given element and show that atoms are more like vibrating drum heads with discrete modes of vibration than like minature solar systems. This point of view is emphasized in Figure 4.3, which shows the standing wave pattern of the electron in the hydrogen atom corresponding to the $n = 3$ state of the Bohr theory.

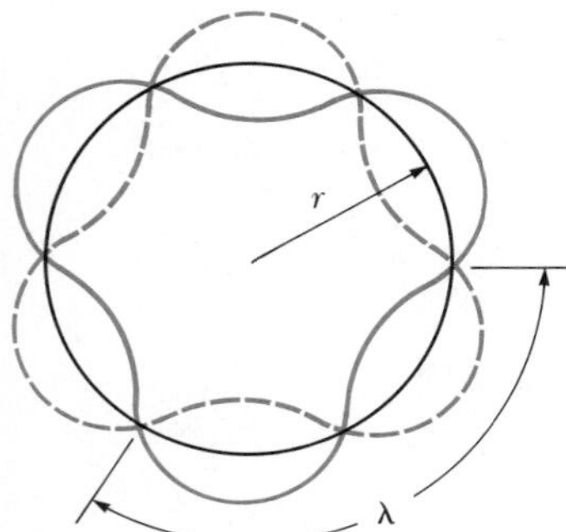

Figure 4.3 Standing waves fit to a circular Bohr orbit. In this particular diagram, three wavelengths are fit to the orbit, corresponding to the $n = 3$ energy state of the Bohr theory.

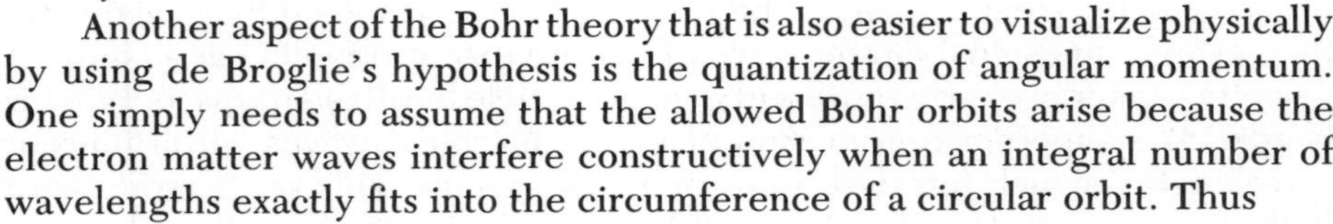

Another aspect of the Bohr theory that is also easier to visualize physically by using de Broglie's hypothesis is the quantization of angular momentum. One simply needs to assume that the allowed Bohr orbits arise because the electron matter waves interfere constructively when an integral number of wavelengths exactly fits into the circumference of a circular orbit. Thus

$$n\lambda = 2\pi r \tag{4.6}$$

where r is the radius of the orbit. From Equation 4.1, we see that $\lambda = h/mv$. Substituting this into Equation 4.6, and solving for mvr, the angular momentum of the particle, gives

$$\boxed{mvr = n\hbar} \tag{4.7}$$

Note that this is precisely the Bohr condition of quantization of angular momentum.

EXAMPLE 4.1 Why Don't We See the "Fuzzy" Edges of a Baseball?

An object will appear "wavelike" if it exhibits interference or diffraction, both of which require scattering objects or apertures of about the same size as the wavelength. A baseball of mass 140 g traveling at a speed of 60 mph (27 m/s) has a wavelength given by

$$\lambda = \frac{h}{mv} = \frac{6.63 \times 10^{-34}\ \text{J}\cdot\text{s}}{(0.140\ \text{kg})(27\ \text{m/s})} = 1.75 \times 10^{-34}\ \text{m}$$

Even a nucleus (whose size is $\sim 10^{-15}$ m) is much too large to diffract this incredibly small wavelength! This explains why all macroscopic objects appear particle-like.

EXAMPLE 4.2 What Size Particles Do Exhibit Diffraction?

A particle of charge q and mass m is accelerated from rest through a potential difference V. (a) Find its de Broglie wavelength.

Solution: When a charge is accelerated from rest through a potential difference V, its gain in kinetic energy $\frac{1}{2}mv^2$ must equal its loss in potential energy qV, since energy is conserved. That is,

$$\tfrac{1}{2}mv^2 = qV$$

Since $p = mv$, we can express this in the form

$$\frac{p^2}{2m} = qV \qquad \text{or} \qquad p = \sqrt{2mqV}$$

Substituting this expression for p into the de Broglie relation $\lambda = h/p$ gives

$$\lambda = \frac{h}{p} = \frac{h}{\sqrt{2mqV}}$$

(b) Calculate λ if the particle is an electron and $V = 50$ V.

Solution: The de Broglie wavelength of the electron accelerated through 50 V is

$$\lambda = \frac{h}{\sqrt{2mqV}}$$

$$= \frac{6.63 \times 10^{-34}\ \mathrm{J \cdot s}}{\sqrt{2(9.11 \times 10^{-31}\ \mathrm{kg})(1.6 \times 10^{-19}\ \mathrm{C})(50\ \mathrm{V})}}$$

$$= 1.74 \times 10^{-10}\ \mathrm{m} = 1.74\ \text{Å}$$

This wavelength is of the order of atomic dimensions and the spacing between atoms in a solid. Such low-energy electrons are normally used in electron diffraction experiments.

Exercise 1 (a) Show that the de Broglie wavelength for an electron accelerated from rest through a *large* potential difference, V, is

$$\lambda = \frac{12.27}{V^{1/2}}\left(\frac{Ve}{2m_0c^2} + 1\right)^{-1/2} \qquad (4.8)$$

where λ is in angstroms and V is in volts. (b) Calculate the percent error introduced when $\lambda = 12.27/V^{1/2}$ is used instead of the correct relativistic expression for 10 MeV electrons.
Answer: (b) 230%

4.2 THE DAVISSON-GERMER EXPERIMENT

The crucial experimental proof that electrons possess a wavelength $\lambda = h/p$ was furnished by the diffraction experiments of Davisson and Germer at the Bell Laboratories in New York City in 1927 (Fig. 4.4).[1] In fact, de Broglie had already suggested in 1924 that a stream of electrons traversing a small aperture should exhibit diffraction phenomena. In 1925 Einstein (again!) was led to the necessity of postulating matter waves from an analysis of fluctuations of a molecular gas. In addition, he noted that a molecular beam should show small but measurable diffraction effects. In the same year, Walter Elsasser, much impressed with de Broglie's and Einstein's ideas on matter waves, pointed out that the slow electron scattering experiments of C. J. Davisson and C. H. Kunsman at the Bell Labs could be explained by electron diffraction.

Clear-cut proof of the wave nature of electrons was obtained in 1927 by the work of Davisson and Germer in the United States and G. P. Thomson (the son of J. J. Thomson) in England. Both cases are intriguing not only for their physics but also for their human interest. The first case was an accidental discovery, and the second involved the discovery of the particle properties of the electron by the father and the wave properties by the son.

The crucial experiment of Davisson and Germer was an offshoot of an attempt to understand the arrangement of atoms on the surface of a nickel sample by elastically scattering a beam of low-speed electrons from a polycrystalline nickel target. A schematic drawing of their apparatus is shown in Figure 4.5. Note that their device allowed the variation of three experimental parameters, electron energy, nickel target orientation, α, and scattering angle, ϕ. Before a fortuitous accident occurred, the results seemed quite pedestrian. For constant electron energies of about 100 eV, the scattered intensity rapidly decreased as ϕ increased. But then someone dropped a flask of liquid air on the glass vacuum system, rupturing the vacuum and oxidizing

Figure 4.4 Clinton J. Davisson and Lester A. Germer (right) at Bell Laboratories in New York City.

[1] Clinton J. Davisson and Lester H. Germer, *Phys. Rev.* **30:** 705, 1927.

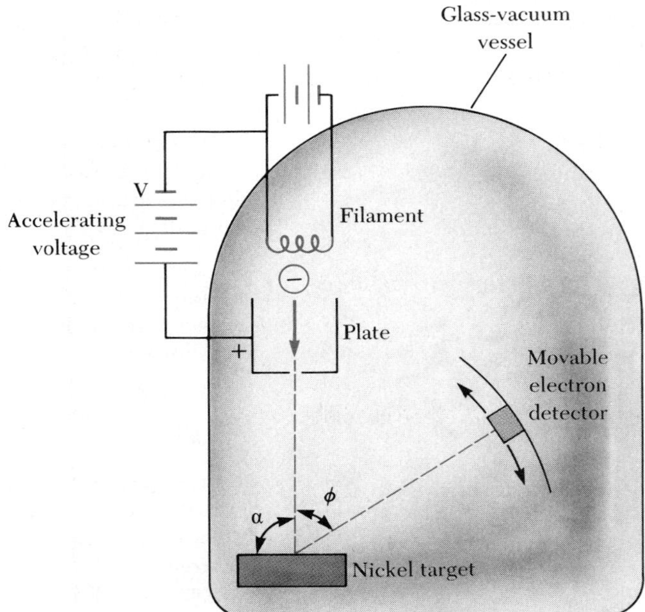

Figure 4.5 A schematic diagram of the Davisson-Germer apparatus.

the nickel target, which had been at high temperature. In order to remove the oxide, the sample was reduced by heating cautiously[2] in a flowing stream of hydrogen. When the apparatus was reassembled, quite different results were found: strong variations in the intensity of scattered electrons with angle were observed, as shown in Figure 4.6. What had happened was that the prolonged heating had annealed the nickel sample, causing large single-crystal regions to develop. These crystalline regions furnished the extended regular lattice needed to observe electron diffraction. Once Davisson and Germer realized that it was the elastic scattering from *single crystals* that produced such unusual results (1925), they initiated a thorough investigation of elastic scattering from large single crystals with predetermined crystallographic orientation. However, even these experiments were not conducted at first as a test of

[2] At present this can be done without the slightest fear of stinks or bangs, as 5% hydrogen 95% argon safety mixtures are commercially available.

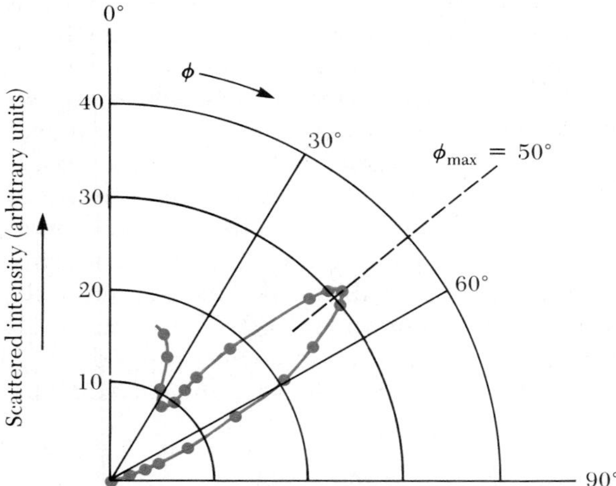

Figure 4.6 A polar plot of scattered intensity versus scattering angle for 54 eV electrons, based on the original work of Davisson and Germer. The scattered intensity is proportional to the distance of the point from the origin.

de Broglie's wave theory. Following discussions with Richardson, Born, and Franck, the experiments and their analysis finally culminated in 1927 in the proof that electrons experience diffraction with a wavelength given by $\lambda = h/p$.

The idea that electrons behave like waves when interacting with the atoms of a crystal is so striking that Davisson and Germer's proof deserves closer scrutiny. In effect they calculated the wavelength of electrons from a simple diffraction formula and compared this result with de Broglie's formula $\lambda = h/p$. Although they tested this result over a wide range of target orientations and electron energies, we will consider in detail only the simple case shown in Figures 4.5 and 4.6 with $\alpha = 90°$, $V = 54.0$ V, and $\phi = 50°$, corresponding to the $n = 1$ diffraction maximum. In order to obtain the de Broglie wavelength for this case, we first obtain the velocity of an electron accelerated through a potential difference V from the conservation of energy relation

$$\tfrac{1}{2}mv^2 = Ve$$

Substituting $v = \sqrt{2Ve/m}$ into the de Broglie relation gives

$$\lambda = \frac{h}{mv} = \frac{h}{\sqrt{2Vem}} \tag{4.9}$$

Thus the wavelength of 54 V electrons is

$$\lambda = \frac{6.63 \times 10^{-34}\ \text{J}\cdot\text{s}}{\sqrt{2(54\ \text{V})(1.60 \times 10^{-19}\ \text{C})(9.11 \times 10^{-31}\ \text{kg})}}$$

$$= 1.67 \times 10^{-10}\ \text{m} = 1.67\ \text{Å}$$

The experimental wavelength may be obtained by considering the nickel atoms to be a reflection diffraction grating, as shown in Figure 4.7. Only the surface layer of atoms is considered because low-energy electrons, unlike x-rays, do not penetrate deeply into the crystal. Constructive interference occurs when the path length difference between two adjacent rays is an integral number of wavelengths or

$$d \sin \phi = n\lambda \tag{4.10}$$

As d was known to be 2.15 Å from x-ray diffraction measurements, Davisson and Germer calculated λ to be

$$\lambda = (2.15\ \text{Å})(\sin 50°) = 1.65\ \text{Å}$$

which is in excellent agreement with the de Broglie formula.

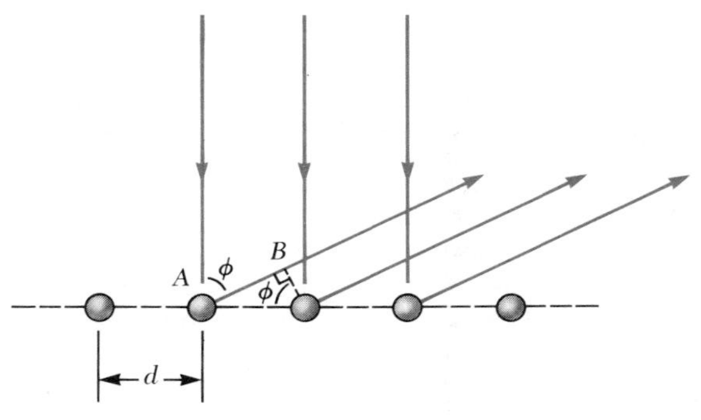

Figure 4.7 Constructive interference from electron matter waves scattered at an angle ϕ.

Figure 4.8 Diffraction of 50 kV electrons from a film of Cu_3Au alloy. The alloy film was 400 Å thick. (Courtesy of the late Dr. L. H. Germer.)

It is interesting to note that while the diffraction lines from low-energy reflected electrons are quite broad (see Fig. 4.6), the lines from high-energy electrons transmitted through metal foils are quite sharp (see Fig. 4.8). This effect occurs because hundreds of atomic planes are penetrated by high-energy electrons and consequently Equation 4.10, which treats diffraction from a surface layer, no longer holds. Instead, the Bragg law, $2d \sin \theta = n\lambda$, applies to high-energy electron diffraction. The maxima are extremely sharp in this case because if $2d \sin \theta$ is not exactly equal to $n\lambda$ there will be no diffracted wave. This occurs because there are scattering contributions from so many atomic planes that eventually the path length difference between the wave from the first plane and some deeply buried plane will be *an odd multiple of* $\lambda/2$, resulting in complete cancellation of these waves (see Problem 22).

If de Broglie's postulate is true for all matter, then any object of mass m has wavelike properties and a wavelength $\lambda = h/mv$. In the years following Davisson and Germer's discovery, experimentalists tested the universal character of de Broglie's postulate by searching for diffraction of other "particle" beams. In subsequent experiments, diffraction was observed for helium atoms (Estermann and Stern in Germany) and hydrogen atoms (Johnson in the United States). Following the discovery of the neutron in 1932, it was shown that neutron beams of the appropriate energy also exhibit diffraction when incident on a crystalline target (Fig. 4.9).

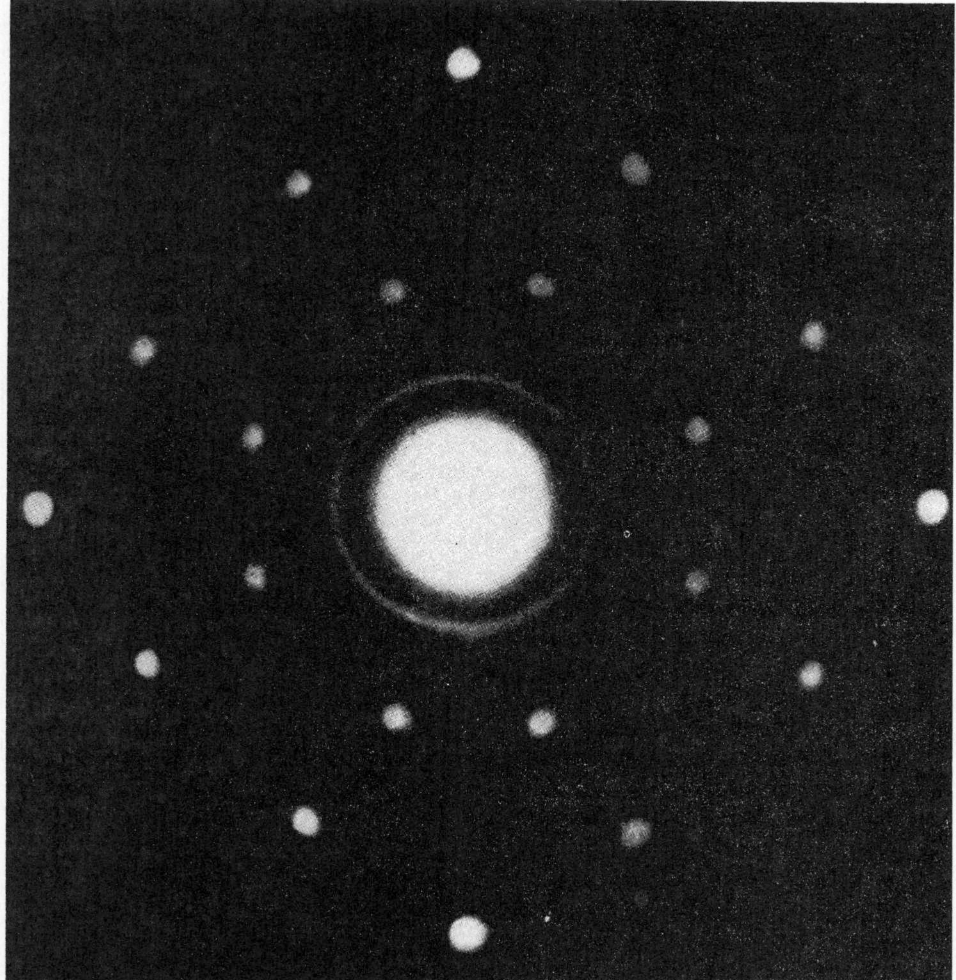

Figure 4.9 Diffraction pattern of neutrons produced by a single crystal of NaCl. (Courtesy of Dr. E. O. Wollan.)

EXAMPLE 4.3 Thermal Neutrons

What kinetic energy in eV should neutrons have if they are to be diffracted from crystals?

Solution: Appreciable diffraction will occur if the de Broglie wavelength of the neutron is of the same order of magnitude as the interatomic distance. Taking $\lambda = 1$ Å, we find

$$p = \frac{h}{\lambda} = \frac{6.63 \times 10^{-34}\ \text{J}\cdot\text{s}}{10^{-10}\ \text{m}} = 6.63 \times 10^{-24}\ \text{kg}\cdot\text{m/s}$$

The kinetic energy is given by

$$K = \frac{p^2}{2m} = \frac{(6.63 \times 10^{-24}\ \text{J}\cdot\text{s})^2}{2(1.66 \times 10^{-27}\ \text{kg})}$$

$$= 1.32 \times 10^{-20}\ \text{J} = 0.0825\ \text{eV}$$

Because the average thermal energy of a particle in thermal equilibrium is $\frac{1}{2}kT$ per degree of freedom, neutrons at room temperature (300 K) possess a kinetic energy of

$$K = \tfrac{3}{2}kT = (1.50)(8.62 \times 10^{-5}\ \text{eV/K})(300\ \text{K})$$

$$= 0.0388\ \text{eV}$$

Thus "thermal neutrons" or neutrons in thermal equilibrium with matter at room temperature possess energies of the right order of magnitude to diffract appreciably from single crystals. Neutrons produced in a nuclear reactor are far too energetic to produce diffraction from crystals, and must be slowed down in a graphite column as they leave the reactor. In the graphite moderator, repeated collisions with carbon atoms ultimately reduce the average neutron energies to the average thermal energy of the carbon atoms. When this occurs, these so-

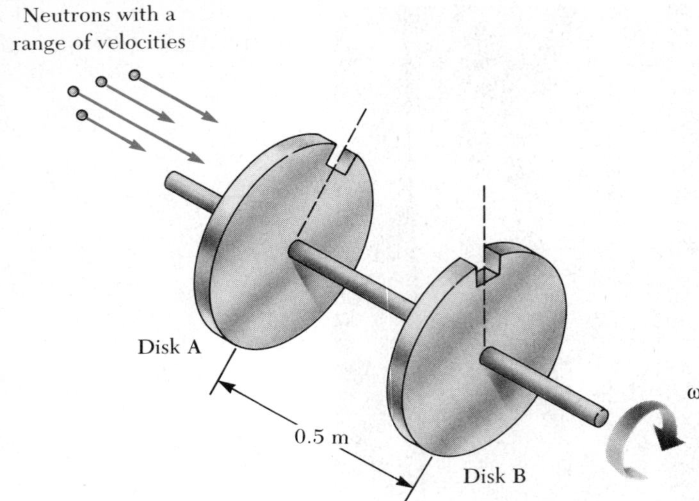

Figure 4.10 A neutron velocity selector. The slot in disk B lags the slot in disk A by 10°.

called "thermalized" neutrons possess a distribution of velocities and a corresponding distribution of de Broglie wavelengths.

Exercise 2 **Monochromatic Neutrons** A beam of neutrons with a single wavelength may be produced by means of a mechanical velocity selector of the type shown in Figure 4.10. (a) Calculate the velocity of neutrons with a wavelength of 1 Å. (b) What rotational speed (in rpm) should the shaft have in order to pass neutrons with wavelength of 1 Å?
Answers: (a) 3.99×10^3 m/s (b) 13 300 rpm

4.3 WAVE GROUPS AND DISPERSION

We now return to the problem of how to construct waves of limited extent (wave groups) in order to represent particles. Actually, waves limited to definite regions of space (pulses, wave packets, or wave groups) are far more common than the ordinary plane wave with *definite wavelength* and infinite extent. Water waves from a stone dropped into a pond, light waves emerging from a briefly opened shutter, a wave generated on a taut rope by a single flip at one end, and a sound wave emitted by a discharging capacitor must all be modeled by wave groups. A wave group consists of a set of waves with *different wavelengths,* and amplitudes and phases chosen so that they interfere constructively over a small region of space. Outside of this region the combination of waves produces a net amplitude that approaches zero rapidly as a result of destructive interference. Perhaps the most familiar physical example in which wave groups arise is the phenomenon of beats. Beating occurs when two sound waves of slightly different wavelength (and hence different frequency) are combined. The resultant sound wave has a frequency equal to the average of the two combining waves and an amplitude that fluctuates or "beats" at a rate given by the difference of the two original frequencies. This case is illustrated in Figure 4.11.

Let us examine this same situation mathematically. Consider a one-dimensional wave propagating in the positive x direction with phase velocity v_p, wavelength λ, frequency f, and amplitude A. This traveling wave may be described by

$$y = A\cos\left(\frac{2\pi x}{\lambda} - 2\pi f t\right) \tag{4.11}$$

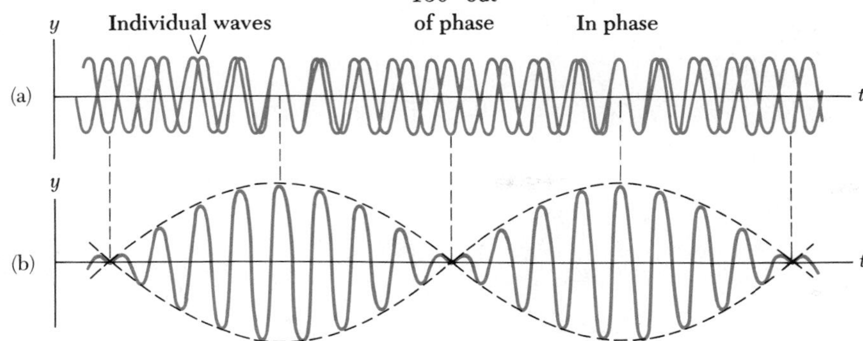

Figure 4.11 Beats are formed by the combination of two waves of slightly different frequencies traveling in the same direction. (a) The individual waves. (b) The combined wave has an amplitude (broken line) that oscillates in time. (From R. Resnick and D. Halliday, *Physics*, New York, Wiley, 1977; by permission of the publisher)

where λ and f are related by

$$v_p = \lambda f \tag{4.12}$$

A more compact form for Equation 4.11 results if we take $\omega = 2\pi f$ (where ω is the angular frequency) and $k = 2\pi/\lambda$ (where k is the wave number). With these substitutions the infinite wave becomes

$$y = A\cos(kx - \omega t) \tag{4.13}$$

with

$$\boxed{v_p = \frac{\omega}{k}} \tag{4.14}$$

Phase velocity

Let us now form the superposition of two waves of equal amplitude but with slightly different wavelengths, frequencies, and phase velocities. The resultant amplitude y is given by

$$y = y_1 + y_2 = A\cos(k_1x - \omega_1t) + A\cos(k_2x - \omega_2t)$$

Using the trigonometric identity

$$\cos a + \cos b = 2\cos\tfrac{1}{2}(a - b)\cdot\cos\tfrac{1}{2}(a + b)$$

we find

$$y = 2A\cos\frac{1}{2}\{(k_2 - k_1)x - (\omega_2 - \omega_1)t\}\cdot\cos\left\{\left(\frac{k_1 + k_2}{2}\right)x - \left(\frac{\omega_1 + \omega_2}{2}\right)t\right\} \tag{4.15}$$

For the case of two waves with *slightly* different values of k and ω, note that $\Delta k = k_2 - k_1$ and $\Delta\omega = \omega_2 - \omega_1$ are small, while $(k_1 + k_2)$ and $(\omega_1 + \omega_2)$ are large. Thus Equation 4.15 may be interpreted as a slowly varying envelope

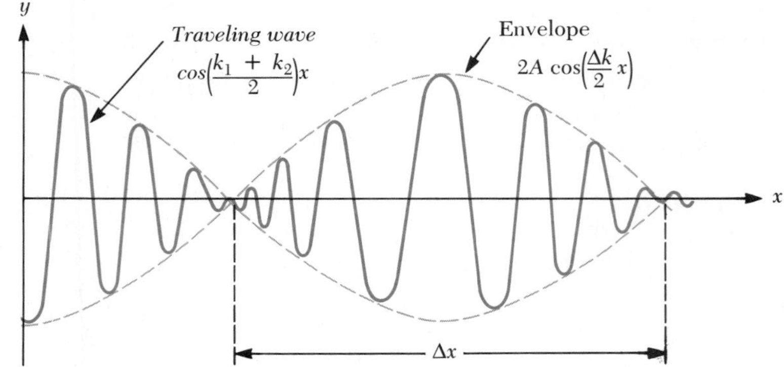

Figure 4.12 Superposition of two waves of slightly different wavelength resulting in primitive wave groups; t has been set equal to zero in Equation 4.15.

$2A \cos \frac{1}{2}\{(k_2 - k_1)x - (\omega_2 - \omega_1)t\}$ modulating a traveling wave with average k and ω values (see Fig. 4.12).

Although our model is primitive and not really confined to a single region of space, it shows several interesting features common to more refined models. For example, the speed of individual traveling waves is different from the envelope speed, so that individual waves actually move through the envelope! (Examine closely the ripples caused by a stone thrown into a pond.) The velocity of either the individual waves or the envelope is given by dividing the appropriate ω term (envelope or individual wave) by the appropriate k term. *For the individual waves* we have

$$v_p = \frac{(\omega_1 + \omega_2)/2}{(k_1 + k_2)/2} \approx \frac{\omega_1}{k_1} = v_1$$

Thus the individual waves move at the phase velocity v_1 of one of the waves (or at v_2 since $v_1 \approx v_2$). However, *the envelope or group moves with a different velocity, or group velocity,* given by

$$v_g = \frac{(\omega_2 - \omega_1)/2}{(k_2 - k_1)/2}$$

In general, the group velocity is given by

$$v_g = \frac{\Delta\omega}{\Delta k} \tag{4.16}$$

The main characteristic of pulses or wave groups is both a limited duration in time, Δt, and a limited extent in space, Δx. In general, it is found that the smaller the spatial width of the pulse, Δx, the larger the range of wavelengths or wavenumbers, Δk, needed to form the pulse. This may be stated mathematically as

$$\Delta x \, \Delta k \approx 1 \tag{4.17}$$

Likewise if the time duration, Δt, of the pulse is small we require a wide spread of frequencies, $\Delta\omega$, to form the group. That is,

$$\Delta t \, \Delta\omega \approx 1 \tag{4.18}$$

In pulse electronics, this latter condition is known as the response time–bandwidth formula. In this situation, $\Delta\omega$ is the range of frequencies that a pulse amplifier must pass (amplify) and Δt is the time-width of the pulse to be amplified.

It is interesting that our simple two-wave model also shows the general principles given by Equations 4.17 and 4.18. If we call (rather artificially) the spatial extent of our group the distance between adjacent minima (labeled Δx in Figure 4.12), we find from the envelope term $2A\cos(\frac{1}{2}\Delta k\, x)$ the condition $\frac{1}{2}\Delta k\, x = \pi$ or

$$\Delta k\, \Delta x = 2\pi \tag{4.19}$$

Here, $\Delta k = k_2 - k_1$ is the range of wavenumbers present. Likewise, if x is held constant and t is allowed to vary in the envelope portion of Equation 4.15, $2A\cos\{\frac{1}{2}(k_2 - k_1)x - \frac{1}{2}(\omega_2 - \omega_1)t\}$, the result is $\frac{1}{2}(\omega_2 - \omega_1)\Delta t = \pi$ or

$$\Delta\omega\, \Delta t = 2\pi \tag{4.20}$$

Thus Equations 4.19 and 4.20 agree with the general principles, respectively, of $\Delta k\, \Delta x \approx 1$ and $\Delta\omega\, \Delta t \approx 1$.

So far we have considered the addition of only two waves with discrete frequencies. In the general case, many waves having slightly different ω and k values must be added to form a packet that is finite over a limited range and zero everywhere else. In this case Equation 4.16 becomes

$$v_g = \left.\frac{d\omega}{dk}\right|_{k_0} \tag{4.21}$$

Group velocity

where k_0 is the central wavenumber of the many waves present. The connection between the group velocity and the phase velocity of the composite waves is easily obtained. Because $\omega = kv_p$, we find

$$v_g = \left.\frac{d\omega}{dk}\right|_{k_0} = v_p|_{k_0} + k\left.\frac{dv_p}{dk}\right|_{k_0} \tag{4.22}$$

where v_p is the phase velocity and is, in general, a function of k or λ. Materials in which the phase velocity varies with wavelength are said to exhibit *dispersion*. An example of a dispersive medium is glass, in which the index of refraction varies with wavelength. Media in which the phase velocity does not vary with wavelength (such as vacuum for electromagnetic waves) are termed nondispersive. The term dispersion arises from the fact that the individual harmonic waves that form a pulse travel at different phase velocities, and cause an originally sharp pulse to change shape and become spread out or dispersed. As an example, dispersion of a laser pulse after traveling 1 km along an optical fiber is shown in Figure 4.13. In a nondispersive medium where all waves have the same velocity, the group velocity is equal to the phase velocity. In a dispersive medium the group velocity can be less than or greater than the phase velocity, depending on the sign of dv_p/dk as shown by Equation 4.22.

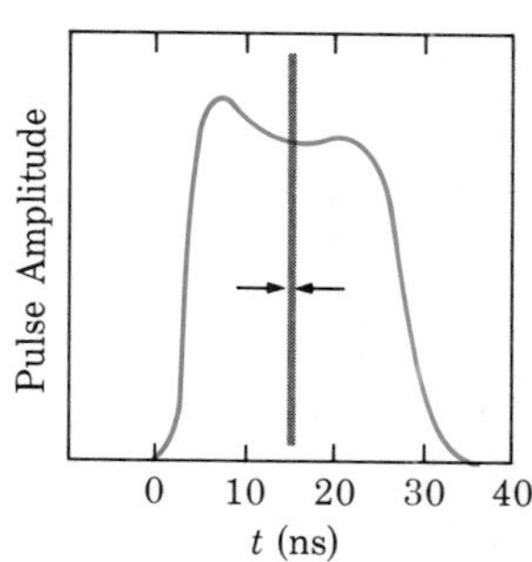

Figure 4.13 Dispersion in a 1-nanosecond laser pulse.

EXAMPLE 4.4 Group Velocity in a Dispersive Medium

In a particular substance the phase velocity of waves doubles when the wavelength is halved. Show that wave groups in this system move at twice the *central* phase velocity.

Solution: From the given information, the dependence of phase velocity on wavelength is

$$v_p = A'/\lambda = Ak$$

for some constants A' and A. From Equation 4.22 we obtain

$$v_g = v_p|_{k_0} + k\left.\frac{dv_p}{dk}\right|_{k_0}$$

$$= Ak_0 + Ak_0 = 2Ak_0$$

Thus,

$$v_g = 2v_p|_{k_0}$$

EXAMPLE 4.5 Group Velocity in Deep Water Waves Newton showed that the phase velocity of deep water waves having wavelength λ is given by

$$v_p = \sqrt{\frac{g\lambda}{2\pi}}$$

where g is the acceleration of gravity, and where the minor contribution of surface tension has been neglected. Show that in this case the velocity of a group of these waves is one-half of the phase velocity of the central wavelength.

Solution: Since $k = 2\pi/\lambda$, we can write v_p as

$$v_p = (g/k)^{1/2}$$

Therefore, we find

$$v_g = v_p|_{k_0} + k\left.\frac{dv_p}{dk}\right|_{k_0} = \left(\frac{g}{k_0}\right)^{1/2} - \frac{1}{2}\left(\frac{g}{k_0}\right)^{1/2}$$

$$= \frac{1}{2}\left(\frac{g}{k_0}\right)^{1/2} \equiv \frac{1}{2}v_p|_{k_0}$$

Electron Wave Groups

We are now in a position to apply our theory of wave groups to electrons. We shall show both the dispersion of de Broglie waves and the satisfying result that the wave group and the particle move at the same velocity. According to de Broglie, individual matter waves have a frequency f and a wavelength λ given by

$$f = \frac{E}{h}$$

and

$$\lambda = \frac{h}{p}$$

where E and p are the relativistic energy and momentum of the particle, respectively. The phase velocity of these matter waves is given by

$$v_p = f\lambda = E/p \tag{4.23}$$

The phase velocity can be expressed as a function of p or k alone by substituting $E = (p^2c^2 + m_0{}^2c^4)^{1/2}$ into Equation 4.23:

$$v_p = c\sqrt{1 + \left(\frac{m_0c}{p}\right)^2} \tag{4.24}$$

The dispersion relation for de Broglie waves can be obtained in the form $v_p = f(k)$ by substituting $p = h/\lambda = \hbar k$ into Equation 4.24. This gives

Phase velocity of matter waves

$$\boxed{v_p = c\sqrt{1 + \left(\frac{m_0c}{\hbar k}\right)^2}} \tag{4.25}$$

Equation 4.25 shows that individual de Broglie waves representing a particle of rest mass m_0 show dispersion even in *empty space* and always travel at different velocities *greater than* or at least equal to c. To obtain the group velocity, we now use

$$v_g = \left[v_p + k\frac{dv_p}{dk}\right]_{k_0}$$

and Equation 4.25. After some algebra, we find

$$v_g = \frac{c}{\left[1 + \left(\frac{m_0c}{\hbar k_0}\right)^2\right]^{1/2}} = \frac{c^2}{v_p|_{k_0}} \tag{4.26}$$

As we have previously shown that $v_p = c^2/v$ (Equation 4.5) where v is the particle velocity, substitution of this expression into Equation 4.26 shows that *the group velocity equals the particle velocity.*

Exercise 3 Verify Equation 4.26.

4.4 THE HEISENBERG UNCERTAINTY PRINCIPLE

In the period 1924–25, Werner Heisenberg (Fig. 4.14), the son of a Greek and Classics professor at the University of Munich, invented a complete theory of quantum mechanics called matrix mechanics. This theory overcame some of the problems with the Bohr theory of the atom, such as the postulate of "unobservable" electron orbits. Heisenberg's formulation was based primarily on measurable quantities such as the transition probabilities for electronic jumps between quantum states. Since transition probabilities depend on the initial and final states, Heisenberg's mechanics used variables labeled by two subscripts. Although at first Heisenberg presented his theory in the form of noncommuting algebra, Max Born quickly realized that this theory could be more elegantly described by matrices. Consequently, Born, Heisenberg, and Pascual Jordan soon worked out a comprehensive theory of matrix mechanics. Although the matrix formulation was quite elegant, it attracted little attention outside of a small group of gifted physicists because it was difficult to apply in specific cases, involved mathematics unfamiliar to most physicists, and was based on rather vague physical concepts.

Although we will investigate this remarkable form of quantum mechanics no farther, we do wish to discuss another of Heisenberg's brilliant discoveries, the uncertainty principle elucidated in a famous paper in 1927. In this paper Heisenberg introduced the notion that it is impossible to determine simultaneously with unlimited precision the position and momentum of a particle. In words we may state the uncertainty principle as follows:

> If a measurement of position is made with precision Δx and a simultaneous measurement of momentum is made with precision Δp, then the

Figure 4.14 Werner Heisenberg (1901–1976) around 1924 (University of Hamburg). Heisenberg obtained his PhD in 1923 at the University of Munich where he studied under Arnold Sommerfeld and became an enthusiastic mountain climber and skier. Later, he worked as an assistant to Max Born at Göttingen and Niels Bohr in Copenhagen. While physicists such as de Broglie and Schrödinger tried to develop pictorialized models of the atom, Heisenberg, with the help of Born and Pascual Jordan, developed an abstract mathematical model called matrix mechanics to explain the wavelengths of spectral lines. The more successful wave mechanics by Schrödinger announced a few months later was shown to be equivalent to Heisenberg's approach. Heisenberg made many other significant contributions to physics including his famous uncertainty principle, for which he received the Nobel Prize in 1932, the prediction of two forms of molecular hydrogen, and theoretical models of the nucleus. During World War II he was director of the Max Planck Institute at Berlin where he was in charge of German research on atomic weapons. Following the war, he moved to West Germany and became director of the Max Planck Institute for Physics at Göttingen.

product of the two uncertainties can never be smaller than the order of $\hbar$. That is,

Momentum-position uncertainty principle

$$\Delta p\,\Delta x \geq \hbar \tag{4.27}$$

In his paper of 1927, Heisenberg was careful to point out that the inescapable uncertainties Δp and Δx do not arise from imperfections in practical measuring instruments. Rather, they arise from the quantum structure of matter itself: from effects such as the partially *unpredictable* recoil of an electron when struck by an *indivisible* light quantum, or the diffraction of light or electrons passing through a small opening. Heisenberg also pointed out that the classical concepts of perfectly defined orbits and predictable momenta based on Newton's laws fail when applied to atomic systems. He did this by using an ingenious set of Gedanken (thought) experiments in which he analyzed different possible measuring devices such as light microscopes and cloud chambers. In essence, he was able to show that the classical belief that the disturbance introduced by a measuring process can be made arbitrarily small could not be extended to the atomic level, where both the irreducible graininess of light and the wave nature of particles manifest themselves.

Although one can show that $\Delta p\,\Delta x \approx \hbar$ emerges in a wide variety of Gedanken experiments such as measuring the position of photons with electrons, electrons with protons, or electrons with a narrow slit, we shall treat only the most famous case introduced by Heisenberg himself—the measurement of an electron's position by means of a light microscope (Fig. 4.15). Since we know that light can scatter from and perturb the electron, let us minimize this effect by considering the scattering of only one light quantum (Fig. 4.16). In order to be collected by the lens, the photon may be scattered through any angle ranging from $-\theta$ to $+\theta$, which consequently imparts to the electron a momentum value ranging from $+(h \sin\theta)/\lambda$ to $-(h \sin\theta)/\lambda$. Thus the uncertainty in the electron's momentum is $\Delta p_e = (2h \sin\theta)/\lambda$. After passing through the lens, the photon lands somewhere on the screen, but the image and consequently the position of the electron is "fuzzy" because the photon is diffracted upon passing through the lens aperture. According to physical optics, the smallest distance Δx between two points in an object that will produce separated images in a microscope is given by $\Delta x = \lambda/(2 \sin\theta)$. Here 2θ is the angle subtended by the objective lens, as shown in Figure 4.16.[3] Multiplying the expressions for Δp and Δx, we find for the electron

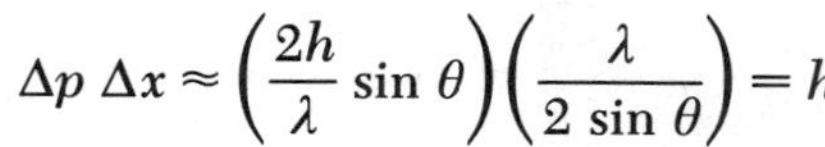

$$\Delta p\,\Delta x \approx \left(\frac{2h}{\lambda}\sin\theta\right)\left(\frac{\lambda}{2\sin\theta}\right) = h$$

in agreement with the uncertainty relation. Note also that this principle is inescapable and relentless! If Δx is reduced by increasing θ, there is an equivalent increase in the uncertainty of the electron's momentum.

Examination of this simple experiment shows several key properties that lead to the uncertainty principle:

1. The indivisible nature of light particles or quanta. (Nothing less than a single photon can be used!)
2. The wave property of light as shown in diffraction.

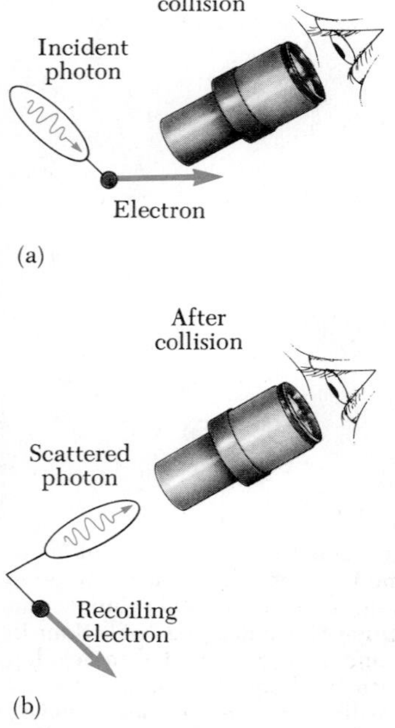

Figure 4.15 A thought experiment for viewing an electron with a powerful microscope (a) The electron is shown before colliding with the photon. (b) The electron recoils (is disturbed) as a result of the collision with the photon.

[3] The resolving power of the microscope is treated clearly in F. A. Jenkins and H. E. White, *Fundamentals of Optics*, 4th ed., McGraw-Hill Book Co., 1976, pp. 332–334.

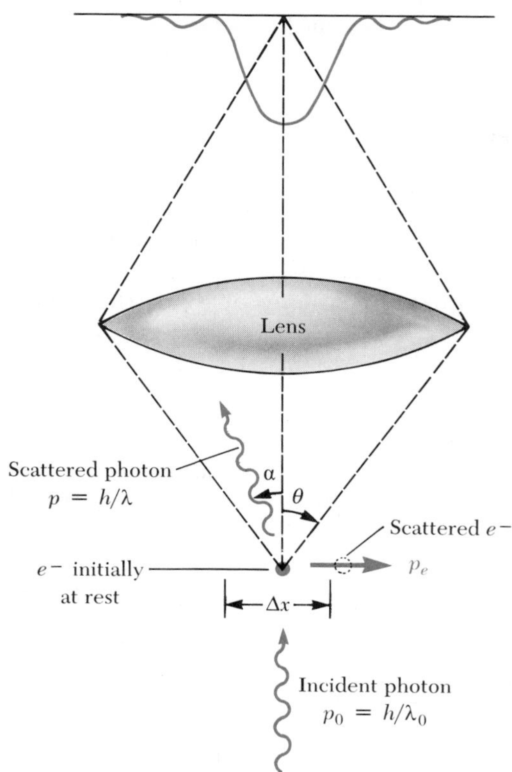

Figure 4.16 The Heisenberg microscope. Conservation of momentum requires $p_{e^-} = (h \sin \alpha)/\lambda$. Because of diffraction by the lens opening, the electron may be anywhere in the region Δx.

3. The impossibility of predicting or measuring the precise classical path of a single scattered photon and hence of knowing the precise momentum transferred to the electron.[4]

Before concluding this section, we should point out that there is another uncertainty principle, which sets a limit on the accuracy with which the energy of a system, ΔE, can be measured if a finite time interval, Δt, is allowed for the measurement. This energy-time uncertainty principle may be stated as

Energy-time uncertainty principle

$$\boxed{\Delta E \, \Delta t \geq \hbar} \tag{4.28}$$

This uncertainty relation is plausible if one considers a frequency measurement of any wave. For example, consider the measurement of a 1000 Hz electrical wave. If our frequency measuring device has a fixed sensitivity of ± 1 cycle, in one second we will measure a frequency of (1000 ± 1) cycles/1 s but in two seconds we will measure a frequency of (2000 ± 1) cycles/2 s. Thus the uncertainty in frequency, Δf, is inversely proportional to Δt, the time during which the measurement is made. This may be stated as

$$\Delta f \, \Delta t \approx 1$$

[4] Attempts to measure the photon's position by scattering electrons from it in a Compton process only serve to make its path to the lens more uncertain.

Since all quantum systems are "wavelike" and can be described by the de Broglie relationship $E = hf$, we may substitute $\Delta f = \Delta E/h$ into the expression above to obtain

$$\Delta E\, \Delta t \approx h$$

in basic agreement with Equation 4.28.

We conclude this section with some examples of the types of calculations that can be done with the uncertainty principle. In the spirit of Fermi or Heisenberg, these "back of the envelope calculations" are surprising for their simplicity and essential description of quantum systems of which the details are unknown.

EXAMPLE 4.6 Locating an Electron
The speed of an electron is measured to have a value of 5.00×10^3 m/s to an accuracy of 0.003 percent. Find the uncertainty in determining the position of this electron.

Solution: The momentum of the electron is

$$p = mv = (9.11 \times 10^{-31}\text{ kg})(5.00 \times 10^3\text{ m/s})$$
$$= 4.56 \times 10^{-27}\text{ kg}\cdot\text{m/s}$$

Because the uncertainty in p is 0.003% of this value, we get

$$\Delta p = 0.00003p = (0.00003)(4.56 \times 10^{-27}\text{ kg}\cdot\text{m/s})$$
$$= 1.37 \times 10^{-31}\text{ kg}\cdot\text{m/s}$$

The uncertainty in position can now be calculated by using this value of Δp and Equation 4.27.

$$\Delta x\, \Delta p \geq \frac{h}{2\pi}$$

$$\Delta x \geq \frac{h}{2\pi\, \Delta p} = \frac{6.63 \times 10^{-34}\text{ J}\cdot\text{s}}{2\pi(1.37 \times 10^{-31}\text{ kg}\cdot\text{m/s})}$$
$$= 0.770 \times 10^{-3}\text{ m} = 0.770\text{ mm}$$

EXAMPLE 4.7 The Width of Spectral Lines
Although an excited atom can radiate at any time from $t = 0$ to $t = \infty$, the average time after excitation at which a group of atoms radiates is called the **lifetime,** τ. (a) If $\tau = 10^{-8}$ s, use the uncertainty principle to compute the line width Δf produced by this finite lifetime.

Solution: We use $\Delta E\, \Delta t \approx \hbar$, where $\Delta E = h\, \Delta f$, and where $\Delta t = 10^{-8}$ s is the average time available to measure the excited state. Thus,

$$\Delta f = \frac{1}{2\pi \times 10^{-8}\text{ s}} = 1.6 \times 10^7\text{ Hz}$$

Note that ΔE is the uncertainty in energy of the excited state. It is also the uncertainty in the energy of the photon emitted by an atom in this state.

(b) If the wavelength of the spectral line involved in this process is 500 nm, find the fractional broadening $\Delta f/f$.

Solution: First, we find the center frequency of this line as follows:

$$f_0 = c/\lambda = \frac{3 \times 10^8\text{ m/s}}{500 \times 10^{-9}\text{ m}} = 6.0 \times 10^{14}\text{ Hz}$$

Hence,

$$\frac{\Delta f}{f_0} = \frac{1.6 \times 10^7\text{ Hz}}{6.0 \times 10^{14}\text{ Hz}} = 2.7 \times 10^{-8}$$

This narrow natural linewidth can be seen with a sensitive interferometer. Usually, however, temperature and pressure effects overshadow the natural linewidth and broaden the line through mechanisms associated with the Doppler effect and collisions.

Exercise 4 Using the nonrelativistic Doppler formula, calculate the Doppler broadening of a 500 nm line emitted by a hydrogen atom at 1000 K. Calculate the broadening by considering the atom to be moving either directly toward or away from an observer with an energy of $\frac{3}{2}kT$.
Answer: 0.0083 nm or 0.083 Å.

EXAMPLE 4.8 Measuring the Momentum of an Atom from the Doppler Shift
Instead of concentrating on position measurements for a particle, let us concentrate on measuring the momentum of an atom in order to see if we can get around the uncertainty principle. (See Fig. 4.17.) Assume that a hydrogen atom in an excited state has a lifetime of $\tau = 10^{-8}$ s and that the wavelength radiated by the atom at rest is 500 nm. (a) If the Doppler shift in frequency of the light emitted from the hydrogen atom is measured to be $\Delta f = 10^{10}$ Hz, find p and Δp for the atom.

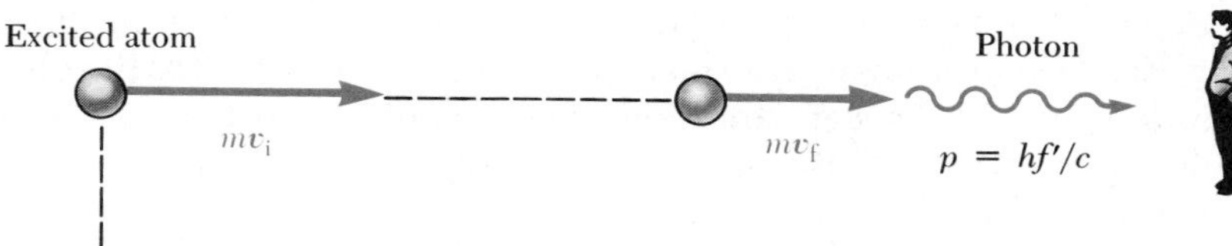

Figure 4.17 Measuring the momentum of an atom from the Doppler shift of emitted radiation.

Solution: The relation between the Doppler shift and velocity, v, is

$$\frac{v}{c} = \frac{\Delta f}{f} = \frac{f' - f}{f}$$

where f is the frequency radiated by the atom at rest and f' is the frequency measured when the atom is moving at the speed v. In this case we have

$$\frac{v}{c} = \frac{\Delta f}{f} = \frac{\Delta f}{c/\lambda} = \frac{(10^{10}\text{ Hz})(500 \times 10^{-9}\text{ m})}{3.0 \times 10^{8}\text{ m/s}}$$

or

$$v = 5000\text{ m/s}$$

Hence, the momentum of the hydrogen atom is

$$p = mv = (1.67 \times 10^{-27}\text{ kg})(5.00 \times 10^{3}\text{ m/s})$$
$$= 8.35 \times 10^{-24}\text{ kg}\cdot\text{m/s}$$

To find the uncertainty in v and hence in p, we differentiate the Doppler shift expression to obtain

$$\frac{dv}{df'} = \frac{c}{f}$$

or

$$dv = \frac{c}{f}\,df'$$

Taking df' to be equal to $1/\tau$ ($\Delta f\,\Delta t \approx 1$ with $\Delta t = \tau$) gives for the minimum uncertainty in v

$$dv = \frac{c}{f\tau}$$

Thus Δp is given by

$$\Delta p = m\,dv = \frac{mc}{f\tau} = \frac{(1.67 \times 10^{-27}\text{ kg})(3.00 \times 10^{8}\text{ m/s})}{(6 \times 10^{14}\text{ Hz})(10^{-8}\text{ s})}$$
$$= 8.35 \times 10^{-26}\text{ kg}\cdot\text{m/s}$$

(b) Estimate the uncertainty in position of the atom as the product of a velocity and a time uncertainty.

Solution: As can be seen in Figure 4.17, the emission of a photon with momentum hf'/c causes an abrupt change Δv in the atom's velocity, where $m\,\Delta v = hf'/c$, or

$$\Delta v = \frac{hf'}{mc}$$

Since the precise time of emission of the photon is uncertain, this introduces an uncertainty in the atom's position. On the average, a photon is emitted between $t = 0$ and $t = 2\tau$ (where τ is the lifetime of the atom), and the atom's position is uncertain by

$$\Delta x = (\Delta v)(2\tau) = \frac{2hf'\tau}{mc}$$
$$= \frac{(2h)(6.00 \times 10^{14}\text{ Hz})(10^{-8}\text{ s})}{(1.67 \times 10^{-27}\text{ kg})(3.00 \times 10^{8}\text{ m/s})}$$
$$= (2.37 \times 10^{25})h = 1.6 \times 10^{-8}\text{ m}$$

(c) Find the value of $\Delta p\,\Delta x$ for this example, and verify that the uncertainty relation is satisfied.

Solution: $\Delta p\,\Delta x = (2.37 \times 10^{25})h \times 8.35 \times 10^{-26} \approx 2h.$

4.5 IF ELECTRONS ARE WAVES, WHAT'S WAVING?

Although we have discussed in some detail the notion of de Broglie matter waves, we have not discussed the precise nature of the field $\Psi(x, y, z, t)$ or wave function which represents the matter waves. We have delayed this discussion because Ψ (Greek letter psi) is rather abstract. Ψ is definitely *not* a measurable disturbance requiring a medium for propagation like a water wave or sound wave. Instead the stuff that is waving requires no medium. Furthermore, Ψ is in general represented by a complex number, and is related to the probability of finding the particle at a given time in a small volume of space. If any of this seems confusing, you should not lose heart, as the nature of the wave function has been confusing people since its invention. It even confused its inventor, Erwin Schrödinger, who incorrectly interpreted $\Psi^*\Psi$ as the electric charge density.[5] The great philosopher of the quantum theory, Bohr, immediately objected to this interpretation. Subsequently, Max Born offered the statistical view of $\Psi^*\Psi$ in late 1926. The confused state of affairs surrounding Ψ at that time was nicely described in a poem by Walter Huckel:

> Erwin with his psi can do
> Calculations quite a few.
> But one thing has not been seen
> Just what does psi really mean?
> (English translation by Felix Bloch)

The currently held view is that a particle is described by a function $\Psi(x, y, z, t)$ called the probability amplitude. The quantity $\Psi^*\Psi = |\Psi|^2$ (wave intensity) represents the probability per unit volume of finding the particle at a time t in a small volume of space. We will treat methods of finding Ψ in much more detail in Chapter 5, but for now all we require is the idea that the probability of finding a particle is directly proportional to the intensity of its matter wave, $|\Psi|^2$.

4.6 THE WAVE-PARTICLE DUALITY

The Description of Electron Diffraction in Terms of Ψ

In this chapter and previous chapters we have seen evidence for both the wave properties and the particle properties of electrons. Historically, the particle properties were first known and connected with a definite mass, a discrete charge, and detection or localization of the electron in a small region of space. Following these discoveries came the confirmation of the wave nature of electrons in scattering at low energy from metal crystals. In view of these results and because of the everyday experience of seeing the world in terms of *either* grains of sand or diffuse water waves, it is no wonder that we are tempted to simplify the issue and ask, "Well, is the electron a wave or a particle?" The answer is that **electrons are very delicate and flexible — they behave like either particles or waves, depending on the kind of experiment performed on them. In any case, it is impossible to simultaneously measure both the wave and particle properties.**[6] The view of Bohr was expressed in an idea known as complementarity. As different as they are, both wave and particle views are needed and they complement each other to fully describe the electron. The view of Feynman[7] is that both electrons and photons behave in their own inimitable way. This is like nothing we have seen before, because we do not live at the very tiny scale of atoms, electrons, and photons.

Perhaps the best way to crystallize our ideas about the wave-particle duality is to consider a "simple" double-slit electron diffraction experiment. This experiment highlights much of the mystery of the wave-particle paradox, shows the impossibility of *measuring simultaneously* both wave and particle properties, and illustrates the use of the wave function, Ψ, in determining

[5] Ψ^* represents the complex conjugate of Ψ. Thus if $\Psi = a + ib$, then $\Psi^* = a - ib$. In exponential form, if $\Psi = Ae^{i\theta}$, then $\Psi^* = Ae^{-i\theta}$. Note that $\Psi^*\Psi = |\Psi|^2$ is real.

[6] Many feel that the elder Bragg's remark, originally made about light, is a more satisfying answer: electrons behave like waves on Mondays, Wednesdays, and Fridays, like particles on Tuesdays, Thursdays, and Saturdays, and like nothing at all on Sundays.

[7] R. Feynman, *The Character of Physical Law*, MIT Press, 1982.

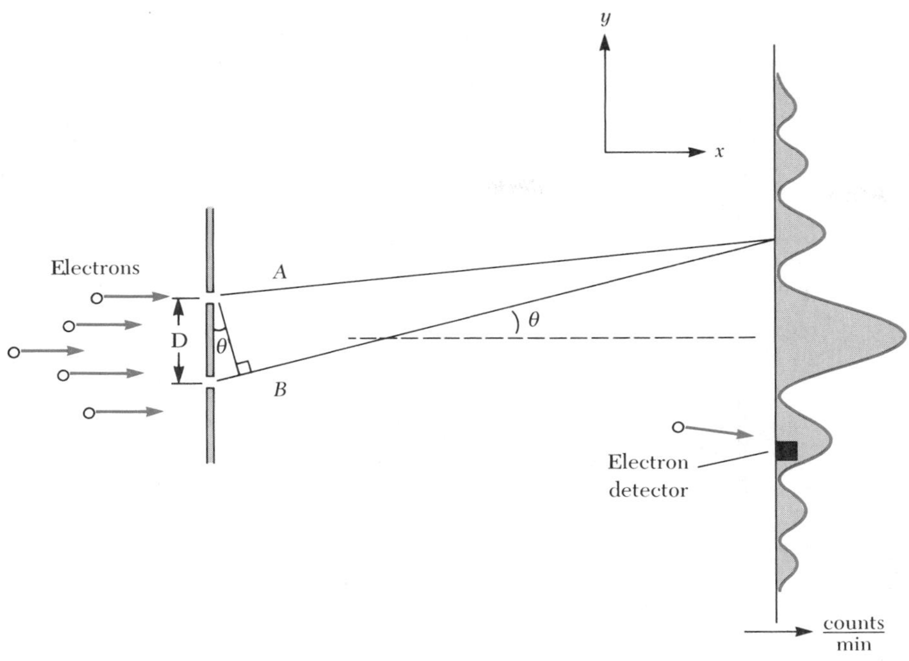

Figure 4.18 Electron diffraction. D is much greater than the individual slit widths and much less than the distance between the slit and detector.

interference effects. A schematic of the experiment with monoenergetic (single wavelength) electrons is shown in Figure 4.18. A parallel beam of electrons falls on a double slit, which has individual openings much smaller than D so that single-slit diffraction effects are negligible. At a distance from the slits much greater than D there is an electron detector capable of detecting individual electrons. It is important to note that the detector always registers discrete particles localized in space and time. In a real experiment this can be achieved if the electron source is weak enough (see Fig. 4.19). *In all cases if the detector collects electrons at different positions for a long enough time, a typical wave*

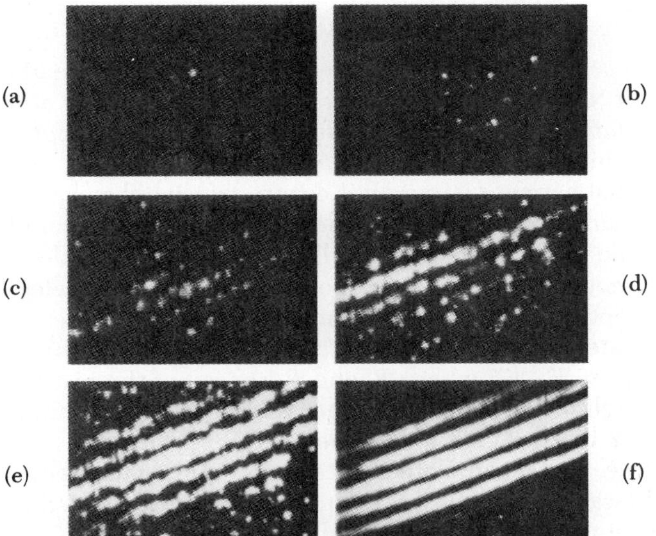

Figure 4.19 Gradual accumulation of interference fringes after the diffraction of increasing number of electrons. (From P. G. Medi, G. F. Missiroli, and G. Pozzi, *Am J. of Phys. 44*, 306, 1976).

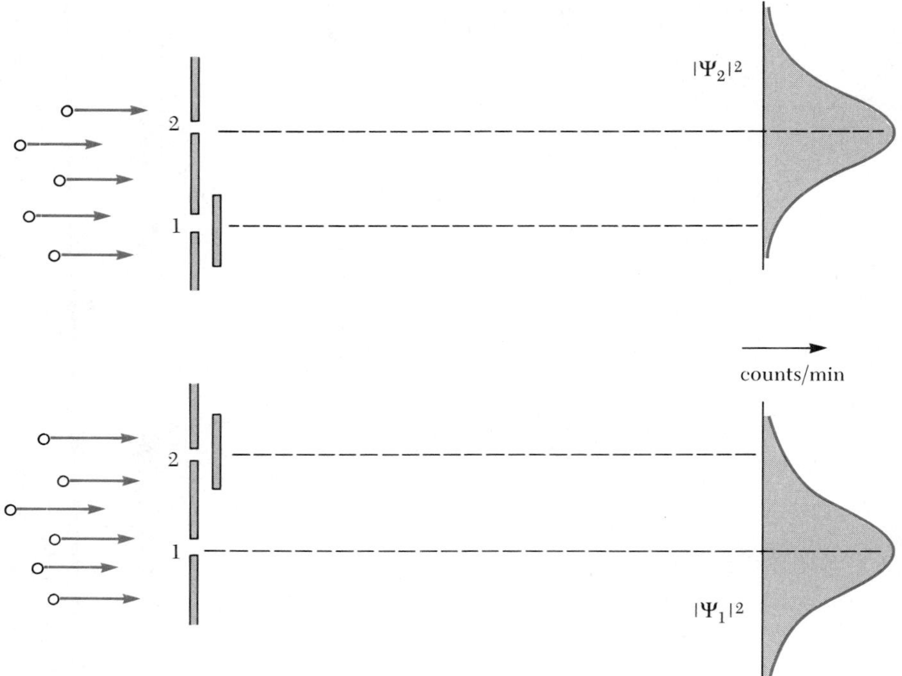

Figure 4.20 The probability of finding electrons at the screen with either the lower or upper slit closed.

interference pattern for the counts/min or probability of arrival of electrons is found. If one imagines a single electron to produce in-phase "wavelets" at the slits, standard wave theory can be used to find the angular separation, θ, of the central probability maximum from its neighboring minimum. The minimum occurs when the path length difference between A and B is half a wavelength or

$$D \sin \theta = \lambda/2$$

As the electron's wavelength is given by $\lambda = h/p_x$, we see that

$$\sin \theta \approx \theta = \frac{h}{2p_x D} \tag{4.29}$$

for small θ. Thus we can see that the dual nature of the electron is clearly shown in this experiment: *while the electrons are detected as particles at a localized spot at some instant of time, the probability of arrival at that spot is determined by finding the intensity of two interfering matter waves.*

But there is more. What happens if one slit is covered during the experiment? In this case one obtains a symmetric curve peaked around the center of the open slit, much like the pattern formed by bullets shot through a hole in armor plate. Plots of the counts per minute or probability of arrival of electrons with the lower or upper slit closed are shown in Figure 4.20. These are expressed as the appropriate square of the absolute value of some wave function, $|\Psi_1|^2 = \Psi_1{}^*\Psi_1$ or $|\Psi_2|^2 = \Psi_2{}^*\Psi_2$, where Ψ_1 and Ψ_2 represent the cases of the electron passing through slit 1 and slit 2, respectively. If an experiment is now performed with slit 2 blocked half of the time and then slit 1 blocked during the remaining time, the accumulated pattern of counts/min is completely different from the case with both slits open. Note in Figure 4.21 that there is no longer a maximum probability of arrival of an electron at $\theta = 0$. In

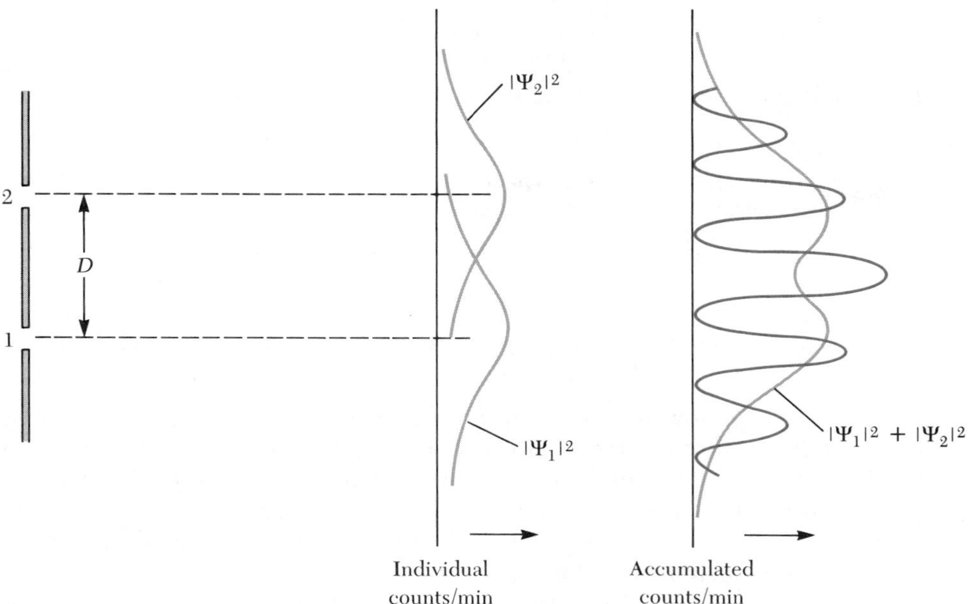

Figure 4.21 Accumulated results from the two-slit electron diffraction experiment with each slit closed half the time. The results with both slits open are shown in color.

fact, *the interference pattern has been lost and the accumulated result is simply the sum of the individual results*. The results shown in Figure 4.21 are easier to understand and more reasonable than the interference effects seen with both slits open. When only one slit is open at a time, we know the electron has the same localizability and indivisibility at the slits as we measure at the detector, since the electron clearly goes through slit 1 or slit 2. Thus the total must be analyzed as the sum of those electrons that come through slit 1, $|\Psi_1|^2$, and those that come through slit 2, $|\Psi_2|^2$. When both slits are open, it is tempting to assume that the electron goes through either slit 1 or slit 2, and that the counts/min are again given by $|\Psi_1|^2 + |\Psi_2|^2$. We know, however, that the experimental results contradict this. Thus our assumption that the electron is localized and goes through only one slit when both slits are open must be wrong (a painful conclusion!). Somehow the electron must be simultaneously present at both slits to exhibit interference.

In order to find the probability of detecting the electron at a particular point on the screen with both slits open, we may say that the electron is in a *superposition state* given by

$$\Psi = \Psi_1 + \Psi_2$$

Thus the probability of detecting the electron at the screen is equal to the quantity $|\Psi_1 + \Psi_2|^2$ and not $|\Psi_1|^2 + |\Psi_2|^2$. Since matter waves that start out in phase at the slits in general travel different distances to the screen (see Fig. 4.18), Ψ_1 and Ψ_2 will possess a relative phase difference ϕ at the screen. Using a phasor diagram (Fig. 4.22) to find $|\Psi_1 + \Psi_2|^2$ immediately yields

$$|\Psi|^2 = |\Psi_1 + \Psi_2|^2 = |\Psi_1|^2 + |\Psi_2|^2 + 2|\Psi_1||\Psi_2| \cos \phi$$

Note that the term $2|\Psi_1||\Psi_2| \cos \phi$ is an interference term that predicts the interference pattern actually observed in this case. For ease of comparison, a summary of the results found in both cases is given in Table 4.1.

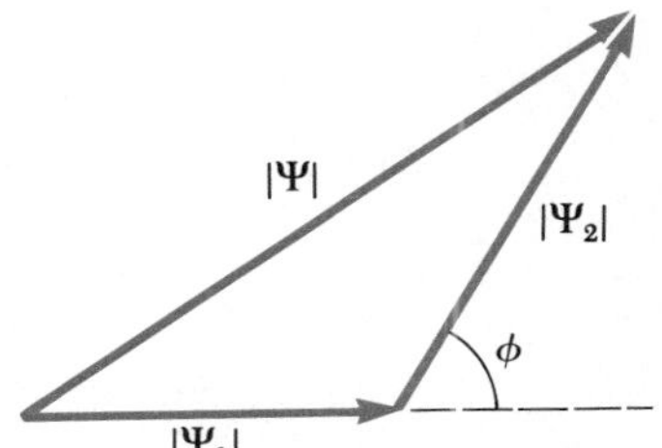

Figure 4.22 Phasor diagram to represent the addition of two complex wave functions, Ψ_1 and Ψ_2.

TABLE 4.1

Case	Wave Function	Counts/Minute at Screen
Electron is measured to pass through slit 1 or slit 2	Ψ_1 or Ψ_2	$\lvert\Psi_1\rvert^2 + \lvert\Psi_2\rvert^2$
No measurements made on electron at slits	$\Psi_1 + \Psi_2$	$\lvert\Psi_1\rvert^2 + \lvert\Psi_2\rvert^2 + 2\lvert\Psi_1\rvert\lvert\Psi_2\rvert\cos\phi$

A Thought Experiment: Measuring Through Which Slit the Electron Passes

Another way to view the electron double-slit experiment is to say that the electron passes through the upper or lower slit only when one *measures* the electron to do so. Once one measures unambiguously which slit the electron passes through (yes, you guessed it . . . here comes the uncertainty principle again . . .), the act of measurement disturbs the electron's path enough to destroy the delicate interference pattern.

Let us look again at our two-slit experiment to see in detail how the interference pattern is destroyed.[8] In order to determine which slit the electron goes through, imagine that a group of particles is placed right behind the slits as shown in Figure 4.23. If we use the recoil of a small particle to determine which slit the electron goes through, we must have the uncertainty in the detecting particle's position, $\Delta y \ll D$. Also, during the collision the detecting particle suffers a change in momentum Δp_y equal and opposite to the change in momentum experienced by the electron, as shown in Figure 4.23. An undeviated electron landing at the first minimum *and producing an interference pattern* has

$$\theta = \frac{p_y}{p_x} = \frac{h}{2p_xD}$$

Thus we require that an electron scattered by a detecting particle have

$$\frac{\Delta p_y}{p_x} \ll \theta = \frac{h}{2p_xD}$$

or

$$\Delta p_y \ll \frac{h}{2D}$$

if the interference pattern is not to be distorted. Since the change in momentum of the scattered electron is equal to the change in momentum of the detecting particle, $\Delta p_y \ll h/2D$ also applies to the detecting particle. Thus we have for the detecting particle

$$\Delta p_y\,\Delta y \ll \frac{h}{2D}\cdot D$$

or

$$\Delta p_y\,\Delta y \ll \frac{h}{2}$$

[8] Although we shall use the uncertainty principle in its standard form, it is worth noting that an alternative statement of the uncertainty principle involves this pivotal double-slit experiment: *It is impossible to design any device to determine through which slit the electron passes that will not at the same time disturb the electron and destroy the interference pattern.*

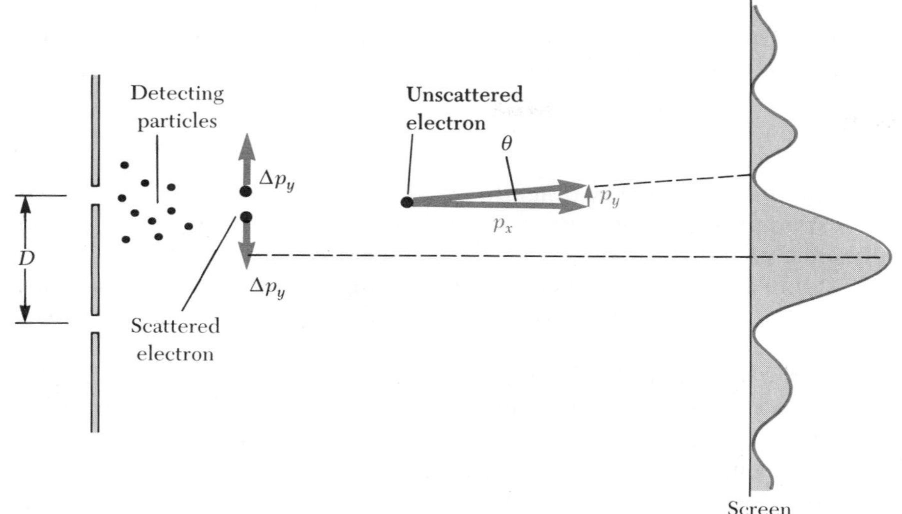

Figure 4.23 A thought experiment to determine through which slit the electron passes.

This is a clear violation of the uncertainty principle. *Hence we get the result that the small uncertainties needed both to see the interference and to know which slit the electron goes through are impossible, since they violate the uncertainty principle.* If Δy is small enough to determine which slit the electron goes through, Δp_y is so large that electrons heading for the first minimum are scattered into adjacent maxima and the interference pattern is destroyed.

Exercise 5 In a real experiment it is likely that some electrons would miss the detecting particles. Thus we would really have two categories of electrons arriving at the detector—those measured to pass through a definite slit and those not observed, or just missed, at the slits. In this case what kind of pattern of counts/min would be accumulated by the detector?
Answer: A mixture of an interference pattern (those not measured) and $|\Psi_1|^2 + |\Psi_2|^2$ (those measured) would result.

4.7 A FINAL NOTE

Most scientists view the world as being made up of distinct and unchanging parts that interact according to strictly deterministic laws of cause and effect. In the classical limit this is fundamentally correct because classical processes involve so many quanta that deviations from the average are imperceptible. At the atomic level, however, we have seen that a given piece of matter (an electron, say) is not a distinct and unchanging part of the universe obeying completely deterministic laws. Such a particle exhibits wave properties when it interacts with a metal crystal and particle properties a short while later when it registers on a Geiger counter. Thus, rather than viewing the electron as a distinct and separate part of the universe with an intrinsic particle nature, we are led to the view that the electron and indeed all particles are changing entities possessing the potential to cycle endlessly between wave and particle behavior. We also find that it is much more difficult to separate the object measured from the measuring instrument at the atomic level, since the type of measuring instrument determines whether wave properties or particle properties are observed.

4.8 SUMMARY

Every lump of matter of mass m and momentum p has wavelike properties with wavelength given by the de Broglie relation

$$\lambda = \frac{h}{p} \tag{4.1}$$

By applying this wave theory of matter to electrons in atoms, de Broglie was able to explain the appearance of integers in the Bohr theory as a natural consequence of interference. In 1927 Davisson and Germer demonstrated directly the wave nature of electrons by showing that low energy electrons were diffracted by single crystals of nickel. In addition, they confirmed Equation 4.1.

While the wavelength of matter waves can be experimentally determined, it is important to understand that they are not just like other waves since their frequency and phase velocity cannot be directly measured. In particular, the phase velocity of an individual matter wave of infinite extent is greater than the velocity of light and varies with wavelength as shown in Equation 4.25:

$$v_p = f\lambda = \left(\frac{E}{h}\right)\left(\frac{h}{p}\right) = c\left[1 + \left(\frac{m_0 c}{\hbar k}\right)^2\right]^{1/2} \tag{4.25}$$

To represent a particle properly, a superposition of matter waves with different wavelengths, amplitudes, and phases must be chosen to interfere constructively over a limited region of space. The resulting wave packet or group can then be shown to travel with the same velocity as the classical particle. In addition, a wave packet localized in a region Δx contains a range of wavenumbers Δk, where $\Delta x\, \Delta k \approx 1$. Since $p = \hbar k$, this implies that there is an uncertainty principle for position and momentum:

$$\Delta p\, \Delta x \geq \hbar \tag{4.27}$$

In quantum mechanics matter waves are represented by a wave function $\Psi(x, y, z, t)$. The probability of finding a particle represented by Ψ in a small volume $dx\, dy\, dz$ at time t is proportional to $|\Psi|^2$. The wave-particle duality of electrons may be seen by considering the passage of electrons through two narrow slits and their arrival at a viewing screen. We find that while the electrons are detected as particles at a localized spot on the screen, the probability of arrival at that spot is determined by finding the intensity of two interfering matter waves.

Although we have seen the importance of matter waves or wave functions in this chapter, we have provided no method of finding Ψ for a given physical system. In the next chapter we shall introduce the Schrödinger wave equation. The solutions to this important differential equation will provide us with the wavefunctions for a given system.

SUGGESTIONS FOR FURTHER READING

1. David Bohm, *Quantum Theory*, Englewood Cliffs, N.J., Prentice-Hall, 1951. Chapters 3 and 6 in this book give an excellent account of wave packets and the wave-particle duality of matter at a more advanced level.
2. Richard Feynman, *The Character of Physical Law*, Cambridge, MA, The M.I.T. Press, 1982, Chapter 6. This monograph is an incredibly lively and readable treatment of the double slit experiment presented in Feynman's inimitable fashion.
3. Banesh Hoffman, *The Strange Story of the Quantum*, New York, Dover Publications, 1959. This short book is a beautifully written nonmathematical discussion of the history of quantum mechanics.

QUESTIONS

1. Is light a wave or a particle? Support your answer by citing specific experimental evidence.
2. Is an electron a particle or a wave? Support your answer by citing some experimental results.
3. An electron and a proton are accelerated from rest through the same potential difference. Which particle has the longer wavelength?
4. If matter has a wave nature, why is this wave-like character not observable in our daily experiences?
5. In what way does Bohr's model of the hydrogen atom violate the uncertainty principle?
6. Why is it impossible to measure the position and velocity of a particle simultaneously with infinite precision?
7. Suppose that a beam of electrons is incident on three or more slits. How would this influence the interference pattern? Would the state of an electron depend on the number of slits? Explain.
8. In describing the passage of electrons through a slit and arriving at a screen, Feynman said that "electrons arrive in lumps, like particles, but the probability of arrival of these lumps is determined as the intensity of the waves would be. It is in this sense that the electron behaves sometimes like a particle and sometimes like a wave." Elaborate on this point in your own words. (For a further discussion of this point, see R. Feynman, *The Character of Physical Law,* Cambridge, Mass., MIT Press, 1982, Chapter 6.)
9. A philosopher once said that "it is necessary for the very existence of science that the same conditions always produce the same results." In view of what has been discussed in this chapter, present an argument showing that this statement is false. How might the statement be reworded to make it acceptable?
10. Do you think that most major experimental discoveries are made by careful planning or by accident? Cite examples.
11. In the case of accidental discoveries, what traits must the experimenter possess in order to capitalize on the discovery?

PROBLEMS

1. Calculate the de Broglie wavelength for a proton moving with a speed of 10^6 m/s.
2. Calculate the de Broglie wavelength for an electron with kinetic energy (a) 50 eV and (b) 50 keV.
3. Calculate the de Broglie wavelength of a 74-kg person who is running at a speed of 5 m/s.
4. The "seeing" ability, or resolution, of radiation is determined by its wavelength. If the size of an atom is of the order of 0.1 nm, how fast must an electron travel to have a wavelength small enough to "see" an atom?
5. An electron and a photon each have energy equal to 50 keV. What are their de Broglie wavelengths?
6. Calculate the de Broglie wavelength of a proton that is accelerated through a potential difference of 10 MV.
7. Show that the de Broglie wavelength of an electron accelerated from rest through a potential difference V is given by $\lambda = 1.226/\sqrt{V}$ nm, where V is in volts.
8. Find the de Broglie wavelength of a ball of mass 0.2 kg just before it strikes the earth after being dropped from a building 50 m tall.
9. An electron has a de Broglie wavelength equal to the diameter of the hydrogen atom. What is the kinetic energy of the electron? How does this energy compare with the ground-state energy of the hydrogen atom?
10. In order for an electron to be confined to a nucleus, its de Broglie wavelength would have to be less than 10^{-14} m. (a) What would be the kinetic energy of an electron confined to this region? (b) On the basis of this result, would you expect to find an electron in a nucleus? Explain.
11. Through what potential difference would an electron have to be accelerated to give it a de Broglie wavelength of 10^{-10} m?
12. A monoenergetic beam of electrons is incident on a single slit of width 0.5 nm. A diffraction pattern is formed on a screen 20 cm from the slit. If the distance between successive minima of the diffraction pattern is 2.1 cm, what is the energy of the incident electrons?
13. A neutron beam with a selected speed of 0.4 m/s is directed through a double slit with a 1-mm separation. An array of detectors is placed 10 m from the slit. (a) What is the de Broglie wavelength of the neutrons? (b) How far off axis is the first zero-intensity point on the detector array? (c) Can we say which slit any particular neutron passed through? Explain.
14. A ball of mass 50 g moves with a speed of 30 m/s. If its speed is measured to an accuracy of 0.1%, what is the minimum uncertainty in its position?
15. A proton has a kinetic energy of 1 MeV. If its momentum is measured with an uncertainty of 5%, what is the minimum uncertainty in its position?
16. We wish to measure simultaneously the wavelength and position of a photon. Assume that the wavelength measurement gives $\lambda = 6000$ Å with an accuracy of one part in a million, i.e., $\Delta\lambda/\lambda = 10^{-6}$. What is the minimum uncertainty in the position of the photon?
17. An air rifle is used to shoot 1-g particles at a speed of 100 m/s through a hole of diameter 2 mm. How far from the rifle must an observer be in order to see the beam spread by 1 cm because of the uncertainty principle? Compare this answer to the diameter of the universe (2×10^{26} m).
18. A two-slit electron diffraction experiment is done with slits of *unequal* widths. When only slit 1 is open, the number of electrons reaching the screen per second is 25 times the number of electrons reaching the screen per second when only slit 2 is open. When both

slits are open, an interference pattern results in which the destructive interference is not complete. Find the ratio of the probability of an electron arriving at an interference maximum to the probability of an electron arriving at an adjacent interference minimum. (Hint: Use the superposition principle.)

19. A beam of electrons is incident on a slit of variable width. If it is possible to resolve a 1% difference in momentum, what slit width would be necessary to resolve the interference pattern of the electrons if their kinetic energy is (a) 0.01 MeV, (b) 1 MeV, and (c) 100 MeV?

20. A photon of energy $E = hc/\lambda$ collides with a particle being studied under a powerful microscope. Assuming that the position of the particle being struck by the photon is uncertain by at least the wavelength λ, (a) show that the uncertainty in the time it takes the photon to move one wavelength is $\Delta t \geq \lambda/c$. (b) If the uncertainty in the energy ΔE of the particle is equal to or greater than the energy of the incident photon, show that the relation $\Delta E\, \Delta t \geq \hbar$ is satisfied. This is another Heisenberg uncertainty relationship, which restricts our ability to know the exact state of a particle.

21. An electron of momentum p is at a distance r from a stationary proton. The electron has a kinetic energy $K = p^2/2m$ and potential energy $U = -ke^2/r$. Its total energy is $E = K + U$. If the electron is bound to the proton to form a hydrogen atom, its average position is at the proton but the uncertainty in its position is approximately equal to the radius r of its orbit. The electron's average momentum will be zero, but the uncertainty in its momentum will be given by the uncertainty principle. Treat the atom as a one-dimensional system in the following: (a) Estimate the uncertainty in the electron's momentum in terms of r. (b) Estimate the electron's kinetic, potential, and total energies in terms of r. (c) The actual value of r is the one that minimizes the total energy, resulting in a stable atom. Find that value of r and the resulting total energy. Compare your answer with the predictions of the Bohr theory.

22. Figure 4.24 shows the top three planes of a crystal with planar spacing d. If $2d \sin\theta = 1.01\lambda$ for the two waves shown, and high-energy electrons of wavelength λ penetrate many planes deep into the crystal, which atomic plane produces a wave that cancels the surface reflection? This is an example of how extremely narrow maxima in high-energy electron diffraction are formed—that is, there are no diffracted beams unless $2d \sin\theta$ is *exactly* equal to an integral number of wavelengths.

23. (a) Show that the formula for Low Energy Electron Diffraction (LEED) when electrons are incident perpendicular to a crystal surface may be written as

$$\sin\phi = \frac{nhc}{d(2m_0c^2k)^{1/2}}$$

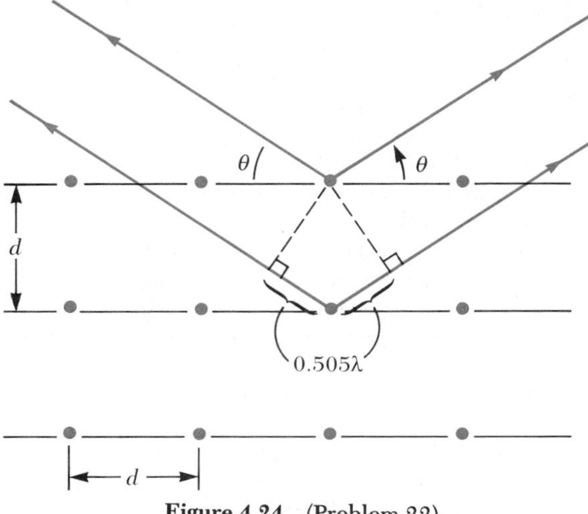

Figure 4.24 (Problem 22)

where n is the order of the maximum, d is the atomic spacing, m_0 is the electron rest mass, K is the electron's kinetic energy, and ϕ is the angle between the incident and diffracted beams. (b) Calculate the atomic spacing in a crystal that has consecutive diffraction maxima at $\phi = 24.1°$ and $\phi = 54.9°$ for 100 eV electrons.

24. *Fourier Integrals.* In order to form a true pulse in space that is zero everywhere outside of a finite range, one must sum an *infinite* number of harmonic waves with continuously varying wavelengths, frequencies, and amplitudes. This sum can be represented by a **Fourier integral,** which is defined as follows:

$$f(x) = (2\pi)^{-1/2} \int_{-\infty}^{+\infty} g(k)e^{ikx}dk$$

where $f(x)$ is a spatially localized wave group, $g(k)$ gives the amplitude of the wave with wave number k, and

$$e^{ikx} = \cos kx + i \sin kx$$

is Euler's compact expression for a harmonic wave. The amplitude distribution, $g(k)$, can be obtained if $f(x)$ is known by using the symmetric formula

$$g(k) = (2\pi)^{-1/2} \int_{-\infty}^{+\infty} f(x)e^{-ikx}dx$$

These equations apply to the case of a spatial pulse at fixed time. However, it is important to note that they are *mathematically identical* to the formulas used to add varying amounts of different frequencies to compose a pulse *in time* at a fixed position. For this second case we have for the strength of a signal $V(t)$ as a function of time:

$$V(t) = (2\pi)^{-1/2} \int_{-\infty}^{+\infty} g(\omega)e^{i\omega t}d\omega$$

where

$$g(\omega) = (2\pi)^{-1/2} \int_{-\infty}^{+\infty} V(t)e^{-i\omega t}dt$$

In this case, $g(\omega)$ is the spectral content of the signal and gives the amount of the harmonic wave with frequency ω present. Assume that a narrow triangular voltage pulse $V(t)$ arises in some type of radar system. (See Fig. 4.25.) (a) Find and sketch the spectral content $g(\omega)$. (b) Show that a relation of the type $\Delta\omega\, \Delta t \approx 1$ holds. (c) If the width of the pulse is $\Delta t = 10^{-9}$ s, what range of frequencies must this system pass if the pulse is to be undistorted?

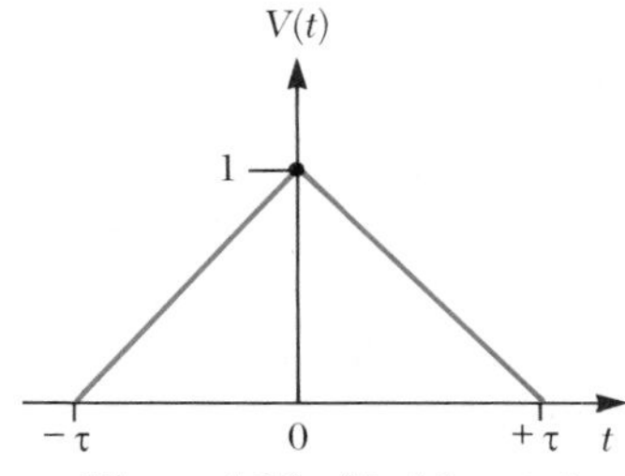

Figure 4.25 (Problem 24)

25. (a) Find and sketch the spectral content of the rectangular pulse of width 2τ shown in Figure 4.26. (b) Show that an uncertainty relation $\Delta\omega\, \Delta t \approx 4\pi$ holds in this case. (c) What range of frequencies is required to compose a 1 μs pulse? a 1 ns pulse?

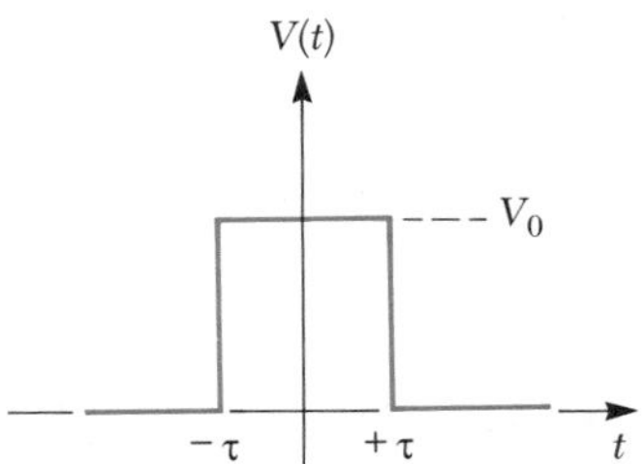

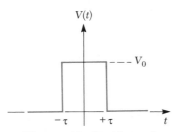

Figure 4.26 (Problem 25)

26. Show that the group velocity for a nonrelativistic free electron is also given by $v_g = p/m = v_0$, where v_0 is the electron's velocity.
27. *A matter wave packet.* (a) Find and sketch the matter wave pulse shape $f(x)$ for a gaussian amplitude distribution $g(k)$ where

$$g(k) = A'e^{-(k-k_0)^2/2(\Delta k)^2}$$

Note that $g(k)$ is peaked at k_0 and has a width Δk. [*Hint:* In order to put $f(x) = (2\pi)^{-1/2} \int_{-\infty}^{\infty} g(k)e^{ikx}dk$ into the standard form $\int_{-\infty}^{\infty} e^{-az^2}dz$, complete the square in k.] (b) By comparing the result for $f(x)$ to the standard form of a gaussian envelope with width Δx, $f(x) = Ae^{-x^2/2(\Delta x)^2}$, show that the width of the matter wave pulse is inversely proportional to the spread of wave numbers composing it: $\Delta x \approx 1/\Delta k$. (c) Express the spread in wave number of the plane waves making up the packet Δk in terms of the spread in momentum of these waves, Δp. Show that $\Delta x\, \Delta k \approx 1$ leads to a momentum-position uncertainty principle.
28. When a pebble is tossed into a pond, a circular wave pulse propagates outward from the disturbance. If you are alert (and it's not a sleepy afternoon in late August), you will see a fine structure in the pulse consisting of surface ripples moving inward through the circular disturbance. Explain this effect in terms of group and phase velocity if the phase velocity of ripples is given by $v_p = \sqrt{2\pi S/\lambda\rho}$ where S is the surface tension and ρ is the density of the liquid.
29. The dispersion relation for free relativistic electron waves is

$$\omega(k) = \sqrt{c^2k^2 + (me^2/\hbar)^2}$$

Obtain expressions for the phase velocity v_p and group velocity v_g of these waves and show that their product is a constant, independent of k. From your result, what can you conclude about v_g if $v_p > c$?

PROBLEMS FOR THE COMPUTER

The following problems require for their solution the use of the program WAVES-1.

30. Display the de Broglie wave of wavenumber $k = 5$ Å^{-1} for a particle in the screen interval -1.5 Å to $+1.5$ Å. Taking the particle to be an electron, track the wave to find its velocity v_p. Report v_p in units of c, and furnish screen dumps of the waveform at two different times to substantiate your result. Compare your measured value with the theoretical prediction $v_p = \omega/k$ for this case.
31. *The Beat Phenomenon.* Display the superposition of two electron waves with wavenumbers $k_1 = 20$ Å^{-1} and $k_2 = 25$ Å^{-1} and identical amplitudes, using the uniform envelope function $a(k) = 1$. Note the corresponding frequencies ω_1 and ω_2. Make suitable measurements on the waveform at subsequent times to determine the phase and group velocities for this case. Compare your measured results with the predictions of Equations 4.14 and 4.16.
32. Construct a wave group using a Gaussian envelope $a(k)$ with 9 wavenumbers in the range 0 to 16 Å^{-1} (steps of 2) and the parameter values $g = 8$, $a = 0.1$. Estimate Δx and Δk for the group, and furnish the appropriate screen dumps to substantiate your claims. Track the wave group for the case where it describes an electron in the screen interval -1.5 Å to $+1.5$ Å. Estimate the group velocity. How much time elapses before the initial localization is destroyed by dispersion? Furnish a screen dump showing the waveform at this time. How do your answers change if the group describes a massless particle (neutrino?) in the same interval?

5

Quantum Mechanics in One Dimension

We have seen that associated with any particle is a matter wave called the *wavefunction*. How this wavefunction affects our description of the particle and its behavior is the subject of *quantum mechanics* or *wave mechanics*. This scheme, developed from 1925 to 1926 by Schrödinger, Heisenberg, and others, makes it possible to understand a host of phenomena involving atoms, molecules, nuclei, and solids. In this and subsequent chapters, we shall describe the basic features of wave mechanics and its application to simple systems. The relevant concepts for particles confined to motion along a line (labeled by x) are developed in the present chapter.

5.1 THE BORN INTERPRETATION

The wavefunction Ψ contains within it all the information that can be known about the particle. That basic premise forms the cornerstone of our investigation: one of our objectives will be to discover how information is extracted from the wavefunction; the other to learn how to obtain this wavefunction for a given system.

The currently held view connects the matter wave Ψ with *probabilities* in the manner first proposed by Max Born in 1925:

Born interpretation of Ψ

The probability that the particle will be found in the infinitesimal interval dx about the point x, denoted by $P(x)\,dx$, is

$$\boxed{P(x)\,dx = |\Psi(x,t)|^2\,dx} \tag{5.1}$$

Therefore, although it is not possible to specify with certainty the location of a particle, it is possible to assign *probabilities* for observing it at any given position. The quantity $|\Psi|^2$, the square of the absolute value of Ψ, represents the *intensity* of the matter wave Ψ and is computed as the product of Ψ with its complex conjugate, that is, $|\Psi|^2 = \Psi^*\Psi$. Notice that Ψ itself is not a measurable quantity; however, $|\Psi|^2$ is measurable and is just the probability per unit length, or **probability density** $P(x)$, for finding the particle at the point x at time t. Because of its relation to probabilities, we insist that $\Psi(x, t)$ be a *single-valued and continuous function of* x and t, so that no ambiguities can arise concerning the predictions of the theory. The wave Ψ also should be *smooth*, a condition that will be elaborated later as it is needed.

Since the particle must be somewhere along the x axis, the sum of the probabilities over all values of x must be 1:

$$\boxed{\int_{-\infty}^{\infty} |\Psi(x,t)|^2\,dx = 1} \tag{5.2}$$

BIOGRAPHY MAX BORN (1882–1970)

Max Born (1882–1970)

Max Born was a German theoretical physicist who made major contributions in many areas of physics including relativity, atomic and solid-state physics, matrix mechanics, the quantum mechanical treatment of particle scattering ("Born approximation"), the foundations of quantum mechanics (Born interpretation of Ψ), optics, and the kinetic theory of liquids. Born received the doctorate in physics from the University of Göttingen in 1907, and he acquired an extensive knowledge of mathematics as the private assistant to the great German mathematician Hilbert. This strong mathematical background proved a great asset when he was quickly able to reformulate Heisenberg's quantum theory in a more consistent way with matrices.

In 1921 Born was offered a post at the University of Göttingen, where he helped build one of the strongest physics centers of the 20th century. This group consisted, at one time or another, of the mathematicians Hilbert, Courant, Klein, and Runge, and the physicists Born, Jordan, Heisenberg, Franck, Pohl, Heitler, Herzberg, Nordheim, and Wigner, among others. In 1926, shortly after Schrödinger's publication of wave mechanics, Born applied Schrödinger's methods to atomic scattering and developed the Born approximation method for carrying out calculations of the probability of scattering of a particle into a given solid angle. This work furnished the basis for Born's startling (in 1926) interpretation of $|\Psi|^2$ as the probability density. For this so-called statistical interpretation of $|\Psi|^2$ he was awarded the Nobel Prize in 1954.

Fired by the Nazis, Born left Germany in 1933 for Cambridge and eventually the University of Edinburgh, where he again became the leader of a large group investigating the statistical mechanics of condensed matter. In his later years Born campaigned against atomic weapons, wrote an autobiography, and translated German humorists into English.

Any wavefunction satisfying Equation 5.2 is said to be *normalized.* Note that normalization is simply a statement that the particle exists at some point at all times. Furthermore, the probability of finding the particle in any finite interval $a \le x \le b$ is

$$P_{ab} = \int_a^b |\Psi(x, t)|^2 \, dx \tag{5.3}$$

That is, the probability is just the area included under the curve of probability density between the points $x = a$ and $x = b$ (Fig. 5.1).

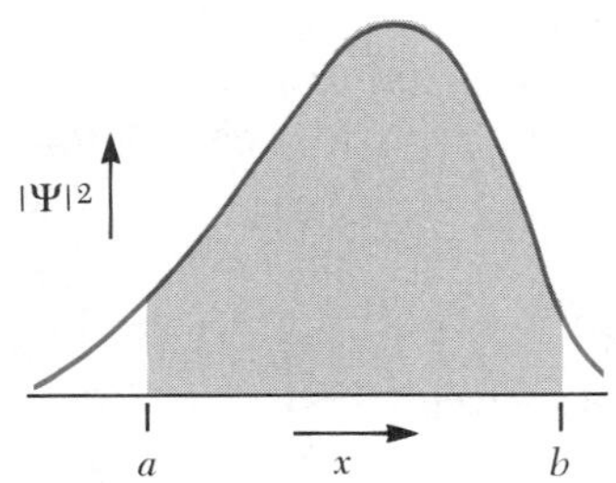

Figure 5.1 The probability for a particle to be in the interval $a \le x \le b$ is the area under the curve from a to b of the probability density function $|\Psi(x, t)|^2$.

EXAMPLE 5.1 Normalizing the Wavefunction
The initial wavefunction of a particle is given as $\Psi(x, 0) = C \exp(-|x|/x_0)$ where C and x_0 are constants. Sketch this function. Find C in terms of x_0 such that $\Psi(x, 0)$ is normalized.

Solution: The given waveform is symmetric, decaying exponentially from the origin in either direction as shown in Figure 5.2. The *decay length* x_0 represents the distance over which the wave amplitude is diminished by the factor $1/e$ from its maximum value $\Psi(0, 0) = C$.

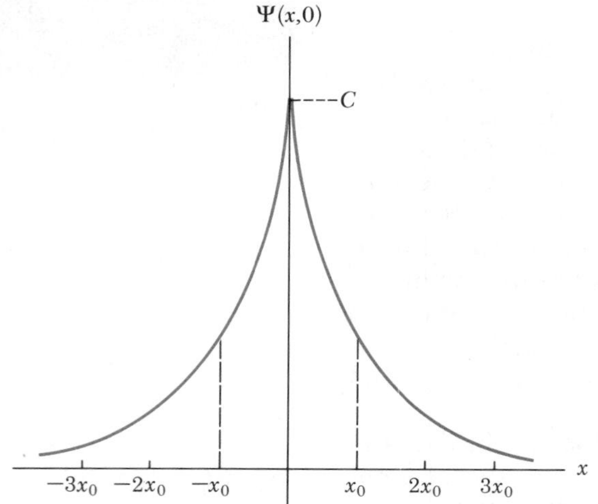

Figure 5.2 (Example 5.1) The symmetric waveform $\Psi(x, 0) = C \exp(-|x|/x_0)$. At $x = \pm x_0$ the wave amplitude is down by the factor $1/e$ from its peak value $\Psi(0, 0) = C$. C is a normalizing constant whose proper value is $C = 1/\sqrt{x_0}$.

The normalization requirement is

$$1 = \int_{-\infty}^{\infty} |\Psi(x, 0)|^2 \, dx = C^2 \int_{-\infty}^{\infty} e^{-2|x|/x_0} \, dx.$$

Since the integrand is unchanged when x changes sign (it is an even function), we may evaluate the integral over the whole axis as twice that over the half axis $x > 0$, where $|x| = x$. Then

$$1 = 2C^2 \int_0^{\infty} e^{-2x/x_0} \, dx = 2C^2 \, (x_0/2) = C^2 \, x_0$$

Thus, we must take $C = 1/\sqrt{x_0}$ for normalization.

EXAMPLE 5.2 Calculating Probabilities
Calculate the probability that the particle in the preceding example will be found in the interval $-x_0 \leq x \leq x_0$.

Solution: The probability in question is the area under the curve of $|\Psi(x, 0)|^2$ from $-x_0$ to $+x_0$, and is obtained by integrating the probability density over the specified interval:

$$P = \int_{-x_0}^{x_0} |\Psi(x, 0)|^2 \, dx = 2 \int_0^{x_0} |\Psi(x, 0)|^2 \, dx$$

where the second step follows because the integrand is an even function, as discussed in Example 5.1. Thus,

$$P = 2C^2 \int_0^{x_0} e^{-2x/x_0} \, dx = 2C^2 \, (x_0/2)\{1 - e^{-2}\}$$

Substituting $C = 1/\sqrt{x_0}$ into this expression gives for the probability $P = 1 - e^{-2} = 0.8647$, or about 86.5%, independent of x_0.

The fundamental problem of quantum mechanics is this: given the wavefunction at some initial instant, say $t = 0$, to find the wavefunction at any subsequent time t. The function $\Psi(x, 0)$ represents the initial information which must be specified; once this is known, however, the initial wave propagates according to prescribed laws of nature.

Since it describes how a given system evolves, quantum mechanics is a dynamical theory much like Newtonian mechanics. There are, of course, important differences. In Newton's mechanics, the state of a particle at $t = 0$ is specified by giving its initial position $x(0)$ and velocity $v(0)$—just two numbers; quantum mechanics demands an entire function $\Psi(x, 0)$—an infinite set of numbers corresponding to the function value at every point x. But both theories describe how this state changes with time when the forces acting on the particle are known. In Newton's mechanics $x(t)$ and $v(t)$ are calculated from Newton's second law; in quantum mechanics $\Psi(x, t)$ must be calculated from another law—*The Schrödinger equation.*

5.2 WAVEFUNCTION FOR A FREE PARTICLE

A free particle is one subject to no force. This special case can be studied using prior assumptions, without recourse to Schrödinger's equation. The development underscores the role of the initial conditions in quantum physics.

The wave number k and frequency ω of free particle matter waves are given by the de Broglie formulas

$$k = p/\hbar \qquad \text{and} \qquad \omega = E/\hbar. \tag{5.4}$$

For nonrelativistic particles ω is related to k as

$$\omega(k) = \frac{\hbar k^2}{2m} \tag{5.5}$$

which follows from the classical connection $E = p^2/2m$ between the energy E and momentum p for a free particle.[1]

For the waveform itself, we should take

Plane wave representation for a free particle

$$\boxed{\Psi_k(x, t) = Ae^{i(kx-\omega t)} = A\{\cos(kx - \omega t) + i \sin(kx - \omega t)\}} \tag{5.6}$$

This is an oscillation with wave number k, frequency ω, and amplitude A. Since the variables x and t occur only in the combination $kx - \omega t$, the oscillation is a *traveling wave*, as befits a free particle in motion. Further, the particular combination expressed by Equation 5.6 is that of a plane wave, for which the probability density $|\Psi|^2 (= A^2)$ is *uniform*. That is, the probability of finding this particle in any interval of the x axis is the same as that for any other interval of equal length, and does not change with time. The plane wave is the simplest traveling waveform with this property—it expresses the reasonable notion that there are no special places for a free particle to be found. The particle's location is completely unknown at $t = 0$ and remains so for all time; however, its momentum and energy are known precisely as $p = \hbar k$ and $E = \hbar\omega$, respectively.

But not *all* free particles are described by the wave of Equation 5.6. For instance, we may establish (by measurement) that our particle initially is in some range Δx about x_0. In that case, $\Psi(x, 0)$ must be a *wave packet* concentrated in this interval, as shown in Figure 5.3a. The plane wave description is

[1] The functional form for $\omega(k)$ was discussed in Section 4.3 for relativistic particles, where $E = \sqrt{(cp)^2 + (m_0c^2)^2}$. In the nonrelativistic case ($v \ll c$) this reduces to $E = p^2/2m_0 + m_0c^2$. The rest energy $E_0 = m_0c^2$ can be disregarded in this case if we agree to make E_0 our energy reference. By measuring all energies from this level, we are in effect setting E_0 equal to zero. Further, in the nonrelativistic case, no distinction is made between the mass m of a particle and its rest mass m_0.

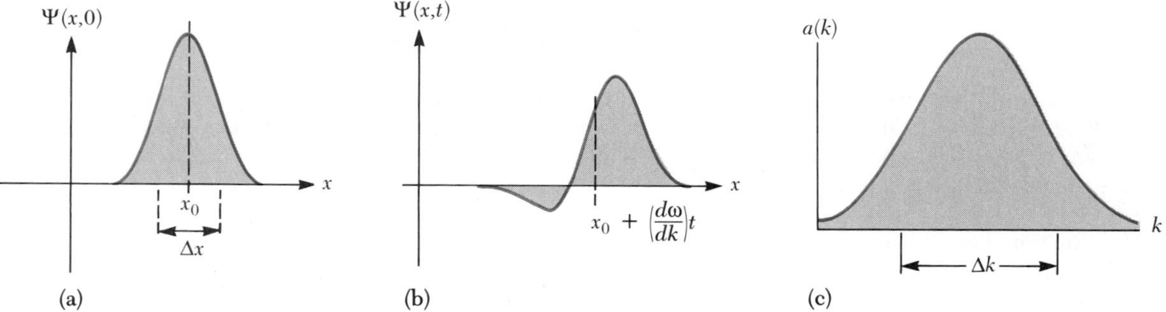

Figure 5.3 (a) A wave packet $\Psi(x, 0)$ formed from a superposition of plane waves. (b) The same wave packet some time t later. Since the plane waves with smaller wave number move at slower speeds, the packet becomes distorted. The body of the packet propagates with the group velocity $d\omega/dk$ of the plane waves. (c) The spectral content $a(k)$ for this packet, indicating the amplitude of each plane wave in the superposition. A narrow wave packet requires a broad spectral content, and vice versa. That is, the widths Δx and Δk are inversely related as $\Delta x\, \Delta k \approx 1$.

inappropriate now since the initial waveshape is not given correctly by $\Psi_k(x, 0) = e^{ikx}$. Instead, a *sum* of plane waves with different wave numbers must be used to represent the packet. Since k is unrestricted, the sum actually is an integral here and we write

Representing a particle with a wave group

$$\Psi(x, 0) = \int_{-\infty}^{\infty} a(k)e^{ikx}\, dk. \tag{5.7}$$

The coefficients $a(k)$ specify the amplitude of the plane wave with wave number k in the mixture, and are chosen to reproduce the initial waveshape. For a given $\Psi(x, 0)$, the required $a(k)$ can be found from the theory of Fourier integrals. We shall not be concerned with the details of this analysis; the essential point is that it can be done for a packet of any shape. If each plane wave comprising the packet is assumed to propagate independently of the others according to Equation 5.6, the packet at any time t is given by

$$\Psi(x, t) = \int_{-\infty}^{\infty} a(k)e^{i(kx-\omega(k)t)}\, dk. \tag{5.8}$$

Notice that the initial data are used only to establish the amplitudes $a(k)$; subsequently, the packet develops according to the evolution of its plane wave constituents. Since each of these constituents moves with a different speed $v_p = \omega/k$ (the phase velocity), we have dispersion (see Section 4.3) and *the packet changes its shape as it propagates* (Fig. 5.3b). The speed of propagation is given by the group velocity $d\omega/dk$ of the plane waves forming the packet. It is also clear that Equation 5.8 no longer describes a particle with precise values of momentum and energy. In order to construct a wave packet (that is, localize the particle), a mixture of wave numbers (hence, particle momenta) is necessary, as indicated by the different $a(k)$. Together, the $a(k)$ furnish the so-called **spectral content** of the packet, which might look like that sketched in Figure 5.3c. The narrower the desired packet $\Psi(x, 0)$, the broader is the function $a(k)$ needed to represent that packet. If Δx denotes the packet width and Δk the extent of the corresponding $a(k)$, one finds that the product always is a number of order unity, that is, $\Delta x\, \Delta k \approx 1$. Together with $p = \hbar k$ this implies an uncertainty principle:

$$\Delta x\, \Delta p \sim \hbar \tag{5.9}$$

In closing, we point out that Equations 5.7 and 5.8 solve the fundamental problem of quantum mechanics for *free* particles subject to *any* initial condition $\Psi(x, 0)$.

EXAMPLE 5.3 Constructing a Wave Packet
Find the wave $\Psi(x, 0)$ that results from taking the function $a(k) = (C\alpha/\sqrt{\pi}) \exp(-\alpha^2k^2)$, where C and α are constants. Estimate the product $\Delta x\, \Delta k$ for this case.

Solution: The function $\Psi(x, 0)$ is given by the integral of Equation 5.7 or

$$\Psi(x, 0) = \int_{-\infty}^{\infty} a(k)e^{ikx}\, dk = \frac{C\alpha}{\sqrt{\pi}} \int_{-\infty}^{\infty} e^{(ikx-\alpha^2k^2)}\, dk$$

To evaluate the integral, we first complete the square in the exponent as

$$ikx - \alpha^2k^2 = -(\alpha k - ix/2\alpha)^2 - x^2/4\alpha^2$$

The second term on the right is constant for the integration over k; to integrate the first term we change variables with the substitution $z = \alpha k - ix/2\alpha$, obtaining

$$\Psi(x, 0) = \frac{C}{\sqrt{\pi}} e^{-x^2/4\alpha^2} \int_{-\infty}^{\infty} e^{-z^2}\, dz$$

The integral now is a standard one whose value is known to be $\sqrt{\pi}$. Then

$$\Psi(x, 0) = Ce^{-x^2/4\alpha^2} = Ce^{-(x/2\alpha)^2}$$

This function $\Psi(x, 0)$, called a **gaussian,** has a single maximum at $x = 0$ and decays smoothly to zero on either side of this point (Fig. 5.4a). The width of this gaussian

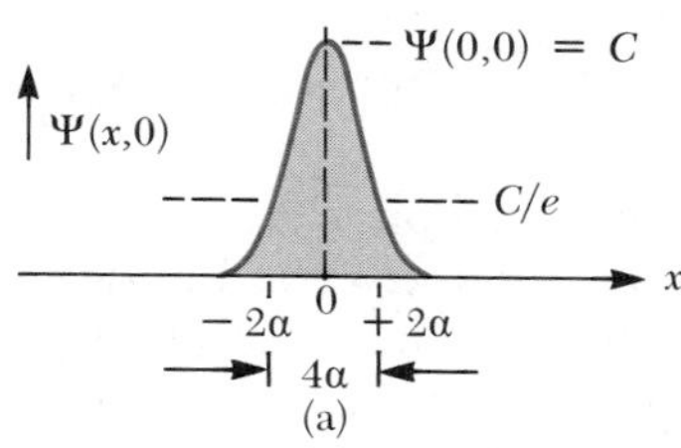

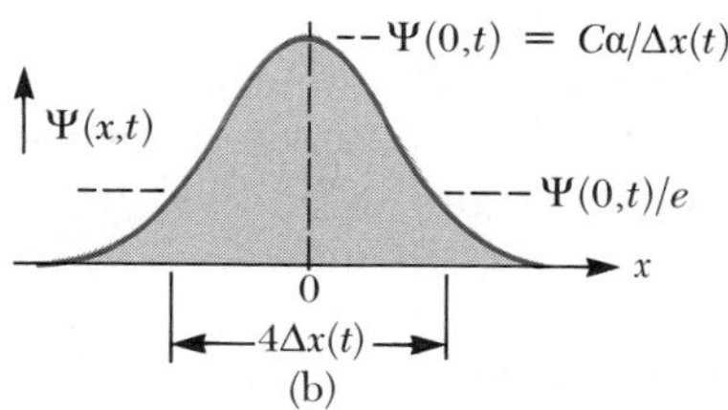

Figure 5.4 (Example 5.3) (a) The gaussian waveform $\Psi(x, 0) = C \exp\{-(x/2\alpha)^2\}$, representing a particle initially localized around $x = 0$. C is the amplitude. At $x = \pm 2\alpha$ the amplitude is down from its maximum value by the factor $1/e$; accordingly, α is identified as the width of the gaussian, $\alpha = \Delta x$. (b) The gaussian waveform of Figure 5.4a at time t. The width has increased to $\Delta x(t) = \sqrt{\alpha^2 + \hbar t/2m}$, and the amplitude is reduced by the factor $\alpha/\Delta x(t)$.

packet becomes larger with increasing α. Accordingly, it is reasonable to identify α with Δx, the initial degree of localization. By the same token, $a(k)$ also is a gaussian, but with amplitude $C\alpha/\sqrt{\pi}$ and width $\frac{1}{2}\alpha$ (since $\alpha^2 k^2 = \{k/2(\frac{1}{2}\alpha)\}^2$). Thus, $\Delta k = \frac{1}{2}\alpha$ and $\Delta x\ \Delta k = \frac{1}{2}$, *independent of* α. The multiplier C is a scale factor chosen to normalize $\Psi(x, 0)$.

Because our gaussian packet is made up of many individual waves all moving with different speeds, the shape of the packet changes over time. In Problem 20 it is shown that the packet *disperses,* its width growing ever larger with the passage of time as

$$\Delta x(t) = \sqrt{\{\Delta x(0)\}^2 + \hbar t/2m}$$

Similarly, the peak amplitude diminishes steadily in order to keep the waveform normalized for all times (Fig. 5.4b). The wave as a whole does not propagate, since for every wave number k present in the wave group there is an equal admixture of the plane wave with the opposing wave number $-k$.

EXAMPLE 5.4 Dispersion of Matter Waves
An atomic electron initially is localized to a region of space 0.1 nm wide (atomic size). How much time elapses before this localization is destroyed by dispersion? Repeat the calculation for a 1-g marble initially localized to 0.1 mm.

Solution: Taking for the initial state a gaussian wave shape, we may use the results of the previous example. In particular, the extent of the matter wave after a time t has elapsed is

$$\Delta x(t) = \sqrt{\{\Delta x(0)\}^2 + \hbar t/2m,}$$

where $\Delta x(0)$ is its initial width. The packet has effectively dispersed when $\Delta x(t)$ becomes appreciable compared to $\Delta x(0)$, say $\Delta x(t) = 10\ \Delta x(0)$. This happens when $\hbar t/2m = 99\{\Delta x(0)\}^2$, or $t = 99(2m/\hbar)\{\Delta x(0)\}^2$.

The electron is initially localized to 0.1 nm ($= 10^{-10}$ m), and its mass is $m = 9.11 \times 10^{-31}$ kg. Thus, the electron wave packet disperses after a time

$$t = 99\left\{\frac{(2)(9.11 \times 10^{-31}\ \text{kg})}{1.055 \times 10^{-34}\ \text{J}\cdot\text{s}}\right\}(10^{-10}\ \text{m})^2$$
$$= 1.71 \times 10^{-14}\ \text{s}$$

The same calculation for a 1-g marble localized to 0.1 mm $= 10^{-4}$ m gives

$$t = 99\left\{\frac{(2)(10^{-3}\ \text{kg})}{1.055 \times 10^{-34}\ \text{J}\cdot\text{s}}\right\}(10^{-4}\ \text{m})^2 = 1.88 \times 10^{25}\ \text{s}$$

or about 5.96×10^{17} years! This is more than ten million times the currently accepted value for the age of the universe. With its much larger mass, the marble does not show the quantum effects of dispersion on any measurable time scale and will, for all practical purposes, remain localized "forever." By contrast, the localization of an atomic electron is destroyed in a time that is very short, even by atomic standards.

5.3 WAVEFUNCTIONS IN THE PRESENCE OF FORCES

For a particle acted upon by a force F, $\Psi(x, t)$ must be found from **Schrödinger's equation:**

$$\boxed{-\frac{\hbar^2}{2m}\frac{\partial^2\Psi}{\partial x^2} + U(x)\Psi = i\hbar\frac{\partial\Psi}{\partial t}} \qquad (5.10)$$

The Schrödinger wave equation

Again, we assume knowledge of the initial wavefunction $\Psi(x, 0)$. In this expression, $i = \sqrt{-1}$ is the imaginary unit and $U(x)$ is the potential energy func-

BIOGRAPHY ERWIN SCHRÖDINGER (1887–1961)

Erwin Schrödinger (1887–1961)

Erwin Schrödinger was an Austrian theoretical physicist best known as the creator of wave mechanics. As a young man he was a good student who liked mathematics and physics, but also Latin and Greek for their logical grammar. He received a doctorate in physics from the University of Vienna. Although his work in physics was interrupted by World War I, by 1920 he had produced important papers on statistical mechanics, color vision, and general relativity, which he at first found quite difficult to understand. Expressing his feelings about a scientific theory in the remarkably open and outspoken way he maintained throughout his life, Schrödinger found general relativity initially "depressing" and "unnecessarily complicated." Other Schrödinger remarks in this vein, with which some readers will enthusiastically agree, are as follows: The Bohr-Sommerfeld quantum theory was "unsatisfactory," "even disagreeable." "I . . . feel intimidated, not to say repelled, by what seem to me the very difficult methods [of matrix mechanics] and by the lack of clarity."

Shortly after de Broglie introduced the concept of matter waves in 1924, Schrödinger began to develop a new relativistic atomic theory based on de Broglie's ideas, but his failure to include electron spin led to the failure of this theory for hydrogen. However, by January of 1926, by treating the electron as a nonrelativistic particle, Schrödinger had introduced his famous wave equation and successfully obtained the energy values and wavefunctions for hydrogen. As Schrödinger himself pointed out, an outstanding feature of his approach was that the discrete energy values emerged from his wave equation in a natural way (as in the case of standing waves on a string), and in a way superior to the artificial postulate approach of Bohr. Another outstanding feature of Schrödinger's wave mechanics was that it was easier to apply to physical problems than Heisenberg's matrix mechanics, since it involved a

tion for the force F; that is, $F = -dU/dx$. Schrödinger's equation is not derivable from any more basic principle, but is one of the *laws* of quantum physics. As with any law, its "truth" must be gauged ultimately by its ability to make predictions in agreement with experiment.

The Schrödinger equation propagates the initial wave forward in time. To see how this works, suppose $\Psi(x, 0)$ has been given. Then the left hand side (LHS) of Schrödinger's equation can be evaluated and Equation 5.10 gives $\partial\Psi/\partial t$ at $t = 0$, the initial rate of change of the waveform. From this we compute the wavefunction a short time δt later as $\Psi(x, \delta t) = \Psi(x, 0) + [\partial\Psi/\partial t]_0\delta t$. This allows the LHS to be re-evaluated, now at $t = \delta t$. With each such repetition, Ψ is advanced another step δt into the future. Continuing the process generates Ψ at any desired later time t. Such repetitious calculations are ideally suited to computers, and the method outlined above may be used to solve Schrödinger's equation numerically.

But how can we obtain an explicit mathematical expression for $\Psi(x, t)$?Returning to the free particle case, we see that the plane waves $\Psi_k(x, t)$ of Equation 5.6 serve a dual purpose: on the one hand, they are the waveforms of particles whose momentum (hence, energy) is known precisely; on the other, they become the building blocks for constructing wavefunctions satisfying any initial condition. The question naturally arises: do analogous functions partial differential equation very similar to the classical wave equation. Intrigued by the remarkable differences in conception and mathematical method of wave and matrix mechanics, Schrödinger did much to hasten the universal acceptance of all of quantum theory by demonstrating the mathematical equivalence of the two theories in 1926.

Although Schrödinger's wave theory was generally based on clear physical ideas, one of its major problems in 1926 was the physical interpretation of the wave function Ψ. Schrödinger felt that the electron was ultimately a wave, Ψ was the vibration amplitude of this wave, and $\Psi^*\Psi$ was the electric charge density. As mentioned in Section 4.5 and Chapter 5, Born, Bohr, Heisenberg, and others pointed out the problems with this interpretation and presented the currently accepted view that $\Psi^*\Psi$ is a probability, and that the electron is ultimately no more a wave than a particle. Schrödinger never accepted this view, but registered his "concern and disappointment" that this "transcendental, almost psychical interpretation" had become "universally accepted dogma."

In 1927 Schrödinger, at the invitation of Max Planck, accepted the chair of theoretical physics at the University of Berlin, where he formed a close friendship with Planck and experienced six stable and productive years. In 1933, disgusted with the Nazis like so many of his colleagues, he left Germany. After several moves reflecting the political instability of Europe, he eventually settled at the Dublin Institute for Advanced Studies. Here he spent 17 happy, creative years working on problems in general relativity, cosmology, and the application of quantum physics to biology. This latter effort resulted in a fascinating short book, *What is Life?*, which induced many young physicists to investigate biological processes with chemical and physical methods. In 1956 he returned home to Austria and to his beloved Tirolean mountains. He died there in 1961.

exist when forces are present? The answer is yes! To obtain them we look for solutions to Schrödinger's equation having the separable form[2]

$$\Psi(x, t) = \psi(x)\phi(t) \tag{5.11}$$

where $\psi(x)$ is a function of x only and $\phi(t)$ is a function of t only. (Note that the plane waves have just this form, with $\psi(x) = e^{ikx}$ and $\phi(t) = e^{-i\omega t}$.) Substituting Equation 5.11 into Equation 5.10 and dividing through by $\psi(x)\phi(t)$ gives

$$-\frac{\hbar^2}{2m}\frac{\psi''(x)}{\psi(x)} + U(x) = i\hbar\frac{\phi'(t)}{\phi(t)}$$

where primes denote differentiation with respect to the arguments. Now the LHS of this equation is a function of x only,[3] and the RHS is a function of t only.

[2] Obtaining solutions to partial differential equations in separable form is called *separation of variables.* Upon separating variables, a partial differential equation in, say, N variables is reduced to N ordinary differential equations, each involving only a single variable. The technique is a general one which may be applied to many (but not all!) of the partial differential equations encountered in science and engineering applications.

[3] Implicitly we have assumed that the potential energy $U(x)$ is a function of x only. For potentials which also depend upon t (e.g., those arising from a time-varying electric field), solutions to Schrödinger's equation in separable form usually do not exist.

Since we can assign any value to x, independently of t, the two sides of the equation can be equal only if each is equal to the same constant, which we call E.[4] This yields two equations determining the unknown functions $\psi(x)$ and $\phi(t)$. The resulting equation for the time function $\phi(t)$ is

$$i\hbar \frac{d\phi}{dt} = E\phi(t) \tag{5.12}$$

This can be integrated immediately to give $\phi(t) = e^{-i\omega t}$ with $\omega = E/\hbar$—the same result obtained for free particles! The equation for the space function $\psi(x)$ is

Wave equation for matter waves in separable form

$$\boxed{-\frac{\hbar^2}{2m}\frac{d^2\psi}{dx^2} + U(x)\psi(x) = E\psi(x)} \tag{5.13}$$

Equation 5.13 is called the *time-independent Schrödinger equation.* Explicit solutions to this equation cannot be written for an arbitrary potential energy function $U(x)$. But whatever its form, $\psi(x)$ must be well-behaved on account of its connection with probabilities. In particular, $\psi(x)$ must be everywhere finite, single-valued, and continuous. Furthermore, $\psi(x)$ must be "smooth," that is, the slope of the wave $d\psi/dx$ also must be continuous wherever $U(x)$ has a finite value.[5]

For free particles we take $U(x) = 0$ in Equation 5.13 (to give $F = -dU/dx = 0$) and find that $\psi(x) = e^{ikx}$ is a solution with $E = \hbar^2k^2/2m$. Thus, for free particles the separation constant E becomes the total particle energy; this identification continues to be valid when forces are present. However, the waveform $\psi(x)$ will change with the introduction of forces, since particle momentum (hence k) no longer is constant in that case.

The separable solutions to Schrödinger's equation describe conditions of particular physical interest. One feature shared by all such wavefunctions is especially noteworthy: since $|e^{-i\omega t}|^2 = e^{+i\omega t}e^{-i\omega t} = e^0 = 1$, we have

$$|\Psi(x, t)|^2 = |\psi(x)|^2. \tag{5.14}$$

This equality expresses the time-independence of all probabilities calculated from $\Psi(x, t)$. For this reason solutions in separable form are called **stationary states.** Thus, *for stationay states the probabilities are static* and can be calculated from the time-independent waveform $\psi(x)$.

5.4 THE PARTICLE IN A BOX

Of the problems involving forces, the simplest is particle confinement. Consider a particle moving along the x axis between the points $x = 0$ and $x = L$, where L is the length of the "box." Inside the box the particle is free; at the

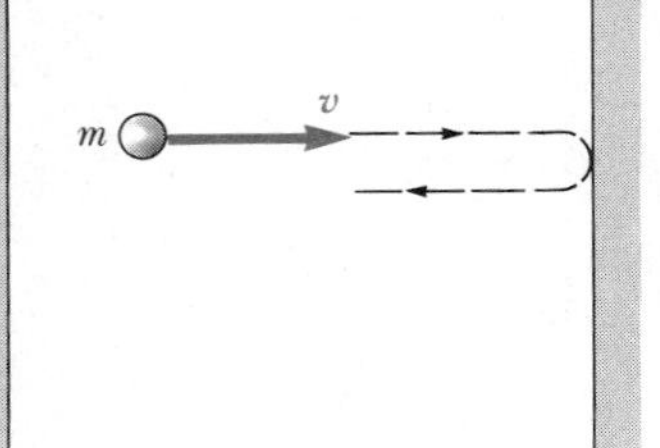

Figure 5.5 A particle of mass m and velocity v bouncing elastically between two impenetrable walls.

[4] More explicitly, changing t cannot affect the LHS since this depends only on x. Since the two sides of the equation are equal, we conclude that changing t cannot affect the RHS either. It follows that the RHS must reduce to a constant. The same argument with x replacing t shows the LHS also must reduce to this same constant.

[5] Upon rearrangement, the Schrödinger equation specifies the curvature of the waveform $d^2\psi/dx^2$ at any point as

$$\frac{d^2\psi}{dx^2} = \frac{2m}{\hbar^2}\{U(x) - E\}\psi(x)$$

It follows that if $U(x)$ is finite at x, the curvature also is finite here and the slope $d\psi/dx$ will be continuous.

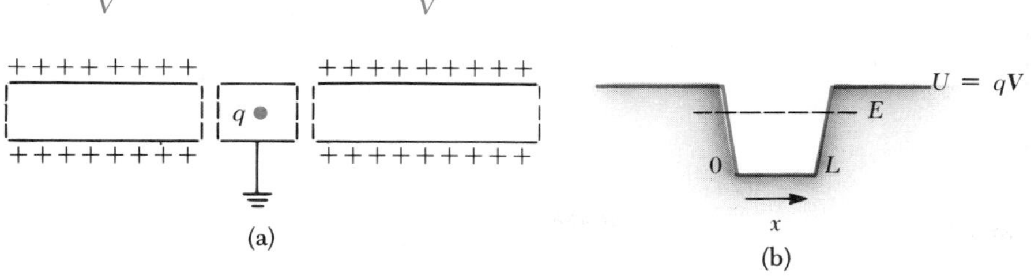

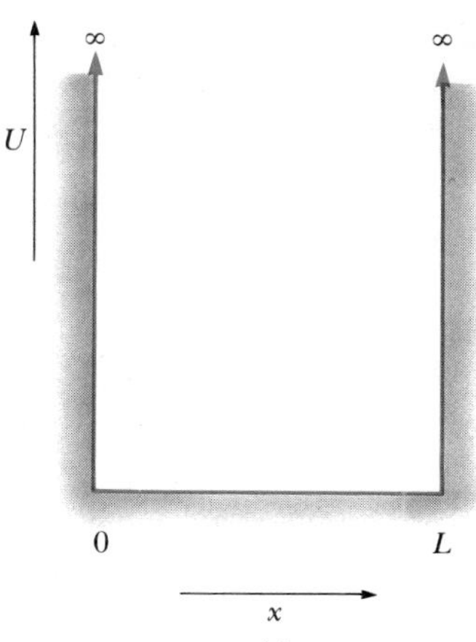

Figure 5.6 (a) Aligned metallic cylinders serve to confine a charged particle. The inner cylinder is grounded, while the outer ones are held at some high electric potential V. A charge q moves freely within the cylinders, but encounters electric forces in the gaps separating them. (b) The potential energy seen by this charge. A charge whose total energy is less than qV is confined to the central cylinder by the strong repulsive forces in the gaps at $x = 0$ and $x = L$. (c) As V is increased and the gaps between cylinders are narrowed, the potential energy approaches that of the infinite square well.

edges, however, it experiences strong forces that serve to contain it. A simple example is a ball bouncing elastically between two inpenetrable walls (Fig. 5.5). A more sophisticated one is a charged particle moving along the axis of aligned metallic tubes held at different potentials, as shown in Figure 5.6a. The central tube is grounded, so that a test charge inside this tube has zero potential energy and experiences no electric force. When both outer tubes are held at a high electric potential V, there are no electric fields within them, but strong repulsive fields arise in the gaps at 0 and L. The potential energy $U(x)$ for this situation is sketched in Figure 5.6b. As V is increased without limit and the gaps are simultaneously reduced to zero, we approach the idealization known as the *infinite square well* or "box" potential (Fig. 5.6c).

From a classical viewpoint, our particle simply bounces back and forth between the confining walls of the box. Its speed remains constant, as does its kinetic energy. Furthermore, classical physics places no restrictions on the values of its momentum and energy. The quantum description is quite different, and leads to the interesting phenomenon of energy quantization.

We are interested in the time-independent waveform $\psi(x)$ of our particle. Since it is confined to the box, the particle can *never* be found outside, which requires ψ to be zero in the exterior regions $x < 0$ and $x > L$. Inside the box $U(x) = 0$ and Equation 5.13 for $\psi(x)$ becomes, after rearrangement,

$$\frac{d^2\psi}{dx^2} = -k^2\psi(x) \qquad \text{with} \qquad k^2 = \frac{2mE}{\hbar^2}$$

Independent solutions to this equation are sin kx and cos kx, indicating that k is the wave number of oscillation. The most general solution is a linear combination of these two,

$$\psi(x) = A \sin kx + B \cos kx \qquad \text{for } 0 < x < L \tag{5.15}$$

This interior wave must match the exterior wave at the walls of the box in order that $\psi(x)$ be continuous everywhere.[6] Thus, we require the interior wave to vanish at $x = 0$ and $x = L$:

$$\begin{aligned} \psi(0) &= B = 0 && \text{[continuity at } x = 0] \\ \psi(L) &= A \sin kL = 0 && \text{[continuity at } x = L] \end{aligned} \tag{5.16}$$

[6] Although $\psi(x)$ must be continuous everywhere, the slope of the wave $\psi'(x)$ is *not* continuous at the walls of the box, where $U(x)$ becomes infinite (cf. footnote 5).

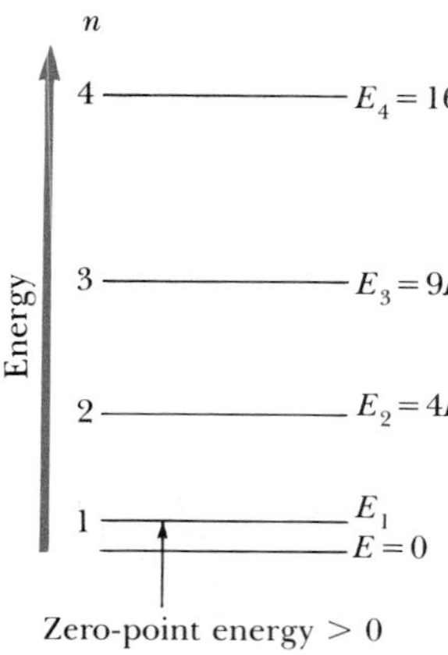

Figure 5.7 Energy level diagram for a particle confined to a one-dimensional box of width L. The lowest allowed energy is E_1 with value $\pi^2\hbar^2/2mL^2$.

The last condition requires that $kL = n\pi$, where n is any positive integer.[7] Since $k = 2\pi/\lambda$, this is equivalent to fitting an integral number of half-wavelengths into the box (see Fig. 5.9). Using Equation 5.16, together with the result that $k = n\pi/L$, we find that the particle energies are *quantized*, being restricted to the values

Allowed energies for a particle in a box

$$E_n = \frac{\hbar^2 k^2}{2m} = \frac{n^2\pi^2\hbar^2}{2mL^2} \qquad n = 1, 2, \ldots \tag{5.17}$$

The lowest allowed energy is given by $n = 1$ and is $E_1 = \pi^2\hbar^2/2mL^2$. This is the **ground state.** Since $E_n = n^2E_1$, the **excited states** for which $n = 2, 3, 4, \ldots$ have energies $4E_1, 9E_1, 16E_1, \ldots$. An energy level diagram is given in Figure 5.7. Notice that $E = 0$ is not allowed; that is, *the particle can never be at rest.* The least energy the particle can have, E_1, is called the **zero-point energy.** This result clearly contradicts the classical prediction, for which $E = 0$ is an acceptable energy, as are all positive values of E.

[7] For $n = 0$ ($E = 0$) Schrödinger's equation requires $d^2\psi/dx^2 = 0$, whose solution is given by $\psi(x) = Ax + B$ for some choice of constants A and B. For this wave to vanish at $x = 0$ and $x = L$, both A and B must be zero, leaving $\psi(x) = 0$ everywhere. In such a case the particle is nowhere to be found, that is, no description is possible when $E = 0$. Also, the inclusion of negative integers $n < 0$ produces no new states, since changing the sign of n merely changes the sign of the wavefunction, leading to the same probabilities as for positive integers.

EXAMPLE 5.5 Energy Quantization for a Macroscopic Object

A small object of mass 1 mg is confined to move between two rigid walls separated by 1 cm. (a) Calculate the minimum speed of the object. (b) If the speed of the object is 3 cm/s, find the corresponding value of n.

Solution: Treating this as a particle in a box, the energy of the particle can only be one of the values given by Equation 5.17, or

$$E_n = \frac{n^2\pi^2\hbar^2}{2mL^2} = \frac{n^2h^2}{8mL^2}$$

The minimum energy results from taking $n = 1$. For $m = 1$ mg ($=10^{-6}$ kg) and $L = 1$ cm ($=10^{-2}$ m) we calculate

$$E_1 = \frac{(6.626 \times 10^{-34}\ \text{J}\cdot\text{s})^2}{8 \times 10^{-10}\ \text{kg}\cdot\text{m}^2} = 5.49 \times 10^{-58}\ \text{J}$$

Since the energy is all kinetic, $E_1 = mv_1{}^2/2$ and the *minimum* speed v_1 of the particle is

$$v_1 = \sqrt{2(5.49 \times 10^{-58}\ \text{J})/(10^{-6}\ \text{kg})} = 3.31 \times 10^{-26}\ \text{m/s}$$

This speed is immeasurably small, so that for practical purposes the object can be considered to be at rest. Indeed, the time required for an object with this speed to move the 1 cm separating the walls is about 3×10^{23} s, or about one million times the present age of the universe! It is reassuring to verify that quantum mechanics applied to macroscopic objects does not contradict our experience.

If, instead, the speed of the particle is $v = 3$ cm/s, then its energy is

$$E = \frac{mv^2}{2} = \frac{(10^{-6}\text{ kg})(3 \times 10^{-2}\text{ m/s})^2}{2} = 4.50 \times 10^{-10}\text{ J}$$

This, too, must be one of the special values E_n. To find which one, we solve for the quantum number n, obtaining

$$n = \frac{\sqrt{8mL^2E}}{h} = \frac{\sqrt{(8 \times 10^{-10}\text{ kg}\cdot\text{m}^2)(4.50 \times 10^{-10}\text{ J})}}{6.626 \times 10^{-34}\text{ J}\cdot\text{s}}$$

$$= 9.05 \times 10^{23}$$

Notice that the quantum number representing a typical speed for this (ordinary size) object is enormous. In fact, the value of n is so large that we would never be able to distinguish the quantized nature of the energy levels. That is, the difference in energy between the two consecutive states, whose quantum numbers are given by $n_1 = 9.05 \times 10^{23}$ and $n_2 = 9.05 \times 10^{23} + 1$, is only about 10^{-33} J, much too small to be detected experimentally. This is another example that illustrates the working of Bohr's correspondence principle, which asserts that quantum predictions must agree with classical results in the limit of large quantum numbers.

EXAMPLE 5.6 Model of an Atom

An atom can be viewed as a number of electrons moving around a positively charged nucleus, where the electrons are subject mainly to the Coulombic attraction of the nucleus (which actually is partially "screened" by the intervening electrons). The potential well that each electron "sees" is sketched in Figure 5.8. Use the model of a particle in a box to estimate the energy (in eV) required to raise an atomic electron from the state $n = 1$ to the state $n = 2$, assuming the atom has a radius of 0.1 nm.

Solution: Taking the length L of the box to be 0.2 nm (the diameter of the atom) and $m = 511$ keV/c^2 for the electron, we calculate

$$E_1 = \frac{\pi^2\hbar^2}{2mL^2}$$

$$= \frac{\pi^2(197.3\text{ eV}\cdot\text{nm}/c)^2}{2(511 \times 10^3\text{ eV}/c^2)(0.2\text{ nm})^2}$$

$$= 9.40\text{ eV}$$

and

$$E_2 = (2)^2E_1 = 4(9.40\text{ eV}) = 37.6\text{ eV}$$

Therefore, the energy that must be supplied to the electron is

$$\Delta E = E_2 - E_1 = 37.6\text{ eV} - 9.40\text{ eV} = 28.2\text{ eV}$$

We could also calculate the wavelength of the photon that would cause this transition by identifying ΔE with the photon energy hc/λ, or

$$\lambda = hc/\Delta E = (1.24 \times 10^3\text{ eV}\cdot\text{nm})/(28.2\text{ eV}) = 44.0\text{ nm}$$

This wavelength is in the far ultraviolet region, and it is interesting to note that the result is roughly correct. Although this oversimplified model gives a good estimate for transitions between lowest-lying levels of the atom, the estimate gets progressively worse for higher-energy transitions.

Exercise 1 Calculate the minimum speed of an atomic electron modeled as a particle in a box with walls that are 0.2 nm apart.
Answer: 182.1 nm/s.

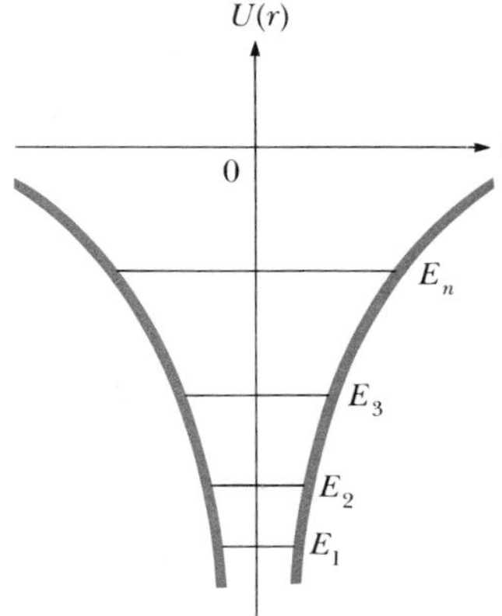

Figure 5.8 (Example 5.6) Model of the potential energy versus r for the one-electron atom.

Returning to the waveforms, we have from Equation 5.15 (with $k = n\pi/L$ and $B = 0$)

$$\boxed{\psi_n(x) = A\sin\left(\frac{n\pi x}{L}\right)} \quad \text{for } 0 < x < L \text{ and } n = 1, 2, \ldots \quad (5.18)$$

Stationary states for a particle in a box

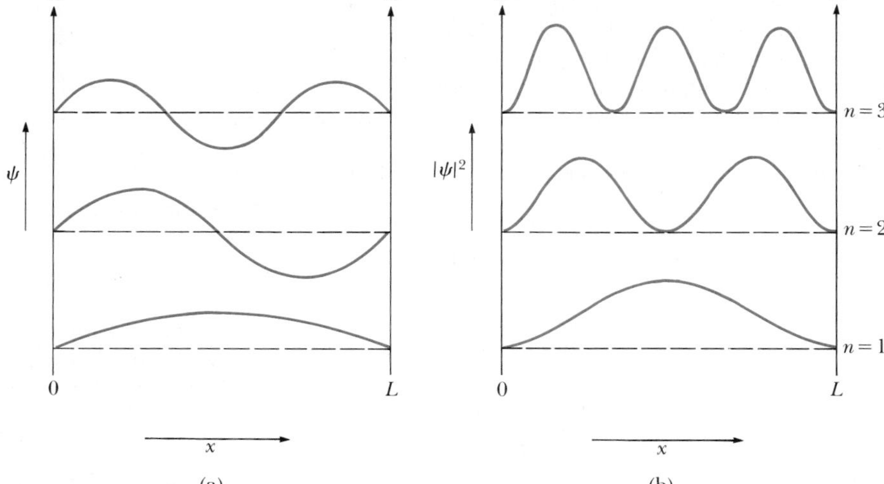

Figure 5.9 The first three allowed stationary states for a particle confined to a one-dimensional box. (a) The wavefunctions for $n = 1$, 2, and 3. (b) The probability distributions for $n = 1$, 2, and 3.

For each value of the **quantum number** n there is a specific waveform $\psi_n(x)$ describing the state of the particle with energy E_n. Figure 5.9 shows plots of ψ_n versus x and of the probability density $|\psi_n|^2$ versus x for $n = 1$, 2, and 3, corresponding to the three lowest allowed energies for the particle. For $n = 1$, the probability of finding the particle is largest at $x = L/2$—this is the *most probable position* for a particle in this state. For $n = 2$, $|\psi|^2$ is a maximum at $x = L/4$ and again at $x = 3L/4$: both points are equally likely places for a particle in this state to be found.

There are also points within the box where it is impossible to find the particle. Again for $n = 2$, $|\psi|^2$ is zero at the midpoint, $x = L/2$; for $n = 3$, $|\psi|^2$ is zero at $x = L/3$ and at $x = 2L/3$, and so on. But how does our particle get from one place to another when there is no probability for its ever being at points in between? It is as if there were *no path at all,* and not just that the probabilities $|\psi|^2$ express our ignorance about a world somehow hidden from view. Indeed, what is at stake here is the very essence of a *particle* as something which gets from one place to another by occupying all intervening positions. The objects of quantum mechanics are not particles, but more complicated things having both particle *and wave* attributes. What is more, these objects are the very "stuff" of nature!

Actual probabilities can be computed only after ψ_n is normalized, that is, we must be sure that all probabilities sum to unity:

$$1 = \int |\psi_n(x)|^2 \, dx = A^2 \int_0^L \sin^2\left(\frac{n\pi x}{L}\right) dx$$

The integral is evaluated with the help of the trigonometric identity $2 \sin^2 \theta = 1 - \cos 2\theta$. Only the first term contributes to the integral, since the cosine integrates to $\sin(2n\pi x/L)$, which vanishes at the limits 0 and L. Thus, normalization requires $1 = A^2L/2$, or

$$A = \sqrt{\frac{2}{L}} \tag{5.19}$$

EXAMPLE 5.7 Probabilities for a Particle in a Box
A particle is known to be in the ground state of an infinite square well with length L. Calculate the probability that this particle will be found in the middle half of the well, that is, between $x = L/4$ and $x = 3L/4$.

Solution: The probability density is given by $|\psi_n|^2$ with $n = 1$ for the ground state. Thus, the probability is

$$P = \int_{L/4}^{3L/4} |\psi_1|^2\,dx = \left(\frac{2}{L}\right)\int_{L/4}^{3L/4} \sin^2(\pi x/L)\,dx$$

$$= \left(\frac{1}{L}\right)\int_{L/4}^{3L/4} \{1 - \cos(2\pi x/L)\}\,dx$$

$$= \left(\frac{1}{L}\right)\left\{\frac{L}{2} - \left(\frac{L}{2\pi}\right)\sin(2\pi x/L)\Big|_{L/4}^{3L/4}\right\}$$

$$= \frac{1}{2} - \left(\frac{1}{2\pi}\right)\{-1 - 1\} = 0.818$$

Notice this is considerably larger than $\frac{1}{2}$, which would be expected for a classical particle that spends equal time in all parts of the well.

Exercise 2 Repeat the calculation of Example 5.7 for a particle in the nth state of the infinite square well, and show that the result approaches the classical value $\frac{1}{2}$ in the limit $n \to \infty$.

5.5 THE FINITE SQUARE WELL (Optional)

The "box" potential is an oversimplification which is never realized in practice. Given sufficient energy, a particle can escape the confines of any well. The potential energy for a more realistic situation—the finite square well—is shown in Figure 5.10, and essentially is that depicted in Figure 5.6b before taking the limit $V \to \infty$. A classical particle with energy E greater than the well height U can penetrate the gaps at $x = 0$ and $x = L$ to enter the outer region. Here it moves freely, but with reduced speed corresponding to a diminished kinetic energy $E - U$.

A classical particle with energy E less than U is permanently bound to the region $0 < x < L$. However, quantum mechanics asserts there is some probability that the particle can be found *outside* this region! That is, the wavefunction generally is nonzero outside the well, and so the probability of finding the particle here also is nonzero. For stationary states, the waveform $\psi(x)$ is found from the time-independent Schrödinger equation. Outside the well where $U(x) = U$, this is

$$\frac{d^2\psi}{dx^2} = \alpha^2\psi(x) \qquad x < 0 \text{ and } x > L$$

where $\alpha^2 = 2m(U - E)/\hbar^2$ is a constant. Since $U > E$, α^2 necessarily is *positive* and the independent solutions to this equation are the *real* exponentials $e^{+\alpha x}$ and $e^{-\alpha x}$. The positive exponential must be rejected in region III where $x > L$ to keep $\psi(x)$ finite as $x \to \infty$; likewise, the negative exponential must be rejected in region I where $x < 0$ to keep $\psi(x)$ finite as $x \to -\infty$. Thus, the exterior wave takes the form

$$\psi(x) = A\,e^{+\alpha x} \qquad \text{for } x < 0$$

$$\psi(x) = B\,e^{-\alpha x} \qquad \text{for } x > L. \tag{5.20}$$

The coefficients A and B are determined by matching this wave smoothly onto the wavefunction in the well interior. Specifically, we require $\psi(x)$ and its first derivative $d\psi/dx$ to be *continuous* at $x = 0$ and again at $x = L$. This can be done only for certain values of E, corresponding to the allowed energies for the bound particle. For these energies, the matching conditions specify the entire waveform except for a multiplicative constant, which then is determined by normalization. Figure 5.11 shows the waveforms and probability densities that result for the three lowest allowed particle energies. Note that in each case the waveforms join smoothly at the boundaries of the potential well.

The fact that ψ is nonzero at the walls *increases* the de Broglie wavelength in the well (compared to that in the infinite well), and this in turn lowers the energy and momentum of the particle. This observation can be used to approximate the allowed

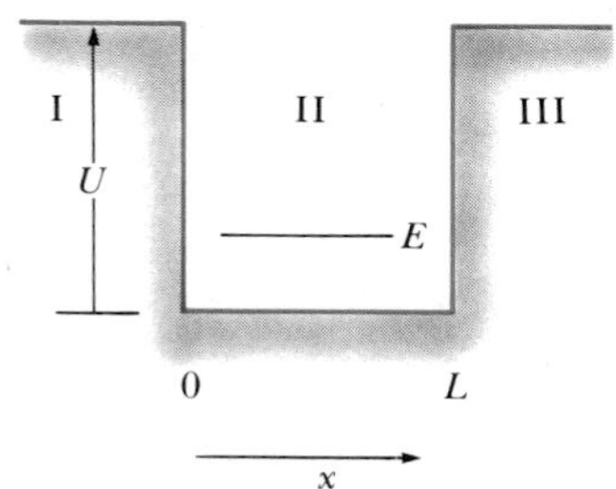

Figure 5.10 Potential energy diagram for a well of finite height U and width L. The energy E of the particle is less than U.

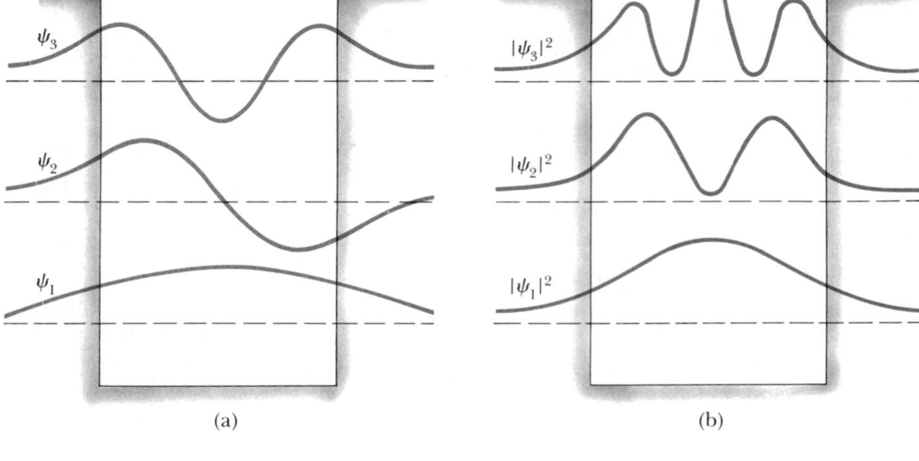

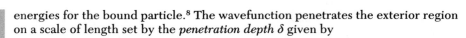

Figure 5.11 (a) Wavefunctions for the lowest three energy states for a particle in a potential well of finite height. (b) Probability densities for the lowest three energy states for a particle in a potential well of finite height.

energies for the bound particle.[8] The wavefunction penetrates the exterior region on a scale of length set by the *penetration depth* δ given by

Penetration depth

$$\delta = \frac{1}{\alpha} = \frac{\hbar}{\sqrt{2m(U-E)}} \tag{5.21}$$

That is, the exterior wave is essentially zero beyond a distance δ on either side of the potential well. If it were truly zero beyond this distance, the allowed energies would be those for an infinite well of length $L + 2\delta$, or (compare Equation 5.17)

Approximate energies for a particle in a well of finite height

$$E_n = \frac{n^2\pi^2\hbar^2}{2m(L+2\delta)^2} \qquad n = 1, 2, \ldots \tag{5.22}$$

The allowed energies for a particle bound to the finite well are given approximately by Equation 5.22 so long as δ is small compared to L. But δ itself is energy-dependent according to Equation 5.21. Thus, Equation 5.22 becomes an *implicit* relation for E which must be solved numerically for a given value of n. The approximation is best for the lowest lying states, and breaks down completely as E approaches U, where δ becomes infinite. From this we infer (correctly) that the number of bound states is limited by the height U of our potential well. Particles with energies E exceeding U are not bound to the well, that is, they may be found with comparable probability in the exterior regions. The case of unbound states will be taken up in the following chapter.

EXAMPLE 5.8 A Bound Electron

Estimate the ground state energy for an electron confined to a potential well of width 0.2 nm and height 100 eV.

Solution: We solve Equations 5.21 and 5.22 together, using an iterative procedure. Since we expect $E \ll U$ (=100 eV), we estimate the decay length δ by first neglecting E to get

$$\delta \approx \frac{\hbar}{\sqrt{2mU}} = \frac{(197.3\ \text{eV}\cdot\text{nm}/c)}{\sqrt{2(511\times 10^3\ \text{eV}/c^2)(100\ \text{eV})}}$$
$$= 0.0195\ \text{nm}$$

[8] This method was reported by S. Garrett in the *American Journal of Physics,* **47:** 195–196, 1979.

Thus, the effective width of the (infinite) well is $L + 2\delta = 0.239$ nm, for which we calculate the ground state energy

$$E = \frac{\pi^2(197.3 \text{ eV}\cdot\text{nm}/c)^2}{2(511 \times 10^3 \text{ eV}/c^2)(0.239 \text{ nm})^2} = 6.58 \text{ eV}.$$

From this E we calculate $U - E = 93.42$ eV and a new decay length

$$\delta \approx \frac{(197.3 \text{ eV}\cdot\text{nm}/c)}{\sqrt{2(511 \times 10^3 \text{ eV}/c^2)(93.42 \text{ eV})}} = 0.0202 \text{ nm}.$$

This, in turn, increases the effective well width to 0.240 nm and lowers the ground state energy to $E = 6.53$ eV. The iterative process is repeated until the desired accuracy is achieved. Since our estimate for E has changed by only 0.05 eV, we may be content with this value. Indeed, another iteration gives the same result to the accuracy reported. This is in good agreement with the actual value, about 6.52 eV for this case.

5.6 THE QUANTUM OSCILLATOR

As a final example of a potential well for which exact results can be obtained, let us examine the problem of a particle subject to a linear restoring force $F = -Kx$. Here x is the displacement of the particle from equilibrium ($x = 0$) and K is the force constant. The corresponding potential energy is given by $U(x) = \frac{1}{2}Kx^2$. The prototype physical system fitting this description is a mass on a spring, but the mathematical description actually applies to any object limited to small excursions about a point of stable equilibrium.

Consider the general potential function sketched in Figure 5.12. The positions a, b, and c all label equilibrium points where the force $F = -dU/dx$ is zero. Further, a and c are examples of *stable* equilibria, while b is *unstable*. The stability of equilibrium is decided by examining the forces in the immediate neighborhood of the equilibrium point. Just to the left of a, for example, $F = -dU/dx$ is positive, that is, the force is directed to the right; conversely, to the right of a the force is directed to the left. Thus, we see that a particle displaced slightly from equilibrium at a encounters a force driving it back to the equilibrium point (restoring force). Similar arguments show that the equilibrium at c also is stable. On the other hand, a particle displaced in either direction from point b experiences a force that drives it further away from equilibrium—an unstable condition. In general, stable and unstable equilibria are marked by potential curves that are concave or convex, respectively, at the equilibrium point. To put it another way, the curvature of $U(x)$ is positive ($d^2U/dx^2 > 0$) at a point of stable equilibrium, and negative ($d^2U/dx^2 < 0$) at a point of unstable equilibrium.

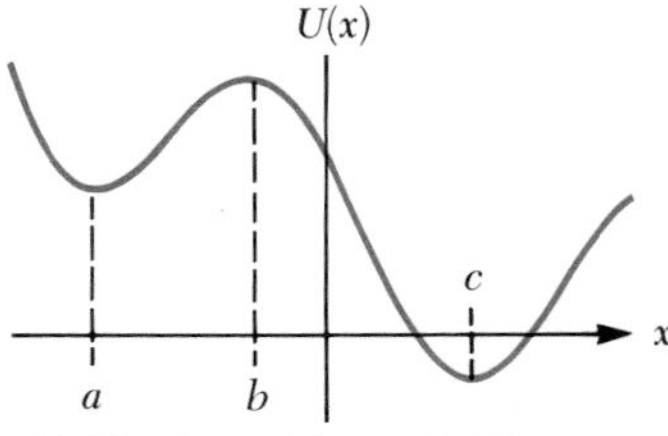

Figure 5.12 A general potential function $U(x)$. The points labeled a and c are positions of stable equilibrium, for which $dU/dx = 0$ and $d^2U/dx^2 > 0$. Point b is a position of unstable equilibrium, for which $dU/dx = 0$ and $d^2U/dx^2 < 0$.

Near a point of stable equilibrium such as a (or c), $U(x)$ can be fit quite well by a parabola:

$$U(x) = U(a) + \tfrac{1}{2}K(x - a)^2 \tag{5.23}$$

Of course, the curvature of this parabola ($=K$) must match that of $U(x)$ at the equilibrium point $x = a$:

$$K = \left.\frac{d^2U}{dx^2}\right|_a \tag{5.24}$$

Further, $U(a)$, the potential energy at equilibrium, may be taken as zero if we agree to make this our energy reference, that is, if we subsequently measure all energies from this level. In the same spirit, the coordinate origin may be

Harmonic approximation to vibrations occurring in real systems

placed at $x = a$, in effect allowing us to set $a = 0$. With $U(a) = 0$ and $a = 0$, Equation 5.23 becomes the spring potential once again; in other words, *a particle limited to small excursions about any stable equilibrium point behaves as if it were attached to a spring with a force constant K prescribed by the curvature of the true potential at equilibrium.* In this way the oscillator becomes a first approximation to the vibrations occurring in many real systems.

The motion of a classical oscillator with mass m is simple harmonic vibration at the angular frequency $\omega = \sqrt{K/m}$. If the particle is removed from equilibrium a distance A and released, it will oscillate between the points $x = -A$ and $x = +A$ (A is the amplitude of vibration), with total energy $E = \frac{1}{2}KA^2$. By changing the initial point of release A, the particle can be given any (nonnegative) energy whatsoever, including zero.

The quantum oscillator is described by the potential energy $U(x) = \frac{1}{2}Kx^2 = \frac{1}{2}m\omega^2x^2$ in Schrödinger's equation. After a little rearrangement we get

$$\frac{d^2\psi}{dx^2} = \frac{2m}{\hbar^2}\left\{\frac{1}{2}m\omega^2x^2 - E\right\}\psi(x) \tag{5.25}$$

as the equation for the stationary states of the oscillator. The mathematical technique for solving this equation is beyond the level of this text. (The exponential and trigonometric forms for ψ employed before will not work here because of the presence of x^2 in the potential.) However, it is instructive to make some intelligent guesses and verify their accuracy by direct substitution. The *ground state* wavefunction should possess the following attributes:

1. ψ should be *symmetric* about the midpoint of the potential well $x = 0$.
2. ψ should be *nodeless*, but approaching zero for $|x|$ large.

Both expectations are derived from our previous experience with the lowest energy states of the infinite and finite square wells, which you might want to review at this time. The symmetry condition (1) requires ψ to be some function of x^2; further, the function must have no zeros (other than at infinity) to meet the nodeless requirement (2). The simplest choice fulfilling both demands is the gaussian form

$$\psi(x) = C_0\, e^{-\alpha x^2} \tag{5.26}$$

for some as-yet-unknown constants C_0 and α. Taking the second derivative of $\psi(x)$ in Equation 5.26 gives (as you should verify)

$$\frac{d^2\psi}{dx^2} = \{4\alpha^2x^2 - 2\alpha\}C_0\, e^{-\alpha x^2} = \{4\alpha^2x^2 - 2\alpha\}\psi(x),$$

which has the same structure as Equation 5.25. Comparing like terms between them, we see that we have a solution provided that both

$$4\alpha^2 = \frac{2m}{\hbar^2}\frac{1}{2}m\omega^2 \quad \text{or} \quad \alpha = \frac{m\omega}{2\hbar} \tag{5.27}$$

and

$$\frac{2mE}{\hbar^2} = 2\alpha = \frac{m\omega}{\hbar} \quad \text{or} \quad E = \tfrac{1}{2}\hbar\omega \tag{5.28}$$

In this way we discover that the oscillator ground state is described by the wavefunction $\psi_0(x) = C_0 \exp(-m\omega x^2/2\hbar)$, and that the energy of this state is $E_0 = \frac{1}{2}\hbar\omega$. The multiplier C_0 is reserved for normalization (see Example 5.9

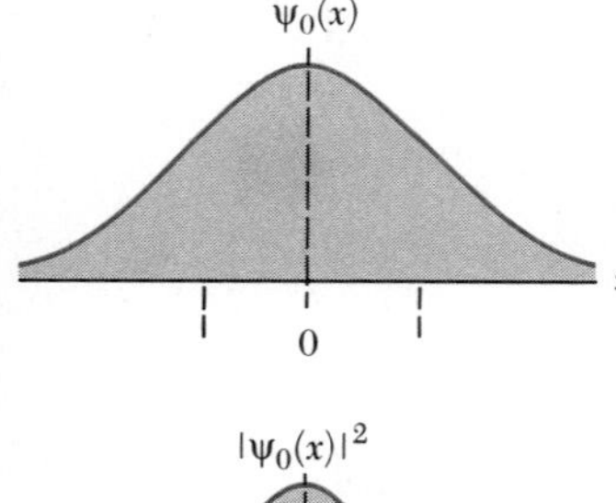

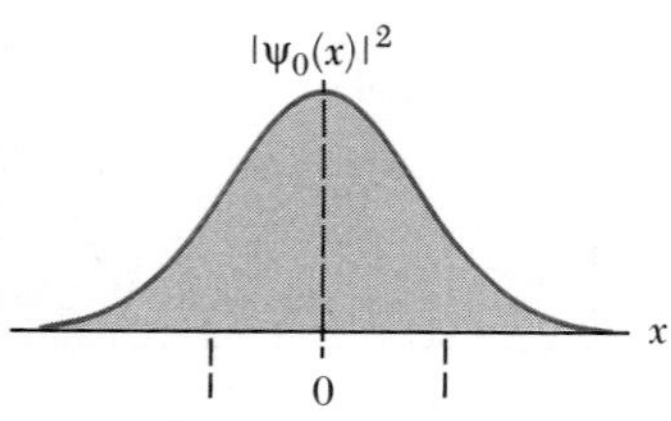

Figure 5.13 (a) Wavefunction for the ground state of a particle in the oscillator potential well. (b) The probability density for the ground state of a particle in the oscillator potential well. The dashed vertical lines mark the limits of vibration for a classical particle with the same energy, $x = \pm A = \pm\sqrt{\hbar/m\omega}$.

below). The ground state wave ψ_0 and associated probability density $|\psi_0|^2$ are illustrated in Figure 5.13. The dashed vertical lines mark the limits of vibration for a classical oscillator with the same energy. Note the considerable penetration of the wave into the classically forbidden regions $x > A$ and $x < -A$. A detailed analysis shows that the particle actually can be found in these nonclassical regions about 16% of the time (see Example 5.11).

EXAMPLE 5.9 Normalizing the Oscillator Ground State Wavefunction

Normalize the oscillator ground state waveform found in the preceding paragraph.

Solution: With $\psi_0(x) = C_0\ e^{-m\omega x^2/2\hbar}$, the integrated probability is

$$\int_{-\infty}^{\infty} |\psi_0(x)|^2\, dx = C_0{}^2 \int_{-\infty}^{\infty} e^{-m\omega x^2/\hbar}\, dx$$

Evaluation of the integral requires advanced techniques. We shall be content here simply to quote the formula

$$\int_{-\infty}^{\infty} e^{-ax^2}\, dx = \sqrt{\frac{\pi}{a}} \qquad a > 0$$

In our case we identify a with $m\omega/\hbar$ and obtain

$$\int_{-\infty}^{\infty} |\psi_0(x)|^2\, dx = C_0{}^2 \sqrt{\frac{\pi\hbar}{m\omega}}$$

Normalization requires this integrated probability to be 1, leading to

$$C_0 = \left(\frac{m\omega}{\pi\hbar}\right)^{1/4}$$

EXAMPLE 5.10 Limits of Vibration for a Classical Oscillator

Obtain the limits of vibration for a classical oscillator having the same total energy as the quantum oscillator in its ground state.

Solution: The ground state energy of the quantum oscillator is $E_0 = \frac{1}{2}\hbar\omega$. At its limits of vibration $x = \pm A$, the classical oscillator has transformed all this energy into elastic potential energy of the spring, given by $\frac{1}{2}KA^2 = \frac{1}{2}m\omega^2A^2$. Therefore

$$\tfrac{1}{2}\hbar\omega = \tfrac{1}{2}m\omega^2A^2 \qquad \text{or} \qquad A = \sqrt{\hbar/m\omega}$$

The classical oscillator vibrates in the interval given by $-A \leq x \leq A$, having insufficient energy to exceed these limits.

EXAMPLE 5.11 The Quantum Oscillator in the Nonclassical Region

Calculate the probability that a quantum oscillator in its ground state will be found outside the range permitted for a classical oscillator with the same energy.

Solution: Since the classical oscillator is confined to the interval $-A \leq x \leq A$, where A is its amplitude of vibration, the question is one of finding the quantum oscillator outside this interval. From the previous example we have $A = \sqrt{\hbar/m\omega}$ for a classical oscillator with energy $\frac{1}{2}\hbar\omega$. The quantum oscillator with this energy is described by the wavefunction $\psi_0(x) = C_0 \exp(-m\omega x^2/2\hbar)$, with $C_0 = (m\omega/\pi\hbar)^{1/4}$ from Example 5.9. The probability in question is found by integrating the probability density $|\psi_0|^2$ in the region beyond the classical limits of vibration, or

$$P = \int_{-\infty}^{-A} |\psi_0|^2\, dx + \int_{A}^{\infty} |\psi_0|^2\, dx$$

From the symmetry of ψ_0, the two integrals contribute equally to P, so

$$P = 2\left(\frac{m\omega}{\pi\hbar}\right)^{1/2} \int_{A}^{\infty} e^{-m\omega x^2/\hbar}\, dx$$

Changing variables from x to $z = \sqrt{m\omega/\hbar}\, x$ and using $A = \sqrt{\hbar/m\omega}$ (corresponding to $z = 1$) leads to

$$P = \frac{2}{\sqrt{\pi}} \int_{1}^{\infty} e^{-z^2}\, dz$$

Expressions of this sort are encountered frequently in probability studies. With the lower limit of integration changed to a variable, say y, the result for P defines the *complementary error function* $\text{erfc}(y)$. Values of the error function may be found in tables. In this way we obtain $P = \text{erfc}(1) = 0.157$, or about 16%.

To obtain excited states of the oscillator, a procedure can be followed similar to that for the ground state. The first excited state should be antisymmetric about the midpoint of the oscillator well ($x = 0$) and display exactly one node. By virtue of the antisymmetry, this node must occur at the origin, so that a suitable trial solution would be $\psi(x) = x \exp(-\alpha x^2)$. Substituting this form into Equation 5.25 yields the same α as before, along with the first excited state energy $E_1 = \frac{3}{2}\hbar\omega$.

Continuing in this manner, we could generate ever higher-lying oscillator states with their respective energies, but the procedure rapidly becomes too laborious to be practical. What is needed is a systematic approach, such as that provided by the method of *power series expansion.*[9] A derivation of this method would take us too far afield, but the result for the allowed oscillator energies is quite simple, and sufficiently important that it be included here:

Energy levels for the harmonic oscillator

$$\boxed{E_n = (n + \tfrac{1}{2})\hbar\omega} \qquad n = 0, 1, 2, \ldots \tag{5.29}$$

The energy level diagram following from Equation 5.29 is given in Figure 5.14. Note the uniform spacing of levels, widely recognized as the hallmark of the oscillator spectrum. The energy difference between adjacent levels is just $\Delta E = \hbar\omega$. In these results we find, for the first time, the quantum justification for Planck's revolutionary hypothesis concerning his *cavity resonators* (see Section 2.2). In deriving his blackbody radiation formula, Planck assumed that these resonators (oscillators), which made up the cavity walls, could possess only those energies that were multiples of $hf = \hbar\omega$. While Planck could not have foreseen the zero-point energy $\hbar\omega/2$, it would make no difference: his resonators still would emit or absorb light energy in the bundles $\Delta E = hf$ necessary to the blackbody spectrum.

The probability densities for some of the oscillator states are plotted in Figure 5.15. The dashed lines, representing the classical probability densities for the same energy, are provided for comparison (see Problem 33 for the calculation of classical probabilities). Note that as n increases, agreement between the classical and quantum probabilities improves, as expected from the correspondence principle.

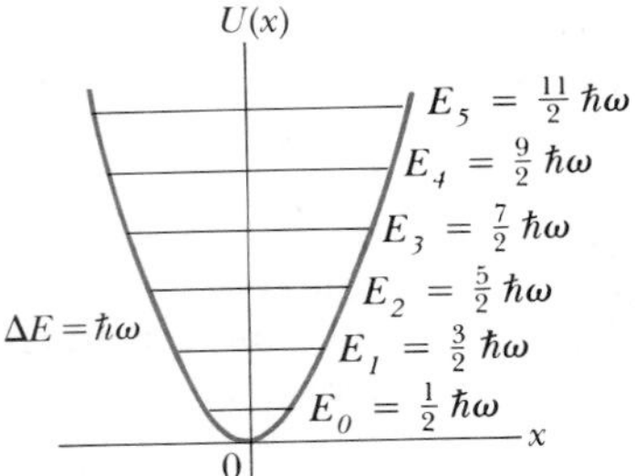

Figure 5.14 Energy level diagram for the quantum oscillator. Note that the levels are equally spaced, with a separation equal to $\hbar\omega$. The ground state energy is equal to E_0.

[9] The method of power series expansion as applied to the problem of the quantum oscillator is developed in any more advanced Quantum Mechanics text. See, for example, E. E. Anderson, *Modern Physics and Quantum Mechanics,* Philadelphia, W. B. Saunders Company, 1971.

EXAMPLE 5.12 Quantization of Vibrational Energy

The energy of a quantum oscillator is restricted to be one of the values $(n + \frac{1}{2})\hbar\omega$. How can this quantization apply to the motion of a mass on a spring, which seemingly can vibrate with any amplitude (energy) whatever?

Solution: The discrete values for the allowed energies of the oscillator would go unnoticed if the spacing between adjacent levels were too small to be detected. At the macroscopic level, a laboratory mass m of, say, 0.01 kg on a spring having force constant $K = 0.1$ N/m (a typical value) would oscillate with angular frequency $\omega = \sqrt{K/m} = 3.16$ rad/s. The corresponding period of vibration is $T = 2\pi/\omega = 1.99$ s. In this case the quantum level spacing is only

$$\Delta E = \hbar\omega = (6.582 \times 10^{-16}\ \text{eV}\cdot\text{s})(3.16\ \text{rad/s})$$
$$= 2.08 \times 10^{-15}\ \text{eV}$$

Such small energies are far below present limits of detection.

At the atomic level, however, much higher frequencies are commonplace. Consider the frequency of revolution for an electron in the first Bohr orbit of hydrogen. From the Bohr theory of Chapter 3, this electron circles the atomic nucleus in an orbit of radius $a_0 = 0.529$ Å with speed $v = (1/137)c$. The angular frequency of revolution is

$$\omega = v/a_0 = (1/137)(3 \times 10^{18}\ \text{Å/s})/(0.529\ \text{Å})$$
$$= 4.14 \times 10^{16}\ \text{rad/s}$$

At such frequencies the quantum of energy $\hbar\omega$ becomes a whopping 27.2 eV!

Exercise 3 The laboratory mass of this example is displaced 1 cm from equilibrium and released from rest. What is the value of the quantum number n for this classical oscillator?

Answer: 4.74×10^{27}.

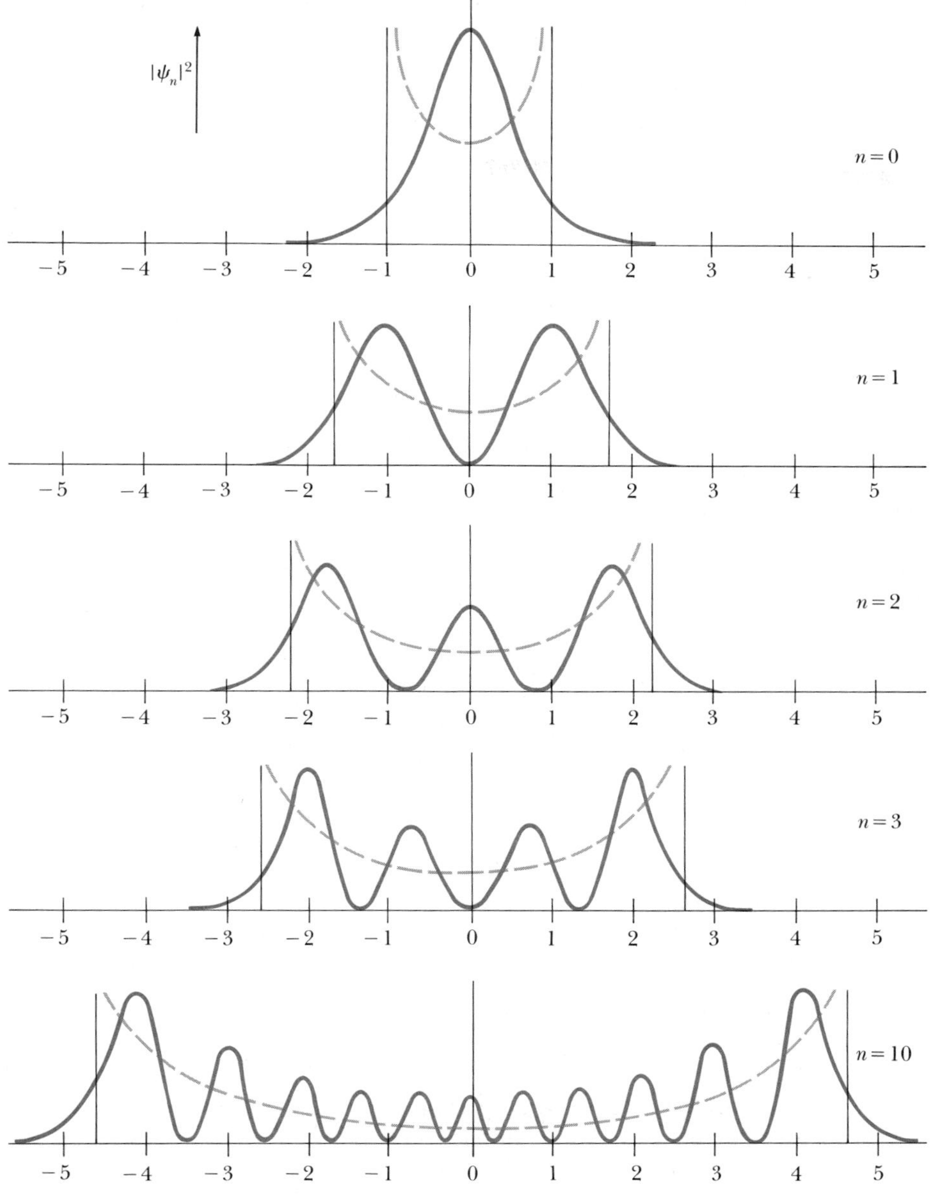

Figure 5.15 Probability densities for a few states of the quantum oscillator. The dashed curves represent the classical probabilities corresponding to the same energies.

5.7 EXPECTATION VALUES

It should be evident by now that two distinct types of dynamical quantities are associated with a given waveform Ψ. One type—like the energy E for the stationary states—is fixed by the quantum number labeling the wave. Therefore, every measurement of this quantity performed on the system described by Ψ yields the *same* value. Quantities such as E are said to be *sharp* to distinguish them from others—like the position x—for which the wavefunction Ψ furnishes only probabilities. We say x is an example of a dynamical quantity which is *fuzzy.* In the following paragraphs we discuss what more can be learned about these "fuzzy" quantities.

Sharp and fuzzy variables

A particle described by the wave Ψ may occupy various places x with probability given by the wave intensity there $|\Psi(x)|^2$. Predictions made this

TABLE 5.1 Hypothetical Data Set for Position of a Particle as Recorded in Repeated Trials

Trial	Position [arbitrary units]	Trial	Position [arbitrary units]	Trial	Position [arbitrary units]
1	$x_1 = 2.5$	7	$x_7 = 8.0$	13	$x_{13} = 4.2$
2	$x_2 = 3.7$	8	$x_8 = 6.4$	14	$x_{14} = 8.8$
3	$x_3 = 1.4$	9	$x_9 = 4.1$	15	$x_{15} = 6.2$
4	$x_4 = 7.9$	10	$x_{10} = 5.4$	16	$x_{16} = 7.1$
5	$x_5 = 6.2$	11	$x_{11} = 7.0$	17	$x_{17} = 5.4$
6	$x_6 = 5.4$	12	$x_{12} = 3.3$	18	$x_{18} = 5.3$

way from Ψ can be tested by making repeated measurements of the particle position. Table 5.1 shows typical results which might be obtained in a hypothetical experiment of this sort. The table consists of 18 entries, each one representing the actual position of the particle recorded in that particular measurement. We see that the entry 5.4 occurs most often (in 3 of the 18 trials); it represents the most probable position based on the data available. The probability associated with this position, again based on the available data, is $3/18 = 0.167$. Of course, these numbers will fluctuate as additional measurements are taken, but they should approach limiting values. The theoretical predictions refer to these limiting values. Obviously, a good test of the theory would require much more data than we have shown.

The information in Table 5.1 also can be used to find the average position of the particle:

$$\bar{x} = (2.5 + 3.7 + 1.4 + \cdots + 5.4 + 5.3)/18 = 5.46$$

This same number can be found in a different way. First, order the table entries by value, starting with the smallest: 1.4, 2.5, 3.3, . . . , 5.4, 6.2, . . . , 8.0, 8.8. Now take each value, multiply by its frequency of occurrence, and sum the results:

$$1.4(1/18) + 2.5(1/18) + \cdots + 5.4(3/18) + 6.2(2/18) + \cdots + 8.8(1/18) = 5.46$$

The two procedures are equivalent, but the latter involves a sum over *ordered values* rather than individual table entries. We may generalize this last expression to include other values for the position of the particle, provided we weight each one by its observed frequency of occurrence (in this case zero). This allows us to write a general prescription to calculate the average particle position from any data set:

$$\bar{x} = \sum xP_x. \tag{5.30}$$

The sum now includes all values of x, each weighted by its frequency or probability of occurrence P_x. Since the possible values of x are distributed continuously over the entire range of real numbers, the sum in Equation 5.30 really should be an integral, and P_x should refer to the probability of finding the particle in the infinitesimal interval dx about the point x; that is, the probability $P_x \rightarrow P(x)\,dx$ where $P(x)$ is the probability density. In quantum mechanics, $P(x) = |\Psi|^2$ and the average value of x, written $\langle x \rangle$, is called the **expectation value.** Then

$$\boxed{\langle x \rangle = \int_{-\infty}^{\infty} x|\Psi(x, t)|^2 \, dx} \qquad (5.31)$$

Average position of a particle

Notice that $\langle x \rangle$ may be a function of time. However, for a stationary state $|\Psi|^2$ is static and $\langle x \rangle$ is independent of t.

In similar fashion we find that the average or expectation value for *any* function of x, say $f(x)$, is

$$\langle f \rangle = \int_{-\infty}^{\infty} f(x)|\Psi|^2 \, dx \qquad (5.32)$$

With $f(x) = U(x)$, Equation 5.32 becomes $\langle U \rangle$, the average potential energy of the particle. With $f(x) = x^2$ the quantum uncertainty in particle position may be found. To see how this is done, we return to Table 5.1 and notice that the entries scatter about the average value. The amount of scatter is measured by the standard deviation σ of the data, defined as

$$\sigma = \sqrt{\frac{\Sigma(x_i - \overline{x})^2}{N}} \qquad (5.33)$$

where N is the number of data points, in this case 18. Writing out the square under the radical gives

$$\frac{\Sigma(x_i)^2}{N} - 2(\overline{x})\frac{\Sigma(x_i)}{N} + (\overline{x})^2 \sum\left(\frac{1}{N}\right) = \overline{(x^2)} - 2(\overline{x})(\overline{x}) + (\overline{x})^2$$
$$= \overline{(x^2)} - (\overline{x})^2$$

and so

$$\sigma = \sqrt{\overline{(x^2)} - (\overline{x})^2}$$

From Equation 5.33 we see that if the standard deviation were zero, all data entries would be identical and equal to the average. In that case the distribution is *sharp;* otherwise the data exhibit some spread (as in Table 5.1) and the standard deviation is greater than zero. In quantum mechanics the standard deviation, written Δx, often is called the *uncertainty* in position. The preceding development implies that the quantum uncertainty in position can be calculated from expectation values as

$$\Delta x = \sqrt{\langle x^2 \rangle - \langle x \rangle^2} \qquad (5.34)$$

The degree to which particle position is fuzzy is given by the magnitude of Δx: Note that the position is *sharp* only if $\Delta x = 0$.

EXAMPLE 5.13 Standard Deviation from Averages
Compute $\overline{(x^2)}$ and the standard deviation for the data given in Table 5.1.

Solution: Squaring the data entries of Table 5.1 and adding the results gives $\Sigma(x_i)^2 = 603.91$. Dividing this by the number of data points, $N = 18$, we find $\overline{(x^2)} = 603.91/18 = 33.55$. Then

$$\sigma = \sqrt{33.55 - (5.46)^2} = 1.93$$

for this case.

EXAMPLE 5.14 Location of a Particle in a Box
Compute the average position $\langle x \rangle$ and the quantum uncertainty in this value, Δx, for the particle in a box, assuming it is in the ground state.

Solution: The possible particle positions within the box are weighted according to the probability density given by $|\Psi|^2 = (2/L)\sin^2(n\pi x/L)$, with $n = 1$ for the ground state. The average position is calculated as

$$\langle x \rangle = \int_{-\infty}^{\infty} x|\Psi|^2 \, dx = (2/L)\int_0^L x \sin^2\left(\frac{\pi x}{L}\right) dx$$

Making the change of variable $\theta = \pi x/L$ (so that $d\theta = \pi\, dx/L$) gives

$$\langle x \rangle = \frac{2L}{\pi^2} \int_0^\pi \theta \sin^2 \theta \, d\theta$$

The integral is evaluated with the help of the trigonometric identity $2 \sin^2 \theta = 1 - \cos 2\theta$, giving

$$\langle x \rangle = \frac{L}{\pi^2} \left\{ \int_0^\pi \theta \, d\theta - \int_0^\pi \theta \cos 2\theta \, d\theta \right\}$$

An integration by parts shows that the second integral vanishes, while the first integrates to $\pi^2/2$. Thus, $\langle x \rangle = L/2$ as expected, since there is equal probability of finding the particle in the left half or the right half of the box.

$\langle x^2 \rangle$ is computed in much the same way, but with an extra factor of x in the integrand. After changing variables to $\theta = \pi x/L$, we get

$$\langle x^2 \rangle = \frac{L^2}{\pi^3} \left\{ \int_0^\pi \theta^2 \, d\theta - \int_0^\pi \theta^2 \cos 2\theta \, d\theta \right\}$$

The first integral evaluates to $\pi^3/3$; the second may be integrated by parts twice to get

$$\int_0^\pi \theta^2 \cos 2\theta \, d\theta = -\int_0^\pi \theta \sin 2\theta \, d\theta = \frac{1}{2} \theta \cos 2\theta \Big|_0^\pi = \pi/2$$

Then

$$\langle x^2 \rangle = \frac{L^2}{\pi^3} \left(\frac{\pi^3}{3} - \frac{\pi}{2} \right) = \frac{L^2}{3} - \frac{L^2}{2\pi^2}$$

Finally, the uncertainty in position for this particle is

$$\Delta x = \sqrt{\langle x^2 \rangle - \langle x \rangle^2} = L \sqrt{\frac{1}{3} - \frac{1}{2\pi^2} - \frac{1}{4}} = 0.181L$$

This is an appreciable figure, amounting to nearly one fifth the size of the box. Consequently, the whereabouts of such a particle are largely unknown. With some confidence, we may assert only that the particle is likely to be in the range $L/2 \pm 0.181L$.

Finally, notice that none of these results depend upon the time, since in a stationary state t enters only through the exponential factor $e^{-i\omega t}$ which cancels when Ψ is combined with Ψ^* in the calculation of averages. Thus, in a stationary state, it is generally true that *all averages, as well as probabilities, are time-independent.*

We have learned how to predict the average position of a particle $\langle x \rangle$, the uncertainty in this position Δx, the average potential energy of the particle $\langle U \rangle$, and so on. But what about the average momentum $\langle p \rangle$ of the particle, or its average kinetic energy $\langle K \rangle$? These could be calculated if $p(x)$, the momentum as a function of x, were known. In classical mechanics, $p(x)$ may be obtained from the equation for the classical path $x(t)$. Differentiating this function once gives the velocity $v(t)$. Then inverting $x(t)$ to get t as a function of x, and substituting this result into $v(t)$, gives $v(x)$ and the desired relation $p(x) = mv(x)$. In quantum mechanics, however, *x and t are independent variables—there is no path,* nor any function connecting p with x! If there were, then p could be found from x using $p(x)$, and both x and p would be known precisely, in violation of the uncertainty principle.

To obtain $\langle p \rangle$ we must use a different approach: we identify the time derivative of the average particle position with the average velocity of the particle. After multiplication by m, this gives the average momentum $\langle p \rangle$:

Average momentum of a particle

$$\langle p \rangle = m \frac{d\langle x \rangle}{dt} \tag{5.35}$$

Equation 5.35 is another of our first principles; it cannot be derived from anything we have said previously. When applied to macroscopic objects where the quantum uncertainties in position and momentum are small, the averages $\langle x \rangle$ and $\langle p \rangle$ become indistinguishable from "the" position and "the" momentum of the object, and Equation 5.35 reduces to the classical definition of momentum. From this viewpoint, we may regard Equation 5.26 as the defining relation for momentum at the microscopic level.

Another expression for $\langle p \rangle$ follows from Equation 5.35 by substituting $\langle x \rangle$ from Equation 5.31 and differentiating under the integral sign. Using Schrödinger's equation to eliminate time derivatives of Ψ and its conjugate Ψ^* gives (after much manipulation!)

$$\langle p \rangle = \int_{-\infty}^{\infty} \Psi^* \left(\frac{\hbar}{i}\right) \frac{\partial \Psi}{\partial x}\, dx \tag{5.36}$$

Exercise 4 Show that $\langle p \rangle = 0$ for *any* state of a particle in a box.

5.8 OBSERVABLES AND OPERATORS (Optional)

An **observable** *is any particle property that can be measured.* The position and momentum of a particle are observables, as are its kinetic and potential energies. Quantum mechanics associates an *operator* with each of these observables. Using this operator, one can calculate the average value of the corresponding observable. An operator here refers to an operation to be performed on whatever function follows it (the *operand*). In this language a constant c becomes an operator, whose meaning is understood by supplying any function $f(x)$ to obtain $cf(x)$. Here the operator c means "multiplication by the constant c." A more complicated operator is d/dx which, after supplying an operand to get df/dx, clearly means "take the derivative with respect to x." Still another example is $(d/dx)^2 = (d/dx)(d/dx)$. Supplying the operand $f(x)$ gives $(d/dx)^2 f(x) = (d/dx)(df/dx) = d^2f/dx^2$. Hence, $(d/dx)^2$ means "take the second derivative with respect to x, that is, take the indicated derivative twice."

The operator concept is useful in quantum mechanics because all expectation values we have encountered so far can be written in the same general form, namely,

$$\langle Q \rangle = \int_{-\infty}^{\infty} \Psi^* [Q] \Psi \, dx \tag{5.37}$$

Operators in quantum mechanics

In this expression, Q is the observable and $[Q]$ is the associated operator. The *order* of terms in Equation 5.27 is important; it indicates that the operand for $[Q]$ always is Ψ. Comparing the general form with that for $\langle p \rangle$ in Equation 5.36 shows that the momentum operator is $[p] = (\hbar/i)(\partial/\partial x)$. Similarly, writing $x|\Psi|^2 = \Psi^* x \Psi$ in Equation 5.31 implies that the operator for position is $[x] = x$. From $[x]$ and $[p]$ the operator for any other observable can be found. For instance, the operator for x^2 is just $[x^2] = [x]^2 = x^2$. For that matter, the operator for potential energy is simply $[U] = U([x]) = U(x)$, meaning that average potential energy is computed as

$$\langle U \rangle = \int_{-\infty}^{\infty} \Psi^* [U] \Psi \, dx = \int_{-\infty}^{\infty} \Psi^* U(x) \Psi \, dx.$$

Still another example is the kinetic energy K. Classically, K is a function of p: $K = p^2/2m$. Then the kinetic energy operator is $[K] = ([p])^2/2m = (-\hbar^2/2m)\partial^2/\partial x^2$, and average kinetic energy is found from

$$\langle K \rangle = \int_{-\infty}^{\infty} \Psi^* [K] \Psi \, dx = \int_{-\infty}^{\infty} \Psi^* \left(-\frac{\hbar^2}{2m}\frac{\partial^2 \Psi}{\partial x^2}\right) dx.$$

To find the average total energy for a particle, we sum the average kinetic and potential energies to get

$$\langle E \rangle = \langle K \rangle + \langle U \rangle = \int_{-\infty}^{\infty} \Psi^* \left\{-\frac{\hbar^2}{2m}\frac{\partial^2}{\partial x^2} + U(x)\right\} \Psi \, dx \tag{5.38}$$

The form of this result suggests that the term in the braces is the operator for total energy. This operator is called the **hamiltonian,** symbolized by $[H]$

$$[H] = -\frac{\hbar^2}{2m}\frac{\partial^2}{\partial x^2} + U(x) \tag{5.39}$$

The designation $[E]$ is reserved for another operator, which arises as follows: Inspection of Schrödinger's equation (Equation 5.10) shows that it can be written

TABLE 5.2 Common Observables and Associated Operators

Observable	Symbol	Associated Operator
position	x	x
momentum	p	$\frac{\hbar}{i}\frac{\partial}{\partial x}$
potential energy	U	$U(x)$
kinetic energy	K	$-\frac{\hbar^2}{2m}\frac{\partial^2}{\partial x^2}$
hamiltonian	H	$-\frac{\hbar^2}{2m}\frac{\partial^2}{\partial x^2}+U(x)$
total energy	E	$i\hbar\frac{\partial}{\partial t}$

neatly as $[H]\Psi = i\hbar\partial\Psi/\partial t$. Using this in Equation 5.38 gives an equivalent expression for $\langle E\rangle$, and leads to the identification of the *energy operator*

$$[E] = i\hbar\frac{\partial}{\partial t}. \tag{5.40}$$

Notice that $[H]$ is an operation involving only the spatial coordinate x, while $[E]$ depends only on the time t. That is, $[H]$ and $[E]$ really are two different operators, but they produce identical results when applied to any solution of Schrödinger's equation. This is because the LHS of Schrödinger's equation is simply $[H]\Psi$, while the RHS is none other than $[E]\Psi$ (cf. Equation 5.10)! Table 5.2 summarizes the observables we have discussed and their associated operators.

Quantum Uncertainty and the Eigenvalue Property

In Section 5.7 we showed how the quantum uncertainty in position Δx could be found from the expectation values $\langle x^2\rangle$ and $\langle x\rangle$. But the argument given there applies to *any* observable, that is, the quantum uncertainty ΔQ for any observable Q is calculated as

Quantum uncertainty for any observable Q

$$\boxed{\Delta Q = \sqrt{\langle Q^2\rangle - \langle Q\rangle^2}} \tag{5.41}$$

Again, if $\Delta Q = 0$, Q is said to be a *sharp* observable, and all measurements of Q yield the *same value*. More often, however, $\Delta Q > 0$ and repeated measurements reveal a distribution of values—as in Table 5.1 for the observable x. In such cases, we say the observable is *fuzzy*, suggesting that, prior to actual measurement, the particle cannot be said to possess a unique value of Q.

In classical physics all observables are sharp. The extent to which sharp observables can be specified in quantum physics is limited by uncertainty principles, such as

$$\Delta x\,\Delta p \geq \hbar/2. \tag{5.42}$$

The uncertainties here are to be calculated from Equation 5.41. Equation 5.42 says that, no matter what the state of the particle, the spread in distributions obtained in measurements of x and of p will be inversely related; when one is small, the other will be large. Alternatively, if the position of the particle is quite "fuzzy," its momentum will be relatively "sharp," and vice versa. The degree to which both may be simultaneously sharp is limited by the size of h. The incredibly small value of h in SI units is an indication that quantum ideas are unnecessary at the macroscopic level.

Despite restrictions imposed by uncertainty principles, some observables in quantum physics may still be sharp. The energy E of all stationary states is one example. In the free particle plane waves of Section 5.2 we have another: the plane wave with wavenumber k

$$\Psi_k(x, t) = e^{i(kx-\omega t)}$$

describes a particle with momentum $p = \hbar k$. Evidently, momentum is a sharp observable for this wavefunction. We find that the action of the momentum operator in this instance is especially simple:

$$[p]\Psi_k(x, t) = \left(\frac{\hbar}{i}\frac{\partial}{\partial x}\right) e^{i(kx-\omega t)} = \hbar k \Psi_k(x, t),$$

that is, *the operation [p] returns the original function multiplied by a constant.* This is an example of an *eigenvalue problem* for the operator $[p]$.[10] The wavefunction Ψ_k is the **eigenfunction** and the constant, in this case $\hbar k$, is the **eigenvalue.** Notice that the eigenvalue is just the sharp value of particle momentum for this wave. This connection between sharp observables and eigenvalues is a general one: *for an observable Q to be sharp, the wavefunction must be an eigenfunction of the operator for Q. Further, the sharp value for Q in this state is the eigenvalue.* In this way the eigenvalue property can serve as a simple *test* for sharp observables, as the following examples illustrate.

Eigenfunctions and eigenvalues

EXAMPLE 5.15 Plane Waves and Sharp Observables

Use the eigenvalue test to show that the plane wave $\Psi_k(x, t) = e^{i(kx-\omega t)}$ is one for which total energy is a sharp observable. What value does the energy take in this case?

Solution: To decide the issue we examine the action of the energy operator $[E]$ on the candidate function $e^{i(kx-\omega t)}$. Since taking a derivative with respect to t of this function is equivalent to multiplying the function by $-i\omega$, we have

$$[E]e^{i(kx-\omega t)} = \left(i\hbar\frac{\partial}{\partial t}\right) e^{i(kx-\omega t)} = \hbar\omega\, e^{i(kx-\omega t)}$$

showing that $e^{i(kx-\omega t)}$ is an eigenfunction of the energy operator $[E]$ and the eigenvalue is $\hbar\omega$. Thus, energy is a sharp observable and has the value $\hbar\omega$ in this state.

It is instructive to compare this result with the outcome found by using the other energy operator $[H]$. The hamiltonian for a free particle is simply the kinetic energy operator $[K]$, since the potential energy is zero in this case. Then

$$[H]e^{i(kx-\omega t)} = \left(-\frac{\hbar^2}{2m}\frac{\partial^2}{\partial x^2}\right) e^{i(kx-\omega t)} = \left(-\frac{\hbar^2}{2m}\right)(ik)^2 e^{i(kx-\omega t)}$$

Again, the operation returns the original function with a multiplier, so that $e^{i(kx-\omega t)}$ also is an eigenfunction of $[H]$. The eigenvalue in this case is $\hbar^2k^2/2m$, which also must be the sharp value of particle energy. The equivalence with $\hbar\omega$ follows from the dispersion relation for free particles (cf. Footnote 1).

Exercise 5 Show that total energy is a sharp observable for *any* stationary state.

[10] The eigenvalue problem for any operator $[Q]$ is $[Q]\psi = q\psi$, that is, the result of the operation $[Q]$ on some function ψ is simply to return a multiple q of the same function. This is possible only for certain special functions ψ, the *eigenfunctions*, and then only for certain special values of q, the *eigenvalues*. Generally, $[Q]$ is known; the eigenfunctions and eigenvalues are found by imposing the eigenvalue condition.

EXAMPLE 5.16 Sharp Observables for a Particle in a Box
Are the stationary states of the infinite square well eigenfunctions of $[p]$? of $[p]^2$? If so, what are the eigenvalues? Discuss the implications of these results.

Solution: The candidate function in this case is any one of the square well wavefunctions $\Psi(x, t) = \sqrt{2/L}\,\sin(n\pi x/L)\, e^{-iE_nt/\hbar}$. Since the first derivative gives $(d/dx)\sin(n\pi x/L) = (n\pi/L)\cos(n\pi x/L)$, we see at once that the operation $[p]$ will *not* return the original function Ψ, and so these are not eigenfunctions of the momentum operator. However, they *are* eigenfunctions of $[p]^2$. In particular, we have $(d^2/dx^2)\sin(n\pi x/L) = -(n\pi/L)^2 \sin(n\pi x/L)$, so that

$$[p]^2\Psi(x, t) = -(\hbar/i)^2(n\pi/L)^2\Psi(x, t) = (n\pi\hbar/L)^2\Psi(x, t)$$

The eigenvalue is the multiplier $(n\pi\hbar/L)^2$. Thus, the squared-momentum (or magnitude of momentum) is sharp for such states, and repeated measurements of p^2 (or $|p|$) for the state labeled by n will give identical results equal to $(n\pi\hbar/L)^2$ (or $n\pi\hbar/L$). By contrast, the momentum itself is not sharp, meaning that different values for p will be obtained in successive measurements. In particular, it is the sign or direction of momentum that is fuzzy, consistent with the classical notion of a particle bouncing back and forth between the walls of the "box."

5.9 SUMMARY

In quantum mechanics, matter waves (or de Broglie waves) are represented by a wavefunction Ψ. The probability that a particle constrained to move along the x axis will be found in an interval dx at time t is given by $|\Psi|^2\, dx$. These probabilities summed over all values of x must be 1 (certainty). That is,

$$\int_{-\infty}^{\infty} |\Psi|^2\, dx = 1 \tag{5.2}$$

This is called the **normalization condition.** Furthermore, the probability that the particle will be found in any interval $a \le x \le b$ is obtained by integrating the **probability density** $|\Psi|^2$ over this interval.

Aside from furnishing probabilities, the wavefunction can be used to find the average value or **expectation value** of any dynamical quantity. The average position of a particle at any time t is given as

$$\langle x \rangle = \int_{-\infty}^{\infty} \Psi^* x \Psi\, dx. \tag{5.31}$$

In general, the average value of any observable Q at time t is found from

$$\langle Q \rangle = \int_{-\infty}^{\infty} \Psi^*[Q]\Psi\, dx, \tag{5.37}$$

where $[Q]$ is the associated operator. The operator for position is just $[x] = x$, while that for particle momentum is $[p] = (\hbar/i)\partial/\partial x$.

The wavefunction Ψ must satisfy the **Schrödinger equation**

$$-\frac{\hbar^2}{2m}\frac{\partial^2\Psi}{\partial x^2} + U(x)\Psi(x, t) = i\hbar\frac{\partial\Psi}{\partial t} \tag{5.10}$$

Separable solutions to this equation, called **stationary states,** are $\Psi(x, t) = \psi(x)e^{-i\omega t}$, with $\psi(x)$ a time-independent wave satisfying the **time-independent Schrödinger equation**

$$-\frac{\hbar^2}{2m}\frac{d^2\psi}{dx^2} + U(x)\psi(x) = E\psi(x) \tag{5.13}$$

The approach of quantum mechanics is to solve Equation (5.13) for ψ and E, given the potential energy $U(x)$ for the system. In doing so, we must require

1. that $\psi(x)$ be continuous,
2. that $\psi(x)$ be finite for all x including $x = \pm\infty$,
3. that $\psi(x)$ be single-valued, and
4. that $d\psi/dx$ be continuous wherever $U(x)$ is finite.

Explicit solutions to Schrödinger's equation can be found for several potentials of special importance. For a free particle the stationary states are the **plane waves** $\psi(x) = e^{ikx}$ of wavenumber k and energy $E = \hbar^2k^2/2m$. The particle momentum in such states is $p = \hbar k$, but the location of the particle is completely unknown. A free particle known to be in some range Δx is described not by a plane wave, but by a **wave packet** or **group,** formed from a superposition of plane waves. The momentum of such a particle is not known precisely, but only to some accuracy Δp which is related to Δx by the **uncertainty principle**

$$\Delta x \, \Delta p \geq \hbar/2$$

For a particle confined to a one-dimensional box of length L, the stationary state waves are those for which an integral number of half-wavelengths can be fit inside, that is $L = n\lambda/2$. In this case the energies are **quantized** as

$$E_n = \frac{n^2\pi^2\hbar^2}{2mL^2} \qquad n = 1, 2, 3, \ldots \tag{5.17}$$

and the wavefunctions within the box are given by

$$\psi_n(x) = \sqrt{\frac{2}{L}} \sin\left(\frac{n\pi x}{L}\right) \qquad n = 1, 2, 3, \ldots \tag{5.18}$$

For the **harmonic oscillator** the potential energy function is $U(x) = \frac{1}{2}m\omega^2x^2$, and the total particle energy is quantized according to the relation

$$E_n = \left(n + \frac{1}{2}\right)\hbar\omega \qquad n = 0, 1, 2, \ldots \tag{5.29}$$

The lowest energy is $E_0 = \frac{1}{2}\hbar\omega$; the separation between adjacent energy levels is uniform and equal to $\hbar\omega$. The wavefunction for the oscillator ground state is

$$\psi_0(x) = C_0\, e^{-\alpha x^2}, \tag{5.26}$$

where $\alpha = m\omega/2\hbar$ and C_0 is a normalizing constant. The oscillator results apply to any system executing small-amplitude vibrations about a point of stable equilibrium. The effective spring constant in the general case is

$$K = m\omega^2 = \left.\frac{d^2U}{dx^2}\right|_a$$

with the derivative of the potential evaluated at the equilibrium point a.

The stationary state waves for any potential share the following attributes:

1. their time dependence is $e^{-i\omega t}$,
2. they yield probabilities which are time-independent,
3. all average values obtained from stationary states are time-independent, and
4. the energy in any stationary state is a **sharp observable,** that is, repeated measurements of particle energy performed on identical systems always yield the same result, $E = \hbar\omega$.

For other observables, such as position, repeated measurements usually yield different results. We say these observables are **fuzzy.** Their inherent "fuzziness" is reflected by the spread in results about the average value, as measured by the standard deviation or **uncertainty.** The uncertainty in any observable Q can be calculated from expectation values as

$$\Delta Q = \sqrt{\langle Q^2 \rangle - \langle Q \rangle^2} \tag{5.41}$$

SUGGESTIONS FOR FURTHER READING

1. L. de Broglie, *New Perspectives in Physics,* New York, Basic Books Inc., 1962. A series of essays by Louis de Broglie on various aspects of theoretical physics and on the history and philosophy of science. A large part of this work is devoted to de Broglie's developing attitudes toward the interpretation of wave mechanics and wave-particle duality.
2. For an in-depth look at the problems of interpretation and measurement surrounding the formalism of quantum theory see M. Jammer, *The Philosophy of Quantum Mechanics,* New York, John Wiley Sons, Inc., 1974.
3. A concise, solid introduction to the basic principles of quantum physics with applications may be found in Chapter 41 of R. Serway's *Physics for Scientists and Engineers with Modern Physics* (2nd ed.), Philadelphia, Saunders College Publishing, 1986.
4. A novel but delightfully refreshing exposition of quantum theory is presented by R. Feynman, R. Leighton, and M. Sands in *The Feynman Lectures on Physics Vol. III, Modern Physics,* Reading, Mass, Addison-Wesley Publishing Co., 1965. This work is more advanced and sophisticated than other introductory texts in the field, though still somewhat below the intermediate level.

QUESTIONS

1. The probability density at certain points for a particle in a box is zero, as seen in Figure 5.9. Does this imply that the particle cannot move across these points? Explain.
2. Discuss the relation between the zero-point energy and the uncertainty principle.
3. Consider a square well with one finite wall and one infinite wall. Compare the energy and momentum of a particle trapped in this well to the energy and momentum of an identical particle trapped in an infinite well with the same width.
4. Sketch the wavefunction and probability density for the infinite square well in the case where n is large. Explain how these results illustrate the correspondence principle.
5. Explain why a wave packet moves with the group velocity rather than with the phase velocity.
6. Consider the relationship between the average value of a quantity and its most probable value. For what kinds of probability distributions are these the same? Explain.
7. According to Section 5.2, a free particle can be represented by any number of waveforms, depending on the values chosen for the coefficients $a(k)$. What is the source of this ambiguity, and how is it resolved?
8. Since the Schrödinger equation can be formulated in terms of operators as $[H]\Psi = [E]\Psi$, is it incorrect to conclude from this the operator equivalence $[H] = [E]$?
9. For a particle in a box, the squared-momentum p^2 is a sharp observable, but the momentum itself is fuzzy. Explain how this can be so, and how it relates to the classical motion of such a particle.
10. A common fallacy among beginning students of quantum mechanics is that a stationary state represents a condition of no motion. Give a simple example for which this is certainly incorrect. What, then, is "stationary" about a stationary state?

PROBLEMS

1. A free electron has a wave function

$$\psi(x) = A \sin(5 \times 10^{10}\, x)$$

where x is measured in m. Find (a) the electron's de Broglie wavelength, (b) the electron's momentum, and (c) the electron's energy in eV.

2. The wave function of a particle is given by

$$\psi(x) = A \cos(kx) + B \sin(kx)$$

where A, B, and k are constants. Show that ψ is a solution of the Schrödinger equation (Eq. 5.13), assuming the particle is free ($U = 0$), and find the corresponding energy E of the particle.

3. In a region of space, a particle with zero energy has a wave function given by

$$\psi(x) = Axe^{-x^2/L^2}$$

(a) Find the potential energy U as a function of x.
(b) Make a sketch of $U(x)$ versus x.

4. A bead of mass 5 g slides freely on a wire 20 cm long. Treating this system as a particle in a one-dimensional box, calculate the value of n corresponding to the state of the bead if it is moving at a speed of 0.1 nm per year (that is, apparently at rest).
5. Show that allowing the state $n = 0$ for a particle in a one-dimensional box violates the uncertainty principle, $\Delta p\, \Delta x \geq \hbar/2$.
6. The nuclear potential that binds protons and neutrons in the nucleus of an atom is often approximated by a square well. Imagine a proton confined in an infinite square well of length 10^{-5} nm, a typical nuclear diameter. Calculate the wavelength and energy associated with the photon that is emitted when the proton undergoes a transition from the first excited state ($n = 2$) to the ground state ($n = 1$). In what region of the electromagnetic spectrum does this wavelength belong?
7. Sketch the wave function $\psi(x)$ and the probability density $|\psi(x)|^2$ for the $n = 4$ state of a particle in a *finite* potential well.
8. An electron is contained in a one-dimensional box of width 0.1 nm. (a) Draw an energy level diagram for the electron for levels up to $n = 4$. (b) Find the wavelengths of *all* photons that can be emitted by the electron in making transitions that would eventually get it from the $n = 4$ state to the $n = 1$ state.
9. Consider a particle moving in a one-dimensional box with walls at $x = -L/2$ and $x = L/2$. (a) Write the wave functions and probability densities for the states $n = 1, n = 2$, and $n = 3$. (b) Sketch the wave functions and probability densities. (*Hint:* Make an analogy to the case of a particle in a box with walls at $x = 0$ and $x = L$.)
10. A ruby laser emits light of wavelength 694.3 nm. If this light is due to transitions from the $n = 2$ state to the $n = 1$ state of an electron in a box, find the width of the box.
11. A proton is confined to moving in a one-dimensional box of width 0.2 nm. (a) Find the lowest possible energy of the proton. (b) What is the lowest possible energy of an electron confined to the same box? (c) How do you account for the large difference in your results for (a) and (b)?
12. A particle of mass m_0 is placed in a one-dimensional box of length L. The box is so small that the particle's motion is *relativistic*, so that $E = p^2/2m_0$ is *not valid.* (a) Derive an expression for the energy levels of the particle. (b) If the particle is an electron in a box of length $L = 10^{-12}$ m, find its lowest possible kinetic energy. By what percent is the nonrelativistic formula for the energy in error?
13. Consider a "crystal" consisting of two nuclei and two electrons as shown in Figure 5.16. (a) Taking into account all the pairs of interactions, find the potential energy of the system as a function of d. (b) Assuming the electrons to be restricted to a one-dimensional box of length $3d$, find the minimum kinetic energy of the two electrons. (c) Find the value of d for which the total energy is a *minimum.* (d) Compare this value of d with the spacing of atoms in lithium, which has a density of 0.53 g/cm³ and an atomic weight of 7. (This type of calculation can be used to estimate the densities of crystals and certain stars.)

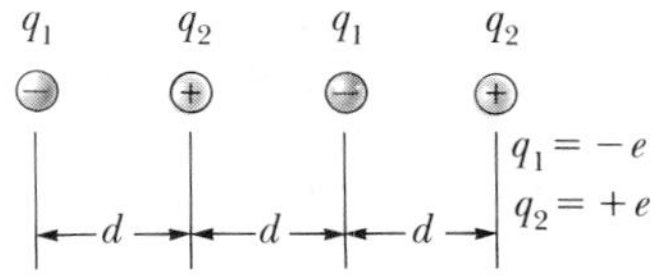

Figure 5.16 (Problem 13)

14. An electron is trapped in an infinitely deep potential well 0.3 nm in width. (a) If the electron is in its ground state, what is the probability of finding it within 0.1 nm of the left-hand wall? (b) Repeat (a) for an electron in the 99th energy state above the ground state. (c) Are your answers consistent with the uncertainty principle?
15. An electron is trapped at a defect in a crystal. The defect may be modeled as a one-dimensional, rigid-walled box of width 1 nm. (a) Sketch the wave functions and probability densities for the $n = 1$ and $n = 2$ states. (b) For the $n = 1$ state, find the probability of finding the electron between $x_1 = 0.15$ nm and $x_2 = 0.35$ nm, where $x = 0$ is the left side of the box. (c) Repeat (b) for the $n = 2$ state. (d) Calculate the energies in eV of the $n = 1$ and $n = 2$ states.
16. A particle is described by the wave function

$$\psi(x) = \begin{cases} A\cos\left(\dfrac{2\pi x}{L}\right) & \text{for } -\dfrac{L}{4} \leq x \leq \dfrac{L}{4} \\ 0 & \text{otherwise} \end{cases}$$

(a) Determine the normalization constant A. (b) What is the probability that the particle will be found between $x = 0$ and $x = L/8$ if a measurement of its position is made?

17. Find the points of maximum and minimum probability density for the nth state of a particle in a one-dimensional box. Check your result for the $n = 2$ state.
18. For any eigenfunction ψ_n of the infinite square well, show that $\langle x \rangle = L/2$ and that

$$\langle x^2 \rangle = \frac{L^2}{3} - \frac{L^2}{2(n\pi)^2}$$

where L is the well dimension.

19. A 1-g marble is constrained to roll inside a tube of length $L = 1$ cm. The tube is capped at both ends. Modeling this as a one-dimensional infinite square well, find the value of the quantum number n if the marble is initially given an energy of 1 mJ. Calculate the *excitation energy* required to promote the marble to the next available energy state.

20. *Spreading of a Gaussian Wave Packet.* The Gaussian wave packet $\Psi(x, 0)$ of Example 5.3 is built out of plane waves according to the spectral function $a(k) = (C\alpha/\sqrt{\pi})\exp(-\alpha^2k^2)$. Calculate $\Psi(x, t)$ for this packet and describe its evolution.

21. *Classical Probabilities.* (a) Show that the classical probability density describing a particle in an infinite square well of dimension L is $P_c(x) = 1/L$. (Hint: The classical probability for finding a particle in dx—$P_c(x)\,dx$—is proportional to the *time* the particle spends in this interval.) (b) Using $P_c(x)$, determine the *classical* averages $\langle x \rangle$ and $\langle x^2 \rangle$ for a particle confined to the well, and compare with the quantum results found in Example 5.14. Discuss your findings in light of the correspondence principle.

22. An electron is described by the wavefunction

$$\psi(x) = \begin{cases} 0 & \text{for } x < 0 \\ Ce^{-x}(1 - e^{-x}) & \text{for } x > 0 \end{cases}$$

where x is in nm and C is a constant. (a) Find the value of C that normalizes ψ. (b) Where is the electron most likely to be found, that is, for what value of x is the probability for finding the electron largest? (c) Calculate $\langle x \rangle$ for this electron and compare your result with its most likely position. Comment on any differences you find.

23. Of the functions graphed in Figure 5.17, which are candidates for the Schrödinger wavefunction of an actual physical system? For those that are not, state why they fail to qualify.

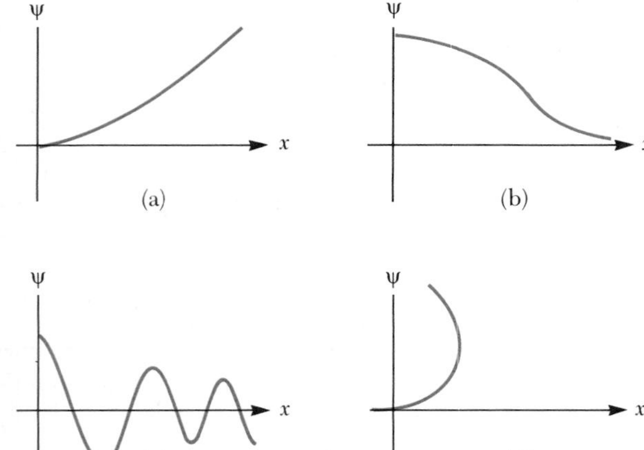

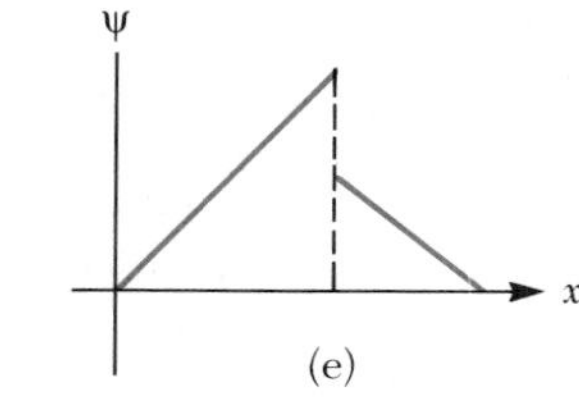

Figure 5.17 (Problem 23)

*24. Consider a particle with energy E bound to a *finite* square well of height U and width $2L$ situated on $-L \le x \le +L$. Since the potential energy is symmetric about the midpoint of the well, the stationary state waves will be either symmetric or antisymmetric about this point. (a) Show that for $E < U$, the conditions for smooth joining of the interior and exterior waves lead to the following equation for the allowed energies of the symmetric waves:

$$k \tan kL = \alpha \qquad \text{(symmetric case)}$$

where $\alpha = \sqrt{(2m/\hbar^2)(U - E)}$, and $k = \sqrt{2mE/\hbar^2}$ is the wave number of oscillation in the interior. (b) Show that the energy condition found in (a) can be rewritten as

$$k \sec kL = \frac{\sqrt{2mU}}{\hbar}$$

Apply the result in this form to an electron trapped at a defect site in a crystal, modeling the defect as a square well of height 5 eV and width 0.2 nm. Write a simple computer program to find the ground state energy for the electron. Give your answer accurate to ± 0.001 eV.

*25. Consider a square well having an infinite wall at $x = 0$ and a wall of height U at $x = L$ (Fig. 5.18). For the case $E < U$, obtain solutions to Schrödinger's equation inside the well ($0 \le x \le L$) and in the region beyond ($x > L$) that satisfy the appropriate boundary conditions at $x = 0$ and $x = \infty$. Enforce the proper matching conditions at $x = L$ to find an equation for the allowed energies of this system. Are there conditions for which no solution is possible? Explain.

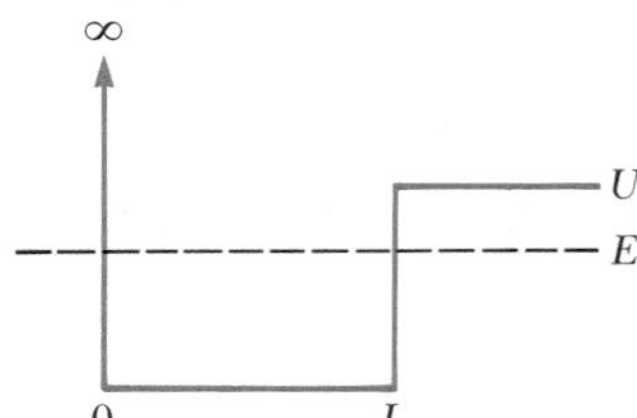

Figure 5.18 (Problem 25)

26. An electron has a wavefunction given by

$$\psi(x) = Ce^{-|x|/x_0}$$

where x_0 is a constant and $C = 1/\sqrt{x_0}$ for normalization (see Example 5.1). For this case, obtain expressions for $\langle x \rangle$ and Δx in terms of x_0. Also calculate the probability that the electron will be found within a standard deviation of its average position, that is, in the range $\langle x \rangle - \Delta x$ to $\langle x \rangle + \Delta x$, and show that this is independent of x_0.

27. Which of the following functions are eigenfunctions of the momentum operator $[p]$? What are the eigenvalues, where appropriate?

(a) $A \sin(kx)$ (c) $A \cos(kx) + iA \sin(kx)$
(b) $A \sin(kx) - A \cos(kx)$ (d) $A\, e^{ik(x-a)}$

28. Calculate $\langle x \rangle$, $\langle x^2 \rangle$, and Δx for a quantum oscillator in its ground state. *Hint:* Use the integral formula

$$\int_0^\infty x^2 e^{-ax^2}\, dx = \frac{1}{4a}\sqrt{\frac{\pi}{a}} \qquad a > 0$$

29. (a) What value do you expect for $\langle p \rangle$ for the quantum oscillator? Support your answer with a symmetry argument rather than a calculation. (b) Energy conservation for the quantum oscillator can be used to relate $\langle p^2 \rangle$ to $\langle x^2 \rangle$. Use this relation, along with the value of $\langle x^2 \rangle$ from Problem 28, to find $\langle p^2 \rangle$ for the oscillator ground state. (c) Evaluate Δp, using the results of (a) and (b).

30. From the results of Problems 28 and 29, evaluate $\Delta x\, \Delta p$ for the quantum oscillator in its ground state. Is the result consistent with the uncertainty principle?

31. The wavefunction

$$\psi(x) = Cx\, e^{-\alpha x^2}$$

also describes a state of the quantum oscillator, provided the constant α is chosen appropriately. (a) Using Schrödinger's equation, obtain an expression for α in terms of the oscillator mass m and the classical frequency of vibration ω. What is the energy of this state? (b) Normalize this wave. (*Hint:* See the integral of Problem 28.)

32. Show that the oscillator energies in Equation 5.29 correspond to the classical amplitudes

$$A_n = \sqrt{\frac{(2n+1)\hbar}{m\omega}}$$

33. Obtain an expression for the probability density $P_c(x)$ of a classical oscillator with mass m, frequency ω, and amplitude A. (*Hint:* See Problem 21 for the calculation of classical probabilities.)

34. *Nonstationary States.* Consider a particle in an infinite square well described initially by a wave that is a superposition of the ground and first excited states of the well:

$$\Psi(x, 0) = C\{\psi_1(x) + \psi_2(x)\}$$

(a) Show that the value $C = 1/\sqrt{2}$ normalizes this wave, assuming ψ_1 and ψ_2 are themselves normalized. (b) Find $\Psi(x, t)$ at any later time t. (c) Show that the superposition is *not* a stationary state, but that the *average* energy in this state is the arithmetic mean $(E_1 + E_2)/2$ of the ground and first excited state energies E_1 and E_2.

35. For the nonstationary state of Problem 34, show that the average particle position $\langle x \rangle$ oscillates with time as

$$\langle x \rangle = x_0 + A \cos(\Omega t)$$

where

$$x_0 = \frac{1}{2}\left\{\int x|\psi_1|^2\, dx + \int x|\psi_2|^2\, dx\right\}$$

$$A = \int x\psi_1^*\psi_2\, dx$$

and $\Omega = (E_2 - E_1)/\hbar$. Evaluate your results for the mean position x_0 and amplitude of oscillation A for an electron in a well 1 nm wide. Calculate the time for the electron to shuttle back and forth in the well once. Calculate the same time classically for an electron with energy equal to the average $(E_1 + E_2)/2$.

PROBLEMS FOR THE COMPUTER

The following problems require for their solution the use of the program EIGEN. Where appropriate, furnish screen dumps to substantiate your answers.

36. What values are used to describe an electron in a square well of height 100 eV and width 2 Å? Using these values, find the ground state energy and waveform for the electron. How many bound states do you think this well will support?

37. For the ground state of the harmonic oscillator, estimate the depth of penetration of the wave into the nonclassical region. Give your result in units of the amplitude of vibration for a classical oscillator with the same energy.

38. Find the two lowest stationary states for a particle trapped in the triangular well. How many nodes does each state have? What symmetry (if any) do these wavefunctions possess?

The following problems require for their solution the use of the program WAVES-2.

39. Display the nonstationary state that results when the ground and first excited states of the infinite well are superimposed with equal amplitudes [uniform $a(n)$]. Where is the particle most likely to be found when $t = 0$? Find the *recurrence time* T, that is, the least time that elapses for the waveform to return to its original shape. (a) Using numbers appropriate to an electron in a well 1 nm wide, compare T with the shuttle time calculated in Problem 35. Where is the electron most likely to be at $t = \frac{1}{2}T$? (b) Find T for a 1-g mass in a well 1 cm wide. Interpret your result on the basis of the (average) energy possessed by this mass.

40. Repeat Problem 39 for the nonstationary state that results when the ground and first excited states of the infinite well are combined with equal amplitude but opposite sign [harmonic $a(n)$ with $\gamma = 1$].

41. Display the nonstationary state that results when the five lowest symmetric (unnormalized) states of the harmonic oscillator ($n = 0, 2, \ldots, 8$) are combined according to a geometric envelope with $\gamma = \frac{1}{4}$. Record the initial width of this state, measured as the interval bounded by the points where the amplitude drops to $\frac{1}{2}$ its maximum value. What happens to this width with the passage of time? Find the recurrence time (see Problem 39) for this state, and relate it to the period of the classical oscillator with the same frequency.

6

Tunneling Phenomena

In this chapter the principles of wave mechanics are applied to particles striking a potential *barrier*. Unlike potential wells which attract and *trap* particles, barriers *repel* them. With no bound states to be discovered, the emphasis shifts to determining whether a particle incident on the barrier is reflected or transmitted.

In the course of this study we shall encounter for the first time a peculiar but most interesting phenomenon called *tunneling*. A purely wave-mechanical effect, tunneling nevertheless is essential to the operation of many modern-day devices, and shapes our world on a scale from atomic all the way up to galactic proportions. The chapter includes a discussion of the role played by tunneling in several phenomena of practical interest, such as field emission, radioactive decay, and the Josephson effect. Finally, the chapter is followed by a very interesting essay on the scanning tunneling microscope, or STM, a remarkable device which uses tunneling to make images of surfaces with resolution comparable to the size of a single atom.

6.1 THE SQUARE BARRIER

The square barrier is represented by a potential energy $U(x)$ that is constant at U in the barrier region, say between $x = 0$ and $x = L$, and zero outside this region. A realization of the square barrier potential using charged hollow cylinders is shown in Figure 6.1a. The outer cylinders are grounded while the central one is held at some positive potential V. For a particle with charge q, the barrier potential is $U = qV$. The charge experiences no force except in the gaps separating the cylinders. The force in the gaps is repulsive, tending to expel a positive charge q from the central cylinder. The potential energy for the idealized case where the gaps have shrunk to zero size is the *square barrier*, sketched in Figure 6.1b.

A classical particle incident on the barrier, say from the left, experiences a retarding force upon arriving in the gap at $x = 0$. Particles with energies E greater than U are able to overcome this force, but suffer a reduction in speed to a value commensurate with their diminished kinetic energy ($= E - U$) in the barrier region. Such particles continue moving to the right with reduced speed until they reach the gap at $x = L$, where they receive a "kick" accelerat-

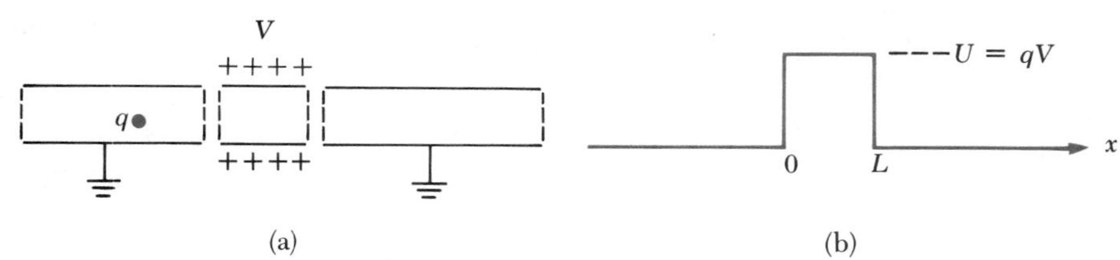

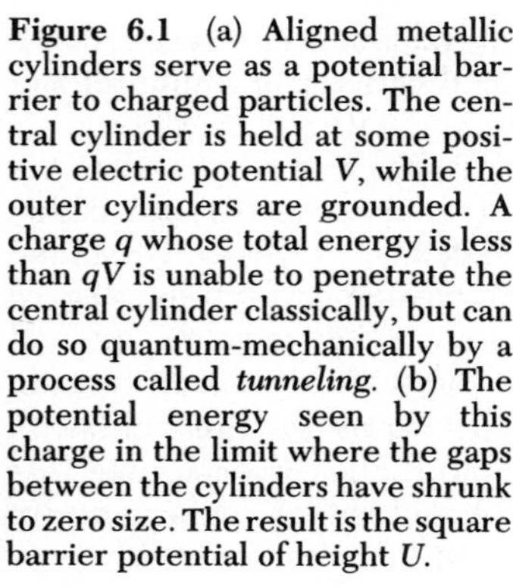
Figure 6.1 (a) Aligned metallic cylinders serve as a potential barrier to charged particles. The central cylinder is held at some positive electric potential V, while the outer cylinders are grounded. A charge q whose total energy is less than qV is unable to penetrate the central cylinder classically, but can do so quantum-mechanically by a process called *tunneling*. (b) The potential energy seen by this charge in the limit where the gaps between the cylinders have shrunk to zero size. The result is the square barrier potential of height U.

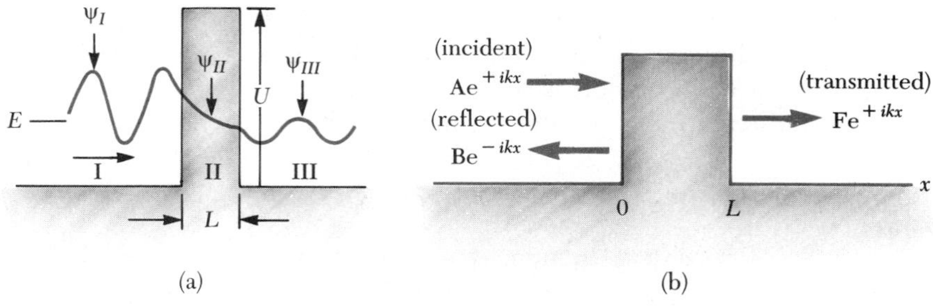

Figure 6.2 (a) A typical stationary state wave for a particle in the presence of a square barrier. The energy E of the particle is less than the barrier height U. Since the wave amplitude is nonzero in the barrier, there is some probability of finding the particle there. (b) Decomposition of the stationary wave into incident, reflected, and transmitted waves.

ing them back up to their original speed. Thus, particles having energy $E > U$ are able to cross the barrier with their speed restored to its initial value. By contrast, particles with energy $E < U$ are turned back (reflected) by the barrier, having insufficient energy to cross, or even penetrate it. In this way the barrier divides the space into classically *allowed* and *forbidden* regions determined by the particle energy: if $E > U$, the whole space is accessible to the particle; for $E < U$ only the interval to the side of the barrier in which the particle originates is accessible—the barrier region itself is forbidden, and this precludes particle motion on the far side as well.

According to quantum mechanics, however, *there is no region inaccessible to our particle, regardless of energy,* since the matter wave associated with the particle is nonzero everywhere. A typical waveform for this case, illustrated in Figure 6.2a, clearly shows the penetration of the wave into the barrier and beyond. This barrier penetration is in complete disagreement with classical physics. The attendant possibility of finding the particle on the far side of the barrier is called *tunneling:* we say the particle has *tunnelled* through the barrier.

The mathematical expression for Ψ to either side of the barrier is easily found. To the left of the barrier the particle is free, so the matter wave here must be the free particle plane waves introduced in Chapter 5:

$$\Psi(x, t) = Ae^{i(kx-\omega t)} + Be^{i(-kx-\omega t)} \tag{6.1}$$

This wave $\Psi(x, t)$ is actually the *sum* of *two* plane waves. Both have frequency ω and energy $E = \hbar\omega = \hbar^2k^2/2m$, but the first moves from left to right (wavenumber k), the second from right to left (wavenumber $-k$). Thus, that part of Ψ proportional to A is interpreted as a wave *incident* on the barrier from the *left;* that proportional to B as a wave *reflected* from the barrier and moving to the *right* (Figure 6.2b). The squared-amplitude, or intensity, of the reflected wave relative to the incident wave intensity is the **reflection coefficient** of the barrier, denoted by R:[1]

$$R = \frac{|B|^2}{|A|^2} \tag{6.2}$$

Reflection coefficient for a barrier

[1] Since the reflection coefficient relates to probabilities, we use the squared-amplitude—not the amplitude itself—of the matter waves in calculating R. The same can be said for the transmission coefficient T.

In wave terminology R is the portion of wave intensity in the reflected beam; in particle language, R becomes the likelihood (probability) that a particle incident on the barrier from the left is reflected by it.

Similar arguments apply to the right of the barrier where, again, the particle is free:

$$\Psi(x, t) = Fe^{i(kx-\omega t)} + Ge^{i(-kx-\omega t)} \tag{6.3}$$

This form for $\Psi(x, t)$ is valid in the range $x > L$, with the term proportional to F describing a wave traveling to the right, and that proportional to G a wave traveling to the left in this region. The latter has *no* physical interpretation for waves incident on the barrier from the *left*, and so is discarded by requiring $G = 0$. The former is that part of the incident wave which is *transmitted* through the barrier. The relative intensity of this transmitted wave is the *transmission coefficient* for the barrier T:

Transmission coefficient for a barrier

$$\boxed{T = \frac{|F|^2}{|A|^2}} \tag{6.4}$$

The transmission coefficient measures the likelihood (probability) that a particle incident on the barrier from the left penetrates to emerge on the other side. Since a particle incident on the barrier is either reflected or transmitted, the probabilities for these events must sum to unity:

$$R + T = 1. \tag{6.5}$$

Equation 6.5 expresses a kind of *sum rule* obeyed by the barrier coefficients. Further, the degree of transmission or reflection will depend upon particle energy. In the classical case $T = 0$ (and $R = 1$) for $E < U$, while $T = 1$ (and $R = 0$) for $E > U$. The wave-mechanical predictions for the functions $T(E)$ and $R(E)$ are more complicated; to obtain them we must examine the matter wave within the barrier.

To find Ψ in the barrier, we look to Schrödinger's equation. Let us consider stationary states $\psi(x)e^{-i\omega t}$ whose energy $E = \hbar\omega$ is below the top of the barrier. This is the case $E < U$ for which no barrier penetration is permitted classically. In the region of the barrier ($0 < x < L$), $U(x) = U$ and the time-independent Schrödinger equation for $\psi(x)$ can be rearranged as

$$\frac{d^2\psi}{dx^2} = \left\{\frac{2m(U-E)}{\hbar^2}\right\}\psi(x)$$

With $E < U$, the term in braces is a positive constant, and solutions to this equation are the *real exponential forms* $e^{\pm\alpha x}$. Since $(d^2/dx^2)e^{\pm\alpha x} = (\alpha)^2 e^{\pm\alpha x}$, we must take the term in braces $= (\alpha)^2$ or

$$\delta = \frac{1}{\alpha} = \frac{\hbar}{\sqrt{2m(U-E)}} \tag{6.6}$$

The parameter δ is the **penetration depth;** it is the distance into the barrier where ψ—hence, the probability of finding the particle—remains appreciable.

The complete waveform in the barrier is then

$$\Psi(x, t) = \psi(x)e^{-i\omega t} = Ce^{-\alpha x - i\omega t} + De^{+\alpha x - i\omega t} \qquad \text{for } 0 < x < L \tag{6.7}$$

The coefficients C and D are fixed by requiring smooth joining of the wave across the barrier edges, that is, both Ψ and $\partial\Psi/\partial x$ must be continuous at $x = 0$

and $x = L$. Writing out the joining conditions using Equations 6.1, 6.3, and 6.7 for Ψ in the regions to the left, to the right, and within the barrier, respectively, gives

$$A + B = C + D$$
$$ikA - ikB = \alpha D - \alpha C$$
$$Ce^{-\alpha L} + De^{+\alpha L} = Fe^{ikL}$$
$$(\alpha D)e^{+\alpha L} - (\alpha C)e^{-\alpha L} = ikFe^{ikL} \tag{6.8}$$

Joining conditions at a square barrier

In keeping with our previous remarks, we have set $G = 0$. Still, there are fewer equations than there are unknowns. This is as it should be, since the amplitude of the incident wave is fixed by experiment and serves only to set the scale for the other amplitudes. That is, doubling the incident wave amplitude simply doubles the amplitudes of the reflected and transmitted waves. Dividing Equations 6.8 through by A furnishes four equations for the ratios B/A, C/A, D/A, and F/A: these equations may be solved by repeated substitution to find B/A, etc. in terms of the barrier height U, the barrier width L, and the particle energy E.

EXAMPLE 6.1 Scattering by a Potential Step

The potential step shown in Figure 6.3 may be regarded as a square barrier in the special case where the barrier width L is infinite. Apply the ideas of this section to discuss the quantum scattering of particles incident from the left on a potential step, in the case where the step height U exceeds the total particle energy E.

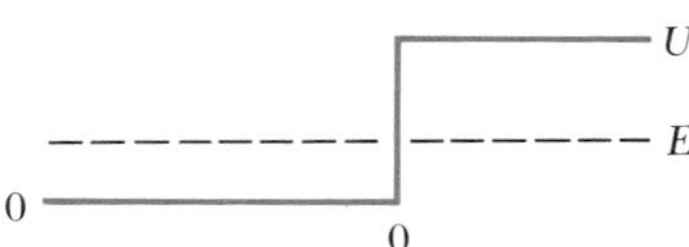

Figure 6.3 (Example 6.1) The potential step of height U may be thought of as a square barrier of the same height in the limit where the barrier width L becomes infinite. All particles incident on the barrier with energy $E < U$ are reflected.

Solution: The waveform everywhere to the right of the origin is the barrier wavefunction given by Equation 6.7. To keep Ψ from diverging for large x we must take $D = 0$, leaving only the decaying wave

$$\Psi(x, t) = Ce^{-\alpha x - i\omega t} \qquad x > 0$$

This must be joined smoothly to the waveform on the left of the origin, given by Equation 6.1:

$$\Psi(x, t) = Ae^{ikx - i\omega t} + Be^{-ikx - i\omega t} \qquad x < 0$$

The conditions for smooth joining at $x = 0$ are

$$A + B = C \qquad \text{and} \qquad ikA - ikB = -\alpha C$$

Solving the second equation for C and substituting into the first (with $\delta = 1/\alpha$) gives $A + B = -ik\delta A + ik\delta B$ or

$$\frac{B}{A} = -\frac{(1 + ik\delta)}{(1 - ik\delta)}$$

The reflection coefficient is $R = |B/A|^2 = (B/A)(B/A)^*$, or

$$R = \left(\frac{1 + ik\delta}{1 - ik\delta}\right)\left(\frac{1 - ik\delta}{1 + ik\delta}\right) = 1$$

Thus, an *infinitely* wide barrier reflects *all* incoming particles with energies below the barrier height, in agreement with the classical prediction. Nevertheless, there is a nonzero wave in the step region since

$$\frac{C}{A} = 1 + \frac{B}{A} = \frac{-2ik\delta}{1 - ik\delta} \neq 0$$

But the waveform for $x > 0$, $\Psi(x, t) = Ce^{-\alpha x - i\omega t}$, is not a *propagating* wave at all, that is, there is no *net transmission* of particles to the right of the step.[2] However, there will be quantum transmission through a barrier of *finite* width, no matter how wide, as illustrated in the next section.

[2] The existence of a barrier wave without propagation is familiar from the optical phenomenon of *total internal reflection* exploited in the construction of beam splitters: light entering a right-angle prism is completely reflected at the hypotenuse face, even though an electromagnetic wave, the *evanescent wave*, penetrates into the space beyond. A second prism brought into near contact with the first can "pick up" this evanescent wave, thereby transmitting and redirecting the original beam. This phenomenon, known as *frustrated total internal reflection*, is the optical analog of tunneling: in effect, photons have tunneled across the gap separating the two prisms.

Wide Barriers

The solution of Equations 6.8 in the general case is algebraically tedious, and the results are often too complicated to be useful. However, a simplifying approximation is possible if the barrier is wide on the scale of length set by δ, that is, if $L \gg \delta$. Because of the great distance to the far edge, the matter wave Ψ for a wide barrier would take on (approximately) the purely decaying waveform of the infinitely wide barrier discussed in Example 6.1. In other words, for a wide barrier the wave growing exponentially in the barrier region should be unimportant. Neglecting the coefficient D of this wave compared to the other amplitudes in the first and second of Equations 6.8 leads to the wide barrier approximation for the transmission coefficient:[3]

Transmission coefficient of a square, wide barrier

$$T(E) = |F/A|^2 \simeq \left(\frac{4k\delta}{1+(k\delta)^2}\right)^2 e^{-2L/\delta} \tag{6.9}$$

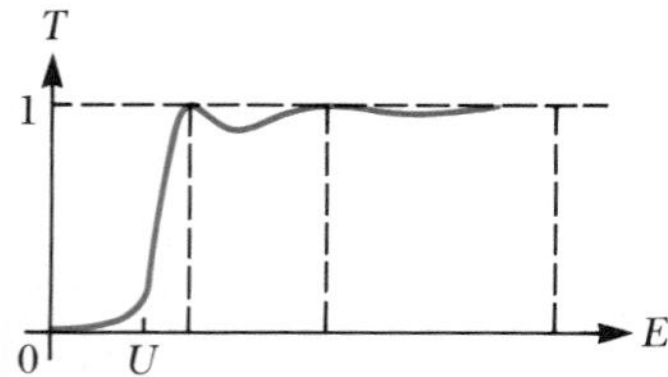

Figure 6.4 A sketch of the transmission coefficient $T(E)$ for a square barrier. The high, wide barrier approximation is accurate only for particle energies E well below the barrier height U. Oscillations in $T(E)$ with E, and transmission resonances where $T(E) = 1$, appear at higher energies, and are further evidence for the wave nature of matter.

The energy dependence of T lies hidden in the parameters k and δ. Reiterating those definitions, we have specifically

$$k^2 = \frac{2mE}{\hbar^2} \quad \text{and} \quad (k\delta)^2 = \frac{E}{U-E} \tag{6.10}$$

A sketch of $T(E)$ for a square barrier is shown in Figure 6.4. The approximation Equation 6.9 holds only for energies E sufficiently below the barrier height U so that $\delta \ll L$ is satisfied; at higher energies we have no recourse but to solve Equations 6.8 without simplification. Notice the oscillations in $T(E)$ for $E > U$ and the *transmission resonances* where $T(E) = 1$ (perfect transmission). This is a wave interference phenomenon, and constitutes further evidence for the wave nature of matter.

[3] The parameter D cannot be ignored in the last two of Equations 6.8 where it multiplies $e^{+\alpha L}$, a term that will be large for wide barriers.

EXAMPLE 6.2 Transmission Coefficient for an Oxide Layer

Two copper conducting wires are separated by an insulating oxide layer (CuO). Modeling the oxide layer as a square barrier of height 10 eV, estimate the transmission coefficient for penetration by 7 eV electrons (a) if the layer thickness is 5 nm, and (b) if the layer thickness is 1 nm.

Solution: From Equation 6.10 we have immediately that $(k\delta)^2 = 7/(10-7) = 2.33$ for this case. Also, using $\hbar c = 197.3$ eV·nm and $m = 511$ keV/c^2 for electrons gives

$$\delta = \frac{\hbar c}{\sqrt{2mc^2(U-E)}} = \frac{(197.3 \text{ eV}\cdot\text{nm})}{\sqrt{2(511\times 10^3 \text{ eV})(3 \text{ eV})}}$$

$$= 0.113 \text{ nm}$$

Since δ is much less than even the 1 nm layer thickness, the wide barrier approximation is applicable. The transmission coefficient from Equation 6.9 is then

$$T \approx \left(\frac{4\sqrt{2.33}}{1+2.33}\right)^2 e^{-2L/(0.113\,\text{nm})} = 3.36e^{-2L/(0.113\,\text{nm})}$$

Substituting $L = 5$ nm, we get

$$T = 3.36e^{-88.73} = 0.979 \times 10^{-38},$$

a fantastically small number! With $L = 1$ nm, however, we find

$$T = 3.36e^{-17.75} = 0.660 \times 10^{-7}.$$

Thus, reducing the layer thickness by a factor of 5 enhances the likelihood of penetration by nearly 31 orders of magnitude!

EXAMPLE 6.3 Tunneling Current Through an Oxide Layer

A 1 mA current of electrons in one of the wires of Example 6.2 is incident on the oxide layer. How much of this current passes through the layer to the adjacent wire if the electron energy is 7 eV and the layer thickness is 1 nm? What becomes of the remaining current?

Solution: Since each electron carries a charge equal to $e = 1.60 \times 10^{-19}$ C, an electron current of 1 mA represents $10^{-3}/(1.60 \times 10^{-19}) = 6.25 \times 10^{15}$ electrons per second impinging on the barrier. Of these, only the fraction T is transmitted, where $T = 0.660 \times 10^{-7}$ from Example 6.2. Thus, the number of electrons per second continuing on to the adjacent wire is

$$(6.25 \times 10^{15})(0.660 \times 10^{-7}) = 4.12 \times 10^{8} \text{ electrons/s}$$

This number represents a transmitted current of

$$(4.12 \times 10^{8}/\text{s})(1.60 \times 10^{-19} \text{ C}) = 6.60 \times 10^{-11} \text{ A} = 66.0 \text{ pA (picoamps)}$$

(Notice that the same transmitted current would be obtained had we simply multiplied the incident current by the transmission coefficient.) The remaining 1 mA − 66.0 pA is reflected at the layer. It is important to note that the *measured* conduction current in the wire on the side of incidence is the *net* of the incident and reflected currents, or again 66.0 pA.

6.2 BARRIER PENETRATION—SOME APPLICATIONS

In actuality, few barriers can be modeled accurately using the square barrier discussed in the preceding section. Indeed, the extreme sensitivity to barrier constants found there suggests that barrier shape will be important in making reliable predictions of tunneling probabilities. The transmission coefficient for a barrier of arbitrary shape, as specified by some potential energy function $U(x)$, can be had from Schrödinger's equation. For high, wide barriers, where the likelihood of penetration is small, a rigorous (though lengthy) treatment yields

$$T(E) \approx \exp\left(-\frac{2}{\hbar}\sqrt{2m}\int\sqrt{U(x)-E}\,dx\right) \tag{6.11}$$

Approximate transmission coefficient of a barrier with arbitrary shape

The integral in Equation 6.11 is to be taken over the *classically forbidden region* where $E < U(x)$. For the square barrier, the integral extends over the region of the barrier $0 < x < L$ where $U(x) = U$, with the result[4]

$$T(E) \approx \exp\left(-\frac{2}{\hbar}\sqrt{2m(U-E)}\,L\right) \tag{6.12}$$

The use of Equation 6.11 in other circumstances is amply illustrated in the remainder of this section where it is applied to several problems of current interest spanning a variety of disciplines.

Field Emission

In field emission, electrons bound to the atoms of a metal are literally "ripped away" by the application of a strong electric field. In this way the metal becomes a source that may be conveniently tapped to furnish electrons for many applications. In the past such *cold cathode emission*, as it was known, was a popular way of generating electrons in vacuum tube circuits, producing less electrical "noise" than hot filament sources, where electrons were "boiled off" by heating the metal to a high temperature. Modern applications include

[4] This result for $T(E)$ differs from that obtained in Section 6.1 using the wide barrier approximation, where the coefficient of the exponential is not unity, but a function of E and U. However, this function is slowly varying and never larger than 4 (for $E = U$), so that both formulas predict the same order-of-magnitude values for T. It is in this spirit that the approximation must be understood.

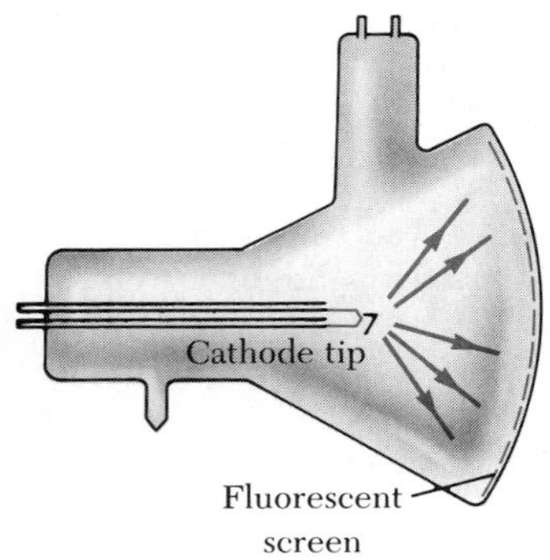

Figure 6.5 Schematic diagram of a field emission microscope. The intense electric field at the tip of the needle-shaped specimen allows electrons to tunnel through the work function barrier at the surface. Since the tunneling probability is sensitive to the exact details of the surface where the electron passes, the number of escaping electrons varies from point to point with the surface condition, thus providing a picture of the surface under study.

the *field emission microscope* (Fig. 6.5) and a related device, the *scanning tunneling microscope* (see the Essay in this chapter), both of which use the escaping electrons to form an image of structural details at the emitting surface.

Field emission is a tunneling phenomenon. Figure 6.6a shows schematically how field emission can be obtained by placing a positively charged plate near the source metal to form, effectively, a parallel plate capacitor. In the gap between the "plates" there is some electric field $\mathcal{E}$ but the electric field inside the metal remains zero due to the shielding by the mobile metal electrons induced at the surface. Thus, an electron in the bulk is virtually free, yet still bound to the metal according to some (constant) potential energy that we denote as $-U$. The total electron energy E, which includes kinetic energy, exceeds this value, but is not larger than zero if our electron is to be bound initially: indeed, $-E$ represents the energy needed to free this electron, a value at least equal to the work function of the metal.

Tunneling model for field emission

Once beyond the surface ($x > 0$), our electron is attracted by the electric force in the gap $F = e\mathcal{E}$, represented by the potential energy $U(x) = -e\mathcal{E}x$. The potential energy diagram is shown in Figure 6.6b, together with the classically allowed and forbidden regions for an electron of energy E. The intersections of E with $U(x)$ at x_1 $(= 0)$ and x_2 $(= -E/e\mathcal{E})$ mark the *classical turning points*, where a classical particle with this energy would be turned around to keep it from entering the forbidden zone. Thus, from a classical viewpoint, an electron initially confined to the metal has insufficient energy to surmount the potential barrier at the surface, and would remain in the bulk forever! It is only by virtue of its wave character that the electron can tunnel through this barrier to emerge on the other side. The probability of such an occurrence is mea-

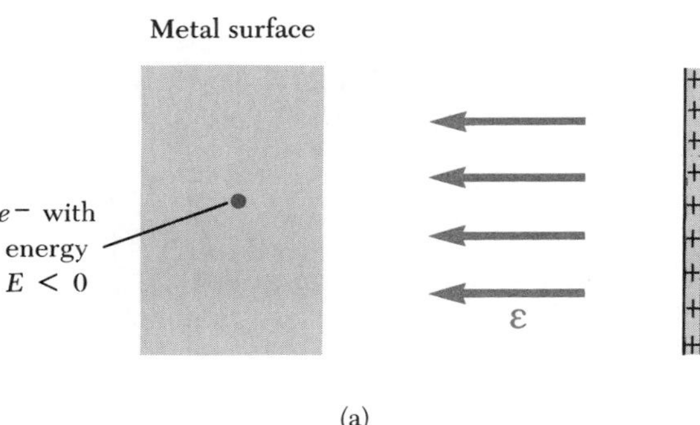

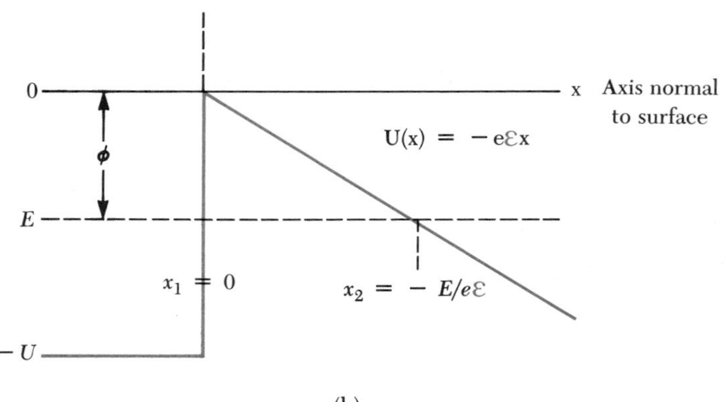

Figure 6.6 (a) Field emission from a metal surface. (b) The potential energy seen by an electron of the metal. The electric field produces the triangular potential barrier shown, through which electrons can tunnel to escape the metal. Turning points at $x_1 = 0$ and $x_2 = -E/e\mathcal{E}$ delineate the classically forbidden region. Tunneling is greatest for the most energetic electrons, for which $-E$ is equal to the work function ϕ of the metal.

sured by the transmission coefficient for the (triangular) barrier depicted in Figure 6.6b.

To calculate $T(E)$ we must evaluate the integral in Equation 6.11 over the classically forbidden region from x_1 to x_2. Since $U(x) = -e\mathcal{E}x$ in this region and $E = -e\mathcal{E}x_2$, we have

$$\int \sqrt{U(x) - E}\, dx = \sqrt{e\mathcal{E}} \int_0^{x_2} \sqrt{x_2 - x}\, dx = -\frac{2}{3}\sqrt{e\mathcal{E}}\, \{x_2 - x\}^{3/2} \Big|_0^{x_2}$$
$$= \frac{2}{3}\sqrt{e\mathcal{E}} \left(\frac{|E|}{e\mathcal{E}}\right)^{3/2}$$

Using this result in Equation 6.11 gives the transmission coefficient for field emission as

$$T(E) \approx \exp\left(\left\{-\frac{4\sqrt{2m}\,|E|^{3/2}}{3e\hbar}\right\}\frac{1}{\mathcal{E}}\right) \tag{6.13}$$

Transmission coefficient for field emission

The sensitivity of T to electron energy in the bulk is evident from this expression. The escape probability is largest for the most energetic electrons, for which $|E| = \phi$, the work function of the metal. It is also apparent that the quantity in curly brackets must have the dimensions of electric field, and represents a characteristic field strength, say $\mathcal{E}_c$, for field emission:

$$\mathcal{E}_c = \frac{4\sqrt{2m}\,|E|^{3/2}}{3e\hbar} \tag{6.14}$$

For $|E| = \phi = 4$ eV, a typical value for many metals, we calculate the field strength to be $\mathcal{E}_c = 5.5 \times 10^{10}$ V/m, a strong field by laboratory standards.

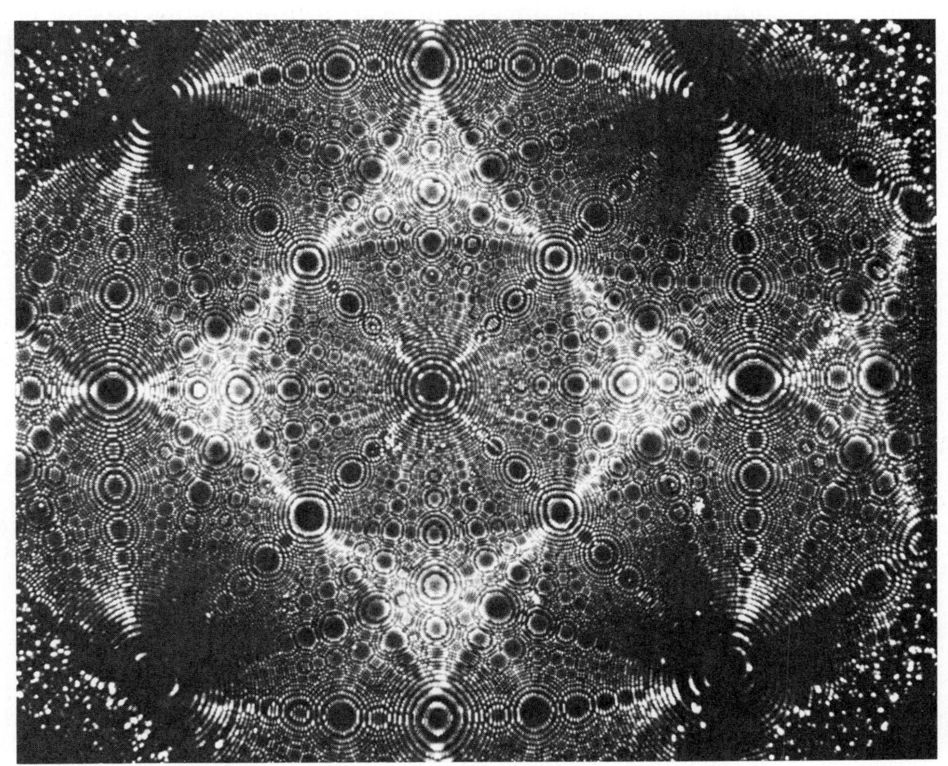

Field ion microscope image of the surface of a platinum crystal with a magnification of 1 000 000×. Individual atoms can be seen on surface layers using this technique. (Courtesy of Prof. T. T. Tsong, The Pennsylvania State University.)

However, measurable emission occurs even with much smaller fields, since the emission *rate* grows with the number of electrons per second that collide with the barrier. This *collision frequency* is quite high for a bulk sample containing something like 10^{22} electrons per cubic centimeter, and values in excess of 10^{30} collisions per second are not uncommon (see Problem 13)! In this way field emission rates on the order of 10^{10} electrons per second (currents of about 1 nA) can be realized with applied fields as small as $\mathcal{E}_c/50$, or about 10^9 V/m.

EXAMPLE 6.4 Tunneling in a Parallel Plate Capacitor

Estimate the leakage current due to tunneling that passes across a parallel plate capacitor charged to a potential difference of 10 kV. Take the plate separation to be $d =$ 0.01 mm and the plate area to be $A = 1.0$ cm^2.

Solution: The number of electrons per second impinging on the plate surface from the bulk is the collision frequency f, about 10^{30} per cm^2 for most metals. Of these, only the fraction given by the transmission coefficient T can tunnel through the potential barrier in the gap to register as a current through the device. Thus, the electron emission rate for a plate of area 1 cm^2 is

$$\lambda = fT(E) = 10^{30} \exp(-\mathcal{E}_c/\mathcal{E})$$

The electric field $\mathcal{E}$ in the gap is 10 kV/0.01 mm $= 10^9$ V/m. Using this and $\mathcal{E}_c = 5.5 \times 10^{10}$ V/m gives for the exponential $\exp\{-55\} = 1.30 \times 10^{-24}$, and an emission rate of $\lambda = 1.30 \times 10^6$ electrons per second. Since each electron carries a charge $e = 1.60 \times 10^{-19}$ C, the tunneling current is

$$I = 2.08 \times 10^{-13} \text{ A} = 0.2 \text{ pA}$$

George Gamow (1904–1969) was born and educated in Russia. After working at various universities in western Europe (including stays with Bohr and Rutherford), he moved to the United States in 1934 where he worked at George Washington University until 1956, and finally at the University of Colorado. In one of his publications regarding cosmology, he stated that the universe's supply of hydrogen and helium were created very quickly, "in less time than it takes to cook a dish of duck and roast potatoes." This comment is characteristic of this interesting physicist who is known as much for his charming popularizations of science and his wonderful sense of humor as for his explanation of alpha decay and theories of cosmology.

Alpha Decay

The decay of radioactive elements with the emission of alpha particles (helium nuclei composed of two protons and two neutrons) was among the longstanding puzzles to which the fledgling wave mechanics was first applied shortly after its inception in 1926. That alpha particles are a disintegration product of such species as radium, thorium, and uranium was well documented as early as 1900, but certain features of this decay remained a mystery, finally unraveled in 1928 in the now-classic works of Gamow, and Gurney and Condon. Their contribution was to recognize that the newly discovered tunnel effect lay behind the two most puzzling aspects of alpha decay:

1. All alpha particles emitted from any one source have nearly the same energy, and, for all known emitters, emerge with kinetic energies in the same narrow range, from about 4 to 9 MeV.
2. In contrast to the uniformity of energies, the half-life of the emitter (time taken for half of the emitting substance to decay) varies over an enormous range — more than twenty orders of magnitude! — according to the emitting element.

For instance, the alphas from thorium emerge with kinetic energy equal to 4.05 MeV, only a little less than half as much as the alphas emitted from polonium (8.95 MeV). Yet the half-life of thorium is 1.3×10^{10} years, compared to only 3.0×10^{-7} seconds for the half-life of polonium!

Gamow attributed this striking behavior to a preformed alpha particle rattling around within the nucleus of the radioactive (parent) element, eventually tunneling through the potential barrier to escape as a detectable decay product (Fig. 6.7a). While inside the parent nucleus the alpha is virtually free, but nonetheless confined to the nuclear potential well by the strong nuclear force. Once outside the nucleus, the alpha particle experiences only the Cou-

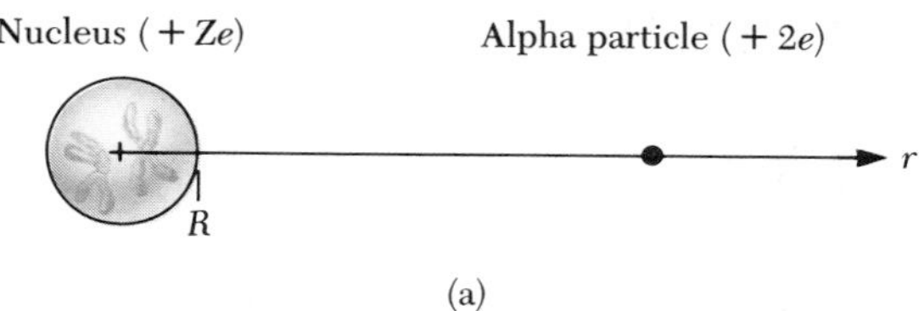

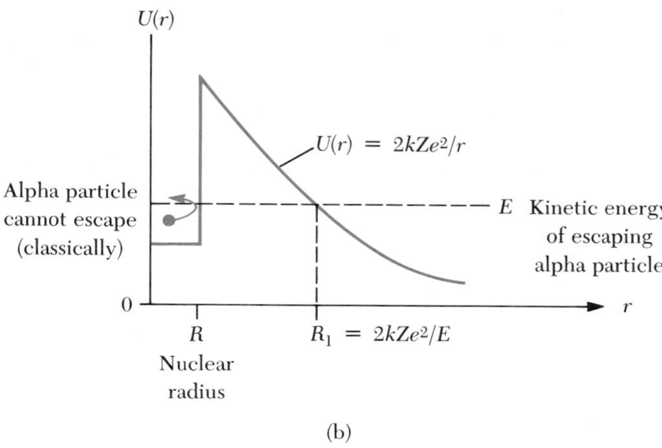

Figure 6.7 (a) Alpha decay of a radioactive nucleus. (b) The potential energy seen by an alpha particle emitted with energy E. R is the nuclear radius, about 10^{-14} m or 10 fm. Alpha particles tunneling through the potential barrier between R and R_1 escape the nucleus to be detected as radioactive decay products.

Tunneling through the Coulomb barrier

lomb repulsion of the emitting (daughter) nucleus. (The nuclear force on the alpha outside the nucleus is insignificant due to its extremely short range, $\sim 10^{-15}$ m.) Figure 6.7b shows the potential energy diagram for the alpha particle as a function of distance r from the emitting nucleus. The nuclear radius R is about 10^{-14} m or 10 fm [note that 1 fm (fermi) $= 10^{-15}$ m] for heavy nuclei;[5] beyond this there is only the energy of Coulomb repulsion $U(r) = kq_1q_2/r$ between the doubly-charged alpha ($q_1 = +2e$) and a daughter nucleus with atomic number Z ($q_2 = +Ze$). Classically, even a 9 MeV alpha particle initially bound to the nucleus would have insufficient energy to overcome the Coulomb barrier ($\sim$30 MeV high) and escape. But the alpha particle, with its wave attributes, may tunnel *through* the barrier to appear on the outside. The total alpha particle energy E inside the nucleus becomes the observed kinetic energy of the emerging alpha once it has escaped. *It is the sensitivity of the tunneling rate to small changes in particle energy which accounts for the wide range of half-lives observed for alpha emitters.*

The tunneling probability and associated decay rate are calculated in much the same way as for field emission, apart from the fact that the barrier shape now is Coulombic, rather than triangular. The tunneling probability for alpha decay is

Transmission coefficient for alpha particles of an unstable nucleus

$$\boxed{T(E) = \exp\left\{-4\pi Z\sqrt{\frac{E_0}{E}} + 8\sqrt{\frac{ZR}{r_0}}\right\}} \tag{6.15}$$

In this expression, $r_0 = \hbar^2/m_\alpha ke^2$ is a kind of "Bohr" radius for the alpha particle. The mass of the alpha particle is $m_\alpha = 7295m_e$, so r_0 has the value

[5] The fermi (fm) is a unit of distance commonly used in nuclear physics.

$a_0/7295 = 7.25 \times 10^{-5}$ Å, or 7.25 fm. The length r_0, in turn, defines a convenient energy unit E_0 analogous to the Rydberg in atomic physics:

$$E_0 = \frac{ke^2}{2r_0} = \left(\frac{ke^2}{2a_0}\right)\left(\frac{a_0}{r_0}\right) = (13.6 \text{ eV})(7295) = 0.0993 \text{ MeV}.$$

To obtain decay rates, $T(E)$ must be multiplied by the number of collisions per second that an alpha particle makes with the nuclear barrier. This collision frequency f is the reciprocal of the transit time for the alpha particle crossing the nucleus, or $f = v/2R$ where v is the speed of the alpha particle inside the nucleus. In most cases f is about 10^{21} collisions per second (see Problem 11). The decay rate λ is then

$$\lambda = fT(E) \approx 10^{21} \exp\{-4\pi Z \sqrt{(E_0/E)} + 8\sqrt{Z(R/r_0)}\}.$$

The reciprocal of λ has dimensions of time, and is related to the half-life of the emitter $t_{1/2}$ as

$$t_{1/2} = \frac{\ln 2}{\lambda} = \frac{0.693}{\lambda} \tag{6.16}$$

EXAMPLE 6.5 Estimating the Half-lives of Thorium and Polonium

Using the tunneling model developed above, estimate the half-lives for alpha decay of the radioactive elements thorium and polonium. The energy of the ejected alphas is 4.05 MeV and 8.95 MeV, respectively, and the nuclear size is about 9 fm in both cases.

Solution: For thorium ($Z = 90$), the daughter nucleus has atomic number $Z = 88$, corresponding to the element radium. Using $E = 4.05$ MeV and $R = 9$ fm, we find for the transmission factor $T(E)$ in Equation 6.15

$$\exp\{-4\pi(88)\sqrt{(0.0993/4.05)} + 8\sqrt{88(9/7.25)}\}$$
$$= \exp\{-89.542\} = 1.29 \times 10^{-39}$$

Taking $f = 10^{21}$ Hz, we obtain for the decay rate the value $\lambda = 2.22 \times 10^{-18}$ alphas per second. The associated half-life is, from Equation 6.16,

$$t_{1/2} = \frac{0.693}{1.29 \times 10^{-18}} = 5.37 \times 10^{17} \text{ s}$$
$$= 1.70 \times 10^{10} \text{ yr}$$

which compares favorably with the actual value for thorium, 1.3×10^{10} yr.

If polonium ($Z = 84$) is the radioactive species, the daughter element is lead, with $Z = 82$. Using for the disintegration energy $E = 8.95$ MeV, we obtain for the transmission factor

$$\exp\{-4\pi(82)\sqrt{0.0993/8.95} + 8\sqrt{82(9/7.25)}\}$$
$$= \exp\{-27.825\} = 8.23 \times 10^{-13}$$

Assuming f is unchanged at 10^{21} collisions per second, we get for this case $\lambda = 8.23 \times 10^{+8}$ alphas per second and a half-life

$$t_{1/2} = \frac{0.693}{8.23 \times 10^{8}} = 8.42 \times 10^{-10} \text{ s}$$

The measured half-life of polonium is 3.0×10^{-7} s.

Given the crudeness of our method, both estimates should be considered satisfactory. Further, the calculations show clearly how a factor of only 2 in disintegration energy leads to half-lives differing by more than 26 orders of magnitude!

The Josephson Junction

An exotic device of recent origin (1962) that operates on the principle of tunneling is the **Josephson junction** named for Brian Josephson, the first one to study theoretically the properties of this system. The device consists of two superconductors separated by a thin insulating oxide layer, 1 to 2 nm thick (Fig. 6.8a). A superconductor is a material that shows a complete absence of resistance to the flow of electric current. In the superconducting state, electrons travel as pairs—the so-called *Cooper pairs*. The Cooper pairs are found with a probability density given by the intensity $|\Psi|^2$ of the matter wave for the

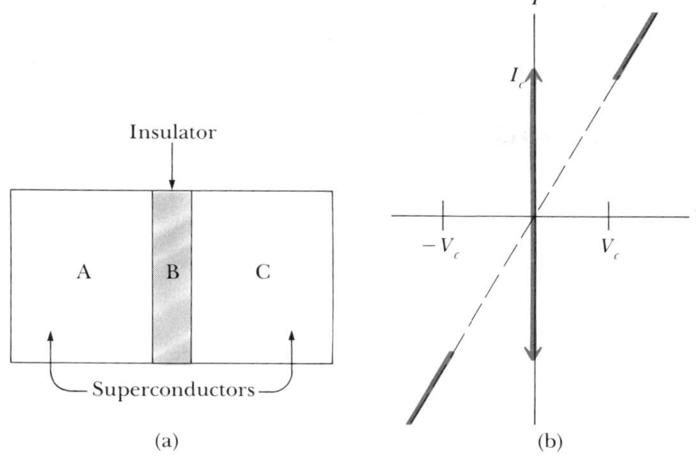

Figure 6.8 (a) A Josephson junction consists of two superconductors, A and B, separated by a very thin insulator, C. (b) The current-voltage characteristic of a Josephson junction. A dc current flows with no voltage drop across the junction up to some critical current I_c: this is the dc Josephson effect. If the voltage exceeds V_c, the current includes an oscillatory component with frequency $\omega = 2eV/\hbar$, and the junction shows some resistance: this is the ac Josephson effect.

superconductor; the phase of the wave Ψ is essentially arbitrary. In the Josephson junction, the Cooper pairs tunnel from one superconductor to the other through the oxide layer. As a result, the phases of the matter waves on either side of the layer are interrelated, and we say that the two superconductors are **phase locked.**

Several effects associated with pair tunneling or phase locking in the Josephson junction have been observed. In the *dc Josephson effect,* a dc current flows across the junction in the absence of electric or magnetic fields. The current is proportional to sin ϕ, where ϕ is the phase difference between the wavefunctions in the two superconductors. In the *ac Josephson effect,* a bias voltage V applied across the junction results in current oscillations at a frequency given by $\omega = 2eV/\hbar$. In effect, when an electron pair passes through the barrier, a photon of energy $\hbar\omega = 2eV$ is either emitted or absorbed. Since a voltage difference of only 1 μV gives a frequency of about 500 MHz, the Josephson junction finds application as a precision voltmeter for very weak signals. Alternatively, with independent and accurate measurements of ω and V, one can obtain a very precise value for e/h. The current-voltage characteristic of a Josephson junction is given in Figure 6.8b. Note that the resistance of the junction is zero for bias voltages between $-V_c$ and $+V_c$. Above V_c, the current has an oscillatory component.

Finally, we note that the application of a magnetic field at right angles to the direction of current through the junction will modify the phase angle ϕ. This feature is exploited in the SQUID or *s*uperconducting *qu*antum *i*nterference *d*evice, which uses two Josephson junctions in parallel for the measurement of very weak magnetic fields. A more detailed discussion of SQUIDs and Josephson junctions is provided in the essay at the end of Chapter 12.

Ammonia Inversion

The "inversion" of the ammonia molecule is another example of tunneling, this time for an entire atom. The equilibrium configuration of the ammonia (NH_3) molecule is shown in Figure 6.9a: the nitrogen atom is situated at the apex of a pyramid whose base is the equilateral triangle formed by the three hydrogen atoms. But this equilibrium is not truly stable; indeed, there is a *second* equilibrium position for the nitrogen atom on the opposite side of the

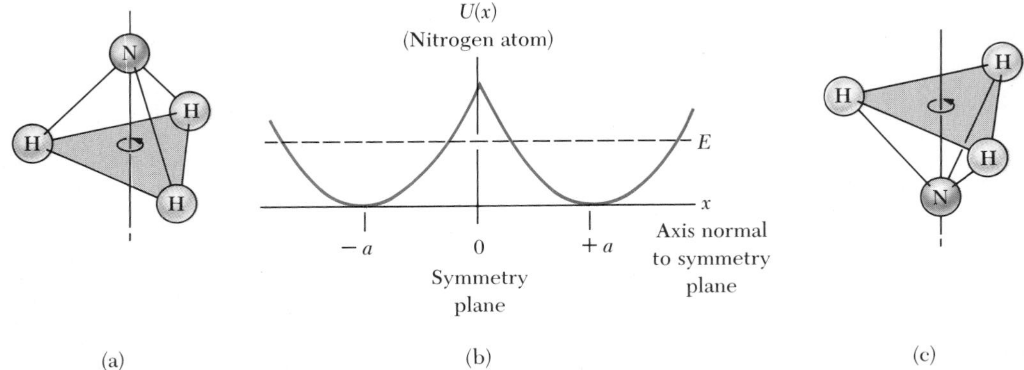

Figure 6.9 (a) The ammonia molecule NH_3. At equilibrium, the nitrogen atom is situated at the apex of a pyramid whose base is the equilateral triangle formed by the three hydrogen atoms. By symmetry, a second equilibrium configuration exists for the nitrogen atom on the opposite side of the plane formed by the hydrogen atoms. (b) The potential energy seen by the nitrogen atom along a line perpendicular to the symmetry plane. The two equilibrium points at $-a$ and $+a$ give rise to the double oscillator potential shown. A nitrogen atom with energy E can tunnel back and forth through the barrier from one equilibrium point to the other, with the result that the molecule alternates between the normal configuration in (a) and the inverted configuration shown in (c).

plane formed by the hydrogen atoms. With its two equilibrium locations, the nitrogen atom of the ammonia molecule constitutes a *double oscillator,* which can be modeled by using the double oscillator potential shown in Figure 6.9b. A nitrogen atom initially located to one side of the symmetry plane will not remain there indefinitely, since there is some probability that it can tunnel through the oscillator barrier to emerge on the other side. When this occurs, the molecule becomes *inverted* (Fig. 6.9c). But the process does not stop there; the nitrogen atom, now on the opposite side of the symmetry plane, has a probability of tunneling *back* through the barrier to take up its original position! The molecule does not just undergo one inversion, but flip-flops repeatedly, alternating between the two classical equilibrium configurations. The "flopping" frequency is fixed by the tunneling rate and turns out to be quite high, about 10^{10} Hz (microwave range of the electromagnetic spectrum)! Notice that the nitrogen atom to one side of the symmetry plane or the other does *not* constitute a stationary state, since the probability for finding it changes with time. Instead, the "flopping" state can be represented by a combination of two stationary states with energies differing by $\hbar\omega$, where ω is the flopping frequency.

Double oscillator representation of the ammonia molecule

Decay of Black Holes

Once inside the event horizon, nothing—not even light—can escape the gravitational pull of a black hole. That was the view held until 1974, when the brilliant British astrophysicist Stephen Hawking proposed that black holes are indeed radiant objects, emitting a variety of particles by a mechanism involving tunneling through the (gravitational) potential barrier surrounding the black hole. The thickness of this barrier is proportional to the size of the black hole, so that the likelihood of a tunneling event initially may be extremely small. However, as the black hole emits particles, its mass and size steadily decrease, making it easier for more particles to tunnel out. In this way emission continues at an ever increasing rate, until eventually the black hole radiates itself out of existence in an explosive climax! Thus, Hawking's scenario leads inexorably to the decay and eventual demise of any black hole. Calculations indicate that a black hole with the mass of our sun would survive against decay by tunneling for about 10^{66} years. On the other hand, a black hole with the mass of only a billion tons and roughly the size of a proton (a *primordial* black hole existing just after the *big bang* origin of the universe), should have almost completely evaporated in the 10 billion years that have elapsed since the time of creation.

6.3 SUMMARY

For potentials representing barriers there are no bound states, only unbound states describing particles scattering from the barrier. When a particle of energy E meets a barrier of height U, part of the incident wave is reflected by the barrier, and part is transmitted through the barrier. Particles incident on the barrier are reflected or transmitted with probability given by the **reflection coefficient** R and **transmission coefficient** T, respectively. These coefficients obey the **sum rule**

$$R(E) + T(E) = 1 \tag{6.5}$$

for any particle energy E. Even when $E < U$, the particle has some nonzero probability of penetrating the barrier. This process, called **tunneling,** is the basic mechanism underlying the phenomena of **field emission,** the **Josephson effect, alpha decay** of radioactive nuclei, particle radiation from **black holes,** and many more. The likelihood of tunneling through a barrier of arbitrary shape $U(x)$ is given approximately as

$$T(E) \approx \exp\left(-\frac{2}{\hbar}\sqrt{2m}\int\sqrt{U(x) - E}\,dx\right) \tag{6.11}$$

The integral in Equation 6.11 is taken over the **classically forbidden** region where $U(x) > E$.

SUGGESTIONS FOR FURTHER READING

1. Many of the topics in this chapter also are treated at about the same level by A. P. French and Edwin F. Taylor in *An Introduction to Quantum Physics,* New York, W. W. Norton and Company, Inc., 1978.
2. Experimental aspects of field emission microscopy and related analytical surface techniques are discussed by Philip F. Kane in Chapter 6 of *Characterization of Solid Surfaces,* New York, Plenum Press, 1974, pp. 133–146.
3. The Josephson effect and its role in determining fundamental physical constants is reviewed in "The Josephson Effect and e/h" by John Clarke in the *American Journal of Physics,* September 1970, pp. 1071–1095. This work also contains a number of other references to this interesting and important topic.
4. For readable accounts of tunneling from black holes, see S. W. Hawking, "The Quantum Mechanics of Black Holes," *Sci. American,* January 1977, pp. 34–40; and J. D. Beckenstein, "Black-hole Thermodynamics," *Physics Today,* January 1980, pp. 24–31. The second title is somewhat more advanced, and contains additional references.

QUESTIONS

1. Consider a particle with energy E scattered from a potential barrier of height $U > E$. How does the amplitude of the reflected wave change as the barrier height is reduced? How does the amplitude of the incident wave change?
2. An electron and a proton of identical energy encounter the same potential barrier. For which is the probability of transmission greatest, and why?
3. In classical physics, only *differences* in energy have physical significance. Discuss how this carries over to

the quantum scattering of particles by examining how adding a constant potential U_0 everywhere affects barrier reflection and transmission coefficients.

4. Suppose a particle with energy E is found within a barrier whose height U is greater than E. Will this particle be found to have a negative kinetic energy? Explain.

5. Explain how a barrier whose width is L might be considered wide for penetration by protons, yet at the same time narrow for penetration by electrons.

6. Discuss the suitability of portraying waveforms and potential barriers on the same graph, as in Figure 6.2a. Can a wave whose crest falls below the top of a square barrier ever penetrate the barrier? Explain.

PROBLEMS

1. A particle incident on the potential step of Example 6.1 with a certain energy $E < U$ is described by the wave

$$\psi(x) = \tfrac{1}{2}\{(1 + i)e^{ikx} + (1 - i)e^{-ikx}\} \qquad \text{for } x \leq 0$$

$$\psi(x) = e^{-kx} \qquad \text{for } x \geq 0$$

(a) Verify by direct calculation that the reflection coefficient is unity in this case. (b) How must k be related to E in order for $\psi(x)$ to solve Schrödinger's equation in the region to the left of the step ($x \leq 0$)? to the right of the step ($x > 0$)? What does this say about the ratio E/U? (c) Evaluate the penetration depth $\delta = 1/k$ for 10 MeV protons incident on this step.

2. Consider the step potential of Example 6.1 in the case where $E > U$. (a) Examine the Schrödinger equation to the left of the step to find the form of the solution in the range $x < 0$. Do the same to the right of the step to obtain the solution form for $x > 0$. Complete the solution by enforcing whatever boundary and matching conditions may be necessary. (b) Obtain an expression for the reflection coefficient R in this case, and show that it can be written in the form

$$R = \frac{(k_1 - k_2)^2}{(k_1 + k_2)^2}$$

where k_1 and k_2 are wave numbers for the incident and transmitted waves, respectively. (c) Evaluate R in the limiting cases of $E \to U$ and $E \to \infty$. Are the results sensible? Explain. (This situation is analogous to the partial reflection and transmission of light striking an interface separating two different media.)

3. Use the results of the preceding problem to calculate the fraction of 25 MeV protons reflected and the fraction transmitted by a 20 MeV step. How do your answers change if the protons are replaced by electrons?

4. An electron beam with kinetic energy 54 eV enters a sharply defined region of *lower* potential where the kinetic energy of the electrons is increased by 10 eV. What fraction of the incident beam will be reflected at the boundary? (This simulates electron scattering at normal incidence from a metal surface, as in the Davisson-Germer experiment.)

5. (a) Show that the probability that a particle of energy $E < U$ will tunnel through the barrier in Figure 6.2 is approximately

$$P \approx e^{-2CL}$$

where

$$C = \frac{\sqrt{2m(U - E)}}{\hbar}$$

and L is the width of the barrier. (b) Give numerical estimates of P for each of the following cases: (1) an electron with $U - E = 0.01$ eV and $L = 0.1$ nm; (2) an electron with $U - E = 1$ eV and $L = 0.1$ nm; (3) an alpha particle ($m = 6.7 \times 10^{-27}$ kg) with $U - E = 10^6$ eV and $L = 10^{-15}$ m; and (4) a bowling ball ($m = 8$ kg) with $U - E = 1$ J and $L = 2$ cm (this corresponds to the ball's getting past a barrier 2 cm wide and too high for the ball to slide over).

6. A beam of electrons is incident on a barrier 5 eV high and 2 nm wide. Write a simple computer program to find what energy the electrons should have if 0.1% of them are to get through the barrier. (Assume the wide barrier result is applicable—check your assumption by comparing δ with L, once the energy is found.)

7. Show that a necessary (but not sufficient) criterion for the applicability of the wide barrier approximation is

$$\frac{2mUL^2}{\hbar^2} \gg 1$$

(The quantity UL^2 is sometimes referred to as the barrier *strength*). Apply this criterion to the case described in Problem 6 and comment on your findings.

8. A model potential of interest for its simplicity is the *delta well*. The delta well may be thought of as a square well of width L and depth S/L in the limit $L \to 0$ (Fig. 6.10). The limit is such that S, the product of the well depth with its width, remains fixed at a finite value known as the *well strength*. The effect of a delta well is to introduce a discontinuity in the slope of the wavefunction at the well site, although the wave itself remains continuous here. In particular, it can be shown that

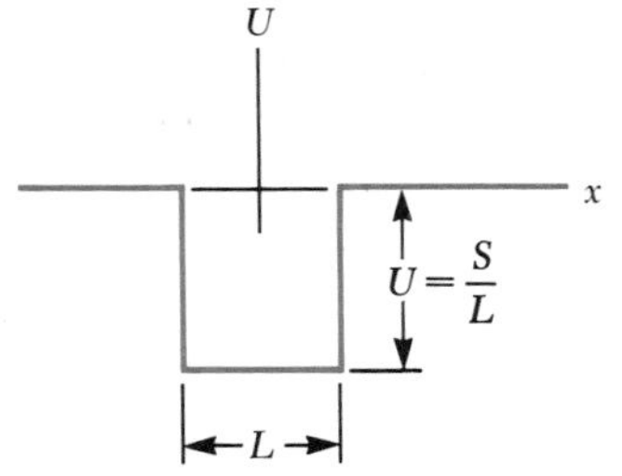

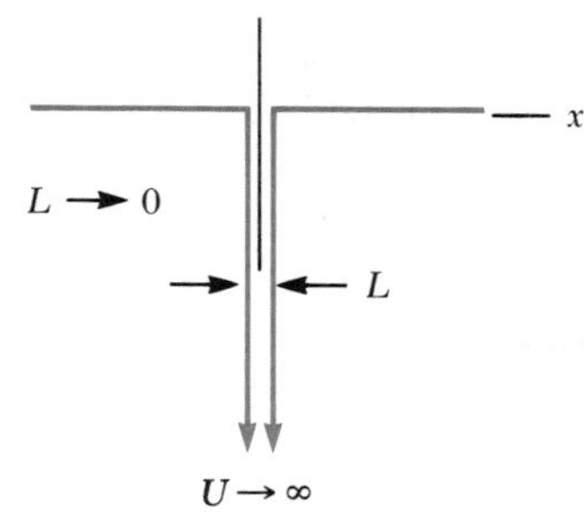

Figure 6.10 (Problem 8)

$$\left.\frac{d\psi}{dx}\right|_{0+} - \left.\frac{d\psi}{dx}\right|_{0-} = -\frac{2mS}{\hbar^2}\,\psi(0)$$

for a delta well of strength S situated at $x = 0$. (a) Solve Schrödinger's equation on both sides of the well ($x < 0$ and $x > 0$) for the case where particles are incident from the left with energy $E > 0$. Note that in these regions the particles are free, so that $U(x) = 0$. (b) Enforce the continuity of ψ and the slope condition at $x = 0$. Solve the resulting equations to obtain the transmission coefficient T as a function of particle energy E. Sketch $T(E)$ for $E \geq 0$. (c) If we allow E to be negative, we find that $T(E)$ diverges for some particular energy E_0. Find this value E_0. (As it happens, E_0 is the energy of a *bound state* in the delta well. The calculation illustrates a general technique, in which bound states are sought among the singularities of the scattering coefficients for a potential well of arbitrary shape.) (d) What fraction of the particles incident on the well with energy $E = |E_0|$ is transmitted and what fraction is reflected?

9. Obtain directly an expression for the reflection coefficient $R(E)$ for the delta well of Problem 8, and verify the sum rule

$$R(E) + T(E) = 1$$

for all particle energies $E > 0$.

10. Consider an alpha particle confined to a thorium nucleus. Model the nuclear potential as a semi-infinite square well with an infinitely high wall at $r = 0$ and a wall of height 30 MeV at the nuclear radius $R = 9$ fm. Use the iterative method of Example 5.8 to estimate the smallest values of energy and velocity permitted for the alpha particle. What conclusion can you draw from the fact that the *ejected* alpha is observed to have a kinetic energy of 4.05 MeV?

11. Estimate the collision frequency of an alpha particle with the nuclear barrier in the tunneling model for the alpha decay of thorium. The daughter nucleus for this case (radium) has $Z = 88$ and a radius of 9 fm. Take for the overall nuclear barrier 30 MeV, measured from the bottom of the nuclear well to the top of the Coulomb barrier (see Figure 6.7).

12. The double oscillator potential of Figure 6.9b is described by the potential energy function

$$U(x) = \tfrac{1}{2}M\omega^2(|x| - a)^2$$

with ω the classical frequency for vibration of a mass M around either of the equilibrium points $x = \pm a$. Evaluate the tunneling integral for barrier penetration by a particle possessing the minimum energy of vibration, $E = \frac{1}{2}\hbar\omega$. (*Hint:* Let $x = a - \sqrt{\hbar/M\omega}\cosh(y)$, where $\cosh(y) = (e^y + e^{-y})/2$ is the hyperbolic cosine function.) Apply this to the ammonia molecule, using the estimate 10^{14} rad/s for ω, and $a = 0.37$ Å. Estimate the collision frequency f for this case and calculate the tunneling rate and its reciprocal, the tunneling time.

13. Verify the claim of Section 6.2 that the electrons of a metal collide with the surface at a rate of about 10^{30} per second. Do this by estimating the collision frequency of electrons in a 1-cm cube of copper metal with one face of the cube surface. Assume that each copper atom contributes one conduction electron to the metal (the chemical valence of copper is +1), and that these conduction electrons move freely with kinetic energy equal to 7 eV.

PROBLEMS FOR THE COMPUTER:

The following problems require for their solutions the use of the program EIGEN. Where appropriate, furnish screen dumps to substantiate your answers.

14. Consider a particle incident on a square barrier of width $3\delta(0)$. For a particle energy which is 0.9 of the barrier height, estimate the transmission and reflection coefficients T and R, and verify the sum rule $T + R = 1$.

15. What program parameters describe an electron incident on a square barrier 10 eV high and 1 Å thick? Estimate the transmission probability for 5 eV electrons incident on the barrier. How does your result change if the particles are protons rather than electrons?

16. Find the ground and first excited states for a double oscillator whose equilibrium points are separated by a distance equal to three times the vibration amplitude for a single oscillator in its ground state. What symmetry, if any, do these wavefunctions exhibit? Repeat your determinations when the equilibrium separation is increased to four times the vibration amplitude. What do you suppose happens when the equilibrium separation approaches infinity?

Essay

THE SCANNING TUNNELING MICROSCOPE

Roger A. Freedman and Paul K. Hansma
Department of Physics, University of California, Santa Barbara

The basic idea of quantum mechanics, that particles have properties of waves and vice versa, is among the strangest found anywhere in science. Because of this strangeness, and because quantum mechanics mostly deals with the very small, it might seem to have little practical application. As we will show in this essay, however, one of the basic phenomena of quantum mechanics — the tunneling of a particle — is at the heart of a very practical device that is one of the most powerful microscopes ever built. This device, the *scanning tunneling microscope* or *STM*, enables physicists to make highly detailed images of surfaces with resolution comparable to the size of a *single atom*. Such images promise to revolutionize our understanding of structures and processes on the atomic scale.

Before discussing how the STM works, we first look at a sample of what the STM can do. An image made by a scanning tunneling microscope of the surface of a piece of gold is shown in Figure 1. You can easily see that the surface is not uniformly flat, but is a series of terraces separated by steps that are only one atom high. Gentle corrugations can be seen in the terraces, caused by subtle rearrangements of the gold atoms.

What makes the STM so remarkable is the fineness of the detail that can be seen in images such as Figure 1. The *resolution* in this image — that is, the size of the smallest detail that can be discerned — is about 2 Å (2×10^{-10} m). For an ordinary microscope, the resolution is limited by the wavelength of the waves used to make the image. Thus an optical microscope has a resolution of no better than 2000 Å, about half the wavelength of visible light, and so could never show the detail displayed in Figure 1. Electron microscopes can have a resolution of 2 Å by using electron waves of wavelength 4 Å or shorter. From the de Broglie formula $\lambda = h/p$, the electron momentum p required to give this wavelength is 3100 eV/c, corresponding to an electron speed $v = p/m = 1.8 \times 10^6$ m/s. Electrons traveling at this high speed would

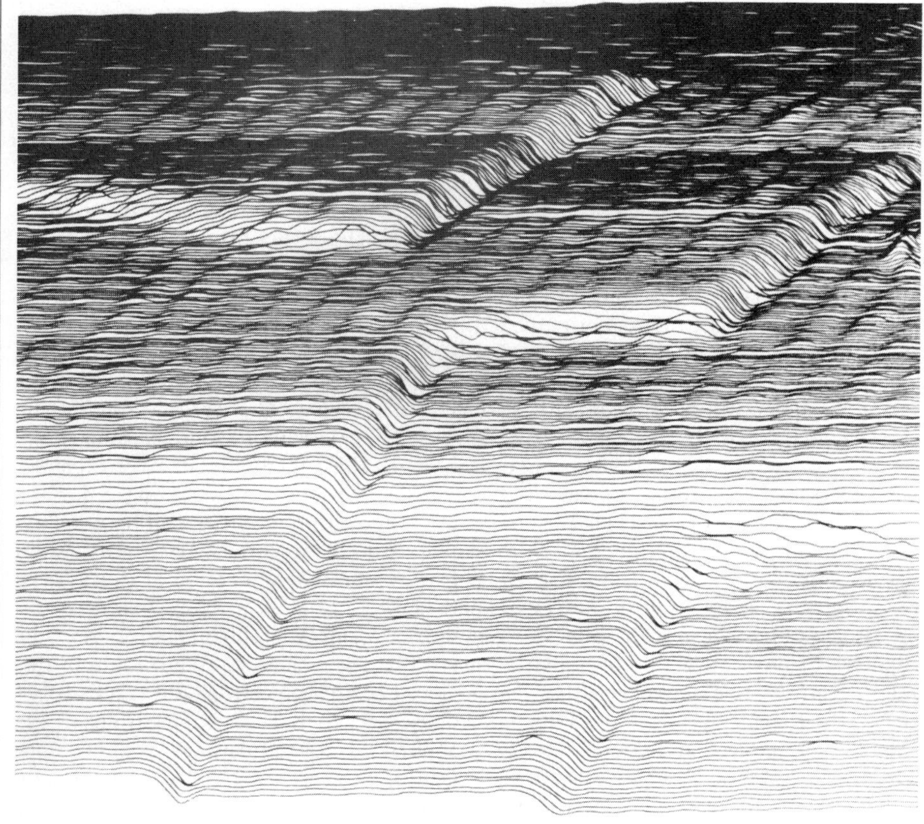

Figure 1 Scanning tunneling microscope image of the surface of crystalline gold. The divisions on the scale are 5 Å. Successive scans are approximately 1.5 Å apart. The figure is from G. Binnig, H. Rohrer, Ch. Gerber, and E. Stoll, *Surface Science* **144:** 321, 1984.

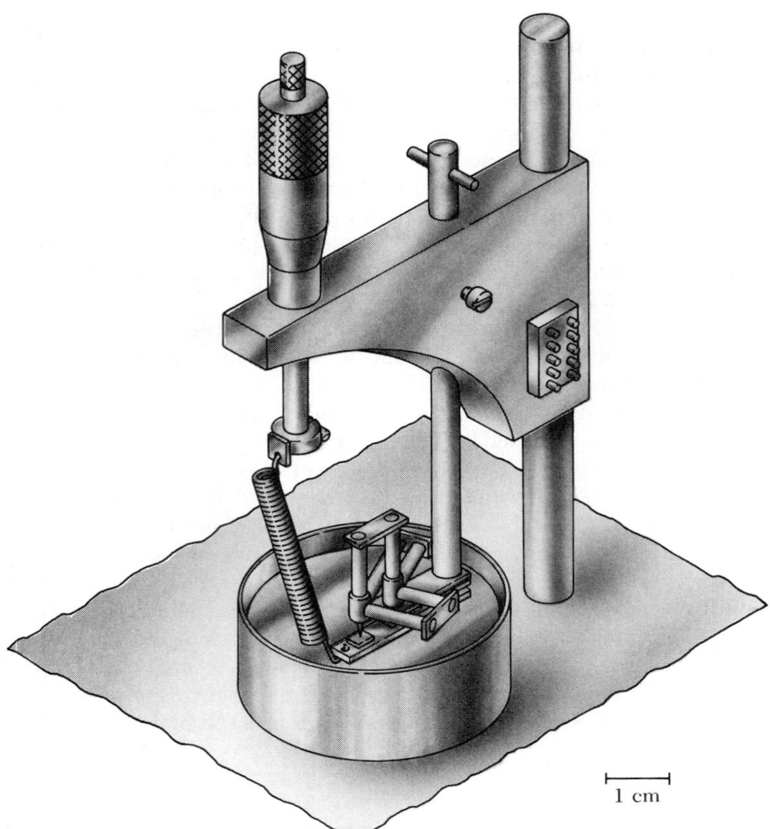

Figure 2 One design for a scanning tunneling microscope (STM). The sample to be studied is mounted on a plate in the cylindrical dish. The probe extends beneath the left tripod. The micrometer attached to the spring is used to position the sample.

penetrate into the interior of the piece of gold in Figure 1, and so would give no information about individual surface atoms.

The image in Figure 1 was made by Gerd Binnig, Heinrich Rohrer, and collaborators at the IBM Research Laboratory in Zurich, Switzerland. Binnig and Rohrer invented the STM and shared the 1986 Nobel Prize in Physics for their work. Such is the importance of this device that unlike most Nobel Prizes, which come decades after the original work, Binnig and Rohrer received their Nobel Prize just six years after their first experiments with an STM.

One design for an STM is shown in Figure 2. The basic idea behind its operation is very simple, as shown in Figure 3. A conducting probe with a very sharp tip is brought near the surface to be studied. Because it is attracted to the positive ions in the surface, an electron in the surface has a lower total energy than would an electron in the empty space between surface and tip. The same thing is true for an electron in the tip. In classical Newtonian mechanics, electrons could not move between the surface and tip because they would lack the energy to escape either material. But because the electrons obey quantum mechanics, they can "tunnel" across the barrier of empty space between the surface and the tip. Let us explore the operation of the STM in terms of the discussion of tunneling in Section 6.2.

For an electron in the apparatus of Figures 2 and 3, a plot of the energy as a function of position would look like Figure 6.6b. The horizontal coordinate in this figure represents electron position. Now L is to be interpreted as the distance between the surface and the tip, so that coordinates less than 0 refer to positions inside the surface material and coordinates greater than L refer to positions inside the tip. The barrier height $U = q\phi$ is the potential energy difference between an electron outside the material and an electron in the material. That is, an electron in the surface or tip has potential energy $-U$ compared to one in vacuum. (We are assuming for the moment that the surface and tip are made of the same material. We will comment on this assmption shortly.) The kinetic energy of an electron in the surface is E so that an amount of energy equal to $(U-E)$ must be given to an electron to remove it from the surface. Thus $(U-E)$ is the work function of an electron in the surface.

For the potential energy curve of Figure 6.6b, one could expect as much tunneling from the surface into the tip as in the opposite direction. In an STM, the direction in which electrons tend to cross the barrier is controlled by applying a voltage between the surface and the tip. With preferential tunneling from the surface into the tip, the tip samples the distribution of electrons in and above the surface. Because of this "bias" voltage, the work functions of surface and tip are different, giving a preferred direction of tunneling. This is also automatically the case if the surface and tip are made of different materials. In addition, the top of the barrier in Figure 6.6b will not be flat but will be tilted to reflect the electric field between the surface and the tip. However, if the barrier energy U is large compared to the difference between the surface and tip work functions, and if the bias voltage is small compared to $\phi = U/q$, we can ignore these complications in our calculations. Then all the results for a square barrier given in Section 6.1 may be applied to an STM. Detailed discussions of the effects that result when these complications are included can be found in the article by Hansma and Tersoff and in the Nobel Prize lecture of Binnig and Rohrer (see Suggestions for Further Reading).

The characteristic scale of length for tunneling is set by the work function $(U-E)$. For a typical value $(U-E) = 4.0$ eV, this scale of length is

$$\delta = \frac{\hbar}{\sqrt{2m(U-E)}} = \frac{\hbar c}{\sqrt{2mc^2(U-E)}}$$

$$= \frac{1.973 \text{ keV}\cdot\text{Å}}{\sqrt{2(511 \text{ keV})(4.0 \times 10^{-3} \text{ keV})}} = 0.98 \text{ Å} \approx 1.0 \text{ Å}$$

The probability that a given electron will tunnel across the barrier is just the transmission coefficient T (Equation 6.4). If the separation L between surface and tip is not small compared to δ, then the barrier is "wide" and we can use the approximate result in Equation 6.9 for T. The current of electrons tunneling across the barrier is simply proportional to T. The tunneling current density can be shown to be

$$j = \frac{e^2V}{4\pi^2 L\delta\hbar} e^{-2L/\delta}$$

In this expression e is the charge of the electron and V is the bias voltage between surface and tip.

We can see from this expression that the STM is very sensitive to the separation L between tip and surface. This is because of the exponential dependence of the tunneling current on L (this is much more important than the $1/L$ dependence). As we saw above, typically $\delta \approx 1.0$ Å. Hence increasing the distance L by just 0.01 Å causes the tunneling current to be multiplied by a factor $e^{-2(0.01 \text{ Å})/(1.0 \text{ Å})} \approx 0.98$; i.e., the current decreases by 2%—a change that is measurable. For distances L greater than 10 Å (that is, beyond a few atomic diameters), essentially no tunneling takes place. This sensitivity to L is the basis of the operation of the STM: monitoring the tunneling current as the tip is scanned over the surface gives a sensitive measure of the topography of the surface. In this way the STM can measure the height of surface features to within 0.01 Å, or approximately one one-hundredth of an atomic diameter.

The STM also has excellent lateral resolution, that is, resolution of features in the plane of the surface. This is because the tips used are *very* sharp indeed, typically only an atom or two wide at their extreme end. Thus the tip samples the surface electrons only in a very tiny region approximately 2 Å wide, and so can "see" very fine detail. You might think that making such tips would be extremely difficult, but in fact it's relatively easy: sometimes just sharpening the tip on a fine grinding stone (or even with fine sandpaper) is enough to cause the tip atoms to rearrange by themselves into an atomically sharp configuration. (If you find this surprising, you're not alone. Binnig and Rohrer were no less surprised when they discovered this.)

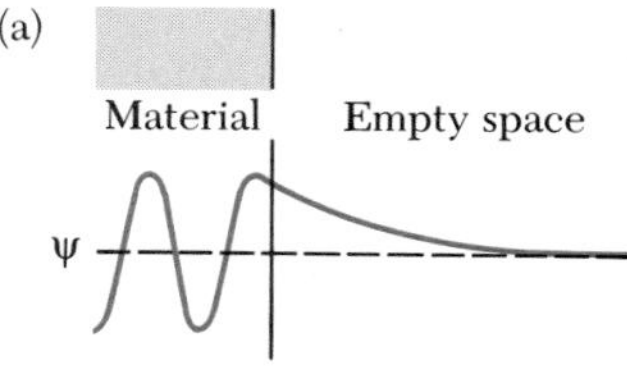

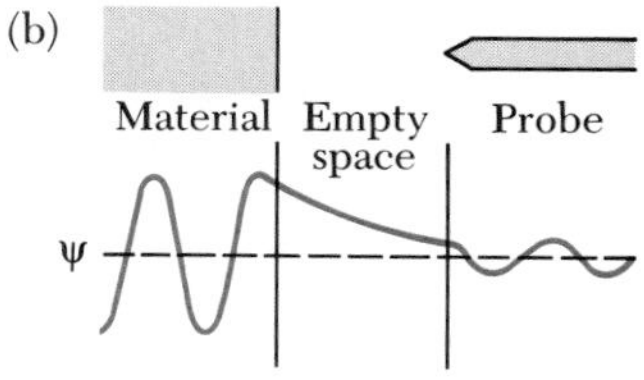

Figure 3 (a) The wavefunction of an electron in the surface of the material to be studied. The wavefunction extends beyond the surface into the empty region. (b) The sharp tip of a conducting probe is brought close to the surface. The wavefunction of a surface electron penetrates into the tip, so that the electron can "tunnel" from surface to tip. Compare this figure to Figure 6.7b.

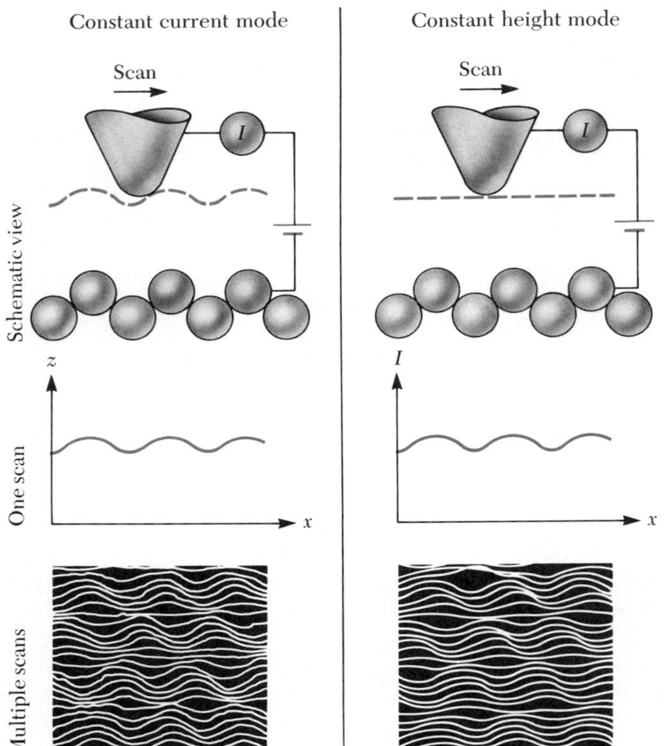

Figure 4 Scanning tunneling microscopes can be operated in either (a) the constant current mode or (b) the constant height mode. The images of the surface of graphite were made by Richard Sonnenfeld at the University of California at Santa Barbara. The constant height mode was first used by A. Bryant, D. P. E. Smith, and C. F. Quate, *Applied Physics Letters* **48:** 832, 1986.

There are two modes of operation for the STM, shown in Figure 4. In the *constant current mode* (Figure 4a), a convenient operating voltage (typically between 2 millivolts and 2 volts) is first established between surface and tip. The tip is then brought close enough to the surface to obtain measurable tunneling current. The tip is then scanned over the surface while the tunneling current I is measured. A feedback network changes the vertical position of the tip, z, to keep the tunneling current constant, thereby keeping the separation L between surface and tip constant. An image of the surface is made by plotting z versus lateral position (x, y). The simplest scheme for plotting the image is shown in the graph below the schematic view. The height z is plotted versus the scan position x. An image consists of multiple scans displaced laterally from each other in the y direction.

The constant current mode was historically the first to be used, and has the advantage that it can be used to track surfaces that are not atomically flat (as in Figure 1). However, the feedback network requires that the scanning be done relatively slowly. As a result, the sample being scanned must be held fixed in place for relatively long times to prevent image distortion.

Alternatively, in the *constant height mode* (Figure 4b), the tip is scanned across the surface at constant voltage and nearly constant height while the current is monitored. In this case the feedback network responds only rapidly enough to keep the average current constant, which means that the tip maintains the same average separation from the surface. The image is then a plot of current I versus lateral position (x, y), as shown in the graph below the schematic. Again, multiple scans along x are displayed laterally displaced in the y direction. The image shows the substantial variation of tunneling current as the tip passes over surface features such as individual atoms.

The constant height mode allows much faster scanning of atomically flat surfaces (100 times faster than the constant current mode), since the tip does not have to be moved up and down over the surface "terrain." This fast scanning means that making an image of a surface requires only a short "exposure time." By making a sequence of

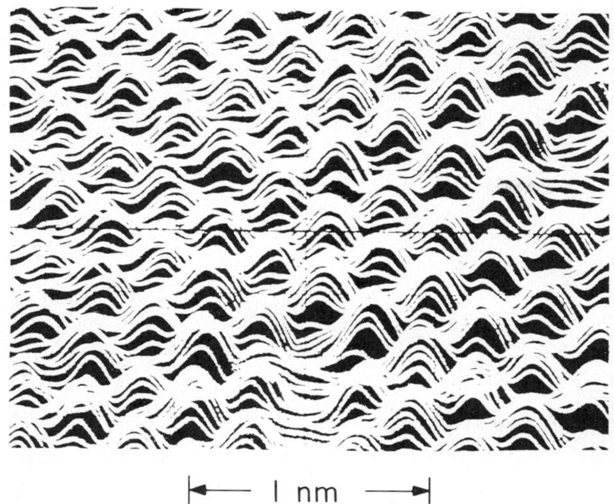

Figure 5 Image of atoms on a surface of tantalum disulfide (TaS_2) immersed in liquid nitrogen. 1 nm = 10^{-9} meter = 10 Å. The figure is from C. G. Slough, W. W. McNairy, R. V. Coleman, B. Drake, and P. K. Hansma, *Physical Review* **B34:** 994, 1986.

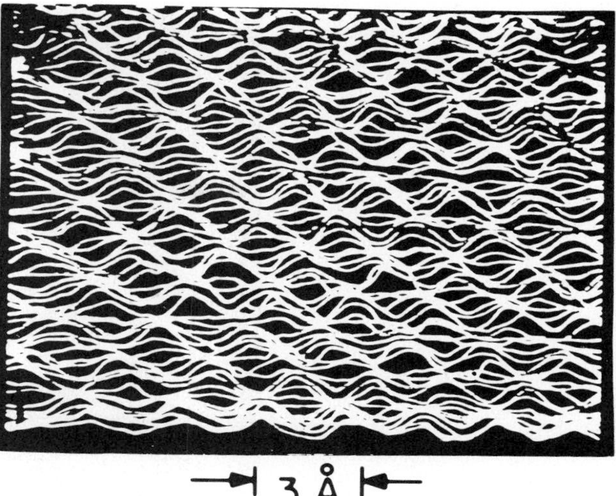

Figure 6 Image of a graphite electrode in an electrolyte used for silver plating. The figure is from R. Sonnenfeld and B. Schardt, *Applied Physics Letters* **49:** 1172, 1986.

such images, researchers may be able to study in real-time processes in which the surfaces rearrange themselves—in effect making an STM "movie."

Individual atoms have been imaged on a variety of surfaces, including those of so-called *layered materials* in which atoms are naturally arranged into two-dimensional layers. Figure 5 shows an example of atoms on one of these layered materials. In this image it is fascinating not only to see individual atoms, but also to note that some atoms are missing. Specifically, there are three atoms missing from Figure 5. Can you find the places where they belong?

Another remarkable aspect of the STM image in Figure 5 is that it was obtained with the surface and tip immersed in liquid nitrogen. While we assumed earlier in this essay that the space between the surface and tip must be empty, in fact electron tunneling can take place not just through vacuum but also through gases and liquids—even water. This seems very surprising since we think of water, especially water with salts dissolved in it, as a conductor. But water is only an *ionic* conductor. For electrons, water behaves as an insulator just as vacuum behaves as an insulator. Thus electrons can flow through water only by tunneling, which makes scanning tunneling microscopy possible "under water."

As an example, Figure 6 shows individual carbon atoms on a graphite surface. It was obtained for a surface immersed in a silver plating solution, which is highly conductive for ions but behaves as an insulator for electrons. (The sides of the conducting probe were sheathed with a nonconductor, so that the predominant current into the probe comes from electrons tunneling into the exposed tip. The design of STM used to make this particular image is the one shown in Figure 2.) Sonnenfeld and Schardt observed atoms on this graphite surface before plating it with silver, after "islands" of silver atoms were plated onto the surface, and after the silver was electrochemically stripped from the surface. Their work illustrates the promise of the scanning tunneling microscope for seeing processes that take place on an atomic scale.

While the original STMs were one-of-a-kind laboratory devices, commercial STMs have recently become available. Figure 7 is an image of a graphite surface in air made with such a commercial STM. Note the high quality of this image and the recognizable rings of carbon atoms. You may be able to see that three of the six carbon atoms in each ring *appear* lower than the other three. All six atoms are in fact at the same level, but the three that appear lower are bonded to carbon atoms lying directly beneath them in the underlying atomic layer. The atoms in the surface layer that

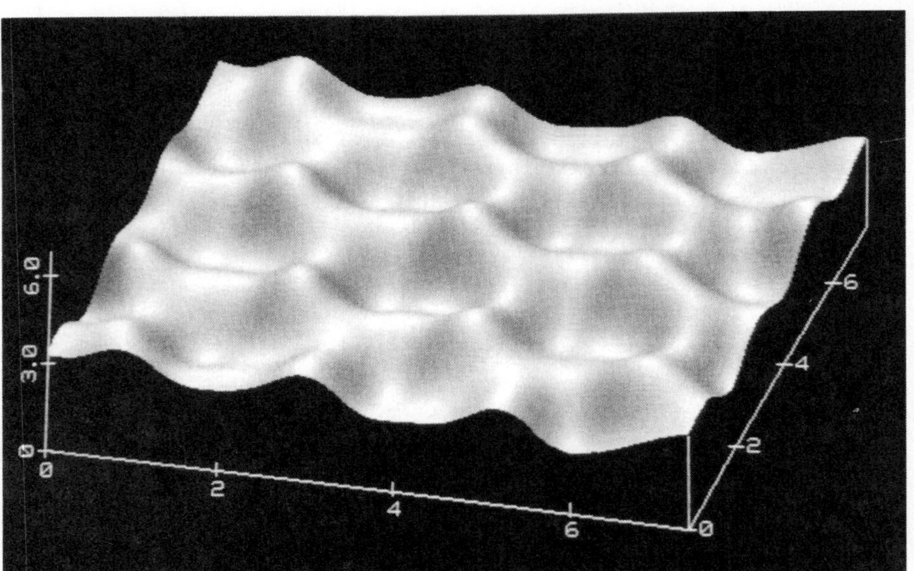

Figure 7 Image of a graphite surface in air, obtained with a commercial STM: the Nanoscope II from Digital Instruments in Goleta, California.

appear higher do not lie directly over subsurface atoms, and hence are not bonded to carbon atoms beneath them. For the higher-appearing atoms, some of the electron density that would have been involved in bonding to atoms beneath the surface instead extends into the space above the surface. This extra electron density makes these atoms appear higher in Figure 7, since what the STM maps is the topography of the electron distribution.

The availability of commercial instruments should speed the use of scanning tunneling microscopy in a variety of applications. These include characterizing electrodes for electrochemistry (while the electrode is still in the electrolyte), characterizing the roughness of surfaces, measuring the quality of optical gratings, and even imaging replicas of biological structures.

Perhaps the most remarkable thing about the scanning tunneling microscope is that its operation is based on a quantum mechanical phenomenon — tunneling — that was well understood in the 1920s, yet the STM itself wasn't built until the 1980s. What other applications of quantum mechanics may yet be waiting to be discovered?

Suggestions for Further Reading

G. Binnig, H. Rohrer, Ch. Gerber, and E. Weibel, *Physical Review Letters* **49:** 57, 1982. The first description of the operation of a scanning tunneling microscope.

G. Binnig and H. Rohrer, *Sci. American,* August 1985, p. 50. A popular description of the STM and its applications.

C. F. Quate, *Physics Today,* August 1986, p. 26. An overview of the field of scanning tunneling microscopy, including insights into how it came to be developed.

P. K. Hansma and J. Tersoff, *Journal of Applied Physics* **61:** R1, 1987. A comprehensive review of the "state of the art" in scanning tunneling microscopy.

G. Binnig and H. Rohrer, *Reviews of Modern Physics* **59:** 615, 1987. The text of the lecture given on the occasion of the presentation of the 1986 Nobel Prize in Physics.

Essay Questions and Problems

1. The density of the earth's atmosphere of air depends on altitude z according to $\rho = \rho_0 \, e^{-z/z_0}$, where z_0, the "scale height" of the atmosphere, is 8430 meters. What is the scale height for the electron "atmosphere" above a conductor? Give both a formula and a numerical estimate.

2. Our discussion of STM operation was based on the assumption that the conducting probe has only a single tip. In fact a probe may have a number of protrusions on its end, each of which acts as a "tip." Hence it might be expected that such a probe would give multiple STM images of the surface, greatly complicating the analysis. Explain why there is in fact no problem with multiple images, provided that the multiple tips differ in proximity to the surface by 20 Å or more.
3. The STM, while using physical concepts that date from the 1920s, was not developed until the 1980s. Suggest some reasons why the STM was not invented half a century earlier.
4. It was stated in the essay that for conventional microscopes, the resolution is limited by the wavelength of the waves used to make the image. To see whether this guideline applies to the STM, estimate the wavelength of an electron in the surface of a conductor and compare your estimate to the vertical resolution (about 0.01 Å) and lateral resolution (about 2 Å). Does the guideline apply to the two resolutions? Why or why not?

7

Quantum Mechanics in Three Dimensions

So far we have shown how quantum mechanics can be used to describe motion in one dimension. Although the one-dimensional case illustrates such basic features of systems as the quantization of energy, we need a full three-dimensional treatment for the applications to atomic, solid state, and nuclear physics that we will meet in later chapters. In this chapter we extend the concepts of quantum mechanics from one to three dimensions, and explore the predictions of the theory for the simplest of real systems — the hydrogen atom.

With the introduction of new degrees of freedom (and the additional coordinates needed to describe them) comes a disproportionate increase in the level of mathematical difficulty. To guide our inquiry, we shall rely on classical insights in helping us identify the observables that are likely candidates for quantization. These must come from the ranks of the so-called sharp observables and, with few exceptions, are the same as the conserved quantities of classical physics.

7.1 PARTICLE IN A THREE-DIMENSIONAL BOX

Let us explore the workings of wave mechanics in three dimensions through the example of a particle confined to a cubic "box." The box has edge length L and occupies the region $0 < x, y, z < L$, as shown in Figure 7.1. A classical particle would rattle around inside the box, colliding with the walls. At each collision, the component of particle momentum normal to the wall is reversed (changes sign), while the other two components of momentum are unaffected (see Fig. 7.2). Thus, the collisions preserve the *magnitude* of each momentum component, in addition to the total particle energy. These four quantities — $|p_x|, |p_y|, |p_z|$, and E — should be among the sharp observables subject to quantization.[1]

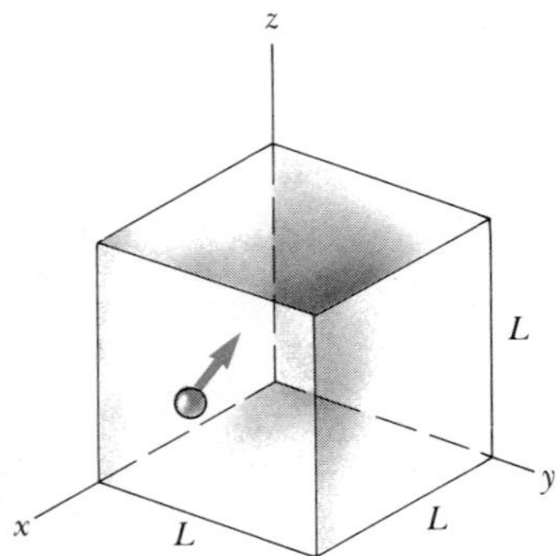

Figure 7.1 A particle confined to move in a cubic box of sides L. Inside the box $U = 0$. The potential energy is infinite at the walls and outside the box.

The matter wave Ψ in three dimensions is a function of $\mathbf{r}$ and t. Again, only the intensity of Ψ is meaningful, and we define the probability density $P(\mathbf{r}, t) = |\Psi(\mathbf{r}, t)|^2$, which is now a *probability per unit volume*. Multiplication by the volume element dV $(= dx\,dy\,dz)$ gives the probability of finding the particle within the volume element dV at the point $\mathbf{r}$ at time t.

Since our particle is confined to the box, the wave function Ψ must be zero outside and at the walls. The wave inside the box is found from Schrödinger's equation,

[1] A sharp observable need not be quantized, that is, limited to discrete values. For example, the free-particle waves in one dimension $e^{i(kx-\omega t)}$ have sharp energy $E = \hbar\omega$ and sharp momentum $p = \hbar k$, but no restrictions apply to either E or p.

$$-\frac{\hbar^2}{2m}\nabla^2\Psi + U(\mathbf{r})\Psi = i\hbar\frac{\partial\Psi}{\partial t} \tag{7.1}$$

Schrödinger equation in three dimensions

We see that $\partial^2/\partial x^2$ in the one-dimensional case is replaced in three dimensions by the **Laplacian,**

$$\nabla^2 = \frac{\partial^2}{\partial x^2} + \frac{\partial^2}{\partial y^2} + \frac{\partial^2}{\partial z^2} \tag{7.2}$$

This is consistent with our belief that the cartesian axes label independent but logically equivalent directions in space. These derivative terms are identified with the particle's kinetic energy along each of the three coordinate axes, while $U(\mathbf{r}) = U(x, y, z)$ is its potential energy. The **stationary states** are solutions to Schrödinger's equation in separable form,

$$\Psi(\mathbf{r}, t) = \psi(\mathbf{r})e^{-i\omega t} \tag{7.3}$$

With this time dependence, the right-hand side of Equation 7.1 reduces to $\hbar\omega\Psi$, leaving $\psi(\mathbf{r})$ to satisfy the **time-independent Schrödinger equation** for a particle whose energy is sharp at the value $E = \hbar\omega$:

$$-\frac{\hbar^2}{2m}\nabla^2\psi(\mathbf{r}) + U(\mathbf{r})\psi(\mathbf{r}) = E\psi(\mathbf{r}) \tag{7.4}$$

The time-independent Schrödinger equation

Since our particle is free inside the box, we take the potential energy $U(\mathbf{r}) = 0$ for $0 < x, y, z < L$. In this case the spatial waveform also is separable; that is, solutions to Equation 7.4 with $U(\mathbf{r}) = 0$ can be found in product form:

$$\psi(\mathbf{r}) = \psi(x, y, z) = \psi_1(x)\psi_2(y)\psi_3(z) \tag{7.5}$$

Substituting Equation 7.5 into Equation 7.4 and dividing every term by the function $\psi(x, y, z)$ gives (for $U(\mathbf{r}) = 0$)

$$-\frac{\hbar^2}{2m\psi_1}\frac{\partial^2\psi_1}{\partial x^2} - \frac{\hbar^2}{2m\psi_2}\frac{\partial^2\psi_2}{\partial y^2} - \frac{\hbar^2}{2m\psi_3}\frac{\partial^2\psi_3}{\partial z^2} = E$$

In this form the independent variables are isolated: the first term on the left depends only on x, the second only on y, and the third only on z. To satisfy the equation *everywhere* inside the cube, each of these terms must reduce to a constant:

$$\begin{aligned} -\frac{\hbar^2}{2m\psi_1}\frac{\partial^2\psi_1}{\partial x^2} &= E_1 \\ -\frac{\hbar^2}{2m\psi_2}\frac{\partial^2\psi_2}{\partial y^2} &= E_2 \\ -\frac{\hbar^2}{2m\psi_3}\frac{\partial^2\psi_3}{\partial z^2} &= E_3 \end{aligned} \tag{7.6}$$

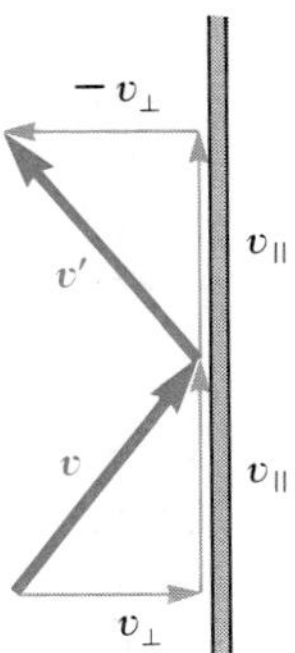

Figure 7.2 Change in particle velocity (or momentum) during collision with a wall of the box. For elastic collision with a smooth wall, the component normal to the wall is reversed, while the tangential components are unaffected.

The stationary states for a particle confined to a cube are obtained from these three separate equations. The energies E_1, E_2, and E_3 are **separation constants,** and represent the energy of motion along the three cartesian axes x, y, and z. Consistent with this identification, the Schrödinger equation requires $E_1 + E_2 + E_3 = E$.

The first of Equations 7.6 is the same as that for the infinite square well in one dimension. Independent solutions to this equation are $\sin k_1x$ and $\cos k_1x$, where $k_1 = \sqrt{2mE_1/\hbar^2}$ is the wavenumber of oscillation. However, only $\sin k_1x$ satisfies the condition that the wave must vanish at the wall $x = 0$. Requiring the wave to vanish also at the opposite wall $x = L$ implies $k_1L = n_1\pi$, where n_1 is any positive integer. In other words, we must be able to fit an integral number of half-wavelengths into our box along the direction marked by x. It follows that the magnitude of particle momentum along this direction must be one of the special values

$$|p_x| = \hbar k_1 = n_1 \frac{\pi\hbar}{L} \qquad n_1 = 1, 2, \ldots$$

Identical considerations applied to the remaining two equations show that the magnitudes of particle momentum in all three directions are similarly quantized:

Allowed values of momentum components for a particle in a box

$$\begin{aligned} |p_x| &= \hbar k_1 = n_1 \frac{\pi\hbar}{L} \qquad && n_1 = 1, 2, \ldots \\ |p_y| &= \hbar k_2 = n_2 \frac{\pi\hbar}{L} \qquad && n_2 = 1, 2, \ldots \\ |p_z| &= \hbar k_3 = n_3 \frac{\pi\hbar}{L} \qquad && n_3 = 1, 2, \ldots \end{aligned} \tag{7.7}$$

Since the momenta are restricted this way, the particle energy (all kinetic) is limited to the following discrete values:

Discrete energies allowed for a particle in a box

$$E = E_1 + E_2 + E_3 = \frac{\pi^2\hbar^2}{2mL^2}\{n_1^2 + n_2^2 + n_3^2\} \tag{7.8}$$

Thus, confining the particle to the cube serves to quantize its momentum and energy according to Equations 7.7 and 7.8. Note that *three quantum numbers are needed to specify the quantum condition, corresponding to the three independent degrees of freedom for a particle in space.* These quantum numbers specify the values taken by the sharp observables for this system.

Collecting the above results, we see that the stationary states for this particle are

$$\begin{aligned} \Psi(x, y, z, t) &= A \sin(k_1x)\sin(k_2y)\sin(k_3z)e^{-i\omega t} \qquad && \text{for } 0 < x, y, z < L \\ &= 0 && \text{otherwise.} \end{aligned} \tag{7.9}$$

The multiplier A is chosen to satisfy the normalization requirement. Example 7.1 shows that $A = (2/L)^{3/2}$ for the ground state, and this result continues to hold for the excited states as well.

EXAMPLE 7.1 Normalizing the Box Wave Functions Find the value of the multiplier A that normalizes the wave of Equation 7.9 having the lowest energy.

Solution: The state of lowest energy is described by $n_1 = n_2 = n_3 = 1$, or $k_1 = k_2 = k_3 = \pi/L$. Since Ψ is nonzero only for $0 < x, y, z < L$, the probability density integrated over the volume of this cube must be unity:

$$1 = \int_0^L dx \int_0^L dy \int_0^L dz\, |\Psi(x, y, z, t)|^2$$

$$1 = A^2 \left\{\int_0^L \sin^2(\pi x/L)\, dx\right\}\left\{\int_0^L \sin^2(\pi y/L)\, dy\right\} \times \left\{\int_0^L \sin^2(\pi z/L)\, dz\right\}$$

Using $2 \sin^2 \theta = 1 - \cos 2\theta$ gives

$$\int_0^L \sin^2(\pi x/L)\, dx = \frac{L}{2} - \frac{L}{4\pi} \sin(2\pi x/L) \Big|_0^L = \frac{L}{2}$$

The same result is obtained for the integrations over y and z. Thus, normalization requires

$$1 = A^2 \left(\frac{L}{2}\right)^3$$

or

$$A = \left(\frac{2}{L}\right)^{3/2}$$

Exercise 1 With what probability will the particle described by the wave function of Example 7.1 be found in the volume $0 < x, y, z < L/4$?

Answer: 0.040, or about 4%.

The ground state, for which $n_1 = n_2 = n_3 = 1$, has energy

$$E_{1,1,1} = \frac{3\pi^2\hbar^2}{2mL^2}$$

There are *three* first excited states, corresponding to the three different combinations of n_1, n_2, and n_3 whose squares sum to 6. That is, we obtain the *same* energy for the three combinations $n_1 = 2$, $n_2 = 1$, $n_3 = 1$, or $n_1 = 1$, $n_2 = 2$, $n_3 = 1$, or $n_1 = 1$, $n_2 = 1$, $n_3 = 2$. The first excited state energy is

$$E_{2,1,1} = E_{1,2,1} = E_{1,1,2} = \frac{6\pi^2\hbar^2}{2mL^2}$$

Note that each of the first excited states is characterized by a different wave function: $\psi_{2,1,1}$ has period L along the x axis and period $2L$ along the y and z axes, while for $\psi_{1,2,1}$ and $\psi_{1,1,2}$ the shortest period is along the y axis and the z axis, respectively.

Whenever different states have the same energy, this energy level is said to be **degenerate.** In the example just described, the first excited level is threefold (or triply) degenerate. This system has degenerate levels because of the high degree of symmetry associated with the cubic shape of the box. The degeneracy would be removed, or *lifted,* if the sides of the box were of unequal lengths (see Example 7.3). In fact, the extent of splitting of the levels increases with the degree of asymmetry.

Figure 7.3 is an energy level diagram showing the first five levels of a particle in a cubic box; Table 7.1 lists the quantum numbers and degeneracies of the various levels.

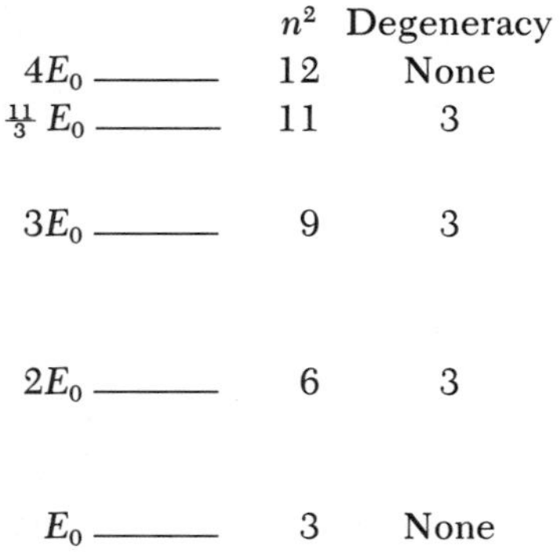

	n^2	Degeneracy
$4E_0$	12	None
$\frac{11}{3} E_0$	11	3
$3E_0$	9	3
$2E_0$	6	3
E_0	3	None

Figure 7.3 An energy level diagram for a particle confined to a cubic box. The lowest, or ground state, energy is $E_0 = 3\pi^2\hbar^2/2mL^2$, and $n^2 = n_1{}^2 + n_2{}^2 + n_3{}^2$. Note that most of the levels are degenerate.

TABLE 7.1 Quantum Numbers and Degeneracies of the Energy Levels for a Particle Confined to a Cubic Box

n_1	n_2	n_3	n^2	Degeneracy
1	1	1	3	None
1	1	2	6	Threefold
1	2	1	6	
2	1	1	6	
1	2	2	9	Threefold
2	1	2	9	
2	2	1	9	
1	1	3	11	Threefold
1	3	1	11	
3	1	1	11	
2	2	2	12	None

Note: $n^2 = n_1{}^2 + n_2{}^2 + n_3{}^2$.

EXAMPLE 7.2 The Second Excited State
Determine the wavefunctions and energy for the second excited level of a particle in a cubic box of edge L. What is the degeneracy of this level?

Solution: The second excited level corresponds to the three combinations of quantum numbers $n_1 = 2, n_2 = 2, n_3 = 1$ or $n_1 = 2, n_2 = 1, n_3 = 2$ or $n_1 = 1, n_2 = 2, n_3 = 2$. The corresponding wave functions inside the box are

$$\Psi_{2,2,1} = A \sin\left(\frac{2\pi x}{L}\right) \sin\left(\frac{2\pi y}{L}\right) \sin\left(\frac{\pi z}{L}\right) e^{-iE_{2,2,1}t/\hbar}$$

$$\Psi_{2,1,2} = A \sin\left(\frac{2\pi x}{L}\right) \sin\left(\frac{\pi y}{L}\right) \sin\left(\frac{2\pi z}{L}\right) e^{-iE_{2,1,2}t/\hbar}$$

$$\Psi_{1,2,2} = A \sin\left(\frac{\pi x}{L}\right) \sin\left(\frac{2\pi y}{L}\right) \sin\left(\frac{2\pi z}{L}\right) e^{-iE_{1,2,2}t/\hbar}$$

The level is threefold degenerate, since each of these wave functions has the same energy,

$$E_{2,2,1} = E_{2,1,2} = E_{1,2,2} = \frac{9\pi^2\hbar^2}{2mL^2}$$

EXAMPLE 7.3 Quantization in a Rectangular Box
Obtain a formula for the allowed energies of a particle confined to a rectangular box with edge lengths L_1, L_2, and L_3. What is the degeneracy of the first excited state?

Solution: For a box having edge length L_1 in the x direction, ψ will be zero at the walls if L_1 is an integral number of half-wavelengths. Thus, the magnitude of particle momentum in this direction is quantized as

$$|p_x| = \hbar k_1 = n_1 \frac{\pi\hbar}{L_1} \qquad n_1 = 1, 2, \ldots$$

Likewise, for the other two directions, we have

$$|p_y| = \hbar k_2 = n_2 \frac{\pi\hbar}{L_2} \qquad n_2 = 1, 2, \ldots$$

$$|p_z| = \hbar k_3 = n_3 \frac{\pi\hbar}{L_3} \qquad n_3 = 1, 2, \ldots$$

The allowed energies are

$$E = (|p_x|^2 + |p_y|^2 + |p_z|^2)/2m$$
$$= \frac{\pi^2\hbar^2}{2m}\left\{\left(\frac{n_1}{L_1}\right)^2 + \left(\frac{n_2}{L_2}\right)^2 + \left(\frac{n_3}{L_3}\right)^2\right\}$$

The lowest energy occurs again for $n_1 = n_2 = n_3 = 1$. Increasing one of the integers by 1 gives the next lowest, or first excited level. If L_1 is the largest dimension, then $n_1 = 2$, $n_2 = 1$, $n_3 = 1$ produces the smallest energy increment, and describes the first excited state. Further, so long as both L_2 and L_3 are not equal to L_1, the first excited level is nondegenerate, that is, there is no other state with this energy. If L_2 or L_3 equals L_1, the level is doubly degenerate; if all three are equal, the level will be triply degenerate. Thus, the higher the symmetry, the more degeneracy we can expect.

7.2 CENTRAL FORCES AND ANGULAR MOMENTUM

The formulation of quantum mechanics in cartesian coordinates is the natural way to generalize from one to higher dimensions, but often it is not the best suited to a given application. For instance, an atomic electron is drawn to the

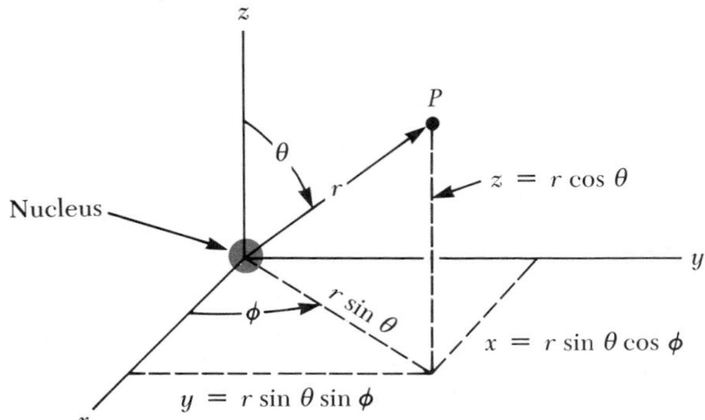

Figure 7.4 The central force on an atomic electron is one directed toward a fixed point, the nucleus. The coordinates of choice here are the spherical coordinates r, θ, ϕ centered on the nucleus.

nucleus of the atom by the Coulomb force between opposite charges. This is an example of a *central force,* that is, one directed toward a fixed *point.* The nucleus is the center of force, and the coordinates of choice here are spherical coordinates r, θ, ϕ centered on the nucleus (Fig. 7.4).

If the central force is conservative, the particle energy (kinetic plus potential) stays constant and E becomes a candidate for quantization. In that case, the quantum states are stationary waves $\psi(\mathbf{r})e^{-i\omega t}$, with $E = \hbar\omega$ being the sharp value of particle energy.

But for central forces, *angular momentum* $\mathbf{L}$ about the force center also is conserved, and we should expect the matter wave ψ to be one for which all three angular momentum components are quantized. This imposes severe constraints on the form of the wave. In fact, these constraints are so severe that it is impossible to find a waveform satisfying all of them at once; that is, *not all components of angular momentum can be known simultaneously!*

The dilemma is reminiscent of our inability to specify simultaneously the position and momentum of a particle. Indeed, if the direction of $\mathbf{L}$ were known precisely, the particle would be confined to the orbital plane (the plane perpendicular to the vector $\mathbf{L}$), and its coordinate in the direction normal to this plane would be accurately known (and unchanging). In that case, however, the particle could have no momentum out of the orbital plane, so that its linear momentum perpendicular to this plane also would be known (to be zero), in violation of the uncertainty principle. The argument just given may be refined to establish an **uncertainty principle for angular momentum:** *it is impossible to specify simultaneously any two components of angular momentum.* Alternatively, if one component of $\mathbf{L}$ is sharp, then the remaining two must be "fuzzy."[2]

Uncertainty principle for angular momentum

Along with E and one component of $\mathbf{L}$, then, what else might be quantized, or sharp, for central forces? A moment's reflection furnishes the answer: with only one component of $\mathbf{L}$ sharp, there is no redundancy in having the magnitude $|\mathbf{L}|$ sharp also. In this way, E, $|\mathbf{L}|$, and one component of $\mathbf{L}$, say L_z, become the sharp observables subject to quantization in the central force problem. Wave functions for which $|\mathbf{L}|$ and L_z are both sharp are the *spherical harmonics.* These functions may be obtained by separating the variables in Schrödinger's equation for a central force. We take the stationary state waveform in spherical coordinates r, θ, ϕ to be

$$\boxed{\psi(\mathbf{r}) = \psi(r, \theta, \phi) = R(r)\,\Theta(\theta)\,\Phi(\phi)} \tag{7.10}$$

Separation of variables for the stationary state wave function

[2] The only instance in which two or more angular momentum components may be known exactly is the trivial case $L_x = L_y = L_z = 0$.

and write Schrödinger's time-independent equation (Equation 7.4) in these coordinates using the spherical coordinate form for the Laplacian:[3]

$$\nabla^2 = \frac{\partial^2}{\partial r^2} + \left(\frac{2}{r}\right)\frac{\partial}{\partial r} + \frac{1}{r^2}\left[\frac{\partial^2}{\partial \theta^2} + \cot\theta\frac{\partial}{\partial\theta} + \csc^2\theta\frac{\partial^2}{\partial\phi^2}\right]$$

Using Equation 7.10 for ψ, and noting that $U(r)$ becomes simply $U(r)$ for a central force seated at the origin, leads to Schrödinger's equation for a central force. After some rearrangement, we arrive at the desired form:

$$\left\{\frac{r^2}{R}\left[\frac{\partial^2 R}{\partial r^2} + \left(\frac{2}{r}\right)\frac{\partial R}{\partial r}\right] - \frac{2mr^2}{\hbar^2}\left[U(r) - E\right]\right\} + \left\{\frac{1}{\Theta}\left[\frac{\partial^2\Theta}{\partial\theta^2} + \cot\theta\frac{\partial\Theta}{\partial\theta}\right] + \csc^2\theta\left\{\frac{1}{\Phi}\frac{\partial^2\Phi}{\partial\phi^2}\right\}\right\} = 0 \tag{7.11}$$

The terms are grouped so that all those involving a single variable appear together, surrounded by curly brackets {. . .}; each such group must reduce to a constant for the equation to hold for every choice of r, θ, and ϕ. The resulting three *separation constants* are interrelated by the original equation, so that only two of them are actually independent. These two are simple forms of $|\boldsymbol{L}|$ and L_z:[4]

$$\frac{1}{\Phi}\frac{\partial^2\Phi}{\partial\phi^2} = -\left(\frac{L_z}{\hbar}\right)^2$$

$$\frac{1}{\Theta}\left[\frac{\partial^2\Theta}{\partial\theta^2} + \cot\theta\frac{\partial\Theta}{\partial\theta}\right] - \csc^2\theta\left(\frac{L_z}{\hbar}\right)^2 = -\frac{|\boldsymbol{L}|^2}{\hbar^2} \tag{7.12}$$

In solving Equations 7.12, we find that acceptable (bounded, single-valued) functions Θ and Φ are possible only if the sharp values $|\boldsymbol{L}|$ and L_z are *quantized* as follows:

Angular momentum and its z-component are quantized

$$\boxed{\begin{aligned}|\boldsymbol{L}| &= \sqrt{\ell(\ell+1)}\hbar \\ L_z &= m_\ell\hbar\end{aligned}} \qquad \begin{aligned}&\ell = 0, 1, 2, \ldots \\ &m_\ell = 0, \pm1, \pm2, \ldots, \pm\ell\end{aligned} \tag{7.13}$$

The **orbital quantum number** ℓ must be a nonnegative integer. For a given value of ℓ, the **magnetic quantum number** m_ℓ is limited to absolute values not larger than ℓ. This limitation is necessary in order that the z-component of angular momentum, L_z, never exceed the magnitude of the vector, $|\boldsymbol{L}|$.

The products $\Theta(\theta)\,\Phi(\phi)$ obtained from Equations 7.12 are the spherical harmonics $Y_\ell^{m_\ell}$; the restrictions in Equations 7.13 are necessary for the spherical harmonics to be well-behaved functions of their arguments. Some spherical harmonics are given in Table 7.2. The constant prefactors are standard and represent a kind of normalization for these functions.[5]

[3] The Laplacian in spherical coordinates is given in any more advanced scientific text, or may be found from the cartesian form by following the arguments of Appendix B.

[4] These identifications are established in Appendix B, along with many other important results from this section. The serious student is referred there for a concise treatment of the quantum central force problem based on the operator methods of Section 5.8.

[5] The normalization is such that the integral of $|Y_\ell^{m_\ell}|^2$ over the surface of a sphere with unit radius is 1.

TABLE 7.2 The Spherical Harmonics $Y_\ell^{m_\ell}(\theta, \phi)$

$$Y_0^0 = \frac{1}{2\sqrt{\pi}}$$

$$Y_1^0 = \frac{1}{2}\sqrt{\frac{3}{\pi}}\cdot\cos\theta$$

$$Y_1^{\pm 1} = \mp\frac{1}{2}\sqrt{\frac{3}{2\pi}}\cdot\sin\theta\cdot e^{\pm i\phi}$$

$$Y_2^0 = \frac{1}{4}\sqrt{\frac{5}{\pi}}\cdot(3\cos^2\theta - 1)$$

$$Y_2^{\pm 1} = \mp\frac{1}{2}\sqrt{\frac{15}{2\pi}}\cdot\sin\theta\cdot\cos\theta\cdot e^{\pm i\phi}$$

$$Y_2^{\pm 2} = \frac{1}{4}\sqrt{\frac{15}{2\pi}}\cdot\sin^2\theta\cdot e^{\pm 2i\phi}$$

$$Y_3^0 = \frac{1}{4}\sqrt{\frac{7}{\pi}}\cdot(5\cos^3\theta - 3\cos\theta)$$

$$Y_3^{\pm 1} = \mp\frac{1}{8}\sqrt{\frac{21}{\pi}}\cdot\sin\theta\cdot(5\cos^2\theta - 1)\cdot e^{\pm i\phi}$$

$$Y_3^{\pm 2} = \frac{1}{4}\sqrt{\frac{105}{2\pi}}\cdot\sin^2\theta\cdot\cos\theta\cdot e^{\pm 2i\phi}$$

$$Y_3^{\pm 3} = \mp\frac{1}{8}\sqrt{\frac{35}{\pi}}\cdot\sin^3\theta\cdot e^{\pm 3i\phi}$$

With the introduction of separation constants, Equation 7.11 simplifies to

$$-\frac{\hbar^2}{2m}\left[\frac{\partial^2 R}{\partial r^2} + \left(\frac{2}{r}\right)\frac{\partial R}{\partial r}\right] + \frac{|\mathbf{L}|^2}{2mr^2}R(r) + U(r)R(r) = ER(r) \qquad (7.14)$$

Radial wave equation

This is the **radial wave equation;** it determines the radial part of our wave ψ and the allowed particle energies E. The reduction from Schrödinger's equation to the radial wave equation represents enormous progress, and it is valid for any central force. Still, the task of solving this equation for a specified potential $U(r)$ is a difficult one, often requiring methods of considerable sophistication.

EXAMPLE 7.4 Orbital Quantum Number for a Stone

A stone with mass 1 kg is whirled in a horizontal circle of radius 1 m with a period of revolution equal to 1 s. What value of orbital quantum number ℓ describes this motion?

Solution: The speed of the stone in its orbit is

$$v = \frac{2\pi R}{T} = \frac{2\pi(1\ \text{m})}{1\ \text{s}} = 6.28\ \text{m/s}$$

The corresponding angular momentum has magnitude

$$|\mathbf{L}| = mvR = (1\ \text{kg})(6.28\ \text{m/s})(1\ \text{m}) = 6.28\ \text{kg}\cdot\text{m}^2/\text{s}$$

But angular momentum is quantized as $\sqrt{\ell(\ell+1)}\hbar$, which is approximately $\ell\hbar$ when ℓ is large. Then

$$\ell \approx \frac{|\mathbf{L}|}{\hbar} = \frac{6.28\ \text{kg}\cdot\text{m}^2/\text{s}}{1.055\times 10^{-34}\ \text{kg}\cdot\text{m}^2/\text{s}} = 5.96\times 10^{34}!$$

Again, we see that macroscopic objects are described by enormous quantum numbers, so that quantization is not evident on this scale.

EXAMPLE 7.5 The Bohr Atom Revisited

Discuss angular momentum quantization in the Bohr model. What orbital quantum number describes the electron in the first Bohr orbit of hydrogen?

Solution: Angular momentum in the Bohr atom is quantized according to the Bohr postulate

$$|\mathbf{L}| = mvr = n\hbar$$

with $n = 1$ for the first Bohr orbit. Comparing this with the quantum mechanical result, Equation 7.13, we see that the two rules are incompatible! The magnitude $|\mathbf{L}|$ can never be an integral multiple of $\hbar$—the smallest nonzero value consistent with Equation 7.13 is $|L| = \sqrt{2}\hbar$ for $\ell = 1$.

In fairness to Bohr, we should point out that the Bohr model makes no distinction between the quantization of $|L|$ and quantization of its components along the coordinate axes. From the classical viewpoint, the coordinate system can always be oriented to align one of the axes, say the z axis, along the direction of $\boldsymbol{L}$. In that case $|L|$ may be identified with $|L_z|$. The Bohr postulate in this form agrees with the quantization of L_z in Equation 7.13, and indicates $\ell = 1$ for the first Bohr orbit! This conflicting result derives from a false assertion, namely, that we may orient a coordinate axis along the direction of $\boldsymbol{L}$. The freedom to do so must be abandoned if the quantization rules of Equation 7.13 are correct! This stunning conclusion is one of the great mysteries of quantum physics, and is implicit in the notion of *space quantization.*

7.3 SPACE QUANTIZATION

For wave functions satisfying Equations 7.12, the orbital angular momentum magnitude $L = |\boldsymbol{L}|$, and L_z, the projection of $\boldsymbol{L}$ along the z axis, are both sharp and quantized according to the orbital and magnetic quantum numbers, respectively. Together, ℓ and m_ℓ specify the orientation of the angular momentum vector $\boldsymbol{L}$. The fact that the direction of $\boldsymbol{L}$ is quantized with respect to an arbitrary axis (the z axis) is referred to as *space quantization.*

Let us look at the possible orientations of $\boldsymbol{L}$ for a given value of orbital quantum number ℓ. Recall that m_ℓ can have values ranging from $-\ell$ to $+\ell$. If $\ell = 0$, then $m_\ell = 0$ and $L_z = 0$. In fact, for this case $|L| = 0$, so that all components of $\boldsymbol{L}$ are zero. If $\ell = 1$, then the possible values for m_ℓ are -1, 0, and $+1$, so that L_z may be $-\hbar$, 0, or $+\hbar$. If $\ell = 2$, m_ℓ can be -2, -1, 0, $+1$, or $+2$, corresponding to L_z values of $-2\hbar$, $-\hbar$, 0, $+\hbar$, or $+2\hbar$, and so on. A vector model describing space quantization for $\ell = 2$ is shown in Figure 7.5a. Note that $\boldsymbol{L}$ can never be aligned with the z axis, since L_z must be smaller than the total angular momentum L. From a three-dimensional perspective, $\boldsymbol{L}$ must lie on the surface of a cone that makes an angle θ with the z axis, as shown in Figure 7.5b. From the figure, we see that θ also is quantized and that its values are specified by the relation

The orientations of L are restricted (quantized)

$$\cos\theta = \frac{L_z}{|\mathbf{L}|} = \frac{m_\ell}{\sqrt{\ell(\ell+1)}} \tag{7.15}$$

Classically θ can take any value; that is, the angular momentum vector $\boldsymbol{L}$ can point in any direction whatsoever. According to quantum mechanics, the

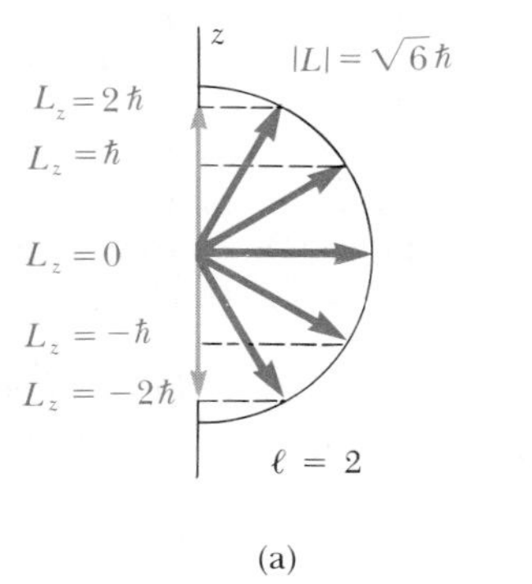

(a)

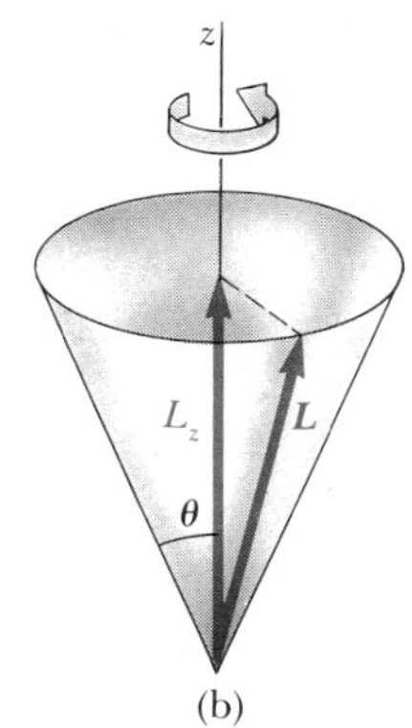

(b)

Figure 7.5 (a) The allowed projections of the orbital angular momentum for the case $\ell = 2$. (b) From a three-dimensional perspective, the orbital angular momentum vector L lies on the surface of a cone. The fuzzy character of L_x and L_y is depicted by allowing L to precess about the z axis, so that L_x and L_y change continually, while L_z maintains the fixed value $m_\ell\hbar$.

possible orientations for L are those consistent with Equation 7.15. Further, these special directions have nothing to do with the forces acting on the particle, provided only that these forces are central so that angular momentum may be conserved. Thus, *the rule of Equation 7.15 does not originate from the law of force but derives from the structure of space itself,* hence the name space quantization.

Figure 7.5 is misleading in showing L with a specific direction, for which all three components L_x, L_y, and L_z are known exactly. As we have mentioned, there is no quantum state for which this condition is true. If L_z is sharp, then L_x and L_y must be "fuzzy." Accordingly, it is more proper to envision the vector L of Figure 7.5b as precessing around the z axis so as to trace out a cone in space. This allows the components L_x and L_y to change continually, while L_z maintains the fixed value $m_\ell \hbar$.[6]

EXAMPLE 7.6 Space Quantization for an Atomic Electron

Consider an atomic electron in the $\ell = 3$ state. Calculate the magnitude $|L|$ of the total angular momentum and the allowed values of L_z and θ.

Solution: With $\ell = 3$, Equation 7.13 gives

$$|L| = \sqrt{3(3+1)}\hbar = 2\sqrt{3}\hbar$$

The allowed values of L_z are $m_\ell \hbar$, with $m_\ell = 0, \pm 1, \pm 2$, and ± 3. This gives

$$L_z = -3\hbar, -2\hbar, -\hbar, 0, \hbar, 2\hbar, 3\hbar$$

Finally, we obtain the allowed values of θ from

$$\cos\theta = \frac{L_z}{|L|} = \frac{m_\ell}{2\sqrt{3}}$$

Substituting the values for m_ℓ gives

$$\cos\theta = \pm 0.866, \quad \pm 0.577, \quad \pm 0.289, \quad \text{and} \quad 0$$

or

$$\theta = \pm 30°, \quad \pm 54.8°, \quad \pm 73.2°, \quad \text{and} \quad 90°$$

Exercise 2 Compare the minimum angles between L and the z axis for the electron of Example 7.6 and for the 1 kg stone of Example 7.4.

Answer: 30 degrees for the electron vs. 2.3×10^{-16} degrees for the stone.

7.4 ATOMIC HYDROGEN

In this section we study the hydrogen atom from the viewpoint of wave mechanics. Its simplicity makes atomic hydrogen the ideal testing ground for comparing theory with experiment. Further, the hydrogen atom is the prototype system for the many complex ions and atoms of the heavier elements. Indeed, our study of the hydrogen atom ultimately will enable us to understand the periodic table of the elements, one of the greatest triumphs of quantum physics.

The object of interest is the orbiting electron, with its mass m and charge $-e$, bound by the force of electrostatic attraction to the nucleus, with its much larger mass M and charge $+Ze$. Z is the atomic number; the choice $Z = 1$ describes the hydrogen atom. Since $M \gg m$, we will assume that the nucleus does not move, but simply exerts the attractive force that binds the electron. This force is the Coulomb force, with its associated potential energy

$$U(r) = \frac{k(+Ze)(-e)}{r} = -\frac{kZe^2}{r} \tag{7.16}$$

where k $(=1/4\pi\epsilon_0)$ is the Coulomb constant.

[6] This precession of the classical vector to portray the inherent fuzziness in L_x and L_y is meant to be suggestive only. In effect, we have identified the quantum averages $\langle L_x \rangle$ and $\langle L_y \rangle$ with their averages over *time* in a classical picture, but the two kinds of averaging in fact are quite distinct.

The stationary state wave functions for hydrogenic atoms

The hydrogen atom constitutes a central force problem; according to Section 7.2, the stationary states for any central force are

$$\boxed{\Psi(r, \theta, \phi, t) = R(r)Y_\ell^{m_\ell}(\theta, \phi)e^{-i\omega t}} \tag{7.17}$$

where $Y_\ell^{m_\ell}(\theta, \phi)$ is a spherical harmonic from Table 7.2. The *radial wave function* $R(r)$ is found from the *radial wave equation* of Section 7.2,

$$-\frac{\hbar^2}{2m}\left[\frac{\partial^2 R}{\partial r^2} + \left(\frac{2}{r}\right)\frac{\partial R}{\partial r}\right] + \frac{|\mathbf{L}|^2}{2mr^2}R(r) + U(r)R(r) = ER(r) \tag{7.18}$$

$U(r)$ is the potential energy of the electron in the field of the nucleus. The remaining terms on the left of Equation 7.18 are associated with the kinetic energy of the electron. The term proportional to $|\mathbf{L}|^2$ is the orbital contribution to kinetic energy, K^{orb}. To see this, consider a particle in circular orbit, where all the kinetic energy is in orbital form (since the distance from the center remains fixed). For such a particle $K^{\text{orb}} = \frac{1}{2}mv^2$ and $|\mathbf{L}| = mvr$. Eliminating the particle speed v, we get

$$K^{\text{orb}} = \frac{m}{2}\left(\frac{|\mathbf{L}|}{mr}\right)^2 = \frac{|\mathbf{L}|^2}{2mr^2}$$

Although it was derived for circular orbits, this result correctly represents the orbital contribution to kinetic energy of a mass m in general motion with angular momentum $\mathbf{L}$.[7]

The derivative terms in Equation 7.18 represent the energy of electron motion toward or away from the nucleus. The leftmost term is just what we should write for the kinetic energy of a matter wave $\psi = R(r)$ associated with motion along the coordinate line marked by r. However, the term involving $\partial R/\partial r$ is mysterious: in fact, the appearance of this term is evidence that *the effective one-dimensional matter wave is not $R(r)$, but $rR(r)$.* In support of our claim we note that

$$\frac{\partial^2(rR)}{\partial r^2} = r\left[\frac{\partial^2 R}{\partial r^2} + \left(\frac{2}{r}\right)\frac{\partial R}{\partial r}\right]$$

Then the radial wave equation written for the *pseudo-wavefunction* $g(r) = rR(r)$ takes the same form as Schrödinger's equation in one dimension,

$$\left(-\frac{\hbar^2}{2m}\right)\frac{\partial^2 g}{\partial r^2} + U_{\text{eff}}(r)g(r) = Eg(r) \tag{7.19}$$

but with an *effective potential energy*

$$U_{\text{eff}} = \frac{|\mathbf{L}|^2}{2mr^2} + U(r) = \ell(\ell + 1)\frac{\hbar^2}{2mr^2} + U(r) \tag{7.20}$$

The intensity of this pseudo-wavefunction $g(r)$ also furnishes probabilities, as described later in this section.

The solution of Equation 7.19 with $U(r) = -kZe^2/r$ (for one-electron or *hydrogenic* atoms) requires methods beyond the scope of this text. Here, as before, we shall present the results and leave their verification to the inter-

[7] For any planar orbit we may write $|\mathbf{L}| = rp_\perp$, where $p_\perp$ is the component of momentum in the orbital plane that is normal to the radius vector $\mathbf{r}$. The orbital part of the kinetic energy is then $K^{\text{orb}} = p_\perp^2/2m = |\mathbf{L}|^2/2mr^2$.

TABLE 7.3 The Radial Wave Functions $R_{n,\ell}(r)$ of Hydrogenic Atoms for $n = 1$, 2, and 3

n	ℓ	$R(r)$
1	0	$\left(\frac{Z}{a_0}\right)^{3/2} 2e^{-Zr/a_0}$
2	0	$\left(\frac{Z}{2a_0}\right)^{3/2} \left(2 - \frac{Zr}{a_0}\right) e^{-Zr/2a_0}$
2	1	$\left(\frac{Z}{2a_0}\right)^{3/2} \frac{Zr}{\sqrt{3}\, a_0} e^{-Zr/2a_0}$
3	0	$\left(\frac{Z}{3a_0}\right)^{3/2} 2\left[1 - \frac{2Zr}{3a_0} + \frac{2}{27}\left(\frac{Zr}{a_0}\right)^2\right] e^{-Zr/3a_0}$
3	1	$\left(\frac{Z}{3a_0}\right)^{3/2} \cdot \frac{4\sqrt{2}}{3} \frac{Zr}{a_0}\left(1 - \frac{Zr}{6a_0}\right) e^{-Zr/3a_0}$
3	2	$\left(\frac{Z}{3a_0}\right)^{3/2} \cdot \frac{2\sqrt{2}}{27\sqrt{5}} \left(\frac{Zr}{a_0}\right)^2 e^{-Zr/3a_0}$

ested reader. Acceptable wave functions can be found only if the energy E is restricted to be one of the following special values:

$$\boxed{E = -\frac{ke^2}{2a_0}\left\{\frac{Z^2}{n^2}\right\}} \qquad n = 1, 2, 3, \ldots \tag{7.21}$$

Allowed energies for hydrogenic atoms

This result is in exact agreement with that found from the simple Bohr theory (see Chapter 3): $a_0 = \hbar^2/mke^2$ is the Bohr radius, 0.529 Å, and $ke^2/2a_0$ is the Rydberg energy, 13.6 eV. The integer n is called the **principal quantum number.** Although n can be any positive integer, the orbital quantum number now is limited to values less than n; that is, ℓ can have only one of the values

$$\boxed{\ell = 0, 1, 2, \ldots, (n-1)} \tag{7.22}$$

Allowed values of ℓ

The cutoff at $(n - 1)$ limits the angular momentum to values consistent with the total energy, as fixed by n. (The magnetic quantum number m_ℓ still is restricted to integer values between $-\ell$ and $+\ell$ so that the z component of angular momentum never exceeds the magnitude of the vector.)

The radial waves $R(r)$ for hydrogenic atoms are products of exponentials with polynomials in r/a_0. These radial wave functions are tabulated as $R_{n,\ell}(r)$ in Table 7.3 for principal quantum numbers up to and including $n = 3$.

For hydrogenic atoms, then, the quantum numbers are n, ℓ, and m_ℓ, associated with the sharp observables E, $|L|$, and L_z, respectively. Notice that the electron energy E depends on n, but not at all on ℓ or m_ℓ. The energy is independent of m_ℓ because of the spherical symmetry of the atom, and this will be true for any central force that varies only with distance r. However, the fact that E also is independent of ℓ is a special consequence of the Coulomb force, and is not to be expected in heavier atoms, say, where the force on any one electron includes the electrostatic repulsion of the remaining electrons in the atom, as well as the Coulombic attraction of the nucleus.

For historical reasons, all states with the same principal quantum number n are said to form a **shell.** These shells are identified by the letters K, L, M, . . . , which designate the states for which $n = 1, 2, 3, \ldots$. Likewise,

TABLE 7.4 Spectroscopic Notation for Atomic Shells and Subshells

n	Shell Symbol	ℓ	Subshell Symbol
1	K	0	s
2	L	1	p
3	M	2	d
4	N	3	f
5	O	4	g
6	P	5	h
. . .		. . .	

states having the same value of both n and ℓ are said to form a **subshell.** The letters s, p, d, f, . . . are used to designate the states for which $\ell = 0, 1, 2, 3, \ldots$. This *spectroscopic notation* is summarized in Table 7.4.

EXAMPLE 7.7 The $n = 2$ Level of Hydrogen
Enumerate all states of the hydrogen atom corresponding to the principal quantum number $n = 2$, giving the spectroscopic designation for each. Calculate the energies of these states.

Solution: When $n = 2$, ℓ can have the values 0 and 1. If $\ell = 0$, m_ℓ can only be 0. If $\ell = 1$, m_ℓ can be -1, 0, or $+1$. Hence, we have a $2s$ state with quantum numbers

$$n = 2, \quad \ell = 0, \quad m_\ell = 0$$

and three $2p$ states for which the quantum numbers are

$$n = 2, \quad \ell = 1, \quad m_\ell = -1$$

$$n = 2, \quad \ell = 1, \quad m_\ell = 0$$

$$n = 2, \quad \ell = 1, \quad m_\ell = +1$$

Because all of these states have the same principal quantum number, $n = 2$, they also have the same energy, which can be calculated from Equation 7.21. For $Z = 1$ and $n = 2$, this gives

$$E_2 = -(13.6 \text{ eV})\{1^2/2^2\} = -3.4 \text{ eV}$$

Exercise 3 How many possible states are there for the $n = 3$ level of hydrogen? For the $n = 4$ level?
Answers: 9 states for $n = 3$, and 16 states for $n = 4$.

Hydrogenic Atoms — Ground State

The ground state of a one-electron atom with atomic number Z, for which $n = 1$, $\ell = 0$, and $m_\ell = 0$, has energy

$$E_1 = -(13.6 \text{ eV})Z^2 \tag{7.23}$$

The wave function for this state is

$$\psi_{1,0,0} = R_{1,0}(r)Y_0^{\,0} = \pi^{-1/2}(Z/a_0)^{3/2}e^{-Zr/a_0} \tag{7.24}$$

The constants are such that ψ is automatically normalized.

Notice that $\psi_{1,0,0}$ does not depend on angle, since it is the product of a radial wave with $Y_0^{\,0} = 1/\sqrt{4\pi}$. In fact, all the $\ell = 0$ waves share this feature; that is, *all s-state waves are spherically symmetric.*

The electron described by the wave of Equation 7.24 is found with a probability per unit volume given by

$$|\psi_{1,0,0}|^2 = \frac{Z^3}{\pi a_0^{\,3}}\, e^{-2Zr/a_0}$$

The probability distribution also is spherically symmetric, as it would be for any s-state wave; that is, the likelihood for finding the electron in the atom is the same at all points equidistant from the center (nucleus). It is convenient to define a **radial probability distribution** with its associated density $P(r)$ such that $P(r)\,dr$ is the probability of finding the electron anywhere in the spherical shell of radius r and thickness dr (Fig. 7.6). The shell volume is its surface area, $4\pi r^2$, multiplied by the shell thickness, dr. Since the probability density $|\psi_{1,0,0}|^2$ is the same everywhere in the shell, we have

$$P(r)\,dr = |\psi|^2 4\pi r^2\,dr$$

or, for the hydrogenic 1s state,

$$P_{1s}(r) = \frac{4Z^3}{a_0{}^3}\, r^2 e^{-2Zr/a_0} \tag{7.25}$$

The same result is obtained for $P_{1s}(r)$ from the intensity of the pseudo-matter wave $g(r) = rR(r)$:

$$\boxed{P(r) = |g(r)|^2 = r^2|R(r)|^2} \tag{7.26}$$

The radial probability density for any state

In fact, *Equation 7.26 gives the correct radial probability density for any state;* for the spherically symmetric s-states this is the same as $4\pi r^2|\psi|^2$, since then $\psi(r) = (1/\sqrt{4\pi})R(r)$.[8]

A plot of the function $P_{1s}(r)$ is presented in Figure 7.7a; Figure 7.7b shows the 1s electron "cloud" $|\psi_{1,0,0}|^2$ from which $P_{1s}(r)$ derives. $P(r)$ may be loosely interpreted as the probability of finding the electron at distance r from the nucleus, irrespective of its angular position. Thus, the peak of the curve in Figure 7.7a represents the most probable distance of the 1s electron from the nucleus. Furthermore, the normalization condition becomes

$$1 = \int_0^\infty P(r)\,dr \tag{7.27}$$

where the integral is taken over all *possible* values of r. The average distance of the electron from the nucleus is found by weighting each possible distance with the probability that the electron will be found at that distance:

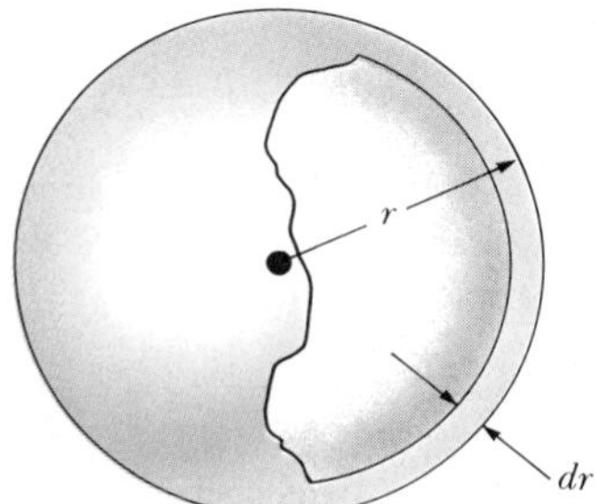

Figure 7.6 $P(r)\,dr$ is the probability that the electron will be found in the volume of a spherical shell with radius r and thickness dr. The shell volume is just $4\pi r^2\,dr$.

[8] From its definition, $P(r)\,dr$ always may be found by integrating $|\psi|^2 = |R(r)|^2|Y_\ell^{m_\ell}|^2$ over the volume of a spherical shell having radius r and thickness dr. Since the volume element in spherical coordinates is $dV = r^2\,dr \sin\theta\, d\theta\, d\phi$, and the integral of $|Y_\ell^{m_\ell}|^2$ over angle is unity (see footnote 3), this leaves $P(r) = r^2|R(r)|^2$.

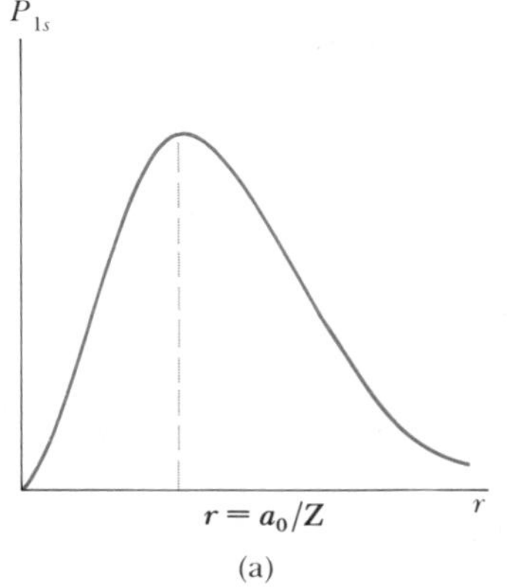

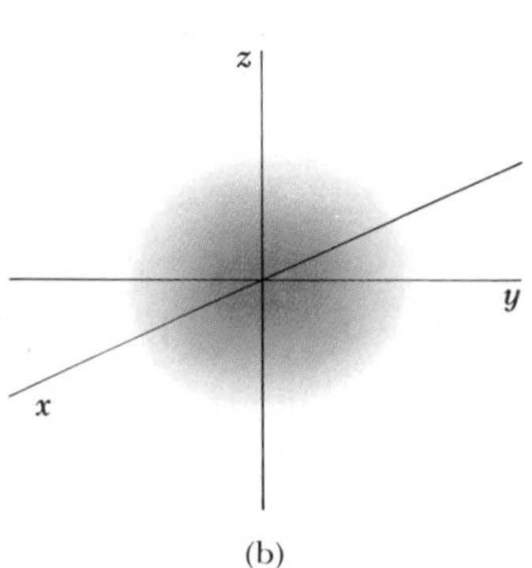

Figure 7.7 (a) The curve $P_{1s}(r)$, representing the probability of finding the electron as a function of distance from the nucleus in a 1s hydrogenic state. Note that the probability takes its maximum value when r equals a_0/Z. (b) The spherical electron "cloud" for a hydrogenic atom in its 1s state. The shading at every point is proportional to the probability density $|\psi_{1s}(r)|^2$.

The average distance of an electron from the nucleus

$$\langle r \rangle = \int_0^\infty rP(r)\,dr \tag{7.28}$$

In fact, the average value of *any* function of distance $f(r)$ is obtained by weighting the function value at every distance with the probability at that distance:

$$\langle f \rangle = \int_0^\infty f(r)P(r)\,dr \tag{7.29}$$

EXAMPLE 7.8 Probabilities for the Electron in Hydrogen

Calculate the probability that the electron in the ground state of hydrogen will be found outside the first Bohr radius.

Solution: The probability is found by integrating the radial probability density for this state, $P_{1s}(r)$, from the Bohr radius a_0 to ∞. Using Equation 7.25 with $Z = 1$ for hydrogen gives

$$P = \frac{4}{a_0^3}\int_{a_0}^\infty r^2 e^{-2r/a_0}\,dr$$

We can put the integral in dimensionless form by changing variables from r to $z = 2r/a_0$. Noting that $z = 2$ when $r = a_0$, and that $dr = (a_0/2)\,dz$, we get

$$P = \frac{1}{2}\int_2^\infty z^2 e^{-z}\,dz = -\frac{1}{2}\{z^2 + 2z + 2\}e^{-z}\Big|_2^\infty = 5e^{-2}$$

This is about 0.677 or 67.7%.

EXAMPLE 7.9 The Electron-Proton Separation in Hydrogen

Calculate the most probable distance of the electron from the nucleus in the ground state of hydrogen, and compare this with the average distance.

Solution: The most probable distance is the value of r that makes the radial probability $P(r)$ a maximum. The slope here is zero, so the most probable value of r is obtained by setting $dP/dr = 0$ and solving for r. Using Equation 7.25 with $Z = 1$ for the $1s$ or ground state of hydrogen, we get

$$0 = \left(\frac{4}{a_0^3}\right)\frac{d}{dr}\{r^2e^{-2r/a_0}\} = \left(\frac{4}{a_0^3}\right)e^{-2r/a_0}\left\{-\frac{2r^2}{a_0} + 2r\right\}$$

The right-hand side is zero for $r = 0$ and for $r = a_0$. Since $P(0) = 0$, $r = 0$ is a minimum of $P(r)$, not a maximum. Thus, the most probable distance is

$$r = a_0$$

The average distance is obtained from Equation 7.28, which in this case becomes

$$\langle r \rangle = \frac{4}{a_0^3}\int_0^\infty r^3 e^{-2r/a_0}\,dr$$

Again introducing $z = 2r/a_0$, we obtain

$$\langle r \rangle = \frac{a_0}{4}\int_0^\infty z^3 e^{-z}\,dz$$

The definite integral on the right is one of a broader class

$$\int_0^\infty z^n e^{-z}\,dz = n!$$

whose value $n! = n(n-1)\ .\ .\ .\ (1)$ is established by repeated integration by parts. Then

$$\langle r \rangle = \frac{a_0}{4}(3!) = \frac{3}{2}a_0$$

The average distance and the most probable distance are not the same, because the probability curve $P(r)$ is not symmetric about the peak distance a_0. Indeed, values of r greater than a_0 are weighted more heavily in Equation 7.28 than values smaller than a_0, so that the average $\langle r \rangle$ actually exceeds a_0 for this probability distribution.

Hydrogenic Atoms—Excited States

There are four first excited states for hydrogenic atoms: $\psi_{2,0,0}$, $\psi_{2,1,0}$, $\psi_{2,1,1}$, and $\psi_{2,1,-1}$. All have the same principal quantum number $n = 2$, hence the same total energy

$$E_2 = -(13.6 \text{ eV}) \frac{Z^2}{4}$$

Accordingly, the first excited level E_2 is fourfold degenerate.

The 2*s* state, $\psi_{2,0,0}$, is again spherically symmetric. Plots of the radial probability density for this and several other hydrogenic states are shown in Figure 7.8. Note that the plot for the 2*s* state has two peaks. In this case, the most probable distance ($\sim 5a_0/Z$) is marked by the highest peak. An electron in the 2*s* state would be much farther from the nucleus (on the average) than an electron in the 1*s* state. Likewise, the most probable distances are even greater for an electron in any of the $n = 3$ states (3*s*, 3*p*, or 3*d*). Observations such as these continue to support the old idea of a shell structure for the atom, even in the face of the uncertainties inherent in the wave nature of matter.

The remaining three first excited states, $\psi_{2,1,1}$, $\psi_{2,1,0}$, and $\psi_{2,1,-1}$, have $\ell = 1$ and make up the 2*p* subshell. These states are not spherically symmetric. All of them have the same radial wave $R_{2,1}(r)$, but they are multiplied by

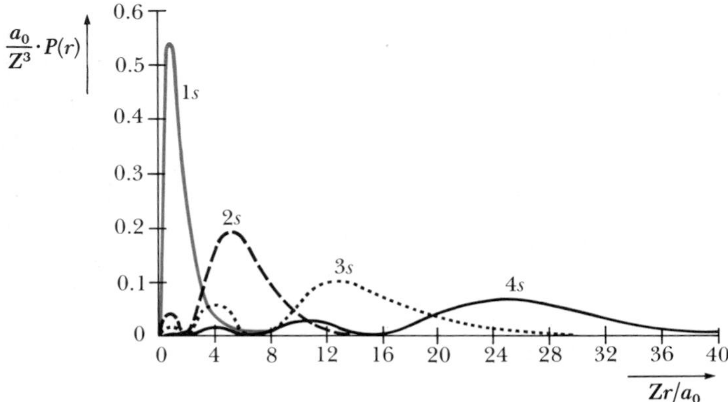

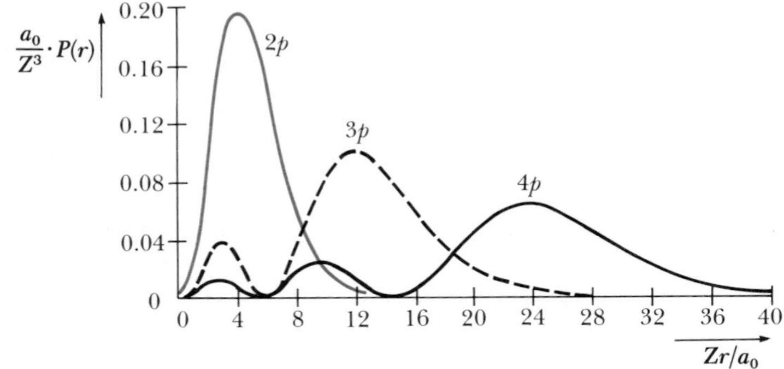

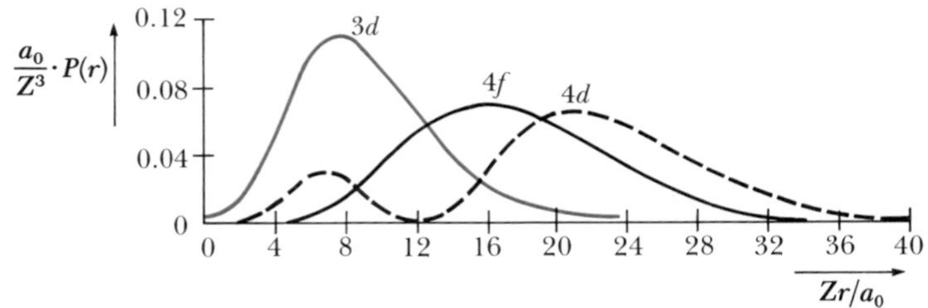

Figure 7.8 The radial probability density function for several states of hydrogenic atoms.

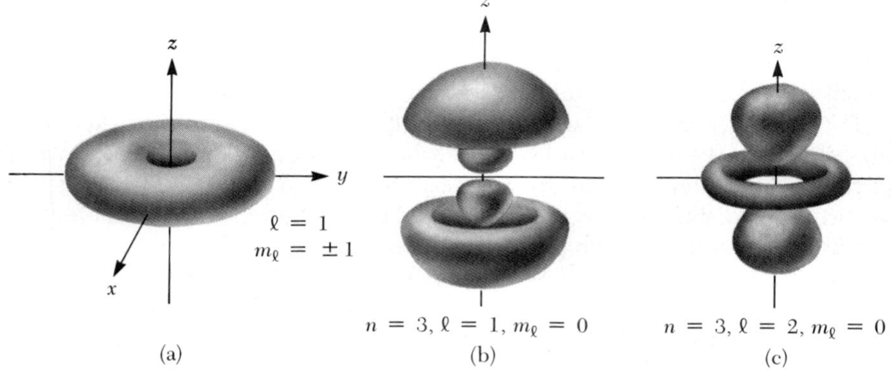

Figure 7.9 (a) The probability density $|\psi_{2,1,1}|^2$ for a hydrogenic $2p$ state. Note the axial symmetry about the z axis. (b) and (c) The probability densities $|\psi(r)|^2$ for several other hydrogenic states. The electron "cloud" is axially symmetric about the z axis for all the hydrogenic states $\psi_{n,\ell,m_\ell}(r)$.

different spherical harmonics and, thus, depend differently on the angles θ and ϕ. For example, the wave function $\psi_{2,1,1}$ is

$$\psi_{2,1,1} = R_{2,1}(r)Y_1^{\,1} = \pi^{-1/2}\left(\frac{Z}{a_0}\right)^{3/2}\left(\frac{Zr}{8a_0}\right)e^{-Zr/2a_0}\sin\theta\, e^{i\phi}$$

Notice, however, that the probability density $|\psi_{2,1,1}|^2$ is independent of ϕ and therefore is symmetric about the z axis (Fig. 7.9a). Since $|e^{im_\ell\phi}|^2 = 1$, the same is true for all the hydrogenic waves ψ_{n,ℓ,m_ℓ}, as suggested by the remaining illustrations in Figure 7.9.

The $\psi_{2,1,0}$ state has distinct directional characteristics, as shown in Figure 7.10a, and is sometimes designated $[\psi_{2p}]_z$ to indicate the preference for the electron in this state to be found along the z axis. Other highly directional states can be formed by combining the waves with $m_\ell = +1$ and $m_\ell = -1$. Thus, the wave functions

$$[\psi_{2p}]_x = \frac{1}{\sqrt{2}}\{\psi_{2,1,1} + \psi_{2,1,-1}\}$$

and

$$[\psi_{2p}]_y = \frac{1}{\sqrt{2}}\{\psi_{2,1,1} - \psi_{2,1,-1}\} \tag{7.30}$$

have large probability densities along the x and y axes, respectively, as shown in Figures 7.10b and 7.10c. The wavefunctions of Equations 7.30, formed as superpositions of first excited state waves with *identical* energies E_2, are themselves stationary states with this same energy; indeed, the three waves

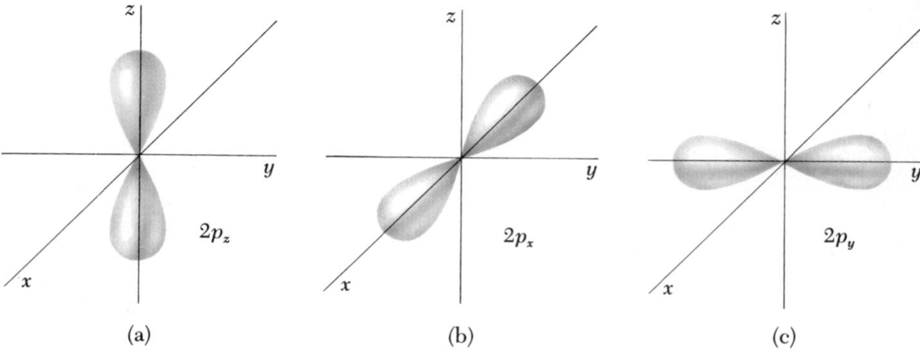

Figure 7.10 (a) Probability distribution for an electron in the hydrogenic $2p_z$ state, described by the quantum numbers $n = 2$, $\ell = 1$, $m_\ell = 0$. (b) and (c) Probability distributions for the $2p_x$ and $2p_y$ states. The three distributions $2p_x$, $2p_y$, and $2p_z$ have the same structure, but differ in their spatial orientation.

$[\psi_{2p}]_z$, $[\psi_{2p}]_x$, and $[\psi_{2p}]_y$ together constitute an equally good description of the $2p$ states for hydrogenic atoms. Wave functions with a highly directional character, such as these, play an important role in chemical bonding, the formation of molecules, and chemical properties.

The next excited level, E_3, is ninefold degenerate; that is, there are nine different wave functions with this energy, corresponding to all possible choices for ℓ and m_ℓ consistent with $n = 3$. Together, these nine states constitute the third shell, with subshells $3s$, $3p$, and $3d$ composed of one, three, and five states, respectively. The wave functions for these states may be constructed from the entries in Tables 7.2 and 7.3. Generally, they are more complicated than their second shell counterparts because of the increasing number of nodes in the radial part of the wave.

The progression to higher lying states leads to still more degeneracy and wave functions of ever-increasing complexity. The nth shell has energy

$$E_n = -(13.6\text{ eV})\frac{Z^2}{n^2}$$

and contains exactly n subshells, corresponding to $\ell = 0, 1, \ldots, n-1$. Within the ℓth subshell there are $2\ell + 1$ orbitals. Altogether, the nth shell contains a total of n^2 states, all with the same energy E_n; that is, *the energy level E_n is n^2 degenerate.* Equivalently, we say the nth shell can hold as many as n^2 electrons, assuming no more than one electron can occupy a given orbital. This argument underestimates the shell capacity by a factor of two, owing to the existence of electron spin, as discussed in the next chapter.

7.5 SUMMARY

In three dimensions, the matter wave intensity $|\Psi(\mathbf{r}, t)|^2$ represents the probability per unit volume for finding the particle at $\mathbf{r}$ at time t. Probabilities are found by integrating this probability density over the volume of interest.

The matter wave itself must satisfy the Schrödinger equation

$$-\frac{\hbar^2}{2m}\nabla^2\Psi + U(\mathbf{r})\Psi = i\hbar\frac{\partial\Psi}{\partial t} \tag{7.1}$$

Stationary states are solutions to this equation in separable form: $\Psi(\mathbf{r}, t) = \psi(\mathbf{r})e^{-i\omega t}$, with $\psi(\mathbf{r})$ satisfying

$$-\frac{\hbar^2}{2m}\nabla^2\psi(\mathbf{r}) + U(\mathbf{r})\psi(\mathbf{r}) = E\psi(\mathbf{r}) \tag{7.4}$$

This is the *time-independent Schrödinger equation,* from which we obtain the time-independent waveform $\psi(\mathbf{r})$ and the allowed values of particle energy E.

For a particle confined to a cubic box whose sides are L, the magnitudes of the components of particle momentum normal to the walls of the box are sharp, as is the particle energy. The sharp momentum values are quantized as

$$\begin{aligned} |p_x| &= n_1\frac{\pi\hbar}{L} \\ |p_y| &= n_2\frac{\pi\hbar}{L} \\ |p_z| &= n_3\frac{\pi\hbar}{L} \end{aligned} \tag{7.7}$$

and the allowed energies are found to be

$$E = \frac{\pi^2\hbar^2}{2mL^2}\{n_1{}^2 + n_2{}^2 + n_3{}^2\} \tag{7.8}$$

The three *quantum numbers* n_1, n_2, and n_3 are all positive integers, one for each degree of freedom of the particle. Many levels of this system are *degenerate,* that is, there is more than one wavefunction with the same energy.

For particles acted upon by a *central force,* the *angular momentum* $\mathbf{L}$ is conserved classically, and is quantized along with particle energy. Wave functions for which the z component L_z and magnitude $|\mathbf{L}|$ of angular momentum are simultaneously sharp are the *spherical harmonics* $Y_\ell^{m_\ell}(\theta, \phi)$ in the spherical coordinate angles θ and ϕ. For any central force, angular momentum is quantized by the rules

$$|\mathbf{L}| = \sqrt{\ell(\ell+1)}\,\hbar$$

and

$$L_z = m_\ell\hbar \tag{7.13}$$

The *orbital quantum number* ℓ must be a nonnegative integer. For a fixed value of ℓ, the *magnetic quantum number* m_ℓ is restricted to integer values lying between $-\ell$ and $+\ell$. Since $|\mathbf{L}|$ and L_z are quantized differently, the classical freedom to orient the z axis in the direction of $\mathbf{L}$ must be abandoned. This stunning conclusion is the essence of *space quantization.* Furthermore, no two components of $\mathbf{L}$, such as L_z and L_y, can be sharp simultaneously. This implies a lower limit to the uncertainty product $\Delta L_z\,\Delta L_y$, and gives rise to an uncertainty principle for the components of angular momentum.

A central force of considerable importance is the force on the electron in a one-electron atom. This is the *Coulomb* force, described by the potential energy $U(r) = -kZe^2/r$, where Z is the atomic number of the nucleus. The allowed energies for this case are given by

$$E = -\frac{ke^2}{2a_0}\left\{\frac{Z^2}{n^2}\right\} \qquad n = 1, 2, \ldots \tag{7.21}$$

This coincides exactly with the results obtained from the Bohr theory. The energy depends only on the *principal quantum number* n. For a fixed value of n, the orbital and magnetic quantum numbers are restricted as

$$\ell = 0, 1, 2, \ldots, n-1$$

$$m_\ell = 0, \pm 1, \pm 2, \ldots, \pm\ell \tag{7.22}$$

All states with the same principal quantum number n form a *shell,* identified by the letters K, L, M, . . . (corresponding to $n = 1, 2, 3, \ldots$). All states with the same values of n and ℓ form a *subshell,* designated by the letters s, p, d, f, . . . (corresponding to $\ell = 0, 1, 2, \ldots$).

The wave functions for an electron in a hydrogenic atom

$$\psi(r, \theta, \phi) = R_{n,\ell}(r)Y_\ell^{m_\ell}(\theta, \phi)$$

depend on the three quantum numbers n, ℓ, and m_ℓ, and are products of spherical harmonics multiplied by radial waves $R_{n,\ell}(r)$. The *pseudo-wavefunction* $g(r) = rR_{n,\ell}(r)$ is analogous to the wave function $\psi(x)$ in one dimension; the intensity of $g(r)$ gives the *radial probability density*

$$P(r) = |g(r)|^2 = r^2|R_{n,\ell}(r)|^2 \tag{7.26}$$

$P(r)\ dr$ is the probability that the electron will be found at a distance between r and $r + dr$ from the nucleus. The *most probable distance* is the one that maximizes $P(r)$, and generally differs from the *average distance* $\langle r \rangle$ calculated as

$$\langle r \rangle = \int_0^\infty rP(r)\, dr \tag{7.28}$$

The most probable values are found to coincide with the radii of the allowed orbits in the Bohr theory.

SUGGESTIONS FOR FURTHER READING

1. For more on the use of angular momentum to simplify the three-dimensional Schrödinger equation in central force applications, see Chapter 11 of *An Introduction to Quantum Physics,* by A. P. French and Edwin F. Taylor, New York: W. W. Norton and Company, Inc., 1978.
2. A good discussion of the radial probability density and its use for the lowest states of hydrogen is contained in Chapter 7 of *Modern Physics,* by Kenneth Krane, New York: John Wiley and Sons, Inc., 1983.
3. Illustrations in perspective of the electron "cloud" $|\psi(r)|^2$ for various hydrogenic states may be found in *Quantum Physics of Atoms, Molecules, Solids, Nuclei,* and *Particles,* by R. Eisberg and R. Resnick, New York: John Wiley and Sons, Inc., 1974.

QUESTIONS

1. Why are three quantum numbers needed to describe the state of a one-electron atom?
2. Compare the Bohr theory with the Schrödinger treatment of the hydrogen atom. Comment specifically on the total energy and orbital angular momentum.
3. How do we know whether a given $2p$ electron in an atom has $m_\ell = 0, +1,$ or -1? What value of m_ℓ characterizes a directed orbital such as $[\psi_{2p}]_x$ of Equation 7.39?
4. For atomic s states, the probability density $|\psi|^2$ is largest at the origin, yet the probability for finding the electron a distance r from the nucleus, given by $P(r)$, goes to zero with r. Explain.
5. For the electron in the ground state of hydrogen—as with many other quantum systems—the kinetic and potential energies are fuzzy observables while their sum, the total energy, is sharp. Explain how this can be so.
6. Discuss the relationship between space quantization and Schrödinger's equation. If the latter were invalid, would space quantization still hold?

PROBLEMS

1. A particle of mass m moves in a three-dimensional box of unequal sides L_1, L_2, and L_3. Find the energies of the six lowest states if $L_1 = L$, $L_2 = 2L$, and $L_3 = 2L$. Which of these are degenerate?
2. An electron moves in a cube whose sides have a length of 0.2 nm. Find values for the energy of (a) the ground state and (b) the first excited state of the electron.
3. A particle of mass m moves in a three-dimensional box with sides L. If the particle is in the third excited level, corresponding to $n^2 = 11$, find (a) the energy of the particle, (b) the combinations of n_1, n_2, and n_3 that would give this energy, and (c) the wave functions for these different states.
4. A particle of mass m moves in a two-dimensional box of sides L. (a) Write expressions for the wave functions and energies as a function of the quantum numbers n_1 and n_2 (assuming the box is in the xy plane). (b) Find the energies of the ground state and first excited state. Is either of these states degenerate? Explain.
5. Assume that the nucleus of an atom can be regarded as a three-dimensional box of width 2×10^{-14} m. If a proton moves as a particle in this box, find (a) the ground-state energy of the proton in MeV and (b) the energies of the first and second excited states. (c) What are the degeneracies of these states?
6. The normalized ground-state wave function for the electron in the hydrogen atom is

$$\psi(r, \theta, \phi) = \frac{1}{\sqrt{\pi}} \left(\frac{1}{a_0}\right)^{3/2} e^{-r/a_0}$$

where r is the radial coordinate of the electron and a_0 is the Bohr radius. (a) Sketch the wave function versus

r. (b) Show that the probability of finding the electron between r and $r + dr$ is given by $|\psi(r)|^2 4\pi r^2\, dr$. (c) Sketch the probability versus r and from your sketch find the radius at which the electron is most likely to be found. (d) Show that the wave function as given is normalized. (e) Find the probability of locating the electron between $r_1 = a_0/2$ and $r_2 = 3a_0/2$.

7. (a) Determine the quantum numbers ℓ and m_ℓ for the He^+ ion in the state corresponding to $n = 3$. (b) What is the energy of this state?

8. (a) Determine the quantum numbers ℓ and m_ℓ for the Li^{2+} ion in the states for which $n = 1$ and $n = 2$. (b) What are the energies of these states?

9. A general expression for the energy levels of one-electron atoms is

$$E_n = -\left(\frac{\mu k^2 q_1{}^2 q_2{}^2}{2\hbar^2}\right)\frac{1}{n^2}$$

where k is the Coulomb constant, q_1 and q_2 are the charges of the two particles, and μ is the reduced mass. The wavelength for the $n = 3$ to $n = 2$ transition of the hydrogen atom is 656.3 nm (visible red light). What are the wavelengths for this same transition in (a) positronium, which consists of an electron and a positron, and (b) singly ionized helium? (*Note:* a positron is a positively charged electron.)

10. If an electron has an angular orbital momentum of 4.714×10^{-34} J·s, what is the orbital quantum number for this state of the electron?

11. Calculate the possible values of the z component of angular momentum for an electron in a d subshell.

12. Consider an electron for which $n = 4$, $\ell = 3$, and $m_\ell = 3$. Calculate the numerical value of (a) the orbital angular momentum and (b) the z component of the orbital angular momentum.

13. The orbital angular momentum of the earth in its motion about the sun is 4.83×10^{31} kg·m²/s. Assuming it is quantized according to Equation 7.24, find (a) the value of ℓ corresponding to this angular momentum and (b) the fractional change in L as ℓ changes from ℓ to $\ell + 1$.

14. Calculate the angular momentum for an electron in (a) the $4d$ state and (b) the $6f$ state of hydrogen.

15. A hydrogen atom is in the 6g state. (a) What is the principal quantum number? (b) What is the energy of the atom? (c) What are the values for the orbital quantum number and the magnitude of the electron's orbital angular momentum? (d) What are the possible values for the orbital magnetic quantum number? For each value, find the corresponding z component of the electron's orbital angular momentum and the angle that the orbital angular momentum vector makes with the z axis.

16. Prove that the nth energy level of an atom has degeneracy equal to n^2.

17. The time-independent Schrödinger equation for the hydrogen atom written in cartesian coordinates is

$$\frac{\partial^2\psi}{\partial x^2} + \frac{\partial^2\psi}{\partial y^2} + \frac{\partial^2\psi}{\partial z^2} = \frac{2m}{\hbar^2}\left(-\frac{ke^2}{r} - E\right)\psi$$

where k is the Coulomb constant and $r = \sqrt{x^2 + y^2 + z^2}$. Show that the wave function given by Equation 7.33 is a solution of the Schrödinger equation with energy

$$E = -\frac{mk^2e^4}{2\hbar^2}$$

18. Suppose that a hydrogen atom is in the 2s state. Taking $r = a_0$, calculate values for (a) $\psi_{2s}(a_0)$, (b) $|\psi_{2s}(a_0)|^2$, and (c) $P_{2s}(a_0)$.

19. The radial part of the wave function for the hydrogen atom in the $2p$ state is given by

$$R_{2p}(r) = Are^{-r/2a_0}$$

where A is a constant and a_0 is the Bohr radius. Using this expression, calculate the average value of r for an electron in this state.

20. A dimensionless number that often appears in atomic physics is the *fine-structure constant* α, given by

$$\alpha = \frac{ke^2}{\hbar c}$$

where k is the Coulomb constant. (a) Obtain a numerical value for $1/\alpha$. (b) In scattering experiments, the "size" of the electron is the *classical electron radius*, $r_0 = ke^2/m_ec^2$. In terms of α, what is the ratio of the Compton wavelength, $\lambda = h/m_ec$, to the classical electron radius? (c) In terms of α, what is the ratio of the Bohr radius, a_0, to the Compton wavelength? (d) In terms of α, what is the ratio of the *Rydberg wavelength*, $1/R$, to the Bohr radius?

21. Obtain the stationary states for a *free* particle in three dimensions by separating the variables in Schrödinger's equation. Do this by substituting the separable form $\Psi(r, t) = \psi_1(x)\psi_2(y)\psi_3(z)\phi(t)$ into the time-dependent Schrödinger equation and dividing each term by $\Psi(r, t)$. Isolate all terms depending only on x from those depending only on y, etc., and argue that four separate equations must result, one for each of the unknown functions ψ_1, ψ_2, ψ_3, and ϕ. Solve the resulting equations. What dynamical quantities are sharp for the states you have found?

22. For a particle confined to a cubic box of dimension L, show that the normalizing factor is $A = (2\pi/L)^{3/2}$, the *same* value for *all* the stationary states. How is this result changed if the box has edge lengths L_1, L_2, and L_3, all of which are different?

23. Calculate the average potential and kinetic energies for the electron in the ground state of hydrogen.

24. Consider a particle of mass m confined to a three-dimensional cube of length L so small that the particle motion is *relativistic*. Obtain an expression for the allowed particle energies in this case. Compute the ground state energy for an electron if L = 10 fm (10^{-5} nm), a typical nuclear dimension.

25. As shown in Example 7.9, the average distance of the electron from the proton in the hydrogen ground state is 1.5 bohrs. For this case calculate Δr, the uncertainty in distance about the average value, and compare it with the average itself. Comment on the significance of your result.

26. Compare the most probable distances of the electron from the proton in the hydrogen $2s$ and $2p$ states with the radius of the second Bohr orbit in hydrogen, $4a_0$.

27. Compute the probability that a $2s$ electron of hydrogen will be found inside the Bohr radius for this state, $4a_0$. Compare this with the probability of finding a $2p$ electron in the same region.

28. Calculate the uncertainty product $\Delta r \, \Delta p$ for the $1s$ electron of a hydrogenic atom with atomic number Z. (*Hint:* Use $\langle p \rangle = 0$ by symmetry and deduce $\langle p^2 \rangle$ from the average kinetic energy, calculated as in Problem 23.)

29. An electron outside a dielectric is attracted to the surface by a force $F = -A/x^2$, where x is the perpendicular distance from the electron to the surface, and A is a constant. Electrons are prevented from crossing the surface, since there are no quantum states in the dielectric for them to occupy. Assume that the surface is infinite in extent, so that the problem is effectively one-dimensional. Write the Schrödinger equation for an electron outside the surface $x > 0$. What is the appropriate boundary condition at $x = 0$? Obtain a formula for the allowed energy levels in this case. (*Hint:* compare the equation for $\psi(x)$ with that satisfied by the pseudo-wavefunction $g(r) = rR(r)$ for hydrogenic atoms.)

PROBLEMS FOR THE COMPUTER:

The following problems require for their solution the use of the program EIGEN3. Where appropriate, furnish screen dumps to substantiate your answers.

30. Find the ground state energy and wavefunction for the spherical well with $U = 4$. How many bound states do you think there are in this case? Can you guess what the results should be in the limit $U \to \infty$?

31. The hydrogen atom can be studied using the defect well with $\beta = 0$. For this case, verify that the next-to-lowest energy for s waves ($\ell = 0$) coincides with the lowest energy for p waves ($\ell = 1$). How do the corresponding waveforms compare? In particular, report the most probable distance for each. Explain these results in terms familiar from your textbook study of atomic hydrogen.

The following problems require for their solution the use of the program ORBS.

32. Display the probability distribution for the electron in the $2s$ state of hydrogen. Describe the symmetry of this distribution. Locate (approximately) those places where the electron is most likely to be found. Do the same for the $3s$ state of hydrogen, and compare the two sets of results.

33. Display and describe the probability distribution for the $3d$ electron in hydrogen. Find the most probable places for this electron and compare your results with that for a $3p$ electron and a $3s$ electron.

8

Atomic Structure

Much of what we have learned about the hydrogen atom with its single electron can be used directly to describe such single-electron ions as He^+ and Li^{+2}, which are hydrogenlike in their electronic structure. However, multielectron atoms such as neutral helium and lithium introduce extra complications that stem from the interactions among the atomic electrons. Thus, the study of the atom inevitably involves us in the complexities of systems consisting of many interacting electrons. In this chapter we will learn some of the basic principles needed to treat such systems effectively, and apply them to describe the physics of electrons in atoms.

Being of like charge and confined to a small space, the electrons of an atom repel each other strongly through the Coulomb force. In addition, we shall discover that the atomic electrons behave like tiny bar magnets, interacting magnetically with each other as well as with any external magnetic field applied to the atom. These magnetic properties derive in part from a new concept—*electron spin*—which will be explored at some length in this chapter.

Another new physical idea, known as the *exclusion principle,* is also presented in this chapter. This principle is extremely important in understanding the properties of multielectron atoms and the periodic table. In fact, the implications of the exclusion principle are almost as far-reaching as those of the Schrödinger equation itself.

8.1 ORBITAL MAGNETISM AND THE NORMAL ZEEMAN EFFECT

An electron orbiting the nucleus of an atom should give rise to magnetic effects, much like those arising from an electric current circulating in a wire loop. In particular, the motion of the charge generates a magnetic field within the atom, and the atom as a whole will be subject to forces and torques when it is placed in an external magnetic field. These magnetic interactions can all be described in terms of a single property of the atom—the magnetic dipole moment.

To calculate the magnetic moment of an orbiting charge, we reason by analogy with the current-carrying loop of wire. The moment $\boldsymbol{\mu}$ of such a loop has magnitude $\mu = iA$, where i is the current and A is the area bounded by the loop. The direction of this moment is perpendicular to the plane of the loop, and its sense is given by a right-hand rule, as shown in Figure 8.1a. This characterization of a current loop as a magnetic dipole implies that its magnetic behavior is similar to that of a bar magnet with its N-S axis directed along $\boldsymbol{\mu}$ (Fig. 8.1b).

For a circulating charge q, the (time-averaged) current is simply q/T where T is the orbital period. Further, A/T is just the area swept out per unit time, and equals the angular momentum L of the orbiting charge divided by

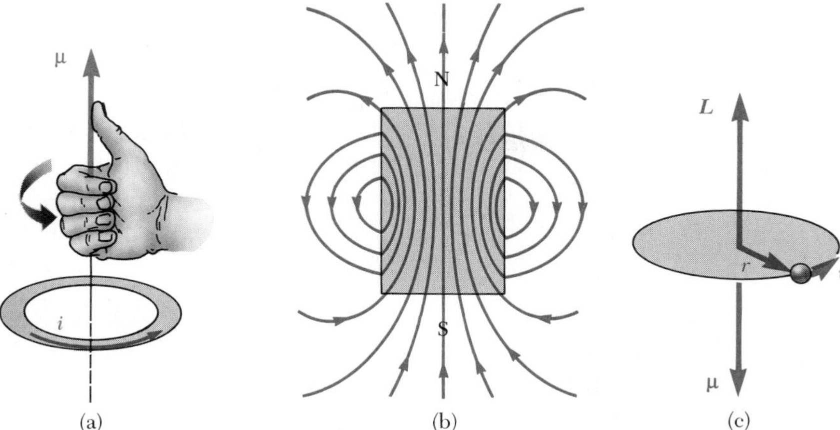

Figure 8.1 (a) The magnetic field in the space surrounding a current-carrying wire loop is that of a magnetic dipole with moment $\boldsymbol{\mu}$ perpendicular to the plane of the loop. The vector $\boldsymbol{\mu}$ points in the direction of the thumb if the fingers of the right hand are curled in the sense of the current i (right-hand rule). (b) The magnetic field in the space surrounding a bar magnet is also that of a magnetic dipole. The dipole moment vector $\boldsymbol{\mu}$ points from the south to the north pole of the magnet. (c) The magnetic moment $\boldsymbol{\mu}$ of an orbiting electron with angular momentum $\boldsymbol{L}$. Since the electron is negatively charged, $\boldsymbol{\mu}$ and $\boldsymbol{L}$ point in opposite directions.

twice the particle mass m.[1] This relation is easily verified for circular orbits, where $L = mvr$, $v = 2\pi r/T$, and $A = \pi r^2$, so that

$$L = m\left(\frac{2\pi r}{T}\right) r = 2m\left(\frac{\pi r^2}{T}\right) = 2m\left(\frac{A}{T}\right) \quad \text{or} \quad \frac{A}{T} = \frac{L}{2m}$$

The same result holds for orbital motion of any kind (see Problem 16), so that $\boldsymbol{\mu} = i\boldsymbol{A}$ becomes

$$\boxed{\boldsymbol{\mu} = \frac{q}{2m}\boldsymbol{L}} \tag{8.1}$$

Magnetic moment of an orbiting charge

for the magnetic moment of an orbiting charge q. Since $\boldsymbol{L}$ is perpendicular to the orbital plane, so too is $\boldsymbol{\mu}$. You may verify that the sense of the vector described by Equation 8.1 is consistent with that expected from the right-hand rule. Thus, the magnetic moment vector is directed along the angular momentum vector, and its magnitude is fixed by the proportionality constant $q/2m$, called the **gyromagnetic ratio.** For electrons, $q = -e$ so $\boldsymbol{\mu}$ actually is opposite the direction of $\boldsymbol{L}$ (Fig. 8.1c).

On the atomic scale, the elemental unit of angular momentum is $\hbar$. It follows that the natural unit for atomic moments is the quantity $e\hbar/2m$, called the **Bohr magneton** and designated by the symbol μ_{B}. Its value in SI units (joules/tesla) is

$$\mu_{\mathrm{B}} = \frac{e\hbar}{2m} = 9.274 \times 10^{-24}\ \mathrm{J/T} \tag{8.2}$$

Bohr magneton

[1] This is one of *Kepler's laws* of planetary motion, later shown by Newton to be a consequence of any central force.

With atomic energies on the scale of electron volts ($1\text{ eV} = 1.60 \times 10^{-19}\text{ J}$) and laboratory fields often specified in gauss ($1\text{ T} = 10^4\text{ G}$), a convenient practical unit for μ_B is eV/G. In these units $\mu_B = 5.788 \times 10^{-9}\text{ eV/G}$.

Since $\boldsymbol{\mu}$ is proportional to $\boldsymbol{L}$, the orbital magnetic moment is subject to space quantization, as illustrated in Figure 8.2. In particular, the z component of the orbital magnetic moment is fixed by the value of the magnetic quantum number m_ℓ as

$$\mu_z = -\frac{e}{2m} L_z = -\mu_B m_\ell \tag{8.3}$$

Just as with angular momentum, the magnetic moment vector should be visualized as precessing about the z axis to preserve this sharp value of μ_z while depicting the fuzziness of the remaining components μ_x and μ_y.

The interaction of an atom with an applied magnetic field depends upon the size and orientation of the atom's magnetic moment. Suppose an external field $\boldsymbol{B}$ is imposed in the direction of the z axis. According to classical electromagnetism, the atom experiences a torque

$$\boldsymbol{\tau} = \boldsymbol{\mu} \times \boldsymbol{B} \tag{8.4}$$

A magnetic moment precesses in a magnetic field

that tends to align its moment with the applied field. Instead of aligning itself with $\boldsymbol{B}$, however, the moment actually precesses around the field direction! This unexpected precession arises because $\boldsymbol{\mu}$ is proportional to the angular momentum $\boldsymbol{L}$. The motion is analogous to that of a spinning top precessing in the earth's gravitational field: the gravitational torque acting to tip it over instead results in precession because of the angular momentum possessed by the spinning top. Returning to the atomic case, since $\boldsymbol{\tau} = d\boldsymbol{L}/dt$, we see from the torque equation that the *change* in angular momentum $d\boldsymbol{L}$ is always perpendicular to both $\boldsymbol{L}$ and $\boldsymbol{B}$. Figure 8.3 depicts the motion (precession) that results. For atoms in a magnetic field this is known as **Larmor precession.**

From the geometry of Figure 8.3, we see that in a time dt the precession angle increases by $d\phi = (L \sin\theta)^{-1}|d\boldsymbol{L}|$. But Equations 8.1 and 8.4 combine to

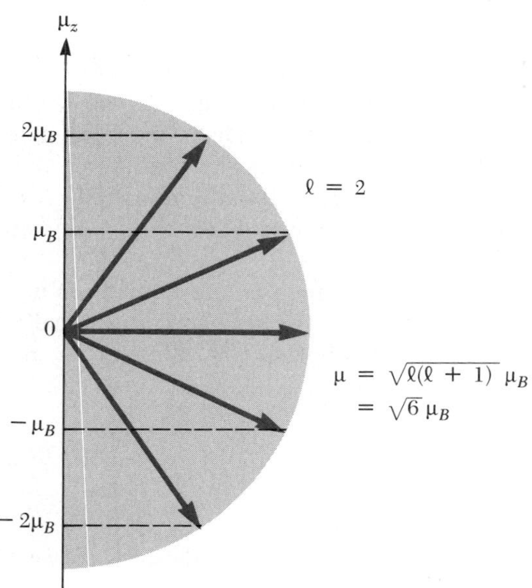

Figure 8.2 The orientations in space and z components of the orbital magnetic moment for the case $\ell = 2$. There are $2\ell + 1 = 5$ different possible orientations.

give $|dL| = |\tau|\ dt = |q/2m|(LB \sin\theta)\ dt$. For electrons, $q = -e$ and the frequency of precession, or *Larmor frequency* ω_L, becomes

$$\omega_L = \frac{d\phi}{dt} = \{L \sin\theta\}^{-1}\frac{|dL|}{dt} = \frac{e}{2m}B \tag{8.5}$$

It is useful to introduce the quantum of energy $\hbar\omega_L$ associated with the Larmor frequency ω_L. This energy is related to the work required to reorient the atomic moment against the torque of the applied field. Remembering that the work of a torque τ to produce an angular displacement $d\theta$ is $dW = \tau\ d\theta$, we have from Equation 8.4

$$dW = -\mu B \sin\theta\ d\theta = d(\mu B \cos\theta) = d(\boldsymbol{\mu}\cdot\mathbf{B})$$

The minus sign signifies that the external torque must *oppose* that produced by the magnetic field $\mathbf{B}$. The work done is stored as orientational potential energy of the dipole in the field. Writing $dW = -dU$, we identify the **magnetic potential energy** U as

The energy of a magnetic moment depends upon its orientation in a magnetic field

$$\boxed{U = -\boldsymbol{\mu}\cdot\mathbf{B}} \tag{8.6}$$

Equation 8.6 expresses the fact that the energy of a magnetic dipole in an external magnetic field $\mathbf{B}$ depends upon its orientation in this field. The magnetic energy is minimal when $\boldsymbol{\mu}$ and $\mathbf{B}$ are aligned; therefore, this alignment is the preferred orientation. Since the possible orientations for $\boldsymbol{\mu}$ are restricted by space quantization, the magnetic energy is quantized accordingly. Taking the z axis along $\mathbf{B}$, and combining Equations 8.1, 8.3, and 8.6, we find

$$U = \frac{e}{2m}\mathbf{L}\cdot\mathbf{B} = \frac{eB}{2m}L_z = \hbar\omega_L m_\ell \tag{8.7}$$

From Equation 8.7 we see that the magnetic energy of an atomic electron depends upon the magnetic quantum number m_ℓ and, hence, is quantized. The total energy of this electron is the sum of its magnetic energy U plus whatever energy it had in the absence of an applied field, say E_0. Thus,

$$E = E_0 + \hbar\omega_L m_\ell \tag{8.8}$$

For atomic hydrogen, E_0 depends only on the principal quantum number n; in more complex atoms the atomic energy also varies according to the subshell label ℓ, as discussed further in Section 8.5.

Unlike energies, the wave functions of atomic electrons are unaffected by the application of a magnetic field. This somewhat surprising result can be understood by recognizing that, according to classical physics, the only effect of the field is to cause (Larmor) precession around the direction of $\mathbf{B}$. For atomic electrons this translates into precession of $\mathbf{L}$ about the z axis. However, such a precession is already implicit in our semiclassical picture of electron orbits in the absence of external fields, as required by the sharpness of L_z while L_x and L_y remain fuzzy. From this viewpoint, the introduction of an applied magnetic field merely transforms this *virtual precession*[2] into a real one at the Larmor frequency!

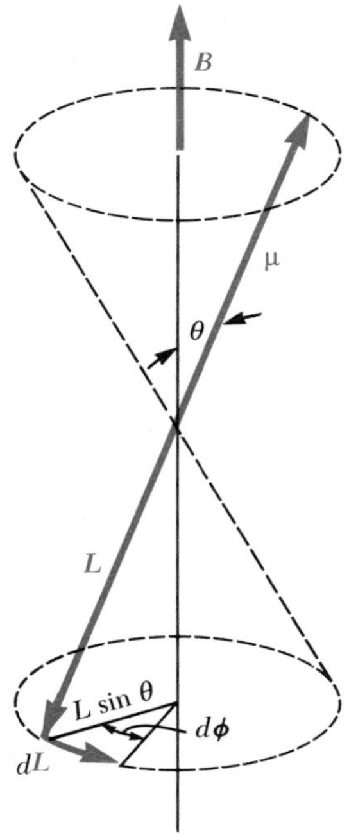

Figure 8.3 Larmor precession of the orbital moment $\boldsymbol{\mu}$ in an applied magnetic field $\mathbf{B}$. Since $\boldsymbol{\mu}$ is proportional to $\mathbf{L}$, the torque of the applied field causes the moment vector $\boldsymbol{\mu}$ to precess around the direction of $\mathbf{B}$ with frequency $\omega_L = eB/2m$.

[2] In zero magnetic field, the precession of the classical vector may be termed virtual (not real) since, even though the same value may not be obtained for L_x (or L_y) in successive measurements, the average value $\langle L_x\rangle$ (or $\langle L_y\rangle$) does not change over time. With $\mathbf{B}$ nonzero, however, it can be shown that $(d^2/dt^2)\langle L_x\rangle = -\omega_L{}^2\langle L_x\rangle$ (and similarly for $\langle L_y\rangle$), indicating that $\langle L_x\rangle$ (and $\langle L_y\rangle$) oscillates at the Larmor frequency ω_L.

EXAMPLE 8.1 Magnetic Energy of the Electron in Hydrogen

Calculate the magnetic energy and Larmor frequency for an electron in the $n = 2$ state of hydrogen, assuming the atom is in a magnetic field of strength $B = 1.00$ T.

Solution: Taking the z axis along $\mathbf{B}$, we calculate the magnetic energy from Equation 8.7 as

$$U = \frac{eB}{2m} L_z = \frac{e\hbar}{2m} Bm_\ell = \hbar\omega_L m_\ell$$

For a 1.0 T ($=10^4$ G) field, the energy quantum $\hbar\omega_L$ has the value

$$\hbar\omega_L = \frac{e\hbar}{2m} B = \mu_B B = (9.27 \times 10^{-24} \text{ J/T})(1.00 \text{ T})$$
$$= 9.27 \times 10^{-24} \text{ J} = 5.79 \times 10^{-5} \text{ eV}$$

With $n = 2$, ℓ can be 0 or 1, and m_ℓ is 0 (twice), ± 1, or -1. Thus, the magnetic energy U can be 0, $+\hbar\omega_L$, or $-\hbar\omega_L$. In such applications, the energy quantum $\hbar\omega_L$ is called the **Zeeman energy.** The Zeeman energy divided by $\hbar$ is the Larmor frequency:

$$\omega_L = \frac{5.79 \times 10^{-5} \text{ eV}}{6.58 \times 10^{-16} \text{ eV}\cdot\text{s}} = 8.80 \times 10^{10} \text{ rad/s.}$$

Evidence for the existence of atomic moments is the appearance of extra lines in the spectrum of an atom that is placed in a magnetic field. Consider a hydrogen atom in its first excited ($n = 2$) state. For $n = 2$, ℓ can have values 0 and 1. The magnetic field has no effect on the state for which $\ell = 0$, since then $m_\ell = 0$. However, for $\ell = 1$, m_ℓ can take values of 1, 0, and -1, and the first excited level is split into three levels by the magnetic field (Figure 8.4). The original (Lyman) emission line is replaced by the three lines depicted in Figure 8.4. The central line appears at the same frequency ω_0 as it would without a magnetic field. This is flanked on both sides by new lines at frequencies $\omega_0 \pm \omega_L$. Thus, the magnetic field splits the original emission line into a triplet of lines. This effect of spectral line splitting by a magnetic field is known as the **Zeeman effect** after its discoverer, Pieter Zeeman.

The normal Zeeman effect

Zeeman spectra of atoms excited to higher states should be more complex, since many more level splittings are involved. For electrons excited to the $n = 3$ state of hydrogen, the expected Zeeman lines and the atomic transitions that give rise to them are shown in Figure 8.5. Accompanying each hydrogen

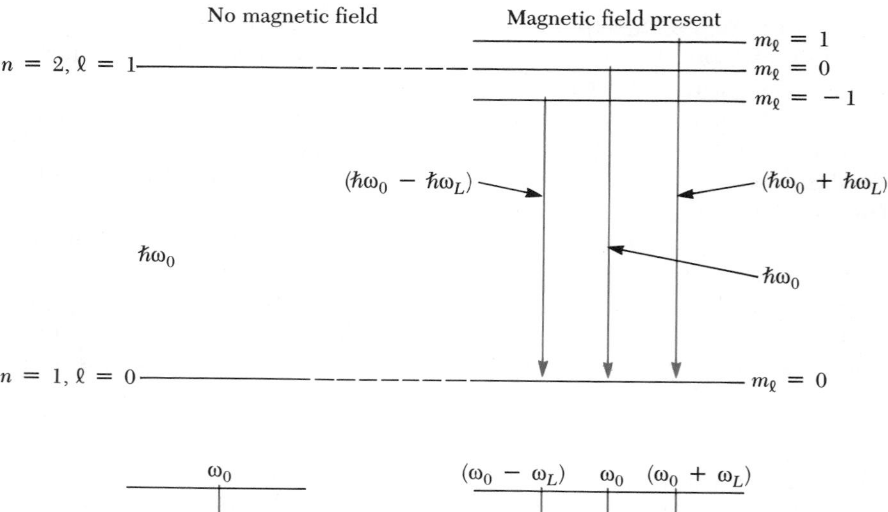

Figure 8.4 Level splittings for the ground and first excited states of a hydrogen atom immersed in a magnetic field $\mathbf{B}$. An electron in one of the excited states decays to the ground state with the emission of a photon, giving rise to emission lines at ω_0, $\omega_0 + \omega_L$, and $\omega_0 - \omega_L$. This is the normal Zeeman effect. When $B = 0$, only the line at ω_0 is observed.

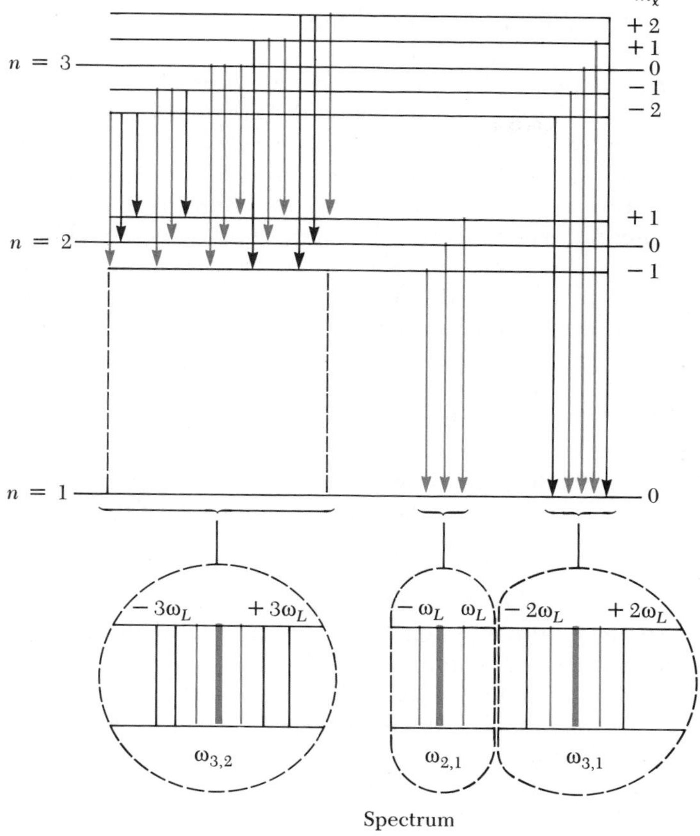

Figure 8.5 Zeeman spectral lines and the underlying atomic transitions that give rise to them for an electron excited to the $n = 3$ state of hydrogen. Because of *selection rules,* only the transitions drawn in color actually occur.

line are anywhere from two to six *satellites* at frequencies removed from the original by multiples of the Larmor frequency. But the observed Zeeman spectrum is not this complicated, owing to **selection rules** that limit the transitions to those for which m_ℓ changes by 0, +1, or −1. The result is that satellites appear at the Larmor frequency *only,* and not at multiples of this frequency. The selection rules express conservation of angular momentum for the *system,* taking into account the angular momentum of the emitted photon. Selection rules are treated in more depth in Chapter 11.

Finally, even the splitting of an emission line into a triplet of equally spaced lines as predicted here, called the *normal Zeeman effect,* frequently is not observed. More commonly, splittings into four, six, or even more unequally spaced lines are seen. This is the *anomalous Zeeman effect,* which has its roots in the existence of electron spin.

8.2 THE SPINNING ELECTRON

The anomalous Zeeman splittings are only one of several phenomena not explained by the magnetic interactions discussed thus far. Another is the observed doubling of many spectral lines referred to as **fine structure.** Both effects are attributed to the existence of a new magnetic moment—the spin moment—that occurs because the electron spins on its axis.

We have seen that the orbital motion of charge gives rise to magnetic effects that can be described in terms of the orbital moment $\boldsymbol{\mu}$ of Equation 8.1.

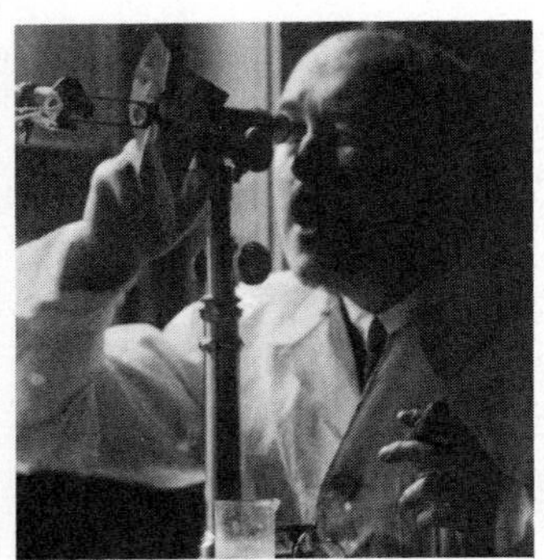

Otto Stern. (Photo courtesy of University of Hamburg, Germany)

BIOGRAPHY OTTO STERN (1888–1969)

Otto Stern was one of the most outstanding experimental physicists of the twentieth century. Born and educated in Germany (PhD in physical chemistry in 1912), he at first worked with Einstein on theoretical issues in molecular theory, in particular applying the new quantum ideas to theories of the specific heat of solids. From about 1920, Stern devoted himself to his real life-work, the development of the molecular-beam method, which enabled him to investigate the properties of free or isolated atoms and culminated in a Nobel Prize in 1945. In this method, a thin stream of atoms is introduced into a high-vacuum chamber where the atoms are free, and hence the properties of individual atoms may be investigated by applying external fields or by some other technique.

Stern first used this method to confirm that silver atoms obey the Maxwell speed distribution. Shortly after, in a series of elegant and difficult experiments with Walter Gerlach, Stern showed that silver atoms obey space quantization and succeeded in measuring the magnetic moment of the silver atom. In the period from 1923 to 1933, Stern directed a remarkably productive molecular beam laboratory at the University of Hamburg. With his students and co-workers he directly demonstrated the wave nature of helium atoms, and measured the magnetic moments of many atoms. Finally, with a great deal of effort, he succeeded in measuring the very small magnetic moments of the proton and deuteron. For these last important fundamental measurements he was awarded the Nobel Prize. In connection with the measurement of the proton's magnetic moment, an interesting story is told by Victor Weisskopf, which should gladden the hearts of experimentalists everywhere:

"There was a seminar held by the theoretical group in Göttingen, and Stern came down and gave a talk on the measurements he was about to finish of the magnetic moment of the proton. He explained his apparatus, but he did not tell us the result. He took a piece of paper and went to each of us saying, 'What is your prediction of the magnetic moment of the proton?' Every theoretician from Max Born down to Victor Weisskopf said, 'Well, of course, the great thing about the Dirac equation is that it predicts a magnetic moment of one Bohr magneton for a particle of spin one-half!' Then he asked us to write down the prediction; everybody wrote 'one magneton.' Then, two months later, he came to give again a talk about the finished experiment, which showed that the value was 2.8 magnetons. He then projected the paper with our predictions on the screen. It was a sobering experience." [1]

In protest over Nazi dismissals of some of his closest co-workers, Stern resigned his post in Hamburg and came to the Carnegie Institute of Technology in the United States in 1933. Here he worked on molecular beam research until his retirement in 1946.

[1] From *Physics in the Twentieth Century, My Life as a Physicist,* selected essays by Victor F. Weisskopf, The MIT Press, 1972.

A rotating charge gives rise to a spin magnetic moment

Similarly, a charged object in rotation produces magnetic effects related to the spin moment $\boldsymbol{\mu}_s$. The spin moment is found by noting that a rotating body of charge can be viewed as a collection of charge elements Δq with mass Δm all rotating in circular orbits about a fixed line, the axis of rotation (Fig. 8.6). To each of these we should apply Equation 8.1 with $\mathbf{L}$ replaced by $\mathbf{L}_i$, the orbital

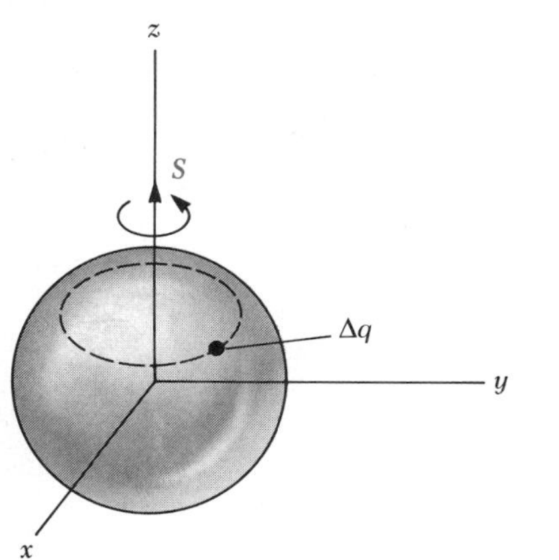

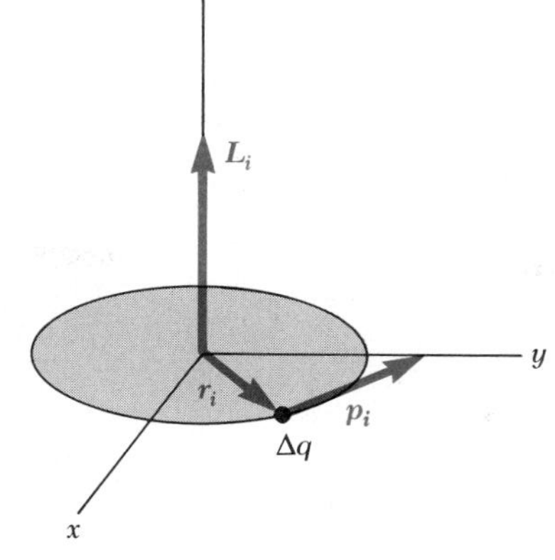

Figure 8.6 A spinning charge q may be viewed as a collection of charge elements Δq orbiting a fixed line, the axis of rotation. The magnetic moments accompanying these orbiting charge elements are summed to give the total magnetic moment of rotation, or spin moment, of the charge q.

angular momentum of the ith charge element. If the charge-to-mass ratio is uniform throughout the body, then $\Delta q/\Delta m$ is the ratio of total charge q to total mass m and we get for the spin moment

$$\mu_s = \frac{q}{2m}\,\Sigma \mathbf{L}_i = \frac{q}{2m}\,\mathbf{S} \tag{8.9}$$

where **S**, the *spin* angular momentum, is the total angular momentum of rotation. The spin angular momentum **S** points along the axis of rotation according to a right-hand rule, as shown in Figure 8.6; its magnitude depends upon the size and shape of the object, as well as its speed of rotation. If the charge-to-mass ratio is not uniform, the gyromagnetic ratio in Equation 8.9 must be multiplied by a dimensionless constant, the **g-factor,** whose value reflects the detailed charge-to-mass distribution within the body. Note that g-factors different from unity are not expected, since a given quantity of charge normally is associated with a proportionate amount of material; that is, the charge-to-mass ratio should be constant.

The existence of a spin moment for the electron was first demonstrated in 1921 in a classic experiment performed by Otto Stern and Walter Gerlach. Electron spin was unknown at that time; the Stern-Gerlach experiment was originally conceived to demonstrate the space quantization associated with orbiting electrons in atoms. In their experiment, a beam of neutral silver atoms was passed through a *nonuniform* magnetic field created in the gap between the pole faces of a large magnet and deposited on glass (Fig. 8.7). A nonuniform field exerts a force on any magnetic moment, so that each atom is deflected in the gap by an amount governed by the orientation of its moment with respect to the direction of inhomogeneity (the z axis), as illustrated in Figure 8.7b. If the moment directions are restricted by space quantization as in Figure 8.2, so too are the deflections. Thus, the atomic beam should split into a number of discrete components, one for each distinct moment orientation present in the beam. This is contrary to the classical expectation that any moment orientation (and hence any beam deflection) is possible, and all would combine to produce a continuous fanning of the atomic beam (Fig. 8.7c).

The Stern-Gerlach experiment produced a staggering result: the silver atomic beam was clearly split—but into only *two* components, not the odd number $(2\ell + 1)$ expected from the space quantization of orbital moments! This is all the more remarkable when we realize that silver atoms in their ground state have no orbital angular momentum ($\ell = 0$), since the outermost electron in silver normally would be in an s state. The result was so surprising that the experiment was repeated in 1927 by Phipps and Taylor with a beam of

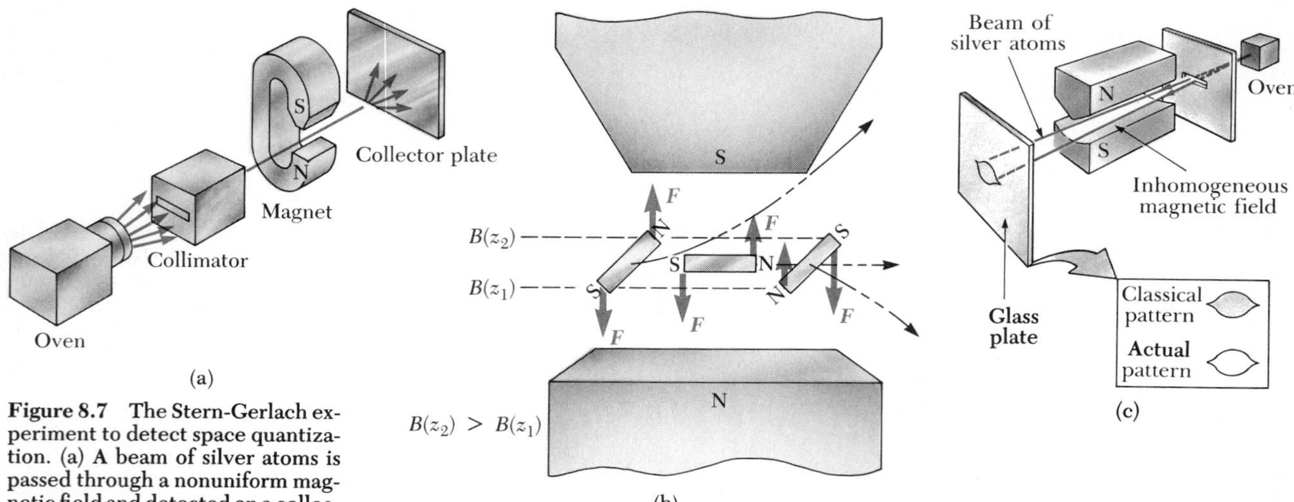

Figure 8.7 The Stern-Gerlach experiment to detect space quantization. (a) A beam of silver atoms is passed through a nonuniform magnetic field and detected on a collector plate. (b) The atoms, with their magnetic moment, are equivalent to tiny bar magnets. In a nonuniform field, each atomic magnet experiences a net force that depends on its orientation. (c) If any moment orientation were possible, a continuous fanning of the beam would be seen at the collector. For space quantization, the fanning is replaced by a set of discrete lines, one for each distinct moment orientation present in the beam.

hydrogen atoms replacing silver, thereby eliminating any uncertainties arising from the use of the more complex silver atoms. The results, however, were unchanged. From these experiments, we are forced to conclude that there is some contribution to the atomic magnetic moment other than the orbital motion of electrons, and that this moment is subject to space quantization.

Our present understanding of the situation dates to the 1925 paper of Samuel Goudsmit and George Uhlenbeck, then graduate students at the University of Leiden. Goudsmit and Uhlenbeck believed that the unknown moment had its origin in the spinning motion of atomic electrons, with the spin angular momentum obeying the same quantization rules as orbital angular momentum. The magnetic moment seen in the Stern-Gerlach experiment is attributed to the spin of the outermost electron in silver. Since all allowed orientations of the spin moment should be represented in the atomic beam, the observed splitting presents a dramatic confirmation of space quantization as applied to electron spin, with the number of components $(2s + 1)$ indicating the value of the **spin quantum number** s.

The spin moment is evidence that the electron is a charge in rotation, although the classical picture of a spinning body of charge given earlier must be adjusted to accommodate the wave properties of matter. The resulting *semiclassical model* of electron spin can be summarized as follows:

Properties of electron spin

1. The spin quantum number s for the electron is $\frac{1}{2}$! This value is dictated by the observation that an atomic beam passing through the Stern-Gerlach magnet is split into just two components $(= 2s + 1)$. Accordingly, there are exactly two orientations possible for the spin axis, described as the up and down spin states of the electron. This is space quantization again, according to the quantization rules for angular momentum[3] as applied to a spin of $\frac{1}{2}$:

$$\boxed{S_z = m_s\hbar} \qquad \text{where } m_s = \tfrac{1}{2} \text{ or } -\tfrac{1}{2} \tag{8.10}$$

The two values $\pm\hbar/2$ for S_z correspond to the two possible orientations for $\mathbf{S}$ shown in Figure 8.8. The value $m_s = +\frac{1}{2}$ refers to the up spin case, some-

[3] For integer angular momentum quantum numbers, the z-component is quantized as $m_s = 0, \pm 1, \ldots \pm s$, which can also be written as $m_s = s, s - 1, \ldots -s$. For $s = \frac{1}{2}$, the latter implies $m_s = \frac{1}{2}$ or $-\frac{1}{2}$.

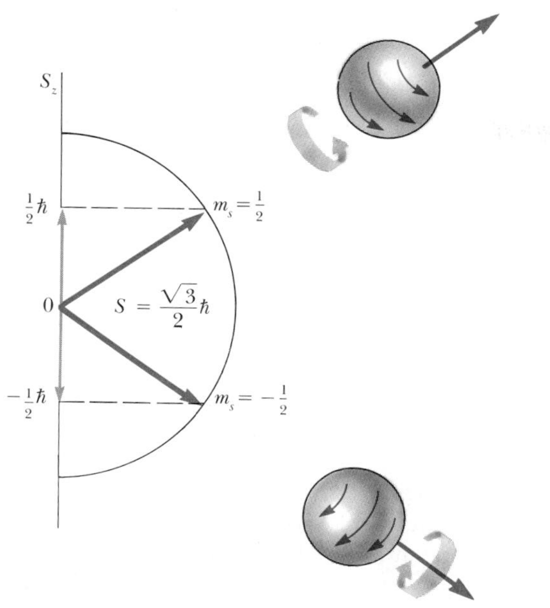

Figure 8.8 The spin angular momentum also exhibits space quantization. This figure shows the two allowed orientations of the spin vector S for a spin $\frac{1}{2}$ particle, such as the electron.

times designated with an up arrow (↑) or simply a plus sign (+). Likewise, $m_s = -\frac{1}{2}$ is the down spin case (↓) or (−). The fact that s has a nonintegral value suggests that spin is not merely another manifestation of orbital motion, as the classical picture implies.

2. The magnitude of the spin angular momentum is

$$|\mathbf{S}| = \sqrt{s(s+1)}\,\hbar = \frac{\sqrt{3}}{2}\hbar \tag{8.11}$$

The spin angular momentum of an electron

and *never changes*! This angular momentum of rotation cannot be changed in any way, but is an intrinsic property of the electron, like its mass or charge. The notion that $|\mathbf{S}|$ is fixed contradicts classical laws, where a rotating charge would be slowed down by the application of a magnetic field owing to the Faraday emf that accompanies the changing field (the diamagnetic effect). Further, if the electron were viewed simply as a spinning ball with angular momentum $\hbar\sqrt{3}/2$ subject to classical laws, parts of the ball near its surface would be rotating with velocities in excess of the speed of light![4] All of this is taken to mean that the classical picture of the electron as a charge in rotation must not be pressed too far; ultimately *the spinning electron is a quantum entity defying any simple classical description.*

3. The spin moment is given by Equation 8.9 with a *g-factor of 2;* that is, the moment is twice as large as would be expected for a body with spin angular momentum given by Equation 8.10. The value $g = 2$ is required by the amount of beam deflection produced by the Stern-Gerlach magnet; the larger the magnetic moment, the greater will be the deflection of the atomic

[4] This follows from the extremely small size of the electron. The exact size of the electron is unknown, but an upper limit of 10^{-6} Å is deduced from experiments in which electrons are scattered from other electrons. According to some current theories, the electron may be a true point object, that is, a particle with zero size!

beam. As already mentioned, any g-factor other than unity is not to be expected classically, although models have been contrived for which $g \neq 1$ (see Problem 19). The factor of 2 is predicted by the relativistic quantum theory of the electron put forth by Paul Dirac in 1929.[5]

With the recognition of electron spin we see that an additional quantum number, m_s, is needed to specify the internal, or spin, state of an electron. Thus, the state of an electron in hydrogen must be described by the four quantum numbers n, ℓ, m_ℓ, and m_s. Further, the total magnetic moment now has orbital and spin contributions:

The total magnetic moment of an electron

$$\boldsymbol{\mu} = \boldsymbol{\mu}_0 + \boldsymbol{\mu}_s = \frac{-e}{2m}\{\mathbf{L} + g\mathbf{S}\} \tag{8.12}$$

Because of the electron g-factor, the total moment $\boldsymbol{\mu}$ is no longer in the same direction as the total (orbital plus spin) angular momentum $\mathbf{J} = \mathbf{L} + \mathbf{S}$. The component of $\boldsymbol{\mu}$ along $\mathbf{J}$ is sometimes referred to as the *effective moment.* When the magnetic field $\mathbf{B}$ applied to an atom is weak, it is the effective moment that determines the magnetic energy of atomic electrons according to Equation 8.6. As we shall discover in Section 8.3, the number of possible orientations for $\mathbf{J}$ and, hence, for the effective moment is even, leading to the even number of spectral lines seen in the anomalous Zeeman effect.

[5] The g-factor for the electron is not exactly 2. The best value to date is $g = 2.00232$. The discrepancy between Dirac's predicted value and the observed value is attributed to the electron interacting with the "vacuum." Such effects are the subject of quantum electrodynamics, developed by Richard Feynman in the early 1950s.

EXAMPLE 8.2 Semiclassical Model for Electron Spin
Calculate the angles between the z axis and the axis of rotation for the spinning electron in the up and down spin states. How should we portray the fuzziness inherent in the x and y components of the spin angular momentum?

Solution: For the electron, the magnitude of the spin angular momentum is $|\mathbf{S}| = \hbar\sqrt{3}/2$, and the z component of spin is $S_z = \pm\hbar/2$. Thus, the spin vector $\mathbf{S}$ is inclined from the z axis at angles given by

$$\cos\theta = \frac{S_z}{|\mathbf{S}|} = \pm\frac{1}{\sqrt{3}}$$

For the up spin state, we take the plus sign and get $\cos\theta = 0.577$ or $\theta = 54.7°$. The down spin orientation is described by the minus sign and gives $\cos\theta = -0.577$, or $\theta = 125.3°$. Since the axis of rotation coincides with the spin vector, these are the angles the rotation axis makes with the z axis.

While S_z is sharp in either the up or down spin orientation, both S_x and S_y are fuzzy. This fuzziness may be depicted by allowing the spin vector to precess about the z axis, as we did for the orbital angular momentum in Chapter 7.

Exercise 1 As it happens, the photon is a spin one particle, that is, $s = 1$ for the photon. Calculate the possible angles between the z axis and the axis of rotation for the photon.
Answers: 45°, 90°, and 135°.

EXAMPLE 8.3 Zeeman Spectrum of Hydrogen Including Spin
Examine the Zeeman spectrum produced by hydrogen atoms initially in the $n = 2$ state when electron spin is taken into account, assuming the atoms to be in a magnetic field of strength $B = 1.0$ T.

Solution: The electron energies now have a magnetic contribution from both the orbital and spin motions. Orienting the z axis along the direction of $\mathbf{B}$, we calculate the magnetic energy from Equations 8.6 and 8.12:

$$U = -\boldsymbol{\mu}\cdot\mathbf{B} = \frac{e}{2m}B\{L_z + gS_z\} = \frac{e\hbar}{2m}B\{m_\ell + gm_s\}$$

The energy $(e\hbar/2m)B$ is the Zeeman energy $\mu_B B$ or $\hbar\omega_L$; its value in this example is

$$\mu_B B = (9.27\times10^{-24}\text{ J/T})(1.0\text{ T}) = 9.27\times10^{-24}\text{ J} = 5.79\times10^{-5}\text{ eV}$$

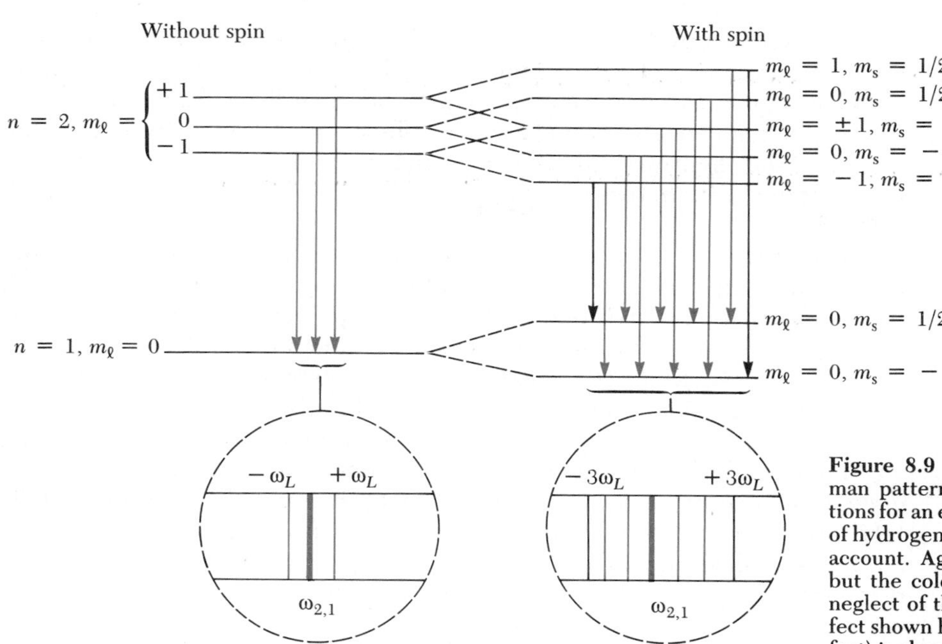

Figure 8.9 (Example 8.3) Predicted Zeeman pattern and underlying atomic transitions for an electron excited to the $n = 2$ state of hydrogen, when electron spin is taken into account. Again, selection rules prohibit all but the colored transitions. Because of the neglect of the spin-orbit interaction, the effect shown here (called the Paschen-Back effect) is observed only in very intense applied magnetic fields.

For the $n = 2$ state of hydrogen, the shell energy is $E_2 = -(13.6 \text{ eV})/2^2 = -3.40$ eV. Since m_ℓ takes the values 0 (twice) and ± 1, there is an orbital contribution to the magnetic energy $U_0 = m_\ell \hbar\omega_L$ that introduces new levels at $E_2 \pm \hbar\omega_L$, as discussed in Example 8.1. The presence of electron spin splits each of these into a pair of levels, the additional (spin) contribution to the energy being $U_s = (gm_s)\hbar\omega_L$ (see Fig. 8.9). Since $g = 2$ and m_s is $\pm\frac{1}{2}$ for the electron, the spin energy in the field $|U_s|$ is again the Zeeman energy $\hbar\omega_L$. Thus, an electron in this shell can have any one of the energies

$$E_2, \quad E_2 \pm \hbar\omega_L, \quad E_2 \pm 2\hbar\omega_L$$

In making a downward transition to the $n = 1$ shell with energy $E_1 = -13.6$ eV, the final state of the electron may have energy $E_1 + \hbar\omega_L$ or $E_1 - \hbar\omega_L$, depending upon the orientation of its spin in the applied field. Thus, the energy of transition may be any one of the following possibilities:

$$\Delta E_{2,1}, \quad \Delta E_{2,1} \pm \hbar\omega_L, \quad \Delta E_{2,1} \pm 2\hbar\omega_L, \quad \Delta E_{2,1} \pm 3\hbar\omega_L$$

Photons emitted with these energies have frequencies

$$\omega_{2,1}, \quad \omega_{2,1} \pm \omega_L, \quad \omega_{2,1} \pm 2\omega_L, \quad \omega_{2,1} \pm 3\omega_L$$

Thus, the spectrum would consist of the original line at $\omega_{2,1}$ flanked on both sides by satellite lines separated from the original by the Larmor frequency, twice the Larmor frequency, and three times this frequency. Notice that the lines at $\omega_{2,1} \pm 2\omega_L$ and $\omega_{2,1} \pm 3\omega_L$ appear solely because of electron spin.

Again, however, the observed pattern is not the predicted one. Selection rules inhibit transitions unless $m_\ell + m_s$ changes by 0, +1, or −1. This has the effect of eliminating the satellites at $\omega_{2,1} \pm 3\omega_L$. Furthermore, the spin moment and the orbital moment of the electron interact with *each other,* a circumstance not recognized in our calculation. Only when this spin-orbit interaction energy is small compared with the Zeeman energy $\hbar\omega_L$ do we observe the spectral lines predicted here. This is the case for the *Paschen-Back effect,* where the magnetic field applied to the atom is intense enough to make $\hbar\omega_L$ the dominant energy.

8.3 THE SPIN-ORBIT INTERACTION AND OTHER MAGNETIC EFFECTS

The existence of both spin and orbital moments for the electron inevitably leads to their mutual interaction. This so-called **spin-orbit interaction** is best understood from the vantage point of the orbiting electron, which "sees" the

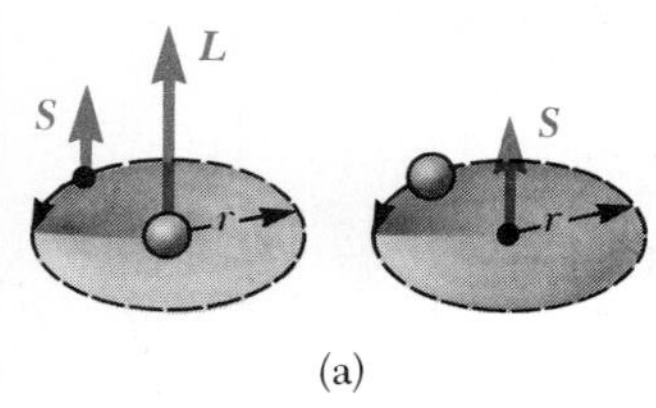

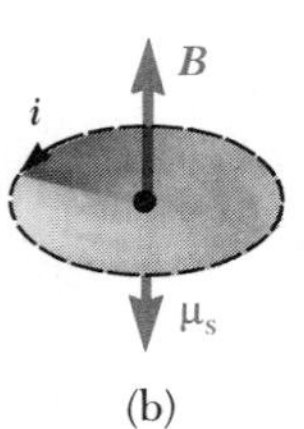

Figure 8.10 (a) An electron with angular momentum $\boldsymbol{L}$ orbiting the nucleus of an atom. In the up-spin orientation shown here, the spin angular momentum $\boldsymbol{S}$ of the electron is essentially aligned with $\boldsymbol{L}$. From the vantage point of the orbiting electron, the nucleus circulates as shown. (b) The apparently circulating nuclear charge is represented by the current i, and causes a magnetic field $\boldsymbol{B}$ at the site of the electron. In the presence of $\boldsymbol{B}$ the electron spin moment $\boldsymbol{\mu}_s$ acquires magnetic energy $U = -\boldsymbol{\mu}_s \cdot \boldsymbol{B}$. The spin moment $\boldsymbol{\mu}_s$ is opposite the spin vector $\boldsymbol{S}$ for the negatively charged electron. The direction of $\boldsymbol{B}$ is given by a right-hand rule: with the thumb pointing in the direction of the current i, the fingers give the sense in which the $\boldsymbol{B}$ field circulates about the wire. The magnetic energy is highest for the case shown, where $\boldsymbol{S}$ and $\boldsymbol{L}$ are "aligned."

atomic nucleus circling *it* (Fig. 8.10). The apparent orbital motion of the nucleus generates a magnetic field at the electron site, and the electron spin moment acquires magnetic energy in this field according to Equation 8.6. This can be thought of as an internal Zeeman effect, with $\boldsymbol{B}$ arising from the orbital motion of the electron itself. The electron has a higher energy when its spin is up, or aligned with $\boldsymbol{B}$, than when its spin is down, or aligned opposite to $\boldsymbol{B}$ (Fig. 8.10b).

The energy difference between the two spin orientations is responsible for the *fine structure doubling* of many atomic spectral lines. For example, the $2p \rightarrow 1s$ transition in hydrogen is split into two lines because the $2p$ level is actually a spin doublet with a level spacing of about 5×10^{-5} eV (Fig. 8.11), while the $1s$ level remains unsplit (there is no orbital field in a state with zero orbital angular momentum). Similarly, the spin-orbit doubling of the sodium $3p$ level gives rise to the well-known sodium D lines discussed in Example 8.4.

The coupling of spin and orbital moments implies that *neither orbital angular momentum nor spin angular momentum is conserved separately.* But total angular momentum $\boldsymbol{J} = \boldsymbol{L} + \boldsymbol{S}$ is conserved, so long as no external torques are present. Consequently, $|\boldsymbol{J}|$ and J_z are sharp observables quantized in the manner we have come to expect for angular momentum:

$$|\boldsymbol{J}| = \sqrt{j(j+1)}\,\hbar$$

$$J_z = m_j\hbar \qquad \text{with } m_j = j, j-1, \ldots, -j \tag{8.13}$$

Permissible values for the total angular momentum quantum number j are

$$j = \ell + s, \quad \ell + s - 1, \quad \ldots, \quad |\ell - s| \tag{8.14}$$

in terms of the orbital (ℓ) and spin (s) quantum numbers. For an atomic electron $s = \frac{1}{2}$ and $\ell = 0, 1, 2, \ldots$, so $j = \frac{1}{2}$ (for $\ell = 0$) and $j = \ell \pm \frac{1}{2}$ (for $\ell > 0$). These results can be deduced from the vector addition model shown in Figure 8.12a. With $j = \frac{1}{2}$, there are only two possibilities for m_j, namely $m_j = \pm\frac{1}{2}$. For $j = \ell \pm \frac{1}{2}$, the number of possibilities $(2j + 1)$ for m_j becomes either 2ℓ or $2\ell + 2$. Notice that the number of m_j values is always *even*, leading to an even number of orientations in the semiclassical model for $\boldsymbol{J}$ (Fig. 8.12b), rather than the odd number predicted for the orbital angular momentum $\boldsymbol{L}$ alone.

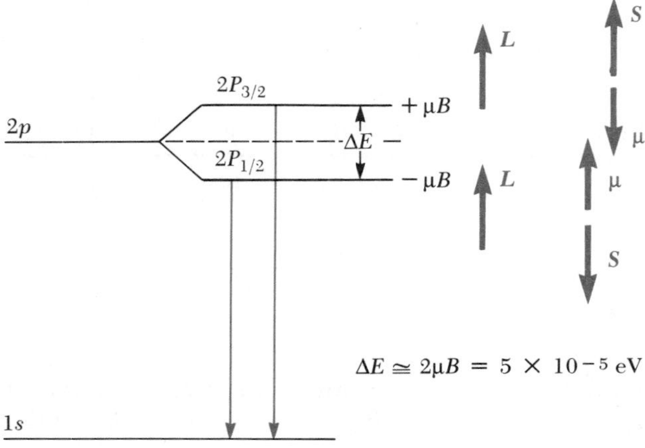

Figure 8.11 The $2p$ level of hydrogen is split by the spin-orbit effect into a doublet separated by the spin-orbit energy $\Delta E = 5 \times 10^{-5}$ eV. The higher energy state is the one for which the spin angular momentum of the electron is "aligned" with its orbital angular momentum. The $1s$ level is unaffected, since no magnetic field arises for orbital motion with zero angular momentum.

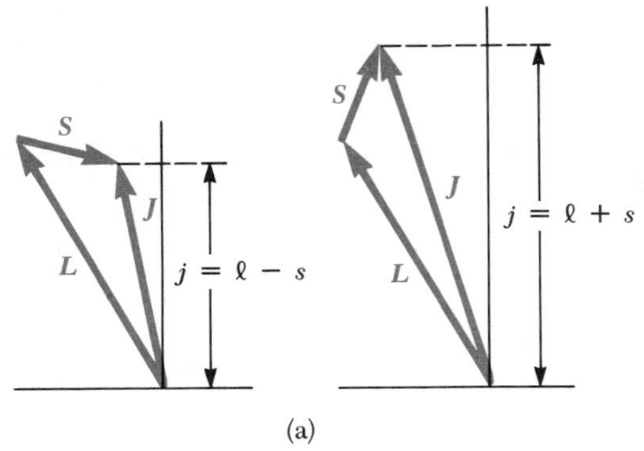

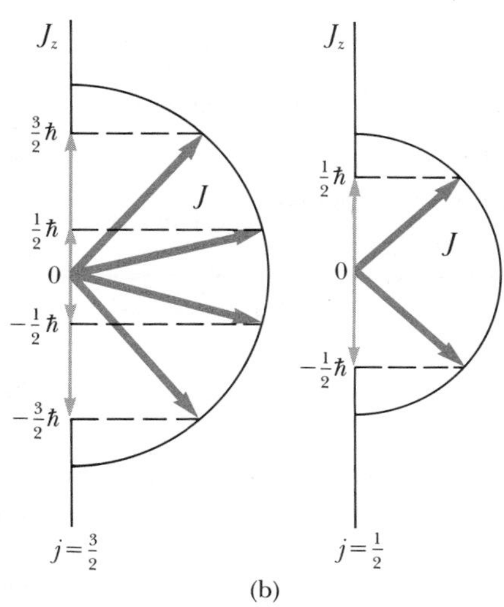

Figure 8.12 (a) A vector model for determining the total angular momentum $\boldsymbol{J} = \boldsymbol{L} + \boldsymbol{S}$ of a single electron. (b) The allowed orientations of the total angular momentum $\boldsymbol{J}$ for the states $j = \frac{3}{2}$ and $j = \frac{1}{2}$. Notice that there are now an even number of orientations possible, not the odd number familiar from the space quantization of $\boldsymbol{L}$ alone.

A common spectroscopic notation is to use a subscript after a letter to designate the total angular momentum of an atomic electron, where the letter itself describes its orbital angular momentum. For example, the notation $1S_{1/2}$ describes the ground state of hydrogen, where the 1 indicates $n = 1$, the S tells us that $\ell = 0$, and the subscript $\frac{1}{2}$ denotes $j = \frac{1}{2}$. Likewise, the spectroscopic notations for the $n = 2$ states of hydrogen are $2S_{1/2}$ ($\ell = 0, j = \frac{1}{2}$), $2P_{3/2}$ ($\ell = 1$, $j = \frac{3}{2}$), and $2P_{1/2}$ ($\ell = 1, j = \frac{1}{2}$). Again, the spin-orbit interaction splits the latter two states in energy by about 5×10^{-5} eV.

Spectroscopic notation extended to include spin

EXAMPLE 8.4 The Sodium Doublet

The famed sodium doublet arises from the spin-orbit splitting of the sodium $3p$ level, and consists of the closely spaced pair of spectral lines at wavelengths of 588.995 nm and 589.592 nm. Show on an energy level diagram the electronic transitions giving rise to these lines, labeling the participating atomic states with their proper spectroscopic designations. From the doublet spacing, deduce the magnitude of the spin-orbit energy.

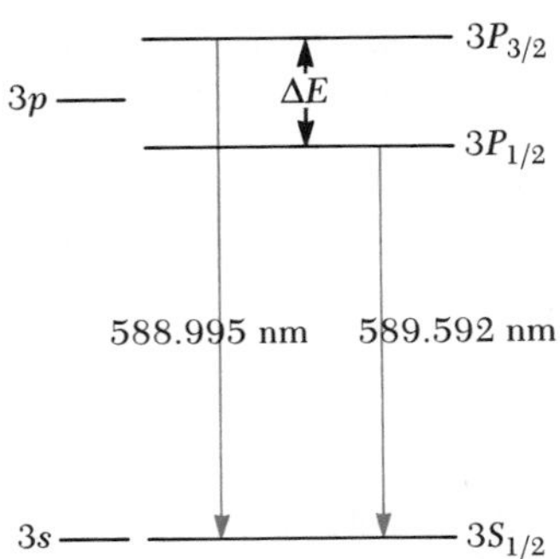

Figure 8.13 (Example 8.4). The transitions $3P_{3/2} \rightarrow 3S_{1/2}$ and $3P_{1/2} \rightarrow 3S_{1/2}$ that give rise to the sodium doublet. The $3p$ level of sodium is split by the spin-orbit effect, while the $3s$ level is unaffected. In the sodium vapor lamp, electrons normally in the $3s$ level are excited to the $3p$ levels by an electric discharge.

Solution: The outer electron in sodium is the first electron to occupy the $n = 3$ shell, and it would go into the lowest energy subshell, the $3s$ or $3S_{1/2}$ level. The next highest levels belong to the $3p$ subshell. The $2(2\ell + 1) = 6$ states of this subshell are grouped into the $3P_{1/2}$ level with two states, and the $3P_{3/2}$ level with four states. The spin-orbit effect splits these levels by the spin-orbit energy. The outer electron, once it is excited to either of these levels by some means (such as an electric discharge in the sodium vapor lamp), returns to the $3S_{1/2}$ level with the emission of a photon. The two possible transitions $3P_{3/2} \rightarrow 3S_{1/2}$ and $3P_{1/2} \rightarrow 3S_{1/2}$ are shown in Figure 8.13. The emitted photons have nearly the same energy, but differ by the small amount ΔE representing the spin-orbit splitting of the initial levels. Since $E = hc/\lambda$ for photons, ΔE is found as

$$\Delta E = \frac{hc}{\lambda_1} - \frac{hc}{\lambda_2} = \frac{hc(\lambda_2 - \lambda_1)}{\lambda_1 \lambda_2}$$

For the sodium doublet, the observed wavelength difference is

$$\lambda_2 - \lambda_1 = 589.592 \text{ nm} - 588.995 \text{ nm} = 0.597 \text{ nm}$$

Using this with $hc = 1240$ eV·nm gives

$$\Delta E = \frac{(1240 \text{ eV}\cdot\text{nm})(0.597 \text{ nm})}{(589.592 \text{ nm})(588.995 \text{ nm})} = 2.13 \times 10^{-3} \text{ eV}$$

Exercise 2 Using the spin-orbit interaction energy calculated in Example 8.4, deduce a value for the magnetic field strength at the site of the orbiting $3p$ electron in sodium.

Answer: $B = 18.38$ T, a large field by laboratory standards.

8.4 EXCHANGE SYMMETRY AND THE EXCLUSION PRINCIPLE

As mentioned earlier, the existence of spin requires that the state of an atomic electron be specified with four quantum numbers. In the absence of spin-orbit effects these could be n, ℓ, m_ℓ, and m_s; if the spin-orbit interaction is taken into account, m_ℓ and m_s are replaced by j and m_j. In either case four quantum numbers are required, one for each of the four degrees of freedom possessed by a single electron.

In those systems where two or more electrons are present, we might expect to describe each electronic state by giving the appropriate set of four quantum numbers. In this connection an interesting question arises, namely, "How many electrons in an atom can have the same four quantum numbers, that is, be in the same state?" This important question was answered by Wolfgang Pauli in 1925 in a powerful statement known as the **exclusion principle.** The exclusion principle states that *no two electrons in an atom can have the same set of quantum numbers.* We should point out that if this principle were not valid, every electron would occupy the 1s atomic state (this being the state of lowest energy), the chemical behavior of the elements would be drastically different, and nature as we know it would not exist!

The exclusion principle follows from our belief that electrons are identical particles—that it is impossible to distinguish one electron from another, even in principle. This seemingly innocuous statement takes on an extra dimension in view of the wave nature of matter, and has far-reaching consequences. To gain an appreciation for this point, let us consider a collision between two electrons, as shown in Figure 8.14. Figures 8.14a and 8.14b depict two distinct events, the scattering effect being much stronger in the latter where the electrons are turned through a larger angle. Each event, however, arises from the same initial condition and leads to the same outcome —both electrons are scattered and emerge at angles θ relative to the axis of incidence. Had we not followed their paths, we could not decide which of the

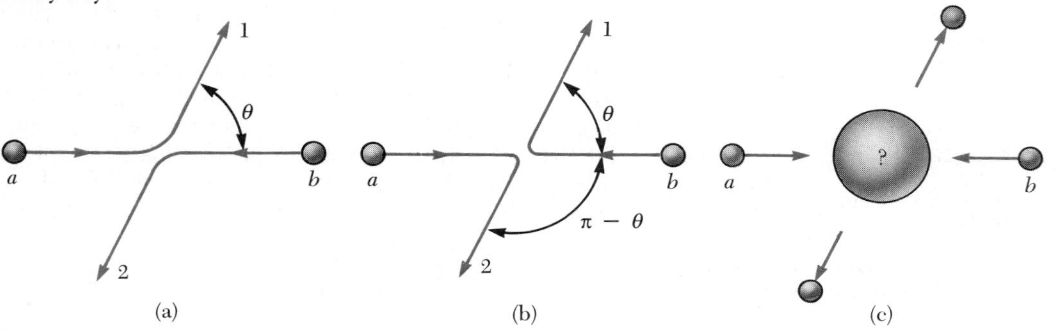

Figure 8.14 The scattering of two electrons as a result of their mutual repulsion. The events depicted in (a) and (b) produce the same outcome for identical electrons, but are nonetheless distinguishable because the path taken by each electron is different in the two cases. In this way, the electrons retain their separate identities during collision. (c) According to quantum mechanics, the paths taken by the electrons are blurred by the wave properties of matter. In consequence, once they have interacted, the electrons cannot be told apart in any way!

BIOGRAPHY WOLFGANG PAULI (1900–1958)

Wolfgang Pauli. (Photo taken by S. A. Goudsmit, AIP Niels Bohr Library)

Wolfgang Pauli was an extremely talented Austrian theoretical physicist who made important contributions in many areas of modern physics. At the age of 21, Pauli gained public recognition with a masterful review article on relativity, which is still considered to be one of the finest and most comprehensive introductions to the subject. Other major contributions were the discovery of the exclusion principle, the explanation of the connection between particle spin and statistics, theories of relativistic quantum electrodynamics, the neutrino hypothesis, and the hypothesis of nuclear spin. An article entitled "The Fundamental Principles of Quantum Mechanics," written by Pauli in 1933 for the *Handbuch der Physik*, is widely acknowledged to be one of the best treatments of quantum physics ever written. Pauli was a forceful and colorful character, well known for his witty and often caustic remarks directed at those who presented new theories in a less than perfectly clear manner. Pauli exerted great influence on his students and colleagues by forcing them with his sharp criticism to a deeper and clearer understanding. Victor Weisskopf, one of Pauli's famous students, has aptly described him as "the conscience of theoretical physics." Pauli's sharp sense of humor was also nicely captured by Weisskopf in the following anecdote:

"In a few weeks, Pauli asked me to come to Zurich. I came to the big door of his office, I knocked, and no answer. I knocked again and no answer. After about five minutes he said, rather roughly, "Who is it? Come in!" I opened the door, and here was Pauli — it was a very big office — at the other side of the room, at his desk, writing and writing. He said, "Who is this? First I must finish calculating." Again he let me wait for about five minutes and then: "Who is that?" "I am Weisskopf." "Uhh, Weisskopf, ja, you are my new assistant." Then he looked at me and said, "Now, you see I wanted to take Bethe, but Bethe works now on the solid state. Solid state I don't like, although I started it. This is why I took you." Then I said, "What can I do for you, sir?" and he said "I shall give you right away a problem." He gave me a problem, some calculation, and then he said, "Go and work." So I went, and after 10 days or so, he came and said, "Well, show me what you have done." And I showed him. He looked at it and exclaimed: "I should have taken Bethe!" [1]

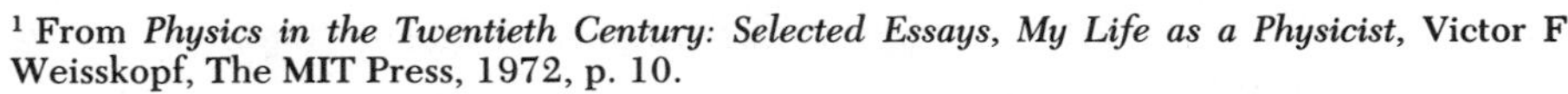

[1] From *Physics in the Twentieth Century: Selected Essays, My Life as a Physicist*, Victor F. Weisskopf, The MIT Press, 1972, p. 10.

two collisions actually occurred, and the separate identities of the electrons would have been lost in the process of collision.

But paths are classical concepts, blurred by the wave properties of matter according to the uncertainty principle. That is, there is an inherent fuzziness to these paths, which blends them inextricably in the collision region, where the electrons may be separated by only a few de Broglie wavelengths. The quantum viewpoint is better portrayed in Figure 8.14c, where the two distinct possibilities (from a classical standpoint) merge into a single quantum event — the scattering of two electrons through an angle θ. The argument makes clear why indistinguishability plays no role in classical physics: all particles, even identical ones, are in fact distinguishable classically through their *paths!* With our acceptance of matter waves, we are forced to conclude that *identical particles cannot be told apart in any way — they are truly indistinguishable.*

Electrons are truly indistinguishable

Inclusion of this remarkable fact into quantum theory leads to the exclusion principle discovered by Pauli.

Let us see how indistinguishability affects our mathematical description of a two-electron system, say the helium atom. Each electron has kinetic energy as well as electrostatic potential energy in the presence of the doubly charged helium nucleus. These contributions are represented in Schrödinger's equation by terms

$$-\frac{\hbar^2}{2m}\nabla_1{}^2\psi + \frac{k(2e)(-e)}{r_1}\psi$$

where $\nabla_1{}^2$ is the Laplacian in this electron's coordinate, r_1. For brevity, let us write the sum of both terms simply as $h(1)\psi$, with the label 1 referring to r_1. For the second electron, we write the same expression, except that r_1 must be replaced everywhere by r_2, the coordinate of the second electron. The stationary states for our two-electron system satisfy Schrödinger's time-independent equation,

$$h(1)\psi + h(2)\psi = E\psi \tag{8.15}$$

The fact that $h(1)$ and $h(2)$ are the *same* but for their arguments reflects the indistinguishability of the two electrons.

Equation 8.15 accounts for the electrons' kinetic and potential energies in the field of the helium nucleus, but ignores their mutual energy of repulsion. In fact, each electron repels the other through the Coulomb force, leading to an *interaction energy* that must be added to the left-hand side of Equation 8.15. For simplicity, we shall ignore this interaction and treat the electrons as independent objects, each unaffected by the other's presence. In Section 8.5 we show how this *independent particle approximation* can be improved to give a better description of reality.

The two-electron wave function will depend upon the coordinates of both particles, $\psi = \psi(r_1, r_2)$, with $|\psi(r_1, r_2)|^2$ representing the probability density for finding one electron at r_1 and the other at r_2. Now, the indistinguishability of electrons requires that a formal interchange of particles produce no observable effects. In particular, all probabilities are unaffected by the interchange, so the wave function ψ must be one for which

$$|\psi(r_1, r_2)| = |\psi(r_2, r_1)|$$

We say that such a wave function exhibits **exchange symmetry.** The wave function itself may be either even or odd under particle exchange. The former is characterized by the property

Exchange symmetry for bosons

$$\boxed{\psi(r_1, r_2) = \psi(r_2, r_1)} \tag{8.16}$$

and describes a class of particles called *bosons.* Photons belong to this class, as do some more exotic particles such as pions. Electrons, as well as protons and neutrons, are examples of *fermions,* for which

Exchange symmetry for fermions

$$\boxed{\psi(r_1, r_2) = -\psi(r_2, r_1)} \tag{8.17}$$

Thus, our two-electron helium wave function must obey Equation 8.17 to account for the indistinguishability of electrons.[6]

[6] It is an experimental fact that integer spin particles are bosons, while half-integer spin particles are fermions. This connection between spin and symmetry under particle exchange can be shown to have a theoretical basis when the quantum theory is formulated so as to conform to the requirements of special relativity.

To recover the Pauli principle, we must examine the wave function more closely. For independent electrons, solutions to Equation 8.15 are easily found. Since each electron "sees" only the helium nucleus, the wave function in each coordinate must be an atomic function of the type discussed in Chapter 7. We denote these atomic functions by ψ_a, where a is a collective label for the *four* quantum numbers n, ℓ, m_ℓ, and m_s (or n, ℓ, j, and m_j if spin-orbit effects are included). The products $\psi_a(r_1)\psi_b(r_2)$ satisfy our equation, since

$$h(1)\psi_a(r_1)\psi_b(r_2) = E_a\psi_a(r_1)\psi_b(r_2)$$

$$h(2)\psi_a(r_1)\psi_b(r_2) = E_b\psi_a(r_1)\psi_b(r_2)$$

E_a and E_b are hydrogenic energies for the states labeled a and b (see Equation 7.21). Thus,

$$[h(1) + h(2)]\psi_a(r_1)\psi_b(r_2) = (E_a + E_b)\psi_a(r_1)\psi_b(r_2) \tag{8.18}$$

and $E = E_a + E_b$ is the total energy of this two-electron state.

Notice that the one-electron energies are simply additive, as we might have anticipated for independent particles. Furthermore, the solution $\psi_a(r_1)\psi_b(r_2)$ describes one electron occupying the atomic state labeled a and the other the state labeled b. But this product is *not* odd under particle exchange, as required for identical fermions. However, you can verify that $\psi_a(r_2)\psi_b(r_1)$ also is a solution to Equation 8.15 with energy $E = E_a + E_b$, corresponding to our two electrons having exchanged states. The antisymmetric combination of these two

$$\psi_{ab}(r_1, r_2) = \psi_a(r_1)\psi_b(r_2) - \psi_a(r_2)\psi_b(r_1) \tag{8.19}$$

does display the correct exchange symmetry, thereby furnishing an acceptable description of the system. From Equation 8.19, however, it is impossible to decide which electron occupies which state—as it should be for identical electrons! Finally, we see that when a and b label the same state ($a = b$), ψ_{ab} is identically zero—the theory allows no solution (description) in this case, in agreement with the familiar statement of the exclusion principle.

EXAMPLE 8.5 Ground State of the Helium Atom
Construct explicitly the two-electron ground state wave function for the helium atom in the independent particle approximation, using the prescription of Equation 8.19. Compare the predicted energy of this state with the measured value, and account in a qualitative way for any discrepancy.

Solution: In the independent particle approximation, each helium electron "sees" only the doubly charged helium nucleus. Accordingly, the ground state wave is constructed from the lowest energy hydrogenic wave functions, with atomic number Z = 2 for helium. These are states for which $n = 1$, $\ell = 0$, and $m_\ell = 0$. Referring to Equation 7.24 of Chapter 7, we find (with Z = 2)

$$\psi_{1,0,0}(r) = \pi^{-1/2}(2/a_0)^{3/2}e^{-2r/a_0}$$

To this orbital function we must attach a spin label (±) indicating the direction of electron spin. Thus, the one-electron state labels a and b in this example are given by $a = (1, 0, 0, +)$, $b = (1, 0, 0, -)$. Since there is no orbital field to interact with the electron spin, the energies of these two states are identical, and are just the hydrogenic levels of Equation 7.21 with $n = 1$ and Z = 2:

$$E_a = E_b = -(2^2/1^2)(13.6\text{ eV}) = -54.4\text{ eV}$$

The antisymmetric two-electron wave function for the ground state of helium is then

$$\psi(r_1, r_2) = \psi_{1,0,0,+}(r_1)\psi_{1,0,0,-}(r_2) - \psi_{1,0,0,-}(r_1)\psi_{1,0,0,+}(r_2)$$

Both terms have the same spatial dependence, but differ as to their spin. The first term of the antisymmetric wave describes electron 1 as having spin up, and electron 2 as having spin down. These spin directions are reversed in the second term. If we introduce the notation $|+, -\rangle$ to describe the two electron spins in the first term, then the second term becomes $|-, +\rangle$, and the total two-electron wave function for the helium ground state can be written

$$\psi(r_1, r_2) = \pi^{-1}(2/a_0)^3e^{-2(r_1+r_2)/a_0}\{|+, -\rangle - |-, +\rangle\}$$

The equal admixture of the spin states $|+, -\rangle$ and $|-, +\rangle$ means the spin of any one of the helium electrons is just as likely to be up as it is to be down. Notice, however, that the spin of the remaining electron is *always* opposite the first. Such spin-spin *correlations* are a direct consequence of the exclusion principle. (The valence electrons of many higher-Z atoms tend to align their spins. This tendency — known as *Hund's Rule* — is another example of spin-spin correlations induced by the exclusion principle.)

The total electronic energy of the helium atom in this approximation is the sum of the one-electron energies E_a and E_b:

$$E = E_a + E_b = -54.4 \text{ eV} - 54.4 \text{ eV} = -108.8 \text{ eV}$$

The magnitude of this number, 108.8 eV, represents the energy (work) required to remove both electrons from the helium atom, in the independent particle model. The measured value is substantially lower, about 79.0 eV, owing to the mutual repulsion of the two electrons. Specifically, it costs less energy — only about 24.6 eV — to remove the first electron from the atom, since the electron left behind *screens* the nuclear charge, making it appear less positive than a bare helium nucleus.

8.5 ELECTRON INTERACTIONS AND SCREENING EFFECTS (Optional)

The preceding discussion of the helium atom exposes a difficulty that arises whenever we treat a system with two or more electrons, namely, handling the effects of electron-electron repulsion. Electrons confined to the small space of an atom are expected to exert strong forces on one another. To ignore these altogether, as in the independent particle model, is simply too crude; to include them exactly is unmanageable, since precise descriptions even for the classical motion in this case are unknown, except through numerical integration. Accordingly, some workable approximation scheme is needed. A most fruitful approach to this problem begins with the notion of an *effective field.*

Any one atomic electron is subject to the Coulomb attraction of the nucleus as well as the Coulomb repulsion of all other electrons in the atom. These influences largely cancel each other, leaving a net effective field with its potential energy $U_{\text{eff}}(r)$. U_{eff} may not be Coulombic — or even spherically symmetric — and may be different for each atomic electron. The success of this approach hinges on how simply and accurately we can model the effective potential. A few of the more obvious possibilities are outlined here.

The outermost or valence electrons of an atom "see" not the bare nucleus, but one shielded or *screened* by the intervening electrons. The attraction is more like that arising from a nucleus with an effective atomic number Z_{eff} somewhat less than the actual number Z, and would be described by

$$U_{\text{eff}}(r) = \frac{k(Z_{\text{eff}}e)(-e)}{r} \tag{8.20}$$

For a Z-electron atom, $Z_{\text{eff}} = Z$ would represent no screening whatever; at the opposite extreme is perfect screening by the $Z - 1$ other electrons, giving $Z_{\text{eff}} = Z - (Z - 1) = 1$. The best choice for Z_{eff} need not even be integral, and useful values may be deduced from *measurements* of atomic ionization potentials (see Example 8.6). Further, the degree of screening depends upon how much time an electron spends near the nucleus, and we should expect Z_{eff} to vary with the shell and subshell labels of the electron in question. In particular, a 4*s* electron is screened more effectively than a 3*s* electron, since its average distance from the nucleus is greater. Similarly, a 3*d* electron is better screened than a 3*s*, or even a 3*p* electron (lower angular momentum implies more eccentric classical orbits, with greater penetration into the nuclear region). The use of a Z_{eff} for valence electrons is appropriate whenever a clear distinction exists between these and inner (core) electrons of the atom, as in the alkali metals.

EXAMPLE 8.6 Z_{eff} for the 3*s* Electron in Sodium

The outer electron of the sodium atom occupies the 3*s* atomic level. The observed value for the ionization energy of this electron is 5.14 eV. From this information, deduce a value of Z_{eff} for the 3*s* electron in sodium. Also give a percentage estimate for the screening effectiveness of the inner (core) electrons in the sodium atom.

Solution: Since the ionization energy, 5.14 eV, represents the amount of energy that must be expended to remove the 3*s* electron from the atom, we infer that the energy of the 3*s* electron in sodium is $E = -5.14$ eV. This should be compared with the energy of a 3*s* electron in a hydrogenic atom with atomic number Z_{eff}, or

$$E = -\frac{Z_{\text{eff}}^2}{3^2}(13.6 \text{ eV})$$

Equating this to -5.14 eV and solving for Z_{eff} gives

$$Z_{\text{eff}} = 3\sqrt{\frac{5.14}{13.6}} = 1.84$$

Since perfect screening by the inner electrons is described by $Z_{\text{eff}} = 1$, the screening effectiveness may be computed as $1/1.84 = 0.54$, or about 54% in this case.

In principle, nuclear shielding can be better described by allowing Z_{eff} to vary continuously throughout the atom in a way that mimics the tighter binding accompanying electron penetration into the core. Two functional forms commonly are used for this purpose. For **Thomas-Fermi screening** we write

$$Z_{\text{eff}}(r) = Z\, e^{-r/a_{\text{TF}}} \tag{8.21}$$

The Thomas-Fermi atom

where a_{TF} is the *Thomas-Fermi screening length.* According to Equation 8.21, Z_{eff} is very nearly Z close to the nucleus ($r \approx 0$), but drops off quickly in the outer region, becoming essentially zero for $r \gg a_{\text{TF}}$. In this way a_{TF} becomes an indicator of atomic size. The Thomas-Fermi model prescribes a_{TF} proportional to $Z^{-1/3}$; the weak variation with Z suggests that all atoms are essentially the same size, regardless of how many electrons they may have. Since the Thomas-Fermi potential is not Coulombic, the one-electron energies that result from the use of Equation 8.21 vary within a given shell; that is, they depend on the principal (n) *and* orbital (ℓ) quantum numbers. The study of these energies and their associated wave functions requires numerical methods, or further approximation. The Thomas-Fermi approximation improves with larger values of Z, and so is especially well suited to describe the outer electronic structure of the heavier elements.

In another approach, nuclear shielding is described by

$$Z_{\text{eff}}(r) = 1 + \frac{b}{r} \tag{8.22}$$

where b is again a kind of screening length. This form is appropriate to the alkali metals, where a lone outer electron is responsible for the chemical properties of the atom. From Equation 8.22, this electron "sees" $Z_{\text{eff}} \approx 1$ for $r \gg b$, and larger values in the core. The special virtue of Equation 8.22 is that it leads to one-electron energies and wave functions that can be found without further approximation. In particular, the energy levels that follow from Equation 8.22 can be shown to be

Quantum defects

$$E_n = -\frac{ke^2}{2a_0}\{n - D(\ell)\}^{-2} \tag{8.23}$$

where $D(\ell)$ is termed the *quantum defect,* since it indicates the departure from the simple hydrogen-atom level structure. As the notation suggests, the quantum defect for an *s* electron differs from that for a *p* or *d* electron, but all *s* electrons have the same quantum defect, regardless of their shell label. Table 8.1 lists some quantum defects deduced experimentally for the sodium atom. Of course, taking $b = 0$ in

TABLE 8.1 Some Quantum Defects for the Sodium Atom

Subshell	*s*	*p*	*d*	*f*
$D(\ell)$	1.35	0.86	0.01	~0

Equation 8.22 causes all quantum defects to vanish, returning us to the hydrogenic level structure discussed in Chapter 7.

The use of a simple Z_{eff}, or the more complicated forms of the Thomas-Fermi or quantum defect method, still results in a U_{eff} with spherical symmetry; that is, the electrons are affected by a *central field*. The **Hartree theory** discards even this feature in order to achieve more accurate results. According to Hartree, the electron "cloud" in the atom should be treated as a classical body of charge distributed with some volume charge density $\rho(r)$. The potential energy of any one atomic electron is then

$$U_{\text{eff}}(r) = -\frac{kZe^2}{r} - \int ke\left[\frac{\rho(r')}{|r - r'|}\right] dV' \tag{8.24}$$

The first term is the attractive energy of the nucleus, and the second term is the repulsive energy of all other atomic electrons. This U_{eff} gives rise to a one-electron Schrödinger equation for the energies E_i and wave functions ψ_i of this, say the ith, atomic electron.

But the Hartree theory is *self-consistent*, in that the charge density ρ due to the other atomic electrons is itself calculated from the electron waves as

Hartree's self-consistent fields

$$\rho(r) = -e\Sigma|\psi_j(r)|^2 \tag{8.25}$$

The sum in Equation 8.25 includes all occupied electron states ψ_j except the ith. In this way the mathematical problem posed by U_{eff} is turned back on itself: we must solve not one Schrödinger equation, but N of them in a single stroke, one for each of the N electrons in the atom! To do this, numerical methods are employed in an iterative solution scheme. An educated guess is made initially for each of the N ground state electron waves. With this, the ρ and U_{eff} for every electron can be computed, and all N Schrödinger equations solved. The resulting wave functions are compared with the initial guesses; if discrepancies appear, the calculation is repeated with the new set of electron waves replacing the old. After several such iterations, agreement is attained between the starting and calculated waves. The resulting N electron wave functions are said to be fully self-consistent. Implementation of the Hartree method is laborious and demands considerable skill, but the results for atomic electrons are among the best available. Indeed, the Hartree and closely related Hartree-Fock methods are the ones frequently called upon today when accurate atomic energy levels and wave functions are required.

8.6 THE PERIODIC TABLE

In principle, it is possible to predict the properties of all the elements by applying the procedures of wave mechanics to each one. However, because of the large number of interactions possible in multi-electron atoms, approximations must be used for all atoms except hydrogen. Nevertheless, the electronic structure of even the most complex atoms can be viewed as a succession of filled levels increasing in energy, with the outermost electrons primarily responsible for the chemical properties of the element.

In the central field approximation, the atomic levels can be labeled by the quantum numbers n and ℓ. From the exclusion principle, the maximum number of electrons in one such *subshell* level is $2(2\ell + 1)$. (If spin-orbit effects are included, each subshell level is actually a doublet composed of two closely spaced sublevels according to the value of $j = \ell \pm \frac{1}{2}$, the quantum number for total angular momentum. The maximum occupancy of each sublevel is $2j + 1$.) The energy of an electron in this level depends primarily on the quantum number n, and to a lesser extent on ℓ (and j). The levels can be grouped conveniently according to the value of n (the shell label), and all those within a group have energies that increase with increasing ℓ. The order of filling the subshell levels with electrons is as follows: once a subshell is filled, the next electron goes into the vacant level that is lowest in energy. This *minimum energy principle* can be understood by noting that if the electron were to occupy a higher level, it would spontaneously decay to a lower one with the emission of energy.

The chemical properties of atoms are determined predominantly by the least tightly bound, or *valence,* electrons, which are in the subshell of highest energy. The most important factors are the occupancy of this subshell and the energy separation between this and the next higher (empty) subshell. For example, an atom tends to be chemically inert if its highest subshell is full and there is an appreciable energy gap to the next higher subshell, since then electrons are not readily shared with other atoms to form a molecule. The quasi-periodic recurrence of similar highest shell structures as Z increases is responsible for the *periodic system* of the chemical elements.

The specification of n and ℓ for each atomic electron is called the **electron configuration** of that atom. We are now in a position to describe the electron configuration of any atom in its ground state:

Hydrogen has only one electron which, in its ground state, is described by the quantum numbers $n = 1, \ell = 0$. Hence, its electron configuration is designated as $1s^1$.

Helium, with its two electrons, has a ground state electron configuration of $1s^2$. That is, both electrons are in the same (lowest energy) subshell, the $1s$. Since two is the maximum occupancy for an s subshell, the subshell (and in this case also the shell) is said to be *closed,* and helium is inert.

Lithium has three electrons. Two of these are assigned to the $1s$ subshell, and the third must be assigned to the $2s$ subshell, since this is slightly lower in energy than the $2p$ subshell. Hence, the electron configuration of lithium is $1s^22s^1$.

With the addition of another electron to make *beryllium,* the $2s$ subshell is closed. The electron configuration of beryllium, with four electrons altogether, is $1s^22s^2$. (However, beryllium is not inert, since the energy gap separating the $2s$ from the next available level—the $2p$—is not very large.)

Boron has a configuration of $1s^22s^22p^1$. (With spin-orbit doubling, the $2p$ electron in boron actually occupies the $2P_{1/2}$ sublevel, corresponding to $n = 2$, $\ell = 1$, and $j = \frac{1}{2}$.)

Carbon has six electrons, and a question arises of how to assign the two $2p$ electrons. Do they go into the same orbital with paired spins ($\uparrow\downarrow$), or do they occupy different orbitals with unpaired spins ($\uparrow\uparrow$)? Experiments show that the energetically preferred configuration is the latter, in which the spins are aligned. This is an illustration of **Hund's rule.** Hund's rule can be partly understood by noting that electrons in the same orbital tend to be closer together, where their mutual repulsion contributes to a higher energy than if they were separated in different orbitals. Some exceptions to this rule do occur in those elements with subshells that are nearly filled or half-filled. The progressive filling of the $2p$ subshell illustrating Hund's rule is shown schematically in Figure 8.15. With *neon,* the $2p$ subshell also becomes closed. The neon atom has ten electrons in the configuration $1s^22s^22p^6$. Since the energy gap separating the $2p$ subshell from the next available level—the $3s$—is quite large, the neon configuration is exceptionally stable and the atom is chemically inert.

A complete list of electron configurations for all the known stable elements is given in Table 8.2. Note that, beginning with *potassium* (Z = 19), the $4s$ subshell starts to fill while the $3d$ level remains empty. Only after the $4s$ subshell is closed to form *calcium* does the $3d$ subshell begin to fill. We infer

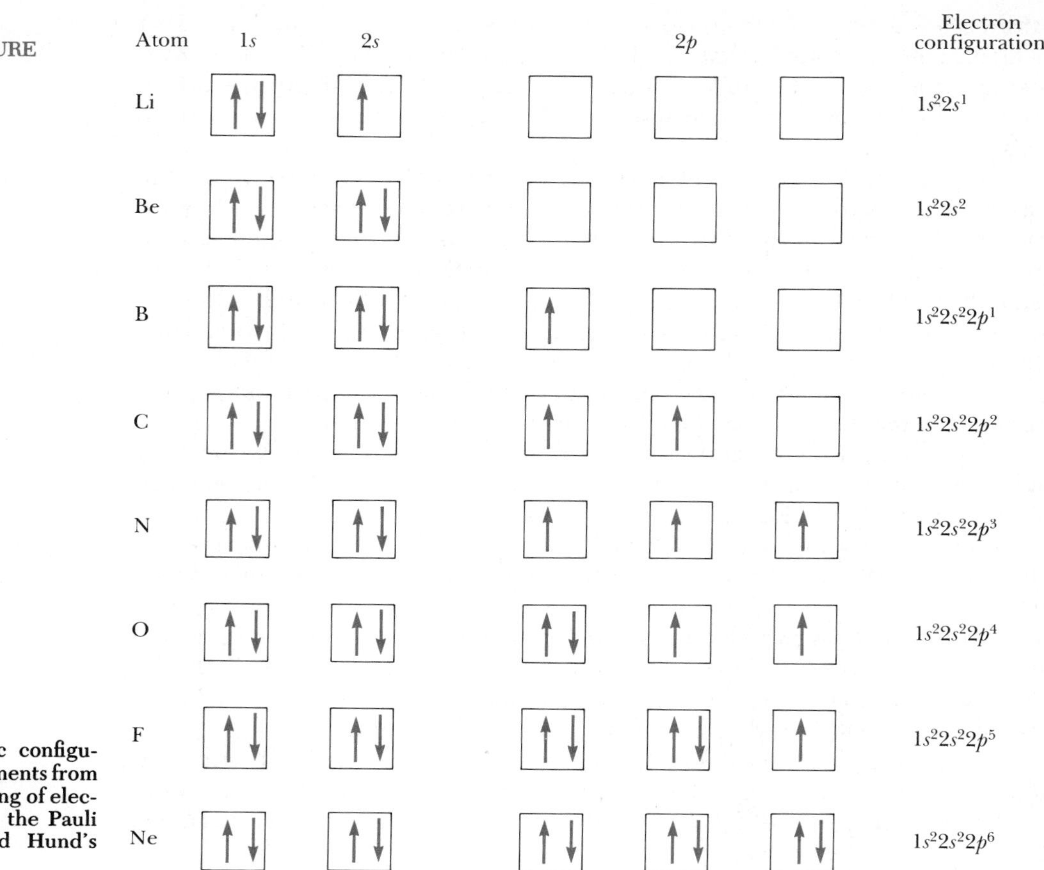

Figure 8.15 Electronic configurations of successive elements from lithium to neon. The filling of electronic states must obey the Pauli exclusion principle and Hund's rule.

that the $3d$ subshell has a higher energy than the $4s$, even though it belongs to a lower-indexed shell. This should come as no surprise, because the energy difference between consecutive shells becomes smaller with increasing n (see the hydrogenic spectrum), while that between subshells is more nearly constant owing to the screening effect discussed in Section 8.5. (In fact, the $3d$ and $4s$ subshells must be very close in energy, as shown by the electron configuration of *chromium,* in which the $3d$ subshell temporarily regains an electron from the $4s$.) The same phenomenon occurs again with *rubidium* ($Z = 37$), where the $5s$ subshell begins to fill at the expense of the $4d$ and $4f$ subshells. Energetically, the electron configurations shown in the table imply the following ordering of subshells with respect to energy:

Ordering of subshells by energy

$$1s < 2s < 2p < 3s < 3p < 4s \sim 3d < 4p < 5s < 4d < 5p < 6s < 4f \sim 5d < 6p < 7s < 6d \sim 5f \ldots$$

The elements from *scandium* ($Z = 21$) to *zinc* ($Z = 30$) form the first *transition series.* These transition elements are characterized by a progressively filling $3d$ subshell, while the outer electron configuration is unchanged at $4s^2$ (except in the case of copper). Consequently, all the transition elements exhibit similar chemical properties. The belated occupancy of inner d subshells is encountered again in the second and third transition series, marked by the progressive filling of the $4d$ and $5d$ subshells, respectively. The second transition series includes the elements *yttrium* ($Z = 39$) to *cadmium* ($Z = 48$); the third contains the elements *lutetium* ($Z = 71$) to *mercury* ($Z = 80$).

TABLE 8.2 Electronic Configurations of the Elements

Z	Symbol	Ground Configuration	Ionization Energy (eV)
1	H	$1s^1$	13.595
2	He	$1s^2$	24.581
3	Li	[He] $2s^1$	5.390
4	Be	$2s^2$	9.320
5	B	$2s^2 2p^1$	8.296
6	C	$2s^2 2p^2$	11.256
7	N	$2s^2 2p^3$	14.545
8	O	$2s^2 2p^4$	13.614
9	F	$2s^2 2p^5$	17.418
10	Ne	$2s^2 2p^6$	21.559
11	Na	[Ne] $3s^1$	5.138
12	Mg	$3s^2$	7.644
13	Al	$3s^2 3p^1$	5.984
14	Si	$3s^2 3p^2$	8.149
15	P	$3s^2 3p^3$	10.484
16	S	$3s^2 3p^4$	10.357
17	Cl	$3s^2 3p^5$	13.01
18	Ar	$3s^2 3p^6$	15.755
19	K	[Ar] $4s^1$	4.339
20	Ca	$4s^2$	6.111
21	Sc	$3d4s^2$	6.54
22	Ti	$3d^2 4s^2$	6.83
23	V	$3d^3 4s^2$	6.74
24	Cr	$3d^5 4s$	6.76
25	Mn	$3d^5 4s^2$	7.432
26	Fe	$3d^6 4s^2$	7.87
27	Co	$3d^7 4s^2$	7.86
28	Ni	$3d^8 4s^2$	7.633
29	Cu	$3d^{10} 4s^1$	7.724
30	Zn	$3d^{10} 4s^2$	9.391
31	Ga	$3d^{10} 4s^2 4p^1$	6.00
32	Ge	$3d^{10} 4s^2 4p^2$	7.88
33	As	$3d^{10} 4s^2 4p^3$	9.81
34	Se	$3d^{10} 4s^2 4p^4$	9.75
35	Br	$3d^{10} 4s^2 4p^5$	11.84
36	Kr	$3d^{10} 4s^2 4p^6$	13.996
37	Rb	[Kr] $5s^1$	4.176
38	Sr	$5s^2$	5.692
39	Y	$4d5s^2$	6.377
40	Zr	$4d^2 5s^2$	6.835
41	Nb	$4d^4 5s^1$	6.881
42	Mo	$4d^5 5s^1$	7.10
43	Tc	$4d^5 5s^2$	7.228
44	Ru	$4d^7 5s^1$	7.365
45	Rh	$4d^8 5s^1$	7.461
46	Pd	$4d^{10}$	8.33
47	Ag	$4d^{10} 5s^1$	7.574
48	Cd	$4d^{10} 5s^2$	8.991
49	In	$4d^{10} 5s^2 5p^1$	5.785
50	Sn	$4d^{10} 5s^2 5p^2$	7.342
51	Sb	$4d^{10} 5s^2 5p^3$	8.639
52	Te	$4d^{10} 5s^2 5p^4$	9.01
53	I	$4d^{10} 5s^2 5p^5$	10.454
54	Xe	$4d^{10} 5s^2 5p^6$	12.127
55	Cs	[Xe] $6s^1$	3.893
56	Ba	$6s^2$	5.210
57	La	$5d6s^2$	5.61
58	Ce	$4f5d6s^2$	6.54
59	Pr	$4f^3 6s^2$	5.48
60	Nd	$4f^4 6s^2$	5.51
61	Pm	$4f^5 6s^2$	
62	Fm	$4f^6 6s^2$	5.6
63	Eu	$4f^7 6s^2$	5.67
64	Gd	$4f^7 5d6s^2$	6.16
65	Tb	$4f^9 6s^2$	6.74
66	Dy	$4f^{10} 6s^2$	6.82
67	Ho	$4f^{11} 6s^2$	
68	Er	$4f^{12} 6s^2$	
69	Tm	$4f^{13} 6s^2$	
70	Yb	$4f^{14} 6s^2$	6.22
71	Lu	$4f^{14} 5d6s^2$	6.15
72	Hf	$4f^{14} 5d^2 6s^2$	7.0
73	Ta	$4f^{14} 5d^3 6s^2$	7.88
74	W	$4f^{14} 5d^4 6s^2$	7.98
75	Re	$4f^{14} 5d^5 6s^2$	7.87
76	Os	$4f^{14} 5d^6 6s^2$	8.7
77	Ir	$4f^{14} 5d^7 6s^2$	9.2
78	Pt	$4f^{14} 5d^8 6s^2$	8.88
79	Au	[Xe, $4f^{14} 5d^{10}$] $6s^1$	9.22
80	Hg	$6s^2$	10.434
81	Tl	$6s^2 6p^1$	6.106
82	Pb	$6s^2 6p^2$	7.415
83	Bi	$6s^2 6p^3$	7.287
84	Po	$6s^2 6p^4$	8.43
85	At	$6s^2 6p^5$	
86	Rn	$6s^2 6p^6$	10.745
87	Fr	[Rn] $7s^1$	
88	Ra	$7s^2$	5.277
89	Ac	$6d7s^2$	6.9
90	Th	$6d^2 7s^2$	
91	Pa	$5f^2 6d7s^2$	
92	U	$5f^3 6d7s^2$	4.0
93	Np	$5f^4 6d7s^2$	
94	Pu	$5f^6 7s^2$	
95	Am	$5f^7 7s^2$	
96	Cm	$5f^7 6d7s^2$	
97	Bk	$5f^8 6d7s^2$	
98	Cf	$5f^{10} 7s^2$	
99	Es	$5f^{11} 7s^2$	
100	Fm	$5f^{12} 7s^1$	
101	Mv	$5f^{13} 7s^2$	
102	No	$5f^{14} 7s^2$	
103	Lw	$5f^{14} 6d7s^2$	
104	Ku	$5f^{14} 6d^2 7s^2$	

Note: The bracket notation is used as a shorthand method to avoid repetition in indicating inner-shell electrons. Thus, [He] represents $1s^2$, [Ne] represents $1s^2 2s^2 2p^6$, [Ar] represents $1s^2 2s^2 2p^6 3s^2 3p^6$, and so on.

Related behavior is seen also with the belated filling of the $4f$ and $5f$ subshells. The *lanthanide series,* stretching from *lanthanum* (Z = 57) to *ytterbium* (Z = 70), is marked by a common $6s^2$ valence configuration, with the added electrons filling out the $4f$ subshell (the energetically close $5d$ levels also are occupied in some instances). The lanthanide elements, or *lanthanides,* also are known as the *rare earths* because of their low natural abundance. *Cerium* (Z = 58), which forms 0.00031 percent by weight of the earth's crust, is the most abundant of the lanthanides.

In the *actinide series* from *actinium* (Z = 89) to *nobelium* (Z = 102), the valence configuration remains $7s^2$, while the $5f$ subshell progressively fills (along with occasional occupancy of the energetically close $6d$).

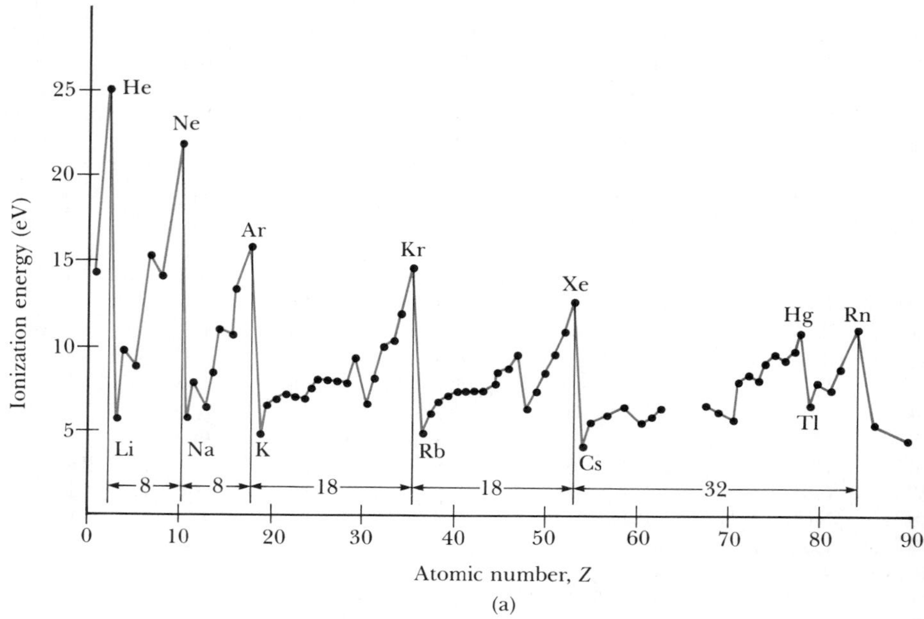

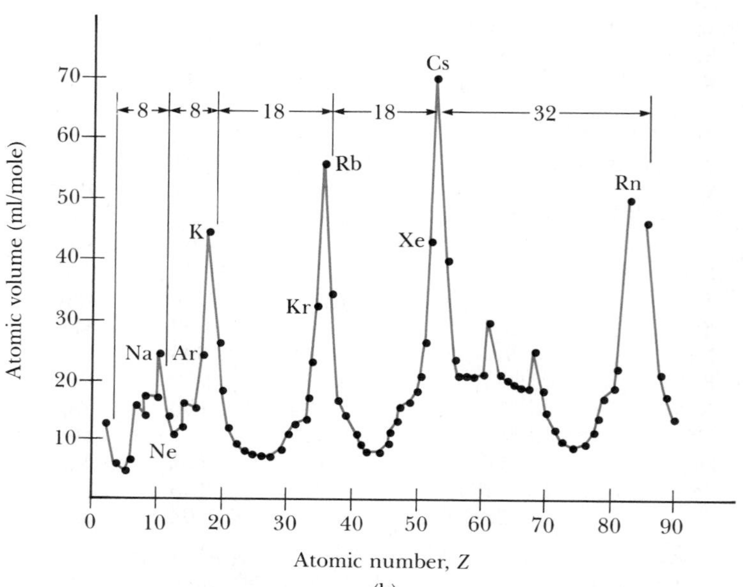

Figure 8.16 (a) Ionization energy of the elements versus atomic number Z. (b) Atomic volume of the elements versus atomic number. The recurring pattern with increasing atomic number Z exemplifies the behavior from which the periodic table gets its name.

Table 8.2 also lists the ionization energies of the elements. The ionization energy for each element is plotted against its atomic number Z in Figure 8.16a. We see that the ionization energy tends to increase within a shell, then drops dramatically as the filling of a new shell begins. The behavior repeats, and it is from this recurring pattern that the periodic table gets its name. A similar repetitive pattern is observed in a plot of the atomic volume per atom versus atomic number (Fig. 8.16b).

The primary features of these plots can be understood from simple arguments. First, the larger nuclear charge that accompanies higher values of Z tends to pull the electrons closer to the nucleus, and binds them more tightly. Were this the only effect, the ionization energy simply would increase and the atomic volume would decrease steadily with increasing Z. But the innermost, or core, electrons screen the nuclear charge, making it less effective in binding the outer electrons. The screening effect varies in a complicated way from one element to the next, but it is most pronounced for a lone electron outside a closed shell, as in the alkali metals (Li, Na, K, Rb, Cs, and Fr). For these configurations the ionization energy dips sharply, only to rise again as the nuclear charge intensifies at higher Z. The variation in ionization energy is mirrored by the behavior of atomic volume, which peaks at the alkali configurations and becomes smaller as the screening effect subsides.

8.7 SUMMARY

The magnetic behavior of atoms is characterized by their *magnetic moment.* The orbital moment of an atomic electron is proportional to its orbital angular momentum:

$$\boldsymbol{\mu} = \frac{-e}{2m}\mathbf{L} \tag{8.1}$$

The constant of proportionality, $-e/2m$, is called the *gyromagnetic ratio.* Since $\mathbf{L}$ is subject to space quantization, so too is the atomic moment $\boldsymbol{\mu}$. Atomic moments are measured in Bohr magnetons, $\mu_B = e\hbar/2m$; the SI value of μ_B is 9.27×10^{-24} J/T.

An atom subjected to an external magnetic field $\mathbf{B}$ experiences a magnetic torque, which results in precession of the moment vector $\boldsymbol{\mu}$ about the field vector $\mathbf{B}$. The frequency of precession is the *Larmor frequency* ω_L given by

$$\omega_L = \frac{eB}{2m} \tag{8.5}$$

Associated with the Larmor frequency is the energy quantum $\hbar\omega_L = \mu_B B$. The ℓth subshell level of an atom placed in a magnetic field is split by the field into $2\ell + 1$ sublevels separated by the Larmor energy $\hbar\omega_L$. This is the *Zeeman effect,* and $\hbar\omega_L$ is also known as the Zeeman energy. The magnetic contribution to the energy of the atom is

$$U = \hbar\omega_L m_\ell \tag{8.7}$$

where m_ℓ is the same *magnetic* quantum number discussed in Chapter 7. Equation 8.7 is a special case of the more general result

$$U = -\boldsymbol{\mu}\cdot\mathbf{B} \tag{8.6}$$

for the magnetic potential energy U of any magnetic moment $\boldsymbol{\mu}$ in an applied field $\mathbf{B}$.

In addition to any orbital magnetic moment, the electron possesses an intrinsic magnetic moment called the *spin* moment $\boldsymbol{\mu}_s$. The spin moment arises from the rotation of the electron on its axis, and is proportional to the angular momentum of rotation, or spin, **S**. The magnitude of the spin angular momentum is

$$|\mathbf{S}| = \frac{\sqrt{3}}{2}\hbar \tag{8.11}$$

corresponding to a *spin* quantum number $s = \frac{1}{2}$, analogous to the orbital quantum number ℓ. The z component of **S** is quantized as

$$S_z = m_s\hbar \tag{8.10}$$

where the *spin magnetic* quantum number m_s is the analog of the orbital magnetic quantum number m_ℓ. For the electron, m_s can be either $+\frac{1}{2}$ or $-\frac{1}{2}$, which describes the up and down spin states, respectively. Equations 8.10 and 8.11 imply that electron spin also is subject to space quantization. This was confirmed experimentally by Stern and Gerlach, who observed that a beam of neutral silver atoms passed through a nonuniform magnetic field was split into two distinct components. The same experiment shows that the spin moment is related to the spin angular momentum by

$$\boldsymbol{\mu}_s = -\frac{e}{m}\mathbf{S}$$

This is *twice* as large as the orbital moment for the same angular momentum. The anomalous factor of 2 is called the *g-factor* of the electron. With the recognition of spin, four quantum numbers—n, ℓ, m_ℓ, and m_s—are needed to specify the state of an atomic electron.

The spin moment of the electron interacts with the magnetic field arising from its orbital motion. The energy difference between the two spin orientations in this orbital field is responsible for the *fine structure doubling* of atomic spectral lines. With the spin-orbit interaction, atomic states are labeled by a quantum number j for the *total* angular momentum $\mathbf{J} = \mathbf{L} + \mathbf{S}$. The value of j is included as a subscript in the spectroscopic notation of atomic states. For example, $3P_{1/2}$ specifies a state for which $n = 3$, $\ell = 1$, and $j = \frac{1}{2}$. For each value of j, there are $2j + 1$ possibilities for the total magnetic quantum number m_j. Thus, the *multiplicity* of the state $3P_{1/2}$ is 2.

Electrons confined to the small space of an atom interact strongly with one another. In particular, the inner or core electrons of an atom screen the nuclear charge seen by the outer or valence electrons. The screening effect is described approximately by introducing an *effective potential* for each electron. More or less sophisticated models for this effective potential are given by the *Thomas-Fermi approximation*, the *quantum defect method*, and *Hartree's self-consistent field*.

The *exclusion principle* states that no two electrons can be in the same quantum state; that is, no two electrons can have the same four quantum numbers. The exclusion principle derives from the notion that electrons are identical particles called *fermions*. Fermions are described by wave functions that are *antisymmetric* in the electron coordinates. Wave functions that are symmetric in the particle coordinates describe another class of objects called *bosons*, to which no exclusion principle applies. All known particles are either fermions or bosons. An example of a boson is the photon.

Using the exclusion principle and the principle of minimum energy, one can determine the electronic configurations of the elements. This serves as a basis for understanding atomic structure and the physical and chemical properties of the elements.

QUESTIONS

1. Why is the direction of the orbital angular momentum of an electron opposite that of its magnetic moment?
2. Why is an inhomogeneous magnetic field used in the Stern-Gerlach experiment?
3. Could the Stern-Gerlach experiment be performed with ions rather than neutral atoms? Explain.
4. Describe some experiments that would support the conclusion that the spin quantum number for electrons can have only the values $\pm\frac{1}{2}$.
5. Discuss some of the consequences of the exclusion principle.
6. Why do lithium, potassium, and sodium exhibit similar chemical properties?
7. From Table 8.2, we find that the ionization energies for Li, Na, K, Rb, and Cs are 5.390, 5.138, 4.339, 4.176, and 3.893 eV, respectively. Explain why these values are to be expected in terms of the atomic structures.
8. Although electrons, protons, and neutrons obey the exclusion principle, some particles that have integral spin, such as photons (spin = 1), do not. Explain.
9. Explain why a photon must have a spin of 1.
10. An energy of about 21 eV is required to excite an electron in a helium atom from the 1*s* state to the 2*s* state. The same transition for the He^+ ion requires about twice as much energy. Explain why this is so.
11. Discuss degeneracy as it applies to a multielectron atom. Can a one-electron atom have degeneracy? Explain.
12. The absorption or emission spectrum of a gas consists of lines that broaden as the density of gas molecules increases. Why do you suppose this occurs?
13. Spin-orbit coupling splits all states except *s* states into doublets. Why are *s* states exceptions to this rule?

PROBLEMS

1. Consider a single-electron atom in the $n = 2$ state. Find all possible values for j and m_j for this state.
2. Find all possible values of j and m_j for a d electron.
3. Give the spectroscopic notation for the following states: (a) $n = 7, \ell = 4, j = \frac{9}{2}$; (b) all the possible states of an electron with $n = 6$ and $\ell = 5$.
4. An electron in an atom is in the $4F_{5/2}$ state. (a) Find the values of the quantum numbers n, ℓ, and j. (b) What is the magnitude of the electron's total angular momentum? (c) What are the possible values for the z component of the electron's total angular momentum?
5. How many different sets of quantum numbers are possible for an electron for which (a) $n = 1$, (b) $n = 2$, (c) $n = 3$, (d) $n = 4$, and (e) $n = 5$? Check your results to show that they agree with the general rule that the number of different sets of quantum numbers is equal to $2n^2$.
6. (a) Write out the electronic configuration for oxygen ($Z = 8$). (b) Write out the values for the set of quantum numbers n, ℓ, m_ℓ, and m_s for each of the electrons in oxygen.
7. Which electronic configuration has a lower energy: [Ar]$3d^4 4s^2$ or [Ar]$3d^5 4s^1$? Identify this element and discuss Hund's rule in this case. (*Note:* The notation [Ar] represents the filled configuration for Ar.)
8. Which electronic configuration has the lesser energy and the greater number of unpaired spins: [Kr]$4d^9 5s^1$ or [Kr]$4d^{10}$? Identify this element and discuss Hund's rule in this case. (*Note:* The notation [Kr] represents the filled configuration for Kr.)
9. Devise a table similar to that shown in Figure 8.15 for atoms with 11 through 19 electrons. Use Hund's rule and educated guesswork.
10. List the possible sets of quantum numbers for electrons in (a) the 3*d* subshell and (b) the 3*p* subshell.
11. Zirconium has two unpaired electrons in the *d* subshell. (a) What are all possible values of J_z for each electron? (b) What are all possible values of J_z for the atom as a whole? (c) Calculate the largest possible value of the total angular momentum for zirconium. (*Hint:* Remember to use the exclusion principle.)
12. In the technique known as electron spin resonance (ESR), a sample containing unpaired electrons is placed in a magnetic field. Consider the simplest situation, that in which there is only one electron and therefore only two possible energy states, corresponding to $m_s = \pm\frac{1}{2}$. In ESR, the electron's spin magnetic moment is "flipped" from a lower energy state to a higher energy state by the absorption of a photon. (The lower energy state corresponds to the case where the magnetic moment μ_s is aligned against the magnetic field, and the higher energy state corresponds to the case where μ_s is aligned with the field.) What is the photon frequency required to excite an ESR transition in a magnetic field of 0.35 T?
13. The force on a magnetic moment μ_z in a nonuniform magnetic field B_z is given by

$$F_z = \mu_z \frac{dB_z}{dz}$$

If a beam of silver atoms travels a horizontal distance of 1 m through such a field and each atom has a speed of 100 m/s, how strong must the field gradient dB_z/dz be in order to deflect the beam 1 mm?
14. (a) Starting with the expression $\boldsymbol{J} = \boldsymbol{L} + \boldsymbol{S}$ for the total angular momentum of an electron, derive an expression for the scalar product $\boldsymbol{L}\cdot\boldsymbol{S}$ in terms of the quantum numbers j, ℓ, and s. (b) Using the fact that $\boldsymbol{L}\cdot\boldsymbol{S} = |\boldsymbol{L}||\boldsymbol{S}| \cos\theta$, where θ is the angle between $\boldsymbol{L}$ and $\boldsymbol{S}$, find

the angle between the electron's orbital angular momentum and spin angular momentum for the following states: (1) $P_{1/2}$, $P_{3/2}$ (2) $H_{9/2}$, $H_{11/2}$.

15. Eight identical, noninteracting particles are placed in a cubical box of sides $L = 0.2$ nm. Find the lowest energy of the system (in eV) and list the quantum numbers of all occupied states if (a) the particles are electrons and (b) the particles have the same mass as the electron but do not obey the exclusion principle.

16. Show that for a mass m in orbit with angular momentum L, the rate at which area is swept out by the orbiting particle is

$$\frac{dA}{dt} = \frac{|L|}{2m}$$

(*Hint:* First show that in its displacement $d\mathbf{r}$ *along the path*, the particle sweeps out an area $dA = \frac{1}{2}|\mathbf{r} \times d\mathbf{r}|$, where $\mathbf{r}$ is the position vector of the particle drawn from some origin.)

17. Consider the original Stern-Gerlach experiment employing an atomic beam of silver, for which the magnetic moment is due entirely to the spin of the single valence electron of the silver atom. Assuming the magnetic field B has magnitude 0.5 T, compute the energy difference in eV of the silver atoms in the two exiting beams.

18. When the idea of electron spin was introduced, the electron was thought to be a tiny charged sphere (today it is considered a point object with no extension in space). Find the equatorial speed under the assumption that the electron is a uniform sphere of radius 3×10^{-6} nm, as early theorists believed, and compare your result to the speed of light, c.

19. Consider a right circular cylinder of radius R, with mass M uniformly distributed throughout the cylinder volume. The cylinder is set into rotation with angular speed ω about its longitudinal axis. (a) Obtain an expression for the angular momentum $\mathbf{L}$ of the rotating cylinder. (b) If charge Q is distributed uniformly over the *curved* surface only, find the magnetic moment $\boldsymbol{\mu}$ of the rotating cylinder. Compare your expressions for $\boldsymbol{\mu}$ and $\mathbf{L}$ to deduce the g-factor for this object.

20. An exotic elementary particle called the omega minus (symbol Ω^-) has spin $\frac{3}{2}$. Calculate the magnitude of the spin angular momentum for this particle, and the possible angles the spin angular momentum vector makes with the z axis. Does the Ω^- obey the Pauli exclusion principle? Explain.

21. *Spin-Orbit Energy in an Atom.* Estimate the magnitude of the spin-orbit energy for an atomic electron in the hydrogen $2p$ state. (*Hint:* From the vantage point of the moving electron, the nucleus circles it in an orbit with radius equal to the Bohr radius for this state. Treat the orbiting nucleus as a current in a circular wire loop and use the result from classical electromagnetism,

$$B = \frac{2k_m\mu}{r^3}$$

for the B field at the center of loop with radius r and magnetic moment μ. Here $k_m = 10^{-7}$ N/A^2 is the magnetic constant in SI units.)

*22. The claim is made in Section 8.5 that a d electron is screened more effectively from the nuclear charge in an atom than is a p electron or an s electron. Give a classical argument based on the definition of angular momentum $\mathbf{L} = \mathbf{r} \times \mathbf{p}$ that indicates that smaller values of angular momentum are associated with orbits of larger eccentricity. Verify this quantum-mechanically by calculating the probability that a $2p$ electron of hydrogen will be found inside the $n = 1$ atomic shell and comparing this with the probability of finding a hydrogen $2s$ electron in this same region. For which is the probability largest, and what effect does this have on the degree of screening? The relevant wave functions may be found in Table 7.3 of Chapter 7.

PROBLEMS FOR THE COMPUTER

The following problems require for their solution the use of the program EIGEN3. Where appropriate, furnish screen dumps to substantiate your answers.

*23. Display the first excited state waveform and energy for the screened Coulomb well with atomic number $Z = 10$ and screening length $\alpha = 1$ bohr. What is the value of the orbital quantum number ℓ for this state? What value of ℓ characterizes the next (second) excited state, and what is the energy of this state? Compare your answers with the corresponding results for hydrogenic atoms ($\alpha \rightarrow \infty$)

*24. Find the ground state of the defect well with $\beta = 1$, and determine the corresponding quantum defect $D(\ell = 0)$ for this case. Repeat the procedure for the next lowest s state (first excited state). How does the quantum defect $D(\ell = 0)$ for the first excited state compare with that found for the ground state? Using similar methods, report the quantum defect for p states $D(\ell = 1)$ in the case $\beta = 1$.

* These problems require that section 8.5 be covered.

9

Statistical Physics

Thermodynamics is based upon macroscopic concepts and observations. For example, one of its postulates is the experimentally observed one-way flow of heat from hot objects to cold ones. In this chapter we shall direct our attention to the explanation of thermodynamic properties by means of a microscopic picture involving large numbers of atomic-sized systems. This microscopic approach, known as *statistical physics* or *statistical mechanics,* uses statistical theory to predict the most probable way that a certain amount of energy may be distributed among a large number of particles that obey either classical or quantum mechanics. Although statistical mechanics cannot predict the thermodynamic behavior of a system with *absolute* certainty, it does give predictions with overwhelming likelihood because it is usually applied to systems containing large numbers of particles ($\sim 10^{23}$), and fluctuations from the average are small.

9.1 THE MAXWELL-BOLTZMANN DISTRIBUTION

Although we introduced in Chapter 2 the Maxwell-Boltzmann probability that a particle will be found with energy E at a temperature T, it is important to review this classical distribution function. We especially want to compare it to the new distribution functions that result when the quantum properties of a system of particles are taken into account.

When Maxwell derived his "gaussian" velocity distribution for gas molecules, he opened the field of statistical mechanics and the treatment of collisions. Basically, he attacked the question of whether it is possible to find an equilibrium distribution of molecular velocities for a gas[1] such that collisions that knock molecules out of the speed interval between v and $v + dv$ (destructive collisions) are compensated for by collisions that knock molecules into this same interval (creative collisions). In 1872, Boltzmann, profoundly impressed with Darwin's ideas on evolution, took Maxwell's work a step farther. He not only wanted to establish the properties of the equilibrium or most probable distribution, but he also wished to describe the evolution in time of the gas toward the Maxwellian distribution—the so-called approach to equilibrium problem. With the use of a time-dependent velocity distribution function $f(v, t)$ and his kinetic equation, he was able to show that a system of particles that starts off with a non-Maxwellian velocity distribution steadily approaches and eventually achieves an equilibrium Maxwellian velocity distribution.

Having briefly discussed the contributions of Maxwell and Boltzmann to statistical mechanics, let us review the underlying assumptions and explicit form of the Maxwell-Boltzmann distribution, $F_{\mathbf{MB}}$. The basic assumptions are:

[1] The velocity distribution gives the number of molecules with velocity between v and $v + dv$.

Assumptions of the Maxwell-Boltzmann distribution

1. The particles are identical in terms of physical properties but distinguishable in terms of position. It will be demonstrated later in this chapter that this assumption is equivalent to the statement that the particle size is small compared to the average distance between particles.
2. *The equilibrium distribution is the most probable way of distributing the particles among various allowed energy states* subject to the constraints of a fixed number of particles and fixed total energy.
3. There is no limit on the fraction of the total number of particles in a given energy state.

In order to make these assumptions more concrete, let us consider the analysis of a simple system of distinguishable particles. In particular, we wish to consider the distribution of a total energy of 8E among six particles and to evaluate the average number of particles with a certain amount of energy. In order to work with a manageable diagram, Figure 9.1a enumerates the 20 possible ways of sharing an energy of 8E between six *indistinguishable* particles. Since we are actually interested in *distinguishable* particles, each of the 20 arrangements can be decomposed into many substates or *microstates,* as shown explicitly for one arrangement in Figure 9.1b. The number of microstates for each arrangement is given in parentheses in Figure 9.1a and may be computed from the relation

$$N_{\mathbf{MB}} = \frac{N!}{n_1!\, n_2!\, n_3! \cdots} \tag{9.1}$$

where N is the number of particles and $n_1, n_2, n_3, \ldots$ are the numbers of particles in *occupied* levels. This result may be understood by arguing that the first energy level may be assigned in N ways, the second in $N-1$ ways, and so on, giving $N!$ in the numerator. The factor in the denominator of Equation 9.1 corrects for indistinguishable arrangements where several particles occupy the same energy level. In order to find the *average number of particles with a particular value of energy,* which is also *the unnormalized probability of occupancy of a given energy state,* we use the formula

$$\overline{n_j} = n_{j_1}p_1 + n_{j_2}p_2 + \cdots \tag{9.2}$$

Here $\overline{n_j}$ is the average number of particles in the jth energy level, n_{j_1} is the number of particles found in the jth level in arrangement 1, n_{j_2} is the number of particles found in the jth level in arrangement 2, p_1 is the probability of observing arrangement 1, p_2 is the probability of arrangement 2, and so on. Using the **basic postulate of statistical mechanics,** that *all microstates are equally likely,* we may go on to calculate the various p's and $\overline{n_j}$'s. For example, since there are 1287 microstates (the sum of all the numbers in the parentheses), and 6 distinguishable ways of obtaining arrangement 1, we see that $p_1 = 6/1287$. Using these ideas and Equation 9.2, we calculate the average number of particles with energy 0 as follows:

$$\begin{aligned}\overline{n_0} = {} & (5)(6/1287) + (4)(30/1287) + (4)(30/1287) + (3)(60/1287) \\ & + (4)(30/1287) + (3)(120/1287) + (2)(60/1287) + (4)(15/1287) \\ & + (3)(120/1287) + (3)(60/1287) + (2)(180/1287) \\ & + (1)(30/1287) + (3)(60/1287) + (2)(90/1287) + (2)(180/1287) \\ & + (1)(120/1287) + (0)(6/1287) + (2)(15/1287) \\ & + (1)(60/1287) + (0)(15/1287) \\ = {} & 2.307\end{aligned}$$

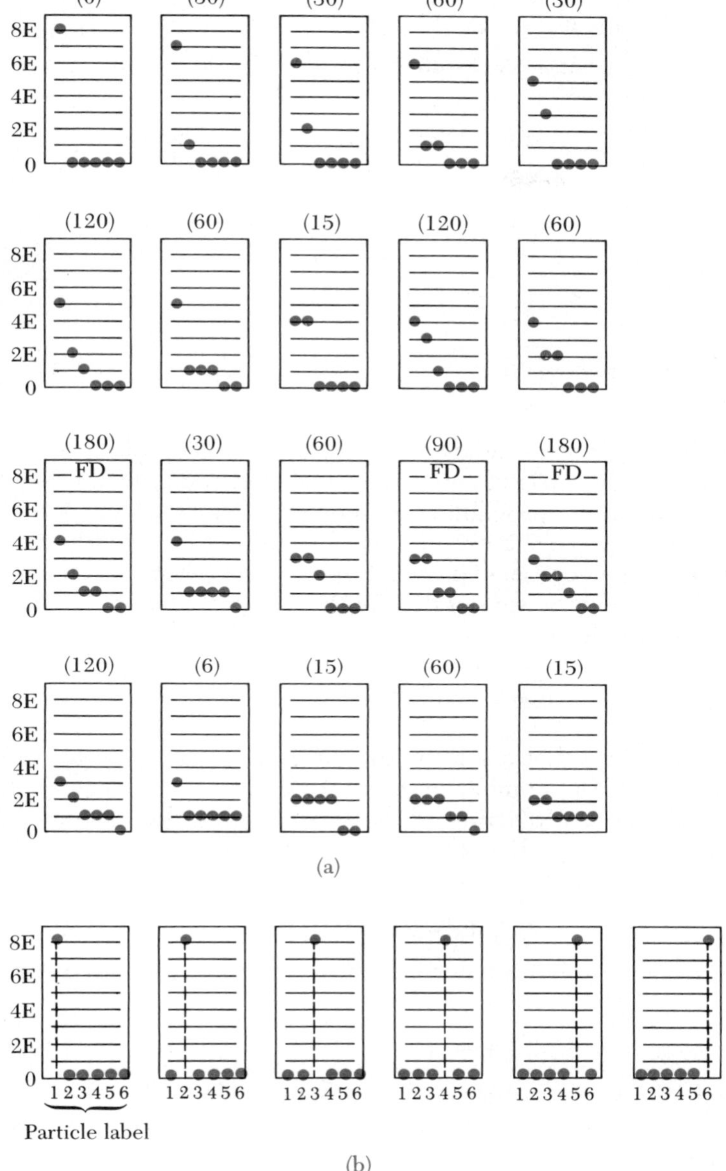

Figure 9.1 (a) The 20 arrangements of six *indistinguishable* particles with a total energy of 8E. (b) The decomposition of the upper left-hand arrangement of part (a) into six distinguishable states for *distinguishable* particles.

It is left as a problem (Problem 1) to show that the average numbers of particles with energies from 1E through 8E are as follows:

$$\overline{n_1} = 1.54 \qquad \overline{n_5} = 0.163$$
$$\overline{n_2} = 1.00 \qquad \overline{n_6} = 0.0699$$
$$\overline{n_3} = 0.587 \qquad \overline{n_7} = 0.0233$$
$$\overline{n_4} = 0.326 \qquad \overline{n_8} = 0.00466$$

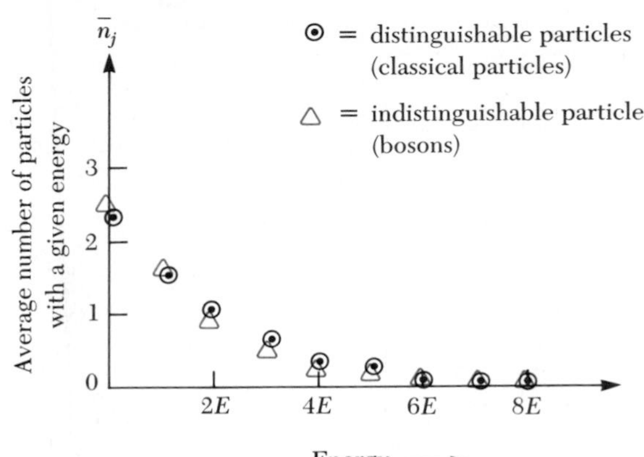

Figure 9.2 The distribution function for an assembly of six particles with a total energy of 8E.

These results, which are plotted in Figure 9.2, show for this simple system, with only six distinguishable particles and a fixed total energy 8E, an approximately exponential decrease in probability with energy. (See Problem 8.)

One may rigorously derive the exponential form of the Maxwell-Boltzmann distribution for a large number of particles by using calculus to maximize the expression for the number of ways of distributing the particles among the allowed energy states (see, for example, reference 1). One finds

Maxwell-Boltzmann distribution

$$\boxed{F_{MB} = Ae^{-E_i/kT}} \tag{9.3}$$

where F_{MB} is the probability that a state with energy E_i is occupied at the absolute temperature T. If the number of states with the same energy E_i is denoted by g_i (g_i is called the *degeneracy* or *statistical weight*), then the number of particles, n_i, with energy E_i is equal to the product of the statistical weight with the probability that the state E_i is occupied, or

$$n_i = g_i F_{MB} \tag{9.4}$$

The parameter A in Equation 9.3 is a normalization constant, which may be determined from

$$\Sigma n_i = N \tag{9.5}$$

where N is the total number of particles in the system.

When the allowed energy states are numerous and closely spaced, the discrete quantities are replaced by continuous functions as follows:

$$g_i \longrightarrow g(E)\,dE$$

$$F_{MB} \longrightarrow Ae^{-E/kT}$$

where $g(E)\,dE$ is the *density of states* or the number of energy states per unit volume in the interval dE. In a similar manner, Equations 9.4 and 9.5 may be replaced as follows:

Number of particles per unit volume with energy between E and $E + dE$

$$n_i = g_i F_{MB} \longrightarrow n(E)\,dE = g(E)F_{MB}(E)\,dE \tag{9.6}$$

$$\Sigma n_i = N \longrightarrow \frac{N}{V} = \int_0^\infty n(E)\,dE = \int_0^\infty g(E)F_{MB}(E)\,dE \tag{9.7}$$

where $n(E)\,dE$ is the number of particles per unit volume with energies between E and $E + dE$. Note that Equations 9.6 and 9.7 may also be used for a system of quantum particles, provided that $g(E)$ and $F_{MB}(E)$ are replaced with the appropriate density of states and quantum distribution functions.

EXAMPLE 9.1 Emission Lines from Stellar Hydrogen
(a) Find the populations of the first and second excited states *relative* to the ground state for atomic hydrogen at room temperature.

Solution: For a gas at ordinary pressures, the atoms maintain the discrete quantum levels of isolated atoms. Recall that the discrete energy levels of atomic hydrogen are given by $E_n = (-13.6/n^2)$ eV and the degeneracy by $g_n = 2n^2$. Thus we have

Ground state	$E_1 = -13.6$ eV	$g_1 = 2$
First excited state	$E_2 = -3.40$ eV	$g_2 = 8$
Second excited state	$E_3 = -1.51$ eV	$g_3 = 18$

Using Equation 9.4 gives

$$\frac{n_2}{n_1} = \frac{g_2 A e^{-E_2/kT}}{g_1 A e^{-E_1/kT}} = \frac{g_2}{g_1} e^{(E_1-E_2)/kT}$$

$$= \frac{8}{2} \exp\{(-10.2 \text{ eV})/(8.617 \times 10^{-5} \text{ eV/K})(300 \text{ K})\}$$

$$= 4e^{-395} \approx 0$$

The ratio of n_3/n_1 will be even smaller. Thus, essentially all atoms are in the ground state at 300 K.

(b) Find the populations of the first and second excited states relative to the ground state for hydrogen heated to 20 000 K in a star.

Solution: When a gas is at very high temperatures (as in a flame, under electric discharge, or in a star), detectable numbers of atoms are in excited states. In this case, $T =$ 20 000 K, $kT = 1.72$ eV, and we find

$$\frac{n_2}{n_1} = \frac{g_2}{g_1} e^{(E_1-E_2)/kT} = 4e^{-10.2/1.72} = 0.0107$$

$$\frac{n_3}{n_1} = \frac{g_3}{g_1} e^{(E_1-E_3)/kT} = 9e^{-12.1/1.72} = 0.00807$$

(c) Find the emission strengths of the spectral lines corresponding to the transitions $E_3 \to E_1$ and $E_3 \to E_2$ relative to $E_2 \to E_1$ at 20 000 K.

Solution: The strength of an emission or absorption line is proportional to the number of atomic transitions per unit time. For particles obeying Maxwell-Boltzmann statistics, the number of transitions per unit time from some initial state (i) to some final state (f) equals the product of the population of the initial state and the probability for the transition i → f.

Note that *the transition rate for particles obeying MB statistics depends only on the initial population, since there are no restrictions on the number of particles in the final state.* Returning to the calculation of the emission strength, S, we find the relative values:

$$\frac{S(3 \longrightarrow 1)}{S(2 \longrightarrow 1)} = \frac{n_3 P(3 \longrightarrow 1)}{n_2 P(2 \longrightarrow 1)}$$

and

$$\frac{S(3 \longrightarrow 2)}{S(2 \longrightarrow 1)} = \frac{n_3 P(3 \longrightarrow 2)}{n_2 P(2 \longrightarrow 1)}$$

In reality, the transition probabilities depend on the wavefunctions of the states involved, but to simplify matters we assume equal probabilities of transition; that is, $P(2 \to 1) = P(3 \to 1) = P(3 \to 2)$. This yields

$$\frac{S(3 \longrightarrow 1)}{S(2 \longrightarrow 1)} = \frac{n_3}{n_2} = \frac{g_3}{g_2} e^{(E_2-E_3)/kT} = \frac{18}{8} e^{-1.89/1.73}$$

$$= 0.75$$

and

$$\frac{S(3 \longrightarrow 2)}{S(2 \longrightarrow 1)} = 0.75$$

If the emission lines are narrow, the measured heights of the 3 → 1 and 3 → 2 lines will be 75% of the height of the 2 → 1 line, as shown in Figure 9.3. For broader lines, the area under the peaks must be used as the experimental measure of emission strength.

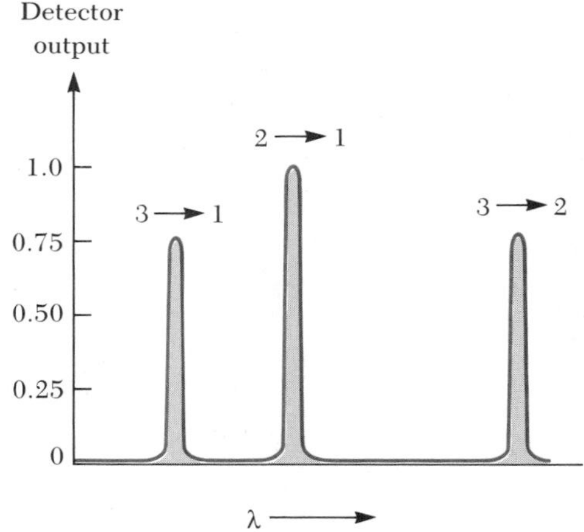

Figure 9.3 The predicted emission spectrum for the 2 → 1, 3 → 1, and 3 → 2 transitions for atomic hydrogen at 20 000 K.

The Maxwell Speed Distribution for Gas Molecules in Thermal Equilibrium at Temperature T

Maxwell's important formula for the number of molecules with *speeds* between v and $v + dv$ in a gas at temperature T may be found by using the Maxwell-Boltzmann distribution in its continuous form (Equations 9.6 and 9.7). In particular, we wish to show that

Maxwell speed distribution for gas molecules

$$n(v)\,dv = \frac{4\pi N}{V}\left(\frac{m}{2\pi kT}\right)^{3/2} v^2 e^{-mv^2/2kT}\,dv$$

where $n(v)\,dv$ is the number of gas molecules per unit volume with speeds between v and $v + dv$, N/V is the total number of molecules per unit volume, m is the mass of a gas molecule, k is Boltzmann's constant, and T is the absolute temperature. This speed distribution function is sketched in Figure 9.4. Note that the v^2 term determines the behavior of the distribution as $v \rightarrow 0$, while the exponential term determines what happens when $v \rightarrow \infty$.

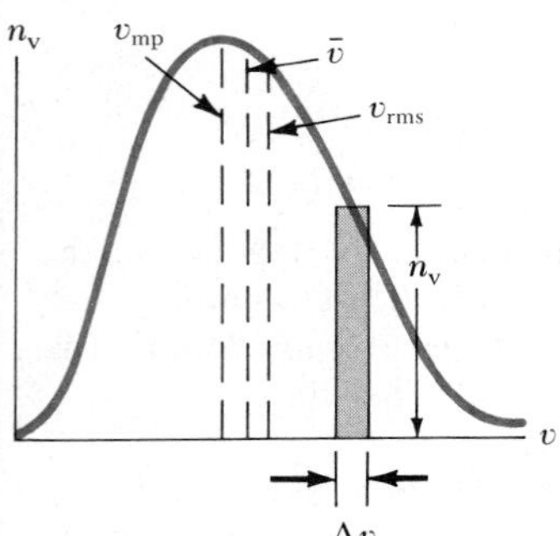

Figure 9.4 The speed distribution of gas molecules at some temperature. The number of molecules in the range Δv is equal to the area of the shaded rectangle, $n(v)\Delta v$. The most probable speed, v_{mp}, the average speed, v, and the root-mean-square speed, v_{rms}, are indicated.

Assuming an ideal gas of point particles such that the energy of each molecule consists only of translational kinetic energy, we have

$$E = \tfrac{1}{2}mv^2$$

for each molecule. Since the gas molecules have translational velocities that are continuously distributed from 0 to ∞, the energy distribution of molecules is also continuous and we may write the number of molecules per unit volume with energy between E and $E + dE$ as

$$n(E)\,dE = g(E)F_{\mathrm{MB}}(E)\,dE = g(E)Ae^{-mv^2/2kT}\,dE$$

In order to find the density of states, $g(E)$, we introduce the concept of **velocity space.** According to this idea, the velocity of each molecule may be represented by a velocity vector with components v_x, v_y, and v_z or by a point in a velocity space with coordinate axes v_x, v_y, and v_z (see Fig. 9.5). From Figure 9.5 we note that the number of states $f(v)\,dv$ with speeds between v and $v + dv$ is proportional to the volume of the spherical shell between v and $v + dv$:

$$f(v)\,dv = C\,4\pi v^2\,dv$$

where C is some constant. Because $E = \frac{1}{2}mv^2$, each speed v corresponds to a single energy E, and the number of energy states, $g(E)\,dE$, with energies between E and $E + dE$ is the same as the number of states with speeds between v and $v + dv$. Thus,

$$g(E)\,dE = f(v)\,dv = C\,4\pi v^2\,dv$$

Substituting this expression for $g(E)\,dE$ into our expression for $n(E)\,dE$, we obtain

$$n(E)\,dE = A\,4\pi v^2 e^{-mv^2/2kT}\,dv$$

where A is some other constant. Since the number of molecules with energy between E and $E + dE$ equals the number of molecules with speed between v

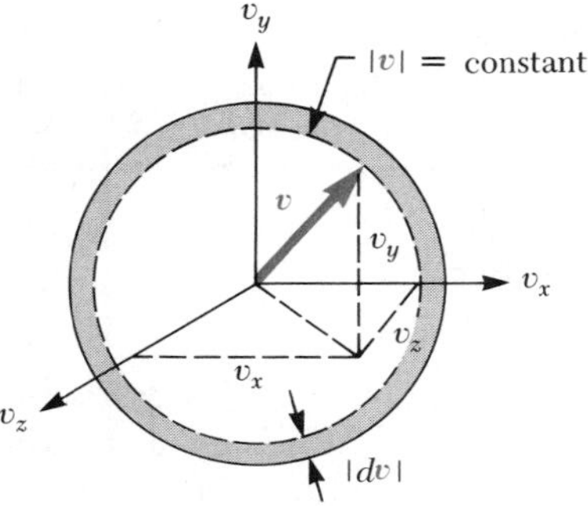

Figure 9.5 Velocity space. The number of states with speeds between v and $v + dv$ is proportional to the volume of a spherical shell with radius v and thickness dv.

and $v + dv$, we may also write

$$n(E)\,dE = n(v)\,dv = A\,4\pi v^2 e^{-mv^2/2kT}\,dv$$

To find A we use the fact that the total number of particles per unit volume is N/V:

$$\frac{N}{V} = \int_0^\infty n(v)\,dv = \int_0^\infty 4\pi A v^2 e^{-mv^2/2kT}\,dv$$

Because

$$\int z^{2j} e^{-az^2}\,dz = \frac{1\cdot 3\cdot 5\cdots(2j-1)}{2^{j+1}a^j}\sqrt{\frac{\pi}{a}} \qquad j = 1, 2, 3, \ldots$$

we find with $j = 1$ and $a = m/2kT$

$$\frac{N}{V} = \frac{(4\pi A)}{2^2\left(\dfrac{m}{2kT}\right)}\sqrt{\frac{\pi 2kT}{m}} = A\left(\frac{2\pi kT}{m}\right)^{3/2}$$

or

$$A = \frac{N}{V}\left(\frac{m}{2\pi kT}\right)^{3/2}$$

Hence we finally obtain Maxwell's famous 1859 result:

$$n(v)\,dv = 4\pi\frac{N}{V}\left(\frac{m}{2\pi kT}\right)^{3/2} v^2 e^{-mv^2/2kT}\,dv$$

In order to find the average speed, $\bar{v}$, indicated in Figure 9.4, we multiply $n(v)\,dv$ by v, integrate over all speeds from 0 to ∞ and divide the result by the total number of molecules per unit volume:

$$\bar{v} = \frac{\displaystyle\int_0^\infty v n(v)\,dv}{N/V} = \frac{4\pi\dfrac{N}{V}\left(\dfrac{m}{2\pi kT}\right)^{3/2}\displaystyle\int_0^\infty v^3 e^{-mv^2/2kT}\,dv}{N/V}$$

Using the definite integral formula

$$\int_0^\infty z^3 e^{-az^2}\,dz = \frac{1}{2a^2}$$

we have

$$\bar{v} = 4\pi\left(\frac{m}{2\pi kT}\right)^{3/2}\left(\frac{1}{2}\right)\left(\frac{2kT}{m}\right)^2 = \sqrt{\frac{8kT}{\pi m}}$$

The root mean square speed may be found by finding the average of v^2, denoted $\overline{v^2}$, and then taking its square root. Consequently, we have

$$\overline{v^2} = \frac{\displaystyle\int_0^\infty v^2 n(v)\,dv}{N/V} = 4\pi\left(\frac{m}{2\pi kT}\right)^{3/2}\int_0^\infty v^4 e^{-mv^2/2kT}\,dv$$

Using the definite integral formula

$$\int_0^\infty z^4 e^{-az^2}\,dz = \frac{3}{8a^2}\sqrt{\frac{\pi}{a}}$$

gives

$$\overline{v^2} = 4\pi\left(\frac{m}{2\pi kT}\right)^{3/2}\left[\frac{3}{8\left(\frac{m}{2kT}\right)^2}\right]\left(\frac{2\pi kT}{m}\right)^{1/2} = \frac{3kT}{m}$$

Since the root mean square speed, v_{rms}, is defined as $v_{\text{rms}} = \sqrt{\overline{v^2}}$, we have

$$v_{\text{rms}} = \sqrt{\frac{3kT}{m}}$$

Note that v_{rms} is *not* the same as the simple average speed, $\bar{v}$, but is about 10% greater as indicated in Figure 9.4. The derivation of the most probable speed, v_{mp}, is left to Problem 5.

EXERCISE 9.1 (a) Show that the formula for the number of molecules with energies between E and $E + dE$ in an ideal gas with absolute temperature T is given *explicitly* in terms of E by

$$n(E)\,dE = \frac{2\pi(N/V)}{(\pi kT)^{3/2}}\,E^{1/2}e^{-E/kT}\,dE$$

(b) Use this result to show that the total energy per unit volume of the gas is given by

$$E_{\text{total}} = \frac{3}{2}\frac{NkT}{V}.$$

9.2 QUANTUM STATISTICS, INDISTINGUISHABILITY, AND THE PAULI EXCLUSION PRINCIPLE

If we reexamine the assumptions that lead to the Maxwell-Boltzmann distribution, keeping the quantum mechanical wave nature of particles in mind, we immediately find a problem with assumption 1. Since particles exhibit wave-like behavior, they are necessarily fuzzy and are not distinguishable when their wave functions overlap. (See Section 8.4, Exchange Symmetry and the Exclusion Principle, for a review of this issue.) If trading molecule A for molecule B no longer counts as a different configuration, then the number of ways a given energy distribution can be realized changes, as does the equilibrium or most probable distribution. Thus the classical Maxwell-Boltzmann distribution must be replaced by a quantum distribution when there is wave function overlap, or when the particle concentration is high. However, the MB distribution is a valid approximation to the correct quantum distribution in the common case of gases at ordinary conditions. Quantum statistics is required only for cases concerning high particle concentrations, such as electrons in a metal[2] or photons in a black body cavity.

It is useful to develop a criterion to determine when the classical distribution is valid. We may say that *the Maxwell-Boltzmann distribution is valid when the average distance between particles, d, is large compared to the de Broglie wavelength, λ, calculated using the average thermal velocity of the particles.* Thus

$$\lambda \ll d \tag{9.8}$$

As $\lambda = h/p$ and the average kinetic energy of a gas is given by $p^2/2m = 3kT/2$, we find

[2] The density of conduction electrons in a metal is several thousand times the number of molecules/cm^3 in a gas at S.T.P.

$$\lambda = \frac{h}{\sqrt{3mkT}} \tag{9.9}$$

where h is Planck's constant, k is Boltzmann's constant, m is the particle mass, and T is the absolute temperature. Since the average spacing between particles is $d = \sqrt[3]{V/N}$, where V is the volume of the system and N is the total number of particles, the condition given by Equation 9.8 becomes

$$\frac{h}{\sqrt{3mkT}} \ll \sqrt[3]{\frac{V}{N}}$$

Thus, the Maxwell-Boltzmann distribution is valid when

Criterion for the validity of Maxwell-Boltzmann statistics

$$\left(\frac{N}{V}\right) \frac{h^3}{(3mkT)^{3/2}} \ll 1 \tag{9.10}$$

As expected, Equation 9.10 shows that the classical Maxwell-Boltzmann distribution holds for low particle concentration and for high particle mass and temperature.

EXAMPLE 9.2 When Can We Use Maxwell-Boltzmann Statistics?
Are Maxwell-Boltzmann statistics valid for (a) hydrogen gas at standard temperature and pressure and (b) conduction electrons in silver at 300 K?

Solution: (a) Under standard conditions, one mole of H_2 gas (corresponding to 6.02×10^{23} molecules) occupies a volume of 22.4×10^{-3} m³. Since we have $m_{H_2} = 3.34 \times 10^{-27}$ kg, $h^3 = 290.9 \times 10^{-102}$ (J·s)³, and $kT = 3.77 \times 10^{-21}$ J, we find

$$\left(\frac{N}{V}\right) \frac{h^3}{(3mkT)^{3/2}} = \left(\frac{6.02 \times 10^{23}}{22.4 \times 10^{-3}\ \text{m}^3}\right) \times \frac{291 \times 10^{-102}\ (\text{J}\cdot\text{s})^3}{[3 \times (3.34 \times 10^{-27}\ \text{kg})(3.77 \times 10^{-21}\ \text{J})]^{3/2}}$$

$$= 3.37 \times 10^{-5}$$

This is much less than one, and from the condition given by Equation 9.10, we conclude that even hydrogen, the lightest gas, is described by Maxwell-Boltzmann statistics.

(b) Silver has a density of 10.5 g/cm³ and a molar weight of 107.9 g. Assuming one free electron per silver atom, the density of free electrons in silver is found to be

$$\frac{10.5\ \text{g/cm}^3}{107.9\ \text{g/mole}} (6.02 \times 10^{23}\ \text{electrons/mole})$$

$$= 5.86 \times 10^{22}\ \text{electrons/cm}^3$$

$$= 5.86 \times 10^{28}\ \text{electrons/m}^3$$

Note that the density of free electrons in silver is about 2000 times greater than the density of hydrogen gas molecules at S.T.P.; that is

$$\frac{(N/V)_{\text{electrons in Ag}}}{(N/V)_{H_2\ \text{at S.T.P.}}} = \frac{5.86 \times 10^{28}\ \text{m}^{-3}}{2.69 \times 10^{25}\ \text{m}^{-3}} = 2180$$

Using $h^3 = 290.9 \times 10^{-102}$ (J·s)³, $m = 9.109 \times 10^{-31}$ kg, and $kT = 4.14 \times 10^{-21}$ J (at $T = 300$ K), we find

$$\left(\frac{N}{V}\right) \frac{h^3}{(3\ mkT)^{3/2}} = (5.86 \times 10^{28}\ \text{m}^{-3}) \times \frac{290.9 \times 10^{-102}\ (\text{J}\cdot\text{s})^3}{[3(9.109 \times 10^{-31}\ \text{kg})(4.14 \times 10^{-21}\ \text{J})]^{3/2}}$$

$$= 1.42 \times 10^4$$

Comparing this result to the condition given by Equation 9.10, we conclude that the Maxwell-Boltzmann distribution *does not hold* for electrons in silver because of the small mass of the electron and because of the high free electron density. We shall see that the correct quantum distribution for electrons is the Fermi-Dirac distribution.

A convenient method for classifying the two quantum distributions, the **Bose-Einstein distribution** and the **Fermi-Dirac distribution,** is in terms of the three assumptions given for the Maxwell-Boltzmann case. To obtain the Bose-Einstein distribution, we simply give up assumption 1 and now assume that the particles are completely indistinguishable. Particles that obey the Bose-

Einstein distribution are called **bosons** and are observed to have *integral spin*. Some examples of bosons are the alpha particle ($S = 0$), the photon ($S = 1$), and the deuteron ($S = 1$). To obtain the Fermi-Dirac distribution, we give up assumptions 1 and 3, and consider indistinguishable particles subject to the constraint that only two particles can occupy a given energy state (corresponding to particles with spin up and down). (See Section 8.4 on the Pauli exclusion principle.) Particles that obey the Fermi-Dirac distribution are called **fermions** and are observed to have *half integral spin*. Some important examples of fermions are the electron, the proton, and the neutron, all with spin 1/2.

In order to see the essential changes in the distribution function introduced by quantum statistics, let us return briefly to our simple system of six particles with a total energy of 8E. *First we consider the Bose-Einstein case, where the particles are indistinguishable and there is no limit on the number of particles in a given energy state.* Figure 9.1a was drawn to represent this situation. Since the particles are indistinguishable, each arrangement is equally likely, so the probability of each arrangement is 1/20. The average number of particles in a particular energy level may be calculated by again using Equation 9.2. The average number of particles in the zero energy level is found to be

$$\begin{aligned}\overline{n_0} = &(5)(1/20) + (4)(1/20) + (4)(1/20) + (3)(1/20) + (4)(1/20)\\ &+ (3)(1/20) + (2)(1/20) + (4)(1/20) + (3)(1/20) + (3)(1/20)\\ &+ (2)(1/20) + (1)(1/20) + (3)(1/20) + (2)(1/20) + (2)(1/20)\\ &+ (1)(1/20) + (0)(1/20) + (2)(1/20) + (1)(1/20) + (0)(1/20)\\ = &\ 49/20 = 2.45\end{aligned}$$

Similarly, we find $\overline{n_1} = 31/20 = 1.55$, $\overline{n_2} = 18/20 = 0.90$, $\overline{n_3} = 9/20 = 0.45$, $\overline{n_4} = 6/20 = 0.30$, $\overline{n_5} = 3/20 = 0.15$, $\overline{n_6} = 2/20 = 0.10$, $\overline{n_7} = 1/20 = 0.05$, and $\overline{n_8} = 0.05$. A plot of these values in Figure 9.2 (shown by the triangle points) shows that the Bose-Einstein distribution gives results generally similar, but not identical, to the Maxwell-Boltzmann distribution. In general, the Bose-Einstein distribution tends to have more particles in the lowest energy levels. At higher energies, the curves come together and both exhibit an exponential fall-off with energy.

In order to illustrate the distinctive shape of the Fermi-Dirac distribution, again consider our simple example of six indistinguishable particles with energy 8E. *Since the particles are fermions, we impose the constraint that no more than two particles can be assigned to a given energy state (corresponding to electrons with spin up and down).* There are only three states (denoted by F.D.) out of the 20 arrangements shown in Figure 9.1a that meet this additional constraint imposed by the Pauli exclusion principle. Since each of these arrangements is equally likely, each has a probability of occurrence of 1/3, and we may again use Equation 9.2 to calculate the average number of fermions in the zero energy level as follows:

$$\overline{n_0} = (2)(1/3) + (2)(1/3) + (2)(1/3) = 2.00$$

Similarly, we find for the average number of fermions with energies of 1E through 8E the following:

$$\overline{n_1} = (2)(1/3) + (2)(1/3) + (1)(1/3) = 1.67$$

$$\overline{n_2} = 3/3 = 1$$

$$\overline{n_3} = 3/3 = 1$$
$$\overline{n_4} = 1/3 = 0.33$$
$$\overline{n_5} = \overline{n_6} = \overline{n_7} = \overline{n_8} = 0$$

When this distribution is plotted, we see that a distinctly different distribution from the Maxwell-Boltzmann or Bose-Einstein curves results. Although it is not entirely clear that the points plotted in Figure 9.6 conform to the smooth curve drawn, consideration of systems with more than six particles proves this to be the case (see Problem 2).

In general, when large numbers of quantum particles are considered, *continuous* distribution functions may be rigorously obtained for both the Bose-Einstein and Fermi-Dirac cases. These functions have the explicit forms:

$$f_{\mathrm{BE}}(E) = \frac{1}{Be^{E/kT} - 1} \tag{9.11}$$

$$f_{\mathrm{FD}}(E) = \frac{1}{Ce^{E/kT} + 1} \tag{9.12}$$

where $f(E)$ is the *probability* (not necessarily normalized to 1) of finding a particle in a particular state of energy E at a given absolute temperature T. As noted earlier, the *number of particles per unit volume* with energy between E and $E + dE$ is given by

$$n(E)\,dE = g(E)f_{\mathrm{BE}}(E)\,dE \tag{9.13}$$

or

$$n(E)\,dE = g(E)f_{\mathrm{FD}}(E)\,dE \tag{9.14}$$

The parameters B and C in Equations 9.11 and 9.12 may be determined from the total number of particles, N, since

$$\left(\frac{N}{V}\right)_{\mathrm{bosons}} = \int_0^\infty g(E)f_{\mathrm{BE}}(E)\,dE \tag{9.15}$$

and

$$\left(\frac{N}{V}\right)_{\mathrm{fermions}} = \int_0^\infty g(E)f_{\mathrm{FD}}(E)\,dE \tag{9.16}$$

In the specific case of photons in a black body cavity, the number of particles is not fixed, since photons are randomly emitted (created) and absorbed (de-

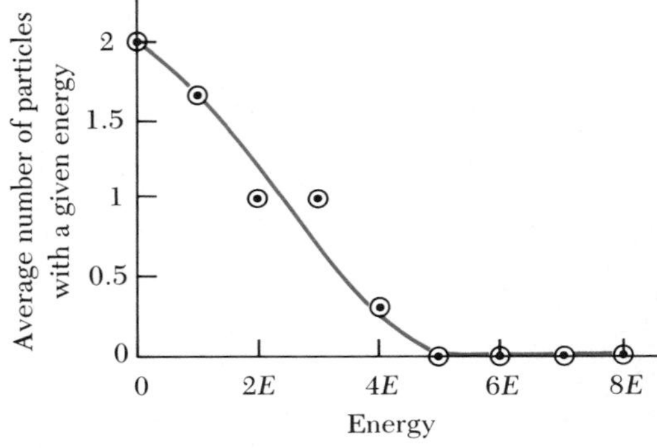

Figure 9.6 The distribution function for six identical particles with total energy 8E constrained so that no more than two particles occupy the same energy state (fermions).

stroyed) by the cavity walls. Thus, Equation 9.15 cannot be used to determine B. In this case, B must be independent of N and can be shown to be 1, so that

Bose-Einstein distribution for photons

$$f(E)_{\text{photons}} = \frac{1}{e^{E/kT} - 1} \tag{9.17}$$

In the case of the Fermi-Dirac distribution, the constant C is often written in an explicitly temperature-dependent form as $C = e^{-E_F/kT}$, where E_F is called the **Fermi energy.** With this substitution, Equation 9.12 changes to the more common form

Fermi-Dirac distribution

$$f_{FD}(E) = \frac{1}{e^{(E-E_F)/kT} + 1} \tag{9.18}$$

This expression shows that the probability of finding an electron having an energy equal to the Fermi energy is exactly 1/2 at any temperature.

A plot comparing the Maxwell-Boltzmann, Bose-Einstein, and Fermi-Dirac distributions as functions of energy at a common temperature of 5000 K is shown in Figure 9.7. Note that for large E, all occupation probabilities decrease to zero as $e^{-E/kT}$. For small values of E, the FD probability saturates at 1, the MB probability constantly increases but remains finite, and the BE probability tends to infinity. The singularity in the Bose-Einstein distribution means that at low enough energies, most of the particles drop into the ground state. When this happens, a new phase of matter with different physical properties can occur. This change in phase for a system of bosons is called a **Bose-Einstein condensation,** and it actually occurs in liquid helium at a temperature of 2.18 K. Below 2.18 K liquid helium becomes a mixture of the normal liquid and a phase with all molecules in the ground state. The ground state phase, called liquid helium II, exhibits many remarkable properties, one being zero viscosity.

Electrons, and other fermions, cannot all occupy the same energy state, so no condensation is possible for fermions. For small values of E such that $E \ll E_F$, the factor $e^{(E-E_F)/kT}$ is small and $f_{FD} \approx 1$. Thus, every energy state at low energy is filled but with only two electrons, as required by the exclusion principle.

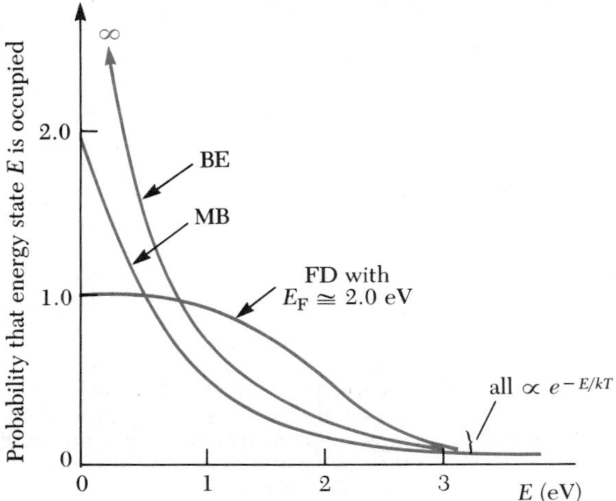

Figure 9.7 A comparison of Maxwell-Boltzmann, Bose-Einstein, and Fermi-Dirac distribution functions at 5000 K.

The history of the discovery of the quantum distributions is interesting. The first quantum distribution to be discovered was the Bose-Einstein function introduced in 1924 by Satyendra Nath Bose, an Indian physicist working in isolation. His paper, which contained a new proof of the Planck formula for blackbody radiation, was sent to Einstein. In this paper, Bose applied the normal methods of statistical mechanics to light quanta but treated the quanta as absolutely indistinguishable. Einstein was impressed by Bose's work and proceeded to translate the paper into German for publication in the *Zeitschrift für Physik*.[3] In order to obtain the quantum theory of the ideal gas, Einstein extended the method to molecules in several papers published in 1924 and 1925.

The exclusion principle

In 1925 Wolfgang Pauli, after an exhaustive study of the quantum numbers assigned to atomic levels split by the Zeeman effect, announced his new and fundamental principle of quantum theory, the exclusion principle: **Two electrons in an atom cannot have the same set of quantum numbers.** In 1926, Enrico Fermi obtained the second type of quantum statistics that occurs in nature by combining Pauli's exclusion principle with the requirement of indistinguishability. Paul Dirac is also credited for this work, since he performed a more rigorous quantum mechanical treatment of these statistics in 1926. The empirical observation that particles with integral spin obey BE statistics and that particles with half-integral spin obey FD statistics was explained much later (in 1940) by Pauli, using relativity and causality arguments.

9.3 AN APPLICATION OF BOSE-EINSTEIN STATISTICS: THE PHOTON GAS

In this section we shall apply BE statistics to the problem of determining the energy density of a photon gas filling a material cavity at temperature T. (In Chapter 2 we discussed the importance of this blackbody problem for quantum physics.) Since the number of photons per unit volume having energy between E and $E + dE$ is $n(E)\,dE = g(E)f(E)\,dE$, the energy density (energy per unit volume) of photons in the range from E to $E + dE$ is

$$u(E)\,dE = E\,n(E)\,dE = \frac{g(E)E\,dE}{e^{E/kT} - 1} \tag{9.19}$$

To complete our calculation, all we need is the factor $g(E)$, the density of states for photons in an enclosure. This important calculation, given in Appendix A, shows that the determination of allowed modes for electromagnetic waves or photons confined to an enclosure depends on the solution vanishing at the boundaries. From Equation A.6, we have for the number of photons per unit volume with frequencies between f and $f + df$

$$N(f)\,df = \frac{8\pi f^2\,df}{c^3} = \frac{8\pi(hf)^2\,d(hf)}{(hc)^3} = \frac{8\pi E^2\,dE}{(hc)^3}$$

using $E = hf$ for photons. Since the number of photons per unit volume with frequencies between f and $f + df$ is equal to the number of photons with energies between E and $E + dE$, we have

$$N(f)\,df = \frac{8\pi E^2\,dE}{(hc)^3} = g(E)\,dE$$

[3] S. N. Bose, *Z. Phys.* **26:** 178, 1924.

Thus we find that the density of states for photons is

$$g(E) = \frac{8\pi E^2}{(hc)^3} \tag{9.20}$$

Substituting Equation 9.20 into Equation 9.19 gives an expression for the energy density:

$$u(E)\,dE = \frac{8\pi}{(hc)^3}\frac{E^3\,dE}{e^{E/kT}-1} \tag{9.21}$$

Converting from photon energy to frequency using $E = hf$ in Equation 9.21, we immediately retrieve the Planck blackbody formula:

$$u(f,T) = \frac{8\pi h}{c^3}\frac{f^3}{e^{hf/kT}-1}$$

EXAMPLE 9.3 Photons in a Box

(a) Find an expression for the number of photons per unit volume in a cavity at temperature T with energies between E and $E + dE$. (b) Find an expression for the total number of photons per unit volume (all energies). (c) Calculate the number of photons/cm³ inside a cavity whose walls are heated to 3000 K. Compare this to a cavity whose walls are at 3 K.

Solution:

(a) $$n(E)\,dE = g(E)f(E)\,dE|_{\text{photons}} = \frac{8\pi E^2\,dE}{(hc)^3(e^{E/kT}-1)}$$

(b) $$N/V = \int_0^\infty n(E)\,dE = \frac{8\pi(kT)^3}{(hc)^3}\int_0^\infty \frac{(E/kT)^2(dE/kT)}{e^{E/kT}-1}$$

or

$$N/V = 8\pi(kT/hc)^3\int_0^\infty \frac{z^2\,dz}{e^z-1}$$

(c) From standard tables,

$$\int_0^\infty \frac{z^2\,dz}{e^z-1} \approx 2.40$$

Therefore,

N/V (at 3000 K) =

$$(8\pi)\left[\frac{(8.62\times10^{-5}\ \text{eV/K})(3000\ \text{K})}{1.24\times10^{-4}\ \text{eV}\cdot\text{cm}}\right]^3(2.40)$$

$$= 5.47\times10^{11}\ \text{photons/cm}^3$$

Likewise, N/V (at 3 K) $= 5.47\times10^2$ photons/cm³. Thus, the photon density decreases by a factor of 10^9 when the temperature drops from 3000 K to 3 K.

9.4 AN APPLICATION OF FERMI-DIRAC STATISTICS: THE FREE-ELECTRON GAS THEORY OF METALS

Since the outer electrons are weakly bound to individual atoms in a metal, we can treat these outer conduction electrons as a gas of fermions trapped within a cavity formed by the metallic surface. Once again, many interesting physical expressions, such as the average energy, Fermi energy, specific heat, and thermionic emission rate, may be derived from the expression for the concentration of electrons with energies between E and $E + dE$:

$$n(E)\,dE = g(E)f_{\text{FD}}(E)\,dE \tag{9.14}$$

Recall that the probability of finding an electron in a particular energy state E is given by the Fermi-Dirac distribution function,

$$f_{\text{FD}}(E) = \frac{1}{e^{(E-E_F)/kT}+1}$$

Plots of this function versus energy are shown in Figure 9.8 for the cases $T = 0$ K and $T > 0$. Note that at $T = 0$ K, we have $f = 1$ for $E < E_F$ and $f = 0$ for

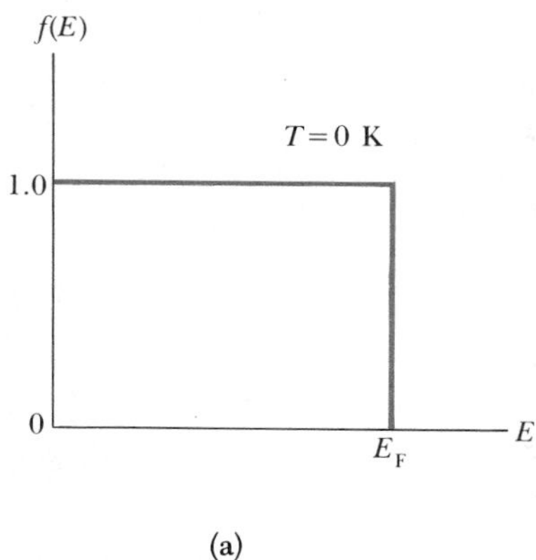

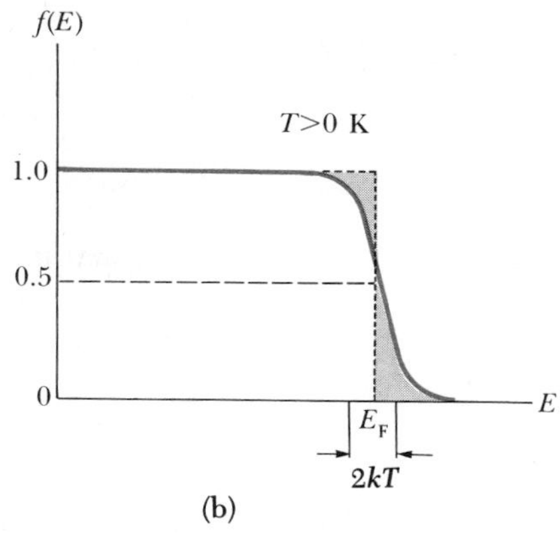

Figure 9.8 A comparison of the Fermi-Dirac distribution functions at (a) absolute zero and (b) finite temperature.

$E > E_F$. Thus all states with energies less than E_F are completely filled and all states with energies greater than E_F are empty. This is in sharp contrast to the predictions of MB and BE statistics, in which all particles condense to a state of zero energy at absolute zero. In fact, far from having zero velocity, a conduction electron in a metal with the cut-off energy E_F has a velocity v_F given by

$$\tfrac{1}{2}m_e v_F^2 = E_F \tag{9.22}$$

where v_F is called the *Fermi velocity*. Substituting a typical value of 5 eV for the Fermi energy yields the remarkable result that electrons at the Fermi level possess velocities of the order of 10^6 m/s at 0 K!

Figure 9.8b shows that as T increases, the distribution rounds off slightly, with states between E and $E - kT$ losing population and states between E and $E + kT$ gaining population. In general, E_F also depends on temperature, but the dependence is weak and we may say that $E_F(T) \approx E_F(0)$ up to several thousand kelvins.

Let us now turn to the calculation of the density of states, $g(E)$, for conduction electrons in a metal. Since the electrons may be viewed as a system of matter waves whose wave functions vanish at the boundaries of the metal, we obtain the same result for electrons as for electromagnetic waves confined to a cavity. In the latter case, we found (see Appendix A) that the density of states with wave number between k and $k + dk$ is given by

$$N(k)\,dk = \frac{k^2\,dk}{2\pi^2} \tag{A.4}$$

In order to apply this expression to electrons in a metal, we must multiply it by a factor of 2 to account for the two allowed spin states of an electron with a given momentum or energy:

$$N(k)\,dk = \frac{k^2\,dk}{\pi^2} \tag{9.23}$$

To obtain $g(E)$ from $N(k)$, we assume nonrelativistic free electrons. Thus

$$E = \frac{p^2}{2m_e} = \frac{\hbar^2 k^2}{2m_e}$$

or

$$k = \left(\frac{2m_e E}{\hbar^2}\right)^{1/2} \tag{9.24}$$

and

$$dk = \frac{1}{2}\left(\frac{2m_e}{\hbar^2}\right)^{1/2} E^{-1/2}\, dE \tag{9.25}$$

Substituting Equations 9.24 and 9.25 into Equation 9.23 yields

Density of states for conduction electrons

$$\boxed{g(E)\, dE = CE^{1/2}\, dE} \tag{9.26}$$

where

$$C = \frac{8\sqrt{2}\pi m_e^{3/2}}{h^3} \tag{9.27}$$

Thus the key expression for the number of electrons per unit volume with energy between E and $E + dE$ becomes

Number of electrons per unit volume with energy between E and $E + dE$

$$\boxed{n(E)\, dE = \frac{CE^{1/2}\, dE}{e^{(E-E_F)/kT} + 1}} \tag{9.28}$$

Figure 9.9 is a plot of Equation 9.28, showing the folding together of an increasing density of states and the decreasing FD distribution. Since

$$N/V = \int_0^\infty n(E)\, dE = C\int_0^\infty \frac{E^{1/2}\, dE}{e^{(E-E_F)/kT} + 1} \tag{9.29}$$

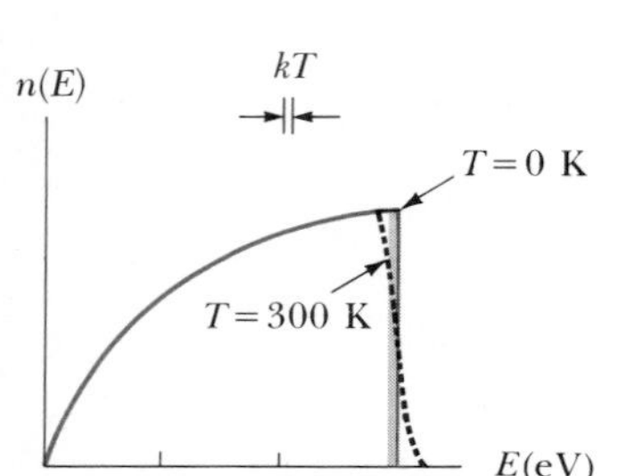

Figure 9.9 The number of electrons with energy between E and $E + dE$. Note that $n(E) = g(E)f_{FD}(E)$.

we can determine the Fermi energy as a function of the electron concentration, N/V. For arbitrary T, Equation 9.29 must be integrated numerically. At $T = 0$ K, the integration is simple since $f(E) = 1$ for $E < E_F$ and is zero for $E > E_F$. Therefore at $T = 0$ K, Equation 9.29 becomes

$$N/V = C\int_0^{E_F} E^{1/2}\, dE = \tfrac{2}{3}CE_F^{3/2} \tag{9.30}$$

Substituting the value of C from Equation 9.27 into Equation 9.30 gives for the Fermi energy at 0 K, $E_F(0)$,

$$E_F(0) = \frac{h^2}{2m_e}\left(\frac{3N}{8\pi V}\right)^{2/3} \tag{9.31}$$

Equation 9.31 shows a gradual increase in $E_F(0)$ with increasing electron concentration. This is expected, because the electrons fill the available energy states, two electrons per state, in accordance with the Pauli exclusion principle up to a maximum energy E_F. Representative values of $E_F(0)$ for various metals calculated from Equation 9.31 are given in Table 9.1. This table also

TABLE 9.1 Calculated Values of Various Parameters for Metals at 300 K Based on the Free-Electron Theory

Metal	Electron Concentration (m^{-3})	Fermi Energy (eV)	Velocity (m/s)	Fermi Temperature (K)
Li	4.70×10^{28}	4.72	1.29×10^6	5.48×10^4
Na	2.65×10^{28}	3.23	1.07×10^6	3.75×10^4
K	1.40×10^{28}	2.12	0.86×10^6	2.46×10^4
Cu	8.49×10^{28}	7.05	1.57×10^6	8.12×10^4
Ag	5.85×10^{28}	5.48	1.39×10^6	6.36×10^4
Au	5.90×10^{28}	5.53	1.39×10^6	6.41×10^4

lists values of the electron velocity at the Fermi level, and the **Fermi temperature, T_F,** defined by

$$T_F \equiv E_F/k \tag{9.32}$$

As a final note, it is interesting that a longstanding puzzle concerning the anomalously small contribution of the conduction electron "gas" to the heat capacity of a solid has a qualitative solution in terms of the Fermi-Dirac distribution. If conduction electrons behaved classically, warming a gas of N electrons from 0 to 300 K would result in an average energy increase of $3kT/2$ for *each* particle, or a total thermal energy per mole, U, given by

$$U = N(\tfrac{3}{2}kT) = \tfrac{3}{2}RT$$

Thus the electronic heat capacity per mole would be given by

$$C_{el} = \frac{\partial U}{\partial T} = \tfrac{3}{2}R$$

An examination of Figure 9.9, however, shows that on heating from 0 K, very few electrons become excited and gain an energy kT. Only a small fraction f within kT of E_F can be excited thermally. The fraction f may be approximated by the ratio of the area of a thin rectangle of width kT and height $N(E_F)$ to the total area under $n(E)$. Thus

$$f \approx \frac{\text{Area of shaded rectangle in Figure 9.9}}{\text{Total area under } n(E)}$$

$$= \frac{(kT)g(E_F)}{C\displaystyle\int_0^{E_F} E^{1/2}\,dE} = \frac{(kT)C(E_F)^{1/2}}{\tfrac{2}{3}CE_F^{\,3/2}} = \frac{3}{2}\frac{kT}{E_F} = \frac{3}{2}\frac{T}{T_F}$$

Since only fN of the electrons gain an energy of the order of kT, the actual total thermal energy gained per mole is

$$\left(\frac{3}{2}\frac{T}{T_F}\right)(N)(kT) = \frac{3}{2}\frac{RT^2}{T_F}$$

From this result, we find that the electronic heat capacity is

$$C_{el} = \frac{\partial U}{\partial T} \approx 3R\frac{T}{T_F}$$

Substituting $T = 300$ K and $T_F = 5 \times 10^4$ K, we find a very small value for the electronic heat capacity at ordinary temperatures:

$$C_{el} \approx 3R\left(\frac{300\text{ K}}{50\,000\text{ K}}\right) = 0.018R$$

Thus, the electrons contribute only $0.018R/1.5R$ or about 1% of the classically expected amount to the heat capacity.

EXAMPLE 9.4 The Fermi Energy of Gold
Calculate (a) the Fermi energy, (b) the Fermi velocity, and (c) the Fermi temperature for gold at 0 K.

Solution: (a) The density of gold is 19.32 g/cm³ and its molar weight is 197 g/mole. Assuming each gold atom contributes one free electron to the Fermi gas, we can calculate the electron concentration as follows:

$$N/V = (19.32\text{ g/cm}^3)\left(\frac{1}{197\text{ g/mole}}\right) \times (6.02 \times 10^{23}\text{ electrons/mole})$$

$$= 5.90 \times 10^{22}\text{ electrons/cm}^3$$

$$= 5.90 \times 10^{28}\text{ electrons/m}^3$$

Using Equation 9.31, we find

$$E_F(0) = \frac{h^2}{2m_e}\left(\frac{3N}{8\pi V}\right)^{2/3}$$

$$= \frac{(6.625 \times 10^{-34}\ \text{J}\cdot\text{s})^2}{2(9.11 \times 10^{-31}\ \text{kg})}\left(\frac{3 \times 5.90 \times 10^{28}\ \text{m}^{-3}}{8\pi}\right)^{2/3}$$

$$= 8.85 \times 10^{-19}\ \text{J} = 5.53\ \text{eV}$$

(b) Since $\frac{1}{2}m_e v_F^2 = E_F$,

$$v_F = \left(\frac{2E_F}{m_e}\right)^{1/2} = \left(\frac{2 \times 5.85 \times 10^{-19}\ \text{J}}{9.11 \times 10^{-31}\ \text{kg}}\right)^{1/2}$$

$$= 1.39 \times 10^6\ \text{m/s}$$

(c) The Fermi temperature is given by

$$T_F = \frac{E_F}{k} = \frac{5.53\ \text{eV}}{8.62 \times 10^{-5}\ \text{eV/K}} = 64\ 000\ \text{K}$$

Thus a gas of classical particles would have to be heated to about 64 000 K to have an average energy per particle equal to the Fermi energy at 0 K!

9.5 SUMMARY

Statistical physics deals with the distribution of a fixed amount of energy among a fixed number of particles that are identical (quantum particles) or distinguishable (classical particles). In most situations, one is not interested in the energies of all the particles at a given instant, but rather in the time average of the number of particles in a particular energy level. The average number of particles in a given energy level is of special interest in spectroscopy because the intensity of radiation emitted or absorbed is proportional to the number of particles in a particular energy state.

For a system described by a continuous distribution of energy levels, the number of particles per unit volume with energy between E and $E + dE$ is given by

$$n(E)\ dE = g(E)f(E)\ dE \tag{9.6}$$

where $g(E)$ is the density of states or the number of energy states per unit volume in the interval dE, and $f(E)$ is the probability that a particle is in the energy state E. The function $f(E)$ is called the *distribution function.*

Three distinct distribution functions are used, depending on whether the particles are distinguishable, and whether there is a restriction on the number of particles in a given energy state:

1. **Maxwell-Boltzmann Distribution (Classical).** The particles are distinguishable and there is no limit on the number of particles in a given energy state.

$$f_{MB}(E) = Ae^{-E/kT}$$

2. **Bose-Einstein Distribution (Quantum).** The particles are indistinguishable and there is no limit on the number of particles in a given energy state.

$$f_{BE}(E) = \frac{1}{Be^{E/kT} - 1} \tag{9.11}$$

3. **Fermi-Dirac Distribution (Quantum).** The particles are indistinguishable, but there can be no more than two particles per energy state.

$$f_{FD}(E) = \frac{1}{e^{(E-E_F)/kT} + 1} \tag{9.18}$$

where E_F is the *Fermi energy.* At $T = 0$ K, all levels below E_F are filled, while all levels above E_F are empty.

At low particle concentrations and high temperature, most systems are best described by Maxwell-Boltzmann statistics. The criterion that determines when the classical Maxwell-Boltzmann distribution is valid is written

$$\left(\frac{N}{V}\right)\frac{h^3}{(3mkT)^{3/2}} \ll 1 \tag{9.10}$$

where N/V is the particle concentration, m is the particle mass, and T is the absolute temperature. For high particle concentration, low particle mass, and modest temperature, there is considerable wave function overlap and quantum distributions must be used to describe systems of indistinguishable particles.

A system of photons in thermal equilibrium at temperature T is adequately described by the Bose-Einstein distribution and a density of states given by

$$g(E) = \frac{8\pi E^2}{(hc)^3} \tag{9.20}$$

Thus, the concentration of photons with energies between E and $E + dE$ is

$$n(E)\,dE = \frac{8\pi E^2}{(hc)^3}\left(\frac{1}{e^{E/kT} - 1}\right) dE$$

Free (conduction) electrons in metals obey the Pauli exclusion principle, and we must use the Fermi-Dirac distribution to treat such a system. The density of states for electrons in a metal is

$$g(E) = \frac{8\sqrt{2}\pi m_e^{3/2}}{h^3} E^{1/2} \tag{9.26}$$

hence the number of electrons per unit volume with energy between E and $E + dE$ is

$$n(E)\,dE = \frac{8\sqrt{2}\pi m_e^{3/2} E^{1/2}}{h^3(e^{(E-E_F)/kT} + 1)}\,dE \tag{9.28}$$

An expression for the Fermi energy as a function of electron concentration may be obtained by integrating Equation 9.28. One finds

$$E_F(0) = \frac{h^2}{2m_e}\left(\frac{3N}{8\pi V}\right)^{2/3} \tag{9.31}$$

where $E_F(0)$ is the Fermi energy at 0 K. The small electronic contribution to the heat capacity of a metal can be explained by noting that only a small fraction of the electrons near E_F gain kT in thermal energy when the metal is heated from 0 K to T K.

SUGGESTIONS FOR FURTHER READING

1. A. Beiser, *Concepts of Modern Physics*, 3rd edition, New York, McGraw-Hill Book Co., 1981.
More advanced treatments of statistical physics may be found in the following books.
2. P. M. Morse, *Thermal Physics*, New York, Benjamin, 1965.
3. C. Kittel, *Thermal Physics*, New York, Wiley, 1969.

QUESTIONS

1. Discuss the basic assumptions of Maxwell-Boltzmann, Fermi-Dirac, and Bose-Einstein statistics. How do they differ, and what are their similarities?
2. Explain the role of the Pauli exclusion principle in describing the electrical properties of metals.

PROBLEMS

1. Verify that for a system of six distinguishable particles with total energy 8E, the average number of particles with energies of 1E, 2E, etc. are: $\overline{n_1} = 1.54$, $\overline{n_2} = 1.00$, $\overline{n_3} = 0.587$, $\overline{n_4} = 0.326$, $\overline{n_5} = 0.163$, $\overline{n_6} = 0.0699$, $\overline{n_7} = 0.0233$, and $\overline{n_8} = 0.00466$.
2. In order to obtain a more clearly defined picture of the Fermi-Dirac distribution, consider a system of 20 Fermi-Dirac particles sharing 94 units of energy. By drawing diagrams like Figure 9.10, show that there are nine different microstates. Using Equation 9.2, calculate and plot the average number of particles in each energy level from 0 to 14E. Locate the Fermi energy at 0 K on your plot from the fact that electrons at 0 K fill all the levels consecutively up to the Fermi energy. (At 0 K the system no longer has 94 units of energy, but has the *minimum* amount of 90E.)

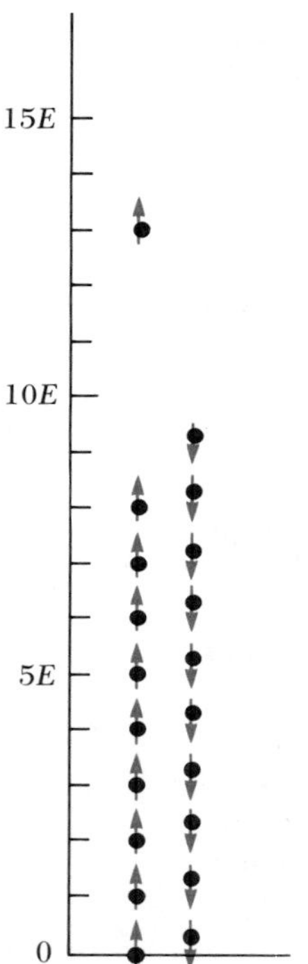

Figure 9.10 (Problem 2) One of the nine equally probable microstates for the 20 FD particles with a total energy of 94E.

3. The Fermi energy of copper at 300 K is 7.05 eV. (a) What is the average energy of a conduction electron in copper at 300 K? (b) At what temperature would the average energy of a molecule of an ideal gas equal the energy obtained in (a)?
4. The Fermi energy of aluminum is 11.63 eV. (a) Assuming that the free electron model applies to aluminum, calculate the number of free electrons per unit volume at very low temperatures. (b) Determine the valence of aluminum by dividing the answer found in part (a) by the number of aluminum atoms per unit volume as calculated from the density and the atomic weight. Note that aluminum has a density of 2.70 g/cm^3.
5. Show that the most probable speed of a gas molcule is given by

$$v_{\text{mp}} = \sqrt{\frac{2kT}{m}}$$

Note that the most probable speed corresponds to the point where the Maxwellian speed distribution curve, $n(v)$, has a maximum.

6. Figure 9.11 shows an apparatus similar to that used by Otto Stern in 1920 to verify the Maxwell velocity distribution. A collimated beam of gas molecules from an oven, *O* is allowed to enter a rapidly rotating cylinder when slit *S* is coincident with the beam. The pulse of molecules created by the rapid rotation of *S* then strikes and adheres to a glass plate detector, *D*. The velocity of a molecule may be detemined from its position on the glass plate (fastest molecules to the right). The number of molecules arriving with a given velocity may be determined by measuring the density of molecules deposited on *D* at a given position. Suppose that the oven contains a gas of bismuth molecules (Bi_2) at 850 K, and that the cylinder is 10 cm in diameter and rotates at 6250 rpm. (a) Find the distance from *A* of the impact points of molecules traveling at $\bar{v}$, v_{rms}, and v_{mp}. (b) Why do you suppose that measurements were originally made with Bi_2 instead of O_2 or N_2?

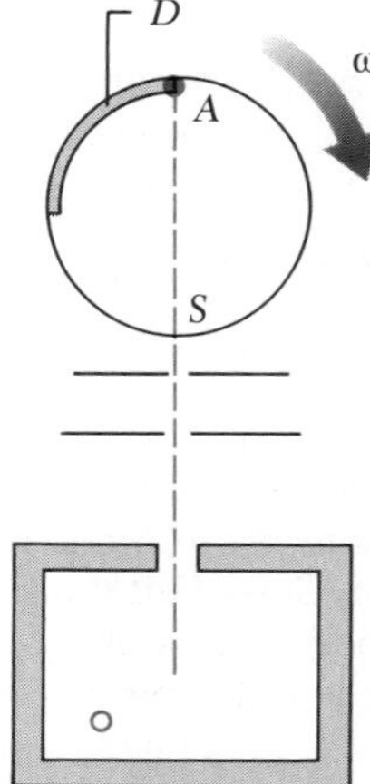

Figure 9.11 (Problem 6) A schematic drawing of an apparatus used to verify the Maxwell velocity distribution.

7. Energy levels known as tunneling levels have been observed from CN^- ions incorporated into KCl crys-

tals. These levels arise from the rotational motion of CN^- ions as they tunnel, at low temperatures, through barriers separating crystalline potential minima. According to one model, the tunneling levels should consist of four equally spaced levels in the far IR with spacing of 12.41×10^{-5} eV (1 cm^{-1}). Assuming that the CN^- ions obey Maxwell-Boltzmann statistics and using Figure 9.12, calculate and sketch the expected appearance of the absorption spectra (strength of absorption vs. energy) at 4 K and 1 K. Assume equal transition probabilities for all allowed transitions, and make the strength of absorption proportional to the peak height. Use $k = 8.62 \times 10^{-5}$ eV/K.

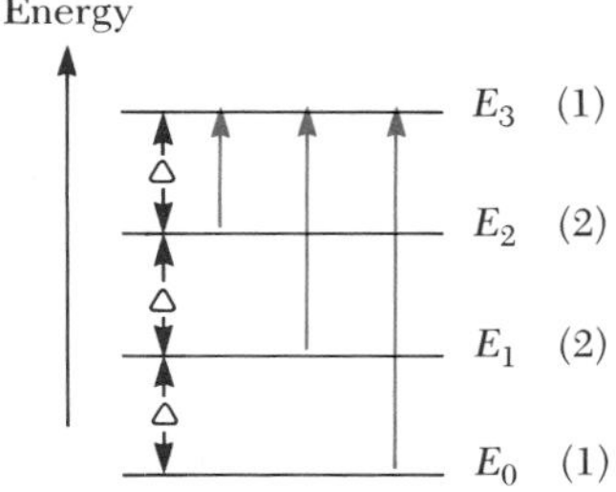

Figure 9.12 (Problem 7) Tunneling energy levels. Allowed transitions are indicated by vertical arrows. The degeneracy of each level is indicated in parentheses.

8. Fit an exponential curve $n = Ae^{-BE}$ to Figure 9.3 to see how closely the system of six distinguishable particles comes to an exponential distribution. Use the values at energies of 0 and 1E to determine A and B.
9. Helium atoms have spin 0 and so are bosons. (a) Must we use the Bose-Einstein distribution at standard temperature and pressure to describe helium gas, or will the Maxwell-Boltzmann distribution suffice? (b) Helium becomes a liquid with a density of 0.145 g/cm³ at 4.2 K and atmospheric pressure. Must the Bose-Einstein distribution be used in this case? Explain.
10. The energy difference between the first excited state of mercury and the ground state is 4.86 eV. (a) If a sample of mercury vaporized in a flame contains 10^{20} atoms in thermal equilibrium at 1600 K, calculate the number of atoms in the $n = 1$ (ground) and $n = 2$ (first excited) states. (Assume the Maxwell-Boltzmann distribution applies and that the $n = 1$ and $n = 2$ states have equal statistical weights.) (b) If the mean lifetime of the $n = 2$ state is τ seconds, the transition probability is $1/\tau$ and the number of photons emitted per second by the $n = 2$ state is n_2/τ, where n_2 is the number of atoms in state 2. If the lifetime of the $n = 2$ state is 100 ns, calculate the power (in watts) emitted by the hot atoms.
11. (a) Find the average energy per photon for photons in thermal equilibrium with a cavity at temperature T. (b) Calculate the average photon energy in electron volts at $T = 6000$ K. *Hint:* Two useful integrals are

$$\int_0^\infty \frac{z^2dz}{e^z - 1} \approx 2.40 \qquad \text{and} \qquad \int_0^\infty \frac{z^3dz}{e^z - 1} = \frac{\pi^4}{15}$$

12. Show that the average kinetic energy of a conduction electron in a metal at 0 K is given by $E = 3E_F/5$. By way of contrast, note that *all* of the molecules in an ideal gas at 0 K have *zero energy! Hint:* Use the standard definition of an average given by

$$\overline{E} = \frac{\int_0^\infty E\, g(E) f_{FD}(E)\, dE}{N/V}$$

13. Although we usually apply Fermi-Dirac statistics to free electrons in a conductor, Fermi-Dirac statistics apply to any system of spin $\frac{1}{2}$ particles, including protons and neutrons in a nucleus. Since protons are distinguishable from neutrons, assume that each set of nucleons independently obeys the Fermi-Dirac distribution and that the number of protons equals the number of neutrons. Using these ideas, estimate E_F and $\overline{E}$ for the nucleons in Zn. (Zn has 30 protons, 34 neutrons, and a radius of 4.8×10^{-15} m.) Are your answers reasonable? Explain.
14. Show that Equation 9.31 can be expressed as

$$E_F = (3.65 \times 10^{-19}) n^{2/3} \text{ eV}$$

where E_F is in eV when n is in electrons/m³.
15. Calculate the probability that a conduction electron in copper at 300 K has an energy equal to 99% of the Fermi energy.
16. Find the probability that a conduction electron in a metal has an energy equal to the Fermi energy at the temperature 300 K.
17. Sodium is a monovalent metal having a density of 0.971 g/cm³ and an atomic weight of 23.0 g/mole. Use this information to calculate (a) the density of charge carriers, (b) the Fermi energy, and (c) the Fermi velocity for sodium.
18. Calculate the energy of a conduction electron in silver at 800 K if the probability of finding the electron in that state is 0.95. Assume that the Fermi energy for silver is 5.48 eV at this temperature.
19. Consider a cube of gold 1 mm on an edge. Calculate the approximate number of conduction electrons in this cube whose energies lie in the range from 4.000 to 4.025 eV.
20. (a) Consider a system of electrons confined to a three-dimensional box. Calculate the ratio of the number of allowed energy levels at 8.5 eV to the number of allowed energy levels at 7.0 eV. (b) Copper has a Fermi energy of 7.0 eV at 300 K. Calculate the ratio of the number of occupied levels at an energy of 8.5 eV to the number of occupied levels at the Fermi energy. Compare your answer with that obtained in part (a).

CALCULATOR/COMPUTER PROBLEMS

21. Consider a system of 10^4 oxygen molecules/cm^3 at a temperature T. Write a program that will enable you to calculate and plot a graph of the Maxwell distribution $n(v)$ as a function of v and T. Use your program to evaluate $n(v)$ for speeds ranging from $v = 0$ to $v = 2000$ m/s (in intervals of 100 m/sec) at temperatures of (a) 300 K and (b) 1000 K. (c) Make graphs of N_v versus v and use the graph at $T = 1000$ K to estimate the number of molecules/cm^3 having speeds between 800 m/s and 1000 m/s at $T = 1000$ K. (d) Calculate and indicate on each graph the rms speed, the average speed, and the most probable speed (see Problem 5).

22. Write a program that will enable you to calculate the Fermi-Dirac distribution function, Equation 9.18, versus energy. Use your program to plot $f(E)$ versus E for (a) $T = 0.2T_F$, and (b) $T = 0.5T_F$, where T_F is the Fermi temperature, defined by Equation 9.32.

23. Copper has a Fermi energy of 7.05 eV at 300 K and a conduction electron concentration of about 8.49×10^{28} m^{-3}. Write programs that will enable you to plot (a) the density of states, $g(E)$, versus E, (b) the particle distribution function, $N(E)$, as a function of energy at $T = 0$ K, and (c) the particle distribution function versus E at $T = 1000$ K. Your energy scales should range from $E = 0$ to $E = 10$ eV.

10

Lasers

10.1 INTRODUCTION

If one can excuse a small pun, we may say that the laser is one of the "flashiest" inventions of 20th century physics. The laser has captured the public interest as much for its theatrical impact as a futuristic weapon in movies as for its myriad uses in medicine, machining, photography, surveying and ranging, isotope separation, communications, and computing. The laser is important not only for its commercial potential, but also because it provides an excellent and interesting example of many of the principles of atoms and radiation in equilibrium. These include blackbody radiation, spontaneous and stimulated emission, thermal equilibrium populations of energy levels, population inversions, and excited state lifetimes. In this chapter we shall develop the theoretical principles of atoms and radiation in thermal equilibrium and the so-called "start-oscillation" condition for a lasing medium. We will then show how these general principles are implemented in actual pulsed and continuous lasers including the ruby crystal laser, gas lasers, and semiconductor lasers.

10.2 ABSORPTION, SPONTANEOUS EMISSION, AND STIMULATED EMISSION

In order to understand the operation of a laser, we must first become familiar with the processes describing the emission and absorption of radiation by atoms. Although it would be stretching the truth to say that Einstein anticipated the laser, this remarkable genius laid down the foundations for the device by his explanation of the processes of atomic absorption, spontaneous emission, and stimulated emission in 1917.[1] Einstein showed that an atom can absorb a photon of energy hf from a radiation field and make a transition from a lower energy state E_1 to a higher state E_2, where $E_2 - E_1 = hf$. Let us set the probability of absorption per unit time per atom equal to $B_{12}u(f)$, where B_{12} is Einstein's coefficient of absorption and $u(f)$ is the energy density per unit frequency in the radiation field. The probability of absorption of radiation is thus assumed to be proportional to the density of radiation. Once excited, an atom in state 2 is observed to have a definite probability per unit time of making a *spontaneous transition* (i.e., one not induced by radiation) back to level 1. The spontaneous transition rate may be characterized by the coefficient A_{21} (transitions/unit time), or by the spontaneous emission lifetime, $t_s = 1/A_{21}$ (time/transition).

Spontaneous transition

In addition to spontaneous emission, which is independent of the radiation density, another kind of emission can occur that is *dependent* on the radiation density. If a photon of energy hf interacts with an atom when it is in

[1] A. Einstein, "Zur Quanten Theorie der Strahlung" (On the Quantum Theory of Radiation), *Phys. Zeit.*, **18:** 121–128, 1917.

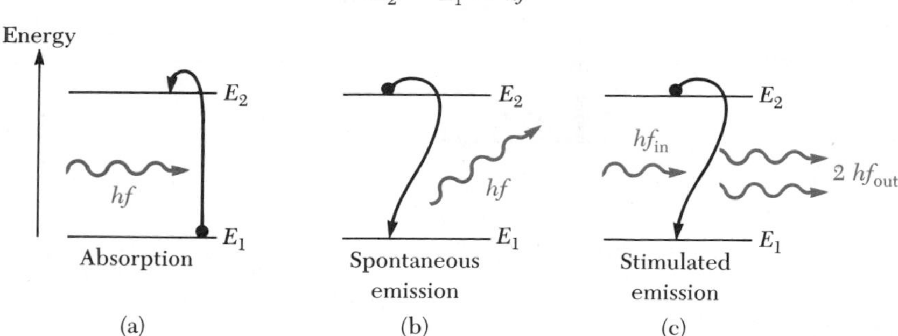

Figure 10.1 The processes of (a) absorption and (b) spontaneous emission. The lifetime of the upper state is t_s and the photon is emitted in a random direction. (c) Stimulated emission. In this process, the emitted photons are in phase with the stimulating photon and all have the same direction of travel.

level 2, the electric field associated with this photon can stimulate or induce atomic emission such that the emitted electromagnetic wave (photon) vibrates in phase with the stimulating wave (photon) and travels in the same direction. Two such photons are said to be *coherent*. We may set the stimulated transition rate per atom equal to $B_{21}u(f)$, where B_{21} is Einstein's coefficient of induced emission. The three processes of absorption, spontaneous emission, and stimulated emission are summarized in Figure 10.1.[2]

Next, following Einstein, consider a mixture of atoms and radiation in thermal equilibrium at temperature T. The populations N_1 and N_2 of the energy levels E_1 and E_2 obey the Boltzmann relation (discussed in Chapter 9).

$$\frac{N_2}{N_1} = e^{-(E_2-E_1)/kT} = e^{-hf/kT} \tag{10.1}$$

This equilibrium ratio of populations represents a case of *dynamic equilibrium* in which the number of atomic transitions per unit time up (from E_1 to E_2) is balanced by the number of spontaneous *and* stimulated transitions per unit time down (from E_2 to E_1). Since the number of atoms making transitions is proportional to the population of the starting level as well as to the transition probability, we may write:

The number of atoms going from $1 \rightarrow 2$ per unit time $= N_1u(f, T)B_{12}$

where $u(f,T)$ is the radiation energy density per unit frequency ($\mathrm{J/m^3 \cdot Hz}$).

The number of atoms going from $2 \rightarrow 1$ per unit time by stimulated emission $= N_2u(f, T)B_{21}$

For spontaneous emission there is no dependence on the radiation field and we merely have:

The number of atoms going from $2 \rightarrow 1$ per unit time spontaneously $= N_2A_{21}$

[2] Although spontaneous emission has no classical analog and is a truly quantum effect (arising from quantization of the radiation field), absorption and stimulated emission do have classical analogs. If the atom-photon interaction is modeled as a charged harmonic oscillator driven by an electromagnetic wave, the classical oscillator can either absorb energy from the wave or give up energy, depending on the relative phases of oscillator and wave.

Since the number of upward transitions per unit time must equal the number of downward transitions per unit time, we have

$$N_1 u(f, T)B_{12} = N_2[B_{21}u(f, T) + A_{21}] \tag{10.2}$$

Using Equation 10.1, this becomes

$$u(f, T) = \frac{A_{21}}{B_{12}e^{hf/kT} - B_{21}} \tag{10.3}$$

Equation 10.3 is an expression for the radiation density per unit frequency of a system of atoms and radiation in equilibrium at temperature T expressed in terms of Einstein coefficients A_{21}, B_{12}, and B_{21}. But Planck's blackbody radiation law (Chapter 2) provides us with an expression for a system of radiation and oscillators in thermal equilibrium:

$$u(f, T) = \frac{8\pi hf^3}{c^3}\frac{1}{(e^{hf/kT} - 1)} \tag{2.10}$$

Comparing Equations 10.3 and 2.10, we find the interesting results

Relation of Einstein coefficients of stimulated emission, absorption, and spontaneous emission

$$\boxed{B_{21} = B_{12} = B} \tag{10.4}$$

and

$$\boxed{\frac{A_{21}}{B} = \frac{8\pi hf^3}{c^3}} \tag{10.5}$$

Equation 10.4 states that any atomic system that has a finite probability per unit time of absorption has an *equal* probability of stimulated emission. Equation 10.5 shows that the process of spontaneous emission dominates the process of stimulated emission at higher frequency. We may use Equation 10.5 to obtain an expression for the induced transition rate, W_i (probability of absorption or emission per unit time per atom), in terms of the measured spontaneous emission lifetime, t_s. Thus

$$W_i = u(f, T)B \tag{10.6}$$

Solving Equation 10.5 for B and substituting in Equation 10.6 yields

$$W_i = \frac{u(f, T)c^3}{8\pi hf^3}A$$

Since $A = 1/t_s$, we find

$$W_i = \frac{u(f, T)c^3}{8\pi hf^3 t_s} \tag{10.7}$$

A more realistic expression than Equation 10.7 takes into account the broadening of the $(2 \rightarrow 1)$ spectral line produced by the finite widths of energy levels 1 and 2. This expression involves the introduction of an atomic line shape function, $g(f)$.[3] Using this concept we may rewrite the expression for

[3] For example, if the atomic line has a Lorentzian shape with a full width at half maximum of Δf centered on f_c, then $g(f)$ has the explicit form

$$g(f) = \frac{\Delta f}{2\pi\left[(f - f_c)^2 + \left(\frac{\Delta f}{2}\right)^2\right]}$$

the induced transition rate in terms of f, t_s, $g(f)$, and the intensity of the radiation field, $I(f)$ (in W/m^2), as

Probability of induced absorption or emission per unit time per atom

$$W_i(f) = \frac{c^2}{8\pi h f^3 t_s} g(f)I(f) \tag{10.8}$$

where $cI(f) = u(f, T)$. This expression for the induced transition rate at the frequency f will prove useful in the derivation of the "start-oscillation" condition for a real lasing medium.

10.3 POPULATION INVERSION AND LASER ACTION

Equation 10.8 shows that there is an equal probability that a photon will cause an upward (absorption) or downward (stimulated emission) atomic transition. When light is incident on a system of atoms in thermal equilibrium, there is usually a net absorption of energy since, according to the Boltzmann distribution, there are more atoms in the ground state than in excited states. However, if one can invert the situation so that there are more atoms in an excited state than in a lower state, a condition called *population inversion* occurs and an amplification of photons can result. Under the proper conditions a single input photon can result in a cascade of stimulated photons, all of which are in phase, traveling in the same direction, and of the same frequency as the input photon. Since the device that does this is a light amplifier, it is called a **laser,** an acronym for light amplification by stimulated emission of radiation.

Although there are many different types of lasers, all lasers have certain essential features. These are:

1. An energy source capable of producing either pulsed or continuous population inversions. In the case of the helium-neon gas laser, the energy source is an electrical discharge that imparts energy by atomic collisions. In the case of crystal lasers, the population inversion is produced by intense flashes of broad-band illumination. The process of excitation with intense illumination is called *optical pumping.*
2. A lasing medium with at least *three* energy levels consisting of a ground state, an intermediate (metastable) state with a relatively long lifetime, t_s, and a high energy pump state (Fig. 10.2). In order to obtain population inversion, t_s must be greater than t_1, the lifetime of the lower or terminal level. Note, too, that amplification cannot be obtained with only two levels because such a system cannot support a population inversion. At most, with extremely intense optical pumping we can increase the population of the

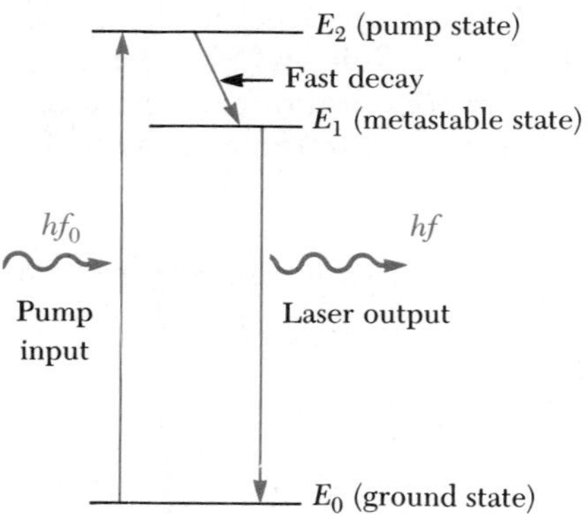

Figure 10.2 Three-level laser system.

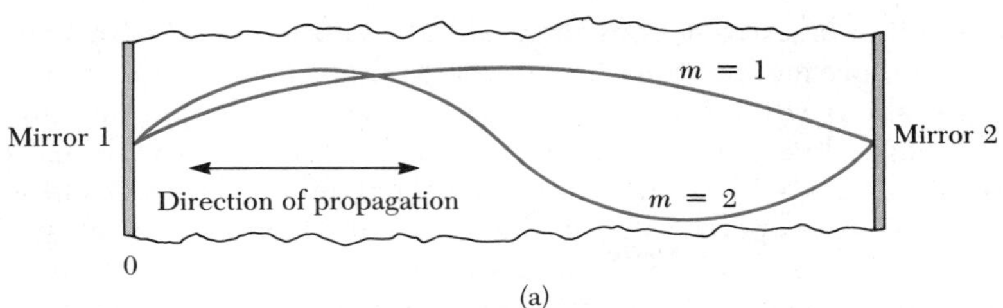

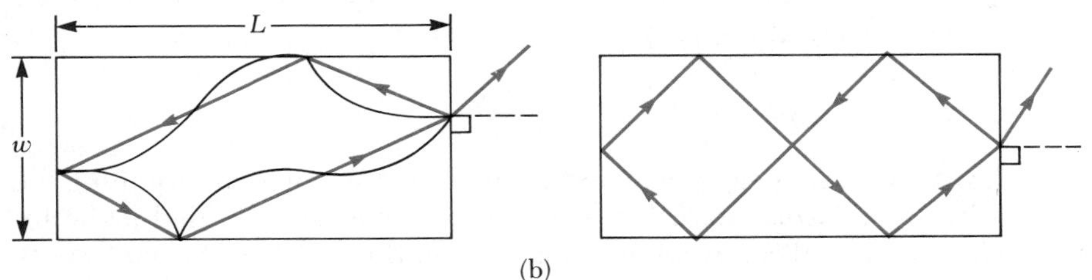

Figure 10.3 Cavity modes in a laser. (a) Longitudinal modes have $L = m\lambda/2$, where m is an integer, λ is the laser wavelength in the laser material, and L is the distance between the two mirrors. (b) Modes with transverse components can exist if both ends and sides of the laser are made reflective. Arrows show direction of propagation of light rays.

upper state until it equals the population of the lower state. Further pumping only serves to excite as many downward transitions as upward transitions, since the probability of absorption is equal to the probability of stimulated emission for a *given* transition. In order to produce a population inversion, energy absorption must occur for a transition *different* from the transition undergoing stimulated emission—thus the need for at least a three-level system.

3. A method of containing the initially emitted photons within the laser so that they can stimulate further emission from other excited atoms. In practice this is usually achieved by placing mirrors at the ends of the lasing medium so that photons make multiple passes through the laser. Thus the laser may be thought of as an optical resonator or oscillator with two opposing reflectors at right angles to the laser beam. The oscillation consists of a plane wave bouncing back and forth between the reflectors. The oppositely traveling plane waves, in turn, generate a highly monochromatic standing wave, which is strongest at resonance when an integral number of half wavelengths just fit between the reflectors (Fig. 10.3a). In order to extract a highly collimated beam from the laser, one of the parallel mirrors is made slightly transmitting so that a small amount of energy leaks out of the cavity.

These three features of lasers lead to several unique characteristics of laser light that make it a much more powerful technological tool than light from ordinary sources. These unique characteristics are high monochromaticity, strong intensity, coherence, and low beam divergence. Typical values of these quantities for a low power laser are shown in Table 10.1. Compared to a

TABLE 10.1 Typical He-Ne Laser Characteristics

Monochromaticity	Beam Power	Beam Diameter at Laser Exit	Beam Divergence
632.81 ± 0.002 nm	10 mW	0.50 mm	1.0 mrad

Irradiance at Laser Exit	Irradiance for a Beam Focused to a 10 μm Diameter
5100 mW/cm^2	1.27×10^7 mW/cm^2

single emission line in a strong conventional light source, the He-Ne laser is about 100 times more monochromatic, generates about 100 times more beam power, and has about 1000 times more exit irradiance (power per unit area). Note that the beam divergence or spread is measured in milliradians (mrad). Thus a 1.0 mrad laser would have a beam diameter of only 1 cm at a distance of 10 m. These unusual properties of laser light have made possible such applications as surgical "welding" of detached retinas, precision surveying and length measurement, a potential source for inducing nuclear fusion reactions, precision cutting of metals and other materials, and telephone communication along optical fibers. These and other interesting applications of lasers are discussed in the essay at the end of this chapter.

Start Oscillation Condition

The model of the laser as an optical resonator is most useful in determining the *start-oscillation* or *lasing condition.* Laser action will be sustained if the increase in the number of coherent photons per pass through the resonant cavity is larger than the decrease caused by losses such as transmission at the ends, scattering from optical inhomogeneities, and absorption by unwanted species. Let us lump all photon loss mechanisms into a single time constant, t_p. This means that the intensity of a specific cavity vibration damps out exponentially according to

$$I = I_0 e^{-t/t_p} \tag{10.9}$$

or that the loss in intensity with time is given by

$$\left(\frac{dI}{dt}\right)_{\text{loss}} = -\frac{I}{t_p} \tag{10.10}$$

The net gain in intensity of a specific cavity vibration produced by stimulated transitions[4] is given as the sum of the gain by stimulated emission from level 2 and loss by stimulated absorption from level 1:

$$\left(\frac{dI}{dt}\right)_{\text{gain}} = (N_2 - N_1)\frac{hfc}{V}W_i \tag{10.11}$$

where V is the laser volume, hfc/V is the intensity contributed per photon, N_2 and N_1 are the populations of levels 2 and 1, and W_i is the stimulated transition rate per atom defined in Equation 10.8. If the degeneracy of level 1, g_1, is not equal to the degeneracy of level 2, g_2, then Equation 10.11 must be corrected as follows:

$$\left(\frac{dI}{dt}\right)_{\text{gain}} = \left[n_2 - n_1\left(\frac{g_2}{g_1}\right)\right] hfcW_i \tag{10.12}$$

where n_2 and n_1 are population densities (atoms/cm³) (see Problem 1). For sustained laser action, the gain in a particular coherent cavity mode must at least be equal to the loss. Thus we have

$$\left(\frac{dI}{dt}\right)_{\text{gain}} \geq \left|\left(\frac{dI}{dt}\right)_{\text{loss}}\right|$$

or

[4] The contribution of spontaneous emission to cavity intensity is omitted since we are only interested in coherent output.

$$\left[n_2 - n_1\left(\frac{g_2}{g_1}\right)\right] hfcW_i \geq \frac{I}{t_p} \tag{10.13}$$

Substituting $W_i = c^2 g(f)I/8\pi hf^3 t_s$ into Equation 10.13 gives the start-oscillation condition as

$$n_2 - n_1\left(\frac{g_2}{g_1}\right) = \frac{8\pi f^2 t_s}{c^3 g(f) t_p} \tag{10.14}$$

As $n_2 - n_1(g_2/g_1)$ has units of atoms/cm^3 and Equation 10.14 has been derived for the threshold of laser action, $n_2 - n_1(g_2/g_1)$ is called the **critical population inversion density** and is denoted as $\Delta N_c/V$. Thus we have

Threshold for laser action, the critical population inversion density

$$\frac{\Delta N_c}{V} = \frac{8\pi f^2 t_s}{c^3 g(f) t_p} \tag{10.15}$$

For a Lorentzian line shape $g(f)$ with a full width at half maximum of Δf centered on f_c we have

$$g(f_c) = \frac{2}{\pi \Delta f} \tag{10.16}$$

Substituting this result into Equation 10.15, we find the critical population density for laser oscillation near the line center to be

$$\boxed{\frac{\Delta N_c}{V} \approx \frac{4\pi^2 f^2 \Delta f}{c^3}\left(\frac{t_s}{t_p}\right)} \tag{10.17}$$

Equation 10.17 is particularly convenient because it gives the required population inversion that must be established in order to have laser action with a given transition of known frequency, width, and lifetime. Equation 10.17 also shows that the only optical cavity property that $\Delta N_c/V$ depends on is the decay time, t_p. Note that c in Equation 10.17 is the speed of light in the laser material.

10.4 PUMPING POWER REQUIREMENTS FOR THREE- AND FOUR-LEVEL LASERS

Perhaps the greatest utility of Equation 10.17 is that it enables one to calculate the minimum pumping power that must be supplied to maintain the critical inversion and produce laser action. Too large a pumping power requirement can easily rule out a given material as a laser medium, particularly since the actual input power is generally greater than the minimum pumping power. The actual input power is greater than the minimum pumping power because energy is supplied by a flash lamp over a broad frequency range and at a higher frequency than the laser output frequency.

The smallest pumping power is required for the broad category of laser materials known as four-level systems. Such systems consist of four energy levels with a separation of the ground state, E_0, and the terminal laser level, E_1, much greater than kT (Fig. 10.4a). The operation of a four-level laser involves excitation to highly unstable levels labeled E_3, which decay almost instantaneously to E_2, the excess energy $E_3 - E_2$ being given off as heat. The level E_2 has a long lifetime, on the order of milliseconds. Because E_1 has a shorter lifetime and essentially zero population (the probability of thermally exciting atoms from E_0 to E_1 is low since $E_1 - E_0 \gg kT$), population inversion of E_2 over E_1 occurs with very little pumping. Quantitatively we have $N_1 \approx 0$ and

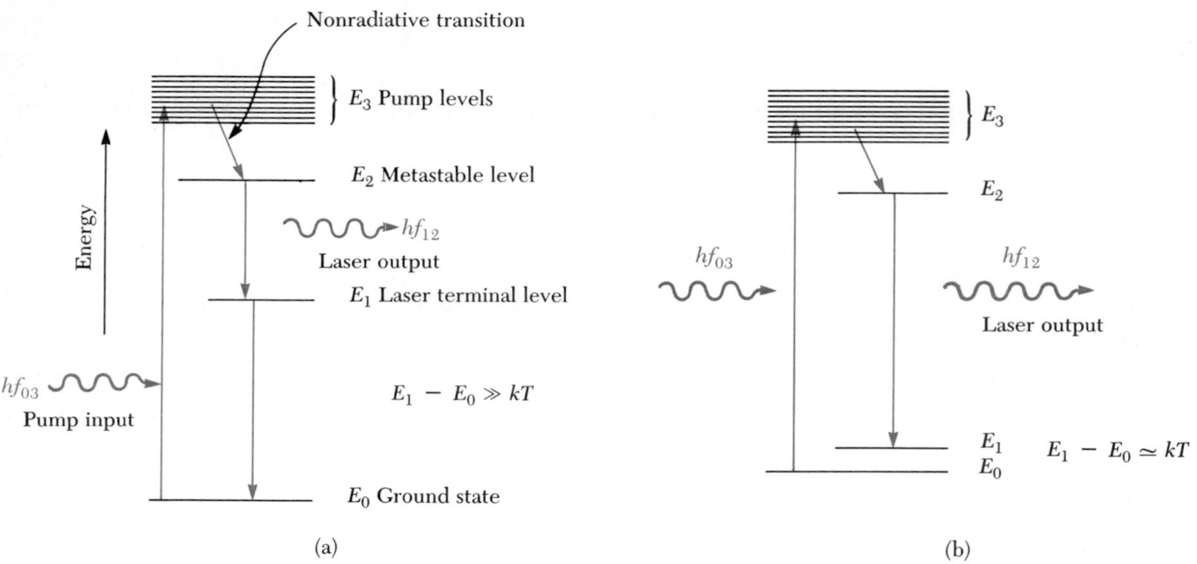

Figure 10.4 (a) A four-level laser system. (b) A three-level laser system.

$N_2 \approx \Delta N_c$ on the threshold of laser action. To maintain $\Delta N_c/V$ atoms per unit volume in level 2 requires a minimum[5] power per unit volume of

$$P = \frac{\Delta N_c}{V}\left(\frac{hf}{t_s}\right) \tag{10.18}$$

Substituting $\frac{\Delta N_c}{V} = \frac{4\pi^2 f^2 \Delta f}{c^3}\left(\frac{t_s}{t_p}\right)$ from Equation 10.17, we have

Minimum pumping power for a four-level laser

$$\boxed{P = \frac{4\pi^2 h f^3 \Delta f}{c^3 t_p}} \tag{10.19}$$

for the minimum power per unit volume that must be supplied to produce laser action in a *four-level system.*

For a three-level laser (Fig. 10.4b), the minimum pumping power is several orders of magnitude greater. For a three-level system in thermal equilibrium, practically all of the optically active atoms in the laser, N, are in level E_1 so that

$$N_1 \approx N \gg \Delta N_c$$

In order to achieve a population inversion of E_2 with respect to E_1, about half the active atoms, $N/2$, must be pumped to E_2. Comparing three- and four-level lasers with equal spontaneous lifetimes and energies, we find for the ratio of minimum pumping powers per unit volume

Comparison of minimum pumping powers for three- and four-level lasers

$$\frac{P_{\text{3-level}}}{P_{\text{4-level}}} \approx \frac{(N/2)hf/t_s V}{\Delta N_c hf/t_s V} = \frac{N/2}{\Delta N_c} \tag{10.20}$$

One more comment concerning the equation $P = \Delta N_c hf/Vt_s$ seems in order. This equation may be interpreted to mean that the pumping source must supply an amount of power equal to that radiated spontaneously in order

[5] Generally more power is required, because the pump levels are considerably above E_2.

to maintain the critical inversion and produce laser action. Since the probability of spontaneous emission increases as (frequency)^3 (Eq. 10.5), the operation of lasers at high frequencies in the vacuum ultraviolet and x-ray regions requires enormous pumping powers and becomes very difficult. Using the relation $P = \Delta N_c hf/Vt_s$ and $t_s \sim 1/f^3$, we find that the *power emitted spontaneously* increases rapidly as f^4.

EXAMPLE 10.1 Lasers at Different Wavelengths
If a laser operating at 1 μm in the IR region emits several milliwatts of spontaneous radiation, calculate the expected spontaneous emission and minimum input power for lasers operating at 1000 Å (vacuum ultraviolet) and 100 Å (soft x-rays).

Solution:

$$P_{\text{spontaneous}} \approx P_{\text{pump}} \propto f^4$$

Thus

$$\frac{P_{\text{spon}}(f_1)}{P_{\text{spon}}(f_2)} = \frac{f_1^4}{f_2^4} = \frac{\lambda_2^4}{\lambda_1^4}$$

$$\frac{P_{\text{spon}}(1000\ \text{Å})}{10^{-3}\ \text{W}} = \frac{(10\,000\ \text{Å})^4}{(1000\ \text{Å})^4} = 10^4$$

or $P_{\text{spon}}(1000\ \text{Å}) = 10$ W.

Strong commercial lamps can produce several hundred milliwatts in selected portions of the visible spectrum. Thus 10 watts of pumping light in the range of 1000 Å is probably difficult but not impossible to obtain. However, when we calculate $P_{\text{spon}}(100\ \text{Å})$ we find a value of 100 000 watts. The supply of this much power in a narrow spectral range necessitates the use of catastrophic pumping, which has apparently actually been achieved with nuclear bomb pumped lasers.[6]

[6] C. A. Robinson, *Aviation Week and Space Technology*, Feb. 23, 1981, p. 25.

10.5 RUBY LASER

Although the first device to produce coherent radiation from stimulated emission (a device called a maser) operated in the microwave region,[7] the first device to produce coherent optical radiation was the ruby laser.[8] A typical experimental set up for operating a pulsed ruby laser at 77 K is shown in Figure 10.5. Ruby consists of a sapphire crystal (Al_2O_3) doped with about 0.05 atomic percent optically active chromium ions ($\sim 10^{19}$ Cr^{3+} ions/cm^3). Ruby crystals are particularly suitable for forming lasers because pure single crystals several inches long with flat polished ends can be easily fabricated. Single crystals are desirable because factors that shorten t_p and increase the minimum pumping power (according to Equation 10.19) may be minimized or controlled in pure single crystals. Two main factors are absorption at f_{12} by atoms other than Cr and scattering by optical inhomogeneities. In addition, ruby meets the requirement of a small atomic linewidth Δf for low pumping power (again see Equation 10.19), especially when cooled to 77 K where the linewidth[9] is only 0.1 cm^{-1}. Finally, chromium in this crystal has appropriate pumping levels in the visible for flash lamp excitation and a metastable state that emits red light (694.3 nm) in a transition to the ground state, as shown in

[7] J. P. Gordon, H. J. Zeiger, and C. H. Townes, "Molecular Microwave Oscillator and New Hyperfine Structure in the Microwave Spectrum of NH_3," *Phys. Rev.* **95:** 282L, 1954.

[8] T. H. Maiman, "Stimulated Optical Radiation in Ruby," *Nature*, **187:** 493, 1960.

[9] The unit cm^{-1} is commonly used in spectroscopy and is a unit of reciprocal wavelength. As $1/\lambda = f/c$, a measurement in cm^{-1} is proportional to frequency or to energy. To convert a measurement in cm^{-1} to frequency in Hz or energy in eV, multiply the measurement by the speed of light, $c = 300 \times 10^{10}$ cm/s or $hc = 1.241 \times 10^{-4}$ eV·cm, respectively.

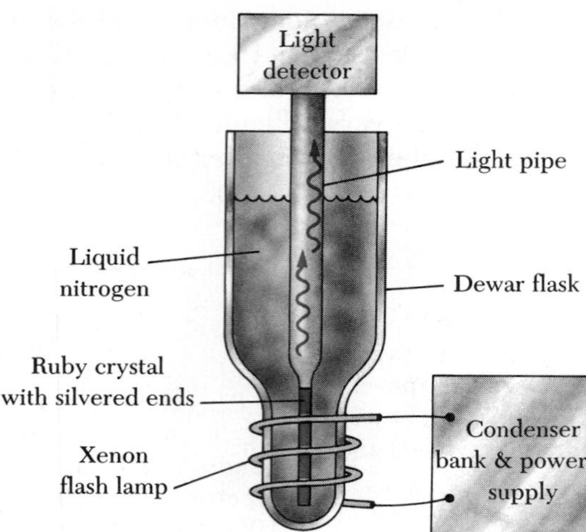

Figure 10.5 Experimental setup for a pulsed ruby laser.

Figure 10.6. Note that ruby is a three-level laser with $^4A_2$, the ground state, acting as the terminal laser level and the level designated ^2E acting as the upper laser level. (The names for the energy levels $^4A_2$ and ^2E are related to the symmetries of the crystal field experienced by the chromium ions, but details of the nomenclature are unimportant for our purposes.)

As noted previously, population inversion is achieved by optical pumping of the ruby rod with a xenon- or mercury-filled flash lamp wrapped around the rod for intimate contact. A very intense pulse of broadband light lasting several milliseconds is obtained by discharging a capacitor through the flash lamp. The flash lamp intensity as a function of *time* is shown dashed in Figure 10.7. Note in this figure that the laser does not operate continuously during the flash lamp pulse but instead emits a series of brief laser spikes, each lasting for several hundred nanoseconds. This occurs because once laser action sets in, the upper level is depopulated at a rate that is fast compared to the pumping rate. Consequently, several tenths of milliseconds are required for the flash lamp to restore population inversion and create the next laser spike.

A laser that allows all of the energy stored in a single flash lamp pulse to be emitted in a single giant laser spike is called a *Q-switched* laser. The power in a

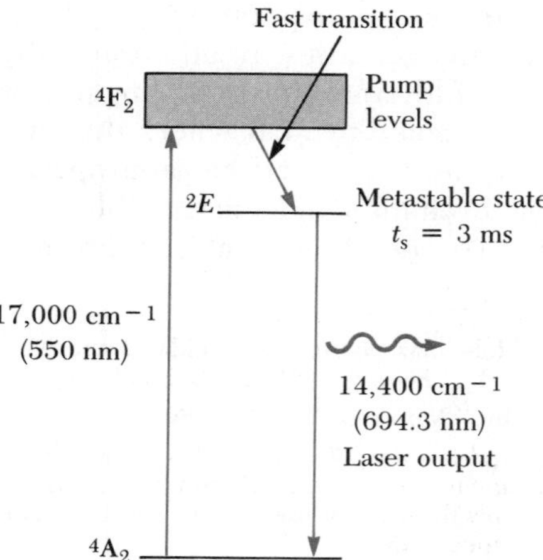

Figure 10.6 Energy levels of chromium in ruby, Al_2O_3.

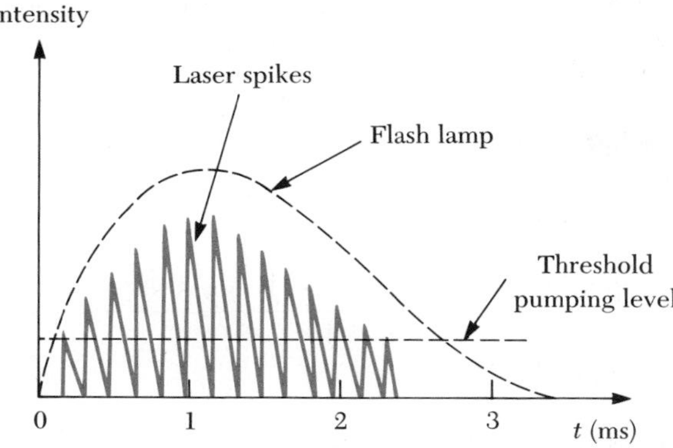

Figure 10.7 Output from a pulsed ruby laser.

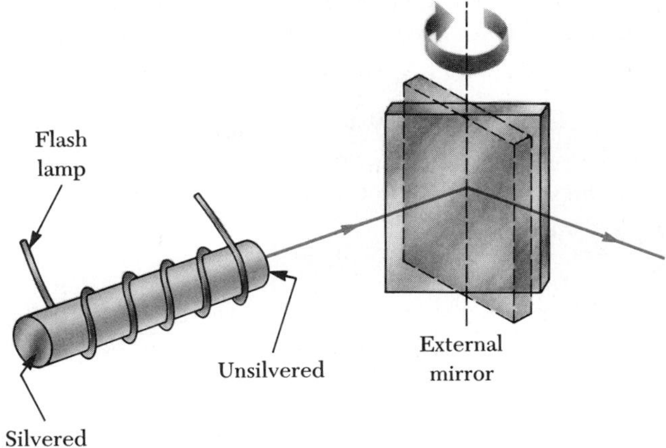

Figure 10.8 A *Q*-switched laser.

laser pulse from such a device is truly phenomenal. For example, a flash lamp operating over several milliseconds can deliver 1 J to the metastable state. If all of this energy is released in a single laser pulse lasting 100 ns, the laser pulse power is (1 J)/(10^{-7} s) = 10 MW! To attain a giant pulse one must suppress stimulated emission until most of the flash lamp energy has been absorbed and a large population inversion has built up. This is achieved by spoiling the resonant nature of the optical cavity until a highly inverted population has built up. In practice, this is accomplished by replacing one of the silvered ends of the ruby rod with a rotating or vibrating mirror, which is phased with the flash lamp pulses so that it is parallel to the other reflector only at the end of the pumping flash (Fig. 10.8). In this way there is no optical resonant cavity and no buildup of photons by multiple reflection until a highly inverted population has been established. The process is called *Q-switching* because *Q*, the quality factor, is a measure of the photon loss from the optical resonant cavity.

Q-switching

EXAMPLE 10.2 Estimation of the Critical Population Inversion and Minimum Pumping Power for the Ruby Laser

Calculate the critical inversion for ruby at room temperature (290 K), using the following data:

$\Delta f = 11 \text{ cm}^{-1}$

$t_s = 3 \times 10^{-3}$ s

$f = 14\,400 \text{ cm}^{-1}$

n = refractive index of ruby = 1.76

E_1 state = $^4A_2 = 0.0 \text{ cm}^{-1}$

E_2 state = $^2E = 14\,400 \text{ cm}^{-1}$

N = chromium ion concentration = 2×10^{19} Cr^{3+}/cm^3

L = ruby length between silvered ends = 10 cm

α = loss of photons per pass from scattering, transmission through imperfectly reflecting mirrors, etc. = 0.02 (2%)

Solution: We wish to use

$$\frac{\Delta N_c}{V} = n_2 - n_1\left(\frac{g_2}{g_1}\right) \approx \frac{4\pi^2 f^2 \Delta f\, t_s}{c^3 t_p}$$

to calculate the inversion. As noted earlier, t_p is the time it takes for the photons passing back and forth between the mirrors to decay or leak away. If the loss per pass, α, is small it takes approximately $1/\alpha$ passes for all of the photons to leak away. Thus t_p is given approximately as

$$t_p = (\text{number of passes for all photons to leak away}) \cdot (\text{time per pass}) = \left(\frac{1}{\alpha}\right)\left(\frac{L}{v}\right)$$

where v = light velocity in the laser material = c/n. Substituting $c = 3 \times 10^{10}$ cm/s and $n = 1.76$ yields $v = 1.70 \times 10^{10}$ cm/s and

$$t_p = \left(\frac{1}{0.02}\right)\left(\frac{10 \text{ cm}}{1.7 \times 10^{10} \text{ cm/s}}\right) = 2.9 \times 10^{-8} \text{ s}$$

In order to obtain units of atoms/cm³ for $\Delta N_c/V$ we must convert f values in cm⁻¹ to Hz. Thus

$$\Delta f = (11\ \text{cm}^{-1})(3.0 \times 10^{10}\ \text{cm/s}) = 3.3 \times 10^{11}\ \text{s}^{-1}$$

$$f = (14\ 400\ \text{cm}^{-1})(3.0 \times 10^{10}\ \text{cm/s}) = 4.32 \times 10^{14}\ \text{s}^{-1}$$

Substituting in Equation 10.17, where we again take v to be the speed of light in the laser material, we have

$$\frac{\Delta N_c}{V} = \frac{4\pi^2 f^2 \Delta f t_s}{v^3 t_p}$$

$$= \frac{(4\pi^2)(4.3 \times 10^{14}\ \text{s}^{-1})^2(3.3 \times 10^{11}\ \text{s}^{-1})(3 \times 10^{-3}\ \text{s})}{(1.70 \times 10^{10}\ \text{cm/s})^3(2.9 \times 10^{-8}\ \text{s})}$$

$$= 5.1 \times 10^{16}\ \text{atoms/cm}^3$$

Hence the critical population *difference* between ^2E and $^4A_2$ levels that must be established for laser action is about 5×10^{16} atoms/cm³. In order to find the *minimum* pumping power for ruby, we note that about half of the chromium atoms must be maintained in the E level, since ruby is a three-level system. Thus the minimum pumping power that must be supplied by the flash lamp is

$$P_{\text{min}} \approx \left(\frac{N}{2}\right)\frac{hf}{t_s}$$

$$= \frac{(1.0 \times 10^{19})(6.63 \times 10^{-34}\ \text{J}\cdot\text{s})(4.3 \times 10^{14}\ \text{s}^{-1})}{3 \times 10^{-3}\ \text{s}}$$

$$\approx 900\ \text{W/cm}^3$$

This is a large amount of energy to supply, particularly when it is recognized that the actual input power to the laser is generally greater than the minimum pumping power.

10.6 RARE EARTH SOLID STATE LASERS

The most common four-level solid state lasers, which are more efficient than three-level systems, consist of rare earth ions incorporated substitutionally into various hosts. Such lasers are efficient, often have output wavelength in the IR, and are capable of supplying extremely high power pulses. For example, the neodymium glass laser (ND : glass) and the neodymium yttrium-aluminum garnet laser (Nd : YAG) have been used in laser fusion schemes.[10] The rare earth ions are particularly suitable for forming lasers because they possess strong visible absorption bands for pumping and sharp fluorescence bands for lasing, which are relatively insensitive to the crystalline electric field of the host crystal. This occurs because the electrons responsible for the optical transitions ($4f$) are shielded from the host's electric field by outer $5s$ and $5p$ shells. A partial listing of rare earth solid state lasers is given in Table 10.2. A more complete listing would contain many more host materials, including plastics and liquids.

[10] These schemes involve hitting microscopic glass beads containing a mixture of deuterium and tritium with multiple laser beams, heating the beads until the hydrogen isotopes undergo nuclear fusion.

TABLE 10.2 Some Rare Earth Solid State Lasers

Laser Material	Output Wavelength (nm)
Nd^{3+} : glass	1060
Nd^{3+} : $CaWO_4$	1063
Yb^{3+} : glass	1015
Gd^{3+} : glass	312.5
Sm^{2+} : CaF_2	708
Tm^{2+} : CaF_2	1116
Ho^{3+} : $CaWO_4$	2050
Er^{3+} : $CaWO_4$	1612
Tm^{3+} : $CaWO_4$	1911
Dy^{2+} : CaF_2	2360
Pr^{3+} : $CaWO_4$	1047

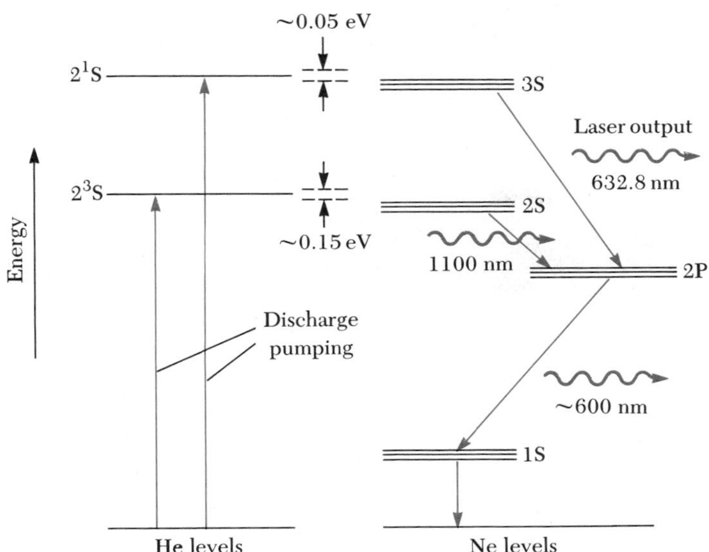

Figure 10.9 Energy level diagram for the He-Ne laser.

10.7 THE HELIUM-NEON GAS LASER AND HOLOGRAPHY

Since most of the light from a flash lamp is of the wrong wavelength to induce pumping, much of the input energy to a solid state laser is wasted and must be dissipated as heat. Thus solid state lasers are generally operated on a pulsed basis to allow the system to cool between pulses. An attractive alternative to pulsed lasers are gas lasers, which may be operated to give a steady and continuous output, called *c*ontinuous *w*ave or cw operation. To avoid the excess heat production accompanying optical pumping, atoms in gas lasers are excited directly by DC or radio frequency electric discharges. The discharge readily occurs in a low pressure gas, and it creates a glowing plasma of excited atoms and ions as electrons and ions collide with neutral atoms.

The most common gas laser is the helium-neon laser. It was also the first gas laser to be constructed.[11] This system employs helium to narrow the broad band of discharge energies into a few sets of neon levels so that preferential pumping of *particular* neon levels occurs. The process works in the following way. Energetic electrons in the discharge excite the more abundant helium atoms to a variety of states, which cascade quickly into two metastable helium states, 2^1S and 2^3S (Fig. 10.9). Helium atoms in these long-lived states then collide with unexcited neon atoms and preferentially excite them into 3*S* and 2*S* neon states, which have almost exactly the same energies as the helium 2^1S and 2^3S states. Thus there is a resonant energy exchange in the collision and a high probability that neon atoms will be pumped to 3*S* and 2*S* states. Decay from these neon states results in strongest laser emission at ~1100 nm (about 10 lines in the IR) and at 632.8 nm in the red part of the visible spectrum, as shown in Figure 10.9. It is interesting that the laser gain has been found to be inversely proportional to the He-Ne tube diameter. This occurs because the 1*S* neon atoms decay to the ground state by losing energy in collisions with the

[11] A. Javan, W. B. Bennet, Jr., and D. R. Herriott, "Population Inversion and Continuous Optical Laser Oscillation in a Gas Discharge Containing a He-Ne Mixture," *Phys. Rev. Lett.* **6**:106, 1961.

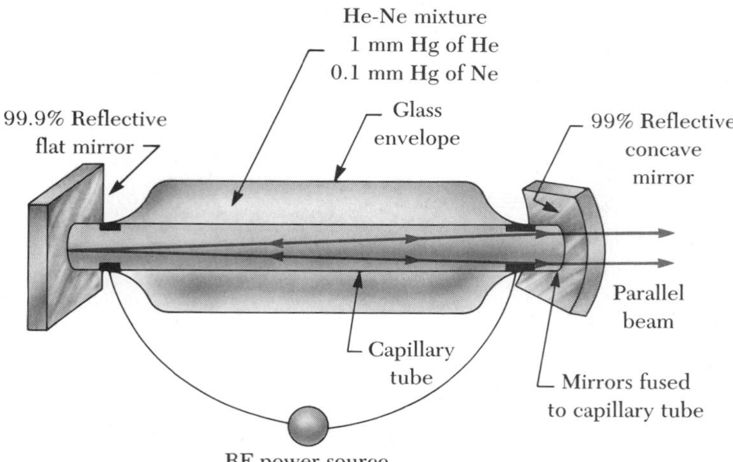

Figure 10.10 A typical He-Ne gas laser.

walls of the containing vessel, and such collisions are more probable the narrower the tube. If the tube has a wide diameter, the 1*S* population can increase to the point that the photons emitted by decaying 2*P* atoms (600 nm) have a high probability of re-exciting 1*S* atoms to the 2*P* state. This decreases the population inversion between 2*P* and 2*S* and between 2*P* and 3*S*, and thus lowers the laser gain.

A sketch of a typical He-Ne laser is shown in Figure 10.10. It consists basically of a glass tube about 0.5 m long, an excitation source for the gas, and end mirrors that are sealed onto the glass capillary tube to form an optical resonant cavity. The use of one flat mirror and one concave mirror allows easy optical alignment of the laser. The divergence in the beam that is created by the concave mirror is corrected by grinding the outer surface of this mirror to

Figure 10.11 A $\frac{3}{8}$-in titanium plate being cut with a CO_2 laser at 90 in/min. (Courtesy of Coherent Radiation.)

act as a converging lens. The ratio of He to Ne concentrations is indicated by the gas pressures shown in Figure 10.10. Direct excitation of neon by discharge electrons is possible, but this excitation does not preferentially populate the $2S$ and $3S$ neon states. In fact, direct discharge excitation of neon tends to destroy population inversion. Thus direct excitation of neon is avoided by making the neon density considerably less than the helium density.

The light output from a typical He-Ne laser is 5 to 10 mW, while the electrical input to the gas discharge is on the order of several watts, giving an efficiency of less than 1%. Several other gas systems are capable of providing much higher continuous light outputs at higher efficiencies. For example, very powerful CO_2 lasers operating in the IR region have been constructed; they can generate thousands of watts at efficiencies of about 10%. Such a laser is shown in operation in Figure 10.11.

EXAMPLE 10.3 A Comparison of Critical Inversion and Minimum Pumping Power in Gas and Solid State Lasers

Let us compare two four-level systems with almost identical emission energies. Consider first the solid state calcium tungstate-neodymium laser ($CaWO_4 : Nd^{3+}$). Typical parameters for this system are:

$$\lambda_0 = 1.06\ \mu\text{m}$$
$$L = 3\ \text{cm}$$
$$n = 1.9$$
$$t_p = \frac{L}{\alpha v} = 3.8 \times 10^{-9}\ \text{s} \qquad (\alpha = 5\%)$$
$$\Delta f = 7\ \text{cm}^{-1}$$
$$t_s = 10^{-4}\ \text{s}$$
$$N = \text{concentration of optically active } Nd^{+3} \text{ ions} = 10^{18}\ \text{cm}^{-3}$$

Solution: Using Equation 10.17 we find:

$$\frac{\Delta N_c}{V} = 4.5 \times 10^{15}\ Nd^{3+}\ \text{ions/cm}^3$$

Using $P = \frac{\Delta N_c}{V}\left(\frac{hf}{t_s}\right)$, we find for the minimum pumping power:

$$P = 10\ \text{W/cm}^3$$

We next calculate the critical population density and minimum pumping power for a He-Ne gas laser operating at 1.15 μm ($2S \rightarrow 2P$ transition), using the following typical parameters:

$$\lambda_0 = 1.15\ \mu\text{m}$$
$$L = 100\ \text{cm}$$
$$n = 1.0$$
$$t_p = \frac{L}{\alpha c} = 3.3 \times 10^{-7}\ \text{s} \ (\alpha = 1\%)$$
$$\Delta f = 9 \times 10^8\ \text{Hz}$$
$$t_s = 10^{-7}\ \text{s}$$

We find

$$\frac{\Delta N_c}{V} = 3.0 \times 10^7\ \text{atoms/cm}^3$$

and

$$P = \frac{\Delta N_c}{V}\left(\frac{hf}{t_s}\right) \approx 10^{-4}\ \text{W/cm}^3$$

The concentration of optically active neon atoms, N, may be found by assuming that neon obeys the ideal gas law and is admitted to the laser at 0.1 mm Hg pressure:

$$N = \frac{6.02 \times 10^{23}\ \text{atoms}}{22.4\ \text{liters}} \times \frac{0.1\ \text{mm}}{(760\ \text{mm})\left(10^3\ \frac{\text{cm}^3}{\text{liter}}\right)}$$
$$= 3.5 \times 10^{15}\ \text{atoms/cm}^3$$

Thus comparing solid and gas lasers at the same energy, we find that the critical population inversion, the minimum pumping power, and the concentration of optically active atoms are all several orders of magnitude higher in the solid. These figures allow us to understand why it is relatively easy to obtain continuous lasing with gas systems but not with solid lasers.

Holography

One of the most unusual and interesting applications of the gas laser is in the production of three-dimensional images of an object in a process called **holography.** Figure 10.12 shows how a hologram is made. Light from the laser is split into two parts by a half-silvered mirror at B. One part of the beam reflects

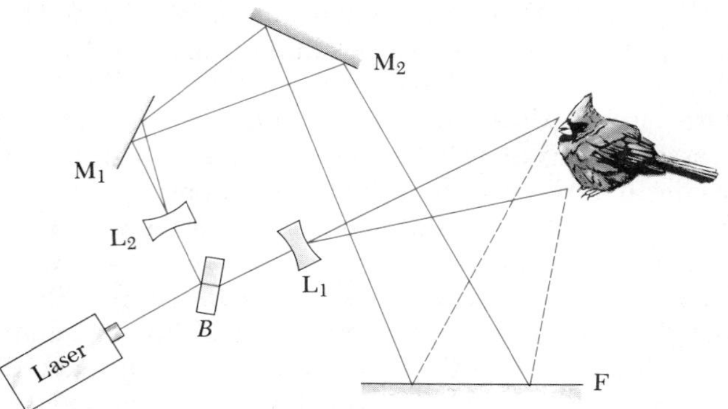

Figure 10.12 Experimental arrangement for producing a hologram.

off the object to be photographed and strikes an ordinary photographic film. The other half of the beam is diverged by lens L_2, reflects from mirrors M_1 and M_2, and finally strikes the film. The two beams overlap to form an extremely complicated interference pattern on the film. Such an interference pattern can be produced only if the phase relationship of the two waves is maintained constant throughout the exposure of the film. This condition is met if one uses light from a laser because such light is coherent (all of the photons in the beam have the same phase). The hologram records not only the intensity of the light scattered from the object (as in a conventional photograph), but also the phase difference between the reference beam and the beam scattered from the object. Because of this phase difference, an interference pattern is formed that produces an image with full three-dimensional perspective.

A hologram is best viewed by allowing coherent light to pass through the developed film as one looks back along the direction from which the beam comes. One sees a three-dimensional image of the object such that as the viewer's head is moved, the perspective changes, as for an actual object. The applications of holography promise to be many and varied. For example, someday your television set may be replaced by one using holography. Several other interesting applications of lasers are discussed in the essay at the end of this chapter.

10.8 SEMICONDUCTOR LASERS

One of the most important light sources for fiber optics communication is the semiconductor laser. These lasers are eminently suited to fiber optics communication because of their small size (maximum dimension several tenths of a millimeter), high efficiency (25 to 30%), simplicity, long lifetime ($p-n$ junctions have estimated working lifetimes of about 100 years), ease of modulation (by controlling the junction current), and fast response (extending well into the GHz range). A typical gallium arsenide (GaAs) $p-n$ junction laser is shown in Figure 10.13 along with nominal operating values. The first semiconductor laser was made of GaAs and operated at 0.842 μm in the IR.[12,13] Shortly thereafter, Holonyak and Bevacqua[14] reported the operation of a visible GaAs

[12] R. N. Hall, G. E. Fenner, J. D. Kingsley, T. J. Soltys, and R. O. Carlson, "Coherent Light Emission from GaAs Junctions," *Phys. Rev. Lett.* **9:** 336, 1962.

[13] M. I. Nathan, W. P. Dumke, G. Burns, F. H. Dill, and G. Lasher, "Stimulated Emission of Radiation from GaAs $p-n$ Junctions," *Appl. Phys. Lett.* **1:** 62, 1962.

[14] N. Holonyak, Jr., and S. F. Bevacqua, *Appl. Phys. Lett.* **1:** 82, 1962.

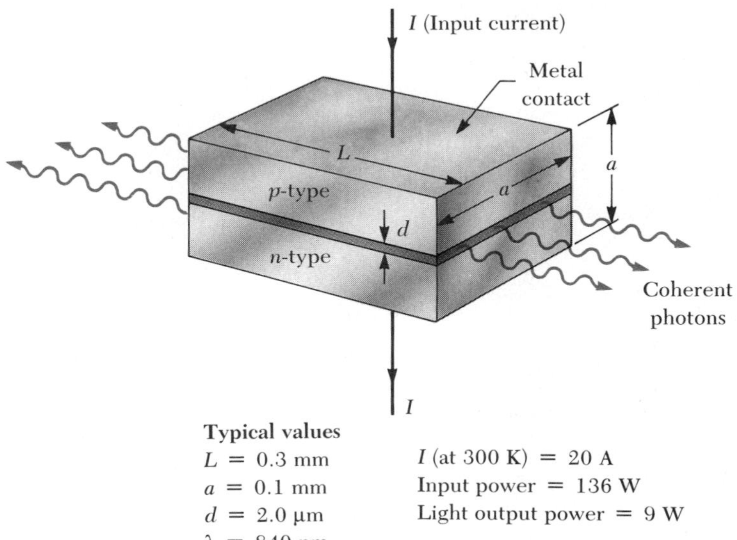

Typical values
$L = 0.3$ mm
$a = 0.1$ mm
$d = 2.0$ μm
$\lambda = 840$ nm

I (at 300 K) = 20 A
Input power = 136 W
Light output power = 9 W

Figure 10.13 A gallium arsenide *p-n* junction laser.

$p-n$ junction laser, and the laser was on its way to becoming an important part of semiconductor device technology.

Semiconductor lasers are similar to other lasers in that they require a population inversion between quantum levels in order to produce laser action. They are notably different from other lasers in that (a) discrete energy levels in conventional lasers become *energy bands* in semiconductors, and (b) optical and discharge pumping in conventional lasers are replaced by *injection pumping* in semiconductors. This consists of passing large forward currents through the laser diode and injecting electrons and holes into p and n regions, respectively, where they recombine and emit radiation. Although we will treat the physics of the solid state more systematically and deal with the laser diode in Chapter 12, let us briefly sketch the main points needed to understand semiconductor lasers.

Energy bands in semiconductors

In semiconductors the discrete energy levels found in isolated atoms widen into bands because of the perturbations caused by neighboring atoms (Fig. 10.14). We are basically interested in the two outermost energy bands, known as the valence (v) and conduction (c) bands. In a semiconductor like silicon in thermal equilibrium at room temperature, the valence band is filled with electrons and the conduction band, where electrons can move freely, contains no electrons. Since the two levels are separated by a forbidden energy region or gap, $E_g = 1.1$ eV, very few electrons possess enough thermal energy at room temperature to cross the gap and carry a current in the material. Hence silicon, like other semiconductors, is normally a poor conductor of electricity. However, light, heat, or electrical energy added to a semiconductor can promote electrons into the conduction band and increase the conductivity of a semiconductor. The hole (vacancy) left by the electron in the v-band plays an equally important role in conduction in a semiconductor. One can say that is has a physical existence, since it behaves like a positive charge moving opposite in direction to the conduction electron. In fact, the hole actually completes the current path through the semiconductor. This process of the creation of an electron-hole pair is called *pair generation*. An electron in the c-band can also "fall into" a hole and recombine with it, emitting light of band gap energy in the process. Both of these important processes are indicated

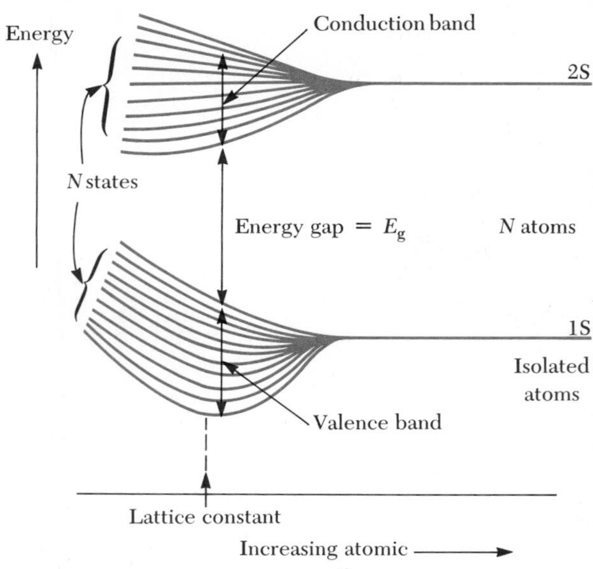

Figure 10.14 Formation of energy bands with decreasing atomic separation.

schematically in Figure 10.15. If the thermal equilibrium situation of empty c-band and full v-band is disturbed by intense optical or electrical excitation, a semiconductor population inversion can be set up, with the v-band empty of electrons (full of holes) to a depth X and the c-band full of electrons to a height Y. As can be seen in Figure 10.16, once a few spontaneously emitted photons with energy E_g induce stimulated emission at E_g, stimulated emission will eventually far exceed absorption since photons with energy E_g lack sufficient energy to be absorbed. Eventually the dominance of the rate of stimulated emission over the rate of absorption will lead to line narrowing and laser action.

A Final Note on Other Types of Semiconductor Lasers

For simplicity we have described only the basic GaAs injection laser. To leave the reader with the impression that this is the most important laser would be a serious distortion of the truth, for the present field of semiconductor lasers is enormously rich and varied. While the GaAs laser is a single-homojunction

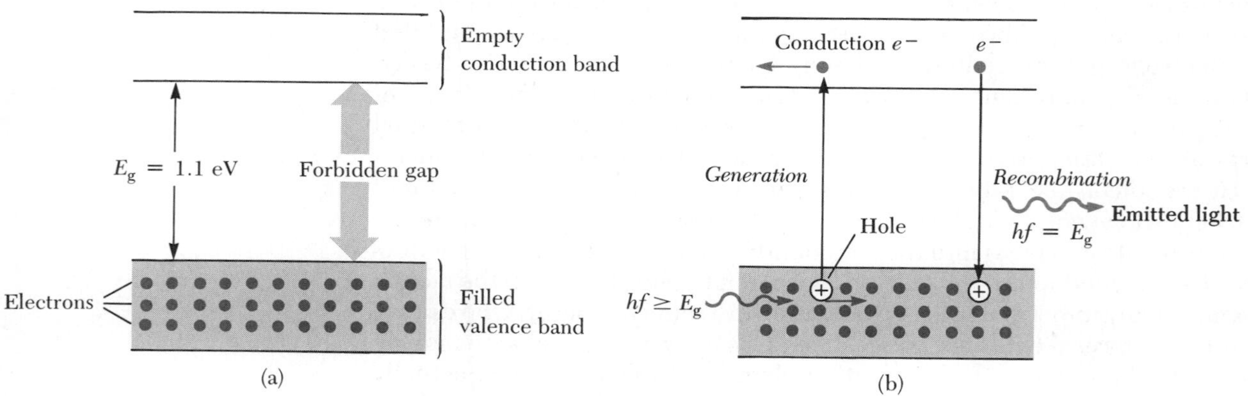

Figure 10.15 (a) Energy bands in pure silicon at 0 K. (b) Generation and recombination of electron-hole pairs in silicon.

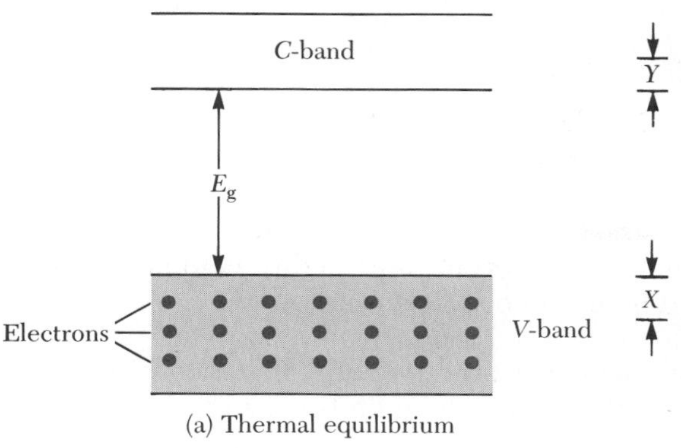

(a) Thermal equilibrium

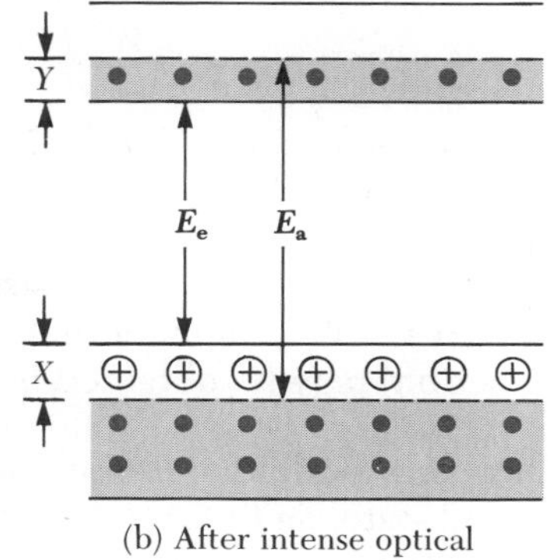

(b) After intense optical or electrical pumping

Figure 10.16 A population inversion in a semiconductor. Note that E_e, the emission energy, is equal to E_g and that E_a, the absorption energy, is equal to $E_g + X + Y$. The absorption energy is $E_g + X + Y$ and not just $E_g + X$ because electrons obey the Pauli exclusion principle. Thus electrons must absorb enough energy to occupy states above Y.

laser (one junction fabricated from a single compound), many current lasers are double-heterojunction (DH) lasers such as the PbTe-PbSnTe and the GaAs-AlGaAs lasers.[15] When two junctions are formed in the laser, the injected carriers are confined to a narrower region so that a lower threshold current is required for laser action, permitting continuous laser operation at room temperature. In addition, by varying the amount of Sn or Al in the ternary alloy, the band gap energy and, consequently, the laser wavelength may be varied. Also, GaAs-AlGaAs multilayer junctions are relatively easy to fabricate since AlGaAs layers may be grown epitaxially on GaAs layers. DH lasers can be formed with many other compounds including quaternary alloys, and tunable semiconductor lasers can be made whose emission wavelength may be varied by changing the applied pressure or magnetic field. These techniques have resulted in semiconductor lasers that span the broad range from 0.3 to 40 μm. These and other developments in this field, which is a truly fascinating combination of optics, material science, and device physics, are treated more extensively in some of the references cited at the end of this chapter.

10.9 SUMMARY

The processes of atomic absorption, spontaneous emission, and stimulated emission are crucial for an understanding of laser action. By considering a group of atoms and radiation in thermal equilibrium and balancing the rate of upward transitions against the rate of downward transitions, one can show that

$$B_{21} = B_{12} = B \tag{10.4}$$

and

$$\frac{A_{21}}{B} = \frac{8\pi hf^3}{c^3} \tag{10.5}$$

where B_{12} is Einstein's coefficient of absorption, B_{21} is Einstein's coefficient of stimulated emission, and $A_{21} = 1/t_s$ is the probability of spontaneous emission per unit time per atom.

The *induced* transition rate, $W_i(f)$, or the probability of absorption or stimulated emission per unit time per atom, is given by

[15] PbSnTe and AlGaAs are called ternary (three-component) alloys. They are crystalline solid solutions with Sn and Al randomly replacing some Pb and Ga atoms, respectively.

$$W_i(f) = \frac{c^2}{8\pi h f^3 t_s} g(f)I(f) \tag{10.8}$$

where f is the frequency of the incident radiation, t_s is the spontaneous emission lifetime, $g(f)$ is the atomic line shape function, and $I(f)$ is the intensity of the incident radiation.

Amplification of the light from a particular atomic transition and generation of an intense, monochromatic, coherent, and highly collimated beam (laser action) may be obtained if certain conditions are met:

1. The lasing medium must contain at least three energy levels consisting of a ground state, an intermediate state with a long lifetime, and a high energy pump state.
2. There must be an electrical or optical energy source capable of "pumping" atoms into excited states faster than they leave, so that a population inversion is produced.
3. There must be a method of confining the first wave of emitted photons within the laser so that they can stimulate further emission.

An expression for the minimum population inversion required to start laser action may be derived by setting the gain in intensity within the laser cavity from stimulated emission equal to the loss from absorption and other factors. One finds

$$\frac{\Delta N_c}{V} = n_2 - n_1\left(\frac{g_2}{g_1}\right) = \frac{8\pi f^2 t_s}{c^3 g(f) t_p} \tag{10.15}$$

where $\Delta N_c/V$ is the critical population inversion density (atoms/cm^3), n_1 and n_2 are the concentrations of atoms in the lower and upper laser levels, respectively, g_1 and g_2 are the degeneracies of these levels, t_s is the spontaneous emission lifetime of the upper laser level, $g(f)$ is the line shape function, t_p is the time constant for the decay of laser intensity from all photon loss mechanisms, and c is the speed of light in the laser material. In practice, t_p may be estimated from the loss per pass, α, the length per pass, L, and the speed of light in the laser, c, by using

$$t_p = \left(\frac{1}{\alpha}\right)\left(\frac{L}{c}\right)$$

If one assumes that the laser transition has a Lorentzian line shape with a full width at half maximum of Δf, Equation 10.15 takes the particularly simple form

$$\frac{\Delta N_c}{V} \approx \frac{4\pi^2 f^2 \Delta f}{c^3}\left(\frac{t_s}{t_p}\right) \tag{10.17}$$

A four-level laser requires less input power than a three-level laser because the lower laser level in a four-level system is essentially unpopulated. This means that only the critical population density $\Delta N_c/V$ must be maintained in the upper laser level. Using this fact yields a minimum pumping power per unit volume for a four-level system given by

$$P = \frac{\Delta N_c}{V}\left(\frac{hf}{t_s}\right) \tag{10.18}$$

or, using Equation 10.17,

$$P = \frac{4\pi^2 hf^3 \Delta f}{c^3 t_p} \qquad (10.19)$$

Actual lasers include solid state crystalline lasers such as the ruby laser, gas lasers like the He-Ne laser, and semiconductor lasers such as the GaAs laser. Although large, pure single crystals of ruby are easy to fabricate, a large pumping power is required to induce lasing in ruby, and consequently this laser must be operated on a pulsed basis. Gas lasers offer an attractive alternative to crystalline solid state lasers because they may be operated continuously. Continuous operation is possible because of the modest pumping power requirements of gas lasers, which have small concentrations of optically active atoms and low critical population inversion densities. The most important light source for optical signal processing is the semiconductor laser. This laser is essentially a $p-n$ junction operated under strong forward bias. Typical advantages of these lasers are small size, high efficiency, simplicity, ease of modulation, and fast response.

SUGGESTIONS FOR FURTHER READING

1. D. R. Herriott, "Applications of Laser Light," *Sci. American,* September 1968, 141–156.
2. C. K. N. Patel, "High-Power Carbon Dioxide Lasers," *Sci. American,* August 1968, 22–23.
3. A. L. Schawlow, "Laser Light," *Sci. American,* September 1968, 120–126.
4. Special Issue on Applications of Lasers in Research, *Physics Today,* vol. 30, no. 5, May 1977.
5. D. C. O'Shea, W. R. Callen, and W. T. Rhodes, *Introduction to Lasers and their Applications,* Reading, MA, Addison-Wesley, 1977.
6. B. G. Streetman, *Solid State Electronic Devices,* Englewood Cliffs, NJ, Prentice-Hall, 1980.

More advanced treatments of lasers may be found in:

7. J. I. Pankove, *Optical Processes in Semiconductors,* New York, Dover Publications, 1971.
8. S. M. Sze, *Physics of Semiconductor Devices,* New York, John Wiley & Sons, 2nd edition, 1981.

QUESTIONS

1. Radiative emission by an atom can be spontaneous or stimulated. Distinguish between these two processes, identifying the characteristic features of each. Describe the role(s) each one plays in the operation of the ruby laser.
2. Why must a lasing medium possess at least three energy levels?
3. Since the light from semiconductor lasers can easily be modulated, these lasers are of great use in lightwave communication systems. Show that laser communication systems possess a great advantage over microwave communication systems in terms of the amount of information that can be carried. *Hint:* If each speech channel requires a bandwidth of 5 kHz, calculate the number of channels carried by a laser beam of 10^{16} Hz and by a microwave beam of 10^{10} Hz.
4. Explain how low temperature operation of a laser can greatly increase its efficiency in at least two ways.
5. Discuss the main criteria that must be met to achieve laser action in a three-level system.
6. Why is it necessary to use a pumping light source in a laser?
7. What are the three fundamental ways in which light (photons) interacts with matter (atoms)? Give a brief explanation of each.

PROBLEMS

1. Assume that energy levels E_1 and E_2 are degenerate with degeneracies of g_1 and g_2. Starting with the correct form of Equation 10.1, derive Equation 10.12,

$$\left(\frac{dI}{dt}\right)_{\text{gain}} = \left[n_2 - n_1\left(\frac{g_2}{g_1}\right)\right] hfcW_i$$

Hint: Follow the development in Equations 10.1 through 10.12, balance the numbers of upward and downward transitions to show that $B_{21} = (g_1/g_2)B_{12}$, and calculate separate expressions for the induced emission rate and absorption rate.

2. The energy level scheme of uranium in CaF_2 is shown

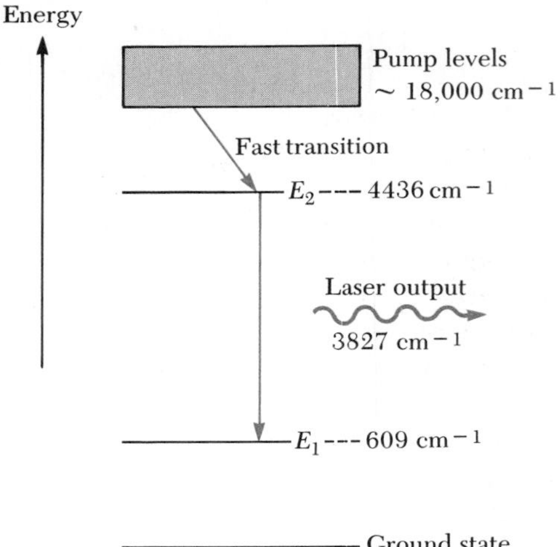

Figure 10.17 (Problem 2) Energy levels of uranium in calcium fluoride, CaF_2. Note that for the state E_2, $t_s = 130\ \mu s$. Note: $1\ cm^{-1} = 1.241 \times 10^{-4}$ eV.

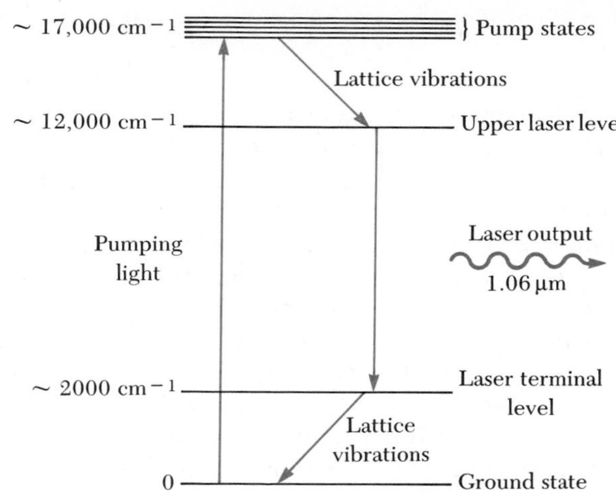

Figure 10.18 (Problem 5) Neodymium/glass four-level laser.

in Figure 10.17. (a) Calculate the populations of the upper and lower laser levels, E_2 and E_1, relative to the ground state at 300 K. (b) Recalculate the populations at 77 K and show that this system behaves like a four-level laser at 77 K. (c) Find the critical inversion density and minimum pumping power for this system. Use $L = 3$ cm, $n = 1.4$, $\alpha = 0.05$, $\Delta f = 20\ cm^{-1}$ at 77 K, and $t_s = 1.3 \times 10^{-4}$ s. (d) Does this laser emit in the UV, visible, or IR portion of the spectrum? (e) Compare the pumping power of this laser to the pumping power required for ruby. Why is the pumping power so much less for uranium in CaF_2?

3. A pulsed laser is constructed with a ruby crystal as the active element. The crystal (Al_2O_3) contains Cr^{3+} impurities, which substitute for Al^{3+} ions and give ruby its red color. Suppose laser light is emitted from an inverted Cr^{3+} three-level system in 50 ns pulses. The ruby rod contains a total of 3.0×10^{19} Cr^{3+} ions. If the wavelength of the laser light emitted is 694.4 nm, find (a) the energy of one emitted photon in eV, (b) the total energy available per laser pulse, assuming all of the Cr^{3+} ions are in the upper laser state, and (c) the power per pulse. (d) Why do you suppose the power per pulse is much less than this in practice?
4. *How monochromatic is a laser?* The natural line width emitted by a collection of Cr^{3+} ions in ruby in thermal equilibrium at 290 K is about 0.4 nm for the 694.4 nm line. This is the *thermal line width.* Show that the emission from a ruby laser is much more monochromatic than 694.4 ± 0.2 nm if laser emission is confined to a single cavity mode. Take the length of the ruby between polished faces to be 10 cm, $n = 1.8$, and $dn/d\lambda \approx 0$.
5. The neodymium glass laser is a popular high-power solid state laser capable of both continuous and pulsed operation. It consists of Nd^{3+} ions incorporated into a glass matrix and has the energy level structure shown in Figure 10.18. Typical parameters for this laser are:

$$1/\lambda = 9398.50\ cm^{-1}$$
$$L = 5\ cm$$
$$n = 1.8$$
$$\alpha = 7\%$$
$$\Delta f = 15\ cm^{-1}$$
$$t_s = 10^{-4}\ s$$
$$N = 10^{18}\ Nd^{3+}/cm^3$$

(a) Does this laser need to be operated at 77 K in order to act as a four-level system? (b) Calculate the critical inversion density. (c) Calculate the minimum pumping power. (d) Comparing to typical values for the *crystal* $CaWO_4 : Nd^{3+}$ laser (Example 10.4), can you account for the larger loss/pass and line width in the neodymium *glass* laser?
6. In a four-level laser using the praseodymium ion, the lasing transition occurs to a level that is 0.042 eV above the ground state. What is the ratio of the number of ions in this level to the number in the ground state at $T = 300$ K, with no exciting radiation present? What is this ratio at 77 K? (Assume equal statistical weights.) What bearing do these results have on the efficiency of this four-level laser?
7. *A comparison of the external pumping powers of gas and solid state lasers.* Further insight into why gas lasers can be operated continuously, compared to the pulsed operation necessary for solid state lasers, can be obtained by calculating the external pumping power. In general, there is incomplete absorption of the incident pumping power by the optically active atoms. If we let F equal the fraction of the incident power absorbed, the external pumping power, P_e, is related to the minimum pumping power, P, by

$$P_e = \frac{P}{F} = \frac{\Delta N_c}{V}\left(\frac{hf}{t_s F}\right)$$

Because F is proportional to the concentration of the absorbing medium, N, we may write

$$P_e \propto \frac{\Delta N_c}{V}\left(\frac{hf}{t_s N}\right)$$

Using this result, calculate the *ratio* of external pumping powers for the lasers described in Example 10.3.

8. *More than you ever wanted to know about the ruby laser.* The detailed energy level structure and degeneracies for the Cr^{3+} ion in ruby are shown in Figure 10.19. (a) Assuming that *all* of the Cr ions are in the three lowest levels, and using the values of ruby from Example 10.2, calculate the populations of these levels. [*Hint:* $n(^4A_2)$ and $n(\overline{E})$ are related by the critical population inversion density found in Example 10.2, and $n(\overline{E})$ and $n(2\overline{A})$ are related by the Maxwell-Boltzmann statistics.] (b) Determine the minimum pumping power, assuming that it depends on the *total* population in $\overline{E}$ and $2\overline{A}$. (These levels cannot be pumped separately). Note that the degeneracies of the lowest three states are $g(^4A_2) = 4$, $g(\overline{E}) = 2$, $g(2\overline{A}) = 2$.

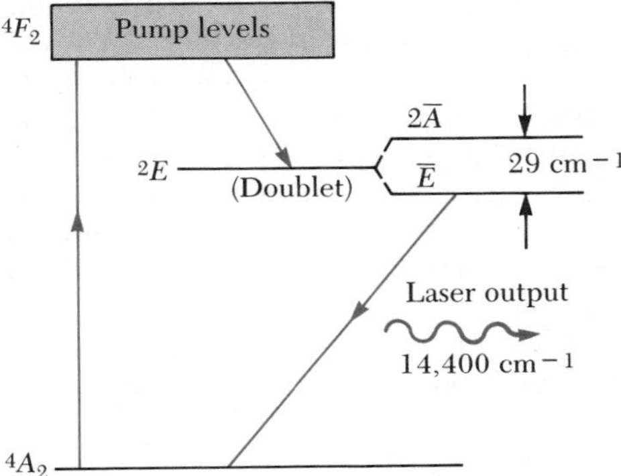

Figure 10.19 (Problem 8) Energy-level diagram for Cr^{3+} in ruby (Al_2O_3) giving the degeneracies of pertinent levels.

9. *Cavity modes in lasers.* Under high resolution and at threshold current, the emission spectrum from a GaAs laser is seen to consist of many sharp lines, as shown in Figure 10.20. Although one line generally dominates at higher currents, let us consider the origin and spacing of the multiple lines or modes shown in this figure. Recall that standing waves or resonant modes corresponding to the sharp emission lines will be formed when an integral number m of half-wavelengths fits

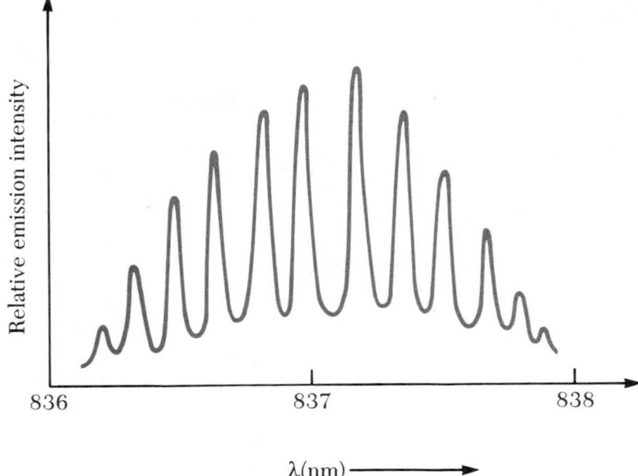

Figure 10.20 (Problem 9) High resolution emission spectrum of the GaAs laser operated just below the laser threshold (2 K).

between the cleaved GaAs surfaces. If L is the distance between cleaved faces, n is the index of refraction of the semiconductor, and λ is the wavelength in air, we have

$$m = \frac{L}{\lambda/2n}$$

(a) Show that the wavelength separation between adjacent modes (the change in wavelength for the case $\Delta m = +1$) is given by

$$|\Delta\lambda| = \frac{\lambda^2}{2L\left(n - \lambda \dfrac{dn}{d\lambda}\right)}$$

(*Hint:* Since m is large and we want the small change in wavelength that corresponds to $\Delta m = 1$, we can consider m to be a continuous function of λ and differentiate $m = 2Ln/\lambda$).

(b) Using the following typical values for GaAs, calculate the wavelength separation between adjacent modes. Compare your results to Figure 10.20.

$$\left.\begin{aligned} \lambda &= 837 \text{ nm}, \\ 2L &= 0.60 \text{ mm} \\ n &= 3.58 \\ dn/d\lambda &= 3.8 \times 10^{-4}\ (\text{nm})^{-1} \end{aligned}\right\} \text{at } \lambda = 837 \text{ nm}$$

(c) Estimate the mode separation for the He-Ne gas laser using $\lambda = 632.8$ nm, $2L = 0.6$ m, $n \approx 1$, and $dn/d\lambda = 0$. On the basis of this calculation and the result of part (b), what is the controlling factor in mode separation in both solids and gases?

Essay

LASERS AND THEIR APPLICATIONS

Isaac D. Abella
The University of Chicago

LASERS

The enormous growth of laser technology mentioned in Section 10.1 has stimulated a broad range of scientific and engineering applications that exploit some of the unique properties of laser light. These properties derive from the distinctive way laser light is produced in contrast to the generation of ordinary light. Laser light originates from atoms, ions, or molecules through a process of *stimulated emission* of radiation. The active laser medium is contained in an enclosure or *cavity,* which organizes the normally random emission process into an intense directional, monochromatic and coherent wave. The end mirrors provide the essential optical feedback, which selectively builds up the stimulating wave along the tube axis. However, in an ordinary sodium vapor street lamp, for example, the atoms spontaneously emit in random directions and at irregular times, over a broad spectrum, resulting in the isotropic illumination of incoherent light.

Currently operating laser systems use a variety of gases, solids, and liquids as their working laser substances. These devices are designed to emit either continuous or pulsed monochromatic beams, and operate over a broad range of the optical spectrum (ultraviolet, visible, infrared) with output powers from milliwatts (10^{-3} W) to megawatts (10^{6} W). The particular application determines the choice of laser system, wavelength, power level, or other relevant variables, since no one laser has all the desirable properties.

As noted in Section 10.3, several conditions must be satisfied for a laser to operate successfully.

The requirement for population inversion, that is, more atoms in a particular excited state than in a lower state, essentially means that energy must be supplied from outside the system. Otherwise, atoms would eventually radiate, and develop an increasing probability for absorbing light. Finally, they fall to the lowest energy state and stop emitting altogether. Therefore, all laser systems must be connected to external energy sources, usually electrical, as required by conservation of energy (see Figure 1). For example, we can energize the atoms in a gas laser medium by bombarding them with electrons (so-called "electrical pumping"). We can also supply energy to lamps, whose light populates excited states by photon absorption ("optical pumping") for those solids or liquids that do not conduct electric charge. These pumping mechanisms tend to have low efficiency (ratio of laser energy output to electric energy supplied), typically a few percent, with the balance discharged as heat into cooling water or circulating air.

Controlling the electrical input into the laser system allows a variable laser energy output, which may be important in many applications. Thus, the argon ion laser system can emit up to about 10 watts in the green optical beam by adjustment of the electric current in the argon gas, which in turn controls the degree of population inversion.

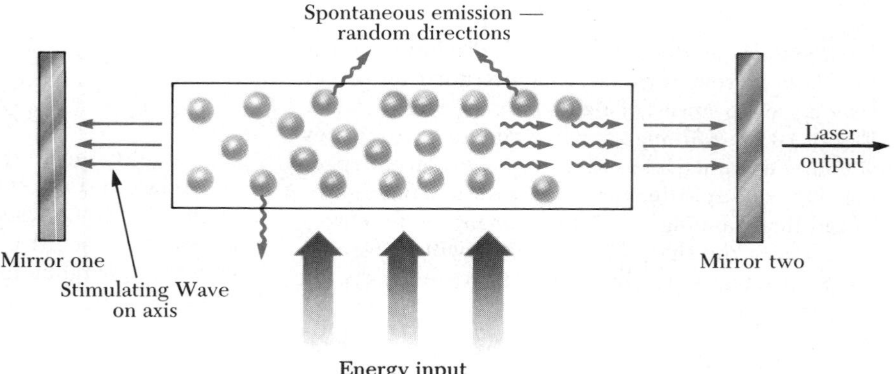

Figure 1 A schematic of a laser design. The tube contains atoms, which represent the active medium. An external source of energy (optical, electrical, etc.) is needed to "pump" the atoms to excited energy states. The parallel end mirrors provide the feedback of the stimulating wave.

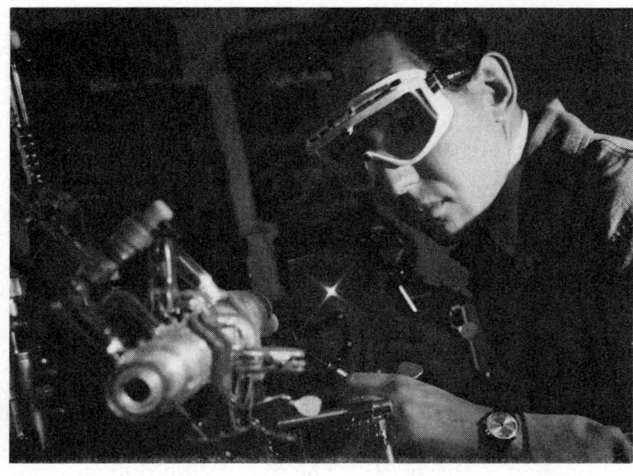

Figure 2 A scientist using a laser to study excited states in molecules. (Courtesy of Laser Precision Corp.)

Chemical lasers operate without direct electrical input. Several highly reactive gases are mixed in the laser chamber, and the energy released in the ensuing reaction populates the excited levels in the molecules. In this case, the reactants need to be resupplied for the laser to operate for any length of time.

Some laser systems have fluid media, containing dissolved dye molecules. The dye lasers are usually pumped to excited levels by an external laser. The advantage of this arrangement is that dye lasers can be continuously "tuned" over a wide range of wavelengths, using prisms or gratings, whereas the pump source has a fixed wavelength. Color variability is important for those cases where the laser is directed at materials whose absorption depends on wavelength. Thus, the laser can be tuned into exact coincidence with selected energy states. For example, blood does not absorb red light to any great extent, which excludes red laser light from most surgical applications on blood-rich tissue.

A variety of laser systems are in general use today. They include the 1 mW helium-neon laser, usually operating in the red at 632.8 nm (although yellow and green beams are available); the argon ion laser, which operates in the green or blue and can emit up to 10 watts; the carbon dioxide gas laser, which emits in the infrared at 10 μm and can produce several hundred watts; and the neodymium YAG laser, a powerful solid-state optically-pumped system that emits at 1.06 μm in either continuous or pulsed mode. The recently perfected diode junction laser emits in the near infrared and operates by passage of current through a semiconductor material. The recombination radiation (see Section 10.8) is essentially direct conversion of electrical energy to laser light and is a very efficient process. The diodes can emit up to 5 watts and can be used to energize other laser materials.

APPLICATIONS

We shall describe several applications that serve to illustrate the wide variety of laser utilization. First, lasers are being used to make precision long-range distance measurements (rangefinding). It has become important, for astronomical and geophysical purposes, to measure as precisely as possible the distances from various points on the surface of the earth to a point on the moon's surface. To facilitate this, the Apollo astronauts set up a compact array, a 0.5 m square of reflector prisms, on the moon. This array allows laser pulses directed from an earth station to be retro-reflected to the same station. Using the known speed of light and the measured round-trip travel time of a 1 ns pulse, one can determine the earth-moon distance, 380 000 km, to a precision of better than 10 cm. Such information would be useful, for example, in making more reliable earthquake predictions and for learning more about the motions of the earth-moon system. This technique requires a high-power pulsed laser for its success, since the burst of photons must return to a collecting telescope on earth with sufficient power to be detected. Variants of this method are also used to measure the distances to inaccessible points on the earth.

The low-power helium-neon laser is the basis for a technical innovation seen in many supermarkets. The laser beam can be focused with a lens to a very small bright spot, which is then reflected from an oscillating mirror producing a swiftly moving dot image. If this spot is scanned over the product identification bar-code printed on supermarket products, the variation of the reflected light can be detected and decoded. This scanner has improved speed and accuracy at the checkout counter. The spot of light must be small enough to resolve the different widths of the individual bars and bright enough to be "seen" in reflection by the optical detector below the counter. Since this laser is operated in public, the power must be low enough to be safe for any reasonable use of the system. This puts stringent limits on the type of laser used in this application.

Similarly, a laser (light emitting) diode is used to decode the digital information on the compact audio disc, the so-called CD. On the compact disc, the music has been digitized as pits and grooves embedded in the plastic. The fluctuating reflection of the weak laser spot from the disc surface is detected by a photocell and decoded by digital-to-analog circuits. The high resolution of the laser beam allows the CD player to reproduce music with with extremely high fidelity, without the noise or hiss associated with regular long-playing records or magnetic tape. There are also video versions of the laser disc. In all of these decoding applications, the essential laser properties that are exploited are the accurate focusing of the beam, the monochromaticity (to be able to operate in the presence of background illumination), and enough power that the diffusely reflected light can be observed. High-power lasers could damage the disc and are not employed.

The amount of information stored on a compact audio disc as digital data is estimated to be about 1 gigabyte (10^9 characters), which is enormously high density data storage. By way of comparison, the data storage capacity on magnetic "floppy" discs in personal computers is typically 800 kilobytes (8×10^5 characters). Developments are under way to transfer this optical storage technology to computer disc drives having both read and write capability. The compact disc as a read-only device with prerecorded data requires very little modification to be used with a computer, and is already in use to store encyclopedia or dictionary volumes. However, the ability to alter or erase an optical disc would require a combination of optical and magnetic methods.

Novel medical applications use the fact that different laser wavelengths can be absorbed in specific biological tissues. A widespread eye condition, glaucoma, is manifested by a high fluid pressure in the eye, which can lead to destruction of the optic nerve. A simple laser operation (iridectomy) can "burn" a tiny hole in a clogged membrane, relieving the destructive pressure. Along the same lines, a serious side effect of diabetes is the formation of weak blood vessels (neovascularization), which often leak blood into extremities. When this occurs in the eye, vision deteriorates (diabetic retinopathy), leading to blindness in diabetic patients. It is now possible to direct the green light from the argon ion laser through the clear eye lens and eye fluid, focus on the retina edges, and photo-coagulate the leaky vessels. These procedures have greatly reduced blindness in patients with glaucoma and diabetes.

Laser surgery is now a practical reality. Infrared light at 10 μm from a carbon dioxide laser can cut through muscle tissue, primarily by heating and evaporating the water contained in cellular material. Laser power of about 100 W is required in this technique. The advantage of the "laser knife" over conventional methods is that laser radiation cuts and coagulates at the same time, leading to substantial reduction of blood loss. In addition, the technique virtually eliminates cell migration, which is very important in tumor removal. Furthermore, a laser beam can be trapped in fine glass-fiber light-guides (endoscopes) by means of total internal reflection. The light fibers can be introduced through natural orifices, conducted around internal organs, and directed to specific interior body locations, eliminating the need for massive surgery. For example, bleeding tissues in the gastrointestinal tract can be optically cauterized by a fiberoptic endoscope inserted through the mouth.

Finally, we describe an application to biological and medical research. It is often important to isolate and collect unusual cells for study and growth. A laser cell-separator exploits the fact that specific cells can be tagged with fluorescent dyes. All of the cells in a sample are dropped from a tiny charged nozzle, and they are laser scanned for the dye tag. If the detector is triggered by the correct light-emitting tag, a small voltage is applied to parallel plates and the falling electrically charged cell is deflected into a collection beaker. This is an efficient method for extracting the proverbial needles from the haystack.

Essay Problems

1. (a) In the case of the earth-moon laser rangefinder, what is the round-trip time for a laser pulse? (b) What precision in timing is required, that is, how small a time change needs to be detectable, to be able to measure the distance to an error of 10 cm? What effect does the earth's atmosphere have?
2. A laser beam of wavelength $\lambda = 600$ nm is directed at the moon from a laser tube of 1 cm diameter. Does the beam spread at all? What is the diameter of the "spot" on the moon's surface? Does your answer demand a good strategy for successful lunar-array illumination ? How would you do it?
3. Estimate how much chemical reagent is required to produce a chemical laser of 1 kW output, in moles/second. Pick a reasonable exothermic reaction rate in kcal/mole. Estimate how many kilograms of each of hydrogen and chlorine gas would be needed to make an HCl laser operate for an hour at 1 kW output.

11

Molecular Structure

Except for the inert gases, elements generally combine to form chemical compounds whose basic unit is the *molecule*, an aggregate of individual atoms joined by chemical bonds. The physical and chemical properties of molecules derive from their constituent atoms—their arrangement, the manner and degree to which they interact, and their individual electronic structures.

The properties of molecules can be studied experimentally by examining their spectra. As with atoms, a molecule can emit or absorb photons, with accompanying electronic transitions among the allowed energy levels of the molecule. The resulting emission or absorption spectrum is different for each molecule, and acts as a sort of fingerprint of its electronic structure.

But molecules also emit or absorb energy in ways not found in atoms. Molecules can rotate, storing energy in the form of kinetic energy of rotation; and they can vibrate, and so possess energy of vibration. As we shall discover, both the rotational and vibrational energies are quantized, and so give rise to their own unique spectra. It follows that molecular spectra are vastly more complicated than atomic spectra, but also carry a good deal more information. In particular, the vibration-rotation spectrum tells us how the individual atoms that form the molecule are arranged, and the strength of their interaction.

In this chapter, we shall describe the bonding mechanisms in molecules, the various modes of molecular excitation, and the radiation emitted or absorbed by molecules. In the course of our study, we shall encounter the quantum origins of the chemical bond, and discover why some atoms bond to form a molecule while others do not. Central to this inquiry are the roles played by the exclusion principle and tunneling, both nonclassical ideas that underscore the importance of wave mechanics to the study of molecular structure.

11.1 BONDING MECHANISMS—A SURVEY

Two atoms combine to form a molecule because of a net attractive force between them when their separation is greater than their equilibrium separation in the molecule. Furthermore, the total energy of the stable bound molecule is *less* than the total energy of the separated atoms.

Fundamentally, the bonding mechanisms in a molecule are primarily due to electrostatic forces between atoms (or ions). When two atoms are separated by an infinite distance, the force between them is zero, as is the electrostatic potential energy of the system. As the atoms are brought closer together, both attractive and repulsive forces act. At very large separations, the dominant forces are attractive in nature. For small separations, repulsive forces between like charges begin to dominate. The potential energy of the system can be positive or negative, depending on the separation between the atoms.

The total potential energy of the system can be approximated by the expression

$$U = -\frac{A}{r^n} + \frac{B}{r^m}$$

where r is the internuclear separation, A and B are constants associated with the attractive and repulsive forces, and n and m are small integers. Figure 11.1 presents a sketch of the total potential energy versus internuclear separation. Note that the potential energy for large separations is negative, corresponding to a net attractive force. At the equilibrium separation, the attractive and repulsive forces just balance and the potential energy has its minimum value.

A complete description of the binding mechanisms in molecules is a highly complex problem because it involves the mutual interactions of many particles. In this section, we shall discuss some simplified models in the following order of decreasing bond strength: the ionic bond, the covalent bond, the hydrogen bond, and the van der Waals bond.

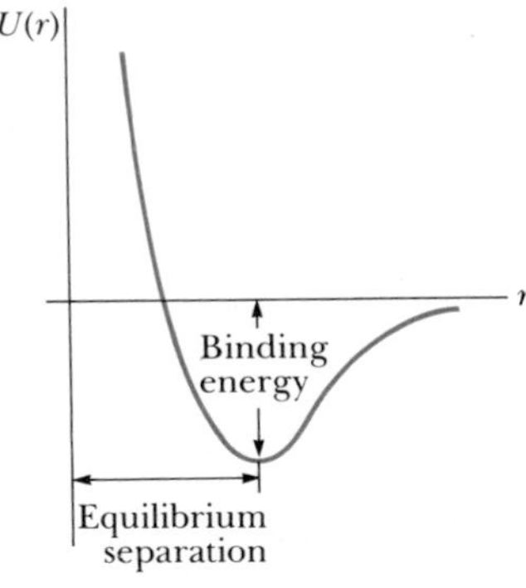

Figure 11.1 The total particle energy as a function of the internuclear separation for a system of two atoms.

Ionic Bonds

Ionic bonds are fundamentally due to the Coulomb attraction between oppositely charged ions. A familiar example of an ionically bonded molecule is sodium chloride, NaCl, which is common table salt. Sodium, which has an electronic configuration $1s^2 2s^2 2p^6 3s$, becomes more stable when it gives up its $3s$ valence electron to form a Na^+ ion. The energy required to ionize the atom to form Na^+ is 5.1 eV. Chlorine, which has an electronic configuration $1s^2 2s^2 2p^5$, is one electron short of the closed-shell structure of argon. Because closed-shell configurations are energetically more favorable, the Cl^- ion is more stable than the neutral Cl atom. The energy released when an atom takes on an electron is called the **electron affinity.** For chlorine, the electron affinity is 3.7 eV. Therefore, the energy required to combine Na^+ and Cl^- to form NaCl is $5.1 - 3.7 = 1.4$ eV.

The total energy versus the internuclear separation for NaCl is shown in Figure 11.2. Note that the total energy of the molecule has a minimum value of -4.2 eV at the equilibrium separation of about 0.24 nm. The energy required to separate the NaCl molecule into neutral sodium and chlorine atoms, called the **dissociation energy,** is equal to 4.2 eV.

When the two ions are brought closer than 0.24 nm, the electrons in the closed shells begin to overlap, which results in a repulsion between the closed

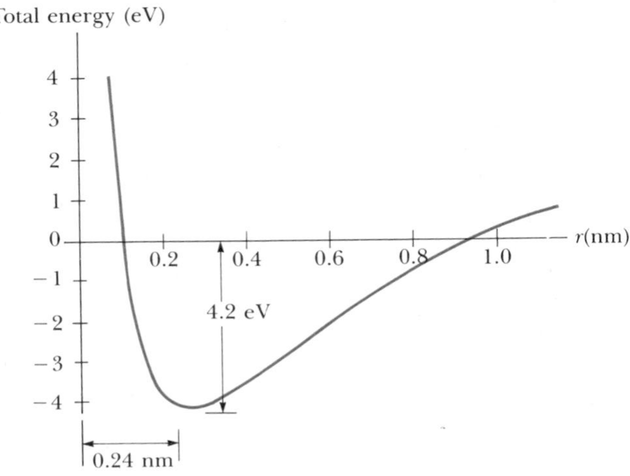

Figure 11.2 Total energy versus the internuclear separation for the NaCl molecule. The energy required to separate the NaCl molecule into neutral atoms of Na and Cl is 4.2 eV.

shells. When the ions are far apart, core electrons from one ion do not overlap with those of the other ion. However, as they are brought closer together, the core-electron wave functions begin to overlap. Because of the exclusion principle, some electrons must occupy a higher-energy state.

Covalent Bonds

A covalent bond between two atoms can be visualized as the sharing of electrons supplied by one or both atoms that form the molecule. Many diatomic molecules such as H_2, F_2, and CO owe their stability to covalent bonds. In the case of the H_2 molecule, the two electrons are equally shared between the nuclei, and form a so-called molecular orbital. The two electrons are more likely to be found between the two nuclei, hence the electron density is large in this region. The formation of the molecular orbital from the *s* orbitals of the two hydrogen atoms is shown in Figure 11.3. Because of the exclusion principle, the two electrons in the ground state of H_2 must have anti-parallel spins. It is interesting to note that if a third H atom is brought near the H_2 molecule, the third electron would have to occupy a higher energy quantum state because of the exclusion principle, which is an energetically unfavorable situation. Hence, the H_3 molecule is not stable and does not form.

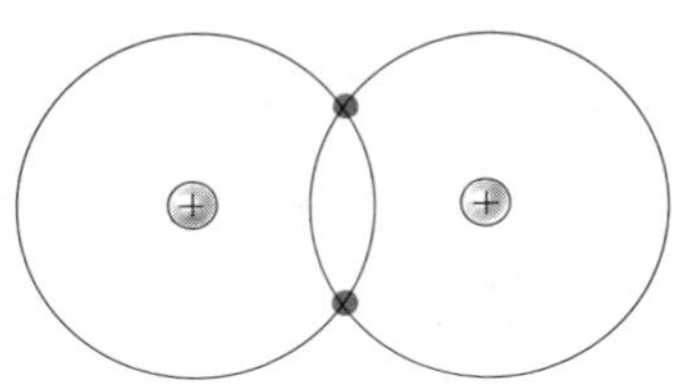

Figure 11.3 The covalent bond formed by the two 1*s* electrons of the H_2 molecule.

More complex stable molecules such as H_2O, CO_2, and CH_4 are also formed by covalent bonds. Consider methane, CH_4, a typical organic molecule shown schematically in the electron sharing diagram of Figure 11.4a. Note that covalent bonds are formed between the carbon atom and each of the four hydrogen atoms. The spatial electron distribution of the four covalent bonds is shown in Figure 11.4b. The four hydrogen nuclei are at the corners of a regular tetrahedron, with the carbon nucleus at the center.

The Hydrogen Bond

Because hydrogen has only one electron, it is expected to form a covalent bond with only one other atom. However, in some molecules, hydrogen forms a different type of bond between two atoms or ions, called a **hydrogen bond.** One example of a hydrogen bond shown in Figure 11.5 is the hydrogen disul-

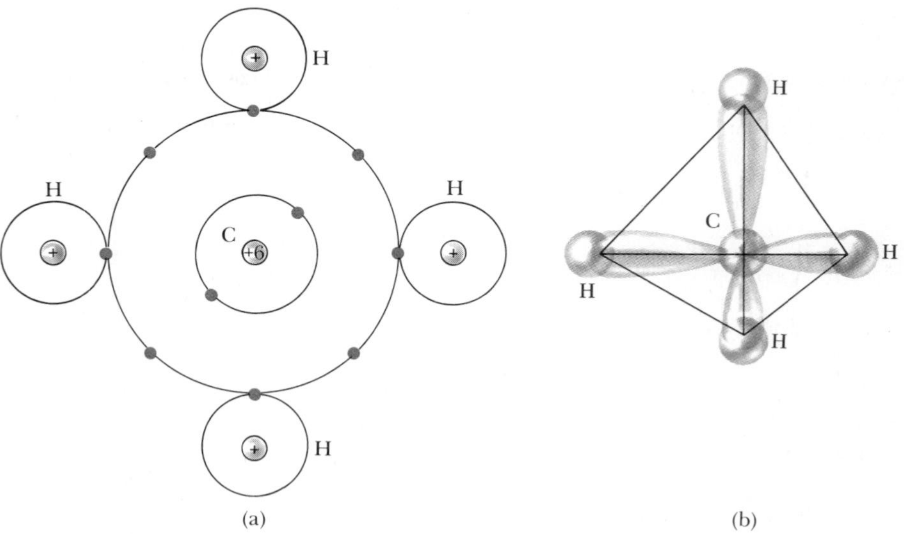

Figure 11.4 (a) Diagram of the four covalent bonds in the CH_4 molecule. (b) The electron distribution in the four covalent bonds of the CH_4 molecule. The carbon atom is at the center of a tetrahedron with hydrogen atoms at its corners.

fide ion, $(HF_2)^-$. The two negative fluorine ions are bound by the positively charged proton between them. This is a relatively weak chemical bond, with a binding energy of about 0.1 eV. Although the hydrogen bond is weak, it is the mechanism responsible for linking giant biological molecules and polymers. For example, in the case of the famous DNA molecule that has a double helix structure, hydrogen bonds link the turns of the helix.

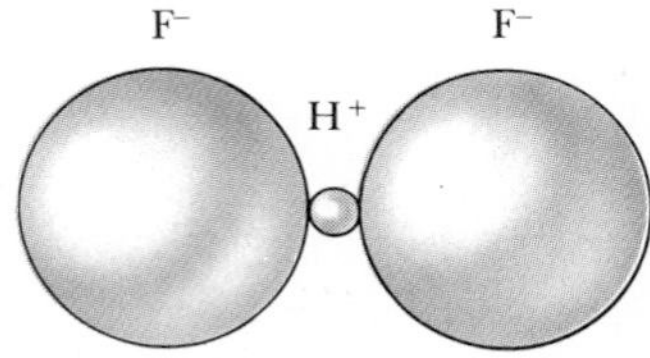

Figure 11.5 Hydrogen bonding in the $(HF_2)^-$ molecular ion. The two negative fluorine ions are bound by the positively charged proton between them.

Van der Waals Bonds

If two molecules are some distance apart, they are attracted to each other by electrostatic forces. Likewise, atoms that do not form ionic or covalent bonds are attracted to each other by electrostatic forces. For this reason, at sufficiently low temperatures where thermal excitations are negligible, substances will condense into a liquid and then solidify (with the exception of helium, which does not solidify at atmospheric pressure). The weak electrostatic attractions between molecules are called **van der Waals forces.**

There are actually three types of van der Waals forces, which we shall briefly describe. The first type, called the *dipole-dipole force,* is an interaction between two molecules each having a permanent electric dipole moment. For example, polar molecules such as HCl and H_2O have permanent electric dipole moments and attract other polar molecules. In effect, one molecule interacts with the electric field produced by another molecule. The attractive force turns out to be proportional to $1/r^7$.

The second type of van der Waals force is a *dipole-induced force* in which a polar molecule having a permanent electric dipole moment *induces* a dipole moment in a nonpolar molecule. This attractive force between a polar and a nonpolar molecule also varies as $1/r^7$.

The third type of van der Waals force is called the *dispersion force.* The dispersion force is an attractive force that occurs between two nonpolar molecules. In this case, the interaction results from the fact that although the average dipole moment of a nonpolar molecule is zero, the average of the square of the dipole moment is nonzero because of charge fluctuations. Consequently, two nonpolar molecules near each other tend to be correlated so as to produce an attractive force, which is the van der Waals force.

11.2 MOLECULAR ROTATION AND VIBRATION

The various means by which a molecule can store energy internally can be divided into several categories: (1) electronic, accompanied by electron rearrangement among the allowed energy levels of the molecule; (2) rotational, corresponding to rotation of the molecule about its center of mass; and (3) vibrational, represented by the vibration of the molecule's constituent atoms. Thus, the total internal energy of a molecule can be written in the form[1]

$$E = E_{\text{el}} + E_{\text{rot}} + E_{\text{vib}}$$

The electronic contribution to the energy of a molecule is difficult to determine because each electron interacts with all the others and with the atomic nuclei forming the molecule. The allowed electronic levels of a molecule will

[1] To these *internal* modes we should add for completeness the translational mode, describing the motion of the molecule's center of mass through space. However, since the quantization of translational energy is unrelated to internal structure, the translational mode is unimportant for the interpretation of molecular spectra.

be taken up in Section 11.4. The remaining two contributions to molecular energy—rotation and vibration—are far easier to describe, and are discussed below.

Molecular Rotation

Let us consider the rotation of a molecule about a fixed axis. We shall confine our discussion to a diatomic molecule, although the same ideas can be extended to polyatomic molecules. As shown in Figure 11.6a, the diatomic molecule has only two rotational degrees of freedom, corresponding to rotations about the y and z axes, that is, the axes perpendicular to the molecular axis.[2]

The energy of a rotating molecule is all kinetic. Let m_1 and m_2 denote the atomic masses, with speeds v_1 and v_2. For a molecule in rotation, the speeds v_1 and v_2 are interrelated. In terms of the angular velocity of rotation ω, we have

$$v_1 = \omega r_1 \qquad \text{and} \qquad v_2 = \omega r_2$$

where r_1 is the distance of m_1 from the axis of rotation, and similarly for r_2 (cf. Fig. 11.6a). The angular momentum of rotation is

$$L = m_1 v_1 r_1 + m_2 v_2 r_2 = \{m_1 r_1^2 + m_2 r_2^2\}\omega = I\omega$$

The quantity in curly brackets { . . . } is the *moment of inertia*, denoted I. With this identification, the energy of rotation is

$$E_{\text{rot}} = \tfrac{1}{2}m_1 v_1^2 + \tfrac{1}{2}m_2 v_2^2 = \tfrac{1}{2}I\omega^2$$

[2] The excitation energy for rotations about the molecular axis is so large that such modes are not observable. This follows because nearly all the molecular mass is concentrated within nuclear dimensions of the rotation axis, giving a negligibly small moment of inertia about the internuclear line (see Problem 11).

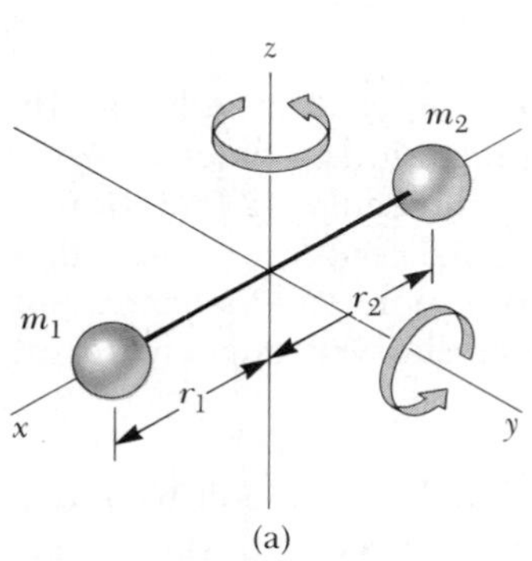

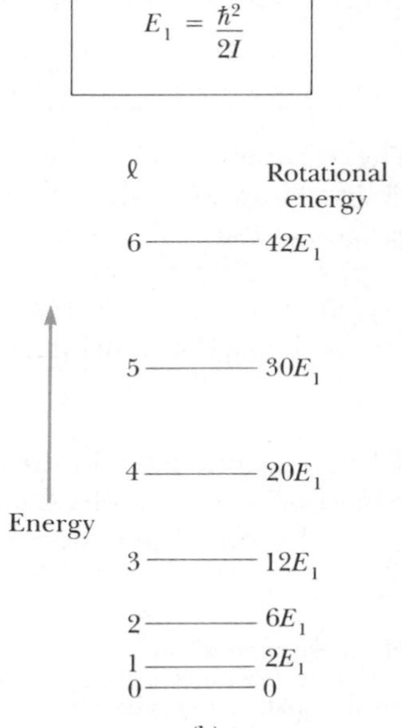

Figure 11.6 (a) A diatomic molecule oriented along the x axis has two rotational degrees of freedom, corresponding to rotation about the y and z axes. (b) Allowed rotational energies of a diatomic molecule as calculated using Equation 11.5.

Eliminating ω from the preceding equations gives the simple result

$$E_{\text{rot}} = \frac{L^2}{2I} \tag{11.1}$$

Rotational energy and angular momentum

Comparing Equation 11.1 with the kinetic energy of a particle in translation, $p^2/2m$, we see that the moment of inertia I of the molecule measures its resistance to rotation in the same way that the mass m of a single particle measures its resistance to translation. However, the value for I depends upon the rotation axis. For the important case where the axis of rotation passes through the center of mass,[3] we have $m_1r_1 = m_2r_2$, and r_1 and r_2 can be written in terms of the atomic separation $R_0 = r_1 + r_2$ as

$$R_0 = \left(\frac{m_2}{m_1} + 1\right) r_2 = \left(1 + \frac{m_1}{m_2}\right) r_1$$

Then the moment of inertia about the center of mass, I_{cm}, becomes

$$I_{\text{cm}} = \left(\frac{m_1 m_2}{m_1 + m_2}\right) R_0{}^2 = \mu R_0{}^2 \tag{11.2}$$

where μ is the **reduced mass** of the molecule,

$$\mu = \frac{m_1 m_2}{m_1 + m_2} \tag{11.3}$$

Unlike moment of inertia, which is a *property* of the molecule, angular momentum L is a *dynamical variable;* in the transition to quantum mechanics, L^2 becomes quantized as discussed in Chapter 7. The correct quantization rule is

$$L^2 = \ell(\ell + 1)\hbar^2 \qquad \ell = 0, 1, 2, \ldots \tag{11.4}$$

which, in turn, restricts the energy of rotation to be one of the values

$$\boxed{E_{\text{rot}} = \frac{\hbar^2}{2I_{\text{cm}}}\ell(\ell + 1)} \qquad \ell = 0,1,2, \ldots \tag{11.5}$$

Allowed energies for rotation

In the context of molecular rotation, the integer ℓ is called the **rotational quantum number.** Thus, we see that *the rotational energy of the molecule is quantized and depends on the moment of inertia of the molecule.* The allowed rotational energies of a diatomic molecule are sketched in Figure 11.6b. These results apply also to polyatomic molecules provided the appropriate generalization of I_{cm} is used.

The spacing between adjacent rotational levels can be calculated from Equation 11.5:

$$\Delta E = E_\ell - E_{\ell-1} = \frac{\hbar^2}{2I_{\text{cm}}}\{\ell(\ell + 1) - (\ell - 1)\ell\} = \frac{\hbar^2}{I_{\text{cm}}}\ell \tag{11.6}$$

where ℓ is the quantum number of the higher energy state. In going from one rotational state to the next, the molecule loses (or gains) energy ΔE. The loss

[3] Since there is no axis about which the molecule is constrained to rotate, the correct axis for calculating rotational energy is one passing through the center of mass. Energy associated with rotation about any other axis would include some energy of translation, as well as energy of rotation.

TABLE 11.1 Microwave Absorption Lines for Several Rotational Transitions of the CO Molecule

Rotational Transition	Wavelength of Absorption Line (m)	Frequency of Absorption Line (Hz)
$\ell=0 \rightarrow \ell=1$	2.60×10^{-3}	1.15×10^{11}
$\ell=1 \rightarrow \ell=2$	1.30×10^{-3}	2.30×10^{11}
$\ell=2 \rightarrow \ell=3$	8.77×10^{-4}	3.46×10^{11}
$\ell=3 \rightarrow \ell=4$	6.50×10^{-5}	4.61×10^{11}

(or gain) is accompanied by photon emission (or absorption) at the frequency $\omega = \Delta E/\hbar$. Thus, photons should be observed at the frequencies $\omega_0 = \hbar/I$, $2\omega_0$, $3\omega_0$, ... These predictions are in excellent agreement with experiment.[4] The wavelengths and frequencies for the absorption spectrum of the CO molecule are given in Table 11.1. The frequencies lie in the *microwave range* of the electromagnetic spectrum, as is typical of the rotational spectra of all molecules. From the data, one can deduce the moment of inertia and the bond length of the molecule, as will be shown in the following example.

EXAMPLE 11.1 Rotation of the CO Molecule
The $\ell = 0$ to $\ell = 1$ rotational transition of the CO molecule occurs at a frequency of 1.15×10^{11} Hz. (a) Use this information to calculate the moment of inertia of the molecule about its center of mass.

Solution: From Equation 11.6, we see that the energy difference between the $\ell = 0$ and $\ell = 1$ rotational levels is $\hbar^2/I_{cm}$. Equating this to the energy of the absorbed photon, we get

$$\frac{\hbar^2}{I_{cm}} = \hbar\omega$$

The angular frequency ω of the absorbed radiation is

$$\omega = 2\pi f = 2\pi(1.15 \times 10^{11}\text{ Hz}) = 7.23 \times 10^{11}\text{ rad/s}$$

so that I_{cm} becomes

$$I_{cm} = \frac{\hbar}{\omega} = \frac{1.055 \times 10^{-34}\text{ J}\cdot\text{s}}{7.23 \times 10^{11}\text{ rad/s}} = 1.46 \times 10^{-46}\text{ kg}\cdot\text{m}^2$$

(b) Calculate the bond length of the molecule.

Solution: Equation 11.2 can be used to calculate the bond length once the reduced mass of the molecule is found. Since the carbon and oxygen atomic masses are 12 u and 16 u, respectively, the reduced mass of the CO molecule is, from Equation 11.3,

$$\mu = \frac{(12\text{ u})(16\text{ u})}{12\text{ u} + 16\text{ u}} = 6.857\text{ u} = 1.14 \times 10^{-26}\text{ kg}$$

where the conversion $1\text{ u} = 1.66 \times 10^{-27}$ kg has been used. Then

$$R_0 = \sqrt{\frac{I_{cm}}{\mu}} = \sqrt{\frac{1.46 \times 10^{-46}\text{ kg}\cdot\text{m}^2}{1.14 \times 10^{-26}\text{ kg}}} = 1.13 \times 10^{-10}\text{ m}$$

$$= 0.113\text{ nm}$$

This example illustrates the immense power of spectroscopic measurements to determine molecular properties!

Molecular Vibration

A molecule is a flexible structure whose atoms are bonded together by what can be considered "effective springs." If disturbed, the molecule can vibrate, taking on vibrational energy. This energy of vibration may be altered if the molecule is exposed to radiation of the proper frequency.

[4] These simple statements must be refined when molecular vibration is taken into account, as discussed in Section 11.3.

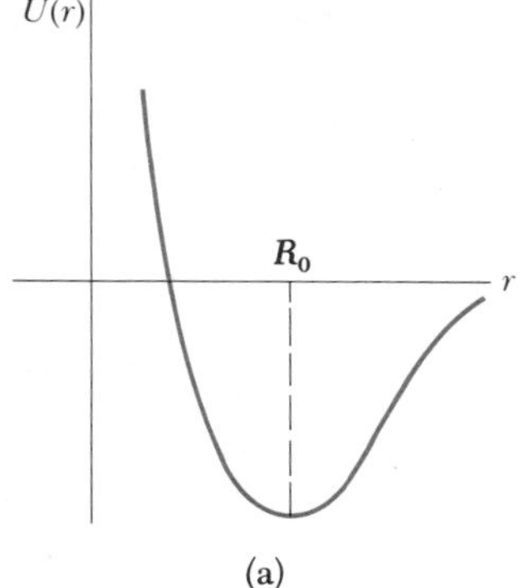

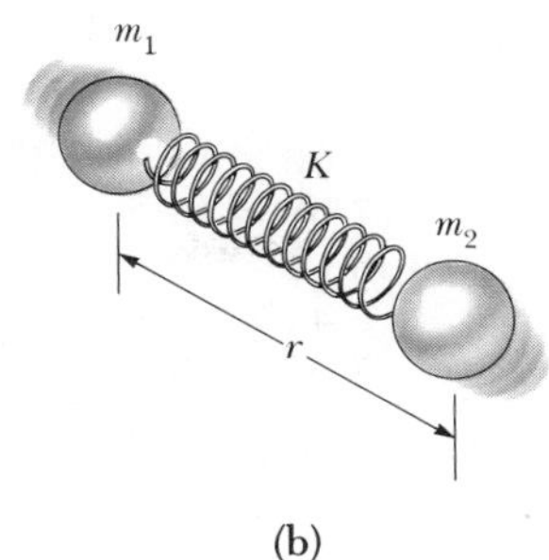

Figure 11.7 (a) A plot of the potential energy of a diatomic molecule versus atomic separation. The parameter R_0 is the equilibrium separation of the atoms. (b) Model of a diatomic molecule whose atoms are bonded by an effective spring of force constant K. The fundamental vibration is along the molecular axis.

Consider a diatomic molecule. The potential energy $U(r)$ versus atomic separation r for such a molecule is sketched in Figure 11.7a. The equilibrium separation of the atoms is denoted there by R_0; for small displacements from equilibrium, the atoms vibrate, as if connected by a spring with unstretched length R_0 and force constant K (Figure 11.7b).[5] Atomic displacements in the direction of the molecular axis give rise to oscillations along the line joining the atoms. For these *longitudinal* vibrations, the system is effectively one-dimensional, with the coordinates of each atom measured along the molecular axis.

Harmonic approximation to molecular vibration

We denote by ξ_1 and ξ_2 the displacements from equilibrium of m_1 and m_2, respectively. In terms of these displacements, the effective spring is stretched a net amount $\xi_1 - \xi_2$, and the elastic energy of the two-atom pair is

$$U = \tfrac{1}{2}K(\xi_1 - \xi_2)^2$$

The kinetic energy of the pair is $(p_1{}^2/2m_1) + (p_2{}^2/2m_2)$, but some of this represents translational energy for the molecule as a whole. To isolate the vibrational component, we examine the problem in the center-of-mass (CM) frame, where the CM of the molecule is at rest. In the CM, the atomic momenta are equal in magnitude but oppositely directed, that is, $p_2 = -p_1$, and the kinetic energy (now of vibration only) is simply

$$KE_{\text{vib}} = p_1{}^2\left(\frac{1}{2m_1} + \frac{1}{2m_2}\right) = \frac{p_1{}^2}{2\mu}$$

where μ is again the reduced mass defined in Equation 11.3.

The preceding relations describe a one-dimensional oscillator with vibration coordinate $\xi = \xi_1 - \xi_2$. The Schrödinger equation for this case is

$$-\frac{\hbar^2}{2\mu}\frac{d^2\psi}{d\xi^2} + \tfrac{1}{2}K\xi^2\psi(\xi) = E_{\text{vib}}\psi(\xi) \tag{11.7}$$

which is that for the quantum oscillator studied in Chapter 5, with x there replaced by ξ. The allowed energies of vibration are the oscillator levels (cf. Equation 5.29),

$$\boxed{E_{\text{vib}} = (v + \tfrac{1}{2})\hbar\omega} \qquad v = 0, 1, 2, \ldots \tag{11.8}$$

Allowed energies for vibration.

[5] In Section 5.6 of Chapter 5 we showed that the effective force constant K is given by the curvature of $U(r)$ evaluated at the equilibrium separation R_0; that is, $K = \partial^2 U/\partial r^2|_{R_0}$.

TABLE 11.2 Fundamental Vibrational Frequencies and Effective Force Constants for Some Diatomic Molecules

Molecule	Frequency (Hz), $v = 0$ to $v = 1$	Force Constant (N/m)
HF	8.72×10^{13}	970
HCl	8.66×10^{13}	480
HBr	7.68×10^{13}	410
HI	6.69×10^{13}	320
CO	6.42×10^{13}	1860
NO	5.63×10^{13}	1530

From G. M. Barrows, *The Structure of Molecules,* New York, W. A. Benjamin, 1963.

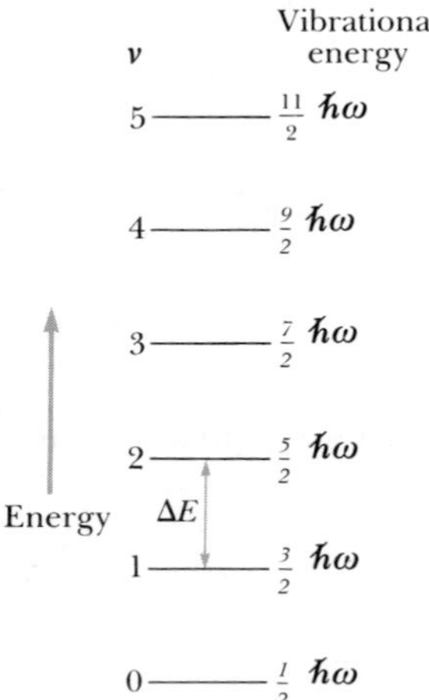

Figure 11.8 Allowed vibrational energies of a diatomic molecule, where ω is the fundamental frequency of vibration given by $\omega = \sqrt{K/\mu}$. Note that the spacings between adjacent vibrational levels are equal.

where v is an integer called the **vibrational quantum number.** The frequency ω is the classical frequency of vibration, and is related to the force constant K by

$$K = \mu\omega^2 \tag{11.9}$$

In the lowest vibrational state, $v = 0$, we find $E_{\text{vib}} = \frac{1}{2}\hbar\omega$, the so-called *zero point energy.* The accompanying vibration—the zero point motion—is always present, even if the molecule is not excited. From Equation 11.8 we see also that the energy difference between any two successive vibrational levels is the same and is given by $\Delta E_{\text{vib}} = \hbar\omega$. A typical value for ΔE_{vib} can be found from the entries in Table 11.2, and turns out to be about 0.26 eV.

The vibrational energies of a diatomic molecule are plotted in Figure 11.8. Normally, more molecules are in the lowest energy state because the thermal energy at ordinary temperatures ($kT \approx 0.025$ eV) is insufficient to excite the molecule to the next available vibrational state. However, electromagnetic radiation can stimulate transitions to the first excited level. Such a transition would be accompanied by the absorption of a photon to conserve energy. Once excited, the molecule can return to the lower vibrational state by emitting a photon of the same energy. For molecular vibrations, photon absorption and emission occur in the *infrared* region of the spectrum. Absorption frequencies for the $v = 0$ to $v = 1$ transitions of several diatomic molecules are listed in Table 11.2, together with the effective force constants K calculated from Equation 11.9. Since larger force constants describe stiffer springs, K indicates the strength of the molecular bond. Notice that the CO molecule, which is bonded by several electrons, is much more rigid than such single-bonded molecules as HCl.

EXAMPLE 11.2 Vibration of the CO Molecule

The CO molecule shows a strong absorption line at the frequency 6.42×10^{13} Hz. (a) Calculate the effective force constant for this molecule.

Solution: The absorption process is accompanied by a molecular transition from the $v = 0$ to the $v = 1$ vibrational level. Since the energy difference between these levels is $\Delta E_{\text{vib}} = \hbar\omega$, the absorbed photon must have carried this much energy. It follows that the photon frequency is just $\Delta E_{\text{vib}}/\hbar = \omega$, the frequency of the CO oscillator! From the information given, we calculate

$$\omega = 2\pi f = 2\pi(6.42 \times 10^{13}) = 4.03 \times 10^{14} \text{ rad/s}$$

Using 12 u and 16 u for the carbon and oxygen atomic masses, we find $\mu = 6.857 \text{ u} = 1.14 \times 10^{-26}$ kg, as in Example 11.1. Then

$$K = \mu\omega^2 = (1.14 \times 10^{-26})(4.03 \times 10^{14})^2$$
$$= 1.86 \times 10^3 \text{ N/m}$$

Thus, infrared spectroscopy furnishes useful information on the elastic properties (bond strengths) of molecules.

(b) What is the classical amplitude of vibration for a CO molecule in the $v = 0$ vibrational state?

Solution: The total vibrational energy for the molecule is $E_{vib} = \frac{1}{2}\hbar\omega$. At maximum displacement, the CO molecule has transformed all this into elastic energy of the spring,

$$\tfrac{1}{2}\hbar\omega = \tfrac{1}{2}KA^2$$

where A is the vibration amplitude. Using $K = \mu\omega^2$ and $\mu = 1.14 \times 10^{-26}$ kg, we get

$$A = \left(\frac{\hbar}{\mu\omega}\right)^{1/2} = \left\{\frac{1.055 \times 10^{-34}}{(1.14 \times 10^{-26})(4.03 \times 10^{14})}\right\}^{1/2}$$

$$= 4.79 \times 10^{-12}\ \text{m} = 0.00479\ \text{nm}$$

Comparing this with the bond length of 0.113 nm, we see that the vibration amplitude is only about 4% of the bond length.

11.3 MOLECULAR SPECTRA

In general, an excited molecule will rotate and vibrate simultaneously. To a first approximation, these motions are independent, with total energy given by the sum of Equations 11.5 and 11.8:

Rotation-vibration spectrum of a molecule

$$E_{\text{rot-vib}} = \frac{\hbar^2}{2I_{cm}}\{\ell(\ell+1)\} + (v + \tfrac{1}{2})\hbar\omega \qquad (11.10)$$

Together, the levels prescribed by Equation 11.10 constitute our approximation to the *rotation-vibration spectrum* of the molecule. For each allowed value of vibrational quantum number v, there is a complete rotational spectrum, corresponding to $\ell = 0, 1, 2, \ldots$ Since successive rotational levels are separated by energies much smaller than the vibrational energy $\hbar\omega$, the rotation-vibration spectrum of a typical molecule appears as shown in Figure 11.9. Each level represents a possible internal condition for the molecule, and is indexed by the *two* quantum numbers ℓ and v specifying the state of rotation and vibration, respectively. Normally, the molecule would take on the configuration with lowest energy, in this case the one designated by $v = 0$ and $\ell = 0$. External influences, such as temperature or the presence of electromagnetic radiation, can change the molecular condition, resulting in a transition from one rotation-vibration level to another. The attendant change in molecular energy must be compensated by absorption or emission of energy in some other form. When electromagnetic radiation is involved, the transitions—referred to as *optical transitions*—are subject to other conservation laws as well, since photons carry both momentum and energy.

Any optical transition between molecular levels with energies E_1 and E_2 must be accompanied by photon emission or absorption at the frequency

$$f = |E_2 - E_1|/h \qquad \text{or} \qquad \Delta E = \pm hf \qquad (11.11)$$

Since hf is the photon energy, this is energy conservation for the system of molecule plus photon. Equation 11.11 expresses a kind of *resonance* between the two; unless photons of the correct frequency (energy) are available, no transition is possible. Similar restrictions apply to other quantities that we know must be conserved in the process of transition. In particular, the initial and final states for an optical transition also must differ by exactly one angular momentum unit:

$$|\ell_2 - \ell_1| = 1 \qquad \text{or} \qquad \Delta\ell = \pm 1 \qquad (11.12)$$

The inference to be drawn from Equation 11.12 is that the photon carries angular momentum in the amount of $\hbar$, that is, the photon is a spin one particle with spin quantum number $s = 1$. Equation 11.12 then expresses angular

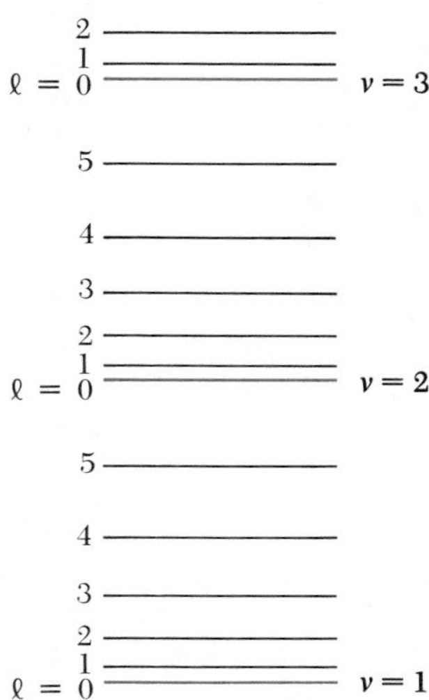

Figure 11.9 The rotation-vibration levels for a typical molecule. Note that the vibrational levels are separated by much larger energies, so that a complete rotational spectrum can be associated with each vibrational level.

momentum conservation for the system molecule plus photon. Equations 11.11 and 11.12 are the **selection rules** for optical transitions.

For the lower vibrational levels, there is also a restriction on the vibrational quantum number ν:

$$|\nu_2 - \nu_1| = 1 \qquad \text{or} \qquad \Delta\nu = \pm 1 \tag{11.13}$$

Equation 11.13 reflects more the harmonic character of the interatomic force than any photon property; indeed, for higher vibrational energies the concept of an effective spring joining the atoms is inaccurate, and Equation 11.13 ceases to be valid.

Selection rules greatly restrict the number of photon frequencies or wavelengths observed in molecular spectra, since transitions are prohibited unless all rules are obeyed simultaneously.[6] For instance, a pure rotational transition normally would not be observed, since this requires $\Delta\nu = 0$ in violation of Equation 11.13. In the same way, a pure vibrational transition ($\Delta\ell = 0$) is forbidden, and we conclude that optical transitions usually involve both molecular vibration and rotation.

The spectrum of a particular molecule can be predicted by considering a collection of such molecules, initially undisturbed. At ordinary temperatures there is insufficient thermal energy to excite any but the $\nu = 0$ vibrational mode, although the molecules will be in various states of rotation. Since a pure rotational transition is forbidden, optical *absorption* must result from transitions in which ν increases by one unit while ℓ either increases or decreases, also by one unit (Figure 11.10*a*). Thus, the molecular absorption spectrum consists of two sequences of lines, represented by $\Delta\ell = \pm 1$, with $\Delta\nu = +1$ for both cases. The energies of the absorbed photons are readily calculated from Equation 11.10:

$$\begin{aligned} \Delta E &= \hbar\omega + \frac{\hbar^2}{2I_{cm}} 2(\ell + 1) \qquad \ell = 0, 1, \ldots \qquad (\Delta\ell = +1) \\ \Delta E &= \hbar\omega - \frac{\hbar^2}{2I_{cm}} 2\ell \qquad \ell = 1, 2, \ldots \qquad (\Delta\ell = -1) \end{aligned} \tag{11.14}$$

Here ℓ is the rotational quantum number of the initial state. The first of Equations 11.14 generates a series of equally spaced lines at frequencies *above* the characteristic vibration frequency ω, and the second generates a series *below* this frequency. Adjacent lines are separated in frequency by the fundamental unit $\hbar/I_{cm}$. Notice that ω itself is excluded, since ℓ cannot be zero if the transition is one for which ℓ decreases ($\Delta\ell = -1$). Figure 11.10b shows the expected frequencies in the absorption spectrum for the molecule; these same frequencies appear in the emission spectrum.

The actual absorption spectrum of the HCl molecule is shown in Figure 11.11 for comparison. The similarities are evident, and reinforce the validity of our model. However, one peculiarity is apparent: each line in the HCl spectrum is split into a doublet. This doubling occurs because the sample is a mixture of two chlorine isotopes, ^{35}Cl and ^{37}Cl, whose different masses give rise to two distinct values for I_{cm}. Notice, too, that not all spectral lines appear with the same intensity, because even the allowed transitions occur at differ-

[6] Transitions may occur in violation of the selection rules. However, these are rare enough that the corresponding spectral line intensity would fall well below the normal level for detection.

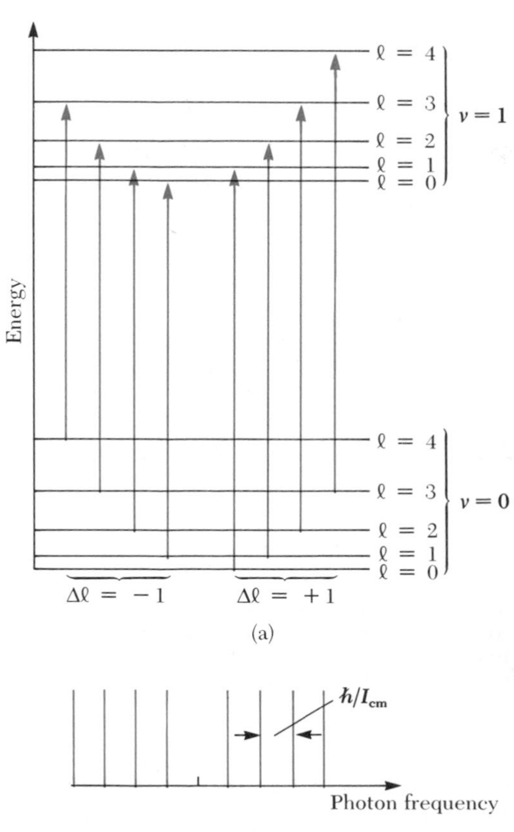

Figure 11.10 (a) Absorptive transitions between the $v = 0$ and $v = 1$ vibrational states of a diatomic molecule. The transitions obey the selection rule $\Delta \ell = \pm 1$, and fall into two sequences, those for which $\Delta \ell = +1$ and those for which $\Delta \ell = -1$. The transition energies are given by Equation 11.14. (b) Expected lines in the optical absorption spectrum of a molecule. The lines on the right side of center correspond to transitions in which ℓ changes by $+1$, while the lines to the left of center correspond to transitions for which ℓ changes by -1. These same lines appear in the emission spectrum.

ent *rates* (number of photons absorbed per second). Transition rates are governed chiefly by the populations of the initial and final states, and these depend upon the degeneracy of the levels as well as the temperature of the system.

Finally, several refinements to our simple model deserve mention. As we have already suggested, the atoms of a molecule exert complicated forces on

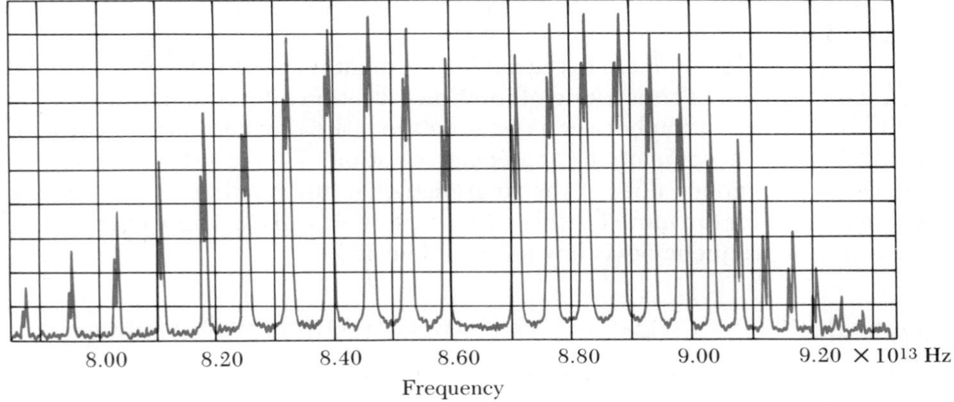

Figure 11.11 The absorption spectrum of the HCl molecule. Each line is split into a doublet because chlorine has two isotopes, ^{35}Cl and ^{37}Cl, which have different nuclear masses.

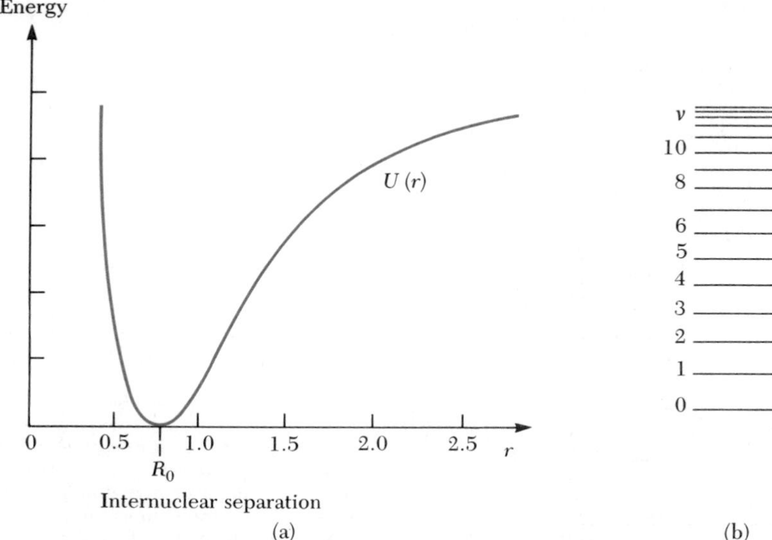

Figure 11.12 (a) Potential energy $U(r)$ for the Morse oscillator. In the equilibrium region around R_0, $U(r)$ is harmonic, but it rises sharply as the atoms are brought closer together. (b) Allowed energies of vibration for the Morse oscillator. Notice that the separation between adjacent levels *decreases* with increasing energy.

Anharmonic effects

one another. These forces are harmonic only near equilibrium. More vibrational energy implies larger vibration amplitude, and with it a breakdown of the harmonic approximation—the effective spring must give way to a more accurate representation of the true interatomic force. One way to do this consists of replacing the harmonic potential $\frac{1}{2}K\xi^2$ in Equation 11.7 with the *Morse potential* shown in Figure 11.12a. Near equilibrium ($r = R_0$) the Morse potential is harmonic; further away, $U(r)$ mimics the asymmetry characteristic of interatomic forces, becoming zero when the atoms are widely separated ($r \rightarrow \infty$) but rising sharply as the atoms come close together. In consequence, oscillations take place about an average position $\langle r \rangle$ that *increases* with the energy of vibration. This is the origin of *thermal expansion*, where the increased vibrational energy results from raising the temperature of the sample.

Unlike the harmonic oscillator, the interval separating successive levels of the Morse oscillator diminishes steadily at higher energies, as illustrated in Figure 11.12b.

As a second refinement to the model, we mention briefly the phenomenon of *bond stretching*. This is a kind of rotation-vibration interaction, wherein the equilibrium separation between two atoms lengthens in response to the centrifugal forces accompanying molecular rotation. The result is that I_{cm} increases with higher values of the rotational quantum number ℓ, the increase being greater for the more flexible structures, such as HCl, whose characteristic vibration frequency ω is low (cf. Problem 12).

11.4 ELECTRON SHARING AND THE COVALENT BOND (Optional)

In this section we examine the allowed energies and wavefunctions for electrons in a molecule. There are two aspects to the problem: one deals with the complexity of electron motion in the field of several nuclei, the other with the efect of all other electrons on the motion of any one. The two may be divided by first treating a one-electron molecular ion such as H_2^+, and subsequently examining the effects of adding one more electron to give the neutral hydrogen molecule H_2.

The Hydrogen Molecular Ion

The hydrogen molecular ion H_2^+ consists of one electron in the field of two protons (Fig. 11.13). The equilibrium separation of the protons in H_2^+ is about 0.1 nm. The electron is drawn to each proton by the force of Coulomb attraction, resulting in a total potential energy for the electron

$$U(\mathbf{r}) = -\frac{ke^2}{|\mathbf{r}|} - \frac{ke^2}{|\mathbf{r}-\mathbf{R}|} \tag{11.15}$$

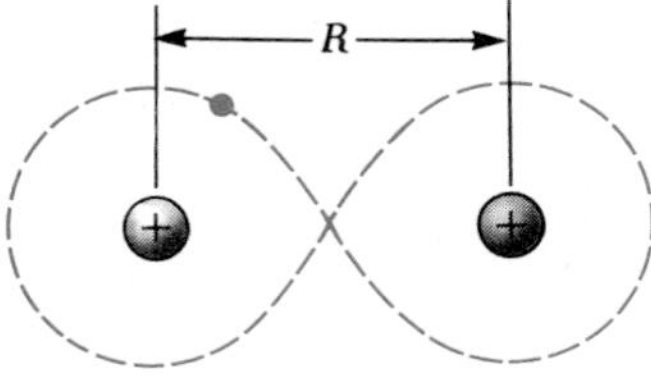

Figure 11.13 The hydrogen molecular ion H_2^+. The lone electron is attracted to both protons by the electrostatic force between opposite charges. The equilibrium separation of the protons in H_2^+ is about 0.1 nm.

The stationary state wavefunctions and energies for the electron in H_2^+ are solutions to the time-independent Schrödinger equation with the potential $U(\mathbf{r})$ of Equation 11.15.

As a guide, it is useful to consider the problem in the two limiting cases $R \to 0$ and $R \to \infty$, for which the results are already known. As $R \to 0$, the two protons coalesce to form a nucleus of helium; consequently, the wavefunctions and energies in this *united atom limit* are those of He^+, given by the formulas of Chapter 7 with atomic number $Z = 2$. In particular, the ground and first excited states are both spherically symmetric waves with energies of −54.4 eV and −13.6 eV, respectively. In the *separated atom limit* $R \to \infty$, the electron sees only the field of a single proton, and the levels are those of atomic hydrogen, −13.6 eV for the ground state and −3.4 eV for the first excited state. Energies for these extreme cases can be marked on the **correlation diagram** of Figure 11.14. The spatial degeneracy of these levels is important, and is given by the numbers in parentheses. Notice especially that the ground state of the united atom is nondegenerate, but the ground state for the separated atoms is doubly degenerate. The extra degeneracy stems from the presence of the second proton in H_2^+: in the separated atom limit, the electron can occupy a hydrogenic orbital around *either* proton. The same argument shows that *all of the atomic levels in the separated atom limit have an extra twofold degeneracy,* corresponding to hydrogenic wavefunctions centered at the site of either proton. The completed correlation diagram for H_2^+ includes the energy levels at any proton separation R.

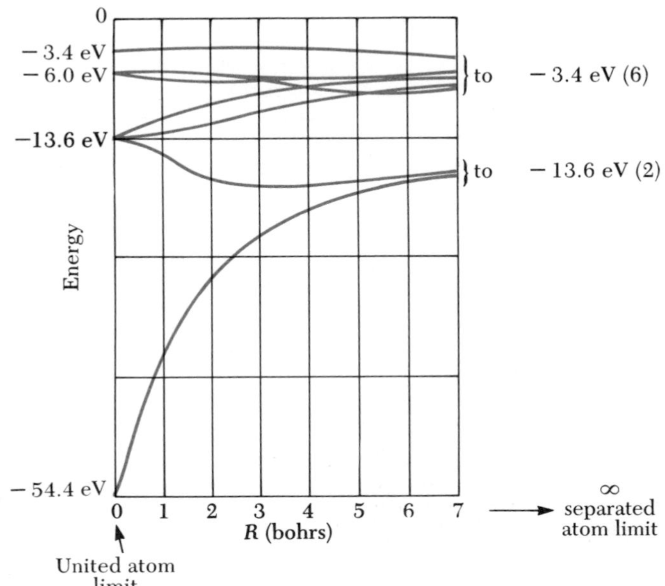

Figure 11.14 Correlation diagram for H_2^+, showing the two lowest energy levels as a function of proton separation. At $R = 0$ the levels are those of the united atom (ion) He^+, while at $R = \infty$ the levels are those of neutral H. The spatial degeneracy of the various levels is given by the numbers in parentheses.

To find out what happens at intermediate values of R, consider first the case where R is finite but large, so that the influence of the other proton, though weak, cannot be ignored. This appearance of a second attractive force center means that the electron will not remain with its parent proton indefinitely. Given sufficient time, the electron inevitably will *tunnel* through the intervening potential barrier to occupy an atomic orbital around the far proton. Some time later, the electron will tunnel back to the original proton. In principle, the situation is no different from the inversion of the ammonia molecule discussed in Section 6.2: in effect, H_2^+ is a double oscillator, for which stationary states are not waveforms concentrated at a single force center, but divided equally between them. To be sure, the tunneling time increases with proton separation, and may be very long indeed—something like 1 s for protons 1 nm apart (an eternity by atomic standards!). But the time itself is inconsequential, since we are interested here in the stationary states that have, so to speak, "all the time in the world" to form.

From the symmetry of H_2^+, the tunneling probability from one proton to the other must be the same in either direction, and so we expect the electron to spend equal amounts of time in the vicinity of each. It follows that the stationary state wave is an equal mixture of atomic-like wave functions centered on each proton as, for example,

Molecular orbitals from atomic wavefunctions

$$\boxed{\psi_+(\mathbf{r}) = \psi_a(\mathbf{r}) + \psi_a(\mathbf{r} - \mathbf{R})} \tag{11.16}$$

The subscript a on the right is a collective label for all the quantum numbers needed to specify an atomic state; for the ground state of atomic hydrogen, a is (1,0,0), meaning $n = 1$, $\ell = 0$, and $m_\ell = 0$. Equation 11.16 describes the *symmetric* combination of two atomic wave functions centered at $\mathbf{r} = 0$ and $\mathbf{r} = \mathbf{R}$. The wave function and probability density given by $\psi_+(\mathbf{r})$ are sketched in Figure 11.15a and b; both are symmetric about the molecular center located at $\mathbf{r} = \mathbf{R}/2$, or halfway between the two protons.

However, we could also have written the antisymmetric form

$$\boxed{\psi_-(\mathbf{r}) = \psi_a(\mathbf{r}) - \psi_a(\mathbf{r} - \mathbf{R})} \tag{11.17}$$

This again leads to a symmetric probability distribution, but one with a *node* at the molecular center (Fig. 11.15c and d). In fact, Equations 11.16 and 11.17 *both*

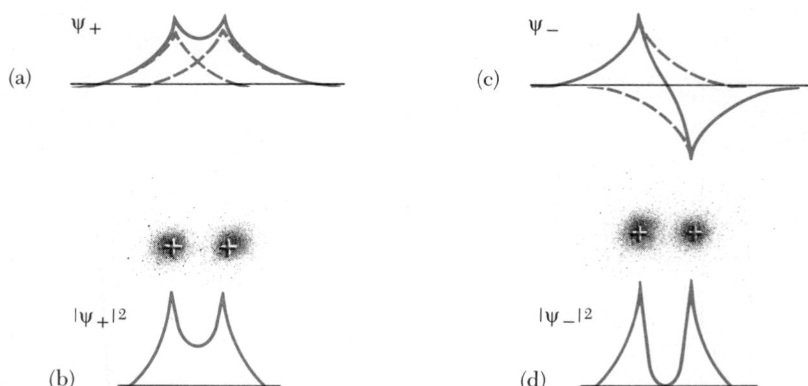

Figure 11.15 The wave function (a) and probability density (b) for the approximate molecular wave ψ_+ formed from the symmetric combination of atomic orbitals centered at $r = 0$ and $r = R$. (c) and (d): Wave function and probability density for the approximate molecular wave ψ_- formed from the antisymmetric combination of these same orbitals.

approximate true molecular wave functions when R is large, but they describe states having distinctly different energies. In particular, ψ_- has a somewhat higher energy than ψ_+, since the electron in the antisymmetric state is more confined, being relegated largely to the vicinity of one or the other nucleus. As with the particle in a box, this greater degree of confinement comes at the expense of increased kinetic energy. Only in the limit $R \to \infty$ do the two energies merge to form the (doubly degenerate) atomic level of the separate atoms. As $R \to 0$, continuity requires the lower of the two energies to approach -54.4 eV, the ground state energy of the united atom He^+; similarly, the higher energy goes over to the first excited state energy of He^+, -13.6 eV.

We have come upon the quantum origins of the covalent bond. The energetically preferred state ψ_+ is referred to as the **bonding orbital,** since the electron in this state spends much of its time in the space between the two protons, shuttling to and fro and acting as a kind of "glue" that holds the molecule together. The more energetic state ψ_- is the **antibonding orbital,** and is largely ineffective in promoting molecular stability. Notice that a bonding-antibonding orbital pair is associated with *every* orbital of the separated atoms, not just with the ground state. The bonding tendency of the electron in H_2^+ is explored further in the paragraphs that follow.

EXAMPLE 11.3 Normalizing a Molecular Wavefunction

Normalize the molecular wavefunction $\psi_\pm$ of Equations 11.16 and 11.17, assuming the atomic wavefunctions ψ_a are already normalized.

Solution: The normalization integral for $\psi_\pm$ is

$$\int |\psi_\pm|^2 dV = \int |\psi_a(\mathbf{r})|^2 dV + \int |\psi_a(\mathbf{r}-\mathbf{R})|^2 dV$$

$$\pm \int \psi_a^*(\mathbf{r})\psi_a(\mathbf{r}-\mathbf{R})dV \pm \int \psi_a(\mathbf{r})\psi_a^*(\mathbf{r}-\mathbf{R})dV$$

Since the atomic wavefunctions are already normalized, the first two integrals on the right are both unity. If R is large, the last two integrals can be neglected, since the integrand for each is the product of one atomic function centered at $\mathbf{r} = 0$ with another centered at $\mathbf{r} = \mathbf{R}$. The resulting product is small everywhere provided R is sufficiently large, and these integrals—known as *overlap integrals*—are very nearly zero. Thus, in the lowest approximation

$$\int |\psi_\pm|^2 dV \simeq 2$$

and approximately normalized wavefunctions can be had by multiplying the waves of Equations 11.16 and 11.17 by the factor $1/\sqrt{2}$.

In a more refined treatment, the overlap integrals would have to be evaluated. For the hydrogen ground state, the atomic wavefunction is

$$\psi_a(r) = (1/\sqrt{\pi})a_0^{-3/2}\, e^{-r/a_0}$$

With this, the first overlap integral, denoted Δ, can be shown to be (see Appendix D)

$$\int \psi_a^*(\mathbf{r})\psi_a(\mathbf{r}-\mathbf{R})dV = \Delta(\mathbf{R}) = \left(1 + R + \frac{R^2}{3}\right) e^{-R}$$

In writing this result, we have substituted simply R for R/a_0, so that all distances are now to be measured in bohrs. The other overlap integral is the same but for complex conjugation, which in the present case does not alter the result:

$$\int \psi_a(\mathbf{r})\psi_a^*(\mathbf{r}-\mathbf{R})dV = \Delta^*(\mathbf{R}) = \Delta(\mathbf{R})$$

With overlaps included, we find

$$\int |\psi_\pm|^2 dV \approx 2 \pm 2\Delta,$$

and the proper normalizing factor for the wave $\psi_\pm$ becomes $1/\sqrt{2(1 \pm \Delta)}$.

At the observed interatomic separation for H_2^+, about $R = 2$ bohrs, Δ is 0.586, but diminishes rapidly with R to become negligible beyond about 5 bohrs. This rapid falloff of Δ with R is characteristic of overlap integrals, and stems from the exponential behavior of atomic wavefunctions. Notice, too, that Δ is the same for $\boldsymbol{R}$ in any direction, that is, $\Delta(\boldsymbol{R})$ depends only on the proton separation $R = |\boldsymbol{R}|$. This is a consequence of the spherical symmetry of the atomic functions, and would not arise had we done the calculation using atomic p states, for example.

The energy of the bonding orbitals in H_2^+ can be estimated from Schrödinger's equation. For the electron in H_2^+ this is

$$-\frac{\hbar^2}{2m}\nabla^2\psi + U(\boldsymbol{r})\psi = E\psi,$$

with $U(\boldsymbol{r})$ the potential energy of the electron in the field of the two protons, as in Equation 11.15. Let us multiply every term by ψ^* and integrate over the whole space. Then the RHS becomes just E (assuming ψ is normalized),

$$E\int |\psi|^2 dV = E,$$

leaving for the LHS an expression which we can use to compute the value of E:

$$E = \int \psi^*\left[-\frac{\hbar^2}{2m}\nabla^2 - \left\{\frac{ke^2}{|\boldsymbol{r}|} + \frac{ke^2}{|\boldsymbol{r}-\boldsymbol{R}|}\right\}\right]\psi\, dV \qquad (11.18)$$

If ψ is the true molecular wavefunction, this equation furnishes the exact value of particle energy in the state ψ; if ψ is an approximate waveform, Equation 11.18 yields an approximation to E.

Bonding orbital of H_2^+

Substituting for ψ the symmetric combination ψ_+ (suitably normalized according to Example 11.3) gives a first approximation to the energy of the bonding orbitals in H_2^+. For the case where ψ_a refers to the hydrogen atom ground state, the molecular energy E_+ can be shown to be

$$E_+ = -1 - \frac{2}{1+\Delta}\left\{\frac{1}{R} - \frac{1}{R}(1+R)e^{-2R} + (1+R)e^{-R}\right\} \qquad (11.19)$$

with Δ the overlap integral of Example 11.3:

$$\Delta = \left(1 + R + \frac{R^2}{3}\right)e^{-R}$$

These expressions for E_+ and Δ are given in *atomic units*, where the rydberg ($= -13.6$ eV) is adopted as the unit of energy to go along with the bohr unit of length.

In the limit of large R, $\Delta \rightarrow 0$ and E_+ approaches the energy of an isolated hydrogen atom $E_a = -1$ Ryd, as it should. As R decreases, E_+ becomes more and more negative, finally reaching -3 Ryd at $R = 0$ where the nuclei coalesce. (The correct value, which is -4 Ryd for the ground state of He^+, is not reproduced owing to the approximation inherent in our use of Equation 11.16 for the molecular wavefunction.) Since E_+ decreases steadily with R, the molecule appears to be unstable, and should collapse to $R = 0$ under the bonding tendency of the orbiting electron. But this overlooks the Coulomb energy of the protons, which must be included to obtain the total molecular energy. Two protons separated by a distance R repel each other with energy ke^2/R, or $2/R$ Ryd. The Coulomb repulsion of the protons offsets the bonding attraction of the electron to stabilize the molecule at

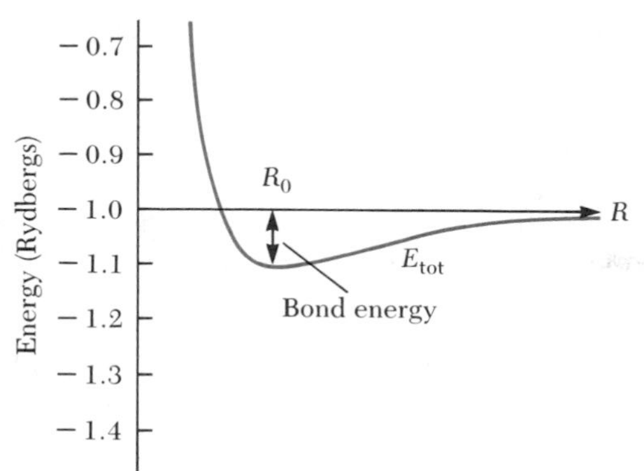

Figure 11.16 The total molecular energy for the bonding orbital of H_2^+, as given by the approximate wave function ψ_+ of Equation 11.16. The predicted bond length occurs at the point of stable equilibrium, around $R = 2.5$ bohrs. The predicted bond energy is about 1.77 eV.

that separation for which the total energy is minimal. The total molecular energy given by our model is sketched as a function of R in Figure 11.16. The minimum comes at $R_0 = 2.49$ bohrs, or 0.132 nm, and agrees reasonably well with the observed *bond length* for H_2^+, 0.106 nm.

At the equilibrium separation of 2.49 bohrs, we find $E_+ = -2.13$ Ryd, and a total molecular energy of $E_+ + 2/R_0 = -1.13$ Ryd. The negative of this, 1.13 Ryd or 15.37 eV, represents the work required to separate the molecule into its constituents, and is called the **dissociation energy.** The measured dissociation energy for H_+ is about 16.3 eV. The **bond energy,** or the work required to separate H_2^+ into H and H^+, is the difference between the molecular energy at $R = \infty$ and at equilibrium. Our model predicts a bond energy of 15.37 eV − 13.6 eV = 1.77 eV, which is somewhat less than the actual value of 2.65 eV. The difference can be attributed to our use of Equation 11.16 for the molecular wavefunction, an approximation which is best for larger values of R.

To obtain the energy of the antibonding orbitals for H_2^+, we replace ψ in Equation 11.18 with the antisymmetric wave ψ_- of Equation 11.17 (again, normalized as in Example 11.3). With ψ_a the hydrogen ground state, we find for the energy of the lowest antibonding orbital in H_2^+

$$E_- = -1 - \frac{2}{1-\Delta}\left\{\frac{1}{R} - \frac{1}{R}(1+R)e^{-2R} - (1+R)e^{-R}\right\} \qquad (11.20)$$

Antibonding orbital of H_2^+

For large separations, E_- coincides with E_+ at −1 Ryd. With decreasing R, however, E_- steadily *increases,* and always lies *above* E_+. In particular, E_- for any finite value of R is *higher* than the energy of $H + H^+$ in isolation, and no stable molecular ion will be formed in this state. Inclusion of the nuclear repulsion only enhances the instability. This is the antibonding effect for H_2^+.

While it must be admitted that E_- seriously overestimates the true molecular energy for small values of R, the conclusion reached vis-a-vis the instability of this state is, nonetheless, correct. The true antibonding orbital energy approaches −1 Ryd in the united atom limit and exhibits a broad, shallow minimum near R = 2 bohrs (see Figure 11.14). However, equilibrium cannot be sustained when the Coulomb repulsion of the nuclei is included, that is, the curve of total molecular energy shows no minimum for the antibonding state.

The Hydrogen Molecule

The addition of one more electron to H_2^+ gives the hydrogen molecule, H_2. The second electron provides even more "glue" for covalent bonding and should produce a stronger bond than that seen in H_2^+. Indeed, the bond energy for H_2 is 4.5 eV, compared to 2.65 eV for H_2^+. (Naively, we might have expected the bond energy to double with the number of electrons, but the electrons repel each other

—an antibonding effect.) The second electron also results in a measured bond length for H_2 equal to 0.074 nm, which is noticeably shorter than the 0.1 nm found for the single-bonded molecular ion H_2^+.

EXAMPLE 11.4 The Hydrogen Molecular Bond

Estimate the bond length and the bond energy of H_2, assuming that each electron moves independently of the other, and is described by the approximate waveform of Equation 11.16.

Solution: If the electrons do not interact, the energy of each must be the bonding energy E_+ of Equation 11.19. The total energy of the molecule in this *independent particle model* is then

$$E_{\text{tot}} = \frac{2}{R} + 2E_+$$

$$= \frac{2}{R} - 2 - \frac{4}{1+\Delta}\left\{\frac{1}{R} - \frac{1}{R}(1+R)e^{-2R} + (1+R)e^{-R}\right\}$$

By trial and error, we discover that the minimum energy occurs at about $R_0 = 1.44$ bohr or 0.076 nm, which becomes the bond length predicted by our model. The molecular energy at this separation is $E_0 = -3.315$ Ryd. Comparing this with the total energy of the separated hydrogen atoms, -2 Ryd, leaves for the bond energy 1.315 Ryd, or 17.88 eV. Due to our neglect of electron repulsion, this model grossly exaggerates the bond energy, although it does get the bond length about right. Even the latter agreement is probably fortuitous, however, since Equation 11.16 is not expected to be accurate even for noninteracting electrons at nuclear separations as small as 1.5 bohr.

So far we have not recognized that the two electrons in H_2, being identical fermions, are subject to the exclusion principle. If both electrons occupy the bonding orbital of H_2—as we have been assuming—*their spins must be opposite.* If their spins were parallel, one electron would be forced into the antibonding orbital, where it has higher energy. Figure 11.17 shows the electronic energy of the molecule for the two cases. These curves were obtained from laborious calculations

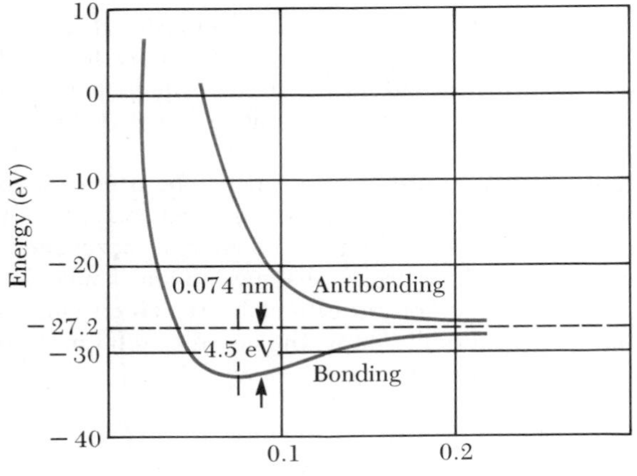

Figure 11.17 Total molecular energy for the bonding and antibonding orbitals of H_2. For both electrons to be in the bonding orbital, their spins must be opposite (↑ ↓). Since the energy of the antibonding orbital exceeds that of the isolated H atoms, no stable molecule can be formed in this state.

which account in an approximate way for the effect of electron-electron repulsion. The results imply that no stable molecule is formed in the parallel spin case.

Identical considerations suggest that three hydrogen atoms will not bond to form a stable molecule. With three electrons, at least two would have their spins aligned, thereby forcing one of the electrons into an antibonding orbital. The resulting configuration evidently has more energy than $H_2 + H$ in isolation, and so is unstable against the formation of an H_2 molecule and a separate H atom. In this way, we conclude correctly that the molecule H_3 does not exist.

11.5 BONDING IN COMPLEX MOLECULES (Optional)

The bonds we found in H_2^+ and H_2 are known as **sigma** type molecular bonds, denoted σ. They arise from the overlap of atomic *s* states, and are characterized by an electron density that is axially symmetric about the line joining the two atoms. Bonding between atoms having more electrons often involves the overlap of *p* states or other atomic orbitals, and leads to other types of molecular bonds with their own distinctive characteristics.

Consider the nitrogen molecule, N_2, the simplest molecule to exhibit bonds other than the σ type. The electron configuration of the nitrogen atom is $1s^22s^22p^3$. The molecular bond in N_2 is due primarily to the three valence electrons in the $2p$ subshell; the inner *s* electrons are too tightly bound to their parent atoms to participate in the sharing necessary for bond formation. The $2p$ subshell is made up of three atomic orbitals ($m_\ell = 1, 0, -1$), and each of these gives rise to a bonding and an antibonding orbital for the molecule. The N_2 molecule has six valence electrons to accommodate, three from each N atom. With two electrons in each of the three bonding orbitals, the N_2 molecule is especially stable, with a bond energy of 9.8 eV and a bond length of 0.11 nm.

Sigma and pi bonds for N_2

Of the three bonding-antibonding orbital pairs in N_2, however, only one is a sigma bond. Recall that atomic *p* states are lobed structures with highly directional characteristics. The electron density in the p_z orbital ($m_\ell = 0$) is concentrated along the z axis; similarly, the p_x and p_y orbitals are marked by electron densities along the x and y axes, respectively (Fig. 11.18). (The p_x and p_y orbitals are formed from the $m_\ell = \pm 1$ states according to the prescription of Equation 7.39 in Chapter 7.) When two N atoms are brought together, one of the three axes—say the z axis—will become the molecular axis for N_2 in order to maximize atomic overlaps and produce the strongest bond. This is the sigma bond, because it has axial symmetry. However, the p_x orbitals also overlap to form a different kind of bond—the **pi bond** (π)—which has a plane of symmetry in the nodal plane of the *p* orbitals. The same is true of the p_y orbitals. Figure 11.19 illustrates the different kinds of bonds in the N_2 molecule. The pi bonds are weaker than the sigma bond because they involve less electron overlap.

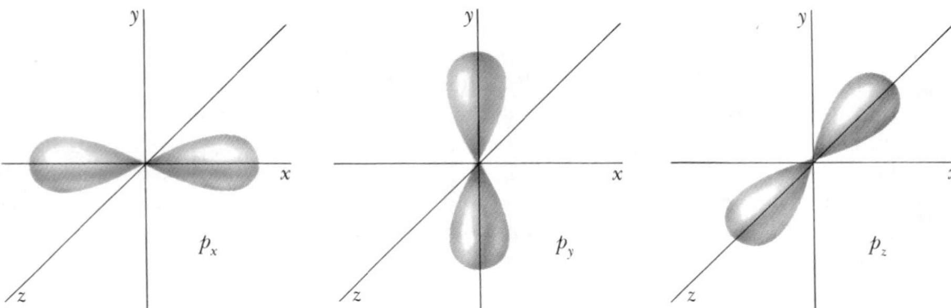

Figure 11.18 Electron distributions for the atomic $2p_x$, $2p_y$, and $2p_z$ orbitals. The former are obtained by rotating the distribution for the $2p_z$ orbital into the x and y axes, respectively.

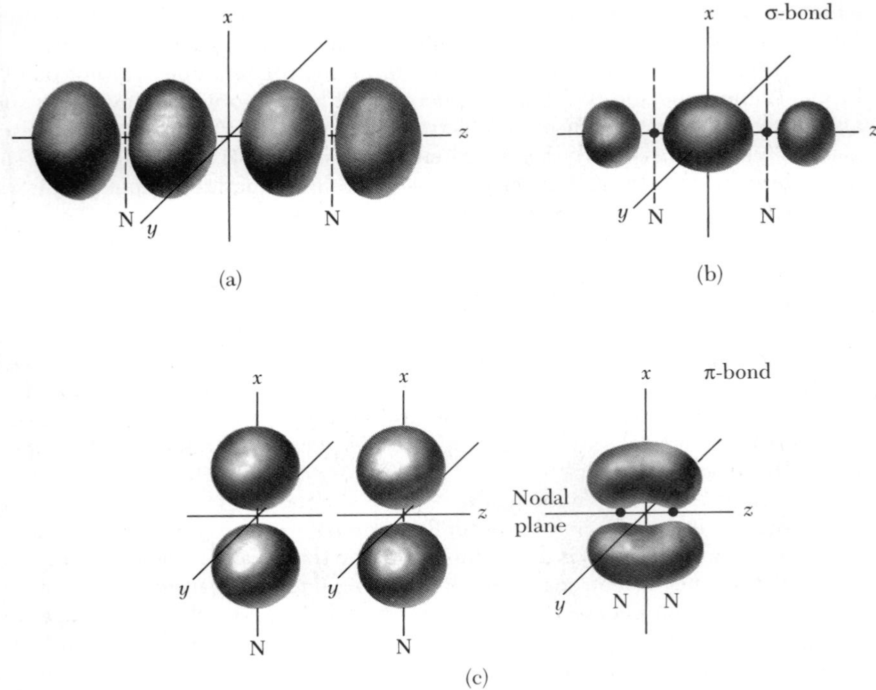

Figure 11.19 (a) and (b): Formation of a sigma bond in N_2 from the overlap of the $2p_z$ orbitals on adjacent N atoms. (c) Formation of a pi bond by overlap of the $2p_x$ orbitals on adjacent N atoms. A similar bond is formed by overlap of the $2p_y$ orbitals.

Bonding in heteronuclear molecules

So far we have been discussing only molecules formed from like atoms. These *homonuclear* molecules exemplify the pure covalent bond. The joining of two different atomic species to form a *heteronuclear* molecule produces *polar covalent* bonds. The hydrogen fluoride molecule, HF, is a good example. The F atom has nine electrons in the configuration $1s^22s^22p^5$. Of the five $2p$ electrons, four completely fill two of the $2p$ orbitals, leaving one $2p$ electron to be shared with the H atom (the filled orbitals are especially stable, and do not significantly affect the molecular bonding in HF). When the two atoms are brought together, the $2p$ orbital of F overlaps the $1s$ orbital of H to form a bonding-antibonding orbital pair for the HF molecule. The result is an ***s-p* bond,** with both electrons occupying the more stable bonding orbital. The *s-p* bond in HF is *polar,* since the shared electrons spend more time in the vicinity of the F atom due to its high electronegativity. Equivalently, the bond is partly ionic, and the HF molecule shows a permanent *electric dipole moment.* The bond energy for HF is 5.90 eV, and the bond length is 0.092 nm.

The case of HF suggests correctly that purely covalent bonds are found only in homonuclear molecules; heteronuclear molecules always form bonds with some degree of ionicity, as measured by a dipole moment. The water molecule, H_2O, is another example of an *s-p* bonded structure with a dipole moment, as is the ammonia molecule, NH_3. In NH_3, each of the three $2p$ electrons in the N atom forms an *s-p* bond with an H atom. Since the p_x, p_y, and p_z atomic orbitals are directed along mutually perpendicular axes, we would expect the three *s-p* bonds to be at right angles. In fact, the measured *bond angle* in NH_3 is somewhat larger—about 107°—because of the electrostatic repulsion of the H nuclei.

In closing, we mention briefly the bonds formed by the carbon atom. These bonds result from ***s-p* hybridization,** a concept that accounts for the almost endless variety of organic compounds. The C atom has six electrons in the configuration

$1s^2 2s^2 2p^2$. We might conclude that only the two $2p$ electrons are prominent in molecular bonding and that carbon is divalent. However, the existence of hydrocarbons like CH_4 (methane) shows that the C atom shares all four of its second-shell electrons, suggesting that the binding energy of the $2s$ electrons in the carbon atom is not much different from that of the $2p$ electrons. But most surprising is the fact that *all four bonds are equivalent!* There are not two *s-p* bonds and two *s-s* bonds as we might have anticipated, but four structurally identical molecular bonds. These bonds arise from the overlap of the $1s$ orbital in H with atomic orbitals in C formed from a *mixture* of carbon $2s$ and $2p$ orbitals. In CH_4 these mixed or *hybrid orbitals* are represented by wavefunctions such as

$$\psi = \psi_{2s} + [\psi_{2p}]_x + [\psi_{2p}]_y + [\psi_{2p}]_z \qquad (11.21)$$

Other combinations arise by subtracting, rather than adding, the individual wavefunctions, but all mix one $2s$ and three $2p$ atomic orbitals to give four *sp^3 hybrids.* In other carbon compounds, hybridization may involve only one or two of the $2p$ orbitals; these are described as sp and sp^2 hybrids, respectively. It is this complexity that gives rise to the rich variety of organic materials.

11.6 SUMMARY

Two or more atoms may combine to form molecules because of a net attractive force between them. The resulting molecular bonds are classified according to the bonding mechanisms, and are of the following types:

1. **Ionic bonds.** Certain molecules form ionic bonds because of the Coulomb attraction between oppositely charged ions. Sodium chloride (NaCl) is one example of an ionically bonded molecule.
2. **Covalent bonds.** The covalent bond in a molecule is formed by the sharing of valence electrons of its constituent atoms. For example, the two electrons of the H_2 molecule are equally shared between the nuclei.
3. **Hydrogen bonds.** This type of bonding corresponds to the attraction of two negative ions by an intermediate hydrogen atom (a proton).
4. **Van der Waals bonds.** This is a weak electrostatic bond between atoms that do not form ionic or covalent bonds. It is responsible for the condensation of inert gas atoms and nonpolar molecules into the liquid phase.

The energy of a molecule in a gas consists of contributions from the electrical interactions, the translation of the molecule, rotations, and vibrations. The **allowed values of the rotational energy** of a diatomic molecule are given by

$$E_{\text{rot}} = \frac{\hbar^2}{2I_{\text{cm}}} \ell(\ell + 1) \qquad \ell = 0, 1, 2, \ldots \qquad (11.5)$$

where I_{cm} is the *moment of inertia* of the molecule and ℓ is an integer called the *rotational quantum number.* Successive rotational levels are separated by energies carried by photons in the microwave range of the electromagnetic spectrum.

The allowed values of the vibrational energy of a diatomic molecule are given by

$$E_{\text{vib}} = (v + \tfrac{1}{2})\hbar\omega \qquad v = 0, 1, 2, \ldots \qquad (11.8)$$

where v is the *vibrational quantum number.* The quantity ω is the classical frequency of vibration, and is related to μ, the *reduced mass* of the molecule, and K, the *force constant* of the effective spring bonding the molecule, by

$\omega = \sqrt{K/\mu}$. Successive vibrational levels are separated by energies carried by photons in the infrared range of the electromagnetic spectrum.

The internal state of motion of a molecule is some combination of rotation and vibration. Any change in the molecular condition is described as a transition from one rotation-vibration level to another. When accompanied by the emission or absorption of photons, these are called **optical transitions.** Besides conserving energy, optical transitions must conform to the **selection rules**

$$\Delta\ell = \pm 1 \quad \text{and} \quad \Delta v = \pm 1 \qquad (11.12, 11.13)$$

It follows that neither a pure rotational transition ($\Delta v = 0$) nor a pure vibrational transition ($\Delta\ell = 0$) is observable. In the most common case, the absorption spectrum of a molecule consists of *two* sequences of lines, corresponding to $\Delta\ell = \pm 1$, with $\Delta v = +1$ for both sequences. Measurements made on such spectra can be used to determine the length and strength of the molecular bonds.

The electronic states of a molecule are classified as **bonding** or **antibonding.** A bonding state is one for which the electron density is large in the space between the nuclei. An electron in a bonding orbital can be thought of as shuttling rapidly from one nucleus to the other, drawing them together as a kind of "glue." The bonding state is energetically preferred over the antibonding state, where the electron spends most of its time in the neighborhood of one or the other nucleus. In a diatomic molecule composed of like atoms (homonuclear molecule), each atomic orbital gives rise to a bonding-antibonding orbital pair for the molecule.

Molecular bonds are further classified according to their symmetry. A **sigma bond** results from the overlap of atomic *s* states, and is characterized by an electron density that is *axially symmetric* about the internuclear line. The overlap of atomic *p* states may give rise to a sigma bond or a **pi bond,** depending on the relative orientation of the overlapping *p* orbitals. In a pi bond the electron density has a *plane of symmetry* in the nodal plane of the *p* orbitals. Pi bonds involve less overlap, and generally are weaker than sigma bonds.

An ***s*-*p* bond** is formed from the overlap of atomic *s* and *p* orbitals, and is the type of bond found in certain heteronuclear molecules like H_2O and NH_3. These bonds are **polar,** meaning that the electron spends more time in the vicinity of one nucleus than the other, resulting in a permanent *dipole moment* for the molecule. In carbon compounds such as CH_4, we find **hybridized bonds.** These bonds result from the overlap of the 1*s* state of H with a *hybrid orbital* of C, represented as a mixture of carbon 2*s* and 2*p* orbitals. Hybridization in its many forms accounts for the almost endless variety of organic compounds.

SUGGESTIONS FOR FURTHER READING

1. Molecular bonding is discussed extensively at an introductory level in Chapter 24 of J. Brackenridge and R. Rosenberg, *The Principles of Physics and Chemistry,* New York: McGraw-Hill Book Company, Inc., 1970. This work also includes a number of nice illustrations of molecular bonds in various compounds.
2. For an in-depth treatment of the bonding in H_2^+ and the H_2 molecule, see J. C. Slater, *Quantum Theory of Molecules* and *Solids, Volume 1,* New York: McGraw-Hill Book Company, Inc., 1963.
3. An elaborate discussion of chemical bonding in organic compounds may be found in Chapter 13 of K. Krane, *Modern Physics,* New York: John Wiley and Sons, Inc., 1983.

QUESTIONS

1. List three ways (modes) in which a diatomic molecule can store energy internally. Which of the three modes is easiest to excite, that is, requires the least energy? Which requires the most energy?
2. How do the effective force constants for the molecules listed in Table 11.2 compare with those found for typical laboratory springs? Comment on the significance of your findings.
3. Describe what hybridization is, why it occurs, and what it has to do with molecular bonding. Give an explicit example of where it occurs.
4. Discuss the mechanisms responsible for the different types of bonds that can occur to form stable molecules.
5. Explain the role of the Pauli exclusion principle in determining the stability of molecules, using H_2 and the (hypothetical) species H_3 as examples.
6. Discuss the relationship between tunneling and the covalent bond. From the viewpoint of possible chemical species, what would be the similarities and differences between the world in which we live and one devoid of tunneling?
7. Distinguish between the dissociation energy of a molecule and its bond energy. Which of the two is expected to be greater?
8. In treating the rotational levels of a diatomic molecule, why do we ignore rotation about the internuclear line? Can you think of a case involving a triatomic molecule where similar considerations might apply?

PROBLEMS

1. Use the data in Table 11.2 to calculate the reduced mass of the NO molecule. Then compute a value for μ using Equation 11.3. Compare the two results.
2. The CO molecule undergoes a rotational transition from the $\ell = 1$ level to the $\ell = 2$ level. Using Table 11.1, calculate the values of the reduced mass and the bond length of the molecule. Compare your results with those of Example 11.1.
3. Use the data in Table 11.2 to calculate the maximum amplitude of vibration for (a) the HI molecule and (b) the HF molecule. Which molecule has the weaker bond?
4. The $\ell = 5$ to $\ell = 6$ rotational absorption line of a diatomic molecule occurs at a wavelength of 1.35 cm (in the vapor phase). (a) Calculate the wavelength and frequency of the $\ell = 0$ to $\ell = 1$ rotational absorption line. (b) Calculate the moment of inertia of the molecule.
5. The HF molecule has a bond length of 0.092 nm. (a) Calculate the reduced mass of the molecule. (b) Plot the potential energy versus internuclear separation in the vicinity of $r = 0.092$ nm.
6. The HCl molecule is excited to its first rotational energy level, corresponding to $\ell = 1$. If the distance between its nuclei is 0.1275 nm, what is the angular velocity of the molecule about its center of mass?
7. The $v = 0$ to $v = 1$ vibrational transition of the HI molecule occurs at a frequency of 6.69×10^{13} Hz. The same transition for the NO molecule occurs at a frequency of 5.63×10^{13} Hz. Calculate (a) the effective force constant and (b) the amplitude of vibration for each molecule. (c) Explain why the force constant of the NO molecule is so much larger than that of the HI molecule.
8. Consider the HCl molecule, which consists of a hydrogen atom of mass 1 u bound to a chlorine of mass 35 u. The equilibrium separation between the atoms is 0.128 nm, and it requires 0.15 eV of work to increase or decrease this separation by 0.01 nm. (a) Calculate the four lowest rotational energies (in eV) that are possible, assuming the molecule rotates rigidly. (b) Find the molecule's "spring constant" and its classical frequency of vibration. (*Hint:* Recall also that $U = \frac{1}{2}Kx^2$.) (c) Find the two lowest vibrational energies and the classical amplitude of oscillation corresponding to each of these energies. (d) Determine the longest wavelength radiation that the molecule can emit in a pure rotational transition and in a pure vibrational transition.
9. The hydrogen molecule comes apart (dissociates) when it is excited internally by 4.5 eV. Assuming that this molecule behaves exactly like a harmonic oscillator with classical frequency $\omega = 8.277 \times 10^{14}$ rad/s, find the vibrational quantum number corresponding to its 4.5-eV dissociation energy.
10. As a model for a diatomic molecule, consider two identical point masses m connected by a rigid massless rod of length R_0. Suppose that this molecule rotates about an axis perpendicular to the rod through its midpoint. Use the Bohr quantization rule for angular momentum to obtain the allowed rotational energies in this approximation. Compare your result with the correct quantum mechanical treatment of this model.
11. For the model of a diatomic molecule described in Problem 10, derive an expression for the minimum energy required to excite the molecule into rotation about the *internuclear line*, that is, about the axis joining the two masses. Assume that the masses are uni-

form spheres of radius r. Apply your result to the hydrogen molecule, H_2, by taking m equal to the proton mass and r equal to the nuclear size, about 10 fm.

12. The rotational motion of molecules has an effect on the equilibrium separation of the nuclei, a phenomenon known as *bond stretching*. To model this effect, consider a diatomic molecule with reduced mass μ, oscillator frequency ω_0, and internuclear separation R_0 when the angular momentum is zero. The effective potential energy for nonzero values of ℓ is then (see Section 7.4)

$$U_{\text{eff}} = \tfrac{1}{2}\mu\omega_0{}^2(r - R_0)^2 + \ell(\ell + 1)\frac{\hbar^2}{2\mu r^2}$$

(a) Minimize the effective potential $U_{\text{eff}}(r)$ to find an equation for the equilibrium separation of the nuclei, R_ℓ, when the angular momentum is ℓ. Solve this equation approximately, assuming $\ell \ll \mu\omega_0 R_0{}^2/\hbar$. (b) Near the corrected equilibrium point, R_ℓ, the effective potential again is nearly harmonic, and can be written approximately as

$$U_{\text{eff}} \approx \tfrac{1}{2}\mu\omega^2(r - R_\ell)^2 + U_0$$

Find expressions for the corrected oscillator frequency ω_ℓ and the energy offset U_0 by matching U_{eff} and its first two derivatives at the equilibrium point R_ℓ. Show that the fractional change in frequency is given by

$$\frac{\Delta\omega}{\omega_0} \approx \frac{3\ell(\ell + 1)\hbar^2}{2\mu^2\omega_0{}^2R_0{}^4}$$

13. As an alternative to harmonic interactions, the *Morse potential*

$$U(r) = U_0\{1 - e^{-\alpha(r-R_0)}\}^2 - U_0$$

can be used to describe the vibrations of a diatomic molecule. The parameters R_0, U_0, and α are chosen to fit the data for a particular atom pair. (a) Show that R_0 is the equilibrium separation, and that the potential energy at the equilibrium point is $U(R_0) = -U_0$. (Further away, $U(r)$ mimics the asymmetry characteristic of interatomic forces, as shown in Figure 11.12a.) (b) Show that near equilibrium ($r \approx R_0$) the Morse potential is harmonic, with force constant $K = m\omega^2 = 2U_0\alpha^2$. (c) The lowest vibrational energy for the Morse oscillator can be shown to be

$$E_{\text{vib}} = -U_0 + \tfrac{1}{2}\hbar\omega - \frac{(\hbar\omega)^2}{16U_0}$$

Obtain from this an expression for the dissociation energy of the molecule. (d) Apply the results of parts (b) and (c) to deduce the Morse parameters U_0 and α for the hydrogen molecule. Use the experimental values 573 N/m and 4.52 eV for the effective spring constant and dissociation energy, respectively. (The measured value of R_0 for H_2 is 0.074 nm.)

14. The allowed energies of vibration for the Morse oscillator can be shown to be

$$E_{\text{vib}} = -U_0 + (v + \tfrac{1}{2})\hbar\omega - (v + \tfrac{1}{2})^2\frac{(\hbar\omega)^2}{4U_0}$$

where $v = 0, 1, \ldots$

Obtain from this an expression for the interval separating successive levels of the Morse oscillator, and show that this interval diminishes steadily at higher energies, as illustrated in Figure 11.12b. From your result, deduce an upper limit for the vibrational quantum number v. What is the largest vibrational energy permitted for the Morse oscillator?

15. A one-dimensional model for the electronic energy of a diatomic homonuclear molecule is described by the potential well and barrier shown in Figure 11.20. In the simplest case, the barrier width is shrunk to zero ($w \to 0$) while the barrier height increases without limit ($U \to \infty$) in such a way that the product Uw approaches a finite value S called the barrier strength. Such a barrier—known as a delta function barrier—can be shown to produce a discontinuous slope in the wave function at the barrier site $L/2$ given by

$$\left.\frac{d\psi}{dx}\right|_{L/2+} - \left.\frac{d\psi}{dx}\right|_{L/2-} = \frac{2mS}{\hbar^2}\psi(L/2)$$

Solve the wave equation in the well subject to this condition to obtain expressions for the energies of the ground state and first excited state as functions of the barrier strength S. Examine carefully the limits $S \to 0$ and $S \to \infty$ and comment on your findings.

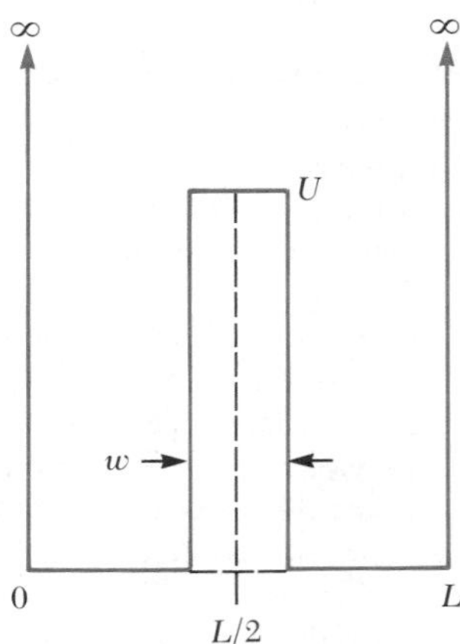

Figure 11.20 (Problem 11.15).

*16. In Section 11.4 it is stated that the approximate electronic energy for the bonding state of $H_2{}^+$ as a function of the internuclear separation R is (in atomic units)

$$E_+ = -1 - \frac{2}{1+\Delta}\left\{\frac{1}{R} - \frac{1}{R}(1+R)e^{-2R} + (1+R)e^{-R}\right\}$$

To this we add the Coulomb energy of the two protons, $2/R$, to get the total energy of this molecular ion, E_{tot}. (a) Write a simple computer program to evaluate E_{tot} for any given value of R. Use your program to verify that the equilibrium separation of the protons in H_2^+ is $R_0 = 2.49$ bohrs, according to this model. (b) Use this model to predict a value for the effective spring constant that governs the vibrations of the H_2^+ molecular ion, and compare your result with the values reported in Table 11.2. (*Hint:* use the well-known finite difference approximation to the second derivative,

$$\frac{d^2f}{dx^2} \approx (1/\Delta x)^2 \{f(x+\Delta x) - 2f(x) + f(x-\Delta x)\}$$

where Δx is a small increment.)

*17. Repeat the calculations of Problem 16 for the case of the neutral hydrogen molecule, H_2. Take for the molecular energy of H_2

$$E_{tot} = \frac{2}{R} + 2E_+$$

where E_+ is the bonding energy of H_2^+ given in Problem 16. Compare your result for the effective spring constant with the experimental value for H_2, about 573 N/m.

PROBLEMS FOR THE COMPUTER

The following problems require for their solution the use of the program EIGEN3. Where appropriate, furnish screen dumps to substantiate your answers.

18. Find the ground state wavefunction and energy for the Morse oscillator using the parameters $b = 6$ and $U = 17$ (as befits the H_2 molecule). What is the physical significance of the coordinate r in this application, and what is its most probable value, expressed in nm, for the H_2 molecule? (See Problem 9 for the pertinent information about H_2.)

*These problems require that Section 11.4 be covered.

19. Find the ground state energy ($\ell = 0$) for the isotopic oscillator with parameter value $b = 6$. Compare your result with E_0, the ground state energy of the one-dimensional oscillator. Is there any discrepancy? Should there be? How do the excited state energies of the isotropic oscillator compare with their one-dimensional counterparts? Explain.

20. Consider the higher angular momentum states of the isotropic oscillator described in Problem 19. Locate the equilibrium point R_ℓ for p waves ($\ell = 1$) and for d waves ($\ell = 2$). In what units do your answers appear? Find the p state with the lowest energy. Compare the energy of this state with the ground state energy E_0 of the one-dimensional oscillator and report the discrepancy as ΔE. Do the same for the d state with lowest energy. Compare your results for ΔE in the two cases with the predictions of Problem 12b. (Note that $\Delta E/E_0$ is the same as $\Delta\omega/\omega_0$.)

The following problems require for their solution the use of the program WAVES-2.

21. Display the four lowest lying states of the divided well with barrier strength $S = 10$. What are the energies of these states? What symmetry (if any) do these wavefunctions exhibit? How many nodes are present for each state, not counting those at the edges of the well? (The divided well serves as a simple model for the electronic states of a diatomic homonuclear molecule —see Problem 15.)

22. Display the nonstationary state that results when the ground and first excited states of the divided well with $S = 10$ are superimposed with equal amplitudes [uniform $a(n)$]. Where is the particle most likely to be found at $t = 0$? Track this wave to see how it evolves. How much time elapses before the particle can be said to have crossed the divider? Report this time for the case of an electron in a well 1 nm wide, and a 1-g mass in a well 1 cm wide. Comment on the significance of your findings.

12

The Solid State

12.1 INTRODUCTION

Matter in the solid state has been a subject of enormous fascination from the beginnings of civilization. Primitive people were attracted to the solid state by its beauty, as shown in the form of radiant, symmetric gemstones, and by its utility as shown in metal tools. These two forces of *utility* and *beauty* are present in physics today with undiminished strength. Industrial applications have made solid state physics the largest subfield in physics, as evidenced by the number of pages allocated in physics journals and the number of physicists employed in this field. The beautiful symmetry and regularity of crystalline solids have both stimulated and allowed rapid progress to be made in the physics of crystalline solids in the 20th century. It is also interesting to note that although rapid theoretical progress has occurred with the most random (gases) and the most regular (crystalline solids) atomic arrangements, much less has been done with liquids and amorphous (irregular) solids until quite recently. The relatively recent rush of interest in low-cost amorphous materials has been driven by applications such as solar cells, memory elements, fiber optic waveguides, and xerography.

In this chapter we first describe how molecules combine to form crystalline and amorphous solids. We shall then introduce one of the simplest classical models of conductors, the free electron gas model, in order to gain physical insight into the processes of electrical and thermal conduction. Next we introduce the quantum theory of metals in order to explain the deficiencies of the classical model. We shall describe the band theory of solids in order to explain the differences between insulators, conductors, and semiconductors, and follow this with a brief discussion of *p-n* junctions and semiconductor devices. We conclude the chapter with an introduction to the exciting field of superconductivity, which has received great attention in the last few years following the discovery of high-temperature superconductors.

12.2 BONDING IN SOLIDS

A crystalline solid consists of a large number of atoms arranged in a regular array, forming a periodic structure. The bonding schemes for molecules described in Chapter 11 are also appropriate in describing the bonding mechanisms in solids. For example, the ions in the NaCl crystal are ionically bonded, while the carbon atoms in the diamond structure form covalent bonds. Another type of bonding mechanism is the metallic bond, which is responsible for the cohesion of copper, silver, sodium, and other metals.

Ionic Solids

Many crystals form by ionic bonding, where the dominant interaction is the Coulomb interaction between the ions. Consider the NaCl crystal shown in Figure 12.1, where each Na^+ ion has six nearest-neighbor Cl^- ions, and each

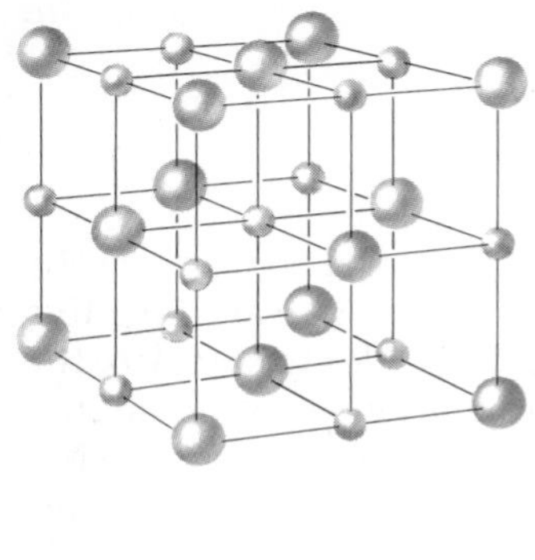

(a)

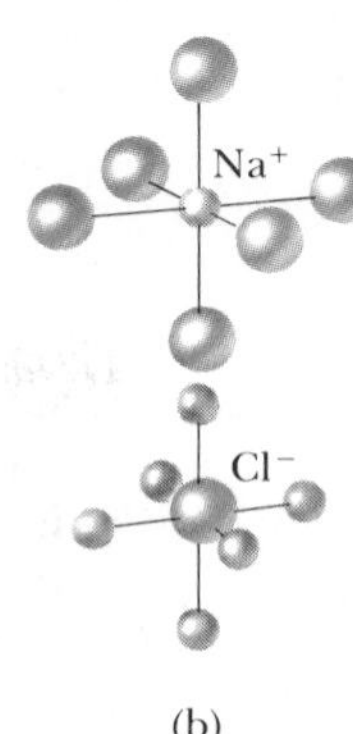

(b)

Figure 12.1 (a) The crystal structure of NaCl. (b) In the NaCl structure, each positive sodium ion is surrounded by six negative chlorine ions, while each chlorine ion is surrounded by six sodium ions.

Cl^- ion has six nearest-neighbor Na^+ ions. Each Na^+ ion is attracted to the six Cl^- ions. The corresponding attractive potential energy is given by $-6ke^2/r$, where r is the Na^+-Cl^- separation and k is the coulomb constant, $(k = 1/4\pi\epsilon_0)$. In addition, there are 12 Na^+ ions at a distance of $\sqrt{2}r$ from the Na^+, which produce a weaker repulsive force on the Na^+ ion. Furthermore, beyond these 12 Na^+ ions one finds more Cl^- ions that produce an attractive force, and so on. The net effect of all these interactions is a resultant electrostatic potential energy given by

$$U_{\text{attractive}} = -\alpha k \frac{e^2}{r} \tag{12.1}$$

where α is a constant called the *Madelung constant.* The value of α depends on the strength of the interaction and the specific crystal structure. For example, $\alpha = 1.7476$ for the NaCl structure. When the atoms are brought close together, the subshells resist overlap because of the exclusion principle, which introduces a repulsive potential energy given by B/r^m, where m is of the order of 10. The total potential energy is therefore of the form

$$U_{\text{total}} = -\alpha k \frac{e^2}{r} + \frac{B}{r^m} \tag{12.2}$$

A plot of the total potential energy versus ion separation is shown in Figure 12.2. The potential energy has its minimum value U_0 at the equilibrium sepa-

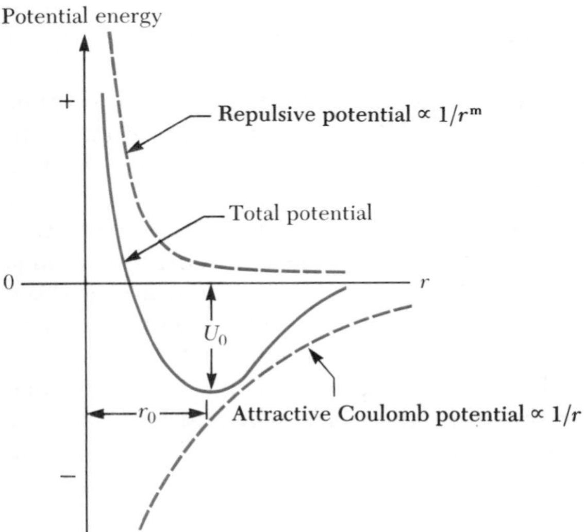

Figure 12.2 Potential energy versus ion separation for an ionic solid. U_0 is the ionic cohesive energy and r_0 is the equilibrium separation between ions.

ration, when $r = r_0$. It is left as a problem (Problem 1) to show that B may be eliminated and that the energy U_0 is given by

Ionic cohesive energy

$$U_0 = -\alpha k \frac{e^2}{r_0}\left(1 - \frac{1}{m}\right) \tag{12.3}$$

The energy U_0, called the **ionic cohesive energy** of the solid, represents the energy required to separate the solid into a collection of positive and negative ions. Its value for NaCl is -7.84 eV per Na^+-Cl^- pair. In order to calculate the **atomic cohesive energy,** which is the binding energy relative to the neutral atoms, we must keep in mind that one gains 5.14 eV in going from Na^+ to Na, and one must supply 3.61 eV in order to convert Cl^- to Cl. Thus the atomic cohesive energy of NaCl is

$$-7.84 \text{ eV} + 5.14 \text{ eV} - 3.61 \text{ eV} = -6.31 \text{ eV}$$

Ionic crystals have the following general properties:

1. They form relatively stable and hard crystals.
2. They are poor electrical conductors because there are no available free electrons.
3. They have high vaporization temperatures.
4. They are transparent to visible radiation but absorb strongly in the infrared region. This occurs because the electrons form such tightly bound shells in ionic solids that visible radiation does not contain sufficient energy to promote electrons to the next allowed shell. The strong infrared absorption (at 20 to 150 μm) occurs because the vibrations of the more massive ions have a lower natural frequency and experience resonant absorption in the low-energy infrared region.
5. They are usually quite soluble in polar liquids such as water. The water molecule, which has a permanent electric dipole moment, exerts an attractive force on the charged ions, which breaks the ionic bonds and dissolves the solid.

Covalent Crystals

As we found in Chapter 11, the covalent bond is very strong and comparable to the ionic bond. Solid carbon, in the form of diamond, is a crystal whose atoms are covalently bonded. Because carbon has the electron configuration $1s^2 2s^2 2p^2$, it lacks four electrons with respect to a filled shell ($2p^6$). Hence, two carbon atoms have a strong attraction for each other, with a cohesive energy of 7.3 eV.

In the diamond structure, each atom is covalently bonded to four other carbon atoms located at four corners of a cube, as shown in Figure 12.3a. In order to form such a configuration of bonds, the electron of each atom must be promoted to the configuration $1s^2 2s^2 2p^3$, which requires an energy of about 4 eV. The crystal structure of diamond is shown in Figure 12.3b. Note that each carbon atom forms covalent bonds with four nearest-neighbor atoms. The basic structure of diamond is called *tetrahedral* (each carbon atom is at the center of a regular tetrahedron), and the angle between the bonds is 109.5°. Other crystals such as silicon and germanium have similar structures.

The properties of some covalent solids are given in Table 12.1. Note that the cohesive energies are larger than for ionic solids, which accounts for the hardness of covalent solids. Diamond is particularly hard, and has an extremely high melting point (about 4000 K). In general, covalently bonded

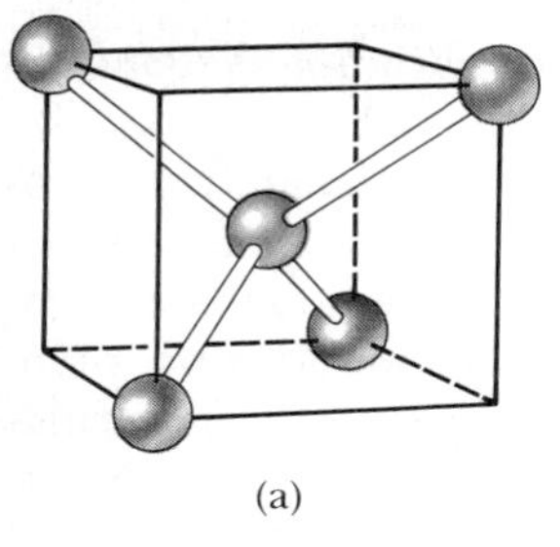

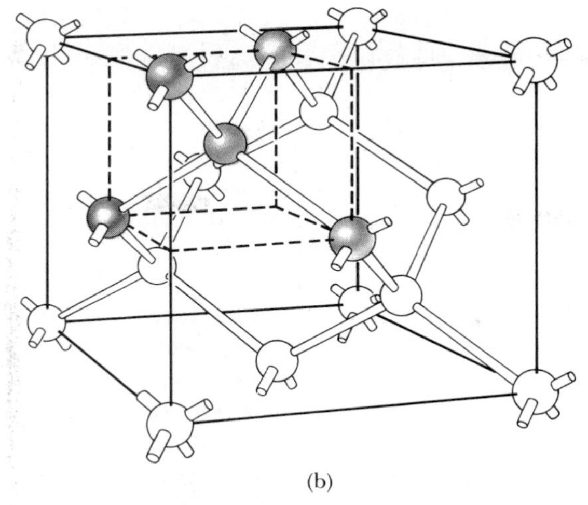

Figure 12.3 (a) Each carbon atom in diamond is covalently bonded to four other carbons and forms a tetrahedral structure. (b) The crystal structure of diamond, showing the tetrahedral bond arrangement. (After W. Shockley, *Electrons and Holes in Semiconductors,* New York, Van Nostrand, 1950.)

solids are very hard, have large bond energies and high melting points, are good insulators, and are transparent to visible light.

Metallic Solids

Metallic bonds are generally weaker than ionic or covalent bonds. The valence electrons in a metal are relatively free to move throughout the material. There are a large number of such mobile electrons in a metal, typically one or two electrons per atom. The metal structure can be viewed as a lattice of positive ions surrounded by a "gas" of nearly free electrons (Fig. 12.4). The binding mechanism in a metal is the attractive force between the positive ions and the electron gas.

Metals have a cohesive energy in the range of 1 to 3 eV, which is smaller than the cohesive energies of ionic or covalent solids. Since visible photons also have energies in this range, light interacts strongly with the free electrons in metals. Hence visible light is absorbed and re-emitted quite close to the surface of a metal, which accounts for the shiny nature of metallic surfaces. In addition to the high electrical conductivity of metals produced by the free electrons, the nondirectional nature of the metallic bond allows many different types of metallic atoms to be dissolved in a host metal in varying amounts. The resulting solid solutions or alloys may be designed to have particularly useful structural properties, such as high strength and low density, since these properties tend to change rather slowly and controllably with alloy composition.

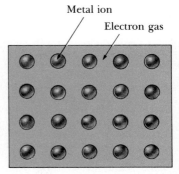

Metal ion
Electron gas

Figure 12.4 Free electron model of a metal.

TABLE 12.1 The Cohesive Energies of Some Covalent Solids

Crystal	Cohesive Energy (eV)
C (diamond)	7.37
Si	4.63
Ge	3.85
InAs	5.70
SiC	12.3
ZnS	6.32
CuCl	9.24

Molecular Crystals

A fourth class of binding can occur even when electrons are not available to participate in bond formation, as in the cases of saturated organic molecules (CH_4) and inert gas atoms with closed electron shells. The weak electric forces at work in this type of bonding are known as **van der Waals forces** and are due to the attractive force between electric dipoles. Some simple molecules containing hydrogen, such as water, have permanent dipole moments (due to the built-in separation of positive and negative centers of charge), and form relatively strong van der Waals bonds known as **hydrogen bonds.** Ice, for example, is hydrogen bonded with a cohesive energy of about 0.52 eV/molecule or 12 kcal/mole. (Note that 1 eV/molecule = 23 kcal/mole.) Two molecules without permanent dipole moments also experience fluctuation induced dipole-dipole attractions, but the van der Waals forces are weaker in this case. Many organic solids with comparatively low melting and boiling points as well as inert gas crystals are held together by such weak forces. Two typical cases are solid methane, with a cohesive energy of about 0.10 eV/molecule, and solid argon, with a cohesive energy of 0.078 eV/molecule and a melting point of 84 K.

Amorphous Solids

The apparent simplicity and regularity of crystalline solids suggest that it will be most profitable to focus attention on perfect crystals. This is indeed the direction we shall follow in most of this chapter. However, real crystals are far from perfect, since they contain irregularities (dislocations) and impurities that can have a profound effect on their strength, conductivity, and other properties. This gives us some motivation for a brief discussion of crystal defects and the nature of amorphous solids.

It is well known that crystals of a given material are formed when the liquid form of that material is cooled sufficiently. For example, water forms crystals when cooled to 0°C. In general, more perfect crystals are formed if a liquid is cooled slowly *(annealing)*, allowing the molecules to relax gradually to states having minimum potential energy. Conversely, rapid cooling of a liquid *(quenching)* causes many dislocations in the resulting lattice structure. If the cooling is rapid enough, almost any liquid will form an amorphous solid (or glass) having no long-range order, as shown in Figure 12.5. Although an amorphous solid has no long-range order, there is extensive short-range order, in that bond lengths and bond angles for nearest neighbors are nearly the same throughout the solid. Even metals can be made to solidify in an amorphous

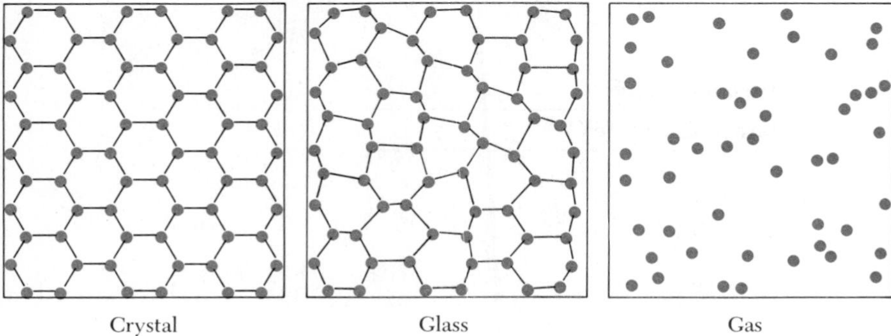

Figure 12.5 Two-dimensional sketches of the atoms in a crystal, glass, and gas. Note that the crystal has equal bond lengths and angles, the glass has a distribution of nearly equal bond lengths and angles, and the gas has a completely random spatial distribution of atoms. (After R. Zallen, *The Physics of Amorphous Solids*, New York, John Wiley and Sons, 1983.)

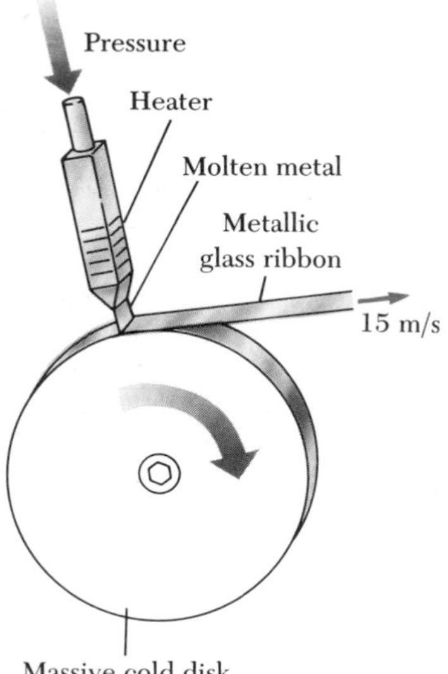

Figure 12.6 Preparing an amorphous metal ribbon by melt spinning.

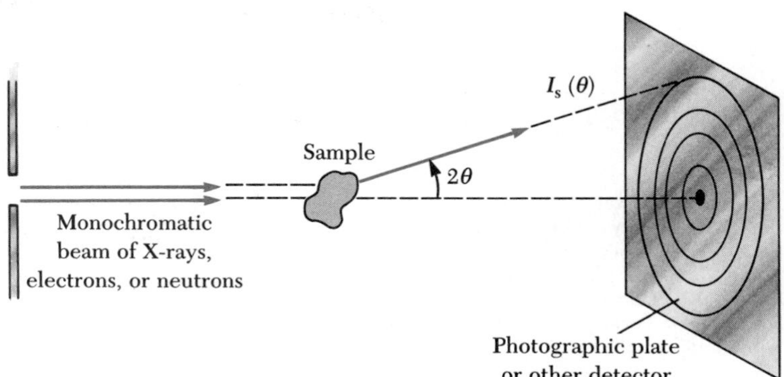

Figure 12.7 Basic experimental setup for determining the distribution of atoms in a sample. (After R. Zallen, *The Physics of Amorphous Solids,* New York, John Wiley and Sons, 1983.)

form (called a *metallic glass*) if the metal in the liquid state can be cooled by 1000 K in about one millisecond. One clever technique for achieving cooling rates of 10^6 K/s, known as *melt spinning,* is illustrated in Figure 12.6.

The main experimental techniques for determining the distribution of atoms in crystalline and amorphous solids are x-ray, electron, and neutron diffraction. The primary experimental data are measurements of the scattered intensity $I_s(\theta)$, as sketched in Figure 12.7. Figure 12.8 shows actual photographic records of the electron intensity scattered from crystalline and amorphous iron samples. Note that the regular spacing of atoms over many atomic separations in the crystalline sample leads to many sharp diffraction lines. This is in contrast to the metallic glass sample, in which the distribution of atomic separations and bond angles over one or two atomic separations leads to fewer, more diffuse lines.

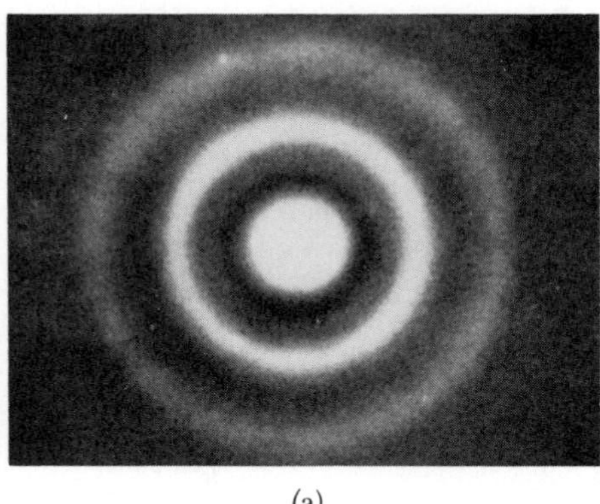

(a)

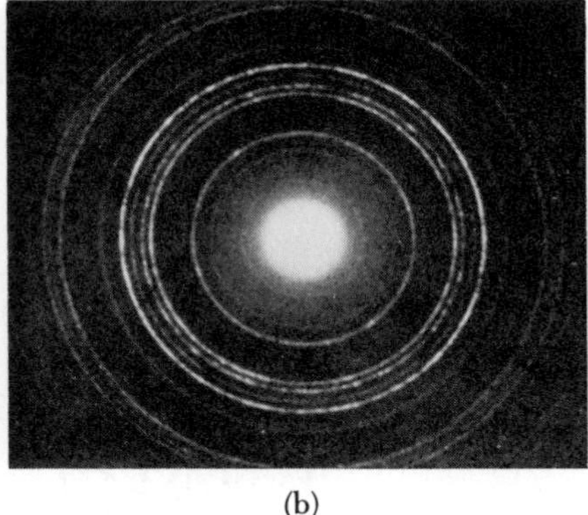

(b)

Figure 12.8 Electron diffraction patterns of (a) amorphous iron and (b) crystalline iron. (From T. Ichikawa, *Phys. Status, Solidi* A19:707, 1973.)

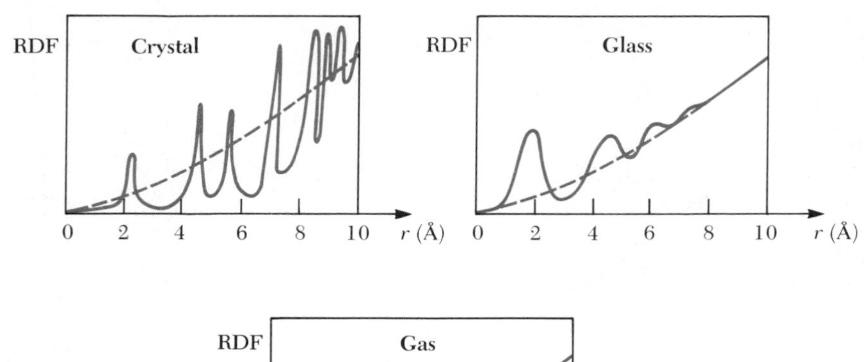

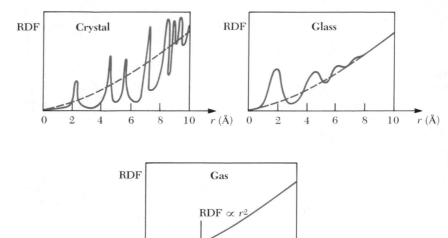

Figure 12.9 Representative radial distribution functions (RDF) for a substance in crystalline, amorphous, and gas phases. (After R. Zallen, *The Physics of Amorphous Solids,* New York, John Wiley and Sons, 1983.)

The actual distribution of atoms in space, which is described by the *radial distribution function,* $\rho(r)$, may be obtained by taking the Fourier transform of $I_s(\theta)$.[1] Note that $\rho(r)\,dr$ is the probability of finding neighboring atoms between r and $r + dr$ from any atom chosen as the origin. Thus we expect $\rho(r)$ to consist of many sharp maxima for crystals, because their structures consist of extremely regular shells of nearest neighbor atoms, next-nearest neighbor atoms, and so forth. On the other hand, we expect fewer and broader maxima for glasses because of their short-range order and distribution of bond lengths. Schematic drawings illustrating these differences in the radial distribution function (RDF) for the crystal, glass, and gas phases are shown in Figure 12.9. Note that the radial distribution functions for the crystal and glass phases are superimposed on a background equivalent to the gas RDF, $\rho_{\text{gas}}(r) = Ar^2$, where A is a constant. The RDFs for both crystal and glass approach the gas RDF for large r values because the distribution of atoms becomes effectively

[1] See, for example, R. Zallen, *The Physics of Amorphous Solids,* Chapter 2, New York, John Wiley and Sons, 1983.

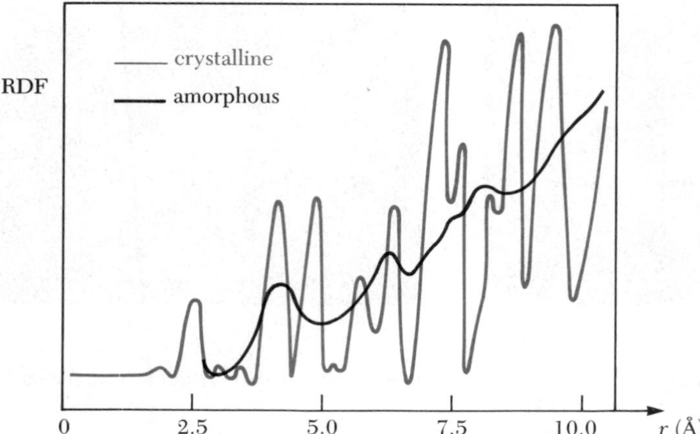

Figure 12.10 RDFs for amorphous and crystalline germanium calculated from x-ray scattering. (From R. J. Temkin, W. Paul, and G. A. N. Connell, *Adv. Phys.* 22:581, 1973.)

TABLE 12.2 Examples of Technological Applications of Amorphous Solids

Type	Material	Use
Oxide glass	$(SiO_2)_{0.8}(Na_2O)_{0.2}$	Transparent window glass
Oxide glass	$(SiO_2)_{0.9}(GeO_2)_{0.1}$	Ultratransparent optical fibers
Organic polymer	Polystyrene	Strong, low-density plastics
Chalcogenide glass	Se, As_2Se_3	Photoconductive films used for xerography
Amorphous silicon	$Si_{0.9}H_{0.1}$	Solar cells
Metallic glass	$Fe_{0.8}B_{0.2}$	Ferromagnetic low-loss ribbons used as transformer cores

continuous over large distances and smooths out to an average value of $\bar{n}$ atoms per unit volume. Since the volume of a spherical shell is $4\pi r^2\,dr$, the number of atoms between r and $r + dr$ in the gas phase is given by $\rho_{gas}(r)\,dr = \bar{n}(4\pi r^2\,dr)$. Thus $\rho_{gas}(r) = Ar^2$, where $A = 4\pi\bar{n}$.

Actual RDFs for crystalline and amorphous germanium obtained from x-ray diffraction measurements are shown in Figure 12.10. Note that the amorphous RDF merges with the smeared background after about four oscillations, indicating only short range order. However, the crystalline RDF is still sharply varying after 14 oscillations. Also, note that the crystalline RDF peaks are not as sharp as might be expected, but are broadened by thermal and zero-point motions of the atoms about their lattice positions.

Amorphous solids have many useful and interesting physical properties such as extremely high optical transparency, extremely high strength, and low density. You may wish to read reference 1 for more information about amorphous solids. We conclude our brief introduction to this rapidly expanding field with Table 12.2, which lists some of the technological applications of amorphous solids.

12.3 CLASSICAL FREE ELECTRON MODEL OF METALS

Shortly after Thomson's discovery of the electron, Drude and Thomson proposed the free electron theory of metals. According to this theory, the physical properties of a metal may be explained by modeling the metal as a classical gas of conduction electrons moving freely through a fixed lattice of ion cores. This picture of a highly mobile electron "fluid" was used with considerable success by Thomson, Drude, and Lorentz to explain the high electrical and thermal conductivities of metals shown in Table 12.3. In particular, the model predicts the correct functional form of Ohm's law and the connection between the electrical and thermal conductivities of a metal known as the Wiedemann-Franz relation.[2] However, the model does not accurately predict the experimental values of electrical and thermal conductivities when classical electronic mean free paths are used in the calculations. In fact, we shall see that the shortcomings of the classical model can be remedied only by taking into account the wave nature of the electron. This involves replacing the incorrect Maxwell-Boltzmann rms velocity for electrons with the Fermi velocity (see Section 9.4) as well as replacing the classical mean free path of electrons with a

[2] The Wiedemann-Franz relation states that the ratio of thermal conductivity to electrical conductivity for metals is proportional to the temperature and that the value of the proportionality constant is independent of the metal considered.

TABLE 12.3 Thermal Conductivity, K, and Electrical Conductivity, σ, of Selected Substances at Room Temperature

Substance	K in $W \cdot m^{-1}K^{-1}$	σ in $(\Omega \cdot m)^{-1}$
Copper	390	59×10^6
Aluminum	210	35×10^6
Iron	63	10×10^6
Steel	50	1.4×10^6
Nichrome	14	0.9×10^6
Quartz	13	
NaCl	7.0	$<10^{-4}$

quantum mean free path that is hundreds to thousands of times the interatomic distance. This remarkably long mean free path occurs because the electron is a wave, and as such is able to pass freely though a perfect lattice, scattering only when impurity atoms or other deviations from crystalline regularity are encountered.

Ohm's Law

Ohm's law was first established as an experimental result applicable to a wide range of metals and semiconductors. It simply states that the current density in a material is directly proportional to the applied electric field. That is,

Ohm's law

$$J = \sigma E \tag{12.4}$$

where J is the current density (A/m^2), σ is the electrical conductivity of the material ($\Omega^{-1}m^{-1}$), and E is the electric field (V/m). Rather than viewing Equation 12.4 simply as a proportionality between applied field and resulting current, it is more instructive to interpret this equation as a definition of σ. Equation 12.4 tells us that a single constant, σ, which depends on the material and temperature *but not on applied field or voltage*, completely characterizes the electrical conduction in an ohmic solid.

Students often have the incorrect notion that Ohm's law is universal in nature. Before discussing a system where Ohm's law is valid, let us briefly consider one case where it does not hold, a low-density gas of electrons in a vacuum. In this case, shown in Figure 12.11a, the electrons accelerate uniformly across the region containing an applied electric field (where there are no ion cores to impede the motion), and have a velocity that steadily increases with x. The velocity as a function of x is given by

$$v^2 = \frac{2eE}{m_e} x = \frac{2Ve}{m_e d} x \tag{12.5}$$

Since the magnitude of the current density J (the charge crossing a unit area per second) is just nev as shown in Figure 12.11b (where n is the electron concentration), we have

$$J(x) = ne\left(\frac{2eE}{m_e} x\right)^{1/2} \tag{12.6}$$

If we now attempt to define a conductivity σ by writing Equation 12.6 in the form of Ohm's law, we have

$$J(x) = \sigma(x, E)E \tag{12.7}$$

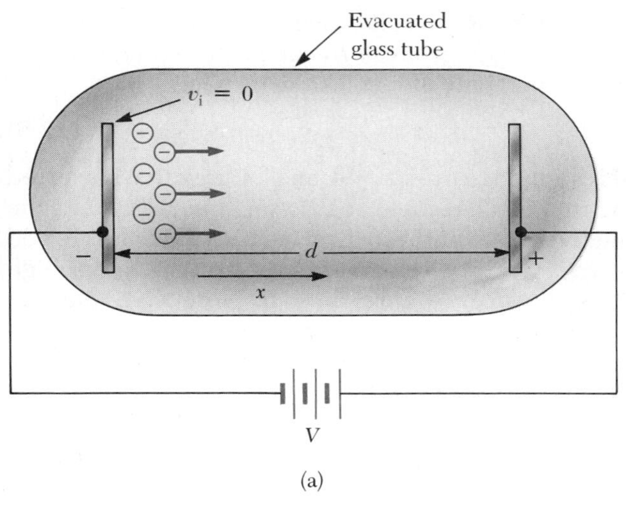

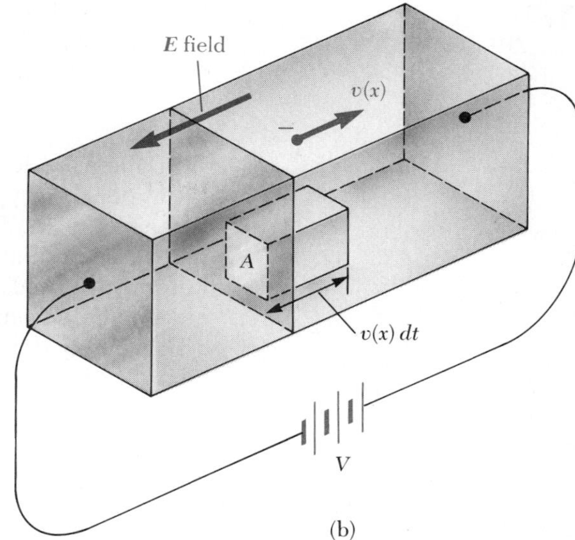

Figure 12.11 (a) Acceleration of electrons in a vacuum. (b) Connection between current density and drift velocity. The charge that passes through A in time dt is the charge contained in the small parallelopiped. Since this charge is $neAv(x)dt$, we find

$$J = \frac{neAv(x)dt}{Adt} = nev(x).$$

where

$$\sigma(x, E) = ne^{3/2}\left(\frac{2x}{m_e E}\right)^{1/2}$$

In this case we find that σ is not a constant for a given material but is a *function of position and electric field or voltage.* Thus Ohm's law is not valid for electrons in a vacuum!

Now let us return to the case of interest and show that Ohm's law does hold for a gas of free electrons in a solid. In the absence of an electric field, the electrons in a metal move randomly along straight-line trajectories, which are constantly interrupted by collisions with lattice ions (see Fig. 12.12a). The root mean square speed at room temperature is fairly high ($\sim 10^5$ m/s) and may be calculated from the classical equipartition theorem:

$$\tfrac{1}{2}m_e\overline{v^2} = \tfrac{3}{2}kT$$

or

$$v_{\text{rms}} = \sqrt{\overline{v^2}} = \left(\frac{3kT}{m_e}\right)^{1/2} \tag{12.8}$$

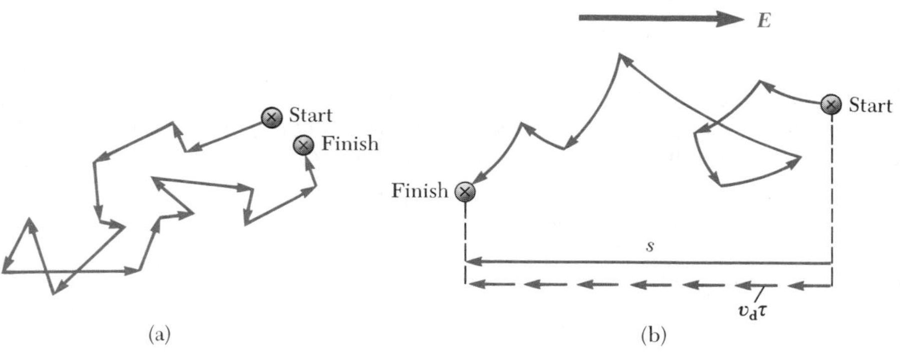

Figure 12.12 (a) Random successive displacements of an electron in a metal without an applied electric field. (b) A combination of random displacements and displacements produced by an external electric field. The net effect of the electric field is to add together multiple displacements of length $v_d\tau$ opposite the field direction.

Each step of the motion (the path between collisions) shown in Figure 12.12a is a "free path"; the average free path or *mean free path, L,* is related to the *mean free time,* τ (the average time between collisions), and v_{rms} by

$$L = v_{rms}\tau \tag{12.9}$$

When an electric field is applied to the sample, an electric force is exerted during each step, resulting in a displacement that is small compared to the mean free path. The cumulative effect of these displacements may be viewed in terms of a small average *drift velocity,* v_d, superimposed on the rather high random thermal velocity, as shown in Figure 12.12b. Note that when an electric field is applied the trajectories become parabolic.

We may derive an expression for v_d by assuming that the total displacement of an electron after p collisions is given by

$$s = \frac{a}{2}(t_1^2 + t_2^2 + t_3^2 + \cdots t_p^2) \tag{12.10}$$

where t_1, t_2, etc. are the successive times between collisions and a is the acceleration (eE/m_e) produced by the electric field. Note that we have ignored the random initial velocities of the electron in Equation 12.10, since these average to zero. We may write Equation 12.10 in terms of averages as

$$s = \frac{a}{2}(p)(\overline{t^2}) \tag{12.11}$$

Since the average value of t^2 is $(\overline{t^2}) = 2\tau^2$ (see Problem 10), Equation 12.11 becomes

$$s = ap\tau^2$$

or

$$s = \frac{eE}{m_e}p\tau^2 \tag{12.12}$$

Comparing Equation 12.12 to the expression for the displacement in terms of the drift velocity, $s = pv_d\tau$, we find

$$\boxed{v_d = \frac{eE\tau}{m_e}} \tag{12.13}$$

As noted earlier, the current density is nev_d, so the expression for the current density becomes

$$\boxed{J = nev_d = \frac{ne^2\tau}{m_e}E} \tag{12.14}$$

The proportionality of J to E given by Equation 12.14 shows that the classical free electron model predicts the observed Ohm's law dependence of J on E. In this case the conductivity, *which is independent of E,* is given by

$$\sigma = \frac{ne^2\tau}{m_e}$$

or, using $\tau = L/\overline{v}$,

$$\sigma = \frac{ne^2L}{m_e\overline{v}} \tag{12.15}$$

TABLE 12.4 Electrical Conductivity of Metals at 300 K

Substance	Measured σ in $(\Omega\cdot\mathrm{m})^{-1}$
Copper	59×10^6
Aluminum	35×10^6
Sodium	22×10^6
Iron	10×10^6
Mercury	1.0×10^6

Substituting the Maxwell-Boltzmann rms thermal velocity into Equation 12.15 for $\bar{v}$ yields

$$\sigma=\frac{ne^2L}{(3kTm_e)^{1/2}} \tag{12.16}$$

Measured values of σ for various metals are listed in Table 12.4 for comparison to calculations using Equation 12.16. Since the resistivity, ρ, is the reciprocal of the conductivity, Equation 12.16 may also be written

Classical expressions for conductivity and resistivity

$$\rho=\frac{(3kTm_e)^{1/2}}{ne^2L} \tag{12.17}$$

EXAMPLE 12.1 The Classical Free Electron Model of Conductivity in Solids

(a) Show that the rms thermal velocity of electrons at 300 K is many orders of magnitude higher than the drift velocity, v_d. In order to find v_d, assume that a copper wire with a cross section of 2 mm × 2 mm can safely carry a current of 10 A. (b) Estimate τ, the average time between collisions for copper at room temperature, assuming that the mean free path is the interatomic distance, 2.6 Å. (c) Calculate the conductivity of copper at room temperature as a test of the classical free electron gas theory, and compare this to the measured value found in Table 12.4.

Solution: (a) By Equation 12.8,

$$v_{\mathrm{rms}}=\left(\frac{3kT}{m_e}\right)^{1/2}=\left[\frac{3(1.38\times10^{-23}\ \mathrm{J/K})(300\ \mathrm{K})}{9.11\times10^{-31}\ \mathrm{kg}}\right]^{1/2}$$

$$=1.17\times10^5\ \mathrm{m/s}$$

To calculate v_d we use $J=nev_d$, or $v_d=J/ne$. The number of free (conduction) electrons per cm^3, n, in copper is given by

$$n=\left(\frac{1\ \text{free electron}}{\text{atom}}\right)\left(\frac{6.02\times10^{23}\ \text{atoms}}{\text{mole}}\right)$$

$$\times\left(\frac{8.96\ \mathrm{g}}{\mathrm{cm}^3}\right)\left(\frac{1\ \text{mole}}{63.5\ \mathrm{g}}\right)$$

$$=8.49\times10^{22}\ \text{electrons/cm}^3$$

Thus

$$v_d=\frac{J}{ne}=\frac{(10\ \mathrm{A})/(4\times10^{-6}\ \mathrm{m}^2)}{(8.49\times10^{28}\ \mathrm{m}^{-3})(1.6\times10^{-19}\ \mathrm{C})}$$

$$=1.8\times10^{-4}\ \mathrm{m/s}$$

(b) Using Equation 12.9,

$$\tau=\frac{L}{v_{\mathrm{rms}}}=\frac{2.6\times10^{-10}\ \mathrm{m}}{1.2\times10^5\ \mathrm{m/s}}=2.2\times10^{-15}\ \mathrm{s}$$

Thus, in this model the electrons experience several hundred trillion collisions per second!

(c) Equation 12.16 gives

$$\sigma=\frac{ne^2L}{(3kTm_e)^{1/2}}=5.3\times10^6\ (\Omega\cdot\mathrm{m})^{-1}$$

From Table 12.4 we see that this value of the conductivity is about 10 times smaller than the measured value!

Although the classical electron gas model does predict Ohm's law, we see from Example 12.1 that it results in conductivity values that are incorrect by an order of magnitude. Worse yet, the *measured resistivity* of most metals is

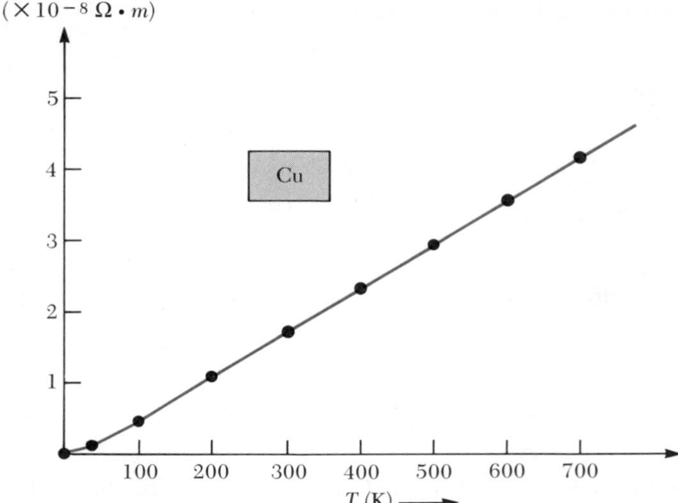

Figure 12.13 The resistivity of pure copper as a function of temperature.

found to be proportional to the absolute temperature (see Fig. 12.13), yet the classical electron gas model predicts a much weaker dependence of ρ on T. From Equation 12.17, we see that the classical model incorrectly predicts that resistivity should be proportional to the square root of the absolute temperature. In Section 12.4 we will give a different model of electron scattering, which explains the linear dependence of ρ on T.

Classical Free Electron Theory of Heat Conduction

The thermal conductivity, K, is defined in a similar way to electrical conductivity, σ. Whereas for electrical conductivity we had current density equal to conductivity times voltage gradient, or

$$J = -\sigma \frac{\Delta V}{\Delta x} \tag{12.18}$$

we now have thermal current density (watts/m^2) equal to thermal conductivity times temperature gradient, or

$$\frac{\Delta Q}{A\Delta t} = -K \frac{\Delta T}{\Delta x} \tag{12.19}$$

Here ΔQ is the heat energy conducted through cross sectional area A in time Δt between two planes with a temperature gradient of $\Delta T/\Delta x$. Since good electrical conductors are also good conductors of heat, it is natural to assume that the highly mobile electron gas is responsible for transporting charge as well as heat energy through the metal via a random collision process. A remarkable triumph of the classical free electron model was to show that if free electrons were responsible for both electrical and thermal conduction in metals, then the ratio of K/σ should be a universal constant, the *same for all metals*, and *dependent only on the absolute temperature*. Since such an unusually simple connection between K and σ had already been experimentally observed (the Wiedemann-Franz law), this prediction confirmed that the motion of the electron gas was basically responsible for both electrical and thermal conductivity.

Since the Wiedemann-Franz law is such a remarkably simple and powerful result, it bears consideration in a bit more detail. The kinetic theory of gases

TABLE 12.5 Experimental Lorentz Numbers $K/\sigma T$ in Units of 10^{-8} W·Ω/K²*

Metal	0°C	100°C
Ag	2.31	2.37
Au	2.35	2.40
Cd	2.42	2.43
Cu	2.23	2.33
Ir	2.49	2.49
Mo	2.61	2.79
Pb	2.47	2.56
Pt	2.51	2.60
Sn	2.52	2.49
W	3.04	3.20
Zn	2.31	2.33

* From C. Kittel, *Introduction to Solid State Physics*, 2nd edition, New York, John Wiley and Sons, 1965.

may be applied to the free electron gas to calculate the new flux of thermal energy (W/m²) carried by electrons moving from a region at temperature $T + \Delta T$ to a region at temperature T.[3] Using this result, one immediately finds that the thermal conductivity depends simply on the heat capacity per unit volume, C_v, the average speed, $\bar{v}$, and the mean free path, L, of the electrons as follows:

$$K = \tfrac{1}{3}C_v\bar{v}L \tag{12.20}$$

If we assume that the electrons behave classically and obey Maxwell-Boltzmann statistics, the heat capacity per *mole* is found to be $3R/2$ or $3N_Ak/2$, where N_A is Avogadro's number. To convert this to heat capacity per unit volume for use in Equation 12.20, we must multiply by the ratio of the electron density to Avogadro's number. Thus

$$C_v = (\tfrac{3}{2}N_Ak)\left(\frac{n}{N_A}\right) = \tfrac{3}{2}kn$$

and Equation 12.20 becomes

$$K = \frac{kn\bar{v}L}{2} \tag{12.21}$$

If we simply use the interatomic distance for L in Equation 12.21 we will, as in the case of electrical conductivity, again find incorrect values. However, if we form the ratio of K to σ we will avoid the problem of having to assign values to n and L, as these quantities cancel. Thus

$$\frac{K}{\sigma} = \frac{kn\bar{v}L/2}{ne^2L/m_e\bar{v}} = \frac{k(\bar{v})^2m_e}{2e^2}$$

Using v_{rms} for $\bar{v}$ in this expression finally gives the desired classical result:

$$\frac{K}{\sigma} = \frac{3k^2}{2e^2}T \tag{12.22}$$

Wiedemann-Franz law

The ratio $K/\sigma T$ is known as the **Lorentz number,** and according to our classical theory it has the same value $3k^2/2e^2 = 1.1 \times 10^{-8}$ W·Ω/K² *for any metal at any temperature.* Table 12.5 shows that, indeed, the ratio of $K/\sigma T$ is nearly con-

[3] See C. Kittel, *Introduction to Solid State Physics*, 6th edition, Chapter 5, New York, John Wiley and Sons, 1986.

stant from metal to metal and with varying temperatures, but that the value of $3k^2/2e^2$ does not agree precisely with the measured Lorentz numbers. Overall, however, the agreement of the simple classical theory with experiment is good and is taken to constitute strong evidence that the free electron gas accounts for both electrical and thermal conductivity in metals.

12.4 QUANTUM THEORY OF METALS

Although the classical electron gas model broadly describes the electrical and thermal properties of metals, there are notable deficiencies in the predictions of this model — namely in the numerical values of K and σ, in the temperature dependence of σ, and in the prediction of an excess heat capacity for metals. (See Section 9.4.) These deficiencies can be rectified by replacing the Maxwell-Boltzmann distribution with the Fermi-Dirac distribution and by calculating the electron mean free path while explicitly taking the wave nature of the electron into account. In general, quantum mechanical calculations of electron mean free path are complicated, and so we shall rely on qualitative arguments and inferences from measured quantities to give physical insight into the remarkable transparency of metals to conduction electrons.

The Replacement of v_{rms} with v_{F}

The classical treatment of electron collisions given so far technically applies to situations where there is a very small probability that a conduction band state is filled with an electron, such as in a semiconductor. In a metal, however, the electrons obey the exclusion principle and essentially all levels in the conduction band are filled up to the Fermi energy, E_{F}. Because of restrictions imposed by the Pauli principle on the quantum states of scattered electrons, one might anticipate a "bottleneck" in electron transport through a metal because of a lack of empty final states. That this is not the case is illustrated in Figure 12.14, which shows the velocity distribution in two dimensions of conduction electrons in a metal. Essentially all the electrons have velocities within a circle of radius $\sqrt{2E_{\text{F}}/m_{\text{e}}}$ in velocity space. Figure 12.14b shows that the application of an E-field causes all the electron velocities to increase by an increment v_{d}. Since the shift of the electron originally at A creates an empty final state for the electron originally at B, there is no bottleneck. Further, the net effect of the

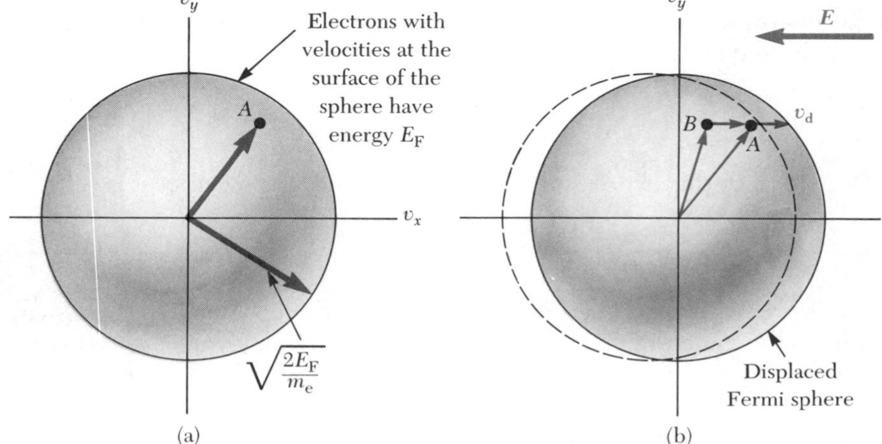

Figure 12.14 (a) Allowed velocity vectors or positions in velocity space of conduction electrons in a metal without an applied electric field. (b) The net effect of an applied electric field is a small displacement of the Fermi sphere, the displaced electrons having a velocity approximately equal to v_{F}.

E-field is to leave an intact core of states for electrons with $E < E_F$ and to produce a displacement of those electrons near the Fermi surface having $v = v_F$. Thus only those electrons with $v = v_F$ are free to move and participate in electrical and thermal conduction, and we can presumably use the classical expressions for σ and K,

$$\sigma = \frac{ne^2L}{m_e\bar{v}} \qquad \text{and} \qquad K = \tfrac{1}{3}C_v\bar{v}L$$

with the Maxwell-Boltzmann thermal velocity $\bar{v}$ replaced by the Fermi velocity, v_F.

Wiedemann-Franz Law Revisited

Let us test the validity of the quantum replacement of $\bar{v}$ with v_F by recalculating $K/\sigma T$ and checking for improved agreement with experiment. First we note that using a Maxwell-Boltzmann molar heat capacity of $3R/2$ for electrons in a metal is incorrect, since conduction electrons obey Fermi-Dirac statistics. As already mentioned in Chapter 9, only a fraction of the electrons within kT of E_F change energy as the temperature changes, leading to a molar heat capacity that is a small fraction of $3R/2$:

$$C \approx 2\left(\frac{kT}{E_F}\right)(\tfrac{3}{2}R) \tag{9.33}$$

An exact calculation is rather complicated[4] but leads to a similar result:

$$C = \frac{\pi^2}{3}\left(\frac{kT}{E_F}\right)(\tfrac{3}{2}R) \tag{12.23}$$

Changing C to a heat capacity per unit volume and substituting into $K = \frac{1}{3}Cv_FL$ yields

$$K = \tfrac{1}{3}\left(\frac{\pi^2}{3}\right)\left(\frac{kT}{E_F}\right)(\tfrac{3}{2}R)\left(\frac{n}{N_A}\right)v_FL$$

Using $R = N_Ak$ and $E_F = \frac{1}{2}m_ev_F^2$, we find

$$K = \frac{\pi^2}{3}\left(\frac{k^2T}{m_ev_F}\right)nL \tag{12.24}$$

Now, forming the Lorentz number $K/\sigma T$ yields

$$\frac{K}{\sigma T} = \frac{(\pi^2/3)(k^2T/m_ev_F)(nL)}{(ne^2L/m_ev_F)T}$$

or

Quantum form of the Wiedemann-Franz law

$$\frac{K}{\sigma T} = \frac{\pi^2k^2}{3e^2} = 2.45 \times 10^{-8}\ \mathrm{W\cdot\Omega/K^2} \tag{12.25}$$

Comparing this result with Table 12.5 shows improved and excellent agreement between theoretical and experimental Lorentz values, thus justifying the replacement of $\bar{v}$ with v_F.

[4] See C. Kittel, op. cit., Chapter 6.

Quantum Mean Free Path of Electrons

Changes from the classical value of electron mean free path arise from the wave properties of the electron. Consider, once again, our friendly copper sample. If we simply use the interatomic distance between copper atoms as the mean free path L for electrons and v_F for $\bar{v}$ in $\sigma = ne^2L/m_e\bar{v}$, we find a value for σ that is about 200 times smaller than the measured value of the conductivity. This discrepancy strongly implies that we are using the wrong value for L and that the scattering sites for electrons are not each ion core, but more widely separated scattering centers. These scattering centers consist of departures from perfect lattice regularity, such as thermal displacements of ions from lattice sites, dislocations, and impurity atoms. Because rigorous quantum calculations of L including these effects are complicated, let us make a rough estimate of L for copper at room temperature by using the measured value of σ. Solving $\sigma = ne^2L/m_e v_F$ for L yields

$$L = \frac{m_e v_F \sigma}{ne^2} \tag{12.26}$$

Using a Fermi energy of 7.05 eV for copper (see Table 9.1) yields

$$v_F = \left(\frac{2E_F}{m_e}\right)^{1/2} = \left(\frac{(2)(7.05 \times 1.6 \times 10^{-19}\ \text{J})}{9.11 \times 10^{-31}\ \text{kg}}\right)^{1/2} = 1.57 \times 10^6\ \text{m/s}$$

Substituting this value into Equation 12.26 gives

$$L = \frac{(9.11 \times 10^{-31}\ \text{kg})(1.57 \times 10^6\ \text{m/s})(5.9 \times 10^7\ \Omega^{-1}\text{m}^{-1})}{(8.49 \times 10^{28}\ \text{electrons/m}^3)(1.60 \times 10^{-19}\ \text{C})^2}$$

$$= 3.9 \times 10^{-8}\ \text{m} = 390\ \text{Å}$$

This is about 150 times the distance between atoms. Using $\tau = L/v_F$, we can also calculate the mean free time between electronic collisions in copper at room temperature to be

$$\tau = \frac{3.90 \times 10^{-8}\ \text{m}}{1.57 \times 10^6\ \text{m/s}} = 2.50 \times 10^{-14}\ \text{s}$$

We can account for the unexpectedly long electron mean free path by taking the wave nature of the electron into account. *Rigorous quantum mechanical calculations show that electron waves with a broad range of energies can pass through a perfect lattice of ion cores unscattered, without resistance, and with an infinite mean free path.* The actual resistance of a metal is due to the random thermal displacements (thermal vibrations) of ions about lattice points and other deviations from a perfect lattice, such as impurity atoms and defects that scatter electron waves. The lack of electron scattering by a perfect lattice can be understood by noting that the electron wave generally travels through the metal unattenuated like light through a transparent crystal. Strong reflections are set up only for limited energies. When this occurs, the electron wave cannot travel freely through the crystal. As we shall see in the next section, these strong reflections occur when the lattice spacing is equal to an integral number of electronic wavelengths, resulting in forbidden energy bands for electrons.

Causes of resistance in a metal

The observed proportionality of resistivity to absolute temperature is the result of scattering of electron waves by lattice ions vibrating thermally about their equilibrium lattice positions. If the ions are modeled as harmonic oscillators, the square of the amplitude of the displacement of the ion from the lattice

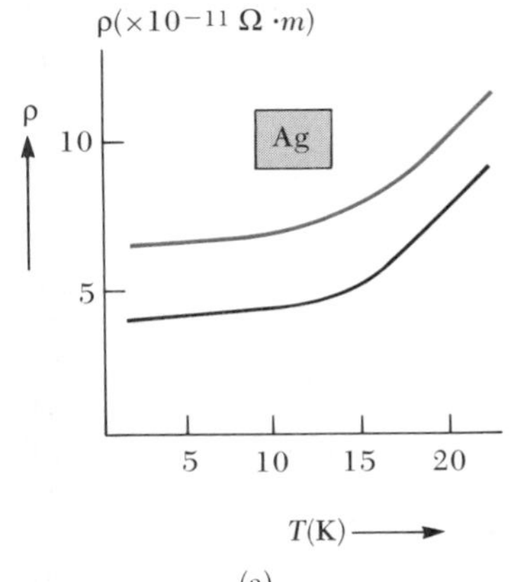

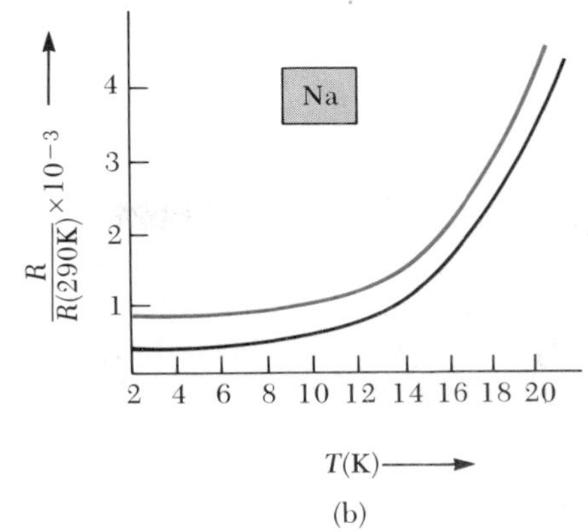

Figure 12.15 (a) Resistivity of silver at low temperature. The colored curve is for a silver sample with a higher concentration of impurity atoms. (From W. J. de Haas, G. J. van den Berg, J. de Boer. *Physica* 2:453, 1935.) (b) Resistance of sodium as a function of temperature. The colored curve is for a sodium sample with a higher concentration of imperfections. (From D. K. C. MacDonald and K. Mendelssohn, *Proc. Roy. Soc. (London)* A202:103, 1950).

site is proportional to T by the equipartition theorem: $A^2 \propto T$. Since the scattering cross section for electron waves is proportional to A^2 and L^{-1} is proportional to the scattering cross section, we have $L^{-1} \propto T$. Substituting $L^{-1} \propto T$ into $\rho = m_e v_F / ne^2 L$, we immediately find that $\rho \propto T$ as observed, in distinction to the classical result that predicted $\rho \propto \sqrt{T}$.

Examination of Figures 12.15a and 12.15b shows that there is also a *temperature-independent* contribution to the resisitivity of a metal, which clearly manifests itself for temperatures of less than 10 K. This *residual resistance*, ρ_i, which remains as $T \rightarrow 0$, varies from sample to sample. It is produced by electron waves scattering from impurities and structural imperfections in a given sample. Note that the parallel nature of the curves shown in Figure 12.15 implies that the *resistivity caused by thermal motion of the lattice*, ρ_L, is independent of the impurity concentration and that ρ_i is independent of temperature. This result is formalized in a result known as **Matthiessen's rule,** which states that the resistivity of a metal may usually be writen in the form

$$\rho = \rho_i + \rho_L \tag{12.27}$$

where ρ_i depends only on the concentration of crystal imperfections, and ρ_L depends only on T.

12.5 BAND THEORY OF SOLIDS

In Sections 12.3 and 12.4 we discussed the classical and quantum models of metals, making the simplifying assumption that the metal consists of a gas of free electrons. Since good insulators have an enormous resistivity compared to metals (a factor of 10^{24} greater), their electronic configurations must be quite different. In fact, most electrons in an insulator are not free, but are involved in ionic or covalent bonds as discussed earlier. Furthermore, in order to completely understand the electronic properties of solids, one must consider the effect of the lattice ions.

In the general case, there are two approaches to the determination of electronic energies in a solid. One is to follow the behavior of energy levels of isolated atoms as they are brought closer and closer to form a solid. The other is to show that energy bands arise when the Schrödinger equation is solved for electrons subject to a periodic potential representing the lattice ions. We shall follow the isolated atom approach here first, since it is simplest and immediately leads to an explanation of the differences between conductors, insulators, and semiconductors. Later in this section we shall explicitly consider the effect of a periodic potential on electron waves to show how energy bands arise.

Isolated Atom Approach to Band Theory

If two identical atoms are very far apart, they do not interact and their electronic energy levels can be considered to be those of isolated atoms. Suppose the two atoms are sodium, each having a 3*s* electron with a specific energy. As the two sodium atoms are brought closer together, their wave functions overlap and the two degenerate isolated 3*s* energy levels are split into two different levels, as shown in Figure 12.16a. We can understand this splitting by considering the appropriate electronic wave functions for the case of widely separated atoms and the case of neighboring atoms. Figure 12.17 shows the isolated atom wave functions ψ_1 and ψ_2 as well as the mixed wave functions $\psi_1 + \psi_2$ and $\psi_1 - \psi_2$ for two atoms close together. Note that an electron in the state $\psi_1 + \psi_2$ has a substantial probability of being found midway between the ion cores, while in the state $\psi_1 - \psi_2$ the probability density vanishes at the midpoint. Since the electron spends part of its time midway between the two attractive ion cores in the state $\psi_1 + \psi_2$, the electron is more tightly bound (has lower energy) in the state $\psi_1 + \psi_2$ than in $\psi_1 - \psi_2$. This leads to the two different 3*s* energy levels shown in Figure 12.16a.

When a large number of atoms are brought together to form a solid, a similar phenomenon occurs. As the atoms are brought close together, the various isolated-atom energy levels begin to split. This splitting is shown in Figure 12.16b for six atoms in close proximity. In this case there are six energy levels corresponding to six different linear combinations of isolated-atom wave functions. Since the width of an energy band (designated ΔE in Figure 12.16) arising from a particular atomic energy level is independent of the number of atoms in a solid, the energy levels are more closely spaced in the case of six atoms than in the case of two atoms.[5] If we extend this argument to a large number of atoms (of the order of 10^{23} atoms/cm^3), we obtain a large number of levels so closely spaced that they may be regarded as a *continuous band* of energy levels (see Fig. 12.16c). In the case of sodium, it is common to refer to the continuous distribution of allowed energy levels as the 3*s* band because they originate from the 3*s* levels of individual sodium atoms. In general, a crystalline solid has a large number of allowed energy bands that arise from different atomic energy levels. Figure 12.18 shows the allowed energy

[5] The width of the energy band depends on the amount of charge distribution overlap between adjacent atoms—hence on the interatomic separation.

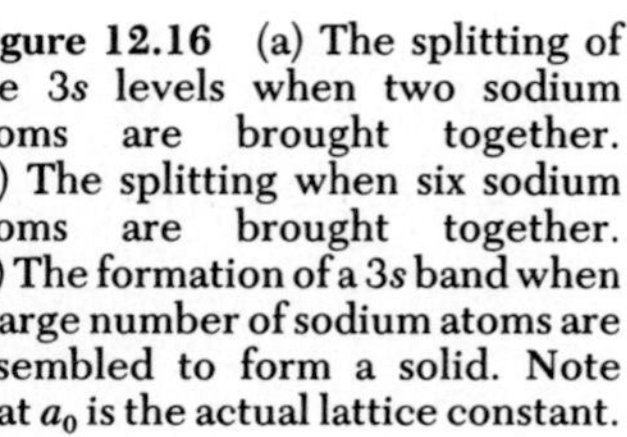
Figure 12.16 (a) The splitting of the 3*s* levels when two sodium atoms are brought together. (b) The splitting when six sodium atoms are brought together. (c) The formation of a 3*s* band when a large number of sodium atoms are assembled to form a solid. Note that a_0 is the actual lattice constant.

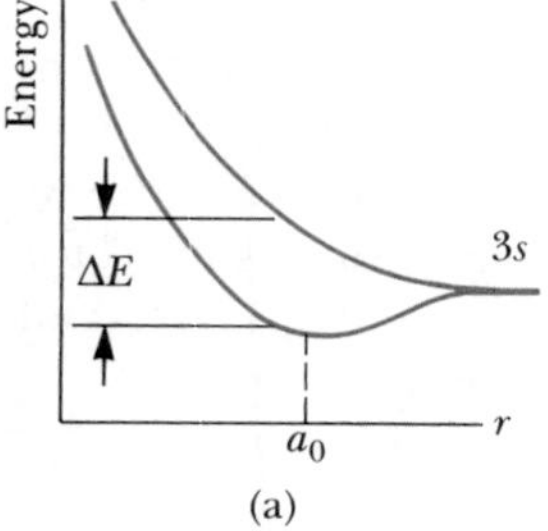

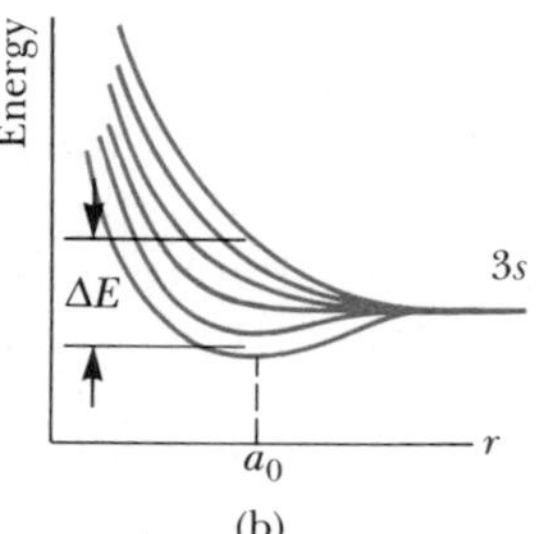

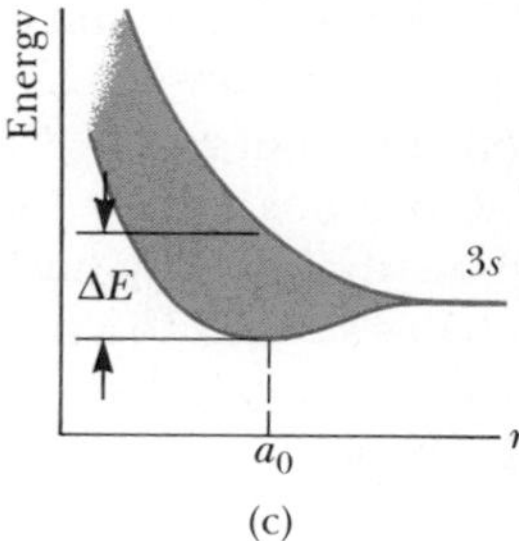

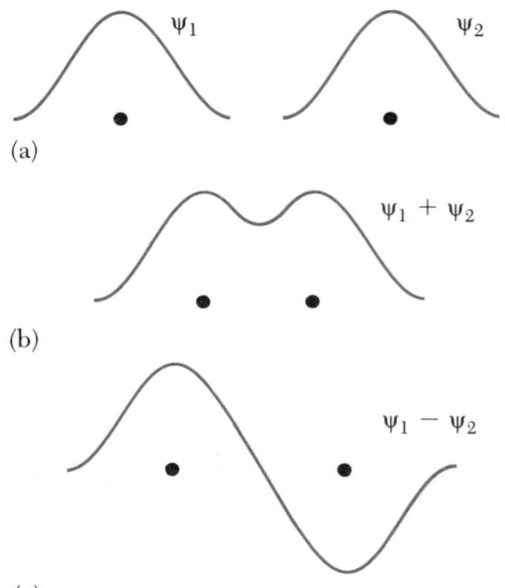

Figure 12.17 (a) Wave functions of electrons bound to two ion cores at large separation. (b) One linear combination of wave functions appropriate for electrons bound to ion cores at small separation is $\psi_1 + \psi_2$. (c) The other is $\psi_1 - \psi_2$.

bands of sodium. Note that energy gaps or forbidden energy bands for electrons occur between the allowed bands.

If the solid contains N atoms, each energy band has N energy levels. The $1s$, $2s$, and $2p$ bands of sodium are each full, as indicated by the gray-shaded areas in Figure 12.18. A level whose orbital angular momentum is ℓ can hold $2(2\ell + 1)$ electrons. The factor of two arises from the two possible electron spin orientations, while the factor $2\ell + 1$ corresponds to the number of possible orientations of the orbital angular momentum. The capacity of each band for a system of N atoms is $2(2\ell + 1)N$ electrons. Hence, the $1s$ and $2s$ bands each contain $2N$ electrons ($\ell = 0$), while the $2p$ band contains $6N$ electrons ($\ell = 1$). Because sodium has only one $3s$ electron and there are a total of N atoms, in the solid, the $3s$ band contains only N electrons and is only half full. The $3p$ band, which is above the $3s$ band, is completely empty.

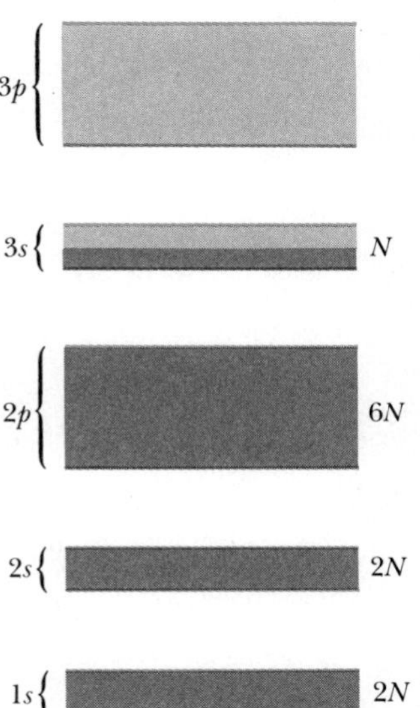

Figure 12.18 The energy bands of sodium. The solid contains N atoms. Note the energy gaps between allowed bands.

Conduction in Metals, Insulators, and Semiconductors

The enormous variation in electrical conductivity of metals, insulators, and semiconductors may be qualitatively explained in terms of energy bands. We shall see that the position and the electronic occupation of the highest band or, at most, of the highest two bands determine the conductivity of a solid.

Metals We can obtain an understanding of electrical conduction in metals by considering the half-filled uppermost $3s$ band of sodium. Figure 12.19 shows this typical half-filled metallic band at $T = 0$ K, where the shaded region represents levels that are filled with electrons. Since electrons obey Fermi-Dirac statistics, all levels below the Fermi energy, E_F, are filled with electrons, while all levels above E_F are empty. In the case of sodium, the Fermi energy lies in the middle of the band. At temperatures greater than 0 K, a few electrons are thermally excited to levels above E_F (as shown by the Fermi-Dirac distribution to the left of Fig. 12.19), but overall there is little change from the 0 K case. *Note, however, that if an electric field is applied to the metal, electrons with energies near the Fermi energy require only a small amount of additional energy from the field to reach nearby empty energy states.* Thus electrons are free to move with only a small applied field in a metal because there are many unoccupied states close to occupied energy states.

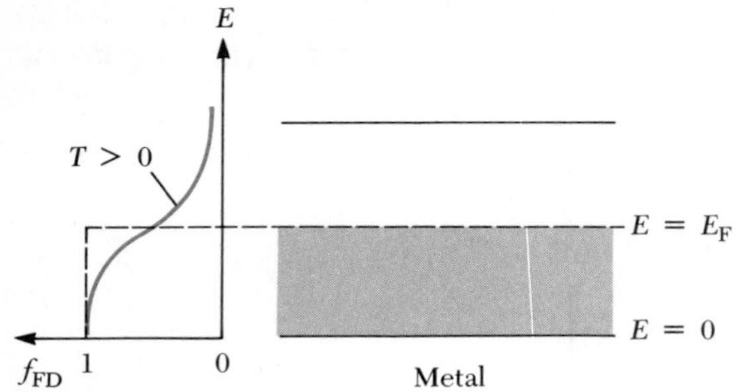

Figure 12.19 A half-filled band of a conductor such as the 3*s* band of sodium. At $T = 0$ K, the Fermi energy lies in the middle of the band. The Fermi-Dirac probability that an energy state E is occupied is shown to the left.

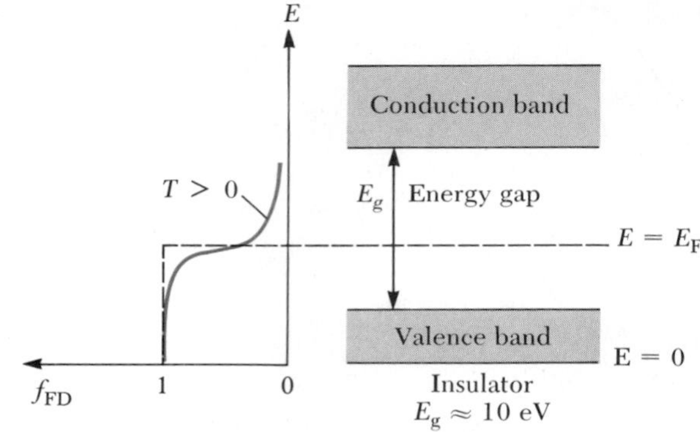

Figure 12.20 An insulator at $T = 0$ K has a filled valence band and an empty conduction band. The Fermi level lies somewhere between these bands.

Insulators Now consider the two highest energy bands of a material having the lower band completely filled with electrons and the higher completely empty at 0 K (see Fig. 12.20). It is common to refer to the separation between the outermost filled and empty bands as the *energy gap*, E_g, of the material. The lower band filled with electrons is called the *valence band*, and the upper empty band is the *conduction band.* The Fermi energy lies somewhere in the energy gap, as shown in Figure 12.20. Since the energy gap for an insulator is large (~ 10 eV) compared to kT at room temperature ($kT = 0.025$ eV at 300 K), the Fermi-Dirac distribution predicts that there will be very few electrons thermally excited into the upper band at normal temperatures. *Thus, although an insulator has many vacant states in the conduction band that can accept electrons, there are so few electrons actually occupying conduction band states that the overall contribution to electrical conductivity is very small, resulting in a high resistivity for insulators.*

Semiconductors Now consider a material that has a much smaller energy gap, on the order of 1 eV. Such materials are called semiconductors. Table 12.6 shows the energy gaps for some representative materials. At $T = 0$ K, all electrons are in the valence band and there are no electrons in the conduction band. Thus semiconductors are *poor* conductors at low temperatures. At ordinary temperatures, however, the situation is quite different. As shown in

TABLE 12.6 Energy Gap Values for Some Semiconductors[*]

Crystal	E_g (eV) 0 K	E_g (eV) 300 K
Si	1.17	1.14
Ge	0.744	0.67
InP	1.42	1.35
GaP	2.32	2.26
GaAs	1.52	1.43
CdS	2.582	2.42
CdTe	1.607	1.45
ZnO	3.436	3.2
ZnS	3.91	3.6

[*] These data were taken from C. Kittel, *Introduction to Solid State Physics*, 6th edition, New York, John Wiley and Sons, 1986.

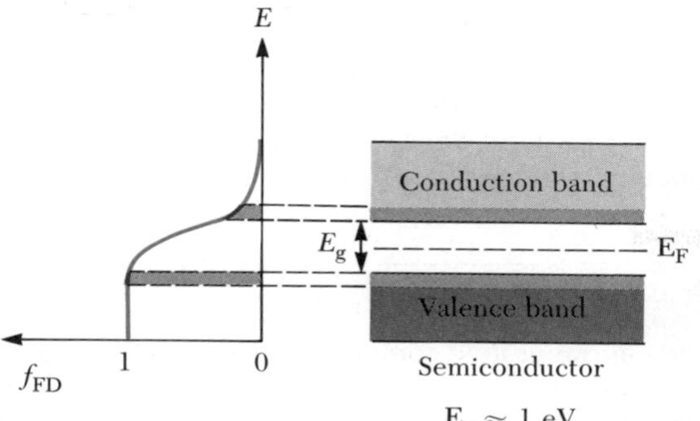

Figure 12.21 The band structure of a semiconductor at ordinary temperatures ($T \approx 300$ K). Note that the energy gap is much smaller than in an insulator and that many electrons occupy states in the conduction band.

Figure 12.21, the populations of the valence and conduction bands are altered. Since the Fermi level, E_F, is located at about the middle of the gap for a semiconductor, and since E_g is small, appreciable numbers of electrons are thermally excited from the valence band to the conduction band. Since there are many empty nearby states in the conduction band, a small applied potential can easily raise the energy of the electrons in the conduction band, resulting in a moderate current. Because thermal excitation across the narrow gap is more probable at higher temperatures, the conductivity of semiconductors depends strongly on temperature and *increases* rapidly with temperature. This contrasts sharply with the conductivity of a metal, which *decreases slowly* with temperature (see Section 12.3).

It is important to point out that there are both negative and positive charge carriers in a semiconductor. When an electron moves from the valence band into the conduction band, it leaves behind a vacant crystal site, or so-called **hole,** in the otherwise filled valence band. This hole (electron-deficient site) appears as a positive charge, $+e$. The hole acts as a charge carrier in the sense that a valance electron from a nearby bond can transfer into the hole, thereby filling it and leaving a hole behind in the electron's original place. Thus the hole migrates through the valence band. In a pure crystal containing only one element or compound, there are equal numbers of conduction electrons and holes. Such combinations of charges are called electron-hole pairs, and a pure semiconductor that contains such pairs is called an *intrinsic semiconductor* (see Fig. 12.22). Note that in the presence of an electric field, the holes move in the direction of the field and the conduction electrons move opposite the field.

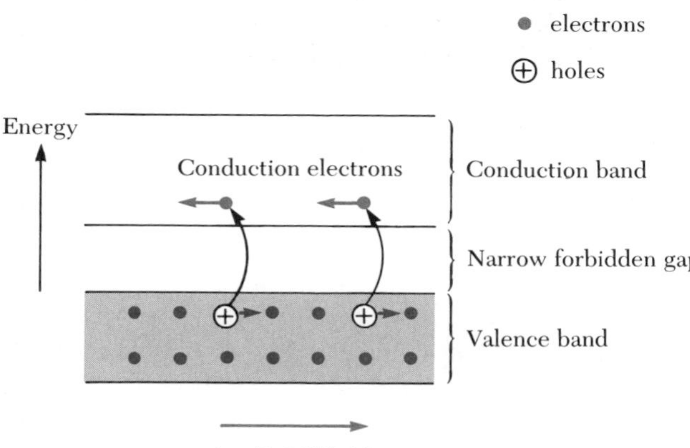

Figure 12.22 An intrinsic semiconductor. Note that the electrons move toward the left and the holes move toward the right when the applied electric field is to the right as shown.

EXAMPLE 12.2

Estimate the electric field strength required to produce conduction in diamond, an excellent insulator at room temperature. Assume a mean free path of 5×10^{-8} m and an energy gap of 7 eV in diamond. Make the simplifying assumption that the electron velocity drops to zero following each collision with an ion core.

Solution: If an electron in diamond is to conduct, it must be supplied with 7 eV of energy from the electric field. Since the electron generally loses most of its energy in each collision, the field must supply 7 eV in a *single* mean free path. Thus for an electron to gain 7 eV in a distance L we require an electric field of

$$E = \frac{V}{L} = \frac{7\ v}{5 \times 10^{-8}\ \mathrm{m}} = 1.4 \times 10^{8}\ \mathrm{V/m}$$

This is an enormous field compared to the field required to produce conduction in metals.

Energy Bands from Electron Wave Reflections

We now wish to consider a completely different approach to understanding the origin of energy bands in solids. This approach involves modifying the free electron wave functions to take into account the scattering of electron waves by the periodic crystal lattice.

Recall that a completely free electron moving in the $+x$ direction is represented by a traveling wave with wave number $k = 2\pi/\lambda$, described by

$$\Psi_{\text{free}} = Ae^{i(kx - \omega t)}$$

According to de Broglie, the wave carries momentum $p = \hbar k$ and energy $E = \hbar\omega$. Further, the energy of the free electron as a function of k is given by

$$E = \frac{p^2}{2m} = \frac{\hbar^2 k^2}{2m}$$

and a plot of E versus k yields a parabola (see Fig. 12.23). We note that allowed energy values are continuously distributed from 0 to infinity and that there are no breaks or gaps in energy at particular k values.

Now let us consider what happens to a traveling electron wave when it passes through a one-dimensional crystal lattice. Let us start with electron waves of very long wavelength or small k and low energy, as in Figure 12.24a. In this case, waves reflected from successive atoms and traveling to the left are all slightly out of phase and cancel out on average. Thus the electron wave does not get reflected and the electron moves through the lattice like a free particle. If, however, we make the electron wavelength shorter and shorter, when we reach the condition $\lambda = 2a$, waves reflected from adjacent atoms will be *in phase* as shown in Figure 12.24b. This occurs because the path length difference is $2a$ for a wave reflected directly from atom 1 compared to a wave that travels from 1 to 2 and then is reflected back to 1, as shown in Figure 12.24c.

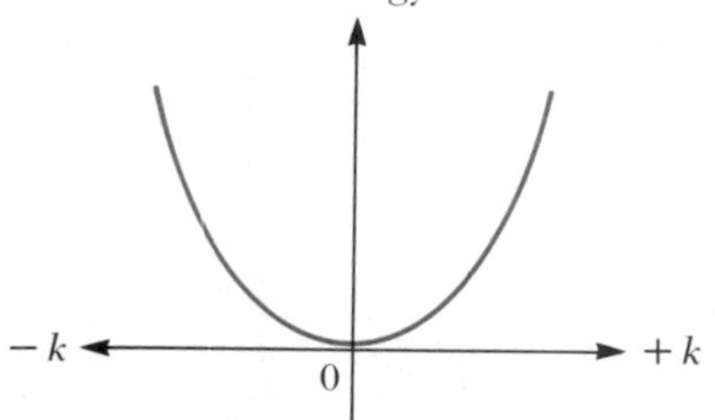

Figure 12.23 Energy versus wave number k for a free electron, where $k = 2\pi/\lambda$.

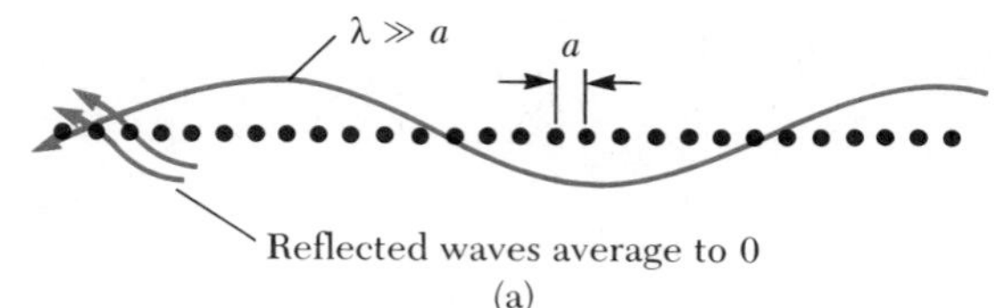

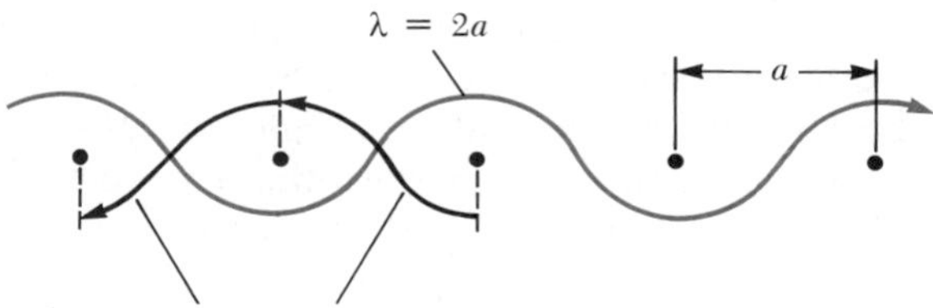

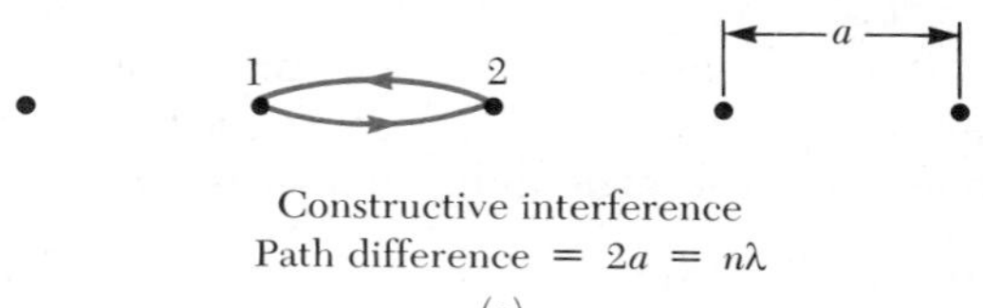

Figure 12.24 (a) Reflection of electron waves when $\lambda \gg a$. (b) Reflection of electron waves when $\lambda = 2a$. (c) Constructive interference of reflected waves.

The longest wavelength at which constructive interference of reflected waves occurs is thus given by

$$2a = 1\lambda$$

In general, constructive interference will also occur at other, shorter wavelengths given by

$$2a = \pm n\lambda \qquad n = 1, 2, 3, \ldots \tag{12.28}$$

where the negative sign arises from reflection from the atom to the left of atom 1. In terms of wave number k, Equation 12.28 predicts strong reflected electron waves when

$$k = \pm\frac{n\pi}{a} \qquad n = 1, 2, 3, \ldots \tag{12.29}$$

Thus, for $k = \pm n\pi/a$ the electron wave functions are not just waves traveling to the right but are composed of equal parts of waves traveling to the right (incident) and to the left (reflected). Since the waves traveling left and right can be added or subtracted, we actually have two different possible standing wave types, denoted Ψ_- and Ψ_+:

$$\Psi_- = Be^{i(kx-\omega t)} - Be^{i(-kx-\omega t)} \tag{12.30}$$

$$\Psi_+ = Be^{i(kx-\omega t)} + Be^{i(-kx-\omega t)} \tag{12.31}$$

where $Be^{i(kx-\omega t)}$ is a wave traveling to the right and $Be^{i(-kx-\omega t)}$ is a wave traveling to the left. Using Euler's identity, $e^{i\theta} = \cos\theta + i\sin\theta$, one can easily show that for $k = \pm\pi/a$, Ψ_- and Ψ_+ take the more useful forms:

$$\Psi_- = \pm 2Bie^{-i\omega t} \sin \frac{\pi x}{a} \tag{12.32}$$

$$\Psi_+ = 2Be^{-i\omega t} \cos \frac{\pi x}{a} \tag{12.33}$$

We now wish to show the key point: Ψ_- *has a slightly higher energy than* Ψ_{free} *and* Ψ_+ *has a slightly lower energy than* Ψ_{free} *at* $k = \pi/a$. *This leads to a discontinuity in energy or a band gap at* $k = \pi/a$. This band gap is shown in Figure 12.25. In order to determine the total energy of standing wave states Ψ_+ and Ψ_-, we must consider both kinetic and potential energies of the electron in these states. The kinetic energy is the same in both states since both have the same momentum, $p = \hbar k = \hbar\pi/a$. Thus, the average kinetic energy of the states Ψ_+ and Ψ_- is

$$\langle K\rangle = \langle\frac{p^2}{2m}\rangle = \frac{\hbar^2k^2}{2m} = \frac{\hbar^2\pi^2}{2ma^2}$$

The potential energies of the two states are different, however, since Ψ_+ and Ψ_- distribute the electrons differently with respect to the positions of the positive ion cores. The distributions of electronic charge for the case of a traveling wave and the two standing waves cases, Ψ_+ and Ψ_-, are shown in Figure 12.26. They are calculated from the probability densities as follows:

$$\Psi^*_{\text{free}}\Psi_{\text{free}} = [Ae^{i(kx-\omega t)}]^*[Ae^{i(kx-\omega t)}] = |A|^2$$

$$\Psi_+{}^*\Psi_+ = \left[2Be^{-i\omega t}\cos\frac{\pi x}{a}\right]^*\left[2Be^{-i\omega t}\cos\frac{\pi x}{a}\right] = 4|B|^2\cos^2\frac{\pi x}{a}$$

$$\Psi_-{}^*\Psi_- = \left[\pm 2Bie^{-i\omega t}\sin\frac{\pi x}{a}\right]^*\left[\pm 2Bie^{-i\omega t}\sin\frac{\pi x}{a}\right] = 4|B|^2\sin^2\frac{\pi x}{a}$$

Note that the traveling wave solution distributes electronic charge uniformly along the lattice. On the other hand, Ψ_- concentrates electronic charge midway between the ions, while Ψ_+ concentrates electronic charge right over the positive ion cores, minimizing the electron's potential energy.[6] Thus, we

[6] Recall that the electron's electrical potential energy is the work done by an external force to move the electron to a particular point. To move the electron to a *positive* ion requires the minimum external work. Thus the electron has the minimum potential energy in this position.

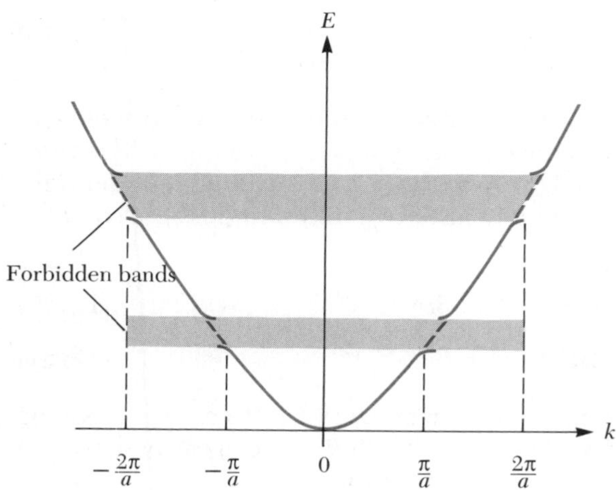

Figure 12.25 Energy versus wave number for a one-dimensional lattice with atomic separation a. The E versus k curve for the free electron case is shown dashed. Note the forbidden bands corresponding to impossible energy states for the electron.

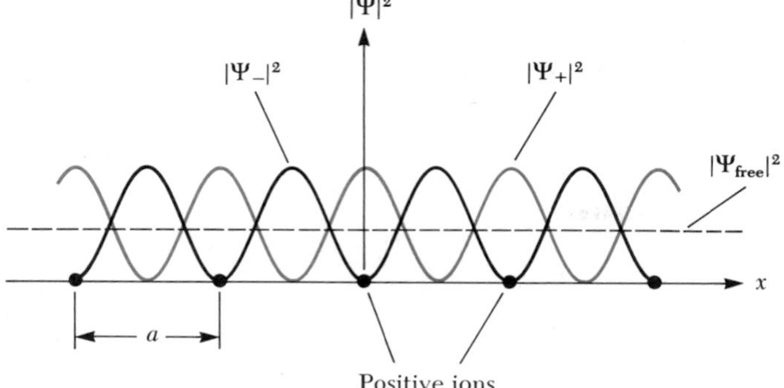

Figure 12.26 Probability densities of standing waves and traveling (free) electron waves in a one-dimensional lattice. Note that the state Ψ_+ places the electron right over the positive ion and consequently has the lowest potential energy.

would expect the average potential energy of the electron in the state Ψ_+ to be somewhat below the free electron energy and the average potential energy in the state Ψ_- to be a bit above the free electron energy. This difference in potential energy between Ψ_+ and Ψ_- leads to a difference in total energy, since the kinetic energies of Ψ_+ and Ψ_- are the same. Thus, the total energy of the state Ψ_+ at $k = \pi/a$ shown in Figure 12.25 lies slightly below the free electron energy parabola, and the total energy of the state Ψ_- lies slightly above the free electron energy. Also, Figure 12.25 shows that the overall effect of the lattice is to introduce forbidden gaps in the free energy continuum at values of $k = \pm n\pi/a$.

12.6 SEMICONDUCTOR DEVICES

The *p-n* Junction

In order to make devices, one must be able to fabricate semiconductors with well defined regions of different conductivity. Both the type (positive or negative) and number of carriers in a semiconductor may be tailored to the needs of a specific device by adding intentional impurities in a process called *doping*. For example, when an atom with five valence electrons, such as arsenic, is added to a semiconductor from Group 4 of the periodic table, four valence electrons participate in the covalent bonds and one electron is left over (see Fig. 12.27a). This extra electron is nearly free and has an energy

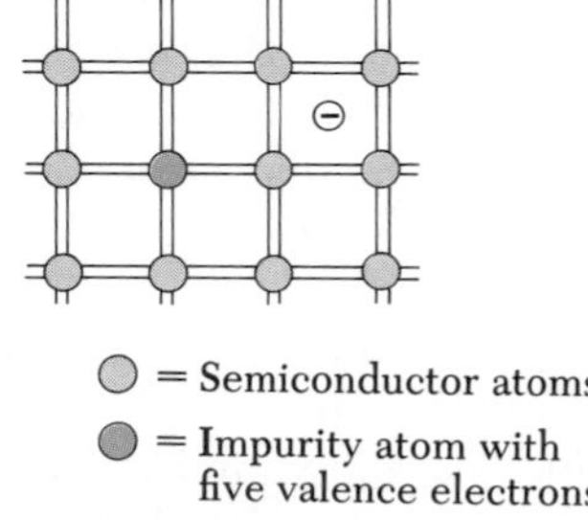

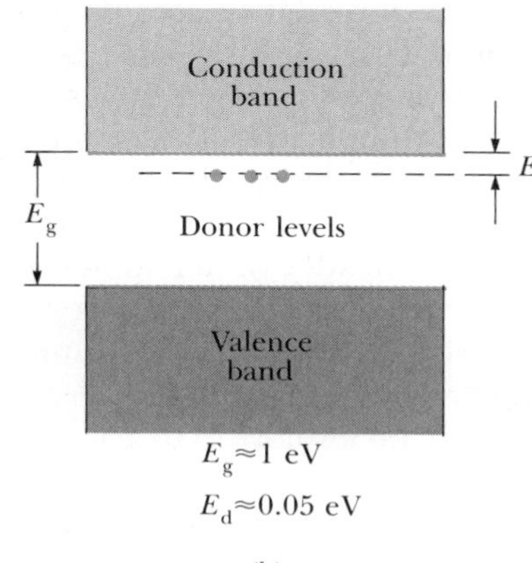

Figure 12.27 (a) Two-dimensional representation of a semiconductor containing a donor atom (colored spot). (b) Energy band diagram of a semiconductor in which the donor levels lie within the forbidden gap, just below the bottom of the conduction band.

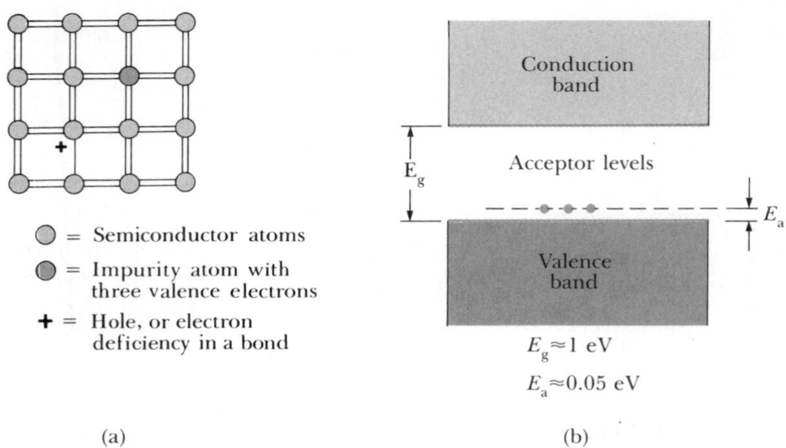

Figure 12.28 (a) Two-dimensional representation of a semiconductor containing an acceptor atom (colored spot). (b) Energy band diagram of a semiconductor in which the acceptor levels lie within the forbidden gap, just above the top of the valence band.

level that lies within the energy gap, just below the conduction band, as shown in Figure 12.27b. Such a pentavalent atom in effect donates an electron to the semiconductor, and is referred to as a *donor atom.* Since the energy spacing between the donor levels and the bottom of the conduction band is very small (typically, about 0.05 eV), a small amount of thermal energy will cause an electron in these levels to move into the conduction band. (Recall that the average thermal energy of an electron at room temperature is about $kT \approx$ 0.025 eV.) Semiconductors doped with donor atoms are called *n-type semiconductors* because the charge carriers are electrons, whose charge is negative.

If the semiconductor is doped with atoms with three valence electrons, such as indium and aluminum, the three electrons form covalent bonds with neighboring atoms, leaving an electron deficiency, or hole, in the fourth bond (see Fig. 12.28a). The energy levels of such impurities also lie within the energy gap, just above the valence band, as indicated in Figure 12.28b. Electrons from the valence band have enough thermal energy at room temperature to fill these impurity levels, leaving behind a hole in the valence band. Because a trivalent atom in effect accepts an electron from the valence band, such impurities are referred to as *acceptors.* A semiconductor doped with trivalent (acceptor) impurities is known as a *p-type semiconductor* because the charge carriers are positively charged holes. When conduction is dominated by acceptor or donor impurities, the material is called an *extrinsic semiconductor.* The typical range of doping densities for *n*- or *p*-type semiconductors is 10^{13} to 10^{19} cm^{-3}.

Now let us consider what happens when a *p*-type semiconductor is joined to an *n*-type semiconductor to form a *p-n junction.* We find that the completed junction consists of three distinct semiconductor regions, as shown in Figure 12.29a: a *p*-type region, a *depletion* region, and an *n*-type region. The depletion region may be visualized to arise when the two halves of the junction are brought together and mobile donor electrons diffuse to the *p* side of the junction, leaving behind immobile positive ion cores. (Conversely, holes dif-

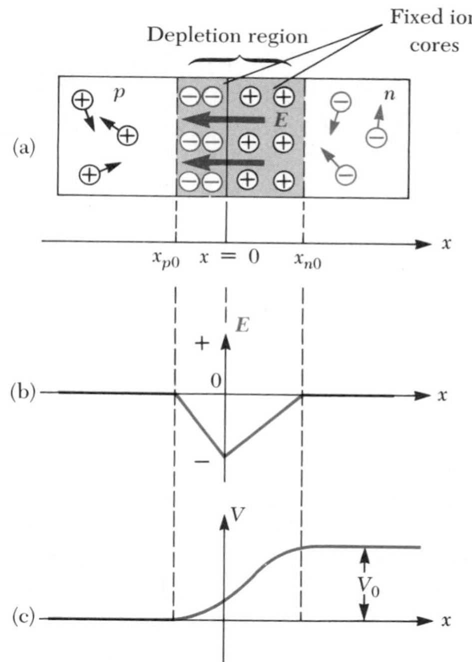

Figure 12.29 (a) Physical arrangement of a *p-n* junction. (b) Built-in electric field versus x for the *p-n* junction. (c) Built-in potential versus x for the *p-n* junction.

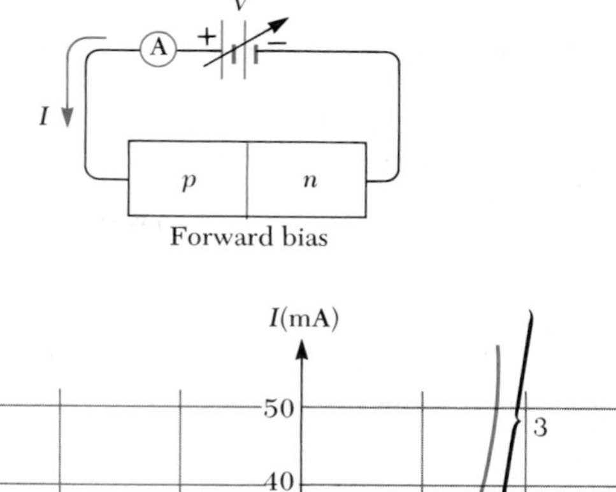

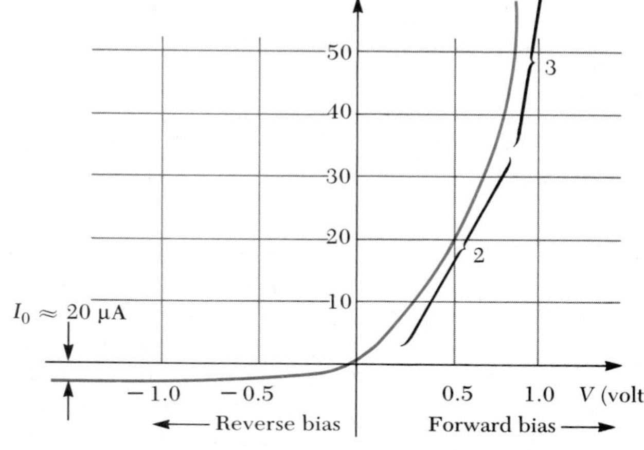

Figure 12.30 The characteristic curve for a real diode.

fuse to the n side and leave a region of fixed negative ion cores.) The region extending several microns from the junction is called the *depletion region* because it is depleted of *mobile* charge carriers. It also contains a built-in electric field on the order of 10^4 to 10^6 V/cm, which serves to sweep mobile charge out of this region and keep it truly depleted. This electric field creates a potential barrier V_0 that prevents the further diffusion of holes and electrons across the junction and insures zero current through the junction when no external voltage is applied.

Perhaps the most notable feature of the *p-n* junction is its ability to pass current in only one direction. Such *diode* action is easiest to understand in terms of the potential diagram in Figure 12.29c. If a positive external voltage is applied to the p side of the junction, the overall barrier is decreased, resulting in a current that increases exponentially with increasing forward voltage or bias. For reverse bias (a positive external voltage applied to the n side of the junction), the potential barrier is increased, resulting in a very small reverse current that quickly reaches a saturation value, I_0, with increasing reverse bias. The current-voltage relation for an ideal diode is given by

$$\boxed{I = I_0(e^{qV/kT} - 1)} \qquad (12.34)$$

where q is the electronic charge, k is Boltzmann's constant, and T is the temperature in kelvins. Figure 12.30 shows a plot of an *I-V* characteristic for a real diode along with a schematic of a diode under forward bias.

EXAMPLE 12.3 Forward and Reverse Currents in a Diode

Calculate the ratio of the forward current to the reverse current for a diode described by Equation 12.34 for an applied voltage of + 1.0 V (forward bias) and − 1.0 V (reverse bias).

Solution:

$$\frac{I_{\text{forward}}}{I_{\text{reverse}}} = \frac{I_0(e^{+qV/kT} - 1)}{I_0(e^{-qV/kT} - 1)}$$

Assuming room temperature operation (300 K), and taking $k = 8.62 \times 10^{-5}$ eV/K, gives

$$\frac{I_f}{I_r} = \frac{I_0(e^{1/(8.6\times10^{-5})(300)} - 1)}{I_0(e^{-1/(8.6\times10^{-5})(300)} - 1)} = \frac{(e^{38.7} - 1)}{(e^{-38.7} - 1)}$$

$$= -6.40 \times 10^{16}!!$$

Light-Emitting and -Absorbing Diodes — LEDs, Lasers, and Solar Cells

As previously noted, light emission and absorption in semiconductors is similar to light emission and absorption by gaseous atoms, except that discrete atomic energy levels must be replaced by bands in semiconductors. As shown in Figure 12.31a, an electron, excited electrically into the conduction band, can easily recombine with a hole (especially if the electron is injected into a p-type region), emitting a photon of band gap energy. Examples of p-n junctions that convert electrical input into light output are light-emitting diodes (LEDs) and injection lasers. Conversely, an electron in the valence band may absorb a photon and be promoted to the conduction band, leaving a hole behind (see Fig. 12.31b). Devices in which light-generated electrons and holes are separated by a junction field and collected as current are called solar cells or photovoltaic devices.

Let us briefly describe the operation of LEDs and injection lasers. Consider the operation of a forward biased diode. Under ordinary forward bias (region 2 in Fig. 12.30), electrons carry the current on the n-side of the junction and holes on the p-side. The concentration of electrons falls by many orders of magnitude as one crosses the depletion region. Relatively few electrons penetrate into the p-side and experience recombination. This rapid falloff in electron (n) and hole (p) concentrations near the junction under ordinary forward bias is shown in Figure 12.32a. If the diode is operated under extreme forward bias (region 3 in Fig. 12.30), electrons and holes are *injected* into and through the depletion region into p and n regions, respectively. This penetration extends several microns from the depletion region (which is no longer depleted, as can be seen in Fig. 12.32b), and results in strong recombination radiation that increases rapidly with forward current density.

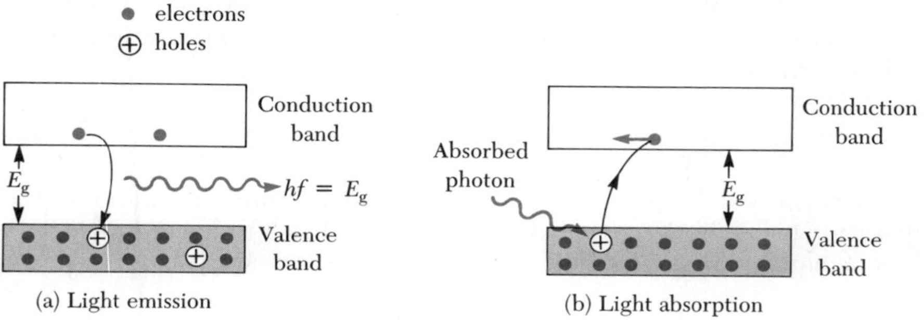

Figure 12.31 (a) Light emission from a semiconductor. (b) Light absorption by a semiconductor.

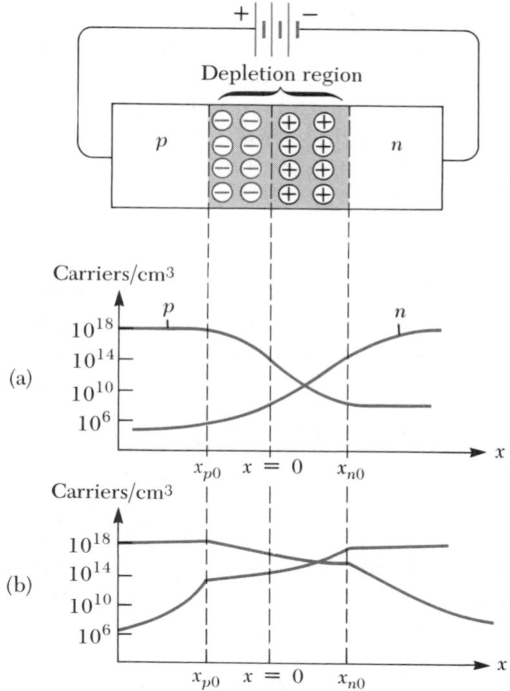

Figure 12.32 Carrier density in a *p-n* junction: (a) The device, (b) normal forward bias, and (c) high current forward bias.

The injection of carriers across the depletion region (electrons into the conduction band on the *p*-side and holes into the valence band on the *n*-side) leads to the formation of a population inversion layer along the plane of the junction. At low current levels the modest population inversion creates a broad emission spectrum, which is dominated by spontaneous emission. LEDs operate at these low current levels. Higher forward current densities create a stronger population inversion, which leads to the dominance of stimulated emission, line narrowing, and lasing, as can be seen in Figures 12.33 and 12.34. The sequence of events occurs as follows. Starting with an inverted population and a spontaneous-emission spectrum of photons, these photons will stimulate further recombinations. The second wave of photons, although

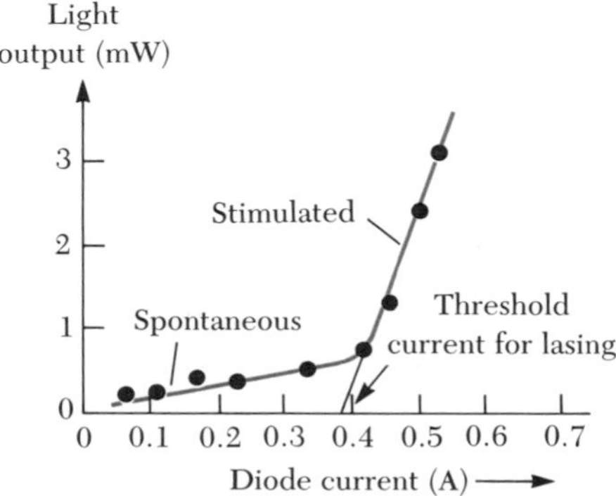

Figure 12.33 Total light output (all wavelengths) versus forward current through a GaAs-AlGaAs laser at room temperature.

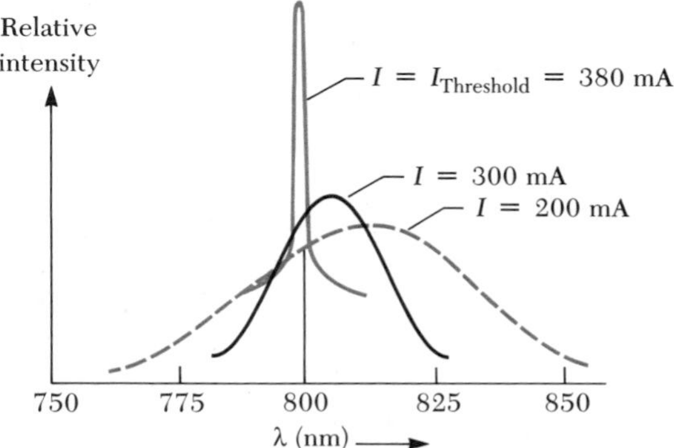

Figure 12.34 Emission spectra from a GaAs-AlGaAs laser as a function of forward current through the laser.

propagating in all directions with random phases, will be more numerous at a given frequency and, therefore, stimulate more transitions at that central frequency than those in the wings of the spectrum. Thus the emission spectrum narrows appreciably with increasing population inversion. Coherent, unidirectional laser light can be obtained by creating a resonant cavity (as with other lasers) that favors the growth of one direction and one phase.

The Junction Transistor

The discovery of the transistor by John Bardeen, Walter Brattain, and William Shockley in 1948 totally revolutionized the world of electronics. For this work, these three men shared a Nobel Prize in 1956. By 1960, the transistor had replaced the vacuum tube in many electronic applications. The advent of the transistor created a multibillion dollar industry that produces such popular devices as pocket radios, handheld calculators, computers, television receivers, and electronic games.

The junction transistor consists of a semiconducting material with a very narrow *n* region sandwiched between two *p* regions. This configuration is called the *pnp transistor.* Another configuration is the *npn transistor,* which consists of a *p* region sandwiched between two *n* regions. Because the operations of the two transistors are essentially the same, we shall describe only the *pnp* transistor.

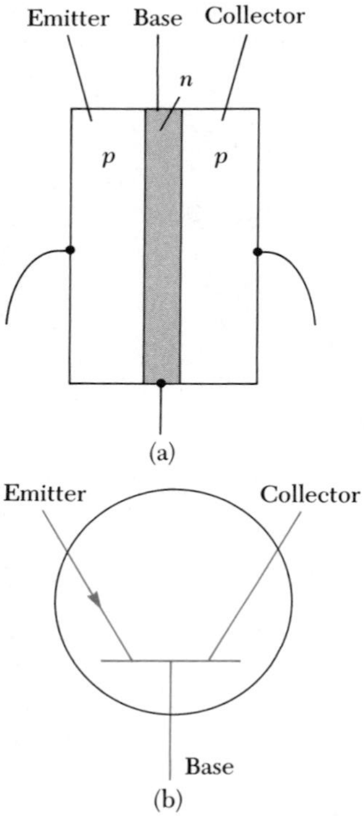

The structure of the *pnp* transistor, together with its circuit symbol, is shown in Figure 12.35. The outer regions of the transistor are called the *emitter* and *collector,* and the narrow central region is called the *base.* Note that the configuration contains two junctions. One junction is the interface between the emitter and the base, and the other is between the base and the collector.

Suppose a voltage is applied to the transistor such that the emitter is at a higher potential than the collector. (This is accomplished with battery V_{ec} in Figure 12.36.) If we think of the transistor as two diodes back to back, we see that the emitter-base junction is forward-biased and the base-collector junction is reverse-biased.

Because the *p*-type emitter is heavily doped relative to the base, nearly all of the current consists of holes moving across the emitter-base junction. Most of these holes do not recombine in the base because it is very narrow. The holes are finally accelerated across the reverse-biased base-collector junction, producing the current I_c in Figure 12.36.

Figure 12.35 (a) The *pnp* transistor consists of an *n* region (base) sandwiched between two *p* regions (the emitter and the collector). (b) The circuit symbol for the *pnp* transistor.

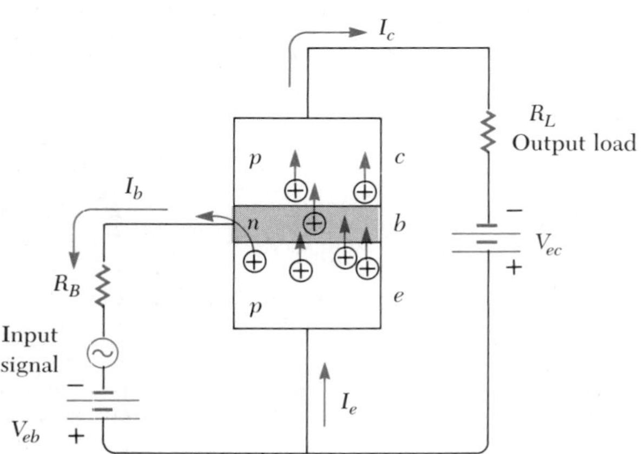

Figure 12.36 A small base current, I_b, controls a large collector current, I_c. Note that $I_e = I_b + I_c$, $I_b \ll I_c$, and $I_e = \beta I_b$, where $\beta \approx 10$ to 100.

Although only a small percentage of holes recombine in the base, those that do limit the emitter current to a small value, because positive charge carriers accumulate in the base and prevent holes from flowing into this region. In order to prevent this limitation of current, some of the positive charge on the base must be drawn off; this is accomplished by connecting the base to a second battery, V_{eb} in Figure 12.36. Those positive charges that are not swept across the collector-base junction leave the base through this added pathway. This base current, I_b, is very small, but a small change in it can significantly change the collector current, I_c. If the transistor is properly biased, the collector (output) current will be directly proportional to the base (input) current and the transistor will act as a current amplifier. This condition may be written

$$I_c = \beta I_b$$

where β, the current gain, is typically in the range from 10 to 100. Thus, the transistor may be used to amplify a small, time-varying signal. The small voltage to be amplified is placed in series with the battery V_{eb} as shown in Figure 12.36. The time-dependent input signal produces a small variation in the base current. This results in a large change in collector current and hence a large change in voltage across the output resistor.

The Integrated Circuit

The integrated circuit, invented independently by Jack Kilby at Texas Instruments in late 1958 and by Robert Noyce at Fairchild Camera and Instrument in early 1959, has been justly called "the most remarkable technology ever to hit mankind." ICs have indeed started a second industrial revolution and are found at the heart of computers, watches, cameras, automobiles, aircraft, robots, space vehicles, and all sorts of communication and switching networks. In simplest terms, an integrated circuit is a collection of interconnected transistors, diodes, resistors, and capacitors fabricated on a single piece of silicon. State of the art chips easily contain several hundred thousand components in a 1 cm^2 area (Fig. 12.37).

It is interesting that integrated circuits were invented partly in an effort to achieve circuit miniaturization and partly in order to solve the interconnection problem spawned by the transistor. In the era of vacuum tubes, power and size considerations of individual components set modest limits on the number of components that could be interconnected in a given circuit. With the advent of the tiny, low-power, highly reliable transistor, design limits on the number of components disappeared and were replaced with the problem of wiring

Figure 12.37 A 32-bit microprocessor integrated circuit containing over 200 000 components is about the size of a dime. (Courtesy of A.T.&T. Bell Laboratories.)

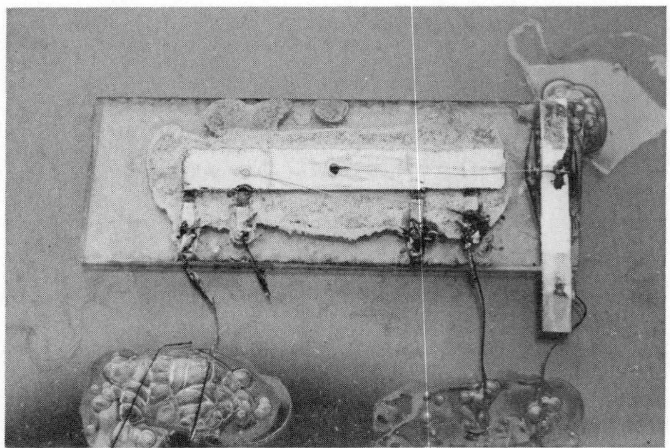

Figure 12.38 Jack Kilby's first integrated circuit, tested on September 12, 1958. (Courtesy of Texas Instruments, Inc.)

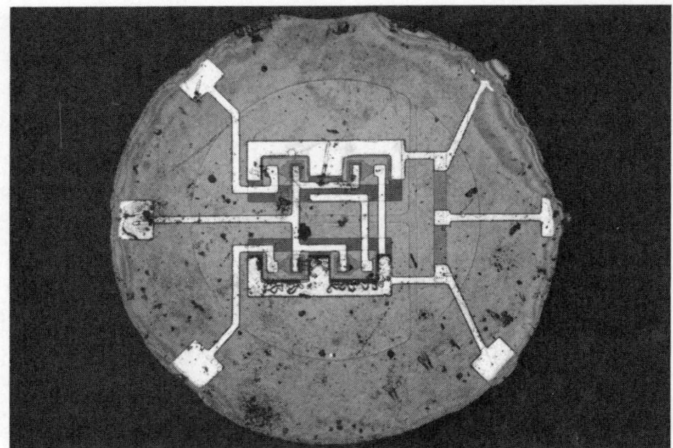

Figure 12.39 One of Robert Noyce's first integrated circuits. Note that the leads connecting different components are built right in as part of the chip. (Courtesy of Fairchild Semiconductor Corporation.)

together hundreds of thousands of components. The magnitude of this problem can be appreciated when one considers that second generation computers (consisting of discrete transistors) contained several hundred thousand components requiring more than a million hand-soldered joints to be made and tested.

Figures 12.38 and 12.39 show pictures of the first ICs containing multiple electronic devices on a single piece of silicon. Noyce's integrated circuit also shows interconnections between devices *fabricated as part of the chip manufacturing process.* In addition to solving the interconnection problem, ICs possess the advantages of miniaturization and fast response, two attributes critical for high speed computers. The fast response is actually also a product of the miniaturization and close packing of components. This is so because the response time of a circuit depends on the time it takes for electrical signals traveling at about 1 ft/ns to pass from one component to another. This time is clearly reduced by the closer packing of components. Additional information on the history, theory of operation, and fabrication of chips may be found in the references at the end of this chapter.

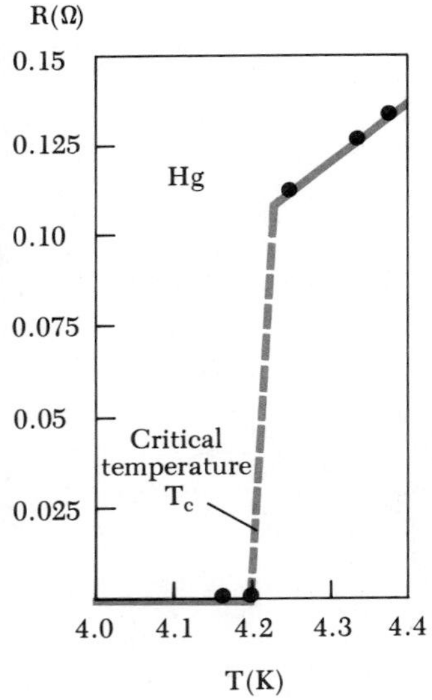

12.7 SUPERCONDUCTIVITY

Superconductors are a class of materials whose *dc* resistivity goes to zero below a certain temperature T_c, called the *critical temperature.* This incredible behavior, called *superconductivity,* was discovered in 1911 by the Dutch physicist H. Kamerlingh Onnes. He and one of his graduate students found that the dc resistance of purified mercury dropped abruptly to zero at a temperature of 4.15 K, as shown in Figure 12.40. Onnes won the Nobel prize in 1913 for the discovery of superconductivity and for the liquefaction of helium. Later, many other elemental metals and compounds were found to exhibit superconductivity.

Figure 12.40 Resistance versus temperature for mercury. The graph follows that of a normal metal above the critical temperature. The dc resistance drops abruptly to zero at the critical temperature, which is 4.15 K for mercury.

Type I Superconductors

The critical temperatures of some superconducting elements, classified as type I superconductors, are given in Table 12.7. The value of the critical temperature is sensitive to chemical composition (impurities), pressure, and

TABLE 12.7 Values for the Critical Temperature and Critical Magnetic Field for Some Type I Superconductors°

Superconductor	T_c(K)	$B_c(0)$ in Tesla
Al	1.180	0.0105
Ga	1.083	0.0058
Hg	4.153	0.0411
In	3.408	0.0281
Pb	7.193	0.0803
Sn	3.722	0.0305
Ta	4.47	0.0829
Ti	0.39	0.010
W	0.015	0.000115
Zn	0.85	0.0054

° The values for the critical field are given for the temperature $T = 0$ K.

crystalline structure. Note that copper, silver, and gold, which are excellent conductors, do not exhibit superconductivity.

When one measures the critical temperature of a superconductor in the presence of a magnetic field, the value of T_c decreases with increasing magnetic field, as indicated in Figure 12.41 for several type 1 superconductors. When the magnetic field exceeds a certain critical field, B_c, the superconducting state is destroyed, and the material behaves as a normal conductor. As you can see in Figure 12.41, the value of the critical field depends on temperature and is a maximum at $T = 0$ K. The maximum current that can be sustained in a superconductor is limited by the value of the critical field. Values for the critical field for type I superconductors are quite low, of the order of 0.1 T, as shown in Table 12.7. For this reason, type I superconductors cannot be used to construct high field magnets.

Because the dc resistance of a superconductor is zero below its critical temperature, once a current is set up in the material, it will persist *without any applied voltage* (which follows from Ohm's law and the fact that $R = 0$.) In fact, steady currents called *supercurrents* have been observed to persist in a superconducting loop for several years with no measurable decay. Recent measurements using nuclear magnetic resonance (see Chapter 13) indicate that the supercurrent will persist for more than 10^5 years! This reaffirms the fact that R is essentially zero.

The magnetic properties of superconductors are as unusual and as difficult to understand as their electrical properties. In 1933 Hans Meissner and Ochsenfeld studied the magnetic behavior of superconductors and found that when a superconductor in the presence of a magnetic field is cooled below T_c,

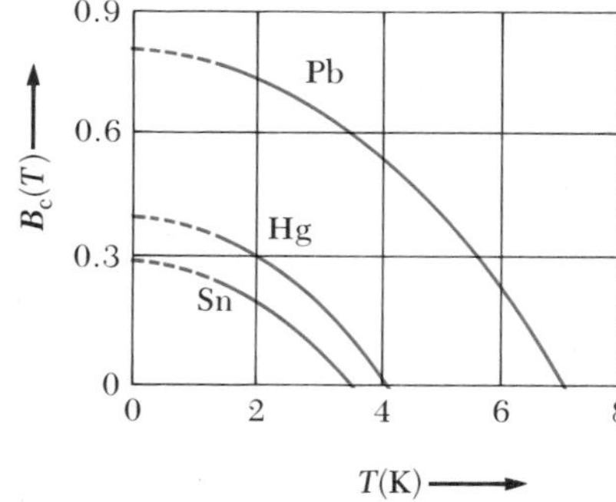

Figure 12.41 Variation of the critical magnetic field with temperature for the elements lead, mercury, and tin. The metal is superconducting at fields and temperatures below its characteristic curve, and is normal above that curve.

Photograph of a small permanent magnet levitated above a disc of the superconductor $YBa_2Cu_3O_7$ which is at a temperature of 77 K. (Photo courtesy of Som Tyagi and Michel Barsoum, Drexel University.)

Meissner effect

all of the magnetic flux is expelled from the interior of the sample. This property of flux expulsion, known as the **Meissner effect,** is described in Figure 12.42a. Since the electrical resistivity, ρ, is zero inside the superconductor, then the electric field inside is also zero, and from Faraday's law, one concludes that the magnetic flux inside cannot change, so $d\boldsymbol{B}/dt = 0$. More important, Meissner and Ochsenfeld also found that $\boldsymbol{B} = 0$ inside a superconductor. That is, a superconductor is both a perfect conductor ($\rho = 0$) and a perfect diamagnet ($\boldsymbol{B} = 0$).

If a magnetic field is applied to a superconductor while it is at a temperature less than T_c, as in Figure 12.42b, persistent surface currents are induced on the superconductor, which produce a magnetic field that exactly cancels the external field everywhere inside the superconductor. (Actually, the magnetic field decreases exponentially from its external value to zero within a

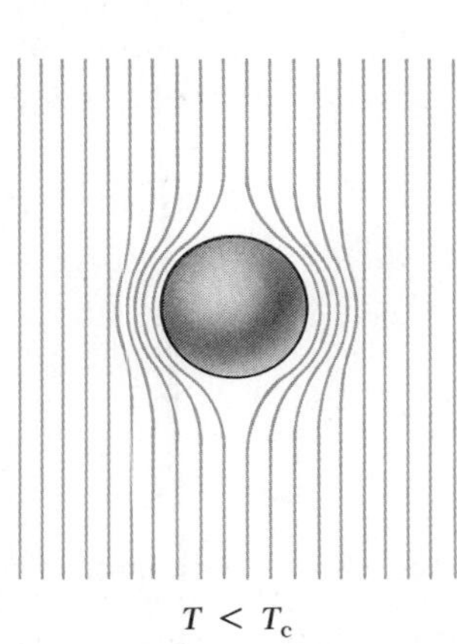

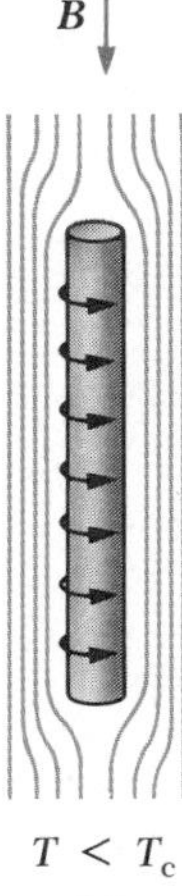

Figure 12.42 (a) The exclusion of magnetic flux from the interior of a superconductor. (b) When an external magnetic field is applied to a superconductor, a surface current is induced which produces a magnetic field that cancels the field inside the superconductor.

depth of about 10^{-8} m.) Likewise, if the magnetic field is applied to the material at $T > T_c$, as in Figure 12.43a, and the sample is cooled to $T < T_c$, surface currents are induced spontaneously to keep the magnetic flux lines out of the sample, as in Figure 12.43b. As you would expect, the surface currents disappear when the external magnetic field is removed.

It is important to emphasize that what we have been describing thus far refers to type I superconductors and for samples in the shape of long cylinders with the applied magnetic field parallel to the cylinder axis. In the superconducting state, and for applied fields less than the critical field, the field does not penetrate the bulk of the sample, and there are only surface currents. Hence the material behaves as a perfect diamagnet. When the applied magnetic field exceeds the critical field, the entire sample becomes normal, the field completely penetrates the sample, and the sample's resistance goes from zero to some value expected for a normal conductor.

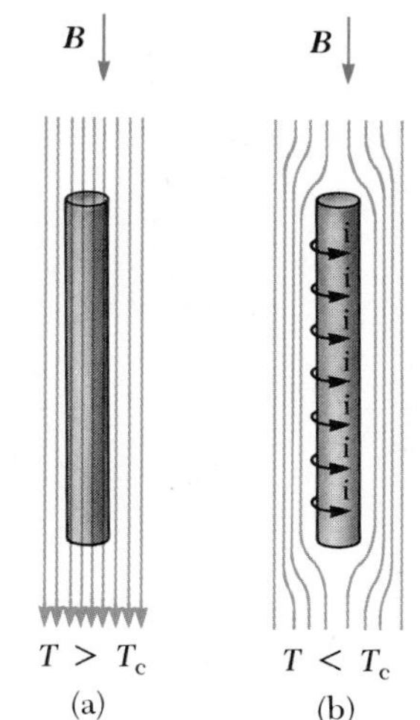

Figure 12.43 (a) When a magnetic field is applied to a conductor above the critical temperature, the field is nonzero inside the conductor. (b) When the sample is cooled to $T < T_c$, surface currents are set up and the flux is excluded from the interior of the superconductor.

Type II Superconductors

By the 1950s it was established that there exists another class of superconductors called type II. In these materials, there are two critical magnetic fields, designated as B_{c1} and B_{c2} in Figure 12.44. When the applied field is less than B_{c1}, there is no flux penetration and the material is entirely superconducting, just as in the case of type I materials. When the applied field exceeds the upper critical field B_{c2}, there is complete flux penetration and the material becomes totally normal, just as for type I materials. However, for fields lying between B_{c1} and B_{c2}, the material is in a *mixed* (or so-called vortex) state. While in this mixed state, the material can have zero resistance and has partial flux penetration. One can view this mixed state as a combination of superconducting regions with filaments of normal material penetrating the sample. Within the filaments, the magnetic flux is finite, and the resistivity is nonzero. Surrounding each filament is a screening current that keeps the magnetic field out of the superconducting regions.

The critical temperatures and the upper critical magnetic fields for several Type II superconductors are given in Table 12.8. Note that the values of the critical fields are very large (of the order of 50 T) compared to the Type I superconductors. For this reason, superconducting magnets requiring very high fields are constructed using these materials.

TABLE 12.8 Values for the Critical Temperature and Upper Critical Magnetic Field for Several Type II Superconductors*

Superconductor	T_c(K)	$B_{c2}(0)$ in Tesla
Nb_3Sn	18.0	24.5
Nb_3Al	18.7	32.4
$Nb_3(AlGe)$	20.7	44
Nb_3Ge	23.2	38
Nb_3Ga	20	35
NbTi	10	15
V_3Ga	14.8	2.08
V_3Si	16.9	2.35
PbMoS	14.4	6.0
NbN	15.7	1.53

* The values of B_{c2} are given for T = 0 K.

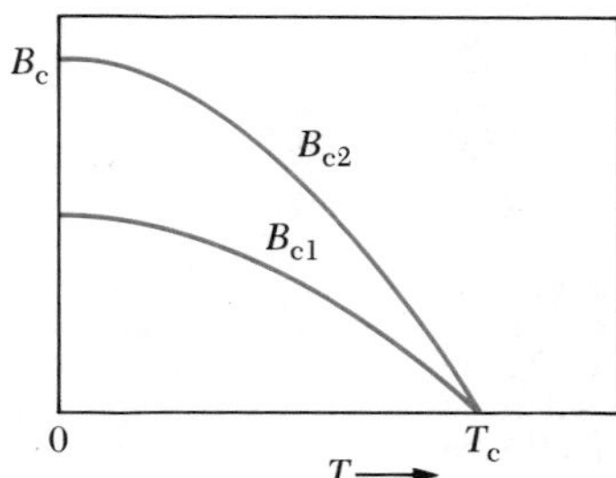

Figure 12.44 The critical fields B_{c1} and B_{c2} as a function of temperature in a type II superconductor.

The BCS Theory

As we described in Section 12.3, the resistivity of a normal metal is due to collisions between the free electrons and the ions of the metal lattice. It is easy to see that the superconducting state could never be explained with this classical model, since the electrons would always suffer some collisions, even in a perfect lattice with no thermal vibrations. The phenomenon of superconductivity can be understood only by using a quantum-mechanical model, where one views the electron wave traveling through the perfect crystal lattice unimpeded in any direction.

A microscopic theory of superconductivity was presented in 1957 by John Bardeen, Leon N. Cooper, and Robert Schrieffer almost 50 years after the discovery of the phenomenon.[7] A central feature of this theory (now known as the BCS theory) is that two electrons in the material are able to form a bound state called a **Cooper pair** if they experience an attractive interaction. This notion at first seem counterintuitive since electrons normally repel one another because of their like charges. However, a net attraction could be obtained if the electrons interact with each other via the lattice as it deforms near an electron. To illustrate this point, Figure 12.45 shows an electron being attracted to nearby positive ions of the lattice. This causes the positive ions to contract towards the electron, producing a small concentration of positive charge in this region. Another electron passing nearby could therefore be attracted to this region. The net effect is a small attractive force between two electrons, with the lattice serving as the mediator.

Cooper pair

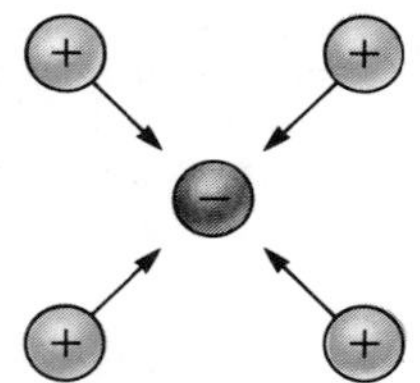

Figure 12.45 Positive ions attracted toward an electron create a region of net positive charge as they move inward.

A Cooper pair in a superconducting material consists of two electrons with equal and opposite momenta and spin, which can be designated as $(p\uparrow)$ and $(-p\downarrow)$. Hence *the pair forms a system with zero total momentum and zero spin.* In other words, the pairs behave as *bosons* all "locked" into the *same* quantum state of zero momentum. In effect, the condensation of all electrons into the same quantum state makes the system behave like a giant quantum mechanical system or "macromolecule." *The condensed state of the Cooper pairs is represented by a single coherent wave function.* In order to reduce the momentum of any single pair by scattering, it would be necessary to reduce the momentum of all the other pairs simultaneously.

The BCS theory has been very successful in explaining the characteristic superconducting properties of zero resistance and flux expulsion. The theory shows that the attractive interaction between electrons in a Cooper pair leads to a ground state separated from excited stated by an energy gap 2Δ, as in Figure 12.46. The energy gap represents the energy needed to break up one of the Cooper pairs, which at $T = 0$ is approximately $kT_c(\approx 1\text{ meV})$. The microscopic calculation in the BCS theory shows that $2\Delta = 3.53\,kT_c$ at $T = 0$ K. As it turns out, the critical magnetic field, thermal properties, and most electromagnetic properties are consequences of the energy gap. Direct experimental evidence for the energy gap is the absorption of electromagnetic radiation by superconductors. Photons with energy greater than the gap energy 2Δ are absorbed by the Cooper pairs and are broken apart.

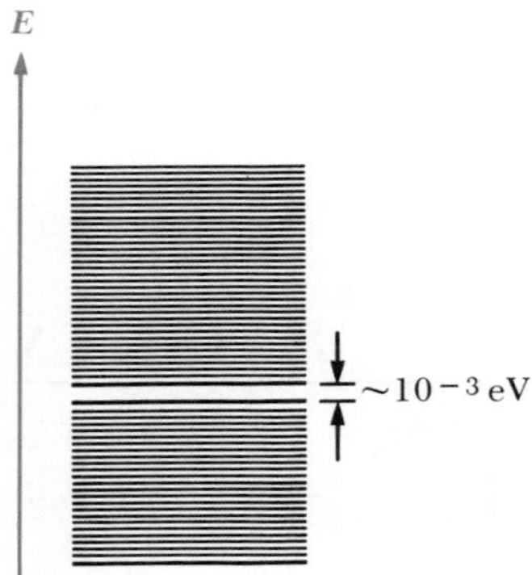

Figure 12.46 Energy bands of a superconductor. The energy gap is very small, of the order of 1 meV, and the Cooper pairs completely fill the band below the energy gap at $T = 0$ K.

As further proof of its success, the BCS theory also predicts that the magnetic flux through a superconducting ring is quantized, and the effective unit of charge is $2e$ rather than e. This has been verified by delicate experiments using small superconducting loops. The basic quantum of flux, Φ_0, has the value $h/2e = 2.0678 \times 10^{-15}\ T\cdot\text{m}^2$.

[7] J. Bardeen, L. N. Cooper, and J. R. Schrieffer, *Phys Rev.* **106:** 162, 1957. These individuals were awarded the Nobel prize in 1972 for their development of the theory of superconductivity.

Finally, some remarkable quantum effects associated with Cooper pairs occur in the Josephson junction described briefly in Chapter 6. Recall that the Josephson junction consists of two superconductors separated by a thin insulating oxide layer (see Fig. 12.47). Although the electrons are not able to pass through the barrier in the classical picture, quantum mechanics allows the pairs to tunnel through the barrier.[8] In the dc Josephson effect, a dc current appears across the function even under zero applied voltage. In the *ac* Josephson effect, a voltage V is applied across the junction, and the current oscillates with a frequency $2eV/h$. For example, a dc voltage of 1 μV produces a frequency of 483.6 MHz. This phenomenon has been used to determine the ratio of e/h to extremely high precision.

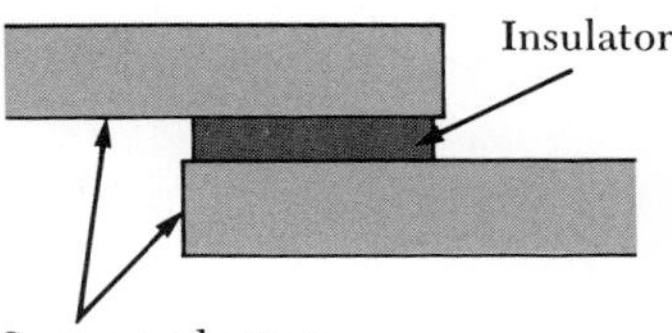

Figure 12.47 A Josephson junction consists of two superconductors separated by a thin insulating layer, usually an oxide about 20 Å thick.

High-Temperature Superconductors — A New Generation

For many years, scientists have searched for new materials that would exhibit superconductivity at higher temperatures, and until recently, the alloy Nb_3Ge had the highest known critical temperature of 23 K. Early in 1986, Georg Bednorz and K. Alex Müller, two scientists at the IBM Research Laboratory in Zurich, made an incredible discovery. They found that an oxide of lanthanum, barium, and copper in the form of a ceramic became superconducting at about 30 K.[9] This discovery took the scientific community by surprise, since such high critical temperatures were not expected, especially in a metallic oxide. By replacing barium with strontium, other researchers raised the value of the critical temperature to about 40 K. Inspired by these developments, scientists worldwide worked to discover materials with even higher T_c values, and research in the superconducting behavior of metallic ceramic oxides accelerated at a tremendous pace. The year 1986 marked the beginning of a new era of high-temperature superconductivity. Bednorz and Müller were awarded the Nobel prize in 1987 (the fastest-ever recognition by the Nobel committee) for their most important discovery. The exceptional interest in these new materials is due to at least four factors pointed out by Müller and Bednorz:

1. The metallic oxides are easy to fabricate and hence can be investigated at smaller laboratories and universities.
2. They have very high T_c values, and very high upper critical magnetic fields, estimated to be in the range of 50–100 T.
3. Their properties and the mechanisms responsible for their superconducting behavior represent a great challenge to theoreticians.
4. They are of considerable technological importance for both current applications and their potential use in inexpensive energy transport.

Early in 1987, research groups at the University of Alabama and the University of Houston announced the discovery of an oxide of yttrium, barium, and copper that became superconducting at about 95 K.[10] The discovery was confirmed in other laboratories around the world, and the compound was soon identified to be $YBa_2Cu_3O_7$. This was another milestone in high T_c superconductivity, since the transition temperature of this compound is above the boiling point of liquid nitrogen (77 K), which is readily available and very inexpensive compared to liquid helium. More recently, even more complex

[8] Brian D. Josephson, an English physicist, and Ivar Giaever, an American physicist, shared the 1973 Nobel prize for the prediction of the tunneling of Cooper pairs through insulators (the Josephson junction) and the tunneling of electrons through insulating barriers, respectively.

[9] J. G. Bednorz and K. A. Müller, Z. *Phys* **B64:** 189, 1986.

[10] M. K. Wu, et al., *Phys. Rev. Letters,* **58:** 908, 1987.

metallic oxides have been investigated, and critical temperatures above 100 K have been observed. Early in 1988, researchers reported the onset of superconductivity at about 120 K in a bismuth-strontium-calcium-copper-oxygen compound and 125 K in a thallium-barium-calcium-copper-oxygen compound. There appears to be a direct correlation between the number of copper-oxygen planes in these compounds and the critical temperature. The critical temperature increases as more copper-oxygen layers are added to the structures. Hence there is good reason to expect that adding more copper-oxygen layers to the complex metallic oxides will raise the critical temperature even higher. It should be noted that the supercurrents are very high in the copper-oxygen planes of the high-T_c materials, and one can view the conduction process as being two-dimensional.

As already mentioned, the discovery of high-T_c superconductivity has attracted worldwide attention in both the scientific community and the business world. Many have termed these developments as creating a "revolution" in physics, comparable to the invention of the transistor in 1948. One can envision many possible applications such as superconducting motors, superfast computers, zero-loss transmission lines, and magnetically levitated trains. However, there are many serious material science problems that might hinder their practical application. For example, the ceramic materials are very brittle and inflexible, making it difficult to fabricate wires. Although the current-carrying capacity of thin films is very large ($\approx 10^{10}$ A/m^2), the capacity is far less than this in bulk samples. While the "old generation" superconductors can easily be integrated into electronic components, the high-T_c materials perform erratically when in contact with other materials.

Finally, it is important to point out that although the conventional BCS theory has been quite successful in explaining the properties of the "old generation" superconductors, theoreticians are still trying to understand the nature of superconductivity in the "new generation" of high-T_c metallic oxides. Many models have been proposed to explain the mechanism(s) responsible for the formation of Cooper pairs and the origin of the band gap in these materials, but it is likely that several years will pass before the problem is resolved. A Superconductivity Supplement is available from the publisher for a more detailed treatment of superconductivity.

12.8 SUMMARY

Bonding mechanisms in solids can be classified in a manner similar to the schemes used to describe bonding of atoms to form molecules. *Ionic solids,* such as NaCl, are formed by the net attractive Coulomb force between positive and negative ion neighbors. Ionic crystals are fairly hard, have high melting points, and have strong cohesive energies of several electron volts. *Covalent solids,* such as diamond, are formed when valence electrons are shared by adjacent atoms to give bonds of great stability. Covalent solids are generally very hard, have high melting points, and possess the highest cohesive energies, ranging up to 12 eV. *Metallic solids* are held together by the metallic bond, which arises from the attractive force between the positive ion cores and the electron "gas." This electron gas acts as a kind of "cement" to hold the ion cores together. Generally, metallic solids are softer and have lower cohesive energies than covalent or ionic solids. The most weakly bound solids are *molecular crystals,* which are held together by van der Waals or dipole-dipole forces. Molecular crystals, such as solid argon, have low melting points and low cohesive energies on the order of 0.1 eV per molecule.

The classical free electron model of metals is generally successful in explaining the gross electrical and thermal properties of metals. According to this theory, a metal is modeled by a gas of conduction electrons moving through a fixed lattice of ion cores. Taking into account the collisions of electrons with the lattice, the classical free electron model predicts that metals will obey Ohm's law, $\mathbf{J} = \sigma\mathbf{E}$, with the electrical conductivity σ given by

$$\sigma = \frac{ne^2L}{(3kTm_e)^{1/2}} \tag{12.16}$$

where n is the concentration of conduction electrons, L is the mean free path, T is the absolute temperature and m_e is the mass of the electron. The classical model may also be applied to heat conduction by electrons in metals to show that K/σ, where K is the thermal conductivity, is a universal constant for all metals and depends only on the absolute temperature (Wiedemann-Franz Law.) The classical model yields for this ratio

$$\frac{K}{\sigma} = \frac{3k^2}{2e^2}\,T \tag{12.22}$$

Although the classical electron gas model broadly describes the electrical and thermal properties of metals, there are notable deficiencies in its predictions—namely in the numerical values of σ and K and in the temperature dependence of σ. These deficiencies may be rectified by taking the wave nature and, hence, the indistinguishability of electrons into account by replacing the Maxwell-Boltzmann distribution with the Fermi-Dirac distribution. When this is done, Equation 12.22 for K/σ takes the new form

$$\frac{K}{\sigma} = \frac{\pi^2k^2}{3e^2}\,T = 2.45 \times 10^{-8}\,T \tag{12.25}$$

and shows excellent agreement with experimental values. The correct quantum expression for the conductivity of a metal is given by

$$\sigma = \frac{ne^2L}{m_e v_F}$$

where v_F is the Fermi velocity and L is the quantum mean free path, *which is usually several hundred times longer than the interatomic distance.* This unusually long mean free path occurs because electron waves can generally pass through a perfect lattice of ions without scattering, and they scatter only from widely spaced crystal imperfections such as thermal displacements of ions and impurity atoms. The observed temperature dependence of σ can also be explained as the result of electron wave scattering from thermally displaced ion cores.

When many atoms are brought together, the individual atomic levels form a set of allowed bands separated by forbidden energy regions or gaps. One can understand the properties of metals, insulators, and semiconductors in terms of the *band theory of solids.* The highest occupied band of a *metal* is partially filled with electrons. Therefore, there are many electrons free to move throughout the metal and contribute to the conduction current. In an *insulator* at $T = 0$ K, the valence band is completely filled with electrons, while the conduction band is empty. The region between the valence band and the conduction band is called the *energy gap* of the material. The energy gap of an insulator is of the order of 10 eV. Because this gap is large compared to kT at ordinary temperatures, very few electrons are thermally excited into the conduction band, which explains the small electrical conductivity of an insulator.

A *semiconductor* is a material with a small energy gap, of the order of 1 eV, and a valence band that is filled at 0 K. Because of this small energy gap, a significant number of electrons can be thermally excited from the valence band into the conduction band as the temperature increases, thereby producing a semiconductor conductivity that *increases* with temperature. The electrical properties of a semiconductor can be modified by adding donor atoms with five valence electrons (such as arsenic) or by adding acceptor atoms with three valence electrons (such as boron). A semiconductor doped with donor atoms is called an *n-type semiconductor,* while one doped with acceptor atoms is called *p-type.* The energy levels of these donor and acceptor atoms fall within the energy gap of the material, the donor levels falling just below the conduction band and the acceptor levels just above the valence band.

When a *p*-type material is joined to an *n*-type material, a *p-n junction* is formed with current-voltage characteristics of a diode. If a *p-n* junction is operated under strong forward bias, electrons can be injected into the *p*-type material where they recombine with holes, emitting photons with band gap energy. This is the principle of operation of the *l*ight *e*mitting *d*iode (LED) and the *injection laser.* If a *p-n* junction absorbs a photon with energy greater than or equal to the band gap, an electron may be promoted to the conduction band (leaving a hole in the valence band), producing a curent. Devices in which light-generated electrons and holes are separated by a built-in junction field and collected as current are called *solar cells.* Two other semiconductor devices that have revolutionized life in the latter half of the 20th century are the *junction transistor* and the *integrated circuit* (IC). The transistor is a small, low power, highly reliable amplifying device that basically consists of two *p-n* junctions placed back to back. The integrated circuit, which is the heart of the small computer, consists of thousands of interconnected transistors, diodes, resistors, and capacitors fabricated on a single chip of silicon. It is interesting that integrated circuits were invented partly in an effort to achieve greater miniaturization and partly in order to solve the interconnection problem spawned by the transistor.

A superconductor is a material whose dc resistance is zero below a temperature T_c, called its *critical temperature.* In addition to being a perfect conductor, ($\rho = 0$), a superconductor is also a perfect diamagnet and expels any magnetic fields that are applied to it ($\boldsymbol{B} = 0$ inside). This is the so-called *Meissner effect.* In a type I superconductor, the material becomes normal when an applied magnetic field exceeds some critical field $\mathbf{B}_{c1}$. A type II superconductor has two critical fields, $\boldsymbol{B}_{c1}$ and $\boldsymbol{B}_{c2}$. When the applied field is less than the lower critical field $\boldsymbol{B}_{c1}$, the material behaves like a type I superconductor. When the field is greater than the upper critical field $\boldsymbol{B}_{c2}$, the material becomes normal. When the field lies between $\boldsymbol{B}_{c1}$ and $\boldsymbol{B}_{c2}$, the material is in a mixed state (part normal and part superconducting). The basic properties of superconductors can be explained in a quantum mechanical model called the BCS theory. In this model, the electrons in the superconductor are attracted to each other and form bound states called *Cooper pairs.* The basic interaction is complex, but arises because of lattice deformations near the electrons.

SUGGESTIONS FOR FURTHER READING

1. R. Zallen, *The Physics of Amorphous Solids,* New York, John Wiley and Sons, 1983. A comprehensive treatment of amorphous solids at a senior undergraduate level. Great care is exercised in introducing new concepts verbally before launching into mathematical arguments.
2. T. R. Reid, *The Chip: How Two Americans Invented the Microchip & Launched a Revolution,* New York, Simon and Schuster, 1984. A popularly written, nonmathematical, extremely interesting chip history.
3. Ben G. Streetman, *Solid State Electronic Devices,* 2nd edition, Englewood Cliffs, N.J., Prentice-Hall, 1980.

A comprehensive treatment of devices at the junior-senior undergraduate level.

4. William Shockley, *Electrons and Holes in Semiconductors*, Melbourne, Fla., Krieger Publishing Co., 1950. The classic "granddaddy" on semiconductors. Although somewhat dated, the book is a marvelous compendium of Shockley's insights and analogies on semiconductor physics. Although the book starts out at a very clear introductory level, about halfway through, the book reaches advanced undergraduate and graduate levels.
5. Charles Kittel, *Introduction to Solid State Physics*, 6th edition, New York, John Wiley and Sons, 1986. A detailed, comprehensive, and up-to-date treatment of solid state physics at the senior/graduate level. Many useful tables of the properties of solids.
6. The September 1977 issue of *Scientific American* is devoted entirely to integrated circuits.

QUESTIONS

1. Explain how the energy bands of metals, semiconductors, and insulators account for these general optical properties: (a) metals are opaque to visible light, (b) semiconductors are opaque to visible light but transparent to infrared (many IR lenses are made of germanium), and (c) many insulators, like diamond, are transparent to visible light.
2. Table 12.6 shows that the energy gaps for semiconductors decrease with increasing temperature. What do you suppose accounts for this behavior?
3. The resistivity of most metals increases with increasing temperature, whereas the resistivity of an intrinsic semiconductor decreases with increasing temperature. What explains this difference in behavior?
4. Discuss the differences among the band structures of metals, insulators, and semiconductors. How does the band structure model enable you to better understand the electrical properties of these materials?
5. Discuss the electrical, physical, and optical properties of ionically bonded solids.
6. Discuss the electrical and physical properties of covalently bonded solids.
7. When a photon is absorbed by a semiconductor, an electron-hole pair is said to be created. Give a physical explanation of this statement, using the energy band model as the basis for your description.
8. Pentavalent atoms such as arsenic are donor atoms in a semiconductor such as silicon, while trivalent atoms such as indium are acceptors. Inspect the periodic table shown on the endpapers, and determine what other elements would be considered as either donors or acceptors.
9. Explain how a *p-n* junction diode operates as a rectifier, a solar cell, and a LED.
10. What are the essential assumptions made in the classical free-electron theory of metals? How does the energy band model differ from the classical free-electron theory in describing the properties of metals?
11. Explain the similarities of electrical conduction and heat conduction in metals.
12. Discuss the differences between crystalline solids, amorphous solids, and gases.
13. Discuss the physical sources of electrical resistance in a metal. Does the resistance depend upon the strong reflections set up when the lattice spacing is equal to an integral number of electronic wavelengths?
14. Does it seem reasonable that the average of the square of a quantity is greater than the square of the average? (Recall from Section 12.3 that $\bar{t} = \tau$ and $\overline{t^2} = 2\tau^2$.)
15. How are the equations governing heat flow and charge flow similar for metals?

PROBLEMS

1. Show that the minimum potential energy of an *ionic* solid is given by
$$U_0 = -\alpha k \frac{e^2}{r_0}\left(1 - \frac{1}{m}\right)$$
where r_0 is the equilibrium separation between ions, α is the Madelung constant, and the repulsive potential energy between ions is B/r^m. U_0 is called the *ionic cohesive energy* and is the energy required to separate to infinity an Na^+Cl^- pair originally at a separation of r_0 in the crystal. $\left(\textit{Hint: } \left.\frac{dU}{dr}\right|_{r=r_0} = 0.\right)$
2. Show that the angle between carbon bonds shown in Figure 12.3a is 109.5°.
3. Calculate the ionic cohesive energy for NaCl, using Equation 12.3. Take $\alpha = 1.7476$, $r_0 = 0.281$ nm, and $m = 8$.
4. LiCl, which has the same crystal structure as NaCl, has $r_0 = 0.257$ nm, and has a measured ionic cohesive energy of 199 kcal/mole. (a) Find the integral value of m that agrees best with the measured value 199 kcal/mole. (b) Is the value of U_0 very sensitive to a change in m? Compute the percent change in U_0 when m increases by unity.

5. Consider a one-dimensional chain of alternating positive and negative ions. Show that the potential energy of an ion in this hypothetical crystal is

$$U(r) = -k\alpha \frac{e^2}{r}$$

where $\alpha = 2 \ln 2$ (the Madelung constant), and r is the interionic spacing. [*Hint:* Make use of the series expansion for $\ln(1 + x)$.]

6. The energy gap for Si at 300 K is 1.14 eV. (a) Find the lowest frequency photon that will promote an electron from the valence band to the conduction band of silicon. (b) What is the wavelength of this photon?

7. From the optical absorption spectrum of a certain semiconductor, one finds that the longest wavelength of radiation absorbed is 1.85 μm. Calculate the energy gap for this semiconductor.

8. (a) Show that the force on an ion in an ionic solid can be written as

$$F = -k\alpha \frac{e^2}{r^2}\left[1 - \left(\frac{r_0}{r}\right)^{m-1}\right]$$

where r_0 is the equilibrium separation and α is the Madelung constant. (*Hint:* Make use of the value of B found in Problem 1.) (b) Imagine that an ion in the solid is displaced a small distance x from r_0, and show that the ion experiences a restoring force $F = -Kx$, where

$$K = \frac{k\alpha e^2}{{r_0}^3}(m-1)$$

(c) Use the result of (b) to estimate the frequency of vibration of an Na^+ ion in NaCl. Take $m = 8$ and use the value $\alpha = 1.7476$. (d) At what wavelength would the Na^+ ion absorb incident radiation? Is the absorption in the UV, visible, or infrared part of the spectrum?

9. (a) Calculate the ionic cohesive energy for KCl, which has the same crystal structure as NaCl. Take $r_0 =$ 0.314 nm and $m = 9$. (b) Calculate the atomic cohesive energy of KCl by using the facts that the ionization energy of potassium is 4.34 eV (that is, Na + 4.34 eV $\rightarrow$ $Na^+ + e$) and that the electron affinity of chlorine is 3.61 eV ($Cl^- + 3.61$ eV $\rightarrow$ Cl + e).

10. Consider a group of N electrons, all of which experience a collision at $t = 0$. One can show that the number of electrons that suffer their next collision between t and $t + dt$ follows an exponentially decreasing distribution. This distribution is given by

$$n(t)\,dt = \frac{Ne^{-t/\tau}}{\tau}\,dt$$

where τ is the mean free time. (a) Show that $\int_0^\infty n(t)\,dt = N$ as expected. (b) Show that this distribution leads to $\bar{t} = \tau$. *Hint:*

$$\bar{t} = \frac{n_1t_1 + n_2t_2 + \cdots + n_ft_f}{N} = \frac{1}{N}\int_0^\infty t\cdot n(t)\,dt$$

(c) Show, similarly, that $\overline{t^2} = 2\tau^2$ as stated in the derivation of Equation 12.12.

11. Silver has a density of 10.5×10^3 kg/m^3 and a resistivity of 1.60×10^{-8} $\Omega\cdot$m at room temperature. (a) On the basis of the classical free electron gas model and assuming that each silver atom contributes one electron to the electron gas, calculate the average time between collisions of the electrons. (b) Calculate the mean free path from τ and the electron's thermal velocity. How does the mean free path compare to the lattice spacing?

12. The Madelung constant may be found by summing an infinite alternating series of terms giving the electrostatic potential energy between an Na^+ ion and its six nearest Cl^- neighbors, its twelve next-nearest Na^+ neighbors, etc. (see Fig. 12.1a). (a) From this expression, show that the first *three* terms of the infinite series for the Madelung constant for the NaCl structure yield $\alpha = 2.13$. (b) Does this infinite series converge rapidly? Calculate the fourth term as a check.

13. The contribution of a single electron or hole to the electric conductivity can be expressed by an important semiconducting quantity called the *mobility*. The mobility, μ, is defined as the particle drift velocity per unit electric field, or, in terms of a formula, $\mu = v_d/E$. Note that mobility describes the ease with which charge carriers drift in an electric field and that the mobility of an electron, μ_n, may be different from the mobility of a hole, μ_p, in the same material. (a) Show that the current density may be written in terms of mobility as $J = ne\mu_nE$, where $\mu_n = e\tau/m_e$. (b) Show that when both electrons and holes are present, the conductivity may be expressed in terms of μ_p and μ_n as $\sigma = ne\mu_n + pe\mu_p$, where n is the electron concentration and p is the hole concentration. (c) If a germanium sample has $\mu_n = 3900$ cm^2/V$\cdot$s, calculate the drift velocity of an electron when a field of 100 V/cm is applied. (d) A pure (intrinsic) sample of germanium has $\mu_n = 3900$ cm^2/V$\cdot$s and $\mu_p = 1900$ cm^2/V$\cdot$s. If the hole concentration is 3.0×10^{13} cm^{-3}, calculate the resistivity of the sample.

14. Gallium arsenide (GaAs) is a semiconductor material of great interest for its high power handling capabilities and fast response time. (a) Calculate the drift velocity of electrons in GaAs for a field of 10 V/cm if the electron mobility is $\mu_n = 8500$ cm^2/V$\cdot$s. (See Problem 13 for the definition of μ.) (b) What percent of the electron's thermal velocity at 300 K is this drift velocity? (c) Assuming an effective electron mass equal to the free electron mass, calculate the average time between electron collisions. (d) Calculate the electronic mean free path.

15. One can roughly calculate the weak binding energy of a donor electron as well as its orbital radius in a semiconductor on the basis of the Bohr theory of the atom. Recall that for a single electron bound to a nucleus of charge Z, the binding energy is given by

$$\frac{ke^2Z^2}{2a_0} = 13.6Z^2 \text{ eV} \qquad (3.35)$$

and the ground state radius by

$$r_1 = a_0/Z \qquad (3.34)$$

For the case of a phosphorus donor atom in silicon, the outermost donor electron is attracted by a nuclear charge of $Z = 1$. However, because the phosphorus nucleus is embedded in the polarizable silicon, the effective nuclear charge seen by the electron is reduced to Z/κ, where κ is the dielectric constant. (a) Calculate the binding energy of a donor electron in Si ($\kappa = 12$) and Ge ($\kappa = 16$). (b) Calculate the radius of the first Bohr orbit of a donor electron in Si and Ge. How does the Bohr orbit radius compare to the nearest neighbor distance in Si and Ge? (Si nearest neighbor = 2.34 Å; Ge nearest neighbor = 2.43 Å.)

16. Assuming that conduction electrons in silver are described by the Fermi free electron gas model with $E_F = 5.48$ eV, repeat the calculations of Problem 11.

17. Sodium is a monovalent metal having a density of 0.971 g/cm³, an atomic weight of 23.0 g/mole, and a resistivity of 4.20 $\mu\Omega\cdot$cm at 300 K. Use this information to calculate (a) the free electron density, (b) the Fermi energy E_F at 0 K, (c) the Fermi velocity v_F, (d) the average time between electronic collisions, (e) the mean free path of the electrons, assuming that E_F at 0 K is the same as E_F at 300 K, and (f) the thermal conductivity. For comparison to (e), the nearest neighbor distance in sodium is 0.372 nm.

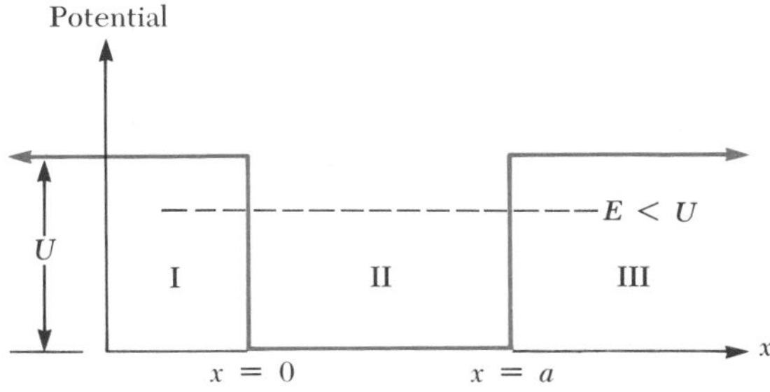

$\Psi_I = Ae^{Kx}$ $\qquad K\hbar = [2m(U - E)]^{1/2}$

$\Psi_{II} = B\cos kx + C\sin kx$ $\qquad k\hbar = (2mE)^{1/2}$

$\Psi_{III} = De^{-Kx}$

Figure 12.48 (Problem 18) Potential and eigenfunctions for a single finite well of depth U, where $E < U$.

COMPUTER PROBLEM

18. The simplest way to model the energy-level splitting that is produced when two originally isolated atoms are brought close together (see Figure 12.16a) is with two finite wells. In this model, the Coulomb potential experienced by the outermost electron in each atom is approximated by a one-dimensional finite square well of depth U and width a. The energy levels for two atoms close together may be found by solving the time-independent Schrödinger equation for a potential consisting of two finite wells separated by a distance b. (a) Start this problem by "warming up" with a solution to the single finite well shown in Figure 12.48. Justify the solutions listed for regions I, II, and III and apply the standard boundary conditions (ψ and

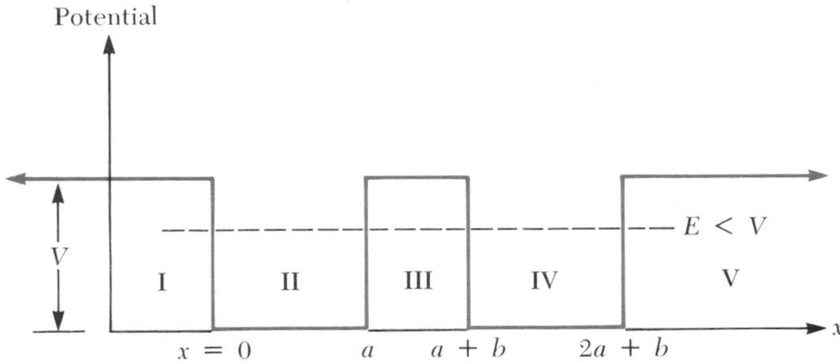

$\Psi_I = Ae^{Kx}$ $\qquad K\hbar = [2m(V - E)]^{1/2}$

$\Psi_{II} = B\cos kx + C\sin kx$ $\qquad K\hbar = (2mE)]^{1/2}$

$\Psi_{III} = De^{Kx} + E'e^{-Kx}$

$\Psi_{IV} = F\cos kx + G\sin kx$

$\Psi_V = He^{-Kx}$

Figure 12.49 (Problem 18) Potential and eigenfunctions for two finite wells. The width of each well is a, and the wells are separated by a distance b.

$d\psi/dx$ continuous at $x = 0$ and $x = a$) to obtain a transcendental equation for the bound state energies. (b) Write a computer program to solve for the bound state energies when an electron is confined to a well with $U = 100$ eV and $a = 1$ Å. You should find two bound states at approximately 19 and 70 eV. (c) Now consider the finite wells separated by a distance b shown in Figure 12.49. Impose the conditions of continuity in ψ and $d\psi/dx$ at $x = 0$ and $x = a$ to obtain

$$\frac{D}{E'} = \frac{2e^{-2Ka}[\cos ka + \frac{1}{2}(K/k - k/K)\sin ka]}{(k/K + K/k)\sin ka}$$

Show that the boundary conditions at $x = a + b$ yield

$$\frac{F}{G} = \frac{(D/E')e^{\beta}[\cos\alpha - (K/k)\sin\alpha] + e^{-\beta}[\cos\alpha + (K/k)\sin\alpha]}{(D/E')e^{\beta}[\sin\alpha + (K/k)\cos\alpha] + e^{-\beta}[\sin\alpha - (K/k)\cos\alpha]}$$

where $\alpha = k(a + b)$ and $\beta = K(a + b)$. The boundary conditions at $x = 2a + b$ yield

$$\frac{F}{G} = \frac{\cos k(2a + b) + (K/k)\sin k(2a + b)}{\sin k(2a + b) - (K/k)\cos k(2a + b)}$$

Thus, the two expressions for F/G may be set equal and the expression for D/E' used to obtain a transcendental equation for the energy E. (d) Write a computer program to solve for the bound state energies when $U = 100$ eV and $a = b = 1$ Å. You should observe twofold splitting of the bound state energies found at 19 and 70 eV for the single well. (The wave functions for this case may be displayed using the Double Well option of the program **EIGEN**.

Essay

SUPERCONDUCTING DEVICES*

Clark A. Hamilton
National Bureau of Standards, Electromagnetic Technology Division

Superconductivity was discovered in 1911 by Dutch physicist Heike Kamerlingh Onnes. In experiments to measure the resistance of frozen mercury, he made the startling discovery that the resistance vanished completely at a temperature a few degrees above absolute zero. Since that time superconductivity has been found in a large number of elemental metals and alloys. The first proposed application of this new phenomenon was to make powerful magnets, since superconductors can carry current with no heat generation. Unfortunately it was soon discovered that the superconductivity in materials such as mercury, tin, and lead is destroyed by only modest magnetic fields. In fact, the conditions under which superconductivity exists in any material can be described by a three-dimensional diagram such as that shown in Figure 1a. The axes are temperature, magnetic field, and current density. The curved surface in this figure represents the transition between normal and superconducting behavior, and its intersections with the three axes are the critical current density (J_c), critical field (H_c), and critical temperature (T_c). The primary focus of superconducting materials research since 1911 has been to expand the volume of superconducting behavior in this diagram. This effort has generated a wide variety of what are now called conventional superconductors with critical temperatures up to 23 K and critical fields up to 40 K (the Earth's magnetic field is 0.00005 T). The primary drawback of these conventional superconductors is the expense and inconvenience of their refrigeration systems. Nevertheless, the fabrication of superconducting magnets for magnetic resonance imaging, accelerator magnets, fusion research, and other applications is a $200 million/year business.

In 1987, Karl Alex Müller and J. Georg Bednorz received the Nobel Prize in Physics for their discovery in 1986 of a new class of higher-temperature superconducting ceramics consisting of oxygen, copper, barium, and lanthanum. Expanding on their work, researchers at the University of Alabama at Huntsville and the University of Houston developed the compound $Y_1Ba_2Cu_3O_x$, which has a transition temperature above 90 K. For the first time superconductivity became possible at liquid nitrogen temperatures. An explosion of research on the new materials has resulted in

* Contribution of the National Bureau of Standards, not subject to copyright.

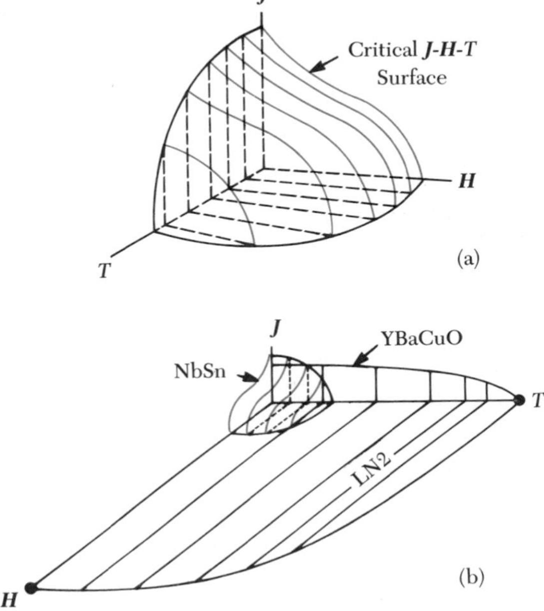

Figure 1 (a) A three-dimensional diagram illustrating the surface separating normal and superconducting behavior as a function of temperature (T), magnetic field (H), and current density (J). (b) A comparison of the J-H-T diagrams for Nb_3Sn and $Y_1Ba_2Cu_3O_x$. (Courtesy of Alan Clark, National Bureau of Standards.)

the discovery of many materials with transition temperatures near 100 K and hints of superconductivity at much higher temperatures. The potential of the new superconductors is illustrated in Figure 1b, which compares the transition parameters of the old and new materials. With their high transition temperatures and critical fields projected to hundreds of tesla, the parameter space where superconductivity exists has been greatly expanded. Unfortunately, the critical current density of the new materials over most of the *H-T* plane is very low. It is expected that this problem will be solved when the new materials can be made in sufficiently pure form.

SUPERCONDUCTIVITY

In 1957 John Bardeen, L. N. Cooper, and J. R. Schrieffer published a theory which shows that superconductivity results from an interaction between electrons and the lattice in which they flow [1, 2, 3]. Coulomb attraction between electrons and lattice ions produces a lattice distortion that can attract other electrons. At sufficiently low temperatures the attractive force overcomes thermal agitation, and the electrons begin to condense into pairs. As the temperature is reduced further, more and more electrons condense into the paired state. In so doing, each electron pair gives up a condensation energy 2Δ. The electron pairs are all described by a common quantum mechanical wave function, and there is thus a long range order among all of the electron pairs in the superconductor. As a result of this long range order, the pairs are able to flow through the lattice without suffering the collisions that lead to resistance in normal conductors. We can model a superconductor as a material in which current is carried by two different electron populations. The first population consists of normal electrons, which behave like the electrons in any normal metal. The second population consists of electron pairs, which behave coherently and are described by a common wavefunction $\Psi(x, t)$. The amplitude of $\Psi(x, t)$ is usually constant throughout the superconductor, but its phase may vary depending on the current. The condensation energy of the electron pairs produces an energy gap between normal and superconducting electrons, which is $2\Delta/e \approx 2$ meV for conventional superconductors. This model is known as the two-fluid model.

JOSEPHSON JUNCTIONS

In addition to the work on large scale applications of superconductivity (primarily magnets), there has been considerable research on small scale superconducting devices. Most of these devices take advantage of the unusual properties of a Josephson junction, which is simply two superconductors separated by an insulating barrier as shown in Figure 2. When the barrier is thin enough, electron pairs can tunnel from

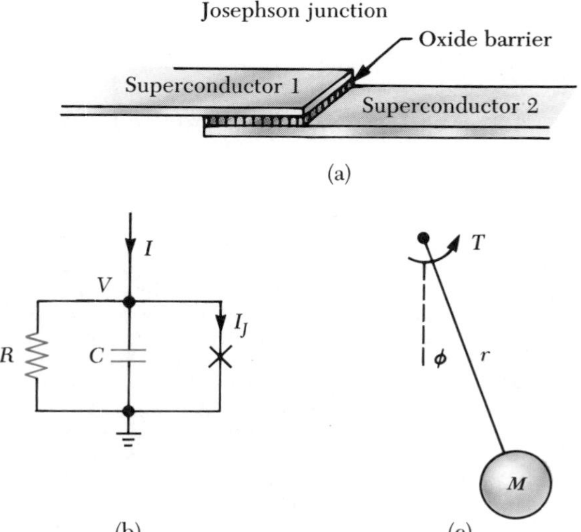

Figure 2 The geometry of a Josephson junction (a), its equivalent circuit (b), and a mechanical analog for the junction behavior.

one superconductor to the other (see Chapter 6). Brian Josephson shared the Nobel Prize in Physics in 1973 for his prediction of the behavior of such a device. The remainder of this essay will discuss the operation of Josephson devices and circuits.

The equations for the current associated with electron pairs across a Josephson junction can be derived by solving two coupled wave equations with a potential energy difference eV depending on the voltage V across the junction [1]. The result is the Josephson equations:

$$I_J = I_0 \sin \phi \tag{1}$$

$$\phi = \frac{2eV}{\hbar} \tag{2}$$

where I_J is the electron pair current through the junction, ϕ is the phase difference between the superconducting wave functions on either side of the barrier, and I_0 is a constant related to the barrier thickness. The full behavior of the junction can be modeled by the equivalent circuit of Figure 2b. The resistance R describes the flow of normal electrons, the X represents a pure Josephson element having a flow of electron pairs, I_J, given by Equations 1 and 2, and C is the normal geometrical capacitance. Thus the total junction current is given by:

$$I = \frac{V}{R} + C\frac{dV}{dt} + I_0 \sin \phi \tag{3}$$

Substituting Equation 2 this becomes

$$I = \frac{C\hbar}{2e}\frac{d^2\phi}{dt^2} + \frac{\hbar}{2eR}\frac{d\phi}{dt} + I_0 \sin \phi \tag{4}$$

Under different drive and parameter conditions, Equation 4 leads to an almost infinite variety of behavior and is the subject of more than 100 papers.

THE PENDULUM ANALOGY

In understanding Equation 4, it is useful to make the analogy between the current in a Josephson junction and the torque applied to a pendulum that is free to rotate 360 degrees in a viscous medium, as shown in Figure 2c. The applied torque, τ, on such a pendulum can be written:

$$\tau = Mr^2\frac{d^2\phi}{dt^2} + k\frac{d\phi}{dt} + Mgr \sin \phi \tag{5}$$

where $k\frac{d\phi}{dt}$ is the damping torque, $Mr^2\frac{d^2\phi}{dt^2}$ is the inertial term, and $Mgr \sin \phi$ is the gravitational restoring torque. Since Equations 4 and 5 have identical form, we can say that the torque on the pendulum is the analog of current in a Josephson junction and the rate of rotation of the pendulum is the analog of voltage across the junction.

We are now in a position to understand the Josephson junction I-V curve shown in Figure 3a. For currents less than I_0, the junction phase assumes a value such that all

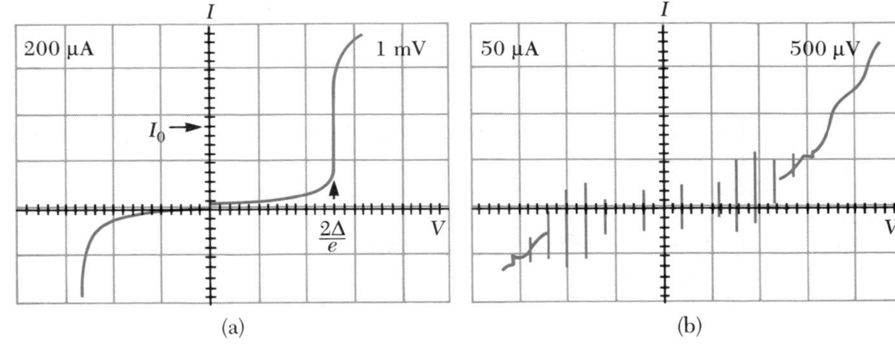

Figure 3 (a) The I-V curve of a Josephson junction. (b) The I-V curve when the junction is exposed to 96 GHz radiation. The quantized voltage levels occur at the values $V_n = nhf/2e$.

of the current is carried by the Josephson element, as described by Equation 1. Except for a small transient, the voltage remains zero. In the analog, the pendulum assumes a fixed angle that balances the torque and the gravitational restoring force. When the torque exceeds the critical value, Mgr, the pendulum rotation accelerates until the viscous damping term can absorb the applied torque. Similarly, when the current applied to a Josephson junction exceeds the critical value I_0, the junction switches to a voltage state in which the junction current oscillates and has an average value determined by the normal state resistance. As the current decreases, the voltage follows the normal state curve until the current reaches a value I_{min}. Below I_{min} the voltage falls abruptly to zero as the junction switches back into the superconducting state. In the analog, with decreasing torque, the pendulum rotation slows down to the point where the applied torque is insufficient to push it over the top, and it then settles into a nonrotating state.

The nonlinearity in the normal state curve for the junction can be explained as follows: For small voltages the resistance is high because there are comparatively few normal state electrons to tunnel across the barrier. At voltages greater than the superconducting energy gap voltage, $2\Delta/e$, there is sufficient energy from the bias source to break superconducting pairs, and the substantial increase in the number of normal carriers causes an abrupt increase in current.

JOSEPHSON VOLTAGE STANDARD

One of the first applications of the Josephson effect comes from the fact that the Josephson current oscillates at a frequency proportional to the applied voltage where the frequency is given by $f = 2eV/h$. Since this relation is independent of all other parameters and has no known corrections, it forms the basis of voltage standards throughout the world. In practice, these standards operate by applying a microwave current through the junction and allowing the Josephson frequency to phase lock to some harmonic of the applied frequency. With the correct choice of design parameters, the only stable operating points of the junction will be at one of the voltages $V = nhf/2e$ where n is an integer. These voltages are very accurately known because frequency can be measured with great accuracy and $2e/h = 483.594$ GHz/mV is a defined value for the purpose of voltage standards. Figure 3b is the I-V curve of a junction that is driven at 96 GHz. About 10 quantized voltage levels span the range from -1 to 1 mV. Each of these levels represents a phase lock between a harmonic of the applied frequency at 96 GHz and the quantum phase difference, ϕ, across the junction. Practical voltage standards use up to 20,000 junctions in series to generate reference voltages up to 10 V.

FLUX QUANTIZATION

Consider a closed superconducting loop surrounding magnetic flux Φ. We know from Faraday's law that any change in Φ must induce a voltage around the loop. Since the voltage across a superconductor is zero, the flux cannot change and is therefore trapped in the loop. There is, however, a further restriction because the phase of the quantum wave function around the loop must be continuous. It can be shown that the variation of the quantum phase along a superconducting wire is proportional to the product of the current and the inductance L of the wire,

$$\Delta\phi = \frac{2\pi LI}{h/2e} \qquad (6)$$

In order for ϕ to be continuous around the loop, $\Delta\phi$ must be a multiple of 2π. It is thus a simple matter to show that the product LI must be quantized in units of $h/2e$. Since LI is just the magnetic flux through the loop, the flux is quantized; the quantity $\Phi_0 = h/2e = 2.07 \times 10^{-15}\ \mathrm{T \cdot m^2}$ is known as the flux quantum.

SQUIDs

One of the most interesting of all superconducting devices is formed by inserting two Josephson junctions in a superconducting loop as shown in Figure 4a [2, 3]. From a circuit viewpoint this looks like two junctions in parallel, and it therefore has an I-V

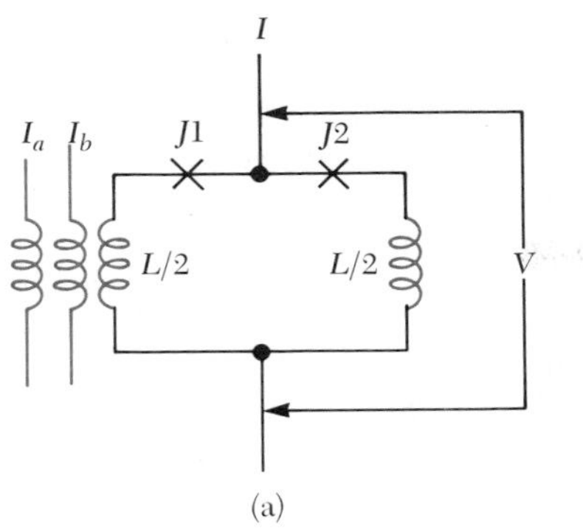

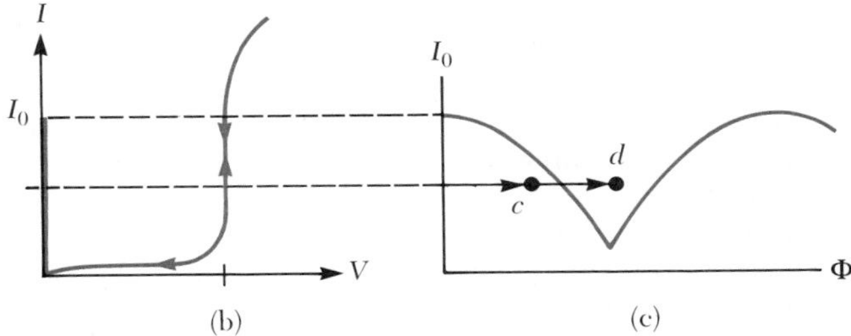

Figure 4 (a) The circuit diagram for a SQUID. (b) Its *I-V* curve, and (c) the dependence of the critical current on applied magnetic flux.

curve like that of a single junction (Fig. 4b). The current through the loop is just the sum of the currents in the two junctions.

$$I = I_{01} \sin \phi_1 + I_{02} \sin \phi_2 \tag{7}$$

The equation for continuity of phase around the loop is

$$\Delta\phi = 2\pi n = \phi_1 - \phi_2 + 2\pi \frac{L(I_1 - I_2)}{\Phi_0} + 2\pi \frac{\Phi}{\Phi_0} \tag{8}$$

When Equations 7 and 8 are solved to maximize I, the total critical current I_{max} is found to be a periodic function of the magnetic flux Φ applied to the loop as shown in Figure 4c. This pattern can be interpreted as the consequence of interference between the superconducting wave functions on either side of the junction. These devices are therefore called Superconducting QUantum Interference Devices or SQUIDs.

The most important application of SQUIDs is the measurement of small magnetic fields. This is done by using a large pick-up loop to inductively couple flux into the SQUID loop. A sensitive electronic circuit senses the changes in critical current I_{max} caused by changing flux in the SQUID loop. This arrangement can resolve flux changes of as little as $10^{-6}\Phi_0$. For a typical SQUID with an inductance $L = 100$ pH, this corresponds to an energy resolution of $\Delta E = \Delta\Phi^2/2L = 2 \times 10^{-32}$ J, which approaches the theoretical limit set by Planck's constant.

A rapidly expanding application of SQUIDs is the measurement of magnetic fields caused by the small currents in the heart and brain. Systems with as many as seven SQUIDs are being used to map these biomagnetic fields. This new nonintrusive tool may one day be important in locating the source of epilepsy and other brain disorders. In such applications it is important to discriminate against distant sources of magnetic noise. This is done by using two pick-up coils wound in opposite directions, so that the SQUID is sensitive only to the *gradient* of the magnetic field.

DIGITAL APPLICATIONS

The possibility of digital applications for Josephson junctions is suggested immediately by the bistable nature of the *I-V* curve in Figure 3a. At a fixed bias current the junction can be in either the zero voltage state or the finite voltage state, representing a binary 0 or 1. In the mid-1970s a major effort was mounted to develop large scale computing systems using Josephson devices [4]. The motivation for this is twofold. First, Josephson junctions can switch states in only a few picoseconds; second, the power dissipation of Josephson devices is nearly three orders of magnitude less than that of semiconductor devices. The low power dissipation results from the low logic levels (0 and 2 mV) of Josephson devices. Dissipation becomes increasingly important in ultra-high speed computers because the need to limit the propagation delay forces the computer into a small volume. As the computer shrinks, it becomes increasingly difficult to remove heat.

A variety of Josephson logic circuits have been proposed. We will consider just one, based on the SQUID circuit of Figure 4a. Suppose that the SQUID carries a current at the level of the lower dashed line in Figure 4. Current in the two input lines (I_a, I_b) moves the bias point horizontally along this line in Figure 4c. For example, a current in either line will move the operating point to (c), while a current in both lines moves the operating point across the threshold curve to (d). When the threshold curve is crossed, the SQUID switches into the voltage state. The output of the device thus is the "AND" function of the two inputs. Other logic combinations can be achieved by adjusting the coupling of the input lines. Memory can be made by using coupled SQUIDs, where one SQUID senses the presence or absence of flux in a second SQUID. Figure 5 is a photograph of two coupled thin film SQUIDs that form a flip-flop circuit. These devices are made by photolithographic processes similar to those used in the semiconductor industry.

Josephson logic and memory chips with thousands of junctions have been built and successfully operated. However, there are still significant problems to be overcome in developing a complete Josephson computer. The greatest obstacle is the dependence of Josephson logic on threshold switching. This results in circuits that operate over only a small range of component parameters. Such circuits are said to have small *margins.* Until a way is found to improve the margins or until fabrication processes with very tight control are developed, the Josephson computer will remain on the drawing boards.

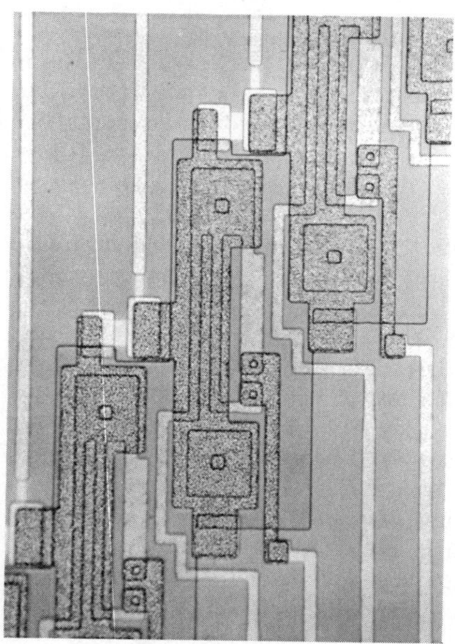

Figure 5 A photograph of a Josephson integrated circuit consisting of two coupled SQUIDs. The minimum feature size is 4 microns.

Suggestions for Further Reading

1. R. P. Feynman, R. B. Leighton and M. Sands, "The Feynman Lectures on Physics," Addison Wesley, Vol. 3, pp. 21.1–21.18, 1966.
2. Donald N. Langenberg, Douglass J. Scalapino, and Barry N. Taylor, "The Josephson Effects," *Sci. American,* May 1966.
3. Donald G. Mcdonald, "Superconducting Electronics," *Physics Today,* February 1981.
4. Juri Matisoo, "The Superconducting Computer," *Sci. American,* May 1980.

Essay Questions and Problems

1. It is well known that superconductors carry dc current without loss. Using the two-fluid model, describe a mechanism that would lead to loss for ac current. [*Hint:* Consider the inertia of the electron pairs.]
2. Starting with the pendulum analog of a Josephson junction, describe a similar analog for an inductor and a SQUID.
3. Using the concept of *gain,* explain why transistor logic has much greater margins than Josephson logic.
4. The basic reason for the reduced power consumption of Josephson logic is the low operating voltage (≈ 2 mV). What fundamental barrier prevents room temperature logic from operating at such low voltage levels? [*Hint:* Consider the fact that logic circuits are by nature nonlinear. How is nonlinearity related to the thermal energy kT/e?]
5. For small values of ϕ, Equations 1 and 2 can be seen to describe an inductance commonly called the Josephson inductance. The resonance of this inductance and the junction capacitance together set an upper limit on the frequency response of many Josephson devices. For a typical junction with $I_0 = 1$ mA and $C = 3$ pF, what is this limit?
6. What is the approximate switching time of a Josephson junction with a critical current of 1 mA and a capacitance of 3 pF?

13

Nuclear Structure

In 1896, the year that marks the birth of nuclear physics, Becquerel discovered radioactivity in uranium compounds. A great deal of activity followed this discovery in an attempt to understand and characterize the nature of the radiation emitted by radioactive nuclei. Pioneering work by Rutherford showed that the emitted radiation was of three types, which he called alpha, beta, and gamma rays. These are classified according to the nature of the electric charge they possess and according to their ability to penetrate matter and ionize air. Later experiments showed that alpha rays are actually helium nuclei, beta rays are electrons, and gamma rays are high-energy photons.

In 1911, Rutherford and his students Geiger and Marsden performed a number of important scattering experiments involving alpha particles. These experiments established that the nucleus of an atom can be regarded as essentially a point mass and point charge and that most of the atomic mass is contained in the nucleus. Furthermore, such studies demonstrated a totally new type of force, the *nuclear force,* which is predominant at distances of less than about 10^{-14} m and zero for large distances. That is, the nuclear force is a short-range force.

Other milestones in the development of nuclear physics include the following events:

1. The observation of nuclear reactions by Cockroft and Walton in 1930, using artificially accelerated particles
2. The discovery of the neutron by Chadwick in 1932
3. The discovery of artificial radioactivity by Joliot and Irene Curie in 1933
4. The discovery of nuclear fission in 1938 by Hahn and Strassmann
5. The development of the first controlled fission reactor in 1942 by Fermi and his collaborators

In this chapter we shall discuss the properties and structure of the atomic nucleus. We start by describing the basic properties of nuclei, followed by a discussion of nuclear forces and binding energy, nuclear models, and the phenomenon of radioactivity. Finally, we discuss nuclear reactions and the various processes by which nuclei decay.

13.1 SOME PROPERTIES OF NUCLEI

All nuclei are composed of two types of particles: protons and neutrons. The only exception to this is the ordinary hydrogen nucleus, which is a single proton. In this section, we shall describe some of the properties of nuclei, such as their charge, mass, and radius. In doing so, we shall make use of the following quantities:

1. The **atomic number,** Z, which equals the number of protons in the nucleus
2. The **neutron number,** N, which equals the number of neutrons in the nucleus

3. The **mass number,** A, which equals the number of nucleons (neutrons plus protons) in the nucleus

It will be convenient for us to have a symbolic way of representing nuclei that will show how many protons and neutrons are present. The symbol to be used is ${}^{A}_{Z}X$, where X represents the chemical symbol for the element. For example, ${}^{56}_{26}Fe$ has a mass number of 56 and an atomic number of 26; therefore it contains 26 protons and 30 neutrons. When no confusion is likely to arise, we shall omit the subscript Z because the chemical symbol can always be used to determine Z.

The nuclei of all atoms of a particular element contain the same number of protons but often contain different numbers of neutrons. Nuclei that are related in this way are called **isotopes.**

Isotopes

The isotopes of an element have the same Z value but different N and A values.

The natural abundances of the various isotopes can differ substantially. For example, ${}^{11}_{6}C$, ${}^{12}_{6}C$, ${}^{13}_{6}C$, and ${}^{14}_{6}C$ are four isotopes of carbon. The natural abundance of the ${}^{12}_{6}C$ isotope is about 98.9%, whereas that of the ${}^{13}_{6}C$ isotope is only about 1.1%. Some isotopes do not occur naturally but can be produced in the laboratory through nuclear reactions. Even the simplest element, hydrogen, has isotopes. They are ${}^{1}_{1}H$, the ordinary hydrogen nucleus; ${}^{2}_{1}H$, deuterium; and ${}^{3}_{1}H$, tritium.

Charge and Mass

The proton carries a single positive charge, equal in magnitude to the charge e on the electron (where $|e| = 1.6 \times 10^{-19}$ C). The neutron is electrically neutral, as its name implies. Because the neutron has no charge, it is difficult to detect.

Nuclear masses can be measured with great precision with the help of the mass spectrometer and the analysis of nuclear reactions. The proton is about 1836 times as massive as the electron, and the masses of the proton and the neutron are almost equal. It is convenient to define, for atomic masses, the **unified mass unit,** u, in such a way that the mass of the isotope ^{12}C is exactly 12 u. That is, the mass of a nucleus (or atom) is measured relative to the mass of an atom of the neutral carbon-12 isotope (the nucleus plus six electrons). Thus, the mass of ^{12}C is defined to be 12 u, where $1 \text{ u} = 1.660559 \times 10^{-27}$ kg. The proton and neutron each have a mass of about 1 u, and the electron has a mass that is only a small fraction of an atomic mass unit:

Ernest Rutherford (1871–1937), an English physicist, was awarded the Nobel prize in 1908 for discovering that atoms can be broken apart by alpha rays and for studying radioactivity. "On consideration, I realized that this scattering backward must be the result of a single collision, and when I made calculations I saw that it was impossible to get anything of that order of magnitude unless you took a system in which the greater part of the mass of the atom was concentrated in a minute nucleus. It was then that I had the idea of an atom with a minute massive center carrying a charge." (Photo courtesy of AIP Niels Bohr Library)

$$\text{Mass of proton} = 1.007276 \text{ u}$$

$$\text{Mass of neutron} = 1.008665 \text{ u}$$

$$\text{Mass of electron} = 0.000549 \text{ u}$$

For reasons that will become apparent later, the mass of the neutron is greater than the combined masses of the proton and the electron.

Because the rest energy of a particle is given by $E = mc^2$, it is often convenient to express the unified mass unit in terms of its energy equivalence. For a proton, we have

$$E = mc^2 = (1.67 \times 10^{-27} \text{ kg})(3 \times 10^8 \text{ m/s})^2$$
$$= 1.50 \times 10^{-10} \text{ J} = 9.38 \times 10^8 \text{ eV} = 938 \text{ MeV}$$

TABLE 13.1 Rest Mass of the Proton, Neutron, and Electron in Various Units

	Mass		
Particle	kg	u	MeV/c^2
Proton	1.6726×10^{-27}	1.007276	938.28
Neutron	1.6750×10^{-27}	1.008665	939.57
Electron	9.109×10^{-31}	5.486×10^{-4}	0.511

Following this procedure, it is found that the rest energy of an electron is 0.511 MeV.

Nuclear physicists often express the unified mass unit in terms of the unit MeV/c^2, where

$$1 \text{ u} \equiv 1.660559 \times 10^{-27} \text{ kg} = 931.50 \text{ MeV}/c^2$$

The rest masses of the proton, neutron, and electron are given in Table 13.1. The masses and some other properties of selected isotopes are provided in Appendix G.

EXAMPLE 13.1 The Unified Mass Unit
Use Avogadro's number to show that the unified mass unit is $1 \text{ u} = 1.66 \times 10^{-27}$ kg.

Solution: We know that 12 kg of ^{12}C contains Avogadro's number of atoms. Avogadro's number, N_A, has the value 6.02×10^{23} atoms/g·mole $= 6.02 \times 10^{26}$ atoms/kg·mole.

Thus, the mass of one carbon atom is

$$\text{Mass of one } ^{12}\text{C atom} = \frac{12 \text{ kg}}{6.02 \times 10^{26} \text{ atoms}} = 1.99 \times 10^{-26} \text{ kg}$$

Since one atom of ^{12}C is defined to have a mass of 12 u, we find that

$$1 \text{ u} = \frac{1.99 \times 10^{-26} \text{ kg}}{12} = 1.66 \times 10^{-27} \text{ kg}$$

The Size of Nuclei

The size and structure of nuclei were first investigated in the scattering experiments of Rutherford, discussed in Section 3.2. In these experiments, positively charged nuclei of helium atoms (alpha particles) were directed at a thin piece of metal foil. As the alpha particles moved through the foil, they often passed near a nucleus of the metal. Because of the positive charge on both the incident particles and the metal nuclei, particles were deflected from their straight-line paths by the Coulomb repulsive force. In fact, some particles were even deflected backwards, through an angle of 180° from the incident direction. These particles were apparently moving directly toward a nucleus, on a head-on collision course.

Rutherford applied the principle of conservation of energy to the alpha particles and nuclei and found an expression for the minimum distance, d, that the particle can approach the nucleus when moving directly toward it before it is turned around by Coulomb repulsion. In such a head-on collision, the kinetic energy of the incoming alpha particle must be converted completely to electrical potential energy when the particle stops at the point of closest approach and turns around (Fig. 13.1). If we equate the initial kinetic energy of the alpha particle to the electrical potential energy of the system (alpha particle plus target nucleus), we have

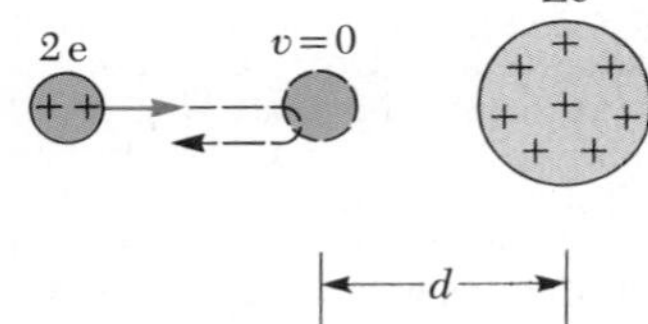

Figure 13.1 An alpha particle on a head-on collision course with a nucleus of charge *Ze*. Because of the Coulomb repulsion between the like charges, the alpha particle will stop instantaneously at a distance *d* from the target nucleus.

$$\tfrac{1}{2}mv^2 = k\frac{q_1q_2}{r} = k\frac{(2e)(Ze)}{d}$$

Solving for d, the distance of closest approach, we get

$$d = \frac{4kZe^2}{mv^2}$$

From this expression, Rutherford found that the alpha particles approached nuclei to within 3.2×10^{-14} m when the foil was made of gold. Thus, the radius of the gold nucleus must be less than this value. For silver atoms, the distance of closest approach was found to be 2×10^{-14} m. From these results, Rutherford concluded that the positive charge in an atom is concentrated in a small sphere, which he called the nucleus, whose radius is no greater than about 10^{-14} m. Because such small lengths are common in nuclear physics, a convenient unit of length that is used is the *femtometer* (fm), sometimes called the **fermi,** defined as

$$1 \text{ fm} \equiv 10^{-15} \text{ m}$$

Since the time of Rutherford's scattering experiments, a multitude of other experiments have shown that most nuclei are approximately spherical and have an average radius given by

$$\boxed{r = r_0 A^{1/3}} \qquad (13.1)$$

where A is the mass number and r_0 is a constant equal to 1.2×10^{-15} m. Because the volume of a sphere is proportional to the cube of its radius, r, it follows from Equation 13.1 that the volume of a nucleus (assumed to be spherical) is directly proportional to A, the total number of nucleons. This suggests that *all nuclei have nearly the same density.* When the nucleons combine to form a compound nucleus, they combine as though they were tightly packed spheres (Fig. 13.2). This fact has led to an analogy between the nucleus and a drop of liquid, in which the density of the drop is independent of its size. We shall discuss the liquid-drop model in Section 13.3.

Figure 13.2 A nucleus can be visualized as a cluster of tightly packed spheres, where each sphere is a nucleon.

EXAMPLE 13.2 The Volume and Density of a Nucleus
Find (a) an approximate expression for the mass of a nucleus of mass number A, (b) an expression for the volume of this nucleus in terms of the mass number, and (c) a numerical value for its density.

Solution: (a) The mass of the proton is approximately equal to that of the neutron. Thus, if the mass of one of these particles is m, the mass of the nucleus is approximately Am.

(b) Assuming the nucleus is spherical and using Equation 13.1, we find that the volume is given by

$$V = \tfrac{4}{3}\pi r^3 = \tfrac{4}{3}\pi r_0{}^3 A$$

(c) The nuclear density can be found as follows:

$$\rho_n = \frac{\text{mass}}{\text{volume}} = \frac{Am}{\frac{4}{3}\pi r_0{}^3 A} = \frac{3m}{4\pi r_0{}^3}$$

Taking $r_0 = 1.2 \times 10^{-15}$ m and $m = 1.67 \times 10^{-27}$ kg, we find that

$$\rho_n = \frac{3(1.67 \times 10^{-27}\text{ kg})}{4\pi(1.2 \times 10^{-15}\text{ m})^3} = 2.3 \times 10^{17}\text{ kg/m}^3$$

Because the density of water is only 10^3 kg/m^3, the nuclear density is about 2.3×10^{14} times as great as the density of water!

Nuclear Stability

Since the nucleus consists of a closely packed collection of protons and neutrons, you might be surprised that it can exist. Because like charges (the protons) in close proximity exert very large repulsive electrostatic forces on each other, these forces should cause the nucleus to fly apart. However, nuclei are stable because of the presence of a force called the **nuclear force.** This

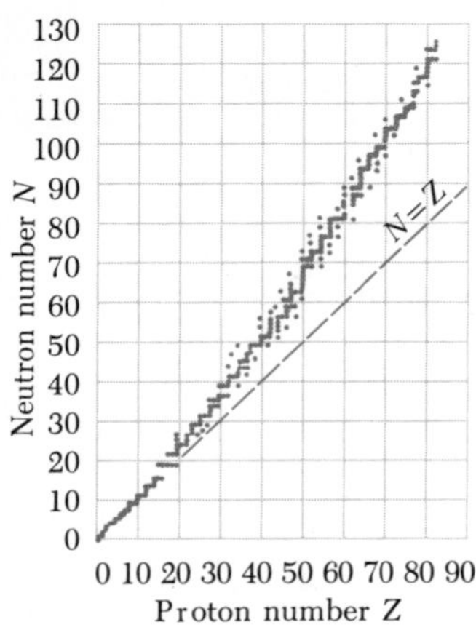

Figure 13.3 A plot of neutron number, N, versus atomic number, Z, for the stable nuclei. The dashed line, corresponding to the condition $N = Z$, is called the *line of stability*.

force, which has a very short range (about 2 fm), is an attractive force that acts between all nuclear particles. The protons attract each other via the nuclear force, and at the same time they repel each other through the Coulomb force. The nuclear force also acts between neutrons and neutrons and between neutrons and protons.

There are about 400 stable nuclei; hundreds of others have been observed, but these are unstable. A plot of N versus Z for a number of stable nuclei is given in Figure 13.3. Note that light nuclei are most stable if they contain equal numbers of protons and neutrons, that is, if $N = Z$. For example, the helium nucleus (two protons and two neutrons) is very stable. Also note that heavy nuclei are more stable if the number of neutrons exceeds the number of protons. This can be understood by noting that, as the number of protons increases, the strength of the Coulomb force increases, which tends to break the nucleus apart. As a result, more neutrons are needed to keep the nucleus stable since neutrons experience only the attractive nuclear force. Eventually, the repulsive forces between protons cannot be compensated by the addition of more neutrons. This occurs when $Z = 83$. Elements that contain more than 83 protons do not have stable nuclei.

It is interesting that most stable nuclei have even values of A. Furthermore, all but eight have Z and N numbers that are both even. In fact, certain values of Z and N correspond to nuclei with unusually high stability. These values of N and Z, called **magic numbers,** are given by

$$\boxed{Z \text{ or } N = 2, 8, 20, 28, 50, 82, 126} \tag{13.2}$$

For example, the alpha particle (two protons and two neutrons), with $Z = 2$ and $N = 2$, is very stable.

Nuclear Spin and Magnetic Moment

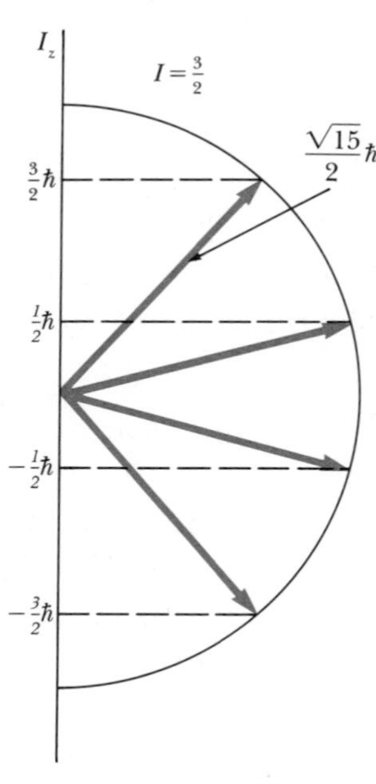

Figure 13.4 The possible orientations of the nuclear spin and its projections along the z axis for the case $I = \frac{3}{2}$.

In Chapter 8, we discussed the fact that an electron has an intrinsic angular momentum associated with its spin. Nuclei, like electrons, also have an intrinsic angular momentum. The magnitude of the *nuclear angular momentum* is $\sqrt{I(I+1)}\hbar$, where I is a quantum number called the *nuclear spin* and may be an integer or half-integer. The maximum value of the angular momentum projected along any direction is $I\hbar$. Figure 13.4 illustrates the possible orientations of the nuclear spin and its projections along the z axis for the case where $I = \frac{3}{2}$.

The nuclear angular momentum has a nuclear magnetic moment associated with it, similar to that of the electron. The *magnetic moment* of a nucleus is measured in terms of the **nuclear magneton** μ_n, a unit of moment defined as

$$\boxed{\mu_n \equiv \frac{e\hbar}{2m_p} = 5.05 \times 10^{-27} \text{ J/T}} \tag{13.3}$$

This definition is analogous to that of the Bohr magneton, μ_B, which corresponds to the spin magnetic moment of a free electron (Section 8.2). Note that μ_n is smaller than μ_B by a factor of about 2000, due to the large difference in masses of the proton and electron.

The magnetic moment of a free proton is not μ_n but $2.7928\ \mu_n$. Unfortunately, there is no general theory of nuclear magnetism that explains this value. Another surprising point is the fact that a neutron also has a magnetic

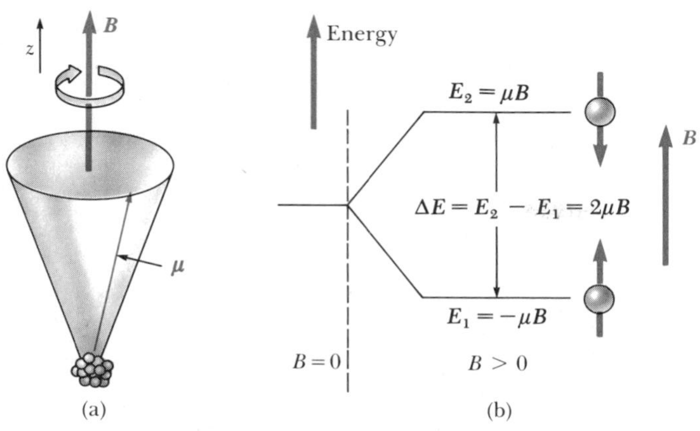

Figure 13.5 (a) When a nucleus having a magnetic moment is placed in an external magnetic field, $\boldsymbol{B}$, the magnetic moment vector $\boldsymbol{\mu}$ precesses about the magnetic field with a frequency that is proportional to the field. (b) A proton, whose spin is $\frac{1}{2}$, can occupy one of two energy states when placed in an external magnetic field. The lower energy state E_1 corresponds to the case where the spin is aligned with the field, and the higher state E_2 corresponds to the case where the spin is opposite the field. The reverse is true for the electrons.

moment, which has a value of $-1.9135\ \mu_n$. The minus sign indicates that its moment is opposite its spin angular momentum.

It is interesting that nuclear magnetic moments (as well as electronic magnetic moments) will precess when placed in an external magnetic field. The frequency at which they precess, called the **Larmor precessional frequency,** ω_p, is directly proportional to the magnetic field. This is described schematically in Figure 13.5a, where the magnetic field is along the z axis. For example, the Larmor frequency of a proton in a magnetic field of 1 T is equal to 42.577 MHz. Recall that the potential energy of a magnetic dipole moment in an external magnetic field is given by $-\boldsymbol{\mu}\cdot\boldsymbol{B}$. When the projection of $\boldsymbol{\mu}$ is along the field, the potential energy of the dipole moment is $-\mu B$, that is, it has its minimum value. When the projection of $\boldsymbol{\mu}$ is against the field, the potential energy is μB and it has its maximum value. These two energy states for a nucleus with a spin of $\frac{1}{2}$ are shown in Figure 13.5b.

Larmor frequency

It is possible to observe transitions between these two spin states using a technique known as **nuclear magnetic resonance.** A dc magnetic field is introduced to align the magnetic moments (Fig. 13.5a), along with a second, weak, oscillating magnetic field oriented perpendicular to $\boldsymbol{B}$. When the frequency of the oscillating field is adjusted to match the Larmor precessional frequency, a torque acting on the precessing moments causes them to "flip" between the two spin states. These transitions result in a net absorption of energy by the spin system, which can be detected electronically. A diagram of the apparatus used in nuclear magnetic resonance is illustrated in Figure 13.6. Note that the absorbed energy is supplied by the generator producing the oscillating magnetic field. The equivalent experiment can be performed on electrons (electron spin resonance), but at much higher frequencies, since electrons have much larger magnetic moments. Nuclear magnetic resonance and electron spin resonance are extremely important methods for studying nuclear and atomic systems and the interactions of these systems with their surroundings. An important medical application of nuclear magnetic resonance is the production of images of various parts of the body (magnetic resonance imaging).

Nuclear magnetic resonance

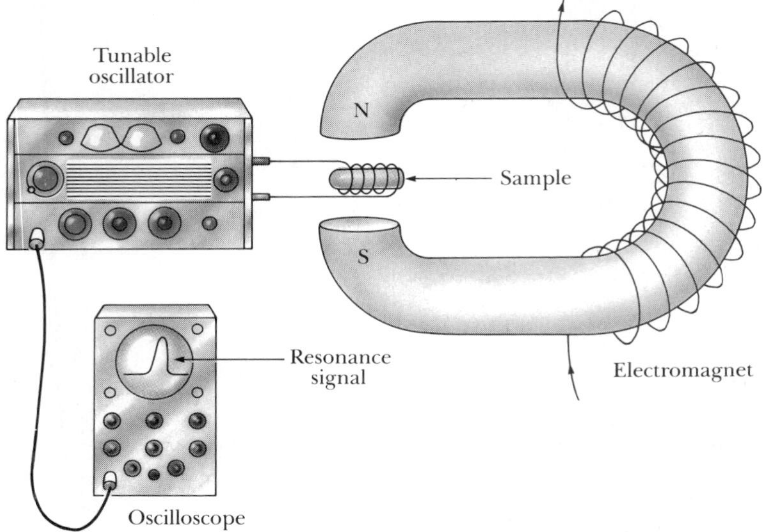

Figure 13.6 An experimental arrangement for nuclear magnetic resonance. The rf magnetic field of the coil, provided by the variable-frequency oscillator, must be perpendicular to the dc magnetic field. When the nuclei in the sample meet the resonance condition, the spins absorb energy from the rf field of the coil, which in turn changes the Q of the circuit containing the coil.

13.2 BINDING ENERGY

The total mass of a stable nucleus is always less than the sum of the masses of its individual nucleons. According to the Einstein mass-energy relationship, if the mass difference, Δm, is multiplied by c^2, we obtain the binding energy of the nucleus. In other words, *the total energy of the bound system (the nucleus) is less than the total energy of the separated nucleons.* Therefore, in order to separate a nucleus into protons and neutrons, energy must be delivered to the system.

In addition to simple radioactive decay, there are two important processes that result in energy release from the nucleus. In **nuclear fission,** a nucleus splits into two or more fragments. In **nuclear fusion,** two or more nucleons combine to form a heavier nucleus.

> In any fission or fusion reaction, the total rest mass of the products is less than that of the reactants.

The decrease in rest mass is accompanied by a release of energy from the system. The following example illustrates this point for the deuteron.

EXAMPLE 13.3 The Binding Energy of the Deuteron Calculate the binding energy of the deuteron, which consists of a proton and a neutron, given that the mass of the deuteron is 2.014102 u.

Solution: We know that the proton and neutron masses are

$$m_p = 1.007825 \text{ u}$$

$$m_n = 1.008665 \text{ u}$$

Note that the masses used for the proton and deuteron in this example are actually those of the neutral atoms. We are able to use atomic masses for these calculations since the electron masses cancel. Therefore,

$$m_p + m_n = 2.016490 \text{ u}$$

To calculate the mass difference, we subtract the deuteron mass from this value:

$$\Delta m = (m_p + m_n) - m_d$$
$$= 2.016490 \text{ u} - 2.014102 \text{ u}$$
$$= 0.002388 \text{ u}$$

Since 1 u corresponds to an equivalent energy of 931.50 MeV (that is, 1 u·c^2 = 931.50 MeV), the mass difference corresponds to the following binding energy:

$$E_b = (0.002388 \text{ u})(931.5 \text{ MeV/u}) = 2.224 \text{ MeV}$$

This result tells us that, in order to separate a deuteron into its constituent parts (a proton and a neutron), it is necessary to add 2.224 MeV of energy to the deuteron. One way of supplying the deuteron with this energy is by bombarding it with energetic particles.

If the binding energy of a nucleus were zero, the nucleus would separate into its constituent protons and neutrons without the addition of any energy, that is, it would spontaneously break apart.

The binding energy of any nucleus represented by ${}^A_Z X$ can be calculated using the following expression:

$$E_b(\text{MeV}) = [Zm_H + Nm_n - M({}^A_Z\text{X})] \times 931.50 \text{ MeV/u} \quad (13.4)$$

Binding energy of a nucleus

where m_H is the mass of the neutral hydrogen atom, $M({}^A_Z\text{X})$ represents the atomic mass of the associated compound nucleus, and the masses are all in unified mass units. Note that *atomic* masses are used in these calculations rather than *nuclear* masses, since tables usually give atomic masses.[1]

It is interesting to examine a plot of the binding energy per nucleon, E_b/A, as a function of mass number for various stable nuclei (Fig. 13.7). Except for the lighter nuclei, the average binding energy per nucleon is about 8 MeV. For the deuteron, the average binding energy per nucleon is found to be $E_b/A = 2.224/2 \text{ MeV} = 1.112 \text{ MeV}$. Note that the curve in Figure 13.7 peaks in the vicinity of $A = 60$. That is, nuclei with mass numbers greater or less than 60 are not as strongly bound as those near the middle of the periodic table. As we shall see later, this fact allows energy to be released in fission and fusion reactions. The curve in Figure 13.7 is slowly varying for $A > 40$, which suggests that the nuclear force saturates. In other words, a particular nucleon can interact with

[1] It is possible to do this because electron masses cancel in such calculations. One exception to this is the β^+ decay process.

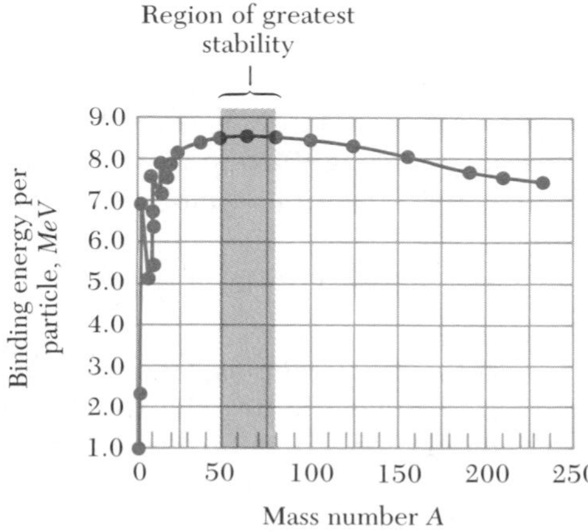

Figure 13.7 A plot of binding energy per nucleon versus mass number for the stable nuclei shown in Figure 13.3.

only a limited number of other nucleons, which can be viewed as being the "nearest neighbors" in the close-packed structure illustrated in Figure 13.2.

The strong nuclear force dominates the Coulomb repulsive force within the nucleus (at short ranges). If this were not the case, stable nuclei would not exist. Morever, the strong nuclear force is nearly independent of charge. In other words, the nuclear forces associated with the proton-proton, proton-neutron, and neutron-neutron interactions are approximately the same, apart from the additional repulsive Coulomb force for the proton-proton interaction.

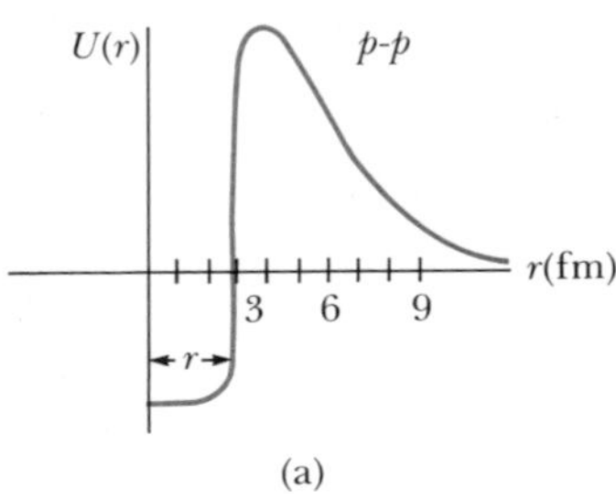

(a)

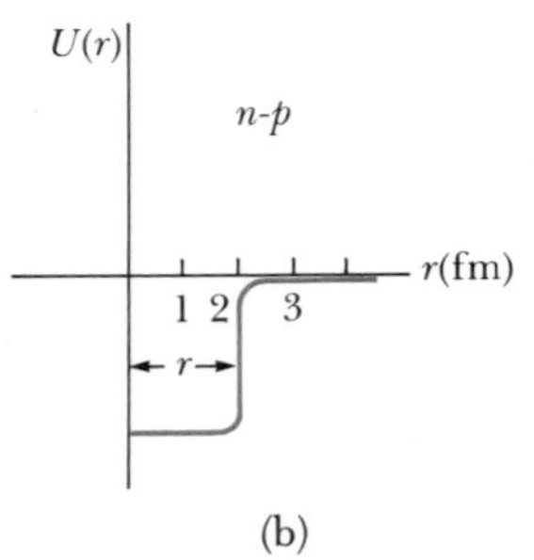

(b)

Figure 13.8 (a) Potential energy versus particle separation for the proton-proton system. (b) Potential energy versus particle separation for the neutron-proton system. The difference in the two curves is due mainly to the large Coulomb repulsion in the case of the proton-proton interaction.

By performing scattering experiments, one can establish that the proton-proton interaction can be represented by the potential energy curve shown in Figure 13.8a. For large values of r, the Coulomb repulsive force is clearly dominant, corresponding to a positive potential energy. When r is reduced to about 3 fm, a break occurs in the curve to negative values of potential energy. That is, as the proton-proton separation approaches the nuclear radius, the attractive nuclear force overcomes the repulsive Coulomb force. In contrast, the proton-neutron interaction can be represented by the diagram shown in Figure 13.8b. In this case, there is no Coulomb force, and so the potential energy is zero at large values of r. However, when the proton-neutron separation is about 2 fm, the nucleons are attracted by the strong nuclear force. Note that the depths of the potential energy minima (corresponding to the nuclear binding energies) are about the same for proton-proton and proton-neutron curves. On inspecting Figure 13.8, we see that the depths of the potentials are approximately equal in the two cases because the interactions are nearly charge-independent. However, the Coulomb repulsion for the proton-proton interaction explains the larger value for r in this case.

13.3 NUCLEAR MODELS

Although the detailed nature of nuclear forces is still not well understood, several phenomenological nuclear models have been proposed, and these are useful in understanding some features of nuclear experimental data and the mechanisms responsible for the binding energy. The models we shall discuss are (1) the liquid-drop model, which accounts for the nuclear binding energy, (2) the independent-particle model, which accounts for the existence of stable isotopes, and (3) the collective model.

Liquid-Drop Model

The **liquid-drop model,** proposed by Bohr in 1936, treats the nucleons as if they were molecules in a drop of liquid. The nucleons interact strongly with each other and undergo frequent collisions as they "jiggle around" within the nucleus. This is analogous to the thermally agitated motion of molecules in a liquid.

The three major effects that influence the binding energy of the nucleus in the liquid-drop model are as follows:

1. The volume effect. Earlier, we showed that the binding energy per nucleon is approximately constant, indicating that the nuclear force exhibits saturation (Fig. 13.7). Therefore, the binding energy is proportional to A and to the nuclear volume. If a particular nucleon is adjacent to n other nucleons and if the binding energy per pair of nucleons is E_b, the binding energy

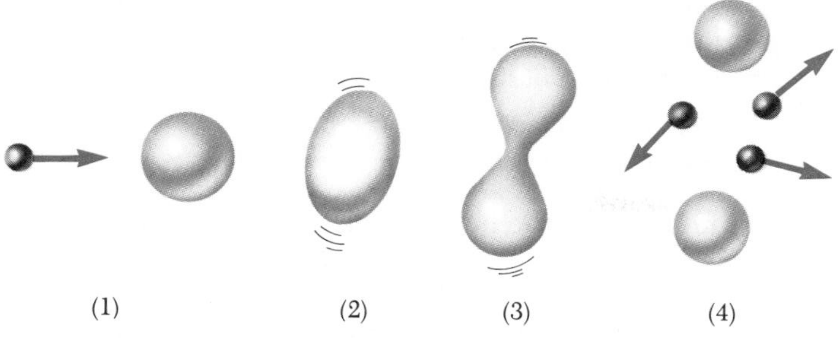

Figure 13.9 Steps leading to fission according to the liquid-drop model of the nucleus.

associated with the volume effect will be C_1A, where $C_1 = nE_b$ and where E_b is approximately $\frac{1}{2}$ MeV for a typical interacting pair of nucleons.

2. The surface effect. Many of the nucleons will be on the surface of the drop, and will have fewer neighbors than those in the interior of the drop. Hence, these surface nucleons will reduce the binding energy by an amount proportional to r^2. Since $r^2 \propto A^{2/3}$ (Eq. 13.1), the reduction in energy can be expressed as $-C_2A^{2/3}$, where C_2 is a constant.
3. The Coulomb repulsion effect. Each proton repels every other proton in the nucleus. The corresponding potential energy per pair of interacting particles is given by ke^2/r, where k is the Coulomb constant. The total Coulomb energy represents the work required to assemble Z protons from infinity to a sphere of volume V. This energy is proportional to the number of protons pairs $Z(Z-1)$ and is inversely proportional to the nuclear radius. Consequently, the reduction in energy that results from the Coulomb effect is $-C_3Z(Z-1)/A^{1/3}$.

Another small effect that contributes to the binding energy in this model is significant only for heavy nuclei having a large excess of neutrons. This effect gives rise to a binding energy term that is proportional to $(A-2Z)^2/A$.

Adding these contributions, we get as the total binding energy

$$E_b = C_1A - C_2A^{2/3} - C_3\frac{Z(Z-1)}{A^{1/3}} - C_4\frac{(A-2Z)^2}{A} \tag{13.5}$$

Semiempirical binding energy formula

Equation 13.5 is often referred to as the **semiempirical binding energy formula.**[2]

The four constants in Equation 13.5 can be adjusted to fit this expression to experimental data, with reasonable results. For nuclei with $A \geq 15$, the constants have the values

$$C_1 = 15.7 \text{ MeV} \qquad C_2 = 17.8 \text{ MeV}$$

$$C_3 = 0.71 \text{ MeV} \qquad C_4 = 23.6 \text{ MeV}$$

Equation 13.5, together with the constants given above, fits the known nuclear mass values very well. However, the liquid drop model does not account for some finer details of nuclear structure, such as stability rules and angular momentum. On the other hand, it does provide a qualitative description of the process of nuclear fission, shown schematically in Figure 13.9. If

[2] For more details on this model, see R. L. Sproull and W. A. Phillips, *Modern Physics*, New York, Wiley, 1980, Chap. 11.

the drop vibrates with a large amplitude (which may be initiated by collision with another particle), it distorts and, under the right conditions, will break apart. We shall discuss the process of fission in Chapter 14.

The Independent-Particle Model

We have mentioned the limitations of the liquid-drop model and its failure to explain some features of nuclear structure. The **independent-particle model,** often called the *shell model,* is based upon the assumption that *each nucleon moves in a well-defined orbit within the nucleus and in an averaged field produced by the other nucleons.* In this model, the nucleons exist in quantized energy states and there are few collisions between nucleons. Obviously, the assumptions of this model differ greatly from those made in the liquid-drop model.

The quantized states occupied by the nucleons can be described by a set of quantum numbers. Since both the proton and the neutron have spin $\frac{1}{2}$, one must apply Pauli's exclusion principle to describe the allowed states (as we did for electrons in Chapter 8). That is, each state can contain only two protons (or two neutrons) with *opposite spin* (Fig. 13.10). The protons have a set of allowed states, and these states differ from those of the neutron because they are different particles. *The proton levels are higher in energy than the neutron levels as a result of the added Coulomb repulsion between protons.*

Using this model of the nucleus, one can begin to understand the stability of certain nuclear states. In particular, if two nucleons are to undergo a collision, the energy of each state after the collision must correspond to one of the allowed states. The collision will simply not occur if that allowed state is already filled by another nucleon. In other words, *a collision between two nucleons can occur only if the process does not violate the exclusion principle.* One can interpret this as an effective force that acts between nucleons to prevent an exchange of energy.

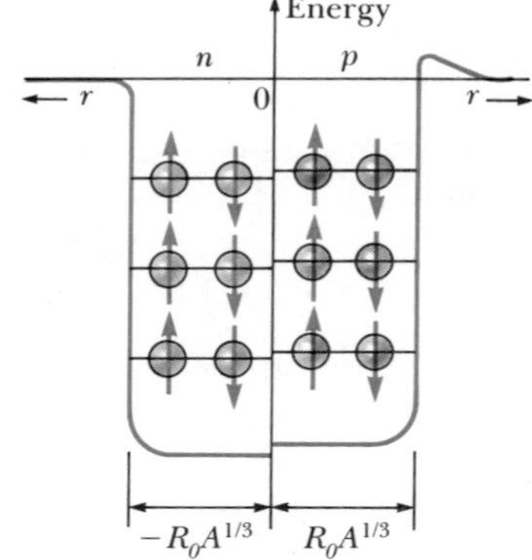

Figure 13.10 A square-well potential containing 12 nucleons. The gray spheres represent protons, and the colored spheres represent neutrons. The energy levels for the protons are slightly higher than those for the neutrons because of the added Coulomb energy in the case of the protons. The difference in the levels increases as Z increases. Note that only two nucleons with opposite spin can occupy a given level, as required by the Pauli exclusion principle.

The independent-particle model is also quite successful in predicting the angular momentum of stable nuclei. In order to properly account for the stable nuclear states, it is necessary to include a large contribution from magnetic effects. In particular, the magnetic moment of each nucleon interacts with the magnetic field produced by its orbital motion.

Finally, it is possible to understand why nuclei containing an even number of protons and neutrons are more stable than others. In fact, there are 160 such stable isotopes (even-even nuclei). Any particular state is filled when it contains two protons (or two neutrons) with opposite spins. *An extra proton or neutron can be added to the nucleus only at the expense of increasing the energy of the nucleus.* This increase in energy leads to a nucleus that is less stable than the original nucleus. A careful inspection of the stable nuclei shows that *the majority have a special stability when their nucleons combine in pairs, which results in a total angular momentum of zero.* This accounts for the large number of high-stability nuclei (those with high binding energies) with the magic numbers given by Equation 13.2.

The Collective Model

A third model of nuclear structure, known as the **collective model,** combines some features of both the liquid-drop model and the independent-particle model. The nucleus is considered to have some "extra" nucleons moving in quantized orbits, in addition to the filled core of nucleons. The extra nucleons are subject to the field produced by the core, as in the independent-particle scheme. Deformations can be set up in the core as a result of a strong interaction between the core and the extra nucleons, thereby setting up vibrational and rotational motions as in the liquid-drop model. The collective model has been very successful in explaining many nuclear phenomena.

13.4 RADIOACTIVITY

In 1896, Becquerel accidentally discovered that uranium salt crystals emit an invisible radiation that can darken a photographic plate even if the plate is covered to exclude light. After several observations of this type under controlled conditions, Becquerel concluded that the radiation emitted by the crystals was of a new type, one that required no external stimulation. This process of spontaneous emission of radiation by uranium was soon to be called **radioactivity.** Subsequent experiments by other scientists showed that other substances were also radioactive. The most significant investigations of this type were conducted by Marie and Pierre Curie. After several years of careful and laborious chemical separation processes on tons of pitchblende, a radioactive ore, the Curies reported the discovery of two previously unknown elements, both of which were radioactive. These were named polonium and radium. Subsequent experiments, including Rutherford's famous work on alpha-particle scattering, suggested that radioactivity was the result of the decay, or disintegration, of unstable nuclei.

Marie Curie (1867–1934), a Polish scientist, shared the Nobel prize in 1903 with her husband, Pierre, and with Becquerel for their work on spontaneous radioactivity and the radiation emitted by radioactive substances. "I persist in believing that the ideas that then guided us are the only ones which can lead to the true social progress. We cannot hope to build a better world without improving the individual. Toward this end, each of us must work toward his own highest development, accepting at the same time his share of responsibility in the general life of humanity."

There are three types of radiation that can be emitted by a radioactive substance: alpha (α) decay, where the emitted particles are ^{4}He nuclei; beta (β) decay, in which the emitted particles are either electrons or positrons; and gamma (γ) decay, in which the emitted "rays" are high-energy photons. A positron is a particle similar to the electron in all respects except that it has a charge of $+e$ (the antimatter twin of the electron). The symbol β^- is used to designate an electron, and β^+ designates a positron.

It is possible to distinguish these three forms of radiation using the scheme described in Figure 13.11. The radiation from a radioactive sample is directed into a region in which there is a magnetic field. Under these conditions, the beam splits into three components, two bending in opposite directions and the third experiencing no change in direction. From this simple observation, one can conclude that the radiation of the undeflected beam carries no charge (the gamma ray), the component deflected upward corresponds to positively

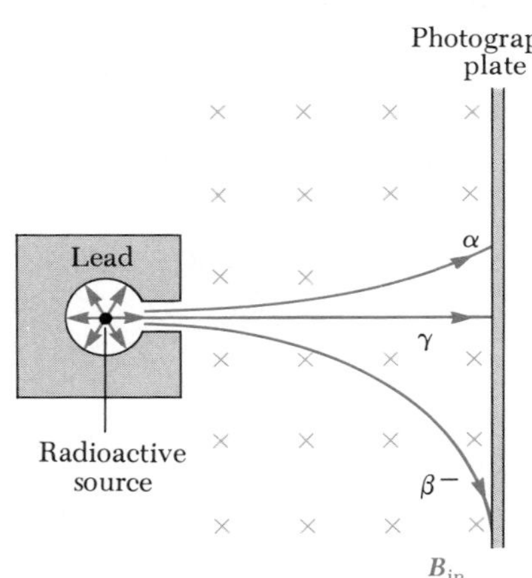

Figure 13.11 The radiation from a radioactive source, such as radium, can be separated into three components by using a magnetic field to deflect the charged particles. The photographic plate at the right records the arrival events.

charged particles (alpha particles), and the component deflected downward corresponds to negatively charged particles (β^-). If the beam includes a positron (β^+), it is deflected upward.

The three types of radiation have quite different penetrating powers. Alpha particles barely penetrate a sheet of paper, beta particles can penetrate a few millimeters of aluminum, and gamma rays can penetrate several centimeters of lead.

The rate at which a particular decay process occurs in a radioactive sample is proportional to the number of radioactive nuclei present (that is, those nuclei that have not yet decayed). If N is the number of radioactive nuclei present at some instant, the rate of change of N is

$$\frac{dN}{dt} = -\lambda N \tag{13.6}$$

where λ is the **decay constant,** or **disintegration constant.** The minus sign indicates that dN/dt is negative, that is, N is decreasing in time.

If we write Equation 13.6 in the form

$$\frac{dN}{N} = -\lambda \, dt$$

we can integrate the expression to give

$$\int_{N_0}^{N} \frac{dN}{N} = -\lambda \int_0^t dt$$

$$\ln\left(\frac{N}{N_0}\right) = -\lambda t$$

or

Exponential decay

$$\boxed{N = N_0 e^{-\lambda t}} \tag{13.7}$$

The constant N_0 represents the number of radioactive nuclei at $t = 0$.

The **decay rate** R $(=|dN/dt|)$ can be obtained by differentiating Equation 13.7 with respect to time. This gives

$$\boxed{R = \left|\frac{dN}{dt}\right| = N_0 \lambda e^{-\lambda t} = R_0 e^{-\lambda t}} \tag{13.8}$$

where $R_0 = N_0\lambda$ is the decay rate at $t = 0$, and $R = \lambda N$. The decay rate of a sample is often referred to as its **activity.** Note that both N and R decrease exponentially with time. The plot of N versus t shown in Figure 13.12 illustrates the exponential decay law.

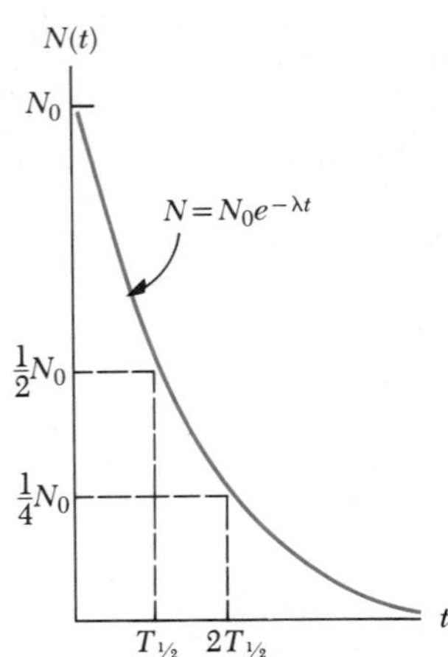

Figure 13.12 Plot of the exponential decay law for radioactive nuclei. The vertical axis represents the number of radioactive nuclei present at any time t, and the horizontal axis is time. The time $T_{1/2}$ is the half-life of the sample.

Another useful parameter used to characterize the decay of a particular nucleus is the **half-life,** $T_{1/2}$.

> The half-life of a radioactive substance is the time it takes half of a given number of radioactive nuclei to decay.

Setting $N = N_0/2$ and $t = T_{1/2}$ in Equation 13.7 gives

$$\frac{N_0}{2} = N_0 e^{-\lambda T_{1/2}}$$

Writing this in the form $e^{\lambda T_{1/2}} = 2$ and taking the natural logarithm of both sides, we get

$$\boxed{T_{1/2} = \frac{\ln 2}{\lambda} = \frac{0.693}{\lambda}} \qquad (13.9)$$

Half-life equation

This is a convenient expression relating the half-life to the decay constant. Note that after an elapsed time of one half-life, there are $N_0/2$ radioactive nuclei remaining (by definition); after two half-lives, half of these will have decayed and $N_0/4$ radioactive nuclei will be left; after three half-lives, $N_0/8$ will be left, and so on. In general, after n half-lives, the number of radioactive nuclei remaining is $N_0/2^n$. Thus, we see that the nuclear decay is independent of the past history of the sample.

The unit of activity is the **curie** (Ci), defined as

$$\boxed{1 \text{ Ci} \equiv 3.7 \times 10^{10} \text{ decays/s}}$$

The curie

This unit was selected as the original activity unit because it is approximately the activity of 1 g of radium. The SI unit of activity is called the **becquerel** (Bq):

$$\boxed{1 \text{ Bq} \equiv 1 \text{ decay/s}}$$

The becquerel

Therefore, $1 \text{ Ci} = 3.7 \times 10^{10}$ Bq. The most commonly used units of activity are the mCi (10^{-3} Ci) and the μCi (10^{-6} Ci).

EXAMPLE 13.4 How Many Nuclei Are Left?

Carbon-14, ${}^{14}_{6}C$, is a radioactive isotope of carbon that has a half-life of 5730 years. If you start with a sample of 1000 carbon-14 nuclei, how many will still be around in 22 920 years?

Solution: In 5730 years, half the sample will have decayed, leaving 500 carbon-14 nuclei remaining. In another 5730 years (for a total elapsed time of 11 460 years), the number will be reduced to 250 nuclei. After another 5730 years (total time 17 190 years), 125 remain. Finally, after four half-lives (22 920 years), only about 62 remain.

It should be noted here that these numbers represent ideal circumstances. Radioactive decay is an averaging process over a very large number of atoms, and the actual outcome depends on statistics. Our original sample in this example contained only 1000 nuclei, certainly not a very large number. Thus, if we were actually to count the number remaining after one half-life for this small sample, it probably would not be 500. If the initial number is very large, the probability of getting extremely close to the predicted number remaining after one half-life increases greatly.

EXAMPLE 13.5 The Activity of Radium

The half-life of the radioactive nucleus ${}^{226}_{88}Ra$ is equal to 1.6×10^3 years. If a sample contains 3×10^{16} such nuclei, determine the activity at this time.

Solution: First, let us calculate the decay constant, λ, using Equation 13.9 and the fact that

$$\begin{aligned} T_{1/2} &= 1.6 \times 10^3 \text{ years} \\ &= (1.6 \times 10^3 \text{ years})(3.15 \times 10^7 \text{ s/year}) \\ &= 5.0 \times 10^{10} \text{ s} \end{aligned}$$

Therefore,

$$\lambda = \frac{0.693}{T_{1/2}} = \frac{0.693}{5.0 \times 10^{10} \text{ s}} = 1.4 \times 10^{-11} \text{ s}^{-1}$$

We can calculate the activity of the sample at $t = 0$ using $R_0 = \lambda N_0$, where R_0 is the decay rate at $t = 0$ and N_0 is the number of radioactive nuclei present at $t = 0$. Since $N_0 = 3 \times 10^{16}$, we have

$$\begin{aligned} R_0 = \lambda N_0 &= (1.4 \times 10^{-11} \text{ s}^{-1})(3 \times 10^{16}) \\ &= 4.2 \times 10^5 \text{ decays/s} \end{aligned}$$

Since $1 \text{ Ci} = 3.7 \times 10^{10}$ decays/s, the activity, or decay rate, at $t = 0$ is

$$R_0 = 11.3\ \mu\text{Ci}$$

EXAMPLE 13.6 The Activity of Carbon

A radioactive sample contains 3.50 μg of pure ${}^{11}_{6}C$, which has a half-life of 20.4 min. (a) Determine the number of nuclei present initially.

Solution: The atomic mass of ${}^{11}_{6}C$ is approximately 11, and therefore 11 g will contain Avogadro's number (6.02×10^{23}) of nuclei. Therefore, 3.50 μg of sample will contain N nuclei, where

$$\frac{N}{6.02 \times 10^{23} \text{ nuclei/mol}} = \frac{3.50 \times 10^{-6} \text{ g}}{11 \text{ g/mol}}$$

or

$$N = 1.92 \times 10^{17} \text{ nuclei}$$

(b) What is the activity of the sample initially and after 8 h?

Solution: Since $T_{1/2} = 20.4$ min $= 1224$ s, the decay constant is

$$\lambda = \frac{0.693}{T_{1/2}} = \frac{0.693}{1224 \text{ s}} = 5.66 \times 10^{-4} \text{ s}^{-1}$$

Therefore, the initial activity of the sample is

$$R_0 = \lambda N_0 = (5.66 \times 10^{-4} \text{ s}^{-1})(1.92 \times 10^{17})$$
$$= 1.09 \times 10^{14} \text{ decay/s}$$

We can use Equation 13.8 to find the activity at any time t. For $t = 8$ h $= 2.88 \times 10^4$ s, we see that $\lambda t = 16.3$ and so

$$R = R_0 e^{-\lambda t} = (1.09 \times 10^{14} \text{ decays/s})e^{-16.3}$$
$$= 9.08 \times 10^6 \text{ decays/s}$$

A listing of activity versus time for this situation is given in Table 13.2.

TABLE 13.2 Activity Versus Time for the Sample Described in Example 13.6

t (h)	R (decays/s)
0	1.09×10^{14}
1	1.42×10^{13}
2	1.85×10^{12}
3	2.41×10^{11}
4	3.15×10^{10}
5	4.10×10^{9}
6	5.34×10^{8}
7	6.97×10^{7}
8	9.08×10^{6}

Exercise 1 Calculate the number of radioactive nuclei remaining after 8 h.
Answer: $N = 1.60 \times 10^{10}$ nuclei.

EXAMPLE 13.7 A Radioactive Isotope of Iodine
A sample of the isotope ^{131}I, which has a half-life of 8.04 days, has a measured activity of 5 mCi at the time of shipment. Upon receipt in a medical laboratory, the activity is measured to be 4.2 mCi. How much time has elapsed between the two measurements?

Solution: We can make use of Equation 13.8:

$$R = R_0 e^{-\lambda t}$$

where R_0 is the initial activity and R is the activity at time t. Writing this in the form

$$\frac{R}{R_0} = e^{-\lambda t}$$

and taking the natural logarithm of each side, we get

$$\ln\left(\frac{R}{R_0}\right) = -\lambda t$$

or

$$(1) \qquad t = -\frac{1}{\lambda}\ln\left(\frac{R}{R_0}\right)$$

To find λ, we can use Equation 13.9:

$$(2) \qquad \lambda = \frac{0.693}{T_{1/2}} = \frac{0.693}{8.04 \text{ days}}$$

Substituting (2) into (1) gives

$$t = -\left(\frac{8.04 \text{ days}}{0.693}\right)\ln\left(\frac{4.2 \text{ mCi}}{5.0 \text{ mCi}}\right) = 2.02 \text{ days}$$

13.5 THE DECAY PROCESSES

As we stated in the previous section, radioactive nuclei spontaneously decay via three processes: alpha decay, beta decay, and gamma decay. Let us discuss these three processes in more detail.

Alpha Decay

If a nucleus emits an alpha particle ($^4_2$He), it loses two protons and two neutrons. Therefore, N decreases by 2, Z decreases by 2, and A decreases by 4. The decay can be written symbolically as

Alpha decay

$$\boxed{^A_Z\text{X} \longrightarrow {}^{A-4}_{Z-2}\text{Y} + {}^4_2\text{He}} \qquad (13.10)$$

where X is called the **parent nucleus** and Y the **daughter nucleus.** As examples, ^{238}U and ^{226}Ra are both alpha emitters and decay according to the schemes

$$^{238}_{92}U \longrightarrow ^{234}_{90}Th + ^{4}_{2}He \quad (13.11)$$

$$^{226}_{88}Ra \longrightarrow ^{222}_{86}Rn + ^{4}_{2}He \quad (13.12)$$

The half-life for ^{238}U decay is 4.47×10^9 years, and the half-life for ^{226}Ra decay is 1.60×10^3 years. In both cases, note that the A of the daughter nucleus is 4 less than that of the parent nucleus. Likewise, Z is reduced by 2. The differences are accounted for in the emitted alpha particle (the ^4He nucleus).

The decay of ^{226}Ra is shown in Figure 13.13. When one element changes into another, as in the process of alpha decay, the process is called *spontaneous decay.* As a general rule, (1) the sum of the mass numbers A must be the same on both sides of the equation and (2) the sum of the charge numbers Z must be the same on both sides of the equation. In addition, the total energy must be conserved. If we call M_X the mass of the parent nucleus, M_Y the mass of the daughter nucleus, and M_α the mass of the alpha particle, we can define the **disintegration energy Q:**

The disintegration energy Q

$$Q = (M_X - M_Y - M_\alpha)c^2 \quad (13.13)$$

Note that Q will be in joules if the masses are in kilograms and c is the usual 3×10^8 m/s. However, when the nuclear masses are expressed in the more convenient unit u, the value of Q can be calculated in MeV using the expression

$$Q = (M_X - M_Y - M_\alpha) \times 931.50 \text{ MeV/u} \quad (13.14)$$

The disintegration energy Q appears in the form of kinetic energy of the daughter nucleus and the alpha particle. The quantity given by Equation 13.13 is sometimes referred to as the Q value of the nuclear reaction. In the case of the ^{226}Ra decay described in Figure 13.13, if the parent nucleus at rest decays, the residual kinetic energy of the products is 4.8 MeV. Most of the kinetic energy is associated with the alpha particle because this particle is much less massive than the recoiling daughter nucleus, ^{222}Rn. That is, since momentum must be conserved, the lighter alpha particle recoils with a much higher speed than the daughter nucleus. Generally, light particles carry off most of the energy in nuclear decays.

Finally, it is interesting to note that if one assumed that ^{238}U (or other alpha emitters) decayed by emitting a proton or neutron, the mass of the decay products would exceed that of the parent nucleus, corresponding to negative Q values. Therefore, such spontaneous decays do not occur.

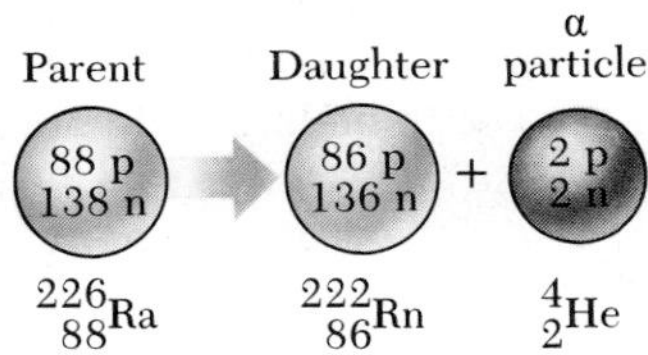

Figure 13.13 Alpha decay of the $^{226}_{88}Ra$ nucleus.

EXAMPLE 13.8 The Energy Liberated When Radium Decays

The ^{226}Ra nucleus undergoes alpha decay according to Equation 13.12. Calculate the Q value for this process. Take the mass of ^{226}Ra to be 226.025406 u, the mass of ^{222}Rn to be 222.017574 u, and the mass of $^{4}_{2}He$ to be 4.002603 u, as found in Appendix G. (Note that atomic mass units rather than nuclear mass units can be used here because the electronic charges cancel in evaluating the mass differences.)

Solution: Using Equation 13.14, we see that

$$\begin{aligned} Q &= (M_X - M_Y - M_\alpha) \times 931.50 \text{ MeV/u} \\ &= (226.025406 \text{ u} - 222.017574 \text{ u} - 4.002603 \text{ u}) \\ &\quad \times 931.50 \text{ MeV/u} \\ &= (0.005229 \text{ u}) \times (931.50 \text{ MeV/u}) = 4.87 \text{ MeV} \end{aligned}$$

It is left as a problem (Problem 57) to show that the kinetic energy of the alpha particle is about 4.8 MeV, whereas the recoiling daughter nucleus has only about 0.1 MeV of kinetic energy.

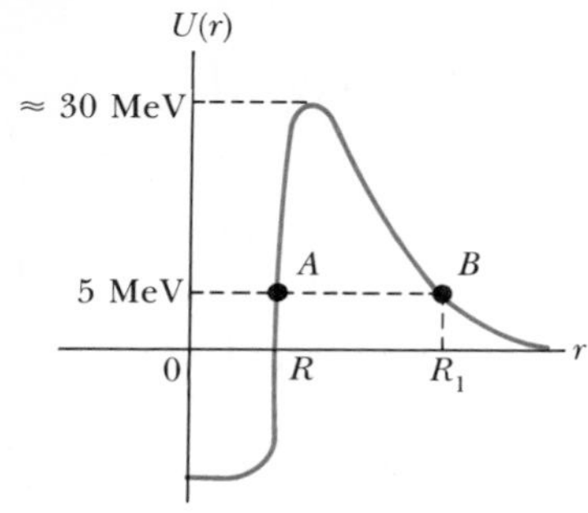

Figure 13.14 Potential energy versus particle separation for the alpha particle–nucleus system. Classically, the energy of the alpha particle is not sufficiently large to overcome the barrier, so the particle should not be able to escape the nucleus.

We now turn to the mechanism of alpha decay. Figure 13.14 is a plot of the potential energy versus distance r from the nucleus for the alpha particle–nucleus system, where R is the range of the nuclear force. The curve represents the combined effects of (1) the Coulomb repulsive energy, which gives the positive peak for $r > R$, and (2) the nuclear attractive force, which causes the curve to be negative for $r < R$. As we saw in Example 13.8, the disintegration energy is about 5 MeV, which is the approximate kinetic energy of the alpha particle, represented by the lower dotted line in Figure 13.14. According to classical physics, the alpha particle is trapped in the potential well. How, then, does it ever escape from the nucleus?

The answer to this question was first provided by Gamow and independently by Gurney and Condon in 1928, using quantum mechanics. Briefly, the view of quantum mechanics is that there is always some probability that the particle can penetrate (or tunnel) through the barrier (Section 6.2). Recall that the probability of locating the particle depends on its wave function ψ and that the probability of tunneling is measured by $|\psi|^2$. Figure 13.15 is a sketch of the

EXAMPLE 13.9 Probability for Alpha Decay

Apply the tunneling methods of Chapter 6 to compute the escape probability from a $^{226}_{88}Ra$ nucleus of an alpha particle with disintegration energy 5 MeV.

Solution: The escape probability is none other than the transmission coefficient $T(E)$ for the Coulomb barrier shown in Figure 13.14. Equation 6.11 of Chapter 6 gives $T(E)$ approximately as

$$T(E) \simeq e^{-(2/\hbar)\sqrt{2m}\int\sqrt{U(r)-E}\,dr}$$

The integral is taken over the *classically forbidden region* where $E < U$. For alpha decay this region is bounded below by the nuclear radius R, and above by $R_1 = 2Zke^2/E$ [from $E = U(R_1) = 2Zke^2/R_1$] (see Fig. 13.14). In this expression, Z is the atomic number of the *daughter* nucleus.

The tunneling integral to be evaluated is

$$\int \sqrt{U(r) - E}\, dr = \sqrt{E}\int_R^{R_1}\sqrt{\frac{R_1}{r} - 1}\, dr$$
$$= R_1\sqrt{E}\int_{R/R_1}^{1}\sqrt{\frac{1}{z} - 1}\, dz$$

An exact value for this integral can be had with some effort, but a useful approximation is found readily by noting that R/R_1 is a small number ($R \sim 10$ fm; $R_1 \sim 50$ fm for $E \sim 5$ MeV). Thus, as a first estimate we set the lower limit to zero and change variables with $z = \cos^2\theta$, obtaining

$$\int_0^1\sqrt{\frac{1}{z} - 1}\, dz = 2\int_0^{\pi/2}\sin^2\theta\, d\theta$$
$$= \int_0^{\pi/2}[1 - \cos 2\theta]\, d\theta = \frac{\pi}{2}$$

To improve upon this we break the original integral into two and approximate the second using $1/z \gg 1$ for z small to get

$$\int_0^1\sqrt{\frac{1}{z} - 1}\, dz - \int_0^{R/R_1}\sqrt{\frac{1}{z} - 1}\, dz \approx \frac{\pi}{2} - \int_0^{R/R_1}\frac{dz}{\sqrt{z}}$$
$$= \frac{\pi}{2} - 2\sqrt{\frac{R}{R_1}}$$

Combining this result with $R_1 = 2Zke^2/E$ gives for the decay probability

$$T(E) = \exp\{-4\pi Z\sqrt{(E_0/E)} + 8\sqrt{Z(R/r_0)}\}.$$

The parameter $r_0 = \hbar^2/M_\alpha ke^2$ is a king of "Bohr" radius for the alpha particle, with the value 7.25 fm, and E_0 is an energy unit analogous to the Rydberg in atomic physics:

$$E_0 = \frac{ke^2}{2r_0} = \frac{14.40\text{ eV}\cdot\text{Å}}{(2)(7.25 \times 10^{-5}\text{ Å})} = 0.0993\text{ MeV}.$$

For the alpha decay of radium, the daughter nucleus is radon with atomic number $Z = 86$ and atomic number $A = 222$. The radius R of the radon nucleus is predicted by Equation 13.1 to be

$$R = (1.2 \times 10^{-5}\text{ fm})(222)^{1/3} = 7.27\text{ fm}$$

Then the decay probability for alpha disintegration at $E = 5$ MeV is

$$\exp\{-4\pi(86)\sqrt{(0.0993/5)} + 8\sqrt{86(7.27/7.25)}\}$$
$$= \exp\{-78.008\} = 1.32 \times 10^{-34}$$

This probability is quite small, but the actual number of disintegrations per second is much larger because of the many collisions the alpha particle makes with the nuclear barrier. This collision frequency f is the reciprocal of the transit time for the alpha particle crossing the nucleus, or $f = v/2R$ where v is the speed of the alpha particle inside the nucleus. In most cases f is about 10^{21} collisions per second (see Problem 6.11), leading to a predicted decay rate $\simeq 10^{-13}$ disintegrations per second for this case.

wave function for a particle of energy E meeting a square barrier of finite height, which approximates the nuclear barrier. Note that the wave function is oscillating both inside and outside the barrier but is greatly reduced in amplitude because of the barrier. As the energy E of the particle is increased, the probability of escaping also increases. Furthermore, the probability increases as the width of the barrier is decreased.

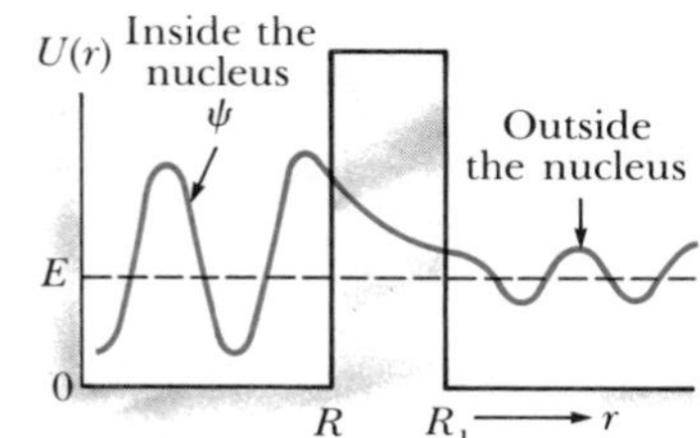

Figure 13.15 The nuclear potential energy is modeled as a square barrier. The energy of the alpha particle is E, which is less than the height of the barrier. According to wave mechanics, the alpha particle has some chance of tunneling through the barrier, as indicated by the finite size of the wave function for $r > R_1$.

Beta Decay

When a radioactive nucleus undergoes beta decay, the daughter nucleus has the same number of nucleons as the parent nucleus but the charge number is changed by 1. Symbolically, the two beta decay processes are

$$ {}^{A}_{Z}\mathrm{X} \longrightarrow {}^{\;\;A}_{Z+1}\mathrm{Y} + \beta^- \tag{13.15}$$

$$ {}^{A}_{Z}\mathrm{X} \longrightarrow {}^{\;\;A}_{Z-1}\mathrm{Y} + \beta^+ \tag{13.16}$$

Again, note that the nucleon number and total charge are both conserved in these decays. As we shall see later, *these processes are not described completely by these expressions.* We shall give reasons for this shortly.

Beta decay

Two typical beta decay processes are

$$ {}^{14}_{6}\mathrm{C} \longrightarrow {}^{14}_{7}\mathrm{N} + \beta^- $$

$$ {}^{12}_{7}\mathrm{N} \longrightarrow {}^{12}_{6}\mathrm{C} + \beta^+ $$

It is important to note that the electron or positron involved in these decays is created within the nucleus as an initial step in the decay process. This is equivalent to saying that, during beta decay, a neutron in the nucleus is transformed into a proton. Indeed, the process $\mathrm{n} \rightarrow \mathrm{p} + \beta^-$ does occur.

Now consider the energy of the system before and after the decay. As with alpha decay, energy must be conserved. Experimentally, one finds that the beta particles are emitted over a continuous range of energies (Fig. 13.16). These results suggest that the decaying nuclei emit electrons with different energies. The kinetic energy of the electrons must be balanced by the decrease in mass of the system, that is, the Q value. However, since all decaying nuclei have the same initial mass, *the Q value must be the same for each decay.* In view of this, why do the emitted electrons have different kinetic energies? The law of conservation of energy seems to be violated! Further analysis shows that, according to the decay processes given by Equations 13.15 and 13.16, the principles of conservation of both angular momentum (spin) and linear momentum are also violated!

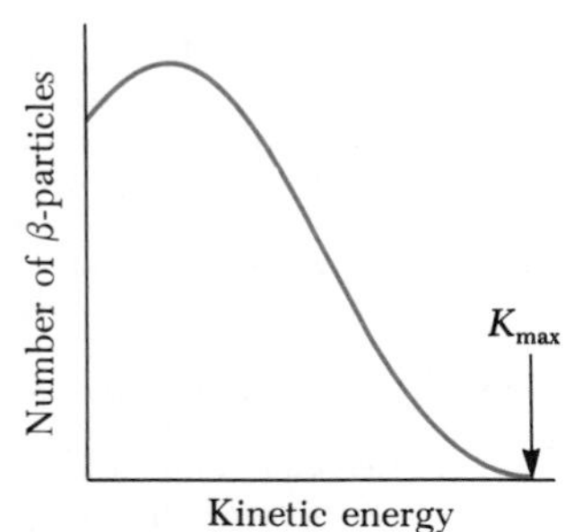

Figure 13.16 A typical beta decay curve. The maximum kinetic energy observed for the beta particle corresponds to the value of Q for the reaction.

After a great deal of experimental and theoretical study, Pauli in 1930 proposed that a third particle must be present to carry away the "missing" energy and momentum. Fermi later named this particle the **neutrino** (little neutral one) since it had to be electrically neutral and have little or no rest mass. Although it eluded detection for many years, the neutrino (symbol ν) was finally detected experimentally in 1956 by Raines and Cowan.

The neutrino has the following properties:

Properties of the neutrino

1. It has zero electric charge.
2. It has a rest mass smaller than that of the electron, and in fact its mass may be zero (although recent experiments suggest that this may not be true).

3. It has a spin of $\frac{1}{2}$, which satisfies the law of conservation of angular momentum.
4. It interacts very weakly with matter and is therefore very difficult to detect.

We can now write the beta decay processes in their correct form:

$$^{14}_{6}\text{C} \longrightarrow {}^{14}_{7}\text{N} + \beta^- + \bar{\nu} \quad (13.17)$$

$$^{12}_{7}\text{N} \longrightarrow {}^{12}_{6}\text{C} + \beta^+ + \nu \quad (13.18)$$

$$\text{n} \longrightarrow \text{p} + \beta^- + \bar{\nu} \quad (13.19)$$

where the symbol $\bar{\nu}$ represents the **antineutrino.**

The positron β^+ emitted in the process given by Equation 13.18 is a particle that is identical to the electron except that it has a positive charge of $+e$. Because it is like the electron in all respects except charge, the positron is said to be an **antiparticle** to the electron. Likewise, the antineutrino is the antiparticle to the neutrino. We shall discuss antiparticles further in Chapter 15. For now, it suffices to say that *a neutrino is emitted in positron decay and an antineutrino is emitted in electron decay.*

Enrico Fermi (1901–1954), an Italian physicist, was awarded the Nobel prize in 1938 for producing the transuranic elements by neutron irradiation and for his discovery of nuclear reactions brought about by slow neutrons. He made many other outstanding contributions to physics including his theory of beta decay, the free electron theory of metals, and the development of the world's first fission reactor in 1942. Fermi was truly a gifted theoretical and experimental physicist. He was also well known for his ability to present physics in a clear and exciting manner. "Whatever Nature has in store for mankind, unpleasant as it may be, men must accept, for ignorance is never better than knowledge."

Carbon Dating

The beta decay of ^{14}C given by Equation 13.17 is commonly used to date old organic samples. Cosmic rays (high-energy particles from outer space) in the upper atmosphere cause nuclear reactions that create ^{14}C. In fact, the ratio of ^{14}C to ^{12}C isotopic abundance in the carbon dioxide molecules of our atmosphere has a constant value of about 1.3×10^{-12}. All living organisms have the same ratio of ^{14}C to ^{12}C because they continuously exchange carbon dioxide with their surroundings. When an organism dies, however, it no longer absorbs ^{14}C from the atmosphere, and so the ratio of ^{14}C to ^{12}C decreases as the result of the beta decay of ^{14}C. It is therefore possible to measure the age of a material by measuring its activity per unit mass due to the decay of ^{14}C. Using carbon dating, samples of wood, charcoal, bone, and shell have been identified as having lived from 1000 to 25 000 years ago. This knowledge has helped us to reconstruct the history of living organisms—including humans—during this time span.

A particularly interesting example is the dating of the Dead Sea Scrolls. This group of manuscripts was first discovered by a shepherd in 1947. Translation showed them to be religious documents, including most of the books of the Old Testament. Because of their historical and religious significance, scholars wanted to know their age. Carbon dating applied to fragments of the scrolls and to the material in which they were wrapped established their age at about 1950 years.

EXAMPLE 13.10 Radioactive Dating

A piece of charcoal of mass 25 g is found in some ruins of an ancient city. The sample shows a ^{14}C activity of 250 decays/min. How long has the tree that this charcoal came from been dead?

Solution: First, let us calculate the decay constant for ^{14}C, which has a half-life of 5730 years.

$$\lambda = \frac{0.693}{T_{1/2}} = \frac{0.693}{(5730 \text{ years})(3.15 \times 10^7 \text{ s/year})}$$

$$= 3.84 \times 10^{-12} \text{ s}^{-1}$$

The number of ^{14}C nuclei can be calculated in two steps. First, the number of ^{12}C nuclei in 25 g of carbon is

$$N(^{12}C) = \frac{6.02 \times 10^{23}\ \text{nuclei/mol}}{12\ \text{g/mol}}(25\ \text{g})$$

$$= 1.25 \times 10^{24}\ \text{nuclei}$$

Assuming that the ratio of ^{14}C to ^{12}C is 1.3×10^{-12}, we see that the number of ^{14}C nuclei in 25 g *before* decay is

$$N_0\ (^{14}C) = (1.3 \times 10^{-12})(1.25 \times 10^{24})$$

$$= 1.63 \times 10^{12}\ \text{nuclei}$$

Hence, the initial activity of the sample is

$$R_0 = N_0\lambda = (1.63 \times 10^{12}\ \text{nuclei})(3.84 \times 10^{-12}\ \text{s}^{-1})$$

$$= 6.26\ \text{decays/s} = 376\ \text{decays/min}$$

We can now calculate the age of the charcoal using Equation 13.8, which relates the activity R at any time t to the initial activity R_0:

$$R = R_0 e^{-\lambda t} \quad \text{or} \quad e^{-\lambda t} = \frac{R}{R_0}$$

Since it is given that $R = 250$ decays/min and since we found that $R_0 = 376$ decays/min, we can calculate t by taking the natural logarithm of both sides of the last equation:

$$-\lambda t = \ln\left(\frac{R}{R_0}\right) = \ln\left(\frac{250}{376}\right) = -0.408$$

$$t = \frac{0.408}{\lambda} = \frac{0.408}{3.84 \times 10^{-12}\ \text{s}^{-1}}$$

$$= 1.06 \times 10^{11}\ \text{s} = 3370\ \text{years}$$

This technique of radioactive-carbon dating has been successfully used to measure the age of many organic relics up to 25 000 years old.

A process that competes with β^+ decay is called **electron capture.** This occurs when a parent nucleus captures one of its own electrons and emits a neutrino. The final product after decay is a nucleus whose charge is $Z - 1$. In general, the electron-capture process can be expressed as

$$^{A}_{Z}X + {}^{0}_{-1}e \longrightarrow {}_{Z-1}^{A}X + \nu \qquad (13.20)$$

Electron capture

In most cases, it is a K-shell electron that is captured, and one refers to this as **K capture.** One example of this process is the capture of an electron by $^{7}_{4}Be$ to become $^{7}_{3}Li$:

$$^{7}_{4}Be + {}^{0}_{-1}e \longrightarrow {}^{7}_{3}Li + \nu$$

Gamma Decay

Very often, a nucleus that undergoes radioactive decay is left in an excited energy state. The nucleus can then undergo a second decay to a lower energy state, perhaps to the ground state, by emitting a photon. Photons emitted in such a de-excitation process are called **gamma rays.** Such photons have very high energy (in the range of 1 MeV to 1 GeV) relative to the energy of visible light (about 1 eV). Recall that the energy of photons emitted (or absorbed) by an atom equals the difference in energy between the two electronic states involved in the transition. Similarly, a gamma ray photon has an energy hf that equals the energy difference ΔE between two nuclear energy states. When a nucleus decays by emitting a gamma ray, it doesn't change, apart from the fact that it ends up in a lower energy state. We can represent a gamma decay process as

$$^{A}_{Z}X^* \longrightarrow {}^{A}_{Z}X + \gamma \qquad (13.21)$$

Gamma decay

where X^* indicates a nucleus in an excited state.

A nucleus may reach an excited state as the result of a violent collision with another particle. However, it is more common for a nucleus to be in an excited state after it has undergone an alpha or beta decay. The following

TABLE 13.3 Various Decay Pathways

Alpha decay	${}^{A}_{Z}X \longrightarrow {}^{A-4}_{Z-2}X + {}^{4}_{2}He$
Beta decay (β^-)	${}^{A}_{Z}X \longrightarrow {}_{Z+1}^{A}X + \beta^- + \bar{\nu}$
Beta decay (β^+)	${}^{A}_{Z}X \longrightarrow {}_{Z-1}^{A}X + \beta^+ + \nu$
Electron capture	${}^{A}_{Z}X + {}_{-1}^{0}e \longrightarrow {}_{Z-1}^{A}X + \nu$
Gamma decay	${}^{A}_{Z}X^* \longrightarrow {}^{A}_{Z}X + \gamma$

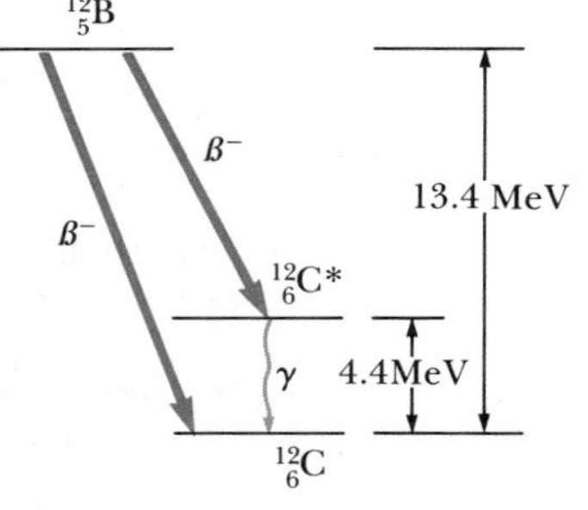

Figure 13.17 The ^{12}B nucleus undergoes β^- decay to two levels of ^{12}C. The decay to the excited level. $^{12}C^*$, is followed by gamma decay to the ground state.

sequence of events represents a typical situation in which gamma decay occurs:

$$^{12}_{5}B \longrightarrow {}^{12}_{6}C^* + {}_{-1}^{0}e + \bar{\nu} \tag{13.22}$$

$$^{12}_{6}C^* \longrightarrow {}^{12}_{6}C + \gamma \tag{13.23}$$

Figure 13.17 shows the decay scheme for ^{12}B, which undergoes beta decay to either of two levels of ^{12}C. It can either (1) decay directly to the ground state of ^{12}C by emitting a 13.4-MeV electron or (2) undergo β^- decay to an excited state of $^{12}C^*$, followed by gamma decay to the ground state. This latter process results in the emission of a 9.0-MeV electron and a 4.4-MeV photon.

The various pathways by which a radioactive nucleus can undergo decay are summarized in Table 13.3.

13.6 NATURAL RADIOACTIVITY

Radioactive nuclei are generally classified into two groups: (1) unstable nuclei found in nature, which give rise to what is called **natural radioactivity,** and (2) nuclei produced in the laboratory through nuclear reactions, which exhibit **artificial radioactivity.**

There are three series of naturally occurring radioactive nuclei (Table 13.4). Each series starts with a specific long-lived radioactive isotope whose half-life exceeds that of any of its descendants. The three natural series begin with the isotopes ^{238}U, ^{235}U, and ^{232}Th, and the corresponding stable end products are three isotopes of lead: ^{206}Pb, ^{207}Pb, and ^{208}Pb. The fourth series in Table 13.4 begins with ^{237}Np and has as its stable end product ^{209}Bi. The element ^{237}Np is a transuranic element (one having an atomic number greater than that of uranium) not found in nature. This element has a half-life of "only" 2.14×10^6 years.

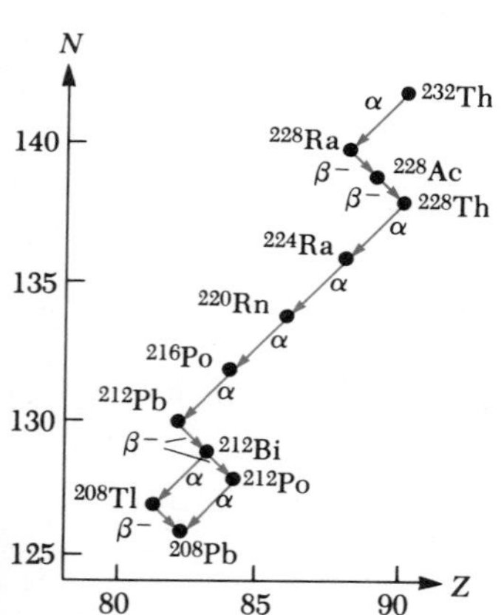

Figure 13.18 Successive decays for the ^{232}Th series.

Figure 13.18 shows the successive decays for the ^{232}Th series. Note that ^{232}Th first undergoes alpha decay to ^{228}Ra. Next, ^{228}Ra undergoes two successive β decays to ^{228}Th. The series continues and finally branches when it reaches ^{212}Bi. At this point, there are two decay possibilities. The end of the decay series is the stable isotope ^{208}Pb.

TABLE 13.4 The Four Radioactive Series

Series		Starting Isotope	Half-Life (years)	Stable End Product
Uranium	Natural	$^{238}_{92}U$	4.47×10^9	$^{206}_{82}Pb$
Actinium	Natural	$^{235}_{92}U$	7.04×10^8	$^{207}_{82}Pb$
Thorium	Natural	$^{232}_{90}Th$	1.41×10^{10}	$^{208}_{82}Pb$
Neptunium		$^{237}_{93}Np$	2.14×10^6	$^{209}_{83}Bi$

The two uranium series are somewhat more complex than the ^{232}Th series. Also, there are several naturally occurring radioactive isotopes, such as ^{14}C and ^{40}K, that are not part of either decay series.

The existence of radioactive series in nature enables our environment to be constantly replenished with radioactive elements that would otherwise have disappeared long ago. For example, because the solar system is about 5×10^9 years old, the supply of ^{226}Ra (whose half-life is only 1600 years) would have been depleted by radioactive decay long ago if it were not for the decay series that starts with ^{238}U, whose half-life is 4.47×10^9 years.

13.7 NUCLEAR REACTIONS

It is possible to change the structure of nuclei by bombarding them with energetic particles. Such collisions, which change the identity or properties of the target nuclei, are called **nuclear reactions.** Rutherford was the first to observe nuclear reactions, in 1919, using naturally occurring radioactive sources for the bombarding particles. Since then, thousands of nuclear reactions have been observed following the development of charged-particle accelerators in the 1930s. With today's advanced technology in particle accelerators and particle detectors, it is possible to achieve particle energies of at least 300 GeV. These high-energy particles are used to create new particles whose properties are helping to solve the mystery of the nucleus.

Consider a reaction in which a target nucleus is bombarded by a particle a, resulting in a nucleus Y and a particle b:

$$\boxed{\mathrm{a} + \mathrm{X} \longrightarrow \mathrm{Y} + \mathrm{b}} \tag{13.24}$$

Nuclear reaction

Sometimes this reaction is written in the more compact form

$$\boxed{\mathrm{X(a, b)Y}}$$

In the previous section, the Q value, or disintegration energy, of a radioactive decay was defined as the energy released as the result of the decay process. Likewise, we define the **reaction energy Q** associated with a nuclear reaction as *the total energy released as the result of the reaction.* More specifically, Q is defined as

$$\boxed{Q = (M_{\mathrm{a}} + M_{\mathrm{X}} - M_{\mathrm{Y}} - M_{\mathrm{b}})c^2} \tag{13.25}$$

Reaction energy Q

As an example, consider the reaction ^{7}Li (p, α)^{4}He, or

$${}^1_1\mathrm{H} + {}^7_3\mathrm{Li} \longrightarrow {}^4_2\mathrm{He} + {}^4_2\mathrm{He}$$

which has a Q value of 17.3 MeV. This reaction was first observed by Cockroft and Walton in 1930 using accelerated charged particles. A reaction such as this, for which Q is positive, is called **exothermic.** After the reaction, this residual energy appears as kinetic energy of (Y, b). That is, the loss in the mass of the system is balanced by an increase in the kinetic energy of the final particles. A reaction for which Q is negative is called **endothermic.** An endothermic reaction will not occur unless the bombarding particle has a kinetic energy greater than Q. The minimum energy necessary for such a reaction to occur is called the **threshold energy.**

Exothermic reaction

Endothermic reaction

Threshold energy

The kinetics of nuclear reactions must obey the law of conservation of linear momentum. That is, the total linear momentum of the system of inter-

TABLE 13.5 *Q* Values for Nuclear Reactions Involving Light Nuclei

Reaction[a]	Measured *Q*-Value (MeV)
$^{2}H(n, \gamma)^{3}H$	6.257 ± 0.004
$^{2}H(d, p)^{3}H$	4.032 ± 0.004
$^{6}Li(p, \alpha)^{3}H$	4.016 ± 0.005
$^{6}Li(d, p)^{7}Li$	5.020 ± 0.006
$^{7}Li(p, n)^{7}Be$	-1.645 ± 0.001
$^{7}Li(p, \alpha)^{4}He$	17.337 ± 0.007
$^{9}Be(n, \gamma)^{10}Be$	6.810 ± 0.006
$^{9}Be(\gamma, n)^{8}Be$	-1.666 ± 0.002
$^{9}Be(d, p)^{10}Be$	4.585 ± 0.005
$^{9}Be(p, \alpha)^{6}Li$	2.132 ± 0.006
$^{10}B(n, \alpha)^{7}Li$	2.793 ± 0.003
$^{10}B(p, \alpha)^{7}Be$	1.148 ± 0.003
$^{12}C(n, \gamma)^{13}C$	4.948 ± 0.004
$^{13}C(p, n)^{13}N$	-3.003 ± 0.002
$^{14}N(n, p)^{14}C$	-0.627 ± 0.001
$^{14}N(n, \gamma)^{15}N$	10.833 ± 0.007
$^{18}O(p, n)^{18}F$	-2.453 ± 0.002
$^{19}F(p, \alpha)^{16}O$	8.124 ± 0.007

From C. W. Li, W. Whaling, W. A. Fowler, and C. C. Lauritsen, *Physical Review* 83:512 (1951).
[a] The symbols n, p, d, α, and γ denote the neutron, proton, deuteron, alpha particle, and photon, respectively.

acting particles must be the same before and after the reaction. This assumes that the only force acting on the interacting particles is their mutual force of interaction; that is, there are no external accelerating electric fields present near the colliding particles.

If a nuclear reaction occurs in which particles a and b are identical, so that X and Y are also necessarily identical, the reaction is called a *scattering event.* If kinetic energy is conserved as a result of the reaction (that is, if $Q = 0$), it is classified as *elastic scattering.* On the other hand, if $Q \neq 0$, kinetic energy is not conserved and the reaction is called *inelastic scattering.* This terminology is identical to that used in dealing with the collision between macroscopic objects.

A list of measured Q values for a number of nuclear reactions involving light nuclei is given in Table 13.5.

In addition to energy and momentum, the total charge and total number of nucleons must be conserved in any nuclear reaction. For example, consider the reaction $^{19}F(p, \alpha)^{16}O$, which has a Q value of 8.124 MeV. We can show this reaction more completely as

$$^{1}_{1}H + ^{19}_{9}F \longrightarrow ^{16}_{8}O + ^{4}_{2}He$$

We see that the total number of nucleons before the reaction $(1 + 19 = 20)$ is equal to the total number after the reaction $(16 + 4 = 20)$. Furthermore, the total charge $(Z = 10)$ is the same before and after the reaction.

13.8 SUMMARY

A nuclear species can be represented by $^{A}_{Z}X$, where A is the **mass number,** which equals the total number of nucleons, and Z is the **atomic number,** which is the total number of protons. The total number of neutrons in a nucleus is the

neutron number N, where $A = N + Z$. Elements with the same Z but different A and N values are called **isotopes.**

Assuming that nuclei are spherical, their radius is given by

$$\boxed{r = r_0 A^{1/3}} \tag{13.1}$$

where $r_0 = 1.2$ fm (1 fm $= 10^{-15}$ m).

Light nuclei are most stable when the number of protons equals the number of neutrons. Heavy nuclei are most stable when the number of neutrons exceeds the number of protons. In addition, most stable nuclei have Z and N values that are both even. Nuclei with unusually high stability have Z or N values of 2, 8, 20, 28, 50, 82, and 126, called **magic numbers.**

Nuclei have an intrinsic spin angular momentum of magnitude $\sqrt{I(I+1)}\hbar$, where I is the **nuclear spin.** The magnetic moment of a nucleus is measured in terms of the **nuclear magneton** μ_n, where

$$\boxed{\mu_n = 5.05 \times 10^{-27} \text{ J/T}} \tag{13.3}$$

The proton has a magnetic moment of 2.7928 μ_n, and the neutron has a magnetic moment of -1.9135 μ_n. When a nuclear moment is placed in an external magnetic field, it precesses about the field with a frequency that is proportional to the field.

The difference between the total mass of the separate nucleons and that of the compound nucleus containing these nucleons, when multiplied by c^2, gives the **binding energy** E_b of the nucleus: $E_b = \Delta mc^2$. We can calculate the binding energy of any nucleus ^A_ZX from the expression

$$\boxed{E_b \text{ (MeV)} = [Zm_H - Nm_n - M(^A_Z\text{X})] \times 931.50 \text{ MeV/u}} \tag{13.4}$$

where all masses are atomic masses, m_H is the mass of the hydrogen atom, and m_n is the neutron mass.

Nuclei are stable because of the **strong nuclear force** between nucleons. This short-range force dominates the Coulomb repulsive force at distances of less than about 2 fm and is nearly independent of charge.

The **liquid-drop model** of nuclear structure treats the nucleons as molecules in a drop of liquid. In this model, the three main contributions influencing the binding energy of the nucleus are the volume effect, the surface effect, and the Coulomb repulsion. Summing such contributions results in the **semiempirical binding energy formula** given by Equation 13.5.

The **independent-particle model** of nuclear structure, or shell model, assumes that each nucleon moves in a well-defined quantized orbit within the nucleus and in an averaged field produced by the other nucleons. The stability of certain nuclei can be explained with this model. Two nucleons will collide only if the energy of each state after the collision corresponds to one of the allowed nuclear states, that is, a state that does not violate the exclusion principle.

The **collective model** of the nucleus combines some features from the liquid-drop model and some from the independent-particle model. It has been very successful in describing various nuclear phenomena.

A radioactive substance can undergo decay by three processes: alpha decay, beta decay, and gamma decay. An alpha particle is the ^{4}He nucleus; a beta particle is either an electron (β^-) or a positron (β^+); a gamma particle is a high-energy photon.

If a radioactive material contains N_0 radioactive nuclei at $t = 0$, the number N of nuclei remaining after a time t has elapsed is given by

$$\boxed{N = N_0 e^{-\lambda t}} \qquad (13.7)$$

where λ is the **decay constant** or **disintegration constant.** The *decay rate,* or *activity,* of a radioactive substance is given by

$$\boxed{R = \left|\frac{dN}{dt}\right| = R_0 e^{-\lambda t}} \qquad (13.8)$$

where $R_0 = N_0\lambda$ is the activity at $t = 0$. The **half-life** $T_{1/2}$, defined as the time it takes half of a given number of radioactive nuclei to decay, is

$$\boxed{T_{1/2} = \frac{0.693}{\lambda}} \qquad (13.9)$$

The unit of activity is the **curie** (Ci), defined as $1\ \text{Ci} \equiv 3.7 \times 10^{10}$ decays/s. The SI unit of activity is the **becquerel** (Bq), defined as 1 decay/s.

Alpha decay can occur because, according to quantum mechanics, some nuclei have barriers that can be penetrated by the alpha particles (the tunneling process). This process is energetically more favorable for those nuclei having a large excess of neutrons. A nucleus can undergo beta decay in two ways. It can emit either an electron (β^-) and an antineutrino ($\overline{\nu}$) or a positron (β^+) and a neutrino (ν). In the electron-capture process, the nucleus of an atom absorbs one of its own electrons (usually from the K shell) and emits a neutrino. In gamma decay, a nucleus in an excited state decays to its ground state and emits a gamma ray.

Nuclear reactions can occur when a target nucleus X is bombarded by a particle a, resulting in a nucleus Y and a particle b:

$$\boxed{\text{a} + \text{X} \longrightarrow \text{Y} + \text{b}} \quad \text{or} \quad \boxed{\text{X(a, b)Y}} \qquad (13.24)$$

The energy released in such a reaction, called the **reaction energy** Q, is given by

$$\boxed{Q = (M_\text{a} + M_\text{X} - M_\text{Y} - M_\text{b})c^2} \qquad (13.25)$$

QUESTIONS

1. Why are heavy nuclei unstable?
2. A proton precesses with a frequency ω_p in the presence of a magnetic field. If the magnetic field intensity is doubled, what happens to the precessional frequency?
3. Explain why nuclei that are well off the line of stability in Figure 13.3 tend to be unstable.
4. Consider two heavy nuclei X and Y with similar mass numbers. If X has the higher binding energy, which nucleus would tend to be more unstable?
5. Discuss the differences between the liquid-drop model and the independent-particle model of the nucleus.
6. If a nucleus has a half-life of 1 year, does this mean that it will be completely decayed after 2 years? Explain.
7. What fraction of a radioactive sample has decayed after two half-lives have elapsed?
8. Explain why the half-lives for radioactive nuclei are essentially independent of temperature.
9. The radioactive nucleus $^{226}_{88}\text{Ra}$ has a half-life of about

1.6×10^3 years. Since the universe is about 5 billion years old, how can you explain why we still can find this nucleus in nature?

10. A free neutron undergoes a beta decay with a half-life of about 15 min. Can a free proton undergo a similar decay?
11. Explain in detail how you can determine the age of an organic sample using the technique of carbon dating.
12. What is the difference between a neutrino and a photon?
13. Use Equations 13.17 to 13.19 to explain why the neutrino must have a spin of $\frac{1}{2}$.
14. If a nucleus such as ^{226}Ra initially at rest undergoes alpha decay, which has more kinetic energy after the decay, the alpha particle or the daughter nucleus?
15. Can a nucleus emit alpha particles with different energies? Explain.
16. Explain why many heavy nuclei undergo alpha decay but do not spontaneously emit neutrons or protons.
17. If an alpha particle and an electron have the same kinetic energy, which undergoes the greater deflection when passed through a magnetic field?
18. If film is kept in a box, alpha particles from a radioactive source outside the box cannot expose the film but beta particles can. Explain.
19. Pick any beta decay process and show that the neutrino must have zero charge.
20. Suppose it could be shown that the cosmic ray intensity was much greater 10 000 years ago. How would this affect the accepted values of the age of ancient samples of once-living matter?
21. Why is carbon dating unable to provide accurate estimates of very old material?
22. Element X has several isotopes. What do these isotopes have in common? How do they differ?
23. Explain the main differences between alpha, beta, and gamma rays.
24. How many protons are there in the nucleus $^{222}_{86}$Rn? How many neutrons? How many electrons are there in the neutral atom?

PROBLEMS

Table 13.6 will be useful for many of these problems. A more complete list of atomic masses is given in Appendix G.

1. Find the radius of (a) a nucleus of $^4_2$He and (b) a nucleus of $^{238}_{92}$U. (c) What is the ratio of these radii?
2. The compressed core of a star formed in the wake of a supernova explosion consists of pure nuclear material and is called a *neutron star.* Calculate the mass of 10 cm^3 of a neutron star.
3. Consider the hydrogen atom to be a sphere of radius equal to the Bohr radius, a_0, given by Equation 3.28, and calculate the approximate value of the ratio of the nucelar density to the atomic density.
4. The Larmor precessional frequency is

$$f = \frac{\Delta E}{h} = \frac{2\mu B}{h}$$

Calculate the radio-wave frequency at which resonance absorption will occur for (a) free neutrons in a magnetic field of 1 T, (b) free protons in a magnetic field of 1 T, and (c) free protons in the earth's magnetic field at a location where the magnetic field strength is 50 μT.

TABLE 13.6 Some Atomic Masses

Element	Atomic Mass (u)	Element	Atomic Mass (u)
$^4_2$He	4.002603	$^{27}_{13}$Al	26.981541
$^7_3$Li	7.016004	$^{30}_{15}$P	29.978310
$^9_4$Be	9.012182	$^{40}_{20}$Ca	39.962591
$^{10}_5$B	10.012938	$^{43}_{20}$Ca	42.958770
$^{12}_6$C	12.000000	$^{56}_{26}$Fe	55.934939
$^{13}_6$C	13.003355	$^{64}_{30}$Zn	63.929145
$^{14}_7$N	14.003074	$^{64}_{29}$Cu	63.929599
$^{15}_7$N	15.000109	$^{93}_{41}$Nb	92.906378
$^{15}_8$O	15.003065	$^{197}_{79}$Au	196.966560
$^{18}_8$O	17.999159	$^{202}_{80}$Hg	201.970632
$^{18}_9$F	18.000937	$^{216}_{84}$Po	216.001790
$^{20}_{10}$Ne	19.992439	$^{220}_{86}$Rn	220.011401
$^{23}_{11}$Na	22.989770	$^{234}_{90}$Th	234.043583
$^{23}_{12}$Mg	22.994127	$^{238}_{92}$U	238.050786

5. (a) Use energy methods to calculate the distance of closest approach for a head-on collision between an alpha particle with an initial energy of 0.5 MeV and a gold nucleus (^{197}Au) at rest. (Assume the gold nucleus remains at rest during the collision.) (b) What minimum initial speed must the alpha particle have in order to approach to a distance of 300 fm?
6. In Example 13.3, the binding energy of the deuteron was calculated to be 2.224 MeV. This corresonds to a value of 1.112 MeV/nucleon. What is the binding energy per nucleon for the heaviest isotope of hydrogen, ^{3}H (called *tritium*)?
7. Using the value of the atomic mass of $^{56}_{26}$Fe given in Table 13.6, find its binding energy. Then compute the binding energy per nucleon and compare your result with Figure 13.7.
8. (a) From Figure 13.8, estimate the electrostatic potential energy due to the Coulomb repulsion between two protons in contact. (b) Compare this value with the rest energy of an electron.
9. A pair of nuclei for which $Z_1 = N_2$ and $Z_2 = N_1$ are called *mirror isobars* (the atomic and neutron numbers are interchangeable). Binding energy measurements on these nuclei can be used to obtain evidence of the charge independence of nuclear forces (that is, proton-proton, proton-neutron, and neutron-neutron forces are approximately equal).Calculate the difference in binding energy for the two mirror nuclei $^{15}_{8}$O and $^{15}_{7}$N.
10. Calculate the binding energy per nucleon for (a) $^{20}_{10}$Ne, (b) $^{40}_{20}$Ca, (c) $^{93}_{41}$Nb, and (d) $^{197}_{79}$Au.
11. Calculate the minimum energy required to remove a neutron from the $^{43}_{20}$Ca nucleus.
12. Two species having the same mass number are known as *isobars*. Calculate the difference in binding energy per nucleon for the isobars $^{23}_{11}$Na and $^{23}_{12}$Mg. How do you account for the difference?
13. The $^{139}_{57}$La isotope of lanthanum is stable. A radioactive isobar of this lanthanum isotope, $^{139}_{59}$Pr, is located above the line of stable nuclei in Figure 13.3 and decays by β^+ emission. Another radioactive isobar of ^{139}La, $^{139}_{55}$Cs, decays by β^- emission and is located below the line of stable nuclei in Figure 13.3. (a) Which of these three isobars has the highest neutron-to-proton ratio? (b) Which has the greatest binding energy per nucleon? (c) Which of the two radioactive nuclei (^{139}Pr or ^{139}Cs) do you expect to be heavier?
14. (a) In the liquid-drop model of nuclear structure, why does the surface-effect term $-C_2A^{2/3}$ have a minus sign? (b) The binding energy of the nucleus increases as the volume-to-surface ratio increases. Calculate this ratio for both spherical and cubical shapes and explain which is more plausible for nuclei.
15. Using the graph in Figure 13.7, estimate how much energy is released when a nucleus of mass number 200 is split into two nuclei, each of mass number 100.
16. Use Equation 13.5 and the given values for the constants C_1, C_2, C_3, and C_4 to calculate the binding energy per nucleon for the isobars $^{64}_{29}$Cu and $^{64}_{30}$Zn.
17. Measurements on a radioactive sample show that its activity decreases by a factor of 5 during a 2-h interval. (a) Determine the decay constant of the radioactive nucleus. (b) Calculate the value of the half-life for this isotope.
18. The half-life of ^{131}I is 8.04 days (a) Calculate the decay constant for this isotope. (b) Find the number of ^{131}I nuclei necessary to produce a sample with an activity of 0.5 μCi.
19. A freshly prepared sample of a certain radioactive isotope has an activity of 10 mCi. After an elapsed time of 4 h, its activity is 8 mCi. (a) Find the decay constant and half-life of the isotope. (b) How many atoms of the isotope were contained in the freshly prepared sample? (c) What is the sample's activity 30 h after it is prepared?
20. Tritium has a half-life of 12.33 years. What percentage of the ^{3}H nuclei in a tritium sample will decay after a period of 5 years?
21. A sample of radioactive material is said to be *carrier-free* when no stable isotopes of the radioactive element are present. Calculate the mass of strontium in a carrier-free 5-mCi sample of ^{90}Sr, whose half-life is 28.8 years.
22. How many radioactive atoms are present in a sample that has an activity of 0.2 μCi and a half-life of 8.1 days?
23. A laboratory stock solution is prepared with an initial activity due to ^{24}Na of 2.5 mCi/ml, and 10 ml of the stock solution is diluted (at $t_0 = 0$) to a working solution whose total volume is 250 ml. After 48 h, a 5-ml sample of the working solution is monitored with a counter. What is the measured activity? (*Note:* 1 ml = 1 milliliter.)
24. Start with Equation 13.8 and find the following useful forms for the decay constant and half-life:

$$\lambda = \frac{1}{t}\ln\left(\frac{R_0}{R}\right) \qquad T_{1/2} = \frac{(\ln 2)t}{\ln(R_0/R)}$$

25. The radioactive isotope ^{198}Au has a half-life of 64.8 h. A sample containing this isotope has an initial activity ($t = 0$) of 40 μCi. Calculate the number of nuclei that will decay in the time interval between $t_1 = 10$ h and $t_2 = 12$ h.
26. Find the energy released in the alpha decay of $^{338}_{92}$U:

$$^{238}_{92}\text{U} \longrightarrow {}^{234}_{90}\text{Th} + {}^{4}_{2}\text{He}$$

You will find the mass values in Table 13.6 useful.

27. The atomic mass values in Appendix G are often replaced by a listing of values of the *mass excess* for each isotope. The mass excess is usually denoted by the symbol Δ and is defined as $\Delta = (M - A)$ when stated in units of u or $\Delta = (M - A)c^2$ when given in units of MeV

or keV. Calculate the mass excess of (a) the neutron and (b) ^{63}Cu.

28. Find the kinetic energy of an alpha particle emitted during the alpha decay of $^{220}_{86}Rn$. Assume that the daughter nucleus, $^{216}_{84}Po$, has zero recoil velocity.

29. Determine which of the following suggested decays can occur spontaneously:
    (a) $^{40}_{20}Ca \longrightarrow {}^{0}_{1}\beta + {}^{40}_{19}K$
    (b) $^{98}_{44}Ru \longrightarrow {}^{4}_{2}He + {}^{94}_{42}Mo$
    (c) $^{144}_{60}Nd \longrightarrow {}^{4}_{2}He + {}^{140}_{58}Ce$

30. Starting with $^{235}_{92}U$, the sequence of decays shown in Fig. 13.19 is observed, ending with the stable isotope $^{207}_{82}Pb$. Enter the correct isotope symbol in each open square.

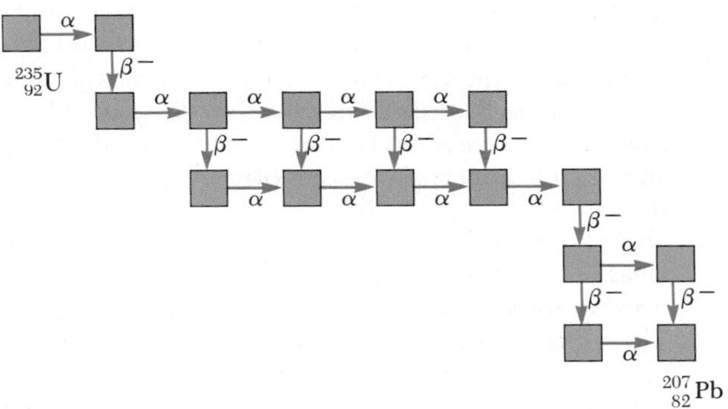

Figure 13.19 (Problem 30).

31. The following reaction, first observed in 1930, led to the discovery of the neutron by Chadwick:

$$^{9}_{4}Be(\alpha, n)^{12}_{6}C$$

Calculate the Q value of this reaction.

32. The following is the first known reaction (achieved in 1934) in which the product nucleus is radioactive:

$$^{27}_{13}Al(\alpha, n)^{30}_{15}P$$

Calculate the Q value of this reaction.

33. There are a few nuclear reactions in which the emerging particle and the product nucleus are identical. One example of this is the reaction

$$^{7}_{3}Li(p, \alpha)^{4}_{2}He$$

Calculate the Q value of this reaction.

34. Show that the following inverse reactions have the same *absolute value* of Q:

$$^{10}_{5}B(\alpha, p)^{13}_{6}C \quad \text{and} \quad ^{13}_{6}C(p, \alpha)^{10}_{5}B$$

35. The nuclear reaction $^{18}O(p, n)^{18}F$ has a Q value of -2.453 ± 0.002 MeV. (a) Write out the reaction. (b) If the atomic mass of ^{18}O is 17.999159 u, what is the atomic mass of ^{18}F?

36. Using the appropriate reactions and Q values from Table 13.5, calculate the masses of ^{8}Be and ^{10}Be in atomic mass units to four decimal places.

37. Copper as it occurs naturally consists of two stable isotopes, ^{63}Cu and ^{65}Cu. What is the relative abundance of the two forms? In your calculations, take the atomic weight of copper to be 63.55 and take the masses of the two isotopes to be 62.95 u and 64.95 u.

38. (a) The first nuclear transmutation was achieved in 1919 by Rutherford, who bombarded nitrogen atoms with alpha particles emitted by the isotope ^{214}Bi. The reaction is

$$^{4}_{2}He + {}^{14}_{7}N \longrightarrow {}^{17}_{8}O + {}^{1}_{1}H$$

What is the Q value of the reaction? What is the threshold energy? (b) The first nuclear reaction utilizing particle accelerators was performed by Cockroft and Walton. In this case, accelerated protons were used to bombard lithium nuclei, producing the following reaction:

$$^{1}_{1}H + {}^{7}_{3}Li \longrightarrow {}^{4}_{2}He + {}^{4}_{2}He$$

Since the masses of the particles involved in the reaction were well known, these results were used to obtain an early proof of the Einstein mass-energy relation. Calculate the Q value of the reaction.

39. (a) One method of producing neutrons for experimental use is based on the bombardment of light nuclei by alpha paticles. In one particular arrangement, alpha particles emitted by plutonium are incident on beryllium nuclei and this results in the production of neutrons:

$$^{4}_{2}He + {}^{9}_{4}Be \longrightarrow {}^{12}_{6}C + {}^{1}_{0}n$$

What is the Q value for this reaction? (b) Neutrons are also often produced by small-particle accelerators. In one design, deuterons (^{2}H) that have been accelerated in a Van de Graaf generator are used to bombard other deuterium nuclei, resulting in the following reaction:

$$^{2}_{1}H + {}^{2}_{1}H \longrightarrow {}^{3}_{2}He + {}^{1}_{0}n$$

Is this reaction exothermic or endothermic? Calculate its Q value.

40. The activity of a sample of radioactive material was measured over 12 h, and the following *net* count rates were obtained at the times indicated:

Time (h)	Counting Rate (counts/min)
1	3100
2	2450
4	1480
6	910
8	545
10	330
12	200

(a) Plot the activity curve on semilog paper. (b) Determine the disintegration constant and half-life of the radioactive nuclei in the sample. (c) What counting rate would you expect for the sample at $t = 0$? (d) Assuming the efficiency of the counting instrument to be 10%, calculate the number of radioactive atoms in the sample at $t = 0$.

41. A by-product of some fission reactors is the isotope $^{239}_{94}Pu$, which is an alpha emitter with a half-life of 24 000 years:

$$^{239}_{94}Pu \longrightarrow {}^{235}_{92}U + \alpha$$

Consider a sample of 1 kg of pure $^{239}_{94}Pu$ at $t = 0$. Calculate (a) the number of $^{239}_{94}Pu$ nuclei present at $t = 0$ and (b) the initial activity in the sample.

42. A piece of charcoal has a mass of 25 g and is known to be about 25 000 years old. (a) Determine the number of decays per minute expected from this sample. (b) If the radioactive background in the counter without a sample is 20 counts/min and we assume 100% efficiency in counting, explain why 25 000 years is close to the limit of dating with this technique.

43. A fission reactor is hit by a nuclear weapon and evaporates 5×10^6 Ci of $^{90}Sr(T_{1/2} = 27.7$ years) into the air. The ^{90}Sr falls out over an area of 10^4 km^2. How long will it take the activity of the ^{90}Sr to reach the agriculturally "safe" level of 2 μCi/m^2?

44. Using a reasonable scale, sketch an energy level diagram for (a) a proton and (b) a deuteron, both in a magnetic field B. (c) What are the absolute values of the changes in energy that accompany the possible transitions between the levels shown in your diagrams?

45. In addition to the radioactive nuclei included in the natural decay series, there are several other radioactive nuclei that occur naturally. One of these is ^{147}Sm, which is 15% naturally abundant and has a half-life of 1.3×10^{10} years. Calculate the number of decays per second per gram (due to this isotope) in a sample of natural samarium. The atomic weight of samarium is 150.4. (Activity per unit mass is referred to as *specific activity*.)

46. (a) Why is the following inverse beta decay forbidden for a free proton?

$$p \longrightarrow n + \beta^+ + \nu$$

(b) Why is the same reaction possible if the proton is bound in a nucleus? For example, the following reaction occurs:

$$^{13}_{7}N \longrightarrow {}^{13}_{6}C + \beta^+ + \nu$$

(c) How much energy is released in the reaction given in (b)? [Take the masses to be $m(\beta^+) = 0.000549$ u, $M(^{13}C) = 13.003355$ u, and $M(^{13}N) = 13.005739$ u.]

47. (a) Find the radius of the $^{12}_{6}C$ nucleus. (b) Find the force of repulsion between a proton at the surface of a $^{12}_{6}C$ nucleus and the remaining five protons. (c) How much work (in MeV) has to be done to overcome this electrostatic repulsion to put the last proton into the nucleus? (d) Repeat (a), (b), and (c) for $^{238}_{92}U$.

48. What activity in disintegrations/min·g would be expected for carbon samples from bones that are 2000 years old? (You should note from Example 13.9 that the ratio of ^{14}C to ^{12}C in living organisms is 1.3×10^{-12}.)

49. Consider a hydrogen atom with the electron in the 1s state. The magnetic field at the nucleus produced by the orbiting electron has a value of 12.5 T. The proton can have its magnetic moment aligned in either of two directions perpendicular to the plane of the electron's orbit. Because of the interaction of the proton's magnetic moment with the electron's magnetic field, there will be a difference in energy between the states with the two different orientations of the proton's magnetic moment. Find that energy difference in eV.

50. Use the Heisenberg uncertainty principle to make a reasonable argument against the hypothesis that free electrons can be present in a nucleus. You should use relativistic expressions for the momentum and energy and include appropriate assumptions and approximations.

51. Supernova explosions are incredibly powerful nuclear reactions that tear apart the cores of relatively massive stars (greater than four times the sun's mass). These blasts are produced by carbon fusion, which requires a temperature of about 6×10^8 K to overcome the strong Coulomb repulsion between carbon nuclei. (a) Estimate the repulsive energy barrier to fusion, using the required ignition temperature for carbon fusion. (In other words, what is the kinetic energy for a carbon nucleus at 6×10^8 K?) (b) Calculate the energy (in MeV) released in each of these "carbon-burning" reactions:

$$^{12}C + {}^{12}C \longrightarrow {}^{20}Ne + {}^{4}He$$

$$^{12}C + {}^{12}C \longrightarrow {}^{24}Mg + \gamma$$

(c) Calculate the energy (in kWh) given off when 2 kg of carbon completely fuses according to the first reaction.

52. A radioactive nucleus with decay constant λ decays to a stable daughter nucleus. (a) Show that the number of daughter nuclei, N_2, increases with time according to the expression

$$N_2 = N_{01}(1 - e^{-\lambda t})$$

where N_{01} is the initial number of parent nuclei. (b) Starting with 10^6 parent nuclei at $t = 0$, with a half-life of 10 h, plot the number of parent nuclei and the number of daughter nuclei as functions of time over the interval 0 to 30 h.

53. Many radioisotopes have important industrial, medical, and research applications. One of these is ^{60}Co, which has a half-life of 5.2 years and decays by the emission of a beta particle (energy 0.31 MeV) and two gamma photons (energies 1.17 MeV and 1.33 MeV). A scientist wishes to prepare a ^{60}Co sealed source that will have an activity of at least 10 Ci after 30 months of use. (a) What is the mimimum initial mass of ^{60}Co required? (b) At what rate will the source emit energy after 30 months?

54. Treat the nucleus as a sphere of uniform volume charge density ρ. (a) Derive an expression for the total energy for such a sphere of charge corresponding to a nucleus of atomic number Z and radius R. (b) Using the result of (a), find an expression for the electrostatic potential energy, U, in terms of the mass number, A, for the case when $N = Z$ (the nucleus is on the line of stability). (c) Evaluate the result of (b) for the nucleus $^{30}_{15}P$.

55. "Free neutrons" have a characteristic half-life of 12 min. What fraction of a group of free neutrons at thermal energy (0.04 eV) will decay before traveling a distance of 10 km?

56. When, after a reaction or disturbance of any kind, a nucleus is left in an excited state, it can return to its normal (ground) state by emission of a gamma-ray photon (or several photons). This process is described by Equation 13.21. The emitting nucleus must recoil in order to conserve both energy and momentum. (a) Show that the recoil energy of the nucleus is given by

$$E_r = \frac{(\Delta E)^2}{2Mc^2}$$

where ΔE is the difference in energy between the excited and ground states of a nucleus of mass M. (b) Calculate the recoil energy of the ^{57}Fe nucleus when it decays by gamma emission from the 14.4-keV excited state. For this calculation, take the mass to be 57 u. (*Hint:* When writing the equation for conservation of energy, use $(Mv)^2/2M$ for the kinetic energy of the recoiling nucleus. Also, assume that $hf \ll Mc^2$ and use the binomial expansion.)

57. The decay of an unstable nucelus by alpha emission is represented by Equation 13.10. The disintegration energy Q given by Equation 13.13 must be shared by the alpha particle and the daughter nucleus in order to conserve both energy and momentum in the decay process. (a) Show that Q and K_α, the kinetic energy of the alpha particle, are related by the expression

$$Q = K_\alpha\left(1 + \frac{M_\alpha}{M}\right)$$

where M is the mass of the daughter nucleus. (b) Use the result of (a) to find the energy of the alpha particle emitted in the decay of ^{226}Ra. (See Example 13.8 for the calculation of Q.)

58. Consider the deuterium-tritium fusion reaction with the tritium nucleus at rest:

$$^2_1H + ^3_1H \longrightarrow ^4_2He + ^1_0n$$

(a) From Equation 13.1, estimate the required distance of closest approach. (b) What is the Coulomb potential energy (in eV) at this distance? (c) If the deuteron has just enough energy to reach the distance of closest approach, what is the final velocity of the deuterium and tritium nuclei in terms of the initial deuteron velocity, v_0? (*Hint:* At this pont, the two nuclei have a common velocity equal to the center-of-mass velocity.) (d) Use energy methods to find the minimum initial deuteron energy required to achieve fusion. (e) Why does the fusion reaction occur at much lower deuteron energies than that calculated in (d)?

59. A neutron star is a nucleus composed entirely of neutrons and has a radius of about 10 km. Such a star is formed when a larger star cools down and collapses under the influence of its own gravitational field. The compression continues to the point where the protons and electrons merge to form neutrons. (a) Assuming that Equation 13.1 is valid, estimate the mass number A and mass M of a neutron star of radius $R = 10$ km. (b) Find the acceleration due to gravity at the surface of such a star. (c) Find the rotational kinetic energy of the star if it rotates about its axis 30 times/s. (Assume that the star is a sphere of uniform density.)

60. The concept of radioactive half-life was described in Section 13.4, and the relationship between $T_{1/2}$ and λ is given by Equation 13.9. Another parameter that is often useful in the description of radioactive processes is the *mean life*, τ. Although the half-life of a radioactive isotope is accurately known, it is not possible to predict the time when any individual atom will decay. The mean life is a measure of the average time of existence of all the atoms in a particular sample. Show that $\tau = 1/\lambda$ (*Hint:* Remember that τ is essentially an average value and use the fact that the number of atoms that decay between t and $t + dt$ is equal to dN. Furthermore, note that these dN atoms have a finite time of existence, equal to t.)

61. Potassium as it occurs in nature includes a radioactive isotope, ^{40}K, which has a half-life of 1.27×10^9 years and a relative abundance of 0.0012%. These nuclei decay by two different pathways: 89% by β^- emission and 11% by β^+ emission. Calculate the total activity in Bq associated with 1 kg of KCl *due to β^- emission.*

62. When the nuclear reaction represented by Equation 13.24 is endothermic, the disintegration energy Q is negative. In order for the reaction to proceed, the incoming particle must have a minimum energy called the *threshold energy*, E_{th}. Some fraction of the energy of the incident particle is transferred to the compound nucleus in order to conserve momentum. Therefore, E_{th} must be greater than Q. (a) Show that

$$E_{th} = -Q\left(1 + \frac{M_a}{M_X}\right)$$

(b) Calculate the threshold energy of the incident alpha particle in the reaction

$$^{4}_{2}He + ^{14}_{7}N \longrightarrow ^{17}_{8}O + ^{1}_{1}H$$

63. During the manufacture of a steel engine component, radioactive iron (^{59}Fe) is included in the total mass of 0.2 kg. The component is placed in a test engine when the activity due to this isotope is 20 μCi. After a 1000-h test period, oil is removed from the engine and found to contain enough ^{59}Fe to produce 800 disintegrations/min per liter of oil. The total volume of oil in the engine is 6.5 liters. Calculate the total mass worn from the engine component per hour of operation. (The half-life for ^{59}Fe is 45.1 days.)

Essay

SYMMETRY IN PHYSICS

Christopher T. Hill
Fermi National Accelerator Laboratory

The concept of symmetry is fundamental to our understanding of the physical world. It is where we can discern true or approximate symmetries, such as those involved in the basic forces of nature, that we profess any real understanding. Where nature displays little or no apparent symmetries, such as in the spectrum of elementary quarks and leptons, we find ourselves most befuddled. Moreover, all thinking in modern theoretical physics is aimed at understanding the possible role of deep mathematical symmetries in nature. The realization of the importance of symmetry to the understanding of the laws of physics is a modern concept, belonging almost entirely to the twentieth century and beginning largely with Einstein and the special theory of relativity. Today symmetry is one of the greatest components of what we mean when we speak of the "beauty of physics."

Mathematicians solve many problems in geometry and topology by turning them into *equivalent algebraic problems.* This approach to understanding symmetry begins with the 19th century French mathematician Galois[1], who in his short, tragic life laid the foundation of what we call "group theory."

Let us think concretely about the symmetries of a very simple geometric object . . . the equilateral triangle. This is the simplest nontrivial example of a system possessing symmetry, and the first time through you may find the mathematics hidden in this simple object to be quite surprising!

It is useful to approach this experimentally (if you can visualize the following manipulations, that's fine; but you are strongly encouraged to perform the experiments yourself). Prepare two cut-outs as in Figure 1 and Figure 2, each featuring an equilateral triangle, both of the same size. The cut-out of Figure 1 has the three axes of symmetry labeled as I, II, and III, while the cut-out of Figure 2 has the vertices labeled as *A*, *B*, and *C*.

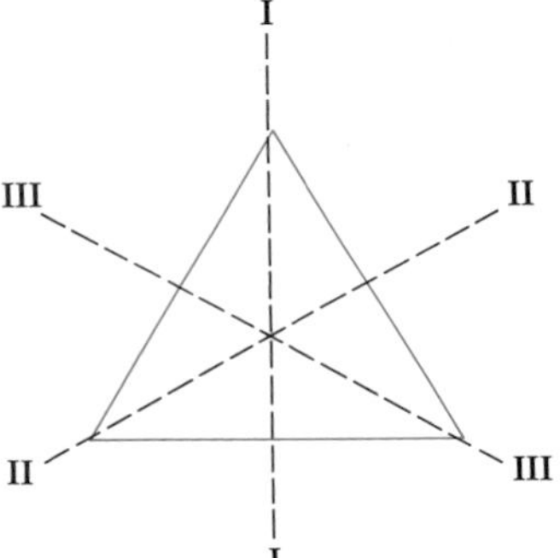

Figure 1 The reference triangle.

The cut-out of Figure 1 is laid down on a table and should be considered to be glued in place; it is a *reference triangle* that serves as a reference grid, or a kind of "coordinate system." Once it is laid in place, we will not move it again. Cut-out 2, on the other hand, is an *experimental triangle;* we will overlay the reference triangle with the experimental triangle. Our problem is to *find all possible* **distinguishable** *ways in which the experimental triangle can be lifted up and brought down on top of the reference triangle.* The vertices of the experimental triangle are labeled to allow us to identify the distinguishable ways in which this can be done.

We shall begin by overlaying the experimental triangle on the reference triangle with the vertices reading *ABC* clockwise around the experimental triangle, and with vertex *A* on axis I. This will be called the **initial position.** We wish to find a way in which we can pick up the experimental triangle and bring it back down on top of the reference triangle so that the vertices read something other than *ABC* clockwise. Each such operation is called a **symmetry operation.** Our problem is to discover all possible distinguishable symmetry operations of the equilateral triangle. How do we proceed?

Perhaps you've considered rotating the experimental triangle until the vertices now read *CAB* clockwise from the top. This certainly corresponds to a symmetry operation, which is a rotation through 120°. We shall designate this first discovery as $\mathbf{Rot}_{120°}$.

Now return to the initial position. What else might we do? It is obvious that a rotation through 240° is another symmetry operation, which yields the result *BCA*. However, it is important to emphasize that we should always return to the initial position before performing the next operation (this is a bit like pressing the **CLEAR** button on a pocket calculator before doing the next calculation). Thus we discover a second distinguishable symmetry operation, which we designate $\mathbf{Rot}_{240°}$.

Are there other symmetry operations? Perhaps you've considered a rotation by −120°. However, we now see that this takes the triangle to *BCA* (from the initial position, of course) and therefore this is not a new operation; i.e., it is not *distinguishable.* We may thus write the equation:

$$\text{Rot}_{-120°} = \text{Rot}_{240°}.$$

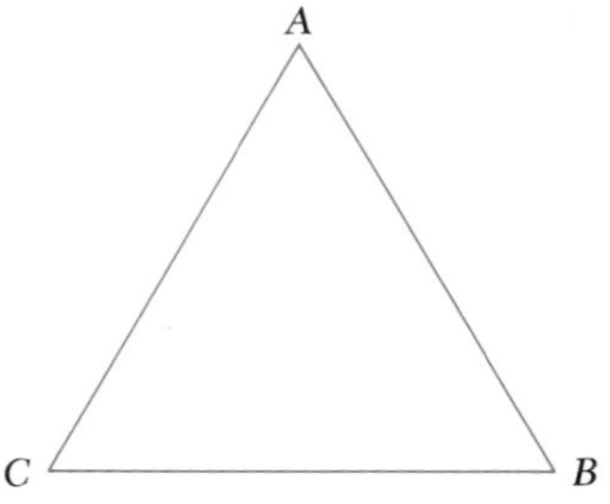

Figure 2 The experimental triangle.

What about a rotation through 360°? We see that it maps the triangle from the initial position *ABC* back to the initial position *ABC*. Consequently, it *is* a symmetry operation, but a very special one. First, it is equivalent to doing nothing at all; as such we shall refer to it as the "do nothing operation," or the *identity operation*. We shall denote it by a boldface **1**. Second, note that the identity element is a symmetry operation of any object; even an amoeba has the identity symmetry. Third, note that a rotation through 360° is equivalent to a rotation through any *integer multiple* of 360°, such as 720° or −360°. All are equivalent to the "do nothing" operation, **1**.

We now have three symmetry operations; are there more? We may consider a *reflection* about one of the three axes of the reference triangle. We begin with the initial position and consider "skewering" the experimental triangle (as if we had a barbecue skewer) along one of the axes of symmetry. For example, skewering along axis I, we then pick up the triangle and flip it and we arrive at the new position, *ACB*. We denote this symmetry operation as a *reflection about axis I* or as $\mathbf{Ref}_{\mathrm{I}}$. Similarly, we return to the initial position and consider the other two reflections (a) about axis II, or $\mathbf{Ref}_{\mathrm{II}}$, which yields the position *BAC*, and (b) about axis III, or $\mathbf{Ref}_{\mathrm{III}}$, which yields the position *CBA*.

We now have a list of six symmetry operations, which are shown in Table 1.

Are there any other symmetry operations? At this point perhaps you've recognized that we have discovered the six possible permutations of three objects, that is, the six permutations of the three vertices of the triangle. That raises an interesting question: *Are the symmetries of all such objects, such as squares, pentagons, hexagons, and cubes, given by the permutations of their vertices?* In fact, the answer is *no*. It doesn't work that way for the square, as we can easily see. Suppose we have a square with vertices labeled *ABCD*. A true symmetry operation of the square is a rotation through 90° that gives *DABC*, which is also a permutation of the vertices. However, is there a symmetry operation that can give the vertex ordering *BACD?* (Think in terms of an experimental square on a cut-out; what would we have to do to the cut-out to get *BACD* starting from *ABCD?*) Clearly, this is not a symmetry operation of the entire square; we would have to *twist* the experimental square to get the vertices into this position, but then the sides would not overlay properly! Thus, while all symmetry operations are indeed permutations, not all permutations are symmetry operations. The equilateral triangle is simpler and it does have only six symmetry operations, the ones we've listed above, which are equivalent (isomorphic) to the permutations of three objects.

Thus far our exercise has been almost trivial, but now we make the great observation of Galois and his colleagues. We now ask whether we can obtain additional symmetry operations by combining two of the operations previously obtained. That is, let us take two of our six operations, say $\mathbf{Rot}_{120°}$ and $\mathbf{Ref}_{\mathrm{II}}$, first perform one of them on the experimental triangle (try $\mathbf{Rot}_{120°}$), and *without returning to the initial position* perform the other operation ($\mathbf{Ref}_{\mathrm{II}}$). We see that if we begin in the initial position, $\mathbf{Rot}_{120°}$ leads to *CAB* and then following with Ref_{II} we obtain the position *ACB*. But *ACB* is not a new position of the triangle; it corresponds to $\mathbf{Ref}_{\mathrm{I}}$, as shown by Table 1. We have therefore discovered an interesting result: *first performing* $\mathbf{Rot}_{120°}$ *and following it by* $\mathbf{Ref}_{\mathrm{II}}$ *yields the* **result** $\mathbf{Ref}_{\mathrm{I}}$.

TABLE 1 The Six Symmetry Operations of the Equilateral Triangle

Notation	Operation	Vertices
1	"do nothing" or identity	*ABC*
$\mathbf{Rot}_{120°}$	rotate by 120°	*CAB*
$\mathbf{Rot}_{240°}$	rotate by 240°	*BCA*
$\mathbf{Ref}_{\mathrm{I}}$	reflect about axis I	*ACB*
$\mathbf{Ref}_{\mathrm{II}}$	reflect about axis II	*BAC*
$\mathbf{Ref}_{\mathrm{III}}$	reflect about axis III	*CBA*

Let us write an equation for this result:

$$\mathbf{Rot}_{120^\circ} \otimes \mathbf{Ref}_{II} = \mathbf{Ref}_{I} \tag{1}$$

Here we have introduced a symbol, ⊗, that represents the action of combining the symmetry operations in the order indicated. It is easily seen that the ⊗ combination of any pair of our symmetry operations (which we also refer to as "elements") produces another of the elements. We say that our set of elements is **closed** under the operation ⊗. Thus, in a sense, combining two symmetry operations is something like *multiplication of numbers.* In this sense the "do nothing operation" is the *identity:*

$$1 \otimes X = X \otimes 1 = X \tag{2}$$

Thus we have made a very important observation: *the symmetry operations form an algebraic system with an operation consisting of performing successive operations.* This algebraic system is called a **group.** The symmetry operations are the analogues of the rational numbers under this group multiplication. We refer to the symmetry operations as **group elements** or simply as **elements.** We present the complete multiplication table of the symmetry group of the equilateral triangle in Table 2.

You should verify this yourself by performing several of the cases with the experimental triangle and reference triangle. Table 2 is to be read like a highway mileage map; if we choose to perform the product $X \otimes Y$, we first find the row labeled by X, then the column labeled by Y, and we look up the corresponding entry in that row and column. There are several important properties of this group multiplication table that are shared by all groups:

- A group is a set of elements and a composition law, ⊗, such that the product of any two elements yields another element in the set.
- Every group has an identity element satisfying Equation 2.
- Each element of the group has a unique inverse element. That is, given an element X, there exists one and only one element, Y (which may even be X itself), such that $X \otimes Y = Y \otimes X = 1$.
- Group multiplication is *associative.* That is, given X, Y, and Z, we have $X \otimes (Y \otimes Z) = (X \otimes Y) \otimes Z$. In words, first perform Y and follow by Z and remember the result (call it W). Now return to the initial position and first do X and follow by W. This result will be the same as having first done X followed by Y then followed by Z. This is the meaning of associativity, and you should carefully think it through to make sure you understand it.
- Each element of the group occurs once and only once in each row and each column of the multiplication table. This can be proved as a theorem from the preceding statements. This is a powerful constraint on the mathematical structure of the group; essentially the group multiplication table forms a kind of "magic square."
- **Group multiplication is not necessarily commutative!** That is, $X \otimes Y$ need not equal $Y \otimes X$!

This last result, that group multiplication is not commutative, is really quite remarkable. Here we have discovered a simple system of six elements with a multiplication law and the system is not even commutative. For example, consider first

TABLE 2 **S_3 Group Multiplication Table**

	1	$\mathbf{Rot}_{120^\circ}$	$\mathbf{Rot}_{240^\circ}$	$\mathbf{Ref}_{I}$	$\mathbf{Ref}_{II}$	$\mathbf{Ref}_{III}$
1	1	$\mathbf{Rot}_{120^\circ}$	$\mathbf{Rot}_{240^\circ}$	$\mathbf{Ref}_{I}$	$\mathbf{Ref}_{II}$	$\mathbf{Ref}_{III}$
$\mathbf{Rot}_{120^\circ}$	$\mathbf{Rot}_{120^\circ}$	$\mathbf{Rot}_{240^\circ}$	1	$\mathbf{Ref}_{III}$	$\mathbf{Ref}_{I}$	$\mathbf{Ref}_{II}$
$\mathbf{Rot}_{240^\circ}$	$\mathbf{Rot}_{240^\circ}$	1	$\mathbf{Rot}_{120^\circ}$	$\mathbf{Ref}_{II}$	$\mathbf{Ref}_{III}$	$\mathbf{Ref}_{I}$
$\mathbf{Ref}_{I}$	$\mathbf{Ref}_{I}$	$\mathbf{Ref}_{II}$	$\mathbf{Ref}_{III}$	1	$\mathbf{Rot}_{120^\circ}$	$\mathbf{Rot}_{240^\circ}$
$\mathbf{Ref}_{II}$	$\mathbf{Ref}_{II}$	$\mathbf{Ref}_{III}$	$\mathbf{Ref}_{I}$	$\mathbf{Rot}_{240^\circ}$	1	$\mathbf{Rot}_{120^\circ}$
$\mathbf{Ref}_{III}$	$\mathbf{Ref}_{III}$	$\mathbf{Ref}_{I}$	$\mathbf{Ref}_{II}$	$\mathbf{Rot}_{120^\circ}$	$\mathbf{Rot}_{240^\circ}$	1

performing $\mathbf{Rot}_{240°}$ and following by $\mathbf{Ref}_{\mathrm{II}}$, that is, calculate $\mathbf{Rot}_{240°} \otimes \mathbf{Ref}_{\mathrm{II}}$. You should obtain the result $\mathbf{Ref}_{\mathrm{III}}$. On the other hand, consider first performing $\mathbf{Ref}_{\mathrm{II}}$ followed by $\mathbf{Rot}_{240°}$; i.e., compute $\mathbf{Ref}_{\mathrm{II}} \otimes \mathbf{Rot}_{240°}$. The result is $\mathbf{Ref}_{\mathrm{I}}$. Summarizing:

$$\mathbf{Rot}_{240°} \otimes \mathbf{Ref}_{\mathrm{II}} = \mathbf{Ref}_{\mathrm{III}}$$

$$\mathbf{Ref}_{\mathrm{II}} \otimes \mathbf{Rot}_{240°} = \mathbf{Ref}_{\mathrm{I}}$$

Thus, although ordinary multiplication is commutative (e.g., $3 \cdot 4 = 4 \cdot 3$), group multiplication need not be. When a group has commutative multiplication it is said to be an **abelian group,** after the mathematician Abel. The general group, such as the equilateral triangle group, is noncommutative, or **nonabelian.**

All possible *continuous* groups (i.e., groups with an infinite number of operations that vary continuously with "angle" parameters, like the rotations of a sphere about a given axis through any angle) were completely classified early in the 20th century by Cartan. Remarkably, only very recently have all possible discrete symmetry groups been classified. This job was made difficult by the existence of certain "sporadic" groups, such as the "monster group" with about 8×10^{53} elements. The classification of the discrete groups constitutes one of the longest and least comprehensible theorems in mathematics [2].

Group mathematics underlies the structure of our physical world. One may wonder how a noncommutative mathematics can have anything at all to do with nature, or physics. Yet it is easily demonstrated. For example, perform two successive rotations on a textbook, first about an imaginary x axis followed by one about an imaginary y axis, and note the book's position. Now return the book to the initial position and perform first the rotation about the y axis followed by a rotation about the x axis. You will find that the book ends up in two different positions. The symmetry group consisting of rotations through 90° is a noncommutative group. The continuous group consisting of all rotations of objects in three dimensions (the full symmetry of a sphere) is thus noncommutative. It is known as $O(3)$, and it governs the physics of angular momentum and spin.

In the mid-1960s [3] it was discovered that the strongly interacting particles could be placed into multiplets of a continuous symmetry group, $SU(3)$. One of the representations of $SU(3)$ has eight components, and is known as an octet (the $SU(3)$ symmetry elements can be represented as 8×8 matrices; the eight members of the octet mix among themselves under an $SU(3)$ transformation). Eight known spin-0 mesons fit into one multiplet, the eight spin-$\frac{1}{2}$ baryons into another, and so on. See Figure 3. There is also a 10-component representation into which the spin-$\frac{3}{2}$ baryonic resonances fit (in fact, one, the Ω^-, was missing at the time $SU(3)$ was discovered and was correctly predicted by the theory). The particles in the multiplets were not degenerate, indicating that the $SU(3)$ symmetry was not exact, but the pattern was clearly established.

The main puzzle was that the smallest representation of $SU(3)$, namely a triplet consisting of three spin-$\frac{1}{2}$ particles with charges $(\frac{2}{3}, -\frac{1}{3}, -\frac{1}{3})$, was not seen directly (in these units the electron charge is -1). These particles are known respectively as the "up" quark, the "down" quark, and the "strange" quark. Today, however, we have compelling evidence that the quarks do exist but are permanently confined by the strong force within the particles we see in the laboratory. All of the mesons are composed of quark and antiquark, while each baryon contains three quarks.

Finally we arrive at one of the most important questions about our physical world. What are the symmetries of the basic laws of physics? What do we mean by a symmetry in a law of physics? Just as in the case of the equilateral triangle, a symmetry operation is an action we perform upon a geometric configuration such that it remains unchanged; we mean now an operation we can perform upon an experiment such that the outcome of the experiment is unchanged. However, the operation must hold for *any conceivable experiment* if it is to qualify as a symmetry of the *laws of physics.*

Let us imagine an effort to map out the symmetries of the laws of physics. We suppose that we have a vast region of perfectly empty space and an enormous amount of time. For example, we take a laboratory into a void in the universe that measures

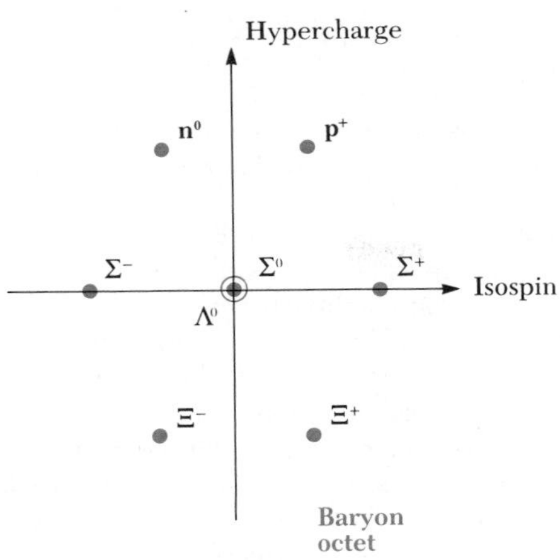

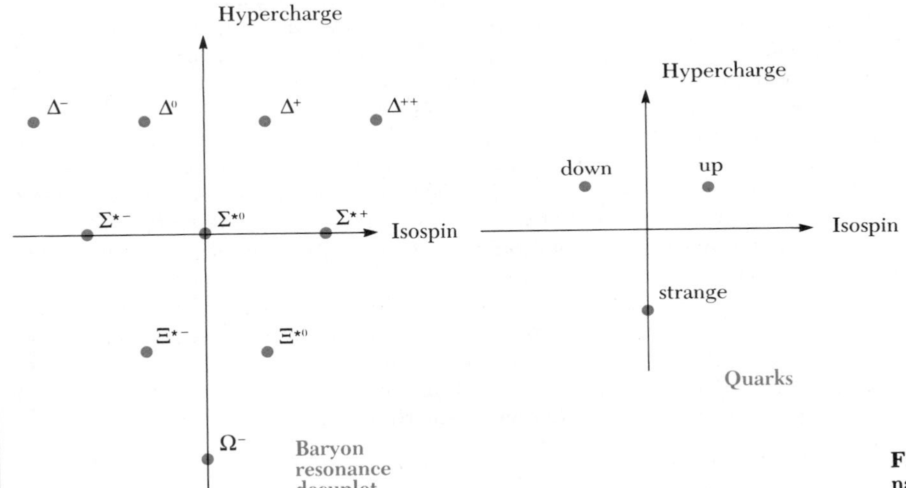

Figure 3 SU(3) baryon octet, resonance decuplet, and quark triplet.

$\approx 10^5$ parsecs on each side. The laboratory can move through the void as we carry out various experiments. We assume that this can be done over enormous time scales. For example, we may measure the following physical quantities:

1. The masses of the electron, proton, mesons, W bosons, and other particles.
2. The electric charges of these particles.
3. The speed of light.
4. Planck's constant.
5. The lifetimes of various particles and nuclear levels.
6. Newton's universal constant of gravitation.

These quantities are measured to enormous precision and the values are plotted against various configurations of the laboratory:

1. The position in space of the laboratory.
2. The orientation in space of the laboratory.
3. The time of the measurement.
4. The velocity of the laboratory.

After many billions of years the results of the study are completed. They may be summarized simply:

> The physical quantities measured by the laboratory have no dependence upon the configuration of the laboratory.

In other words, changing the position, orientation, time, or velocity of the laboratory does not influence the outcome of experiments conducted in the laboratory! *These are symmetries of the laws of physics.*

In fact, such experiments can be performed. For example, we might try to determine whether some basic physical quantity, such as the lifetime of the charged pi meson, depends upon the orientation of the lab. One simply performs the experiment at different times of the day and looks for correlations between time of day and measured lifetime; the rotation of the earth takes care of the reorientation of the laboratory (of course, we must be careful about systematic errors in such an experiment; maybe the power company switches to a different generator in the evening, which somehow contaminates our pion beam and gives us a fake signal.) Such an experiment is remniscent of the famous Michelson-Morley experiment which failed to show any dependence of the speed of light upon the absolute state of motion of the earthbound laboratory through an "ether-filled" space. As the consequence of symmetry is simplicity, this experiment, once properly interpreted by the special theory of relativity, washed away the concept of the ether from physics.

There are other compelling indications from astrophysics that our universe is the same everywhere and for all time. The physical processes occuring in distant stars and more distant galaxies produce the same spectral lines as in laboratory measurements on earth. Such measurements reach back 10 billion years to the early universe where even quasi-stellar objects reveal spectral lines of hydrogen equivalent to those we see today.[1] Also, the measurements are independent of direction in space. In fact, it is hard to imagine such a homogeneous and isotropic universe if the laws of physics themselves are not independent of space, time, direction, and state of motion.

Changing the position, orientation, time, or velocity of the laboratory does not influence the outcome of experiments conducted in the laboratory! In fact, these symmetries form the *Poincaré Group* of space-time symmetries of the laws of nature —the basic space-time symmetry group of the laws of physics. They hold over cosmological as well as microscopic and subnuclear scales. What are the physical consequences of these symmetries? This connection is given by one of the most important theorems in theoretical physics, known as **Nöether's theorem** [4]:

> For every continuous symmetry of the laws of physics there exists a corresponding conservation law.

Since the translational symmetry operations can act in any one of three directions in space, we find that there is a conserved quantity known as **momentum,** which forms a three-component vector. Since temporal symmetry operations act in one direction of time, we find that there exists a conserved quantity known as **energy,** which forms a scalar in Newtonian physics. Relativity unites space and time and in so doing melds energy and momentum together into one quantity called a *four-vector.*

Rotational symmetry operations can be performed in any of three independent directions and thus there exists a vector quantity known as **angular momentum.** In relativity the rotations combine together with the three independent Lorentz transformations, and angular momentum becomes associated with a six-component *tensor.*

Thus Nöether's theorem gives us the remarkable connection between symmetry and conserved physical quantities:

Conserved quantity		Symmetry operation
momentum	$\longleftrightarrow$	space translations
energy	$\longleftrightarrow$	time translations
angular momentum	$\longleftrightarrow$	rotations

[1] Of course, these lines are redshifted due to the general recession of these objects, but the redshift is a universal multiplicative effect and the relative frequency ratios of lines are not affected.

Furthermore, even electric charge, baryon number, quark color, and other conserved quantities are associated with symmetries in a deeper and more abstract manner.

As the arts and music have moved in this century farther away from symmetry, indeed adopting *antisymmetry* as a structural element, it is remarkable that symmetry has been increasingly understood by physicists as fundamental to the formulation of the laws of physics. How many times have we glimpsed an equilateral triangle's simplicity yet missed its inner complexity and logical beauty? So too it is with nature. Perhaps her deepest secrets lie hidden before our very eyes!

Suggestions for Further Reading

1. T. Rothman, "The Short Life of Evariste Galois," *Sci. American*, April 1982.
2. D. Gorenstein, "The Enormous Theorem," *Sci. American*, December 1985.
3. M. Gell-Mann and Y. Neeman, *The Eight-Fold Way*, New York, Benjamin, 1964.
4. L. M. Osen, *Women in Mathematics*, MIT Press, 1974, p. 141.

Essay Questions and Problems

1. Construct the symmetry group of the square and its associated multiplication table. Verify the properties discussed above for the general group.
2. Construct the permutation group of four objects and its associated multiplication table. Is it isomorphic to the square's group?
3. A subgroup of a group is a subset of elements which themselves form a group. Clearly, each subgroup must contain the identity. Can you identify some subgroups of the equilateral triangle group? What is the largest subgroup of the equilateral triangle symmetry group (not counting the entire group itself—each set is a subset of itself; technically we want the largest *proper* subgroup).
4. If the square in the preceding problem is squashed into a rectangle, identify the surviving subgroup (see the preceding problem) which describes the remaining symmetry. (As a preliminary exercise, consider an isosceles triangle in which vertex A is lifted along axis I in Figure 1; what is the resulting symmetry group? Is it a subgroup?)
5. If an infinite floor is tiled with equilateral triangles, we have a *lattice*. Is the symmetry group of the equilateral triangle also a symmetry group of the lattice? Is the full symmetry group of the lattice equivalent to the group of a single triangle? Does a rotation commute with a translation? This problem illustrates how groups enter solid state physics, in which they are of paramount importance.

14

Nuclear Physics Applications

This chapter is concerned primarily with the two means by which energy can be derived from nuclear reactions. These two techniques are fission, in which a nucleus of large mass number splits, or fissions, into two smaller nuclei, and fusion, in which two light nuclei fuse to form a heavier nucleus. In either case, there is a release of energy, which can be used destructively through bombs or constructively through the production of electric power. We shall also examine several devices used to detect radiation.

14.1 COLLISIONS

Since we shall be dealing in this chapter with the interactions between nuclei and matter, it is useful to first introduce a parameter called the *cross section,* which is a measure of the probability that a collision will occur in a nuclear interaction.

When a beam of particles is incident on a thin foil target, not every particle collides with a target nucleus. The probability that a collision will occur depends on the ratio of the *total effective area* of the target nuclei to the total area of the foil. The situation is analogous to that of throwing darts at many inflated balloons attached to a large wall. If the darts are thrown at random in the direction of the wall, and if the balloons are spread out so as not to touch each other, there is some chance that a balloon will be hit on any given throw. Furthermore, if darts are thrown at a rate R_0, the rate R at which balloons will be hit will be less than R_0. In fact, the probability of hitting a balloon will equal R/R_0. The ratio R/R_0 will depend on the number of balloons N on the wall, the area σ of each balloon, and the area A of the wall. Since the total cross-sectional area of the balloons is $N\sigma$, the probability R/R_0 will equal the ratio $N\sigma/A$.

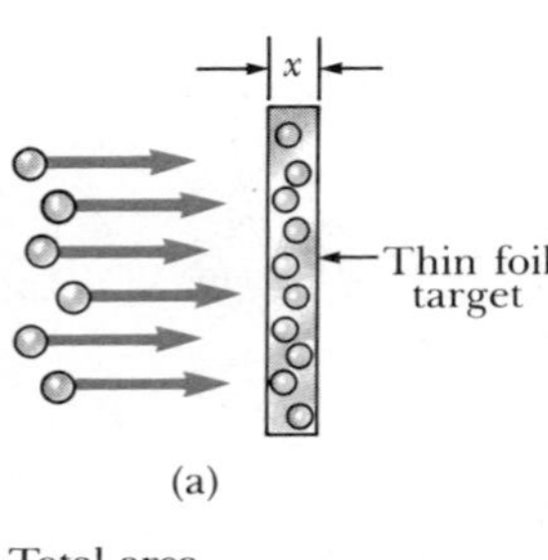

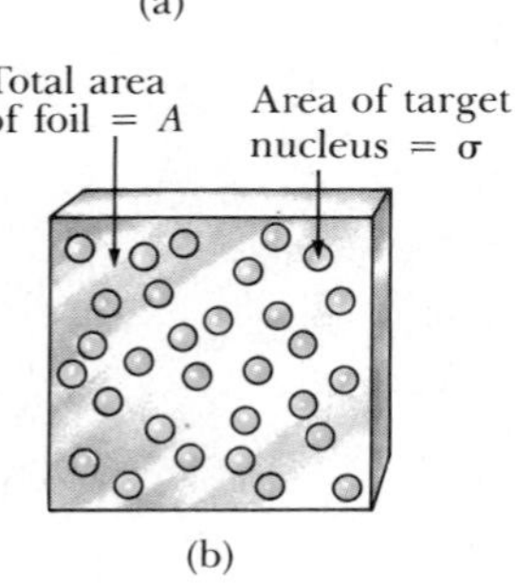

Figure 14.1 (a) A beam of particles incident on a thin foil target of thickness x. The view shown is from the edge of the target. (b) The front view of the target, where the circles represent the target nuclei.

With this analogy, we can now understand the concept of cross section as it pertains to nuclear events. Suppose a beam of particles is incident upon a thin foil target (Fig. 14.1a). Each target nucleus X has an effective area σ, called the **cross section.** You can think of σ as the area of the nucleus that is at right angles to the direction of motion of the bombarding particles. It is assumed that the reaction X(a, b)Y will occur only if the incident particle strikes an area σ. Therefore, the probability that a collision will occur is proportional to σ. That is, the probability increases as the size of the target nuclei increases. In what follows, we shall take the foil thickness to be x and its area to be A. Furthermore, we shall use the following notation:

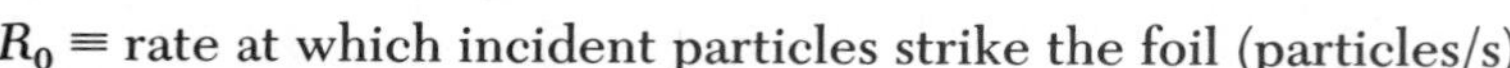

$R_0 \equiv$ rate at which incident particles strike the foil (particles/s)
$R \equiv$ rate at which collision events occur (collisions/s)
$n \equiv$ number of target nuclei per unit volume (particles/m^3), each with cross section σ

Since the total number of target nuclei in the foil is given by $n(Ax)$, the total area exposed to the incident beam must be σnAx. It is assumed that the foil is sufficiently thin so that no nuclei are hidden by others (Fig. 14.1a). The ratio of the rate of collision events to the rate of incident particles, R/R_0, must equal the ratio of the nuclei area σnAx to the total area A of the foil, in analogy to the dart-balloon collision events. That is,

Collision rate is proportional to cross section and target density

$$\frac{R}{R_0} = \frac{\sigma nAx}{A} = \sigma nx \tag{14.1}$$

This result shows that the probability that a collision event will occur is proportional to the nuclei cross section σ, the density of target nuclei n, and the thickness of the target x. A value for σ can therefore be obtained by measuring R, R_0, n, and x and using Equation 14.1.

We can use the same reasoning to arrive at an expression for the number of particles that penetrate (that is, do not interact with) a foil. Suppose that N_0 particles are incident on a foil of thickness dx and that dN is the number of particles that interact with the target nuclei (Fig. 14.2). The ratio of the number of interacting particles to the number of emerging particles, dN/N, equals the ratio of the total target cross section $nA\sigma\, dx$ to the total foil area A. That is,

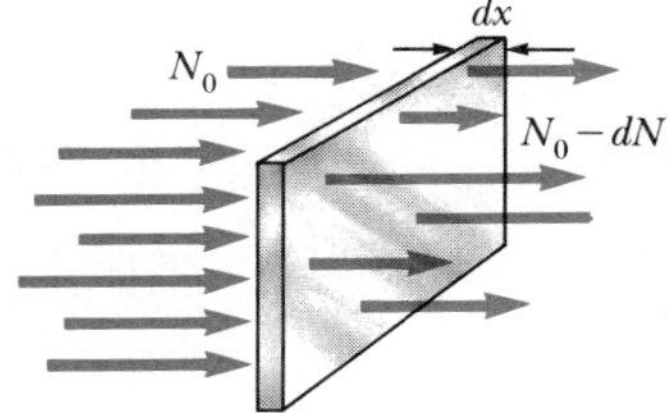

Figure 14.2 If N_0 is the number of particles incident on a target of thickness dx in some time interval, the number emerging from the target is $N_0 - dN$.

$$-\frac{dN}{N} = \frac{nA\sigma\, dx}{A} = n\sigma\, dx$$

The minus sign indicates that particles are being removed from the beam. Integrating this expression and taking $N = N_0$ at $x = 0$, we get

$$\int_{N_0}^{N} \frac{dN}{N} = -n\sigma \int_0^x dx$$

$$\ln\left(\frac{N}{N_0}\right) = -n\sigma x$$

or

$$\boxed{N = N_0 e^{-n\sigma x}} \tag{14.2}$$

Scattering of neutrons

That is, if N_0 is the number of incident particles, the number that emerge from the slab is N. Thus, we see that *the number of particles that penetrate a target decreases exponentially with target thickness.*

Nuclear cross sections, which have dimensions of area, are very small, of the order of the square of the nuclear radius. For this reason, it is common to use the unit **barn** (bn) for measuring nuclear cross sections, where

$$\boxed{1\ \text{bn} \equiv 10^{-28}\ \text{m}^2} \tag{14.3}$$

The barn

In reality, the concept of cross section in nuclear and atomic physics has little to do with the actual geometric area of the target nuclei. The model we have used is simply a convenient one in describing the probability of an event. That is, the probability depends on some effective area of target nuclei. For example, the concept of cross section is also used to describe nuclear reactions, where one introduces a *reaction cross section.* Later, we shall see that, in some events, a nucleus can capture a particle, and one refers to the *capture cross section.*

In general, the total cross section associated with a nuclear or atomic event may be due to a complex combination of interactions. Consider, for example, the process of Rutherford scattering, in which the incident (charged) particles very seldom come close to the target nuclei because a large Coulomb repulsion is present. In this situation, the terminology *scattering cross section* is very common. One observes especially large effective cross sections when using charged incident particles striking targets such as gold. On the other hand, if the incident particles are neutrons, the Coulomb force is not present and the cross sections are much smaller. It is also important to recognize that cross sections depend on the energy of the incident particle.

14.2 INTERACTIONS INVOLVING NEUTRONS

In order to understand the process of nuclear fission and the physics of the nuclear reactor, one must first understand the manner in which neutrons interact with nuclei. As we mentioned, because of their charge neutrality, neutrons are not subject to Coulomb forces. Nevertheless, very slow neutrons can wander through a material and cause nuclear reactions. Since neutrons interact very weakly with electrons, matter appears quite "open" to neutrons. In general, one finds that typical cross sections for neutron-induced reactions increase as the neutron energy decreases. Free neutrons undergo beta decay with a mean lifetime of about 10 min. On the other hand, neutrons in matter are absorbed by nuclei before they decay.

When a fast neutron (energy greater than about 1 MeV) moves through matter, it undergoes many scattering events with the nuclei. In each such event, the neutron gives up some of its kinetic energy to a nucleus. The neutron continues to undergo collisions until its energy is of the order of the thermal energy kT, where k is Boltzmann's constant and T is the absolute temperature. A neutron with this energy is called a **thermal neutron.** At this low energy, there is a high probability that the neutron will be captured by a nucleus, accompanied by the emission of a gamma ray. This **neutron-capture process** can be written

$$\boxed{{}^{1}_{0}\mathrm{n} + {}^{A}_{Z}\mathrm{X} \longrightarrow {}^{A+1}_{Z}\mathrm{X} + \gamma} \tag{14.4}$$

Although we did not indicate it here, the nucleus X is in an excited state X^* for a very short time before it undergoes gamma decay. Also, the product nucleus ${}^{A+1}_{Z}\mathrm{X}$ is usually radioactive and decays by beta emission.

The neutron-capture cross section associated with the above process depends on the nature of the material with which the neutron interacts. For some materials, the cross section is so small that elastic collisions are dominant. Materials for which this occurs are called **moderators** because they slow down (or moderate) the originally energetic neutrons very effectively. Boron, graphite, and water are a few examples of moderator materials.

During an elastic collision between two particles, the maximum kinetic energy is transferred from one particle to the other when they have the same mass. Consequently, a neutron loses all of its kinetic energy when it collides head-on with a proton, in analogy to the collision between a moving and a stationary billiard ball. If the collision is oblique, the neutron loses only part of its kinetic energy. For this reason, materials that are abundant in hydrogen

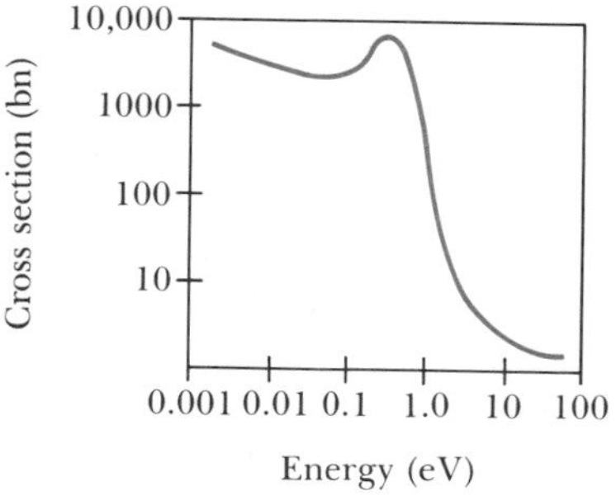

Figure 14.3 The neutron cross section for cadmium. (From H. H. Goldsmith, H. W. Ibser, and B. T. Feld, *Rev. Mod. Phys.* 19:261, 1947.)

atoms with their single-proton nuclei, such as paraffin and water, are good moderators for neutrons.

At some point, many of the neutrons in the moderator become thermal neutrons, which are neutrons in thermal equilibrium with the moderator material. Their average kinetic energy at room temperature is given by

$$K_{av} = \tfrac{3}{2}kT \approx 0.04 \text{ eV}$$

Thermal neutrons have a distribution of velocities, just as the molecules in a container of gas do. A high-energy neutron, whose energy is several MeV, will thermalize (that is, reach K_{av}) in less than 1 ms when incident on a moderator. These thermal neutrons have a very high probability of undergoing neutron capture by the moderator nuclei, since the capture cross section increases with decreasing neutron energy. The variation of the measured cross section versus energy for cadmium is shown in Figure 14.3. The very high cross section below about 0.15 MeV makes cadmium an excellent shielding material against low-energy neutrons.

EXAMPLE 14.1 Neutron Capture by Aluminum

A piece of aluminum foil of thickness 0.3 mm is bombarded by energetic neutrons. The aluminum nuclei undergo neutron capture according to the process $^{27}Al(n, \gamma)^{28}Al$, with a measured capture cross section of 2×10^{-3} bn $= 2 \times 10^{-31}$ m². Assuming the flux of incident neutrons to be 5×10^{12} neutrons/cm²·s, calculate the number of neutrons captured per second by 1 cm² of foil.

Solution: In order to solve this problem, we must first determine the density of nuclei n (which equals the density of atoms). Since the density of aluminum is 2.7 g/cm³ and $A = 27$ for aluminum, we have

$$n = \frac{6.02 \times 10^{23} \text{ nuclei/mol}}{27 \text{ g/mol}} \times 2.7 \text{ g/cm}^3$$

$$= 6.02 \times 10^{22} \frac{\text{nuclei}}{\text{cm}^3} = 6.02 \times 10^{28} \frac{\text{nuclei}}{\text{m}^3}$$

Substituting this value into Equation 14.1, together with the given data for x and σ, gives

$$\frac{R}{R_0} = \sigma n x$$

$$= (2 \times 10^{-31} \text{ m}^2)\left(6.02 \times 10^{28} \frac{\text{nuclei}}{\text{m}^3}\right)(0.3 \times 10^{-3} \text{ m})$$

$$= 3.61 \times 10^{-6}$$

Since the rate of incident particles per unit area is given as $R_0 = 5 \times 10^{12}$ neutrons/cm²·s, we see that

$$R = \left(5 \times 10^{12} \frac{\text{neutrons}}{\text{cm}^2 \cdot \text{s}}\right)(3.61 \times 10^{-6})$$

$$= 1.81 \times 10^{7} \frac{\text{neutrons}}{\text{cm}^2 \cdot \text{s}}$$

Therefore, the number of neutrons captured each second by 1 cm² is only 1.81×10^7, whereas the incident number is 5×10^{12} neutrons. That is, only about 3 neutrons out of 1 million are captured!

14.3 NUCLEAR FISSION

Fission

Nuclear fission occurs when a heavy nucleus, such as ^{235}U, splits, or fissions, into two smaller nuclei. In such a reaction, *the total rest mass of the product is less than the original rest mass.*

Nuclear fission was first observed in 1939 by Otto Hahn and Fritz Strassman, following some basic studies by Fermi. After bombarding uranium ($Z = 92$) with neutrons, Hahn and Strassman discovered among the reaction products two medium-mass elements, barium and lanthanum. Shortly thereafter, Lisa Meitner and Otto Frisch explained what had happened. The uranium nucleus had split into two nearly equal fragments after absorbing a neutron. Such an occurrence was of considerable interest to physicists attempting to understand the nucleus, but it was to have even more far-reaching consequences. Measurements showed that about 200 MeV of energy is released in each fission event, and this fact was to affect the course of human history.

The fission of ^{235}U by slow (low energy) neutrons can be represented by the equation

Fission of ^{235}U

$$\boxed{{}^1_0n + {}^{235}_{92}U \longrightarrow {}^{236}_{92}U^* \longrightarrow X + Y + \text{neutrons}} \tag{14.5}$$

where $^{236}U^*$ is an intermediate state that lasts only for about 10^{-12} s before splitting into X and Y. The resulting nuclei, X and Y, are called **fission fragments.** There are many combinations of X and Y that satisfy the requirements of conservation of mass-energy and charge. In the fission of uranium, there are about 90 different daughter nuclei that can be formed. The process also results in the production of several neutrons, typically two or three. On the average 2.47 neutrons are released per event.

A typical reaction of this type is

$$\boxed{{}^1_0n + {}^{235}_{92}U \longrightarrow {}^{141}_{56}Ba + {}^{92}_{36}Kr + 3\,{}^1_0n} \tag{14.6}$$

The fission fragments, barium and krypton, and the released neutrons have a great deal of kinetic energy following the fission event.

The breakup of the uranium nucleus can be compared to what happens to a drop of water when excess energy is added to it. All the atoms in the drop have energy, but this energy is not great enough to break up the drop. However, if enough energy is added to set the drop into vibration, it will undergo elongation and compression until the amplitude of vibration becomes large enough to cause the drop to break. In the uranium nucleus, a similar process occurs (Fig. 14.4). The sequence of events is as follows:

1. The ^{235}U nucleus captures a thermal (slow-moving) neutron.
2. This capture results in the formation of $^{236}U^*$, and the excess energy of this nucleus causes it to undergo violent oscillations.
3. The $^{236}U^*$ nucleus becomes highly distorted, and the force of repulsion between protons in the two halves of the dumbbell shape tends to increase the distortion.
4. The nucleus splits into two fragments, emitting several neutrons in the process.

Figure 14.5 is a graph of the distribution of fission products versus mass number A. The most probable fission events correspond to nonsymmetric

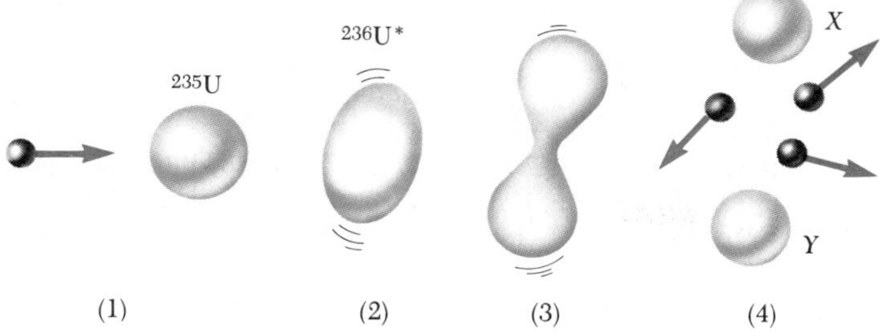

Figure 14.4 The stages in a nuclear fission event as described by the liquid-drop model of the nucleus.

events, that is, events for which the fission fragments have unequal masses. (Symmetric fission events corresponding to $A = 118$ are highly improbable, occurring in only about 0.01% of the cases.) The most probable events correspond to fission fragments with mass numbers of $A \approx 140$ and $A \approx 95$. These fragments, which share the protons and neutrons of the mother nucleus, both fall to the left of the stability line in Figure 13.4. In other words, *fragments that have a large excess of neutrons are unstable.* As a result, the neutron-rich fragments almost instantaneously release two or three neutrons. The remaining fragments are still rich in neutrons and proceed to decay to more stable nuclei through a succession of β^- decays. In the process of such decay, gamma rays are also emitted by nuclei in excited states.

Let us estimate the disintegration energy, Q, released in a typical fission process. From Figure 13.7 we see that the binding energy per nucleon is about 7.6 MeV for heavy nuclei (those having a mass number approximately equal to 240) and about 8.5 MeV for nuclei of intermediate mass. This means that the nucleons in the fission fragments are more tightly bound and therefore have less mass than the nucleons in the original heavy nucleus. This decrease in mass per nucleon appears as released energy when fission occurs. The amount of energy released is (8.5 − 7.6) MeV per nucleon. Assuming a total of 240 nucleons, we find that the energy released per fission event is

$$Q = (240 \text{ nucleons})\left(8.5\,\frac{\text{MeV}}{\text{nucleon}} - 7.6\,\frac{\text{MeV}}{\text{nucleon}}\right) = 220 \text{ MeV}$$

This is indeed a very large amount of energy relative to the amount of energy released in chemical processes. For example, the energy released in the combustion of one molecule of the octane used in gasoline engines is about one millionth the energy released in a single fission event!

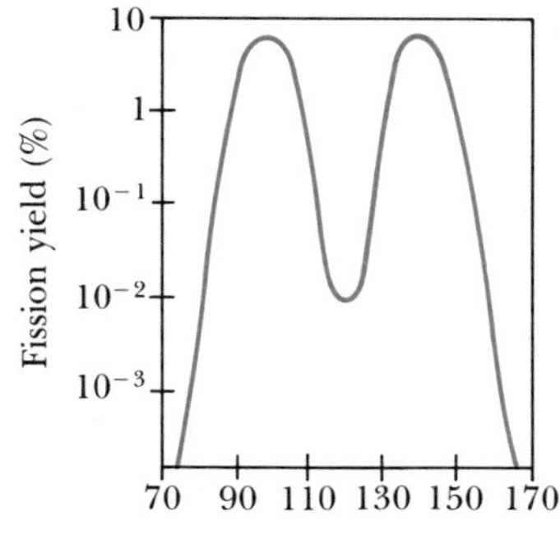

Figure 14.5 The distribution of fission products versus mass number for the fission of ^{235}U with slow neutrons.

EXAMPLE 14.2 The Fission of Uranium

Two other possible ways by which ^{235}U can undergo fission when bombarded with a neutron are (1) by the release of ^{140}Xe and ^{94}Sr as fission fragments and (2) by the release of ^{132}Sn and ^{101}Mo as fission fragments. In each case, neutrons are also released. Find the number of neutrons released in each of these events.

Solution: By balancing mass numbers and atomic numbers, we find that these reactions can be written

$$^{1}_{0}\text{n} + {}^{235}_{92}\text{U} \longrightarrow {}^{140}_{54}\text{Xe} + {}^{94}_{38}\text{Sr} + 2\,{}^{1}_{0}\text{n}$$

$$^{1}_{0}\text{n} + {}^{235}_{92}\text{U} \longrightarrow {}^{132}_{50}\text{Sn} + {}^{101}_{42}\text{Mo} + 3\,{}^{1}_{0}\text{n}$$

Thus, two neutrons are released in the first event and three in the second.

EXAMPLE 14.3 The Energy Released in the Fission of ^{235}U

Calculate the total energy released if 1 kg of ^{235}U undergoes fission, taking the disintegration energy per event to be $Q = 208$ MeV (a more accurate value than the estimate given before).

Solution: We need to know the number of nuclei in 1 kg of uranium. Since $A = 235$, the number of nuclei is

$$N = \left(\frac{6.02 \times 10^{23} \text{ nuclei/mole}}{235 \text{ g/mole}}\right)(10^3 \text{ g})$$

$$= 2.56 \times 10^{24} \text{ nuclei}$$

Hence the total disintegration energy

$$E = NQ = (2.56 \times 10^{24} \text{ nuclei})\left(208 \frac{\text{MeV}}{\text{nucleus}}\right)$$

$$= 5.32 \times 10^{26} \text{ MeV}$$

Since 1 MeV is equivalent to 4.45×10^{-20} kWh, $E = 2.37 \times 10^7$ kWh. This is enough energy to keep a 100-W lightbulb burning for about 30 000 years. Thus, 1 kg of ^{235}U is a relatively large amount of fissionable material.

14.4 NUCLEAR REACTORS

We have seen that, when ^{235}U undergoes fission, an average of about 2.5 neutrons are emitted per event. These neutrons can in turn trigger other nuclei to undergo fission, with the possibility of a chain reaction (Fig. 14.6). Calculations show that if the chain reaction is not controlled (that is, if it does not proceed slowly), it could result in a violent explosion, with the release of an enormous amount of energy, even from only 1 g of ^{235}U. If the energy in 1 g of ^{235}U were released, it would be equivalent to detonating about 20 000 tons of TNT! This, of course, is the principle behind the nuclear bomb, an uncontrolled fission reaction.

Chain reaction

A nuclear reactor is a system designed to maintain what is called a **self-sustained chain reaction.** This important process was first achieved in 1942 by Fermi at the University of Chicago, with natural uranium as the fuel (Fig. 14.7). Most reactors in operation today also use uranium as fuel. Natural uranium contains only about 0.7% of the ^{235}U isotope, with the remaining 99.3% being the ^{238}U isotope. This is important to the operation of a reactor because ^{238}U almost never undergoes fission. Instead, it tends to absorb neutrons, producing neptunium and plutonium. For this reason, reactor fuels must be artificially enriched to contain a few percent of the ^{235}U isotope.

Reproduction constant

Earlier, we mentioned that an average of about 2.5 neutrons are emitted in each fission event of ^{235}U. In order to achieve a self-sustained chain reaction, one of these neutrons, on the average, must be captured by another ^{235}U nucleus and cause it to undergo fission. A useful parameter for describing the level of reactor operation is the **reproduction constant, *K*,** defined as *the average number of neutrons from each fission event that will cause another*

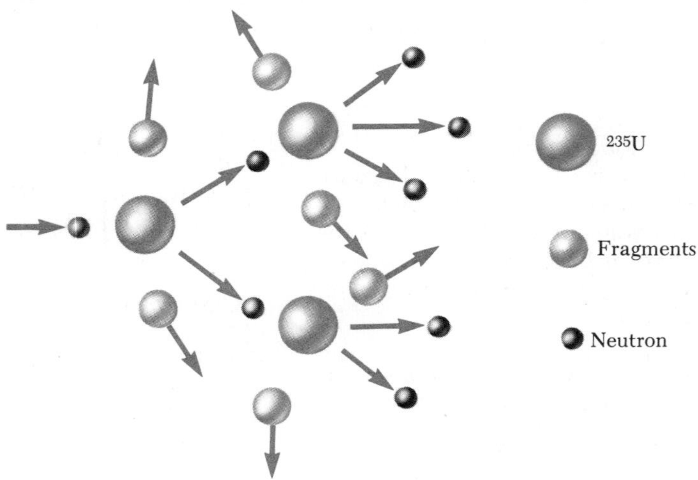

Figure 14.6 A nuclear chain reaction.

Figure 14.7 Sketch of the world's first reactor. Because of wartime secrecy, there are no photographs of the completed reactor. The reactor was composed of layers of graphite interspersed with uranium. A self-sustained chain reaction was first achieved on December 2, 1942. Word of the success was telephoned immediately to Washington with this message: "The Italian navigator has landed in the New World and found the natives very friendly." The historic event took place in an improvised laboratory in the racquet court under the west stands of the University of Chicago's Stagg Field, and the Italian navigator was Fermi.

event. As we have seen, K can have a maximum value of 2.5 in the fission of uranium. However, in practice K is less than this because of several factors, which we shall soon discuss.

A self-sustained chain reaction is achieved when $K = 1$. Under this condition, the reactor is said to be **critical.** When K is less than unity, the reactor is subcritical and the reaction dies out. When K is greater than unity, the reactor is said to be supercritical and a runaway reaction occurs. In a nuclear reactor used to furnish power to a utility company, it is necessary to maintain a value of K close to unity.

The basic design of a nuclear reactor is shown in Figure 14.8. The fuel elements consist of enriched uranium. The function of the remaining parts of the reactor and some aspects of its design will now be described.

Neutron Leakage

In any reactor, a fraction of the neutrons produced in fission will leak out of the core before inducing other fission events. If the fraction leaking out is too large, the reactor will not operate. The percentage lost is large if the reactor is very small because leakage is a function of the ratio of surface area to volume. Therefore, a critical feature of the design of a reactor is to choose the correct surface area-to-volume ratio so that a sustained reaction can be achieved.

Regulating Neutron Energies

Recall that the neutrons released in fission events are very energetic, with kinetic energies of about 2 MeV. It is necessary to slow these neutrons to thermal energies in order to allow them to be captured by other ^{235}U nuclei, since the neutron-capture cross section increases with decreasing energy. As mentioned in Section 14.2, the process of slowing down the energetic neutrons is accomplished by surrounding the fuel with a moderator substance.

In order to understand how neutrons are slowed, consider a collision between a light object and a very massive one. In such an event, the light object rebounds from the collision with most of its original kinetic energy. However, if the collision is between objects whose masses are nearly the same, the incoming projectile will transfer a large percentage of its kinetic energy to the target. In the first nuclear reactor ever constructed, Fermi placed bricks of graphite (carbon) between the fuel elements. Carbon nuclei are about 12 times more massive than neutrons, but after several collisions with carbon nuclei, a neutron is slowed sufficiently to increase its likelihood of fission with ^{235}U. In this design the carbon is the moderator; most modern reactors use water as the moderator.

Neutron Capture

In the process of being slowed down, the neutrons may be captured by nuclei that do not undergo fission. The most common event of this type is neutron capture by ^{238}U. The probability of neutron capture by ^{238}U is very high when the neutrons have high kinetic energies and very low when they have low kinetic energies. Thus the slowing of the neutrons by the moderator serves the dual purpose of making them available for reaction with ^{235}U and decreasing their chances of being captured by ^{238}U.

Control of Power Level

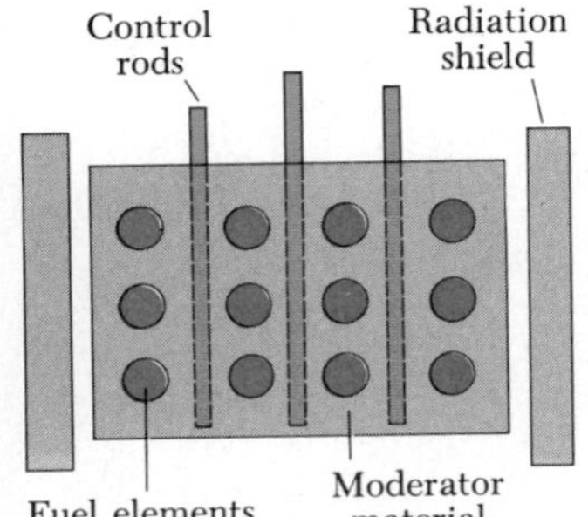

Figure 14.8 Cross-section of a reactor core surrounded by a radiation shield.

It is possible for a reactor to reach the critical stage ($K = 1$) after all the neutron losses described above are minimized. However, a method of control is needed to maintain a K value near unity. If K rises above this value, the heat produced in the runaway reaction would melt the reactor. To control the power level, control rods are inserted into the reactor core (Fig. 14.8). These rods are made of materials, such as cadmium, that are very efficient in absorbing neutrons. By adjusting the number and position of these control rods in the reactor core, the K value can be varied and any power level within the design range of the reactor can be achieved.

A diagram of a pressurized-water reactor is shown in Figure 14.9. This type of reactor is commonly used in electric power plants in the United States. Fission events in the reactor core supply heat to the water contained in the primary (closed) loop, which is maintained at high pressure to keep it from

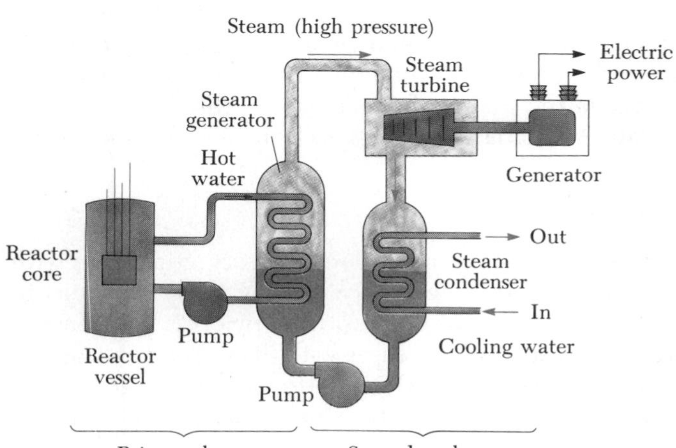

Figure 14.9 Main components of a pressurized-water reactor.

boiling. This water also serves as the moderator. The hot water is pumped through a heat exchanger, and the heat is transferred to the water contained in the secondary loop. The hot water in the secondary loop is converted to steam, which drives a turbine-generator system to create electric power. Note that the water in the secondary loop is isolated from the water in the primary loop in order to avoid contamination of the secondary water and steam by radioactive nuclei from the reactor core.

Safety and Waste Disposal

As more reactors are built around the world, there is justifiable concern about their safety and about their effect on the environment. This is particularly true in light of the 1979 near-disaster at Three Mile Island in Pennsylvania and the more recent tragedy at Chernobyl in the U.S.S.R. The problems of reactor safety are so vast and complex that we can only touch on a few of them here.

One of the inherent dangers in a nuclear reactor is the possibility that the water flow could be interrupted. If this should occur, it is conceivable that the temperature of the reactor could increase to the point where the fuel elements would melt, which would melt the bottom of the reactor and the ground below. This possibility is referred to, appropriately enough, as the "China syndrome." Additionally, the large amounts of heat generated could lead to a high-pressure steam (non-nuclear) explosion that would spread radioactive material throughout the area surrounding the power plant. To decrease the chances of such an event as much as possible, all reactors are built with a backup cooling system that takes over if the regular cooling system fails.

Another problem of concern in nuclear fission reactors is the disposal of radioactive material when the reactor core is replaced. This waste material contains long-lived, highly radioactive isotopes and must be stored over long periods of time in such a way that there is no chance of environmental contamination. At present, sealing radioactive wastes in deep salt mines seems to be the most promising solution.

Other major concerns associated with the proliferation of nuclear power plants are the danger of sabotage at reactor sites and the danger that nuclear fuel (or waste) might be stolen during transport. In some instances, these stolen materials could be used to make an atomic bomb. The difficulties in handling and transporting such highly radioactive materials reduce the possibility of such terrorist activities.

Another consequence of nuclear power plants is thermal pollution. Water from a nearby river is often used to cool the reactor. This raises the temperature of the river downstream from the reactor, which affects living organisms in and adjacent to the river. Another technique to cool the reactor water is to use evaporation towers. In this case, thermal pollution of the atmosphere is a consequence. Clearly, one must be concerned with the detrimental effects of thermal pollution when deciding on the location of nuclear power plants.

We have listed only a few of the problems associated with nuclear power reactors. Among these, the handling and disposal of radioactive wastes appears to be the chief problem. One must, of course, weigh such risks against the problems and risks associated with alternative energy sources.

14.5 NUCLEAR FUSION

Figure 13.7 shows that the binding energy for light nuclei (those having a mass number of less than 20) is much smaller than the binding energy for heavier nuclei. This suggests a possible process that is the reverse of fission. *When two light nuclei combine to form a heavier nucleus, the process is called* **nuclear fusion.** Because the mass of the final nucleus is less than the combined rest masses of the original nuclei, there is a loss of mass accompanied by a release of energy. The following are examples of such energy-liberating fusion reactions:

$$^1_1\text{H} + {}^1_1\text{H} \longrightarrow {}^2_1\text{H} + {}^0_1\text{e} + \nu$$

$$^1_1\text{H} + {}^2_1\text{H} \longrightarrow {}^3_2\text{H} + \gamma \qquad (14.7)$$

This second reaction is followed by either

$$^1_1\text{H} + {}^3_2\text{He} \longrightarrow {}^4_2\text{He} + {}^0_1\text{e} + \nu$$

or

$$^3_2\text{He} + {}^3_2\text{He} \longrightarrow {}^4_2\text{He} + {}^1_1\text{H} + {}^1_1\text{H}$$

These are the basic reactions in what is called the **proton-proton cycle,** believed to be one of the basic cycles by which energy is generated in the sun and other stars with an abundance of hydrogen. Most of the energy production takes place at the interior of the sun, where the temperature is approximately 1.5×10^7 K. As we shall see later, such high temperatures are required in order to drive these reactions, and they are therefore called **thermonuclear fusion reactions.** The hydrogen (fusion) bomb, first exploded in 1952, is an example of an uncontrolled thermonuclear fusion reaction.

Thermonuclear reactions

All of the reactions in Equation 14.7 are exothermic, that is, there is a release of energy. An overall view of the proton-proton cycle is that four protons combine to form an alpha particle, two positrons, and two neutrinos with the release of 25 MeV of energy in the process.

Fusion Reactors

The enormous amount of energy released in fusion reactions suggests the possibility of harnessing this energy for useful purposes here on earth. A great deal of effort is currently under way to develop a sustained and controllable thermonuclear reactor—a fusion power reactor. Controlled fusion is often called the ultimate energy source because of the availability of its source of fuel: water. For example, if deuterium were used as the fuel, 0.12 g of it could be extracted from 1 gal of water at a cost of about four cents. Such rates would make the fuel costs of even an inefficient reactor almost insignificant. An additional advantage of fusion reactors is that comparatively few radioactive by-products are formed. As noted in Equation 14.7, the end product of the fusion of hydrogen nuclei is safe, nonradioactive helium. Unfortunately, a thermonuclear reactor that can deliver a net power output over a reasonable time interval is not yet a reality, and many difficulties must be resolved before a successful device is constructed.

We have seen that the sun's energy is based, in part, upon a set of reactions in which ordinary hydrogen is converted to helium. Unfortunately, the proton-proton interaction is not suitable for use in a fusion reactor because the event requires very high pressures and densities. The process works in the sun only because of the extremely high density of protons in the sun's interior.

The fusion reactions that appear most promising in the construction of a fusion power reactor involve deuterium and tritium, which are isotopes of hydrogen. These reactions are

$$\begin{aligned} {}^2_1\text{H} + {}^2_1\text{H} &\longrightarrow {}^3_2\text{He} + {}^1_0\text{n} \qquad Q = 3.27\ \text{MeV} \\ {}^2_1\text{H} + {}^2_1\text{H} &\longrightarrow {}^3_1\text{H} + {}^1_1\text{H} \qquad Q = 4.03\ \text{MeV} \\ {}^2_1\text{H} + {}^3_1\text{H} &\longrightarrow {}^4_2\text{He} + {}^1_0\text{n} \qquad Q = 17.59\ \text{MeV} \end{aligned} \qquad (14.8)$$

where the Q values refer to the amount of energy released per reaction. As noted earlier, deuterium is available in almost unlimited quantities from our lakes and oceans and is very inexpensive to extract. Tritium, however, is radioactive ($T_{1/2} = 12.3$ years) and undergoes beta decay to ^{3}He. For this reason, tritium does not occur naturally to any great extent and must be artificially produced.

One possible scheme for extracting energy from a fusion reactor is to surround the reactor core with a lithium blanket. Energetic particles, such as neutrons, would be absorbed by the molten lithium, and the thermal energy would then be transferred to a heat exchanger to generate steam. One advantage of this scheme is the formation of tritium from neutron-induced reactions in the lithium blanket; this tritium can be recirculated into the reactor for fuel. The two reactions involved in the breeding of tritium are as follows:

$$\begin{array}{l} \text{n (fast)} + {}^7\text{Li} \longrightarrow {}^3\text{H} + {}^4\text{He} + \text{n (slow)} \\ \text{n (slow)} + {}^6\text{Li} \longrightarrow {}^3\text{H} + {}^4\text{He} + 4.8 \text{ MeV} \end{array} \qquad (14.9)$$

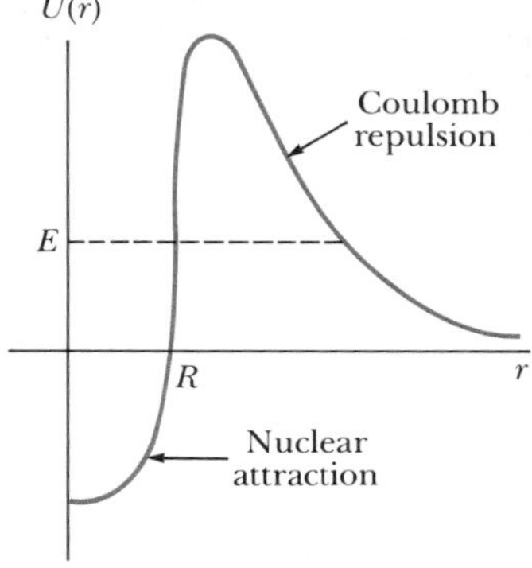

Figure 14.10 The potential energy as a function of separation between two deuterons. The Coulomb repulsive force is dominant at long range, whereas the nuclear attractive force is dominant at short range, where R is of the order of 1 fm.

One of the major problems in obtaining energy from nuclear fusion is the fact that the Coulomb repulsion force between two charged nuclei must be overcome before they can fuse. The potential energy as a function of particle separation for two deuterons (each with charge $+e$) is shown in Figure 14.10. The potential energy is positive in the region $r > R$, where the Coulomb repulsive force dominates, and negative in the region $r < R$, where the strong nuclear force dominates. The fundamental problem then is to give the two nuclei enough kinetic energy to overcome the repulsive force. This can be accomplished by heating the fuel to extremely high temperatures (to about 10^8 K, far greater than the interior temperature of the sun). As you might expect, such high temperatures are not easy to obtain in the laboratory or a power plant. At these high temperatures, the atoms are ionized, and the system consists of a collection of electrons and nuclei, commonly referred to as a **plasma.** The following example illustrates how to estimate the temperature required to achieve the fusion of two deuterons.

High temperatures are required to overcome the large Coulomb barrier

EXAMPLE 14.4 The Fusion of Two Deuterons

The separation between two deuterons must be as little as about 10^{-14} m in order for the attractive nuclear force to overcome the repulsive Coulomb force. (a) Calculate the height of the potential barrier due to the repulsive force.

Solution: The potential energy associated with two charges separated by a distance r is given by

$$U = k\frac{q_1 q_2}{r}$$

where k is the Coulomb constant. For the case of two deuterons, $q_1 = q_2 = +e$, so that

$$U = k\frac{e^2}{r} = \left(9 \times 10^9 \frac{\text{N}\cdot\text{m}^2}{\text{C}^2}\right)\frac{(1.6 \times 10^{-19}\text{ C})^2}{10^{-14}\text{ m}}$$

$$= 2.3 \times 10^{-14}\text{ J} = 0.14\text{ MeV}$$

(b) Estimate the effective temperature required in order for a deuteron to overcome the potential barrier, assuming an energy of $\frac{3}{2}kT$ per deuteron (where in this case k is Boltzmann's constant).

Solution: Since the total Coulomb energy of the pair of deuterons is 0.14 MeV, the Coulomb energy per deuteron is 0.07 MeV $= 1.1 \times 10^{-14}$ J. Setting this equal to the average thermal energy per deuteron gives

$$\tfrac{3}{2}kT = 1.1 \times 10^{-14}\text{ J}$$

where k is equal to 1.38×10^{-23} J/K. Solving for T gives

$$T = \frac{2 \times (1.1 \times 10^{-14}\text{ J})}{3 \times (1.38 \times 10^{-23}\text{ J/K})} = 5.3 \times 10^8\text{ K}$$

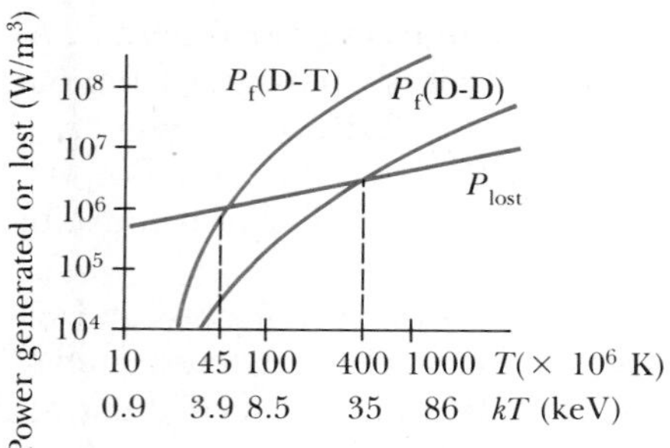

Figure 14.11 The power generated or lost versus temperature for the D-D and D-T fusion reactions. When the generation rate P_f exceeds the loss rate, ignition takes place.

Example 14.4 shows that the deuterons must be heated to about 5×10^8 K in order to achieve fusion. This estimate of the required temperature is too high, however, because the particles in the plasma have a Maxwellian velocity distribution and therefore some fusion reactions will be caused by particles in the high-energy "tail" of this distribution. Furthermore, even those particles that do not have enough energy to overcome the barrier have some probability of penetrating the barrier through tunneling. When these effects are taken into account, a temperature of about 4×10^8 K appears adequate to fuse the two deuterons. As you might imagine, such high temperatures are not easy to attain in the laboratory.

Critical ignition temperature

The temperature at which the power generation rate exceeds the loss rate (due to mechanisms such as radiation losses) is called the **critical ignition temperature.** This temperature for the D-D (deuterium-deuterium) reaction is 4×10^8 K. According to $E = kT$, this temperature is equivalent to about 35 keV. It turns out that the critical ignition temperature for the D-T (deuterium-tritium) reaction is about 4.5×10^7 K, or only 4 keV. A plot of the power generated by fusion versus temperature for the two reactions is shown in Figure 14.11. The straight line in this plot represents the power loss due to the radiation mechanism known as **bremsstrahlung.** This is the principal mechanism of energy loss, in which radiation (primarily x-rays) is emitted as the result of electron-ion collisions within the plasma.[1]

Confinement time

In addition to the high temperature requirements, there are two other critical parameters that determine whether or not a thermonuclear reactor will be successful. These parameters are the **ion density,** n, and the **confinement time,** τ. *The confinement time is the time the interacting ions are maintained at a temperature equal to or greater than the ignition temperature.* Lawson has shown that the density and confinement time must both be large enough to ensure that more fusion energy will be released than is required to heat the plasma. In particular, **Lawson's criterion** states that a net energy output is possible under the following conditions:

Lawson's criterion

$$\begin{aligned} n\tau &\geq 10^{14}\ \text{s/cm}^3 \qquad \text{(D-T)} \\ n\tau &\geq 10^{16}\ \text{s/cm}^3 \qquad \text{(D-D)} \end{aligned} \tag{14.10}$$

A graph of $n\tau$ versus kinetic temperature for the D-T and D-D reactions is given in Figure 14.12.

Lawson's criterion was arrived at by comparing the energy required to heat the plasma with the power generated by the fusion process. The energy

[1] Cyclotron radiation is another loss mechanism; it is especially important in the case of the D-D reaction.

E_h required to heat the plasma is proportional to the ion density n, whereas the energy *generated by the fusion process* is proportional to the product $n^2\tau$. Net energy is produced when the energy generated by the fusion process, E_f, exceeds E_h. This condition leads to Lawson's criterion.[2] The constants involved can be evaluated under suitable conditions. When $E_f = E_h$, a *breakeven* point is reached. As of 1988, no laboratory has reported achieving the breakeven condition.

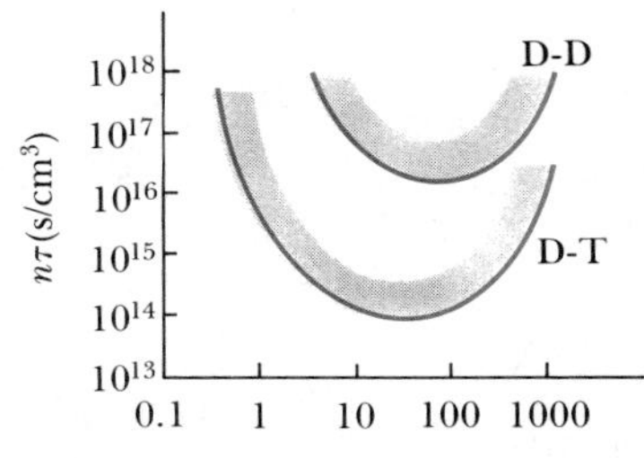

Figure 14.12 The Lawson number $n\tau$ versus kinetic temperature for the D-T and D-D fusion reactions.

In summary, the three basic requirements of a successful thermonuclear power reactor are as follows:

Requirements for a fusion power reactor

1. The plasma temperature must be very high—about 4.5×10^7 K for the D-T reaction and 4×10^8 K for the D-D reaction.
2. The ion density n must be high. It is necessary to have a high density of interacting nuclei to increase the collision rate between particles.
3. The confinement time τ of the plasma must be long. In order to meet Lawson's criterion, the product $n\tau$ must be large. For a given value of n, the probability of fusion between two particles increases as the time of confinement increases.

Current efforts are aimed at meeting Lawson's criterion at temperatures exceeding the critical ignition temperature in one device. Although the minimum plasma densities have been achieved, the problem of confinement time has yet to be solved. How can one confine a plasma at a temperature of 10^8 K for times of the order of 1 s? The two basic techniques under investigation to confine plasmas are magnetic field confinement and inertial confinement. Let us briefly describe these two techniques.

Magnetic Field Confinement

Most fusion-related plasma experiments are using **magnetic field confinement** to contain the plasma. A device called a **tokamak,** first developed in the USSR, is shown in Figure 14.13. The tokamak has a doughnut-shaped geometry (a toroid). A combination of two magnetic fields is used to confine and stabilize the plasma. These are (1) a strong toroidal field B_t, produced by the current in the windings, and (2) a weaker "poloidal" field, produced by the toroidal current I_t. In addition to confining the plasma, the toroidal current is also used to heat it. The resultant confining field is helical, as shown in Figure 14.13. These helical field lines spiral around the plasma and keep it from touching the walls of the vacuum chamber. If the plasma comes into contact with the walls, its temperature is reduced and heavy impurities sputtered from the walls would "poison" it, leading to large power losses. One of the major breakthroughs in the last decade has been in the area of auxiliary heating to reach ignition temperatures. Recent experiments have shown that injecting a beam of energetic neutral particles into the plasma is a very efficient method of heating the plasma to ignition temperatures (5 to 10 keV). RF (radio frequency) heating will probably be needed for reactor-size plasmas.

Table 14.1 lists some characteristics of five major tokamaks operating in the United States, including a major facility at the Princeton Plasma Physics Laboratory. The data in this table represent the best results achieved for each device. The value of $n\tau$ reported by the MIT researchers is about a factor of 3

[2] Note that the Lawson's criterion neglects the magnetic field energy, which is expected to be about 20 times greater than the plasma heat energy. For this reason, it is necessary to have a magnetic energy recovery system or to make use of superconducting magnets.

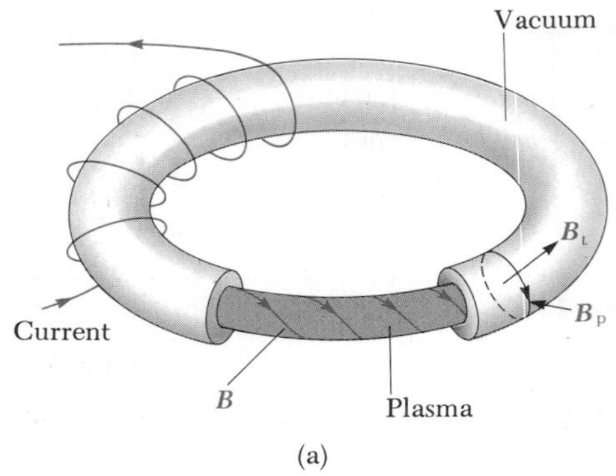

(a)

(b)

Figure 14.13 (a) Schematic diagram of a tokamak used in the magnetic confinement scheme. Note that the total magnetic field $\boldsymbol{B}$ is the superposition of the toroidal field $\boldsymbol{B}_t$ and the poloidal field $\boldsymbol{B}_p$. The plasma is trapped within the spiraling field lines as shown. (b) Photograph of the Princeton TFTR tokamak device, which uses magnetic field confinement. (Courtesy of the Plasma Physics Laboratory at Princeton)

lower than what is required by Lawson's criterion. The tokamak fusion test reactor (TFTR) at Princeton is the largest project in the U.S. fusion program. This device has been designed to use the D-T reaction. Recently the TFTR succeeded in producing plasma temperatures of about 2×10^8 K, more than ten times the temperature at the center of the sun! The corresponding Lawson parameter that was achieved at this temperature was $n\tau \approx 10^{13}$ s/cm^3. Other lower temperature experiments carried out in the TFTR facility reached $n\tau$ values of up to 1.5×10^{14} s/cm^3, close to the goal of 3×10^{14} s/cm^3 (see Table 14.1). In the very near future, researchers hope to demonstrate the breakeven condition. Once this has been achieved, fusion programs in the United States, Europe, Japan, and the Soviet Union (with planned international collaboration) will design and construct large-scale tokamak reactors.

A magnetic confinement technique that has received much attention is the so-called magnetic mirror confinement scheme. The idea of this technique is to trap the plasma in a cylindrical tube by adding additional magnetic field coils at the ends of the tube (Fig. 14.14). The increased fields at the ends of the cylinder serve as magnetic "mirrors" (or a magnetic bottle) for the charged particles, which spiral around the field lines, thereby reducing the leakage problem and increasing the plasma density.

TABLE 14.1 Parameters for Tokamaks in the United States

	n (cm^{-3})	T (keV)	τ (ms)	$n\tau$ (s/cm^3)
TFTR (Princeton)[a]	$\leq 6.5 \times 10^{14}$	10	300–500	$\leq 1.5 \times 10^{14}$
PLT (Princeton)	1.5×10^{14}	7.3	100	1.5×10^{13}
PDX (Princeton)	7×10^{13}	6	90	$\approx 4 \times 10^{12}$
ISX-B (Oak Ridge)	1.3×10^{14}	1.8	40	2×10^{12}
Doublet III (General Atomic)	1.2×10^{14}	≈ 0.7	≈ 70	1×10^{13}
Alcator C (MIT)	7×10^{14}	1.5	35	$> 3 \times 10^{13}$

[a] This is as of August, 1986.

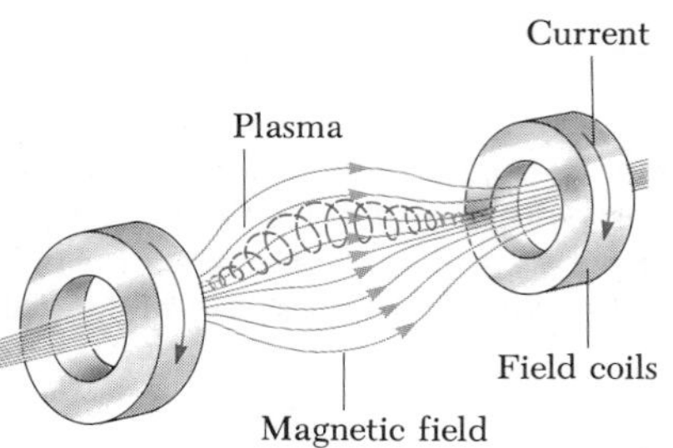

Figure 14.14 Magnetic mirror confinement of a plasma.

Inertial Confinement

The second technique for confining a plasma is called **inertial confinement.** This technique makes use of a target with a very high particle density. In this scheme, the confinement time is very short (typically 10^{-11} to 10^{-9} s), so that, because of their own inertia, the particles do not have a chance to move appreciably from their initial positions.

The laser fusion technique is the most common form of inertial confinement. A small D-T pellet, about 1 mm in diameter, is struck simultaneously by several focused, high-intensity laser beams, resulting in a large pulse of energy (Fig. 14.15). The energy from the laser pulse causes the surface of the fuel pellet to evaporate. The escaping particles produce a reaction force on the core of the pellet, resulting in a strong, inwardly moving, compressive shock wave. This shock wave increases the pressure and density of the core and produces a corresponding increase in temperature. When the temperature of the core reaches ignition temperature, fusion reactions cause the pellet to explode. The process can be viewed as a miniature hydrogen bomb. The SHIVA laser fusion project at the Lawrence Livermore Laboratory uses 20 synchronized laser pulses to deliver a total of 200 kJ of energy to a pellet in less than 10^{-9} s. This corresponds to a power of 2×10^{14} W!

Advantages of fusion

If fusion power can be harnessed, it will offer several advantages: (1) the low cost and abundance of the fuel (deuterium), (2) the absence of weapons-grade material, (3) the impossibility of runaway accidents, and (4) a lesser radiation hazard than with fission. Some of the anticipated problem areas and disadvantages include (1) its as yet unestablished feasibility, (2) the very high proposed plant costs, (3) the possibility of the scarcity of lithium, (4) the limited supply of the helium and superconducting materials that are needed in the tokamak-type reactors, (5) structural damage and induced radioactivity due to neutron bombardment, and (6) the anticipated high degree of thermal pollution. If these basic problems and the engineering design factors can be resolved, nuclear fusion may become a feasible source of energy by the end of this century.

Problem areas and disadvantages of fusion

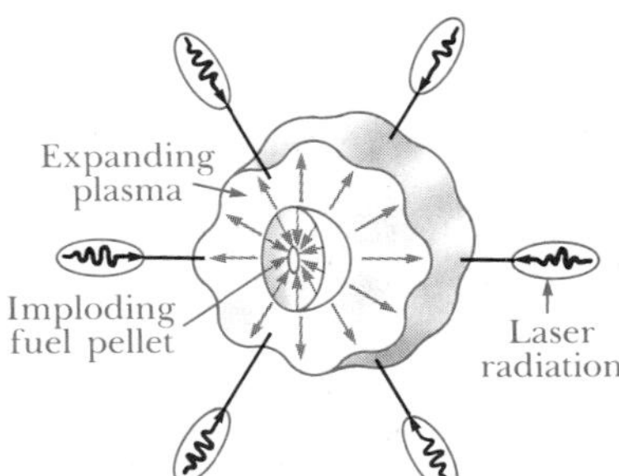

Figure 14.15 In the inertial confinement scheme, a D-T fuel pellet undergoes fusion when struck by several high-intensity laser beams simultaneously.

14.6 THE INTERACTION OF PARTICLES WITH MATTER (Optional)

In this section we shall consider processes involving the interaction between energetic particles and matter. We shall deal mainly with the interaction of charged particles and photons with matter, since these are of primary importance in such factors as shielding characteristics, the design of particle detectors, and biological effects.

Heavy Charged Particles

Range of a particle

A heavy charged particle, such as an alpha particle or proton, moving through a solid, liquid, or gas travels a well-defined distance before coming to rest. This distance is called the **range** of the particle. As the particle passes through the medium, it loses energy primarily through the excitation and ionization of atoms in the medium. Some energy is lost through elastic collisions with nuclei. The highly energetic particle loses its energy in many small increments, and to a good approximation one can treat the problem as a continuous loss of energy. At the end of its range, the particle is left with just thermal energy. The range depends on the charge, mass, and energy of the particle; the density of the medium through which it travels; and the ionization potential and atomic number of the atoms in the medium.

The ranges of alpha particles and protons in air as a function of their energy are plotted in Figure 14.16. Note that for a given energy, the proton has a range about ten times that of the alpha particle. This is because the proton has less charge and does not interact with the medium as strongly as the alpha particle. Furthermore, since the alpha particle is more massive, it travels at a lower average speed. As a result, the slower-moving alpha particle loses its energy more readily through resonance effects, since it has more time to interact with atoms in the medium.

Energy loss rate

The rate of energy loss per unit length $-dE/dx$ (or energy loss rate) versus the energy of the charged particle is shown in Figure 14.17. This **energy loss rate** (also called the *stopping power*) is approximately proportional to the inverse of the energy at low energies ($v \ll c$) and reaches a maximum at some point. At very high energies (as $v \rightarrow c$), the energy loss per unit length is approximately energy-independent. At very low energies, the decrease in energy loss is a result of the fact that the particle is moving too slowly to produce ionization and loses its energy mainly by elastic collisions.

The energy loss rate is approximately proportional to the density of the medium through which the charged particles travel. This is explained by recognizing that the primary energy loss mechanism (ionization) involves the excitation of electrons in the medium, and the density of electrons increases with increasing density of the medium. For example, the range in aluminum of protons having an energy of from 1 to 10 MeV is about 1/1600 of their range in air.

Fission fragments with energies of about 80 MeV have a range of only about 0.02 m in air, which is about 1000 times smaller than the range of protons with the same energy. This difference is due to the larger charge of the fission fragments and to the fact that the energy loss rate is proportional to the square of the charge.

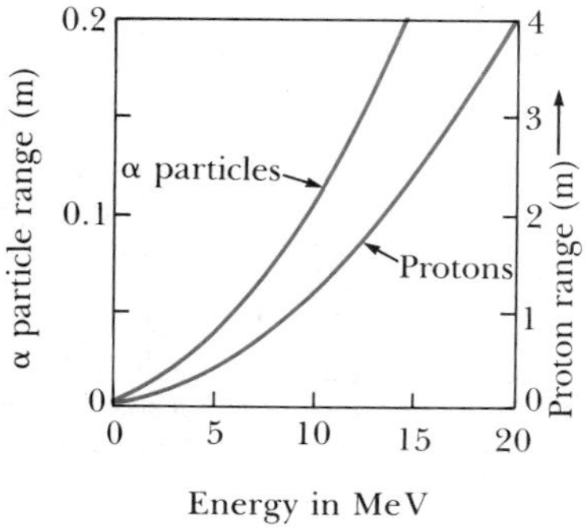

Figure 14.16 The range of alpha particles and protons in air under standard conditions. [From E. Segre (ed.), *Experimental Nuclear Physics*, New York, Wiley, 1953, Vol. 1.]

Electrons

When an electron of energy less than about 1 MeV passes through matter, it loses its energy by the same processes that occur for heavy charged particles. However, the range of the electron is not as well defined. Since an electron is much smaller than a proton, there are large statistical variations in the electron's path length as a result of a phenomenon called **straggling.** It takes only a few large-deflection collisions to stop an electron in matter. Furthermore, electrons scatter much more readily than heavy charged particles with the same energy. As we mentioned earlier, an accelerated charged particle emits electromagnetic radiation called *bremsstrahlung.* This energy loss process is more important for electrons than for heavy charged particles, since electrons undergo greater accelerations when passing through matter. For example, the energy loss rate due to bremsstrahlung for 10-MeV electrons passing through lead is about equal to the loss rate due to ionization.

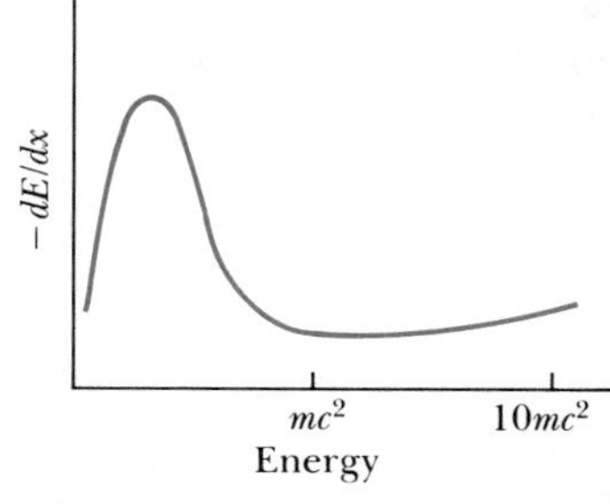

Figure 14.17 The energy loss rate $-dE/dx$ versus the energy of a charged particle moving through a medium.

Photons

Since photons are uncharged, they are not as effective as charged particles in producing ionization or excitation in matter. Nevertheless, photons can be removed from a beam by either scattering or absorption in the medium. Figure 14.18 illustrates the three important processes that contribute to the total absorption of gamma rays in lead:

Gamma-ray absorption

1. At low photon energies (less than about 0.5 MeV), the predominant process that removes photons from a beam is the photoelectric effect.
2. At intermediate photon energies, the predominant process is Compton scattering.
3. At high energies, a process called **pair production** is predominant, in which an electron-positron pair is created as a photon passes near a nucleus in the medium. Since the rest energy of an electron-positron pair is $2mc^2 = 1.02$ MeV (twice the rest energy of an electron), the gamma-ray photon must have at least this much energy to produce a pair. Pair production will be discussed further in Chapter 15.

If a beam of photons is incident on a medium, its intensity decreases exponentially with increasing depth of penetration into the medium. This reduction in intensity is referred to as **attenuation** of the beam. As in the case of neutrons, the intensity in a medium varies according to the relation

Linear absorption coefficient

$$I(x) = I_0 e^{-\mu x} \tag{14.11}$$

Absorption of photons in a medium

where I_0 is the incident photon intensity (measured in photons/m$^2\cdot$s), x is the distance the beam travels in the medium, $I(x)$ is the beam intensity after traveling a distance x, and μ is a parameter called the **linear absorption coefficient** of the medium. This coefficient depends on the energy of the photon as well as on the

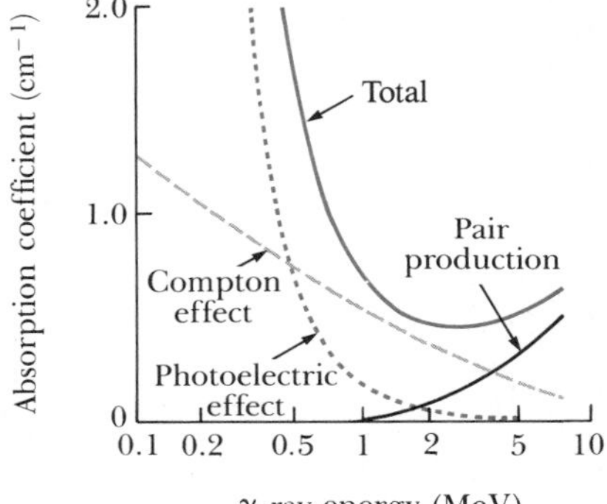

Figure 14.18 The absorption of gamma rays in lead. The three important absorption processes are shown, which add up to give the total absorption constant μ measured in cm^{-1}. (From W. Heitler, *The Quantum Theory of Radiation,* 3rd ed., Oxford, Clarendon Press, 1954.)

TABLE 14.2 Linear Absorption Coefficients for X-rays in Various Media

	Linear Absorption Coefficient (cm^{-1})				
λ (pm)	Air	Water	Aluminum	Copper	Lead
10		0.16	0.43	3.2	43
20		0.18	0.76	13	55
30		0.29	1.3	38	158
40		0.44	3.0	87	350
50	8.6×10^{-4}	0.66	5.4	170	610
60	1.3×10^{-3}	1.0	9.2	286	1000
70	1.95×10^{-3}	1.5	14	430	1600
80	2.73×10^{-3}	2.1	20	625	
90	3.64×10^{-3}	2.8	30	875	
100	4.94×10^{-3}	3.8	41	1200	
150	1.56×10^{-2}	12	124		
200	3.64×10^{-2}	28	275		
250	6.63×10^{-2}	51	524		

properties of the medium. Such variations are shown in Figure 14.18 for gamma rays in lead and in Tables 14.2 and 14.3 for x-rays and gamma rays in various media.

EXAMPLE 14.5 Half-Value Thickness

The half-value thickness of an absorber is defined as that thickness which will reduce the intensity of a beam of particles by a factor of 2. Calculate the half-value thickness for lead, assuming an x-ray beam of wavelength 20 pm (1 pm = 10^{-12} m = 10^{-3} nm).

Solution: The intensity varies with distance traveled in the medium according to Equation 14.11:

$$I(x) = I_0 e^{-\mu x}$$

TABLE 14.3 Linear Absorption Coefficients for Gamma Rays in Various Media

Photon Energy	Linear Absorption Coefficient[a] (cm^{-1})			
(MeV)	Water	Aluminum	Iron	Lead
0.1	0.167	0.432	2.69	59.8
0.15	0.149	0.359	1.43	20.8
0.2	0.139	0.324	1.08	10.1
0.3	0.118	0.278	0.833	3.79
0.4	0.106	0.249	0.722	2.35
0.5	0.0967	0.227	0.651	1.64
0.6	0.0894	0.210	0.598	1.29
0.8	0.0786	0.184	0.525	0.946
1.0	0.0706	0.166	0.468	0.772
1.5	0.0576	0.135	0.380	0.581
2.0	0.0493	0.166	0.332	0.510
3.0	0.0396	0.0953	0.282	0.463
4.0	0.0339	0.0837	0.259	0.470
5.0	0.0302	0.0767	0.247	0.486
6.0	0.0277	0.0718	0.240	0.514
8.0	0.0242	0.0656	0.234	0.532
10.0	0.0221	0.0626	0.236	0.568

[a] Calculated using data in I. Kaplan, *Nuclear Physics*, Reading, MA, Addison-Wesley, 1962, Table 15.6.

In this case, we are looking for a value of x such that $I(x) = I_0/2$. That is, we require that

$$\frac{I_0}{2} = I_0 e^{-\mu x}$$

or

$$e^{-\mu x} = \tfrac{1}{2}$$

Taking the natural logarithm of both sides of this equation gives

$$\mu x = \ln 2$$

or

$$x = \frac{\ln 2}{\mu}$$

Since $\mu = 55\ \text{cm}^{-1}$ for x-rays in lead at a wavelength of 20 pm (Table 14.2),

$$x = \frac{\ln 2}{55\ \text{cm}^{-1}} = \frac{0.693}{55\ \text{cm}^{-1}}$$

$$= 1.26 \times 10^{-2}\ \text{cm} = 0.126\ \text{mm}$$

Hence we conclude that lead is a very good absorber for x-rays.

14.7 RADIATION DAMAGE IN MATTER (Optional)

When radiation passes through matter, it can cause severe damage. The degree and type of damage depend upon several factors, including the type and energy of the radiation and the properties of the medium. For example, metals used in nuclear reactor structures can be severely weakened by high fluxes of energetic neutrons, which often leads to metal fatigue. The damage in such situations is in the form of atomic displacements, often resulting in major alterations in the material properties. Materials can also be damaged by ionizing radiations, such as gamma rays and x-rays. For example, defects called *color centers* can be produced in inorganic crystals such as NaCl by irradiating the crystals with x-rays. One extensively studied color center has been identified as an electron trapped in a Cl^- ion vacancy.

Radiation damage in biological organisms is primarily due to ionization effects in cells. The normal operation of a cell may be disrupted when highly reactive ions or radicals are formed as the result of ionizing radiation. For example, hydrogen and hydroxyl radicals produced from water molecules can induce chemical reactions that may break bonds in proteins and other vital molecules. Furthermore, the ionizing radiation may directly affect vital molecules by removing electrons from their structure. Large doses of radiation are especially dangerous, since damage to a great number of molecules in a cell may cause the cell to die. Although the death of a single cell is usually not a problem, the death of many cells may result in irreversible damage to the organism. Also, cells that do survive the radiation may become defective. These defective cells, upon dividing, can produce more defective cells and lead to cancer.

In biological systems, it is common to separate radiation damage into two categories, *somatic damage* and *genetic damage*. Somatic damage is radiation damage associated with all the body cells except the reproductive cells. Such damage can lead to cancer at high dose rates or seriously alter the characteristics of specific organisms. Genetic damage affects only the reproductive cells of the person exposed to the radiation. Damage to the genes in reproductive cells can lead to defective offspring. Clearly, one must be concerned about the effect of diagnostic treatments such as x-rays and other forms of radiation exposure.

There are several units used to quantify the amount, or *dose*, of any radiation that interacts with a substance.

The roentgen

The **roentgen** (R) is defined as that amount of ionizing radiation that will produce $\frac{1}{3} \times 10^{-9}$ C of electric charge in 1 cm^3 of air under standard conditions.

Equivalently, the roentgen is that amount of radiation that deposits an energy of 8.76×10^{-3} J into 1 kg of air.

For most applications, the roentgen has been replaced by the *rad* (which is an acronym for *radiation absorbed dose*), defined as follows:

The rad

One **rad** is that amount of radiation that deposits 10^{-2} J of energy into 1 kg of absorbing material.

RBE

Although the rad is a perfectly good physical unit, it is not the best unit for measuring the degree of biological damage produced by radiation. This is because the degree of biological damage depends not only on the dose but also on the type of the radiation. For example, a given dose of alpha particles causes about ten times more biological damage than an equal dose of x-rays. The **RBE** (relative biological effectiveness) factor for a given type of radiation is defined as *the number of rad of x-radiation or gamma radiation that produces the same biological damage as 1 rad of the radiation being used.* The **RBE** factors for several types of radiation are given in Table 14.4. Note that the values are only approximate, since they vary with particle energy and the form of the damage.

Finally, the **rem** (roentgen equivalent in man) is defined as the product of the dose in rad and the **RBE** factor:

$$\boxed{\text{dose in rem} \equiv \text{dose in rad} \times \textbf{RBE}} \qquad (14.12)$$

TABLE 14.4 RBE[a] for Several Types of Radiation

Radiation	RBE Factor
X-rays and gamma rays	1.0
Beta particles	1.0–1.7
Alpha particles	10–20
Slow neutrons	4–5
Fast neutrons and protons	10
Heavy ions	20

[a] RBE = relative biological effectiveness

According to this definition, 1 rem of any two radiations will produce the same amount of biological damage. From Table 14.4, we see that a dose of 1 rad of fast neutrons represents an effective dose of 10 rem. On the other hand, 1 rad of beta radiation is equivalent to a dose of about 1 rem.

Low-level radiation from natural sources, such as cosmic rays and radioactive rocks and soil, delivers to each of us a dose of about 0.13 rem/year. The upper limit of radiation dose recommended by the U.S. government (apart from background radiation) is about 0.5 rem/year. Many occupations involve much higher radiation exposures, and so an upper limit of 5 rem/year has been set for combined whole-body exposure. Higher upper limits are permissible for certain parts of the body, such as the hands and forearms. A dose of 400 to 500 rem results in a mortality rate of about 50%. The most dangerous form of exposure is ingestion or inhalation of radioactive isotopes, especially those elements the body retains and concentrates, such as ^{90}Sr. In some cases, a dose of 1000 rem can result from ingesting 1 mCi of radioactive material.

14.8 RADIATION DETECTORS (Optional)

Various devices have been developed for detecting radiation. These devices are used for a variety of purposes, including medical diagnoses, radioactive dating measurements, and the measurement of background radiation.

Geiger counter

The **Geiger counter** (Fig. 14.19) is perhaps the most common device used to detect radiation. It can be considered the prototype of all counters that make use of ionization of a medium as the basic detection process. It consists of a cylindrical metal tube filled with gas at low pressure and a wire along the axis of the tube. The wire is maintained at a high positive potential (about 10^3 V) with respect to the tube. When a high-energy particle or photon enters the tube through a thin window at one end, some of the atoms of the gas become ionized. The electrons removed from the atoms are attracted toward the positive wire, and in the process they ionize other atoms in their path. This results in an avalanche of electrons, which produces a current pulse at the output of the tube. After the pulse is amplified, it can be either

Thin window
Wire electrode
Gas
+
To amplifier and counter
−
Metal tube
(a)

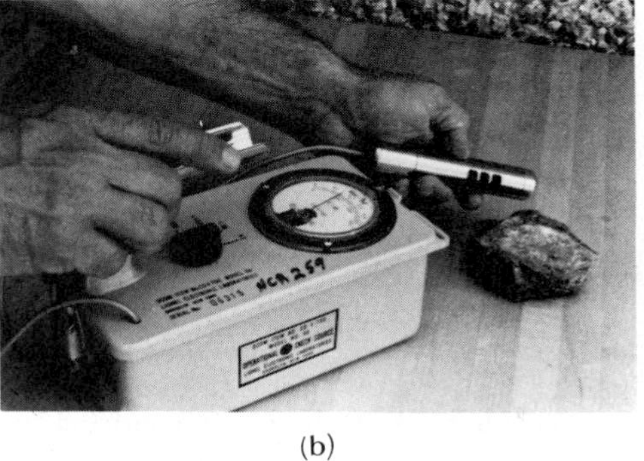

(b)

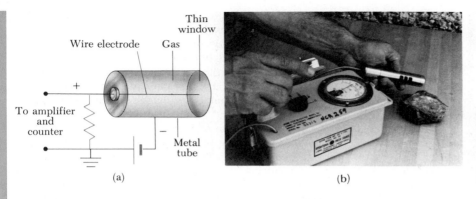

Figure 14.19 (a) Diagram of a Geiger counter. The voltage between the central wire and the metal tube is usually about 1000 V. (b) Using a Geiger counter to measure the activity in a radioactive mineral.

used to trigger an electronic counter or delivered to a loudspeaker, which clicks each time a particle enters the detector.

Semiconductor diode detector

A semiconductor diode detector is essentially a reverse-bias *p-n* junction. Recall from Chapter 12 that a *p-n* junction diode passes current readily when forward-biased and prohibits the flow of current under reverse-bias conditions. As an energetic particle passes through the junction, electrons are excited into the conduction band and holes are formed in the valence band. The internal electric field sweeps the electrons toward the positive (n) side of the junction and the holes toward the negative (p) side. This creates a pulse of current, which can be measured with an electronic counter. In a typical device, the duration of the pulse is about 10^{-7} to 10^{-8} s.

Scintillation counter

A scintillation counter (Fig. 14.20) usually uses a solid or liquid material whose atoms are easily excited by the incoming radiation. These excited atoms emit visible light when they return to their ground state. Common materials used as scintillators are crystals of sodium iodide and certain plastics. If such a material is attached to one end of a device called a **photomultiplier** (PM) tube, the photons emitted by the scintillator can be converted to an electric signal. The PM tube consists of several electrodes, called *dynodes,* whose potentials are increased in succession along the length of the tube as shown in Figure 14.20. The top of the tube contains a photocathode, which emits electrons by the photoelectric effect. As one of these emitted electrons strikes the first dynode, it has sufficient kinetic energy to eject several

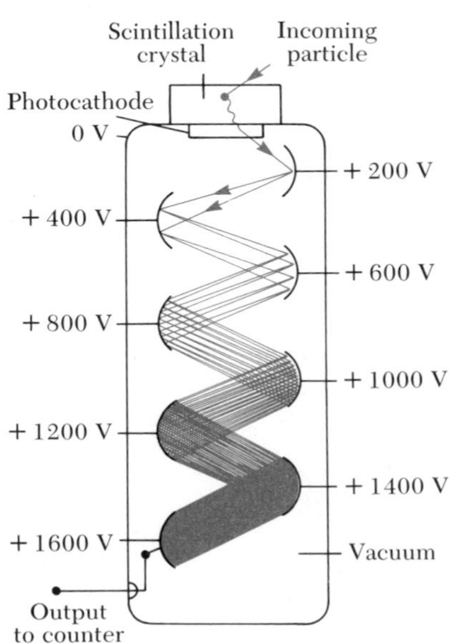

Figure 14.20 Diagram of a scintillation counter connected to a photomultiplier tube.

electrons. When these electrons are accelerated to the second dynode, many more electrons are ejected and an avalanche process occurs. The end result is 1 million or more electrons striking the last dynode. Hence, one particle striking the scintillator produces a sizable electric pulse at the output of the PM tube, and this pulse is in turn sent to an electronic counter. The scintillator device is much more sensitive than a Geiger counter, mainly because of the higher density of the detecting medium. It is especially useful for detecting gamma rays, which interact more weakly with matter than do charged particles.

The devices described so far make use of ionization processes induced by energetic particles. Various devices can be used to view the tracks of charged particles directly. A **photographic emulsion** is the simplest example. A charged particle ionizes the atoms in an emulsion layer. The path of the particle corresponds to a family of points at which chemical changes have occurred in the emulsion. When the emulsion is developed, the particle's track becomes visible. Such devices are common in the film badges used in any environment where radiation levels must be monitored.

Cloud chamber

A **cloud chamber** contains a gas that has been supercooled to just below its usual condensation point. An energetic particle passing through ionizes the gas along its path. These ions serve as centers for condensation of the supercooled gas. The track can be seen with the naked eye and can be photographed. A magnetic field can be applied to determine the signs of the charges as they are deflected by the field.

A device called a **bubble chamber,** invented in 1952 by D. Glaser, makes use of a liquid (usually liquid hydrogen) maintained near its boiling point. Ions produced by incoming charged particles leave bubble tracks, which can be photographed (Fig. 14.21). Because the density of the detecting medium of a bubble chamber is much higher than the density of the gas in a cloud chamber, the bubble chamber has a much higher sensitivity.

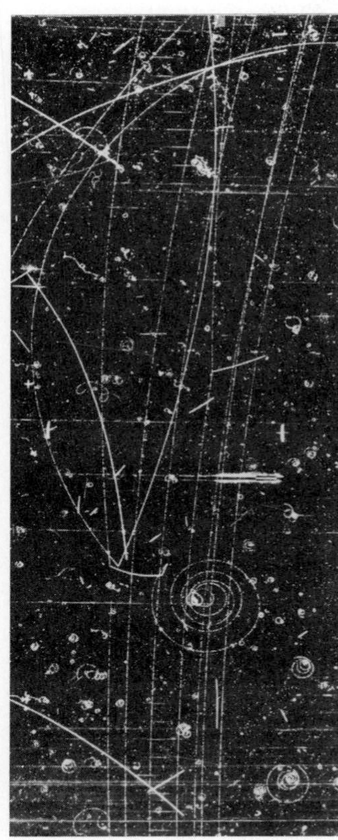

Figure 14.21 Bubble chamber photograph of the production and decay of particles. (Courtesy of Lawrence Radiation Laboratory, University of California, Berkeley, CA)

A **spark chamber** is a counting device that consists of an array of conducting parallel plates. Even-numbered plates are grounded, and odd-numbered plates are maintained at a high potential (about 10 kV). The spaces between the plates contain a noble gas at atmospheric pressure. When a charged particle passes through the chamber, ionization occurs in the gas, resulting in a large surge of current and a visible spark.

14.9 SUMMARY

If N_0 neutrons are incident on a target of thickness x, the number N of neutrons that emerge from the target is

$$N = N_0 e^{-n\sigma x} \tag{14.2}$$

where n is the density of nuclei in the target and σ is the **cross section.** That is, the number of particles that penetrate the target decreases exponentially with increasing thickness. Nuclear cross sections, which have dimensions of area, are usually measured in *barns* (bn), where 1 bn $= 10^{-28}$ m^2.

The cross section for neutrons moving through matter generally increases with decreasing neutron energy. A *thermal neutron* (one whose energy is approximately kT) has a high probability of being captured by a nucleus according to the following **neutron-capture process:**

$${}^1_0\text{n} + {}^A_Z\text{X} \longrightarrow {}^{A+1}_{\ \ Z}\text{X} + \gamma \tag{14.4}$$

Energetic neutrons are slowed down readily in materials called **moderators.** These materials have a small cross section, so that the neutrons lose their energy mainly through elastic collisions.

The process of **nuclear fission** occurs when a very heavy nucleus, such as ^{235}U, splits into two smaller fragments. Thermal neutrons can create fission in ^{235}U by the following process:

$$\boxed{{}^1_0n + {}^{235}_{92}U \longrightarrow {}^{236}_{92}U^* \longrightarrow X + Y + \text{neutrons}} \tag{14.5}$$

where X and Y are the fission fragments and $^{236}U^*$ is a compound nucleus in an excited state. On the average, 2.47 neutrons are released per fission event. The fragments and neutrons have a great deal of kinetic energy following the fission event. The fragments then undergo a series of beta and gamma decays to various stable isotopes. The energy released per fission event is about 208 MeV.

The **reproduction constant** K is defined as the average number of neutrons released from each fission event that will cause another event. In a power reactor, it is necessary to maintain a value of K close to 1. The value of K is affected by such factors as the reactor geometry, the mean energy of the neutrons, and the neutron-capture cross section. Proper design of the reactor geometry is necessary to minimize neutron leakage from the reactor core. Neutron energies are regulated with a moderator material to slow down energetic neutrons and therefore increase the probability of neutron capture by other ^{235}U nuclei. The power level of the reactor is adjusted with control rods made of a material with a large neutron-capture cross section. The value of K can be adjusted by inserting the rods at various depths into the reactor core.

Nuclear fusion is a process in which two light nuclei combine to form a heavier nucleus. A great deal of energy is released in such a process. The major obstacle in obtaining useful energy from fusion is the large Coulomb repulsive force between the charged nuclei at close separations. Sufficient energy must be supplied to the particles to overcome this Coulomb barrier and thereby enable the nuclear attractive force to take over. The temperature required to produce fusion is of the order of 10^8 K. At such high temperatures, all matter is in the form of a **plasma,** which consists of positive ions and free electrons.

In a fusion reactor, the plasma temperature must reach at least the **critical ignition temperature,** which is the temperature at which the power generated by the fusion reactions *exceeds* the power lost in the system. The most promising fusion reaction is the D-T reaction, which has a critical ignition temperature of about 4.5×10^7 K. Two critical parameters involved in fusion reactor design are **ion density** n and **confinement time** τ. The confinement time is the time the interacting particles must be maintained at a temperature equal to or greater than the critical ignition temperature. **Lawson's criterion** states that for the D-T reaction, $n\tau \geq 10^{14}$ s/cm^3.

When energetic particles interact with a medium, they lose their energy by various processes. Heavy particles, such as alpha particles, lose most of their energy by excitation and ionization of atoms in the medium. The particles have a finite *range* in the medium, which depends on the energy, mass, and charge of the particle as well as the properties of the medium. Energetic electrons moving through a medium also lose their energy by excitation and ionization. However, they scatter more readily than heavy particles and therefore undergo fewer collisions before coming to rest.

Photons, which have no charge, are not as effective as charged particles in producing ionization or excitation. However, they can be absorbed in a medium by several processes: (1) the photoelectric effect, predominant at low photon energies; (2) Compton scattering, predominant at intermediate photon energies; and (3) pair production, predominant at high photon energies.

If a beam of photons with intensity I_0 is incident on a medium, the intensity of the beam decreases exponentially with increasing depth of penetration into the medium according to the relation

$$\boxed{I(x) = I_0 e^{-\mu x}} \tag{14.11}$$

where x is the distance traveled in the medium, $I(x)$ is the intensity of the beam in the medium after traveling a distance x, and μ is the **linear absorption coefficient** of the medium.

Devices used to detect radiation are the Geiger counter, semiconductor diode detector, scintillation counter, photographic emulsion, cloud chamber, and bubble chamber.

QUESTIONS

1. Explain the function of a moderator in a fission reactor.
2. Why is water a better shield for neutrons than lead or steel?
3. Discuss the advantages and disadvantages of fission reactors from the point of view of safety, pollution, and resources. Make a comparison with power generated from the burning of fossil fuels.
4. In a fission reactor, nuclear reactions produce heat to drive a turbine generator. How is this heat produced?
5. Why would a fusion reaction produce less radioactive waste than a fission reactor?
6. Why is the temperature required for the D-T fusion less than that needed for the D-D fusion? Estimate the relative importance of Coulomb repulsion and nuclear attraction in each case.
7. What factors make a fusion reaction difficult to achieve?
8. Discuss the similarities and differences between fusion and fission.
9. Discuss the advantages and disadvantages of fusion power from the point of safety, pollution, and resources.
10. Discuss three major problems associated with the development of a controlled fusion reactor.
11. Describe two techniques that are being pursued in an effort to obtain power from nuclear fusion.
12. If two radioactive samples have the same activity measured in Ci, will they necessarily create the same damage to a medium? Explain.
13. Radiation is often used to sterilize such things as surgical equipment and packaged foods. Why do you suppose this works?
14. One method of treating cancer of the thyroid is to insert a small radioactive source directly into the tumor. The radiation emitted by the source can destroy cancerous cells. Very often, the radioactive isotope $^{131}_{53}I$ is injected into the bloodstream in this treatment. Why do you suppose iodine is used?
15. Why should a radiologist be extremely cautious about x-ray doses when treating pregnant women?
16. The design of a PM tube might suggest that any number of dynodes may be used to amplify a weak signal. What factors do you suppose would limit the amplification in this device?

PROBLEMS

1. Consider a slab consisting of two layers of material with thicknesses x_1 and x_2 and target densities n_1 and n_2. If N_0 is the number of particles incident on the first layer of the slab in some time interval, determine the number N that emerge from the second layer in that interval. Assume that the cross section σ is the same for each material. What would you guess is the relationship for three or more layers?
2. The density of the liquid hydrogen target in a bubble chamber is 70 kg/m^3. If 20% of a beam of slow neutrons incident on the bubble chamber has reacted with the hydrogen by the time the beam has traveled 2 m through the hydrogen, what is the cross section, in bn, for the reaction of these slow neutrons with hydrogen atoms?
3. The density of lead is 11.35 g/cm^3, and its atomic weight is 207.2. If 1 cm of lead reduces a beam of 1-MeV gamma rays to 28.65% of its initial intensity, (a) how much lead is required to reduce the beam to 10^{-4}% of its initial intensity? (b) What is the effective cross section of a lead atom for a 1-MeV photon?
4. Neutrons are captured by a cadmium foil. Using the data in Figure 14.3, find (a) the ratio of 10-eV neutrons captured to 1-eV neutrons captured, (b) the ratio of 1-eV neutrons captured to 0.1-eV neutrons captured, and (c) the ratio of 0.1-eV neutrons captured to 0.01-eV neutrons captured. (d) In what range of energies can cadmium be used as an energy selector?
5. The atomic weight of cadmium is 112.41, and its den-

sity is 8.65 g/cm^3. Using Figure 14.3, estimate the attenuation distance of a thermal neutron beam in cadmium. (Note that the attenuation distance is the distance traveled after which the intensity of the beam is reduced to $1/e$ of its initial value, where e is the base of the natural logarithms.)

6. A beam of 100-MeV protons from a Van de Graaf generator is incident on a gold foil having a thickness of 5.1×10^{-5} m. The beam current is 0.1 μA, and the beam has a cross-sectional area of 1 cm^2. If the scattering cross section is 500 bn, calculate (a) the ratio N/N_0, (b) the number per second that pass through the foil, and (c) the number lost from the beam each second by scattering. (Gold has an atomic weight of 197 and a density of 19.3 g/cm^3.)

7. Find the energy released in the following fission reaction:

$$ {}^{1}_{0}\text{n} + {}^{235}_{92}\text{U} \longrightarrow {}^{141}_{56}\text{Ba} + {}^{92}_{36}\text{Kr} + 3({}^{1}_{0}\text{n}) $$

The required masses are

$$ M({}^{1}_{0}\text{n}) = 1.008665 \text{ u} $$
$$ M({}^{235}_{92}\text{U}) = 235.043915 \text{ u} $$
$$ M({}^{141}_{56}\text{Ba}) = 140.9139 \text{ u} $$
$$ M({}^{92}_{36}\text{Kr}) = 91.8973 \text{ u} $$

8. Find the number of ^{6}Li and the number of ^{7}Li nuclei present in 1 kg of lithium. (The natural abundance of ^{6}Li is 7.4%; the remainder is ^{7}Li.)

9. (a) How many grams of ^{235}U must undergo fission to operate a 1000-MW power plant for one day? (b) If the density of ^{235}U is 18.7 g/cm^3, how large a sphere of ^{235}U could you make from this much uranium?

10. In order to minimize neutron leakage from a reactor, the surface area-to-volume ratio should be a minimum for a given shape. For a given volume V, calculate this ratio for (a) a sphere, (b) a cube, and (c) a parallelepiped of dimensions $a \times a \times 2a$. (d) Which of these shapes would have the minimum leakage? Which would have the maximum leakage?

11. It has been estimated that there are 10^9 tons of natural uranium available at concentrations exceeding 100 parts per million, of which 0.7% is ^{235}U. If all the world's energy needs (7×10^{12} J/s) were to be supplied by ^{235}U fission, how long would this supply last? (This estimate of uranium supply was taken from K. S. Deffeyes and I. D. MacGregor, *Scientific American*, January 1980, p. 66.)

12. Two nuclei with atomic numbers Z_1 and Z_2 approach each other with a total energy E. (a) If the minimum distance of approach for fusion to occur is $r = 10^{-14}$ m, find E in terms of Z_1 and Z_2. (b) Calculate the minimum energy for fusion for the D-D and D-T reactions (the first and third reactions in Eq. 14.8).

13. To understand why containment of a plasma is necessary, consider the rate at which a plasma would be lost if it were not contained. (a) Estimate the rms speed of deuterons in a plasma at 10^8 K. (b) Estimate the time such a plasma would remain in a 10-cm cube if no steps were taken to contain it.

14. Of all the hydrogen nuclei in the ocean, 0.0156% are deuterium. The oceans have a volume of 317 million cubic miles. (a) If all the deuterium in the oceans were fused to ${}^{4}_{2}$He, how many joules of energy would be released? (b) Present world energy consumption is about 7×10^{12} W. If consumption were 100 times greater, how many years would the energy calculated in (a) last?

15. It has been pointed out that fusion reactors are safe from explosion because there is never enough energy in the plasma to do much damage. (a) Using the data in Table 14.1, calculate the amount of energy stored in the plasma of the TFTR reactor. (b) How many kg of water could be boiled by this much energy? (The plasma volume of the TFTR reactor is about 50 m^3.)

16. In order to confine a stable plasma, the magnetic energy density in the magnetic field must exceed the pressure $2nkT$ of the plasma by a factor of at least 10. In the following, assume a confinement time $\tau = 1$ s. (a) Using Lawson's criterion, determine the required ion density. (b) From the ignition temperature criterion for the D-T reaction, determine the required plasma pressure. (c) Determine the magnitude of the magnetic field required to contain the plasma.

*17 X-rays of wavelength 25 pm and gamma rays of energy 0.1 MeV have approximately the same absorption coefficient in lead. How do their energies compare?

*18. What is the half-value thickness (Example 14.5) of water to x-rays of wavelength 20 pm? Since the human body is more than 90% water, what does your answer indicate about the use of x-rays as a diagnostic technique?

*19 In a large-scale nuclear attack, typical radiation intensity from radioactive fallout might be 2000 rad in most places. In the following calculations, assume that one third of the radiation is 10-MeV gamma radiation and that the linear absorption coefficient is the same for aluminum and concrete. (a) What thickness (in m) of concrete would be needed to reduce the radiation intensity to 1 rad? (b) If a particular shelter were located at a "hot spot" receiving 100 000 rad, what thickness of concrete would be needed to reduce the radiation intensity to 1 rad?

*20. Assume that an x-ray technician takes an average of eight x-rays per day and receives a dose of 5 rem/year as a result. (a) Estimate the dose in rem per x-ray taken. (b) How does this result compare with low-level background radiation?

*21 In terms of biological damage, how many rad of heavy ions is equivalent to 100 rad of x-rays?

*22. Two workers using an industrial x-ray machine accidentally insert their hands in the x-ray beam for the

* Problems marked with an asterisk relate to optional sections.

same length of time. The first worker inserts one hand in the beam, and the second worker inserts both hands. Which worker received the larger dose in rad?

*23 Calculate the radiation dose in rad supplied to 1 kg of water such that the energy deposited equals (a) the rest energy of the water and (b) its thermal energy. (Assume that each molecule has a thermal energy of kT.)

*24. A person whose mass is 75 kg is exposed to a dose of 25 rad. How many joules of energy are deposited in the person's body?

*25 In a Geiger tube, the voltage between the electrodes is typically 1 kV and the current pulse discharges a 5-pF capacitor. (a) What is the energy amplification of this device for a 0.5-MeV beta ray? (b) How many electrons are avalanched by the initial electron?

*26. In a PM tube, assume that there are seven dynodes with potentials of 100 V, 200 V, 300 V, . . . , 700 V. The average energy required to free an electron from the dynode surface is 10 eV. For each incident electron, how many electrons are freed (a) at the first dynode and (b) at the last dynode? (c) What is the average energy supplied by the tube for each electron? (Assume an efficiency of 100%.)

27. The half-life of tritium is 12 years. If the TFTR fusion reactor contains 50 m^3 of tritium at a density equal to 1.5×10^{14} particles/cm^3, how many Ci of tritium are in the plasma? Compare this with a fission inventory of 4×10^{10} Ci.

28. (a) Estimate the volume of space required to store the radioactive wastes that would be produced in one year if all the annual U.S. electricity production (which is about 2.2×10^{12} kWh/year) came from uranium enriched to 3% ^{235}U. (Assume that the conversion efficiency is 30% and that the waste is in the form of a liquid with a density of 1 g/cm^3.) (b) If the waste could be formed into a cube, what would be the length of the cube's sides?

29. A 1-MeV neutron is emitted in a fission reactor. If it loses one half of its kinetic energy in each collision with a moderator atom, how many collisions must it undergo in order to achieve thermal energy (0.039 eV)?

30. The ^6Li isotope is only about 7.5% naturally abundant. The remaining 92.5% of lithium is ^7Li. It is estimated that there is about 2×10^{13} g of lithium available. If ^6Li is used as a tritium source in fusion reactors, with an energy release of 22 MeV per ^6Li nucleus, estimate the total energy available in D-T fusion. How does this number compare with the world's fossil fuel supply, which is estimated to be about 2.5×10^{23} J?

31. (a) Show that about 2×10^{10} J would be released by the fusion of the deuterons in 1 gal of water. (Note that 1 out of every 6500 water molecules contains a deuteron.) (b) The average energy consumption rate of a person living in the United States is about 10^4 J/s (an average power of 10 kW). At this rate, how long would the energy needs of one person be supplied by the fusion of the deuterons in 1 gal of water? (Assume that the energy released per deuteron is 7.18 MeV.)

32. The densities and atomic weights of the materials in Table 14.3 are as follows:

Substance	Density (g/cm^3)	Atomic Weight
H_2O	1	18
Al	2.7	27
Fe	7.8	55.8
Pb	11.35	207.2

Compute the number of electrons/cm^3 for each material and plot the gamma-ray linear absorption coefficient versus electron density. Draw three graphs corresponding to gamma-ray energies of 0.1, 1.0, and 10 meV. What do you conclude from your graphs?

33. (a) For the second reaction of Equation 14.9, show that the energy produced per event is 4.8 MeV. (b) For the first reaction of Equation 14.9, calculate the minimum energy the fast neutron must have to cause the reaction.

34. Three equally thick layers of aluminum, copper, and lead are used to reduce the intensity of an x-ray beam to one third of its original intensity. The wavelength of the beam is 50 pm. (a) Find the thickness of *each layer* of material. (b) By what fraction would the initial beam intensity be reduced by the lead alone?

35. A particle cannot generally be localized to distances much smaller than its de Broglie wavelength. This means that a slow neutron appears to be larger to a target particle than does a fast neutron, in the sense that the slow neutron will probably be found over a large volume of space. For a thermal neutron at room temperature (300 K), find (a) the linear momentum and (b) the de Broglie wavelength. Compare this effective neutron size with both nuclear and atomic dimensions.

36. Calculate the amount of energy carried off by the ^4_2He nucleus and the neutron in the D-T fusion reaction. (Assume that the momentum is initially zero.) Does this explain why only 20% of the energy produced can be used for critical ignition?

37. (a) Calculate the energy (in kWh) released if 1 kg of ^{239}Pu undergoes complete fission and the energy released per fission event is 200 MeV. (b) Calculate the energy (in MeV) released in the D-T fusion:

$$^2_1H + ^3_1H \longrightarrow ^4_2He + ^1_0n$$

(c) Calculate the energy (in kWh) released if 1 kg of deuterium undergoes fusion. (d) Calculate the energy (in kWh) released by the combustion of 1 kg of coal if each $C + O_2 \longrightarrow CO_2$ reaction yields 4.2 eV. (e) List the advantages and disadvantages of each of these methods of energy generation.

38. A tiny sphere is made of a material that absorbs all the photons incident on it. Many such spheres are embedded randomly in a transparent medium. (a) If the

radius of one of these spheres is b, what is its cross section σ in terms of b for the absorption of photons? (b) If the radius of each sphere is $b = 2 \times 10^{-3}$ m, what is the cross section? (c) If 3×10^4 of these spheres are uniformly embedded in a cylinder of a transparent medium of height 2 m and cross-sectional area 0.5 m^2, and a light of beam intensity 0.75 W/m^2 is incident normally on one end of the cylinder, what is the intensity of the beam of light that emerges from the other end?

39. The sun radiates energy at the rate of 4×10^{23} kW. If the reaction

$$4({}^1_1\text{H}) \longrightarrow {}^4_2\text{He} + 2\beta^+ + 2\nu + \gamma$$

accounted for all the energy released, calculate (a) the number of protons fused per second and (b) the mass transformed into energy per second.

40. Suppose the target in a laser fusion reactor is a sphere of solid hydrogen, with a diameter of 10^{-4} m and a density of 0.2 g/cm^3. Also assume that half of the nuclei are ^{2}H and half are ^{3}H. (a) If 1% of a 200-kJ laser pulse is delivered to this sphere, what temperature will the sphere reach? (b) If all of the hydrogen "burns" according to the D-T reaction, how many J of energy will be released?

41. In a tokamak fusion reactor, suppose a deuteron moves at an angle of 30° to the toroidal magnetic field. Assume that $B_t = 1$ T and $B_p = 0$. (a) Calculate the components of velocity parallel and perpendicular to B_t. (b) What is the radius of the spiral motion for the deuteron? (c) How far does the deuteron travel *along* the magnetic field before it completes one revolution around the magnetic field?

42. The density of cadmium is 8.65 g/cm^3, and its cross section for thermal neutron absorption is 2450 bn. Compare the rates of neutron absorption to neutron decay in cadmium for thermal neutrons at a temperature of 27°C. (*Hint:* The number of neutrons absorbed per second is

$$\left(\frac{dN}{dt}\right)_a = \left(\frac{dN}{dx}\right)_a \frac{dx}{dt} = \left(\frac{dN}{dx}\right)_a v$$

where v is the neutron speed and dN/dx is the number of neutrons lost per meter of travel.)

43. Find the Q value for each of the reactions in the proton-proton cycle (Eq. 14.7) and show that the overall Q value for the cycle is 25.7 MeV.

44. The carbon cycle, first proposed by Bethe in 1939, is another cycle by which energy is released in stars and hydrogen is converted to helium. The carbon cycle requires higher temperatures than the proton-proton cycle. The series of reactions is

$$^{12}\text{C} + {}^1\text{H} \longrightarrow {}^{13}\text{N} + \gamma$$
$$^{13}\text{N} \longrightarrow {}^{13}\text{C} + \beta^+ + \nu$$
$$^{13}\text{C} + {}^1\text{H} \longrightarrow {}^{14}\text{N} + \gamma$$
$$^{14}\text{N} + {}^1\text{H} \longrightarrow {}^{15}\text{O} + \gamma$$
$$^{15}\text{O} \longrightarrow {}^{15}\text{N} + \beta^+ + \nu$$
$$^{15}\text{N} + {}^1\text{H} \longrightarrow {}^{12}\text{C} + {}^4\text{He}$$

(a) If the proton-proton cycle requires a temperature of 1.5×10^7 K, estimate the temperature required for the first step in the carbon cycle. (b) Calculate the Q value for each step in the carbon cycle and the overall energy released. (c) Do you think the energy carried off by the neutrinos is deposited in the star? Explain.

45. In natural silver, the abundance of ^{107}Ag is 51.35% and that of ^{109}Ag is 48.65%. The neutron absorption cross section of ^{107}Ag is 31 bn, and that of ^{109}Ag is 87 bn. The activation products ^{108}Ag and ^{110}Ag decay by beta emission with half-lives of 144 s and 24.5 s, respectively. A silver sample is removed from a fission reactor, and after some delay it is found that the ratio of ^{108}Ag to ^{110}Ag is 20 : 1. How long was the delay? (*Hint:* Assume that the sample has been in the reactor for a time sufficiently long that the decay rate equals the production rate.)

46. An electrical power plant operates on the basis of thermal energy generated in a pressurized water reactor. The electrical power output of the plant is 1 GW, and its efficiency is 30%. The uranium fuel is enriched to 3.0% ^{235}U. (a) Find the total power generated by the reactor. (b) How much power is discharged to the environment as waste heat? (c) Calculate the rate of fission events in the reactor core. (d) Calculate the mass of ^{235}U used up in one year. (e) Using the results from (a), determine the rate at which fuel is converted into energy (in kg/s) in the reactor core and compare your answer with the result from (d).

15

Elementary Particles

In this chapter, we shall examine the properties and classifications of the various known particles and the fundamental interactions that govern their behavior. We shall also discuss the current theory of elementary particles, in which all matter in nature is believed to be constructed from only two families of particles, quarks and leptons. Finally, we shall discuss how clarifications of such models might help scientists understand the evolution of the universe.

15.1 INTRODUCTION

The word "atom" is from the Greek word *atomos,* which means "indivisible." At one time atoms were thought to be the indivisible constituents of matter; that is, they were regarded to be elementary particles. Discoveries in the early part of the 20th century revealed that the atom is not elementary, but has as its constituents protons, neutrons, and electrons. Until 1932, physicists viewed all matter as consisting of only three particles: electrons, protons, and neutrons. With the exception of the free neutron, these particles are very stable. Beginning in 1945, many new particles were discovered in experiments involving high-energy collisions between known particles. These new particles are characteristically very unstable and have very short life times, ranging from 10^{-6} to 10^{-23} s. So far, more than 300 of these unstable, temporary particles have been catalogued.

During the last thirty years, many powerful particle accelerators have been constructed throughout the world, making it possible to observe collisions of particles with great energy under controlled laboratory conditions, so as to reveal the subatomic world in finer detail. Up until the 1960s, physicists were bewildered by the large number and variety of subatomic particles being

An aerial view of the Fermi National Accelerator Laboratory, Batavia, Illinois. The largest circle is the main accelerator. Three experimental lines extend at a tangent from the accelerator. The 16-story twin-towered Wilson Hall is seen at the base of the experimental lines. (Fermilab Photo Dept.)

discovered. They wondered if the particles were like animals in a zoo with no systematic relationship connecting them, or whether a pattern was emerging that would provide a better understanding of the elaborate variety in the subnuclear world. In the last two decades, physicists have made tremendous advances in our knowledge of the structure of matter by recognizing that all particles (with the exception of electrons, photons, and a few related particles) are made of smaller particles called quarks. Thus, protons and neutrons, for example, are not truly elementary but are systems of tightly bound quarks. The quark model has reduced the bewildering array of particles to a manageable number, and has been successful in predicting new quark combinations later found in many experiments.

15.2 THE FUNDAMENTAL FORCES IN NATURE

The key to understanding the properties of elementary particles is to be able to describe the forces between them. All particles in nature are subject to four fundamental forces: strong, electromagnetic, weak, and gravitational.

The **strong force** is very short-ranged and is responsible for the binding of neutrons and protons into nuclei. This force represents the "glue" that holds nucleons together and is the strongest of all the fundamental forces. The strong force is negligible for separations greater than about 10^{-14} m (which is about the size of the nucleus). The **electromagnetic force,** which is about 10^{-2} times the strength of the strong force, is responsible for the binding of atoms and molecules. It is a long-range force between charged particles (Coulomb's law) that decreases in strength as the inverse square of the separation between interacting particles. The **weak force** is a short-range nuclear force that tends to produce instability in certain nuclei. It is responsible for most radioactive decay processes such as beta decay, and its strength is only about 10^{-9} times that of the strong force. (As we shall discuss later, scientists now believe that the weak and electromagnetic forces are two manifestations of a single force called the electroweak force.) Finally, the **gravitational force** is a long-range force that varies as $1/r^2$ and has a strength of only about 10^{-38} times that of the strong force. Although this familiar interaction is the force that holds the planets, stars, and galaxies together, its effect on elementary particles is negligible. The gravitational force is the weakest of all the fundamental forces.

In modern physics, one often describes the interactions between particles in terms of the exchange of field particles or quanta. In the case of the familiar electromagnetic interaction, the field particles are photons. In the language of modern physics, one can say that the electromagnetic force is *mediated* by photons, which are the quanta of the electromagnetic field. Likewise, the strong force is mediated by field particles called *gluons,* the weak force is mediated by particles called the *W* and *Z bosons,* and the gravitational force is mediated by quanta of the gravitational field called *gravitons.* These interactions, their ranges, and their relative strengths are summarized in Table 15.1.

TABLE 15.1 Particle Interactions

Interaction (Force)	Relative Strength	Range of Force	Mediating Field Particle
Strong	1	Short ($\approx$1 fm)	Gluon
Electromagnetic	10^{-2}	Long ($\propto 1/r^2$)	Photon
Weak	10^{-9}	Short ($\approx$1 fm)	$W^{\pm}$, Z
Gravitational	10^{-38}	Long ($\propto 1/r^2$)	Graviton

15.3 POSITRONS AND OTHER ANTIPARTICLES

In the 1920s, the theoretical physicist Paul Adrien Maurice Dirac (1902–1984) developed a version of quantum mechanics that incorporated special relativity. Dirac's theory automatically explained the origin of the electron's spin and its magnetic moment. However, Dirac was faced with a major difficulty in this theory. His relativistic wave equation required solutions corresponding to negative energy states, even for free particles.[1] But if negative energy states existed, one would expect an electron in a state of positive energy to make a rapid transition to one of these states, emitting a photon in the process. Dirac was able to avoid this difficulty by postulating that all negative energy states were filled. Those electrons that occupy the negative energy states are called the "Dirac sea." Electrons in the Dirac sea are not directly observable because the Pauli exclusion principle does not allow them to react to external forces. However, if one of these negative energy states is vacant, leaving a hole in the sea of filled states, the hole can react to external forces and would be observable. (This is analogous to the behavior of a hole in the valence band of a semiconductor.) *The second equivalent intrepretation of this theory was that for every particle, there was also an antiparticle.* The antiparticle would have the same mass as the particle, but their charges would be opposite each other. For example, the electron's antiparticle (called a *positron*) would have a mass of 0.511 MeV/c^2, and a positive charge equal to $+1.6 \times 10^{-19}$ C. Usually we shall designate an antiparticle with a bar over the symbol for the particle. Thus, the positron is denoted by $\bar{\text{e}}$ (although sometimes the notation e^+ is preferred), the antiproton is denoted by $\bar{\text{p}}$, and the antineutrino is denoted by $\bar{\nu}$.

The positron was discovered by Carl Anderson in 1932 (the same year that the neutron was discovered), and in 1936 he was awarded the Nobel prize for his discovery. Anderson made his discovery while examining tracks created by electronlike particles of positive charge in a cloud chamber. (These early experiments used cosmic rays—mostly energetic protons passing through interstellar space—to initiate high-energy reactions of the order of several GeV.) In order to discriminate between positive and negative charges, the cloud chamber was placed in a magnetic field, causing moving charges to follow curved paths. Anderson noted that some of the electronlike tracks deflected in a direction corresponding to a positively charged particle.

Since Anderson's initial discovery, the positron has been observed in a number of experiments. Perhaps the most common process for producing positrons is **pair production.** In this process, a gamma ray with sufficiently high energy collides with a nucleus, and an electron-positron pair is created. Since the total rest energy of the electron-positron pair is $2m_0c^2 = 1.02$ MeV (where m_0 is the rest mass of the electron), the gamma ray must have at least this much energy to create electron-positron pairs. Thus, electromagnetic energy in the form of a gamma ray is transformed into mass in accordance with Einstein's famous relation $E = m_0c^2$. Figure 15.1 shows tracks of electron-positron pairs created by high-energy gamma rays.

A process that is the reverse of pair production can also occur. Under the proper conditions, an electron and positron can combine and annihilate to produce two photons that have a combined energy of at least 1.02 MeV. The reaction can be expressed as

$$\text{e} + \bar{\text{e}} \longrightarrow 2\gamma$$

[1] P. A. M. Dirac, *The Principles of Quantum Mechanics*, 3rd ed., New York, Oxford University Press, Chapter 11, 1947.

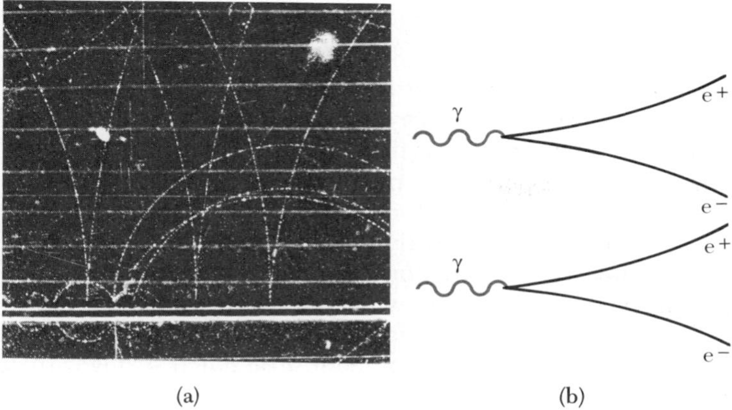

Figure 15.1 (a) Bubble-chamber tracks of electron-positron pairs produced by 300-MeV x-rays striking a lead sheet. (Courtesy of Lawrence Berkeley Laboratory, University of California) (b) Sketch of the pertinent pair-production events. Note that the positrons deflect upward while the electrons deflect downward in an applied magnetic field that is directed into the diagram.

Likewise, a proton-antiproton pair can annihilate to produce two gamma rays; however, this event is very rare.

Practically every known elementary particle has an antiparticle. Among the exceptions are the photon and the neutral pion (π^0). (Note that the π^0 and η^0 are their own antiparticles.) Following the construction of high-energy accelerators in the 1950s, many other antiparticles were discovered. These included the antiproton ($\overline{\text{p}}$) discovered by Emilio Segre and Owen Chamberlain in 1955, and the antineutron ($\overline{\text{n}}$) discovered shortly thereafter.

(a)

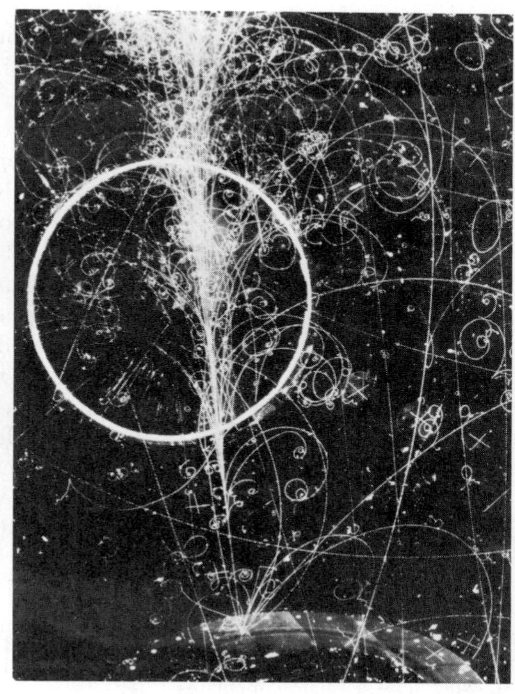

(b)

(Left) The tunnel of the main accelerator at Fermilab. The upper ring of magnets is the 400-GeV accelerator. The lower ring are superconducting magnets for the Tevatron. *(Right)* Photograph of particle interactions in the 15-ft bubble chamber at Fermilab. Such photographs showing particle "tracks" are studied by scanners and experimenters. (Fermilab Photo Dept.)

15.4 MESONS AND THE BEGINNING OF PARTICLE PHYSICS

The structure of matter as viewed by physicists in the mid 1930s was fairly simple. The building blocks of matter were considered to be the proton, the electron, and the neutron. Three other particles were known at the time: the gamma-ray (photon), the neutrino, and the positron. These six particles were considered to be the fundamental constituents of matter. Although the accepted picture of the world was marvelously simple, no one was able to provide an answer to the following important question. If the nucleus of an atom contains many protons in close proximity, which experience strongly repulsive forces due to their like charges, what force holds the nucleus together? Scientists recognized that this mysterious force must be much stronger than anything encountered in nature up to this time.

The first theory to explain the nature of the strong force was proposed in 1935 by the Japanese physicist Hideki Yukawa (1907–1981), an effort that later earned him the Nobel prize. In order to understand Yukawa's theory, it is useful first to recall that *two atoms can form a covalent chemical bond by the exchange of electrons. Similarly, in the modern view of electromagnetic interactions, charged particles interact through the exchange of photons.* Yukawa used this idea to explain the strong force by proposing a new particle whose exchange between nucleons in the nucleus produces the strong force with a range of about 10^{-15} m (the order of the nuclear diameter). Furthermore, he established that the range of the force is inversely proportional to the mass of this new carrier particle, and he predicted that the mass would be about 200 times the mass of the electron. Since the new particle would have a mass between that of the electron and proton, it was called a **meson** (from the Greek *meso,* meaning "middle").

In an effort to substantiate Yukawa's predictions, physicists began an experimental search for the meson by studying cosmic rays that enter the earth's atmosphere from interstellar space. In 1938, Carl Anderson and his collaborators discovered a particle whose mass was 106 MeV, which is about 207 times the mass of the electron. However, subsequent experiments showed that the particle interacted very weakly with matter, and hence could not be the carrier of the strong force. The puzzling situation inspired several theoreticians to propose that there are actually two mesons with slightly different masses. This idea was confirmed in 1947 with the discovery in cosmic rays of the pi meson (π), or simply *pion,* by Cecil Frank Powell (1903–1969) and Guiseppe P. S. Occhialini (1907–). The lighter meson discovered earlier by Anderson, now called a *muon* (μ), has only weak and electromagnetic interaction and plays no role in the strong interaction.

The pion comes in three varieties, corresponding to three charge states: π^+, π^-, and π^0. The π^+ and π^- particles have masses of 139.6 MeV/c^2, while the π^0 has a mass of 135.0 MeV/c^2. Pions and muons are very unstable particles. For example, the π^- first decays into a muon and an antineutrino with a mean lifetime of about 2.6×10^{-8} s. The muon then decays into an electron, a neutrino, and an antineutrino with a mean lifetime of about 2.2×10^{-6} s. The sequence of decays is

$$\pi^- \longrightarrow \mu^- + \bar{\nu}_\mu$$

$$\mu^- \longrightarrow e + \nu_\mu + \bar{\nu}_e \qquad (15.1)$$

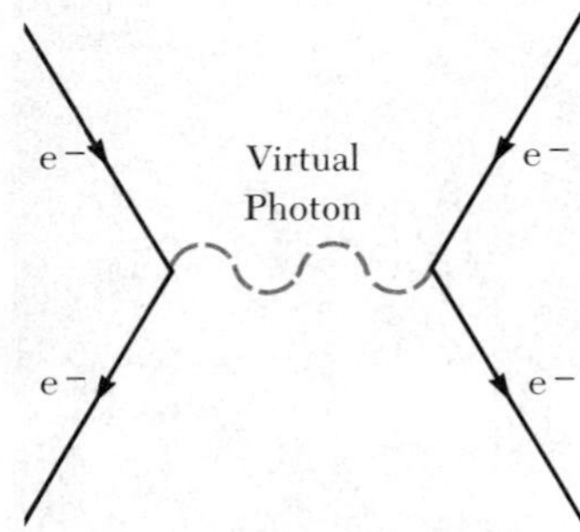

Figure 15.2 Feynman diagram showing how a photon mediates the electromagnetic force between two interacting electrons.

The interaction between two particles can be represented in a simple diagram called a *Feynman diagram,* developed by Richard P. Feynman (1918–1988). Figure 15.2 is a Feynman diagram for the case of the electromagnetic interaction between two electrons. In this simple case, a single photon acts as the carrier of the electromagnetic force between the electrons. The photon transfers energy and momentum from one electron to the other in this interaction. These photons are called *virtual photons* because they can never be detected directly. This is due to the fact that the photon is absorbed by the second electron very shortly after it is emitted by the first electron. The virtual photons violate the law of conservation of energy, but because of the uncertainty principle and the very short lifetime Δt, the photon's energy is less than the uncertainty in its energy, given by $\Delta E \approx \hbar/\Delta t$.

Now consider the pion exchange between a proton and a neutron that mediates the strong force. The Feynman diagram for this interaction is shown in Figure 15.3. One can reason that the energy ΔE needed to create a pion of mass m_π is given by Einstein's equation, $\Delta E = m_\pi c^2$. Again, the very existence of the pion violates energy conservation by an amount ΔE, which is permitted by the uncertainty principle only if this energy is surrendered in a time Δt, where Δt is the time it takes the pion to transfer between nucleons. From the uncertainty principle, we have

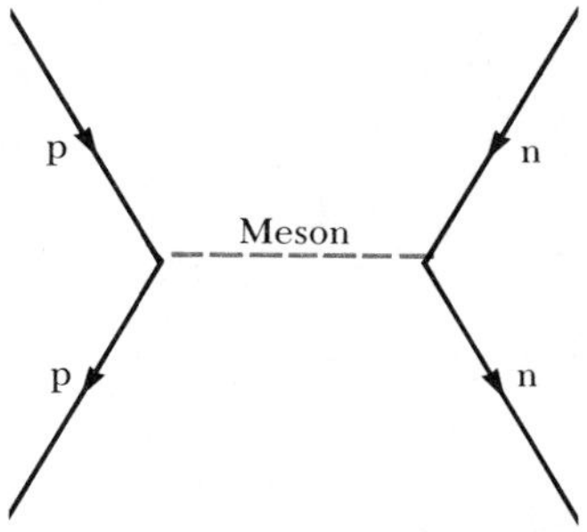

$$\Delta t \approx \frac{\hbar}{\Delta \mathrm{E}} = \frac{\hbar}{\mathrm{m}_\pi \mathrm{c}^2} \tag{15.2}$$

Since the pion cannot travel faster than the speed of light, the maximum distance d it can travel in a time Δt is $c\Delta t$. Using Equation 15.2 and $d = c\Delta t$, we find this maximum distance to be

$$d \approx \frac{\hbar}{\mathrm{m}_\pi \mathrm{c}} \tag{15.3}$$

We know that the range of the strong force is about 1.5×10^{-15} m (see Chapter 13). Using this value for d in Equation 15.3, the rest energy of the pion is calculated to be

$$m_\pi c^2 \approx \frac{\hbar c}{d} = \frac{(1.05 \times 10^{-34}\ \mathrm{J \cdot s})(3 \times 10^8\ \mathrm{m/s})}{1.5 \times 10^{-15}\ \mathrm{m}}$$

$$= 2.1 \times 10^{-11}\ \mathrm{J} \approx 130\ \mathrm{MeV}$$

This corresponds to a mass of 130 MeV/c^2 (about 250 times the mass of the electron) which is in good agreement with the observed mass of the pion.

The concept we have just described is quite revolutionary, and something we have not encountered as yet. In effect, it says that a proton can change into a proton plus a pion, as long as it returns to its original state in a very short time. High-energy physicists often say that a nucleon undergoes "fluctuations" as it emits and absorbs pions. As we have seen, these fluctuations are a consequence of a combination of quantum mechanics (through the uncertainty principle) and special relativity (through Einstein's relation $E = mc^2$).

This section has dealt with the particles that mediate the strong force, namely the pions, and the mediators of the electromagnetic force, photons. The graviton, which is the mediator of the gravitational force, has yet to be observed. The particles that mediate the weak nuclear force are referred to as W^+, W^-, and Z^0. The discovery of the $W^\pm$ and Z^0 particles at CERN was announced in 1983 by Carlo Rubbia (1915–) and his associates, who used a proton-antiproton collider.[2] In this accelerator, protons and antiprotons that

p
n
Meson
p
n

Figure 15.3 Feynman diagram representing a proton interacting with a neutron via the strong force. In this case, the meson mediates the strong force.

[2] Carlos Rubbia, an Italian physicist, and Simon van der Meer, a Dutch physicist, both at CERN, shared the 1984 Nobel Prize in Physics for the discovery of the $W^\pm$ and Z^0 particles and the development of the proton-antiproton collider.

have a momentum of 270 GeV/c undergo head-on collisions with each other. In some of the collisions $W^{\pm}$ and Z^0 particles are produced, which are, in turn, identified by their decay products.

15.5 CLASSIFICATION OF PARTICLES

Hadrons

All particles other than photons can be classified into two broad categories, hadrons and leptons, according to the interactions they experience. Particles that interact through the strong force are called **hadrons.** There are two classes of hadrons, known as **mesons** and **baryons.** These can be classified according to their masses and spins.

Mesons all have zero or integral spins (0 or 1), with masses that lie between the mass of the electron and the mass of the proton. All mesons are known to decay finally into electrons, muons, neutrinos, and photons. The pion is the lightest of known mesons, with a mass of about 140 MeV/c^2 and a spin of 0. Another is the K meson, with a mass of about 500 MeV/c^2 and spin 0.

Baryons, the second class of hadrons, have a mass equal to or greater than the proton mass (hence the name *baryon,* which means "heavy" in Greek), and

TABLE 15.2 A Table of Some Particles and Their Properties

Category	Particle Name	Symbol	Anti-particle	Rest Mass (MeV/c^2)	B	L_e	L_μ	L_τ	S	Lifetime (s)	Principal Decay Modes[a]
Photon	Photon	γ	Self	0	0	0	0	0	0	Stable	
Leptons	Electron	e^-	e^+	0.511	0	+1	0	0	0	Stable	
	Neutrino (e)	ν_e	$\bar{\nu}_e$	0 (?)	0	+1	0	0	0	Stable	
	Muon	μ^-	μ^+	105.7	0	0	+1	0	0	2.20×10^{-6}	$e^-\bar{\nu}_e\nu_\mu$
	Neutrino (μ)	ν_μ	$\bar{\nu}_\mu$	0 (?)	0	0	+1	0	0	Stable	
	Tau	τ^-	τ^+	1784.	0	0	0	−1	0	$<4 \times 10^{-13}$	$\mu^-\bar{\nu}_\mu\nu_\tau$, $e^-\bar{\nu}_e\nu_\tau$
	Neutrino (τ)	ν_τ	$\bar{\nu}_\tau$	0 (?)	0	0	0	−1	0	Stable	
Hadrons											
Mesons	Pion	π^+	π^-	139.6	0	0	0	0	0	2.60×10^{-8}	$\mu^+\nu_\mu$
		π^0	Self	135.0	0	0	0	0	0	0.83×10^{-16}	2γ
	Kaon	K^+	K^-	493.7	0	0	0	0	+1	1.24×10^{-8}	$\mu^+\nu_\mu$, $\pi^+\pi^0$
		K^0_S	$\overline{K}^0_S$	497.7	0	0	0	0	+1	0.89×10^{-10}	$\pi^+\pi^-$, $2\pi^0$
		K^0_L	$\overline{K}^0_L$	497.7	0	0	0	0	+1	5.2×10^{-8}	$\pi^\pm e^\mp\bar{\nu}_e$ $3\pi^0$ $\pi^\pm\mu^\mp\bar{\nu}_\mu$
	Eta	η^0	Self	548.8	0	0	0	0	0	$<10^{-18}$	2γ, 3π
Baryons	Proton	p	$\bar{p}$	938.3	+1	0	0	0	0	Stable	
	Neutron	n	$\bar{n}$	939.6	+1	0	0	0	0	920	$pe^-\bar{\nu}_e$
	Lambda	Λ^0	$\overline{\Lambda}^0$	1115.6	+1	0	0	0	−1	2.6×10^{-10}	$p\pi^-$, $n\pi^0$
	Sigma	Σ^+	$\overline{\Sigma}^-$	1189.4	+1	0	0	0	−1	0.80×10^{-10}	$p\pi^0$, $n\pi^+$
		Σ^0	$\overline{\Sigma}^0$	1192.5	+1	0	0	0	−1	6×10^{-20}	$\Lambda^0\gamma$
		Σ^-	$\overline{\Sigma}^+$	1197.3	+1	0	0	0	−1	1.5×10^{-10}	$n\pi^-$
	Xi	Ξ^0	$\overline{\Xi}^0$	1315	+1	0	0	0	−2	2.9×10^{-10}	$\Lambda^0\pi^0$
		Ξ^-	$\overline{\Xi}^+$	1321	+1	0	0	0	−2	1.64×10^{-10}	$\Lambda^0\pi^-$
	Omega	Ω^-	Ω^+	1672	+1	0	0	0	−3	0.82×10^{-10}	$\Xi^0\pi^{\overline{0}}$, Λ^0K^-

[a] Notations in this column such as $p\pi^-, n\pi^0$ mean two possible decay modes. In this case, the two possible decays are $\Lambda^0 \rightarrow p + \pi^-$ and $\Lambda^0 \rightarrow n + \pi^0$.

their spin is always a noninteger value ($\frac{1}{2}$ or $\frac{3}{2}$). Protons and neutrons are included in the baryon family, as are many other particles. With the exception of the proton, all baryons decay in such a way that the end products include a proton. For example, the Ξ^- hyperon first decays to a Λ^0 hyperon and a negatively charged pion (π^-) in about 10^{-10} s. The Λ^0 then decays to a proton and a π^- in about 3×10^{-10} s.

Today it is believed that hadrons are composed of more elemental units called *quarks*. Later we shall have more to say about the quark model. Some of the important properties of hadrons are listed in Table 15.2.

Leptons

Leptons (from the Greek *leptos* meaning "small" or "light") are a group of particles that participate in the electromagnetic and weak interactions. All leptons have a spin of $\frac{1}{2}$. Included in this group are electrons, muons, and neutrinos, which are all less massive than the lightest hadron. Although hadrons have size and structure, leptons appear to be truly elementary particles with no structure (that is, pointlike).

Quite unlike the situation with hadrons, the number of known leptons is very limited. Currently, scientists believe there only are six leptons (each having an antiparticle)—the electron, the muon, the tau, and a neutrino associated with each of these particles. The τ lepton, discovered in 1975, has a mass about twice that of the proton. (The neutrino associated with the tau has not yet been observed in the laboratory.) We now classify the six known leptons into three groups:

$$\begin{pmatrix} e^- \\ \nu_e \end{pmatrix} \quad \begin{pmatrix} \mu^- \\ \nu_\mu \end{pmatrix} \quad \begin{pmatrix} \tau^- \\ \nu_\tau \end{pmatrix}$$

Although neutrinos are thought to be massless, there is a possibility that they may have a small, nonzero mass. As we shall see later, a firm knowledge of the neutrino's mass could have great significance in cosmological models and predictions of the future of the universe.

15.6 CONSERVATION LAWS

Conservation laws are important in understanding why certain decays or reactions occur and others do not. In general, the laws of conservation of energy, linear momentum, angular momentum, and electric charge provide us with a set of rules that all processes must follow. For example, conservation of electric charge requires that the total charge before a reaction must equal the total charge after the reaction.

A number of new conservation laws are important in the study of elementary particle decays and reactions. Two of these described in this section are the conservation of baryon number and the conservation of lepton number.

Baryon Number

The conservation of baryon number implies that whenever a baryon is created in a reaction or decay, an antibaryon is also created. This can be quantified by assigning a baryon number $B = +1$ to all baryons, $B = -1$ to all antibaryons, and $B = 0$ to all other particles. Thus, the **law of conservation of baryon number** can be stated as follows:

Whenever a nuclear reaction occurs, the sum of the baryon numbers before the process must equal the sum of the baryon numbers after the process.

An equivalent statement is that the net number of baryons remains constant in any process.

Note that if baryon number is absolutely conserved, the proton must be absolutely stable. If it were not for the law of conservation of baryon number, the proton could decay to a positron and a neutral pion. However, such a decay has never been observed. At the present, we can say only that the proton has a lifetime of at least 10^{31} years (which could be compared with the estimated age of the universe, which is about 10^{10} years). In one recent version of the grand unified theory, or GUT, physicists have predicted that the proton is actually unstable. According to this theory, the baryon number (sometimes called the baryonic charge) cannot be absolutely conserved, while electric charge is always conserved.

EXAMPLE 15.1 Checking Baryon Numbers

Determine whether or not each of the following reactions can occur, based on the law of conservation of baryon number.

$$p + n \longrightarrow p + p + n + \bar{p} \quad (1)$$

$$p + n \longrightarrow p + p + \bar{p} \quad (2)$$

Solution: First, let us check reaction (1). Recall that $B = +1$ for baryons and $B = -1$ for antibaryons. Hence the left side of (1) gives a total baryon number of $1 + 1 = 2$. The right side of (1) gives a total baryon number of $1 + 1 + 1 + (-1) = 2$. Hence the reaction can occur, provided the incoming proton has sufficient energy.

Now let us examine reaction (2). The left side of (2) again gives a total baryon number of $1 + 1 = 2$. However, the right side of (2) gives a total number of $1 + 1 + (-1) = 1$. Since the baryon number is not conserved, the reaction cannot occur.

Lepton Number

Recall that there are three varieties of leptons: the electron, the muon, and the recently discovered tau lepton. Each of these is accompanied by a neutrino. There are three separate conservation laws involving lepton numbers, one for each variety of lepton. The **law of conservation of electron-lepton number** states that

the sum of the electron-lepton numbers before a reaction or decay must equal the sum of the electron-lepton numbers after the reaction or decay.

The electron (e^-) and the electron neutrino (ν_e) are assigned a positive electron-lepton number $L_e = +1$, the antileptons e^+ and $\bar{\nu}_e$ are assigned a negative electron-lepton number $L_e = -1$, and all others have $L_e = 0$. For example, consider the decay of the neutron,

Neutron decay

$$n \longrightarrow p + e^- + \bar{\nu}_e$$

Before the decay, the electron-lepton number is $L_e = 0$, while after the decay the electron-lepton number is $0 + 1 + (-1) = 0$. Thus, the electron-lepton number is conserved. It is important to recognize that the baryon number must also be conserved. This can easily be seen by noting that before the decay $B = +1$, while after the decay the baryon number is $+1 + 0 + 0 = +1$.

Similarly, when a decay involves muons, the muon-lepton number, L_μ, is conserved. The μ^- and the ν_μ are assigned positive numbers, $L_\mu = +1$, the antimuons μ^+ and $\bar{\nu}_\mu$ are assigned negative numbers, $L_\mu = -1$, while all others have $L_\mu = 0$. Finally, the tau-lepton number, L_τ, is conserved, and similar assignments can be made for the τ lepton and its neutrino, ν_τ.

EXAMPLE 15.2 Checking Lepton Numbers
Determine which of the following decay schemes can occur on the basis of conservation of lepton number.

$$\mu^- \longrightarrow e^- + \bar{\nu}_e + \nu_\mu \qquad (1)$$

$$\pi^+ \longrightarrow \mu^+ + \nu_\mu + \nu_e \qquad (2)$$

Solution: First let us examine decay (1). Since this decay involves both a muon and an electron, L_μ and L_e must both be conserved. Before the decay, $L_\mu = +1$ and $L_e = 0$. After the decay, $L_\mu = 0 + 0 + 1 = +1$, and $L_e = +1 - 1 + 0 = 0$. Thus, both numbers are conserved, and the decay mode is possible.

Now consider decay (2). Before the decay, $L_\mu = 0$ and $L_e = 0$. After the decay, $L_\mu = -1 + 1 + 0 = 0$, but $L_e = +1$. Thus, the decay is not possible because the electron-lepton number is not conserved.

Exercise 1 Determine whether the decay $\mu^- \rightarrow e^- + \bar{\nu}_e$ can occur.

Answer: No. The muon-lepton number is $+1$ before the decay and is 0 after the decay. Thus, the muon-lepton number is not conserved.

15.7 STRANGE PARTICLES AND STRANGENESS

Many particles discovered in the 1950s were produced by the strong interaction of pions with protons and neutrons in the atmosphere. A group of these particles, namely the K, Λ, and Σ, were found to exhibit unusual properties in their production and decay, and hence were called *strange particles.* One unusual property is that they are always produced in pairs. For example, when a pion collides with a proton, two neutral strange particles are produced with high probability (see Fig. 15.4) following the reaction

$$\pi^- + p \longrightarrow K^0 + \Lambda^0$$

On the other hand, the reaction $\pi^- + p \longrightarrow K^0 + n$ never occurred, even though no known conservation laws were violated and the energy of the pion was sufficient to initiate the reaction. The second peculiar feature of strange particles is that although they are produced by the strong interaction at a high rate, they do not decay into strongly interacting particles at a very high rate as one might expect. Instead, they decay very slowly, which is characteristic of the weak interaction. Their lifetimes are in the range from 10^{-10} s to 10^{-8} s; strongly interacting particles have lifetimes of the order of 10^{-23} s.

In order to explain these unusual properties of strange particles, a new conservation law called conservation of strangeness was introduced, together with a new quantum number S called the **strangeness.** The strangeness numbers for various particles are given in Table 15.2. The production of strange particles in pairs is explained by assigning $S = +1$ to one of the particles and $S = -1$ to the other. The nonstrange particles such as the π mesons, protons, and leptons are assigned strangeness $S = 0$. The **law of conservation of strangeness** can be stated as follows:

> Whenever a nuclear reaction or decay occurs, the sum of the strangeness numbers before the process must equal the sum of the strangeness numbers after the process.

One can explain the slow decay of strange particles by assuming that the strong and electromagnetic interactions obey the law of conservation of strangeness, while the weak interaction does not. Since the decay reaction involves the loss of one strange particle, it violates strangeness conservation, and hence proceeds slowly via the weak interaction.

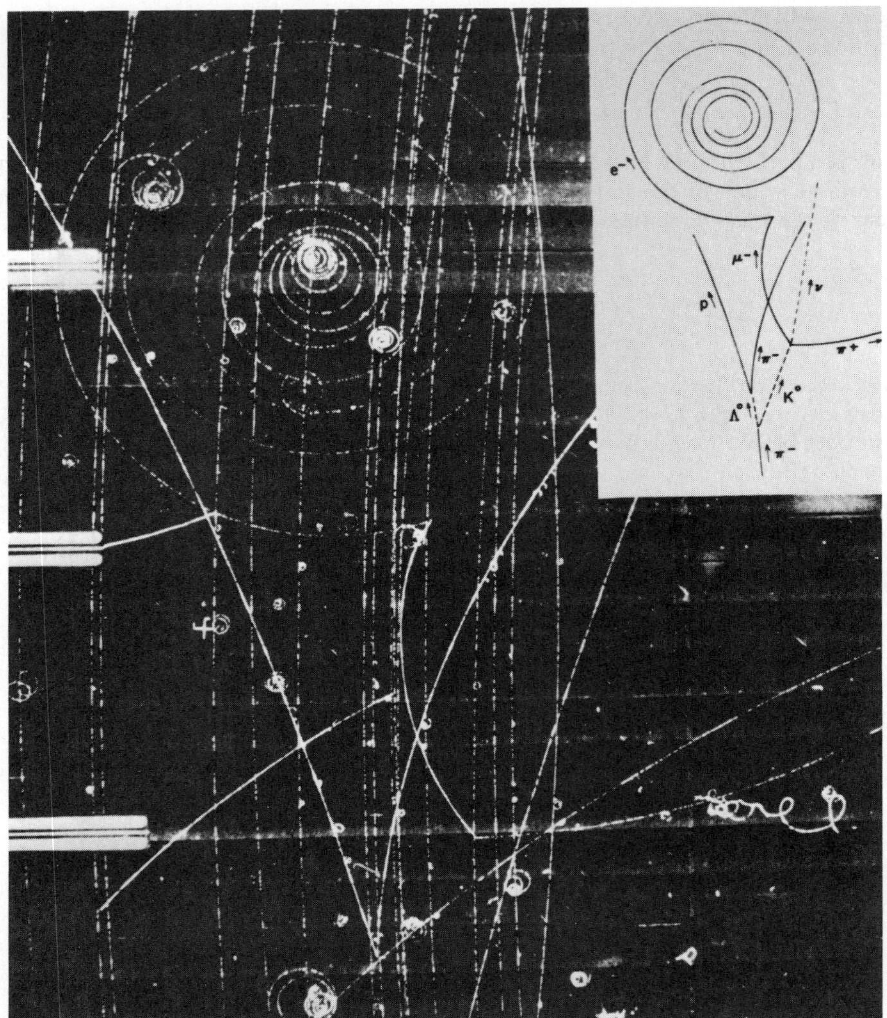

Figure 15.4 This bubble-chamber photograph shows many events, and the inset is a drawing of identified tracks. The strange particles Λ^0 and K^0 are formed (at the bottom) as the π^- interacts with a proton according to $\pi^- + p \rightarrow \Lambda^0 + K^0$. (Note that the neutral particles leave no tracks, as indicated by the dashed lines.) The Λ^0 and K^0 then decay according to $\Lambda^0 \rightarrow \pi^- + p$ and $K^0 \rightarrow \pi^+ + \mu^- + \nu_\mu$. (Courtesy of Lawrence Berkeley Laboratory, University of California, Photographic Services.)

EXAMPLE 15.3 Is Strangeness Conserved?

(a) Determine whether the following reaction occurs, on the basis of conservation of strangeness.

$$\pi^0 + n \longrightarrow K^+ + \Sigma^-$$

Solution: The initial state has a total strangeness $S = 0 + 0 = 0$. Since the strangeness of the K^+ is $S = +1$, and the strangeness of the Σ^- is $S = -1$, the total strangeness of the final state is $+1 - 1 = 0$. Thus, we see that strangeness is conserved and the reaction is allowed.

(b) Show that the following reaction does not conserve strangeness, and hence cannot occur.

$$\pi^- + p \longrightarrow \pi^- + \Sigma^+$$

Solution: In this case, the initial state has a total strangeness $S = 0 + 0 = 0$, while the final state has a total strangeness $S = 0 + (-1) = -1$. Thus strangeness is not conserved, so the reaction does not occur.

Exercise 2 Show that the following observed reaction $p^+ + \pi^- \longrightarrow K^0 + \Lambda^0$ obeys the law of conservation of strangeness.

TABLE 15.3 Properties of Quarks and Antiquarks

Name	Symbol	Spin	Charge	Baryon Number	Strangeness	Charm	Bottomness	Topness
Quarks								
Up	u	$\frac{1}{2}$	$+\frac{2}{3}e$	$\frac{1}{3}$	0	0	0	0
Down	d	$\frac{1}{2}$	$-\frac{1}{3}e$	$\frac{1}{3}$	0	0	0	0
Strange	s	$\frac{1}{2}$	$-\frac{1}{3}e$	$\frac{1}{3}$	−1	0	0	0
Charmed	c	$\frac{1}{2}$	$+\frac{2}{3}e$	$\frac{1}{3}$	0	+1	0	0
Bottom	b	$\frac{1}{2}$	$-\frac{1}{3}e$	$\frac{1}{3}$	0	0	+1	0
Top (?)	t	$\frac{1}{2}$	$+\frac{2}{3}e$	$\frac{1}{3}$	0	0	0	+1
Antiquarks								
Up	$\bar{u}$	$\frac{1}{2}$	$-\frac{2}{3}e$	$-\frac{1}{3}$	0	0	0	0
Down	$\bar{d}$	$\frac{1}{2}$	$+\frac{1}{3}e$	$-\frac{1}{3}$	0	0	0	0
Strange	$\bar{s}$	$\frac{1}{2}$	$+\frac{1}{3}e$	$-\frac{1}{3}$	+1	0	0	0
Charmed	$\bar{c}$	$\frac{1}{2}$	$-\frac{2}{3}e$	$-\frac{1}{3}$	0	−1	0	0
Bottom	$\bar{b}$	$\frac{1}{2}$	$+\frac{1}{3}e$	$-\frac{1}{3}$	0	0	−1	0
Top (?)	$\bar{t}$	$\frac{1}{2}$	$-\frac{2}{3}e$	$-\frac{1}{3}$	0	0	0	−1

15.8 QUARKS—FINALLY

As we have noted, leptons appear to be truly elementary particles because they have no measurable size or internal structure, are limited in number, and do not seem to break down into smaller units. Furthermore, we know that hadrons decay into other hadrons and are large in number. It is important to note that Table 15.2 lists only hadrons that are stable against hadronic decay. Hundreds of others have been discovered, and their properties have been determined. These facts strongly suggest that hadrons cannot be truly elementary, but have some substructure.

The Original Quark Model

In 1963 Murray Gell-Mann and George Zweig independently proposed that the known hadrons have a more elemental substructure. According to their model, all hadrons are composite systems of two or three fundamental constituents called **quarks.** (Gell-Mann borrowed the word "quark" from the passage "Three quarks for Muster Mark" in James Joyce's book *Finnegan's Wake.*) In the original quark model, there were three types of quarks designated by the symbols u, d, and s. These were given the arbitrary names *up, down,* and *sideways* (or now more commonly, *strange*). A most unusual property of quarks is that they have fractional electric charges. The u, d, and s quarks have charges of $+2e/3$, $-e/3$, and $-e/3$, respectively, where e is the charge of an electron. Each quark has a baryon number of $\frac{1}{3}$ and a spin of $\frac{1}{2}$. Furthermore, the u and d quarks have strangeness of 0, while the s quark has strangeness of −1. Other properties of quarks and antiquarks are given in Table 15.3. Note that associated with each quark is an antiquark of opposite charge, baryon number, and strangeness. The compositions of all hadrons known at the time could be completely specified by two simple rules: (1) Mesons consist of one quark and one antiquark, giving them a baryon number of 0 as required. (2) Baryons consist of three quarks. Table 15.4 lists the quark compositions of several mesons and baryons. For example, a π^- meson contains one $\bar{u}$ and one d quark (designated as $\bar{u}d$), with $Q = -2e/3 - e/3 = -e$, $B = -\frac{1}{3} + \frac{1}{3} = 0$, and $S = 0 + 0 = 0$. A proton, on the other hand, contains two u quarks and one d

TABLE 15.4 Quark Compositions of Several Hadrons

Particle	Quark Composition
Mesons	
π^+	$u\bar{d}$
π^-	$\bar{u}d$
K^+	$u\bar{s}$
K^-	$\bar{u}s$
K^0	$d\bar{s}$
Baryons	
p	uud
n	udd
Λ^0	uds
Σ^+	uus
Σ^0	uds
Σ^-	dds
Ξ^0	uss
Ξ^-	dss
Ω^-	sss

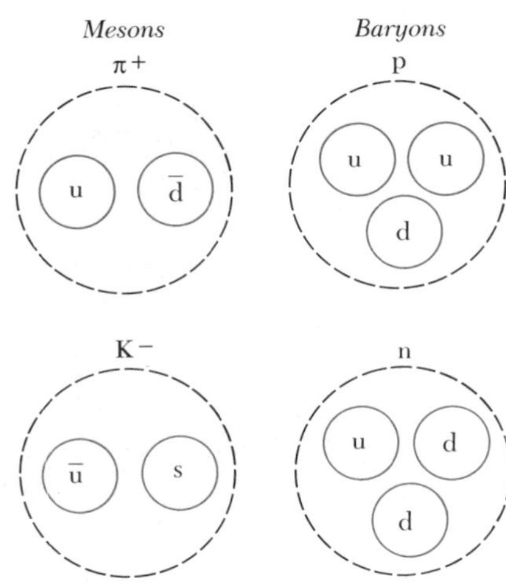

Figure 15.5 Quark compositions of several particles. Note that the mesons on the left contain two quarks, while the baryons on the right contain three quarks.

quark, with $Q = +1$, $B = 1$, and $S = 0$. Note that just two of the quarks, u and d, are contained in all hadrons encountered in ordinary matter (protons and neutrons). The third quark, s, is needed only to construct strange particles with a strangeness number of either $+1$ or -1. For example, the neutral particle Λ^0 has uds as its quark composition, with $Q = 0$, $B = 1$, and $S = -1$. Figure 15.5 is a pictorial representation of the quark compositions of several particles.

Charm and Other Recent Developments

Although the original quark model was highly successful in classifying particles into families, there were some discrepancies between predictions of the model and certain experimental decay rates. Consequently, a fourth quark was proposed by several physicists in 1967. They argued that if there are four leptons (as was thought at the time), then there should also be four quarks because of an underlying symmetry in nature. The fourth quark, designated by c, was given the new property, or quantum number, called **charm.** The *charmed* quark would have a charge $+2e/3$, but its new property of charm would distinguish it from the other three quarks. The new quark would have a charm of $C = +1$, while its antiquark would have a charm of $C = -1$, as indicated in Table 15.3. Charm, like strangeness, would be conserved in

The American physicists Richard Feynman *(left)* and Murray Gell-Mann *(right)* won the Nobel prize in physics in 1965 and 1969, respectively, for their theoretical studies dealing with subatomic particles. (Photo courtesy of Michael R. Dressler)

strong and electromagnetic interactions, but would not be conserved in weak interactions.

In 1974, a new heavy meson called the J/Ψ particle was discovered independently by a group led by Burton Richter at the Stanford Linear Accelerator (SLAC) and another group led by Samuel Ting at the Brookhaven National Laboratory. Richter and Ting were awarded the Nobel Prize in 1976 for their discovery. The particle did not fit into the original three-quark model, but had the properties of a combination of a charmed quark and its antiquark ($c\bar{c}$). Its mass was much greater than those of the other known mesons ($\sim 3100 \text{ MeV}/c^2$) and its lifetime was much longer than those of other strongly decaying particles. Soon thereafter, other related charmed mesons called D^+, D^-, and D^0 were discovered, corresponding to such quark combinations as $c\bar{d}$ and $\bar{c}d$, all of which have large masses and long lifetimes. In 1975, the same group at SLAC that discovered the J/Ψ particle reported strong evidence for the tau (τ) lepton, with a mass of $1784 \text{ MeV}/c^2$. Such discoveries led to more elaborate quark models, and the proposal of two new quarks, named *top* (t) and *bottom* (b). (Some physicists prefer the whimsical names *truth* and *beauty*.) To distinguish these quarks from the old ones, quantum numbers called *topness* and *bottomness* were assigned to these new particles, and are included in Table 15.3. In 1977, researchers at the Fermi National Laboratory under the direction of Leon Lederman reported the discovery of a very massive new meson, Υ, whose composition is considered to be $b\bar{b}$.

At this point, you are probably wondering whether such discoveries will ever end. How many "building blocks" of matter really exist? At present, physicists believe that the fundamental particles in nature include six quarks and six leptons (together with their antiparticles). Some of the properties of these particles are given in Table 15.5.

In spite of many extensive experimental efforts, no isolated quark has ever been observed. Physicists now believe that quarks are permanently confined inside ordinary particles because of an exceptionally strong force that prevents them from escaping. This force, called the "color" force, increases with separation distance (similar to the force of a spring); the properties of this force are discussed in the next section. The great strength of the force between quarks has been described by one author as follows:[3]

> Quarks are slaves of their own color charge, . . . bound like prisoners of a chain gang . . . Any locksmith can break the chain between two prisoners, but no locksmith is expert enough to break the gluon chains between quarks. Quarks remain slaves forever.

TABLE 15.5 The Fundamental Particles and Some of Their Properties

Particle	Rest Energy	Charge
Quarks		
u	360 MeV	$+\frac{2}{3}e$
d	360 MeV	$-\frac{1}{3}e$
c	1500 MeV	$+\frac{2}{3}e$
s	540 MeV	$-\frac{1}{3}e$
t (?)	~100 GeV	$+\frac{2}{3}e$
b	5 GeV	$-\frac{1}{3}e$
Leptons		
e^-	511 keV	$-e$
μ^-	107 MeV	$-e$
τ^-	1784 MeV	$-e$
ν_e	<30 eV	0
ν_μ	<0.5 MeV	0
ν_τ	<250 MeV	0

15.9 THE STANDARD MODEL

Shortly after the concept of quarks was proposed, scientists recognized that certain particles had quark compositions that were in violation of the Pauli exclusion principle. Recall that quarks are fermions with spins of $\frac{1}{2}$, and hence are expected to follow the exclusion principle. One example is the Ω^- (sss) baryon, which contains three s quarks having parallel spins, giving it a total spin of $\frac{3}{2}$. Other examples of baryons that have identical quarks with parallel spins are the Δ^{++} (uuu) and the Δ^- (ddd). To resolve this problem, it was suggested that quarks possess a new property called **color.** This property is similar in many respects to electric charge, except that it occurs in three

[3] Harald Fritzsch, *Quarks, The Stuff of Matter*, London, Allen Lane, 1983.

varieties (of color) called red, green, and blue. Another term that is used to distinguish among the six quarks is **flavor.** The various flavors are those already mentioned (top, bottom, etc.) and each flavor of quark can have three colors. (The whimsical names "color" and "flavor" should not be taken literally.) Of course, *the antiquarks have the colors antired, antigreen, and antiblue.* In order to satisfy the exclusion principle, all three quarks in a baryon must have different colors. A meson consists of a quark of one color and an antiquark of the corresponding anticolor. The result is that baryons and mesons are always colorless (or white). Furthermore, the new property of color increases the number of quarks by a factor of three.

Although the concept of color in the quark model was originally conceived to satisfy the exclusion principle, it also provided a better theory for explaining certain experimental results. For example, the modified theory correctly predicts the lifetime of the π^0 meson. The theory of how quarks interact with each other is called **quantum chromodynamics,** or QCD, because it is similar in its structure to quantum electrodynamics (the theory of interaction between electric charges). In QCD, the quark is said to carry a *color charge,* in analogy to electric charge. The strong force between quarks is often called the *color force.* As mentioned earlier, the strong interaction between hadrons is mediated by massless particles called **gluons** (analogous to photons for the electromagnetic force). According to the theory, there are eight gluons, six of which have color charge. Because of their color charge, quarks can attract each other and form composite particles. When a quark emits or absorbs a gluon, its color changes. For example, a blue quark that emits a gluon may become a red quark, while the red quark that absorbs this gluon becomes a blue quark. The color force between quarks is analogous to the electric force between charges—like colors repel and opposite colors attract. Therefore, two red quarks repel each other, but a red quark will be attracted to an antired quark. The attraction between quarks of opposite color to form a meson ($q\bar{q}$) is indicated in Figure 15.6a. Differently colored quarks also attract each other, but with less intensity than opposite colors of quark and antiquark. For example, a cluster of red, blue, and green quarks all attract each other to form baryons as indicated in Figure 15.6b. Thus, all baryons contain three quarks, each of which has a different color.

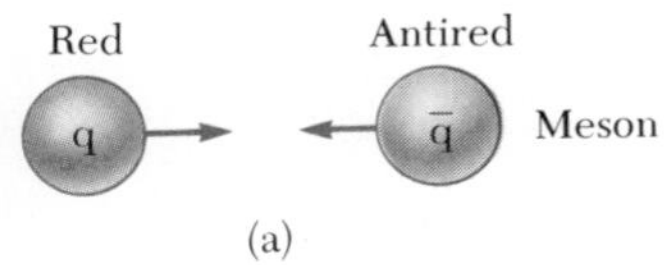

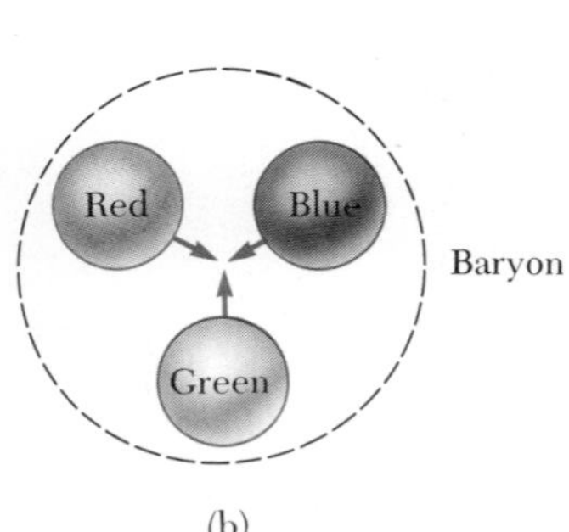

Figure 15.6 (a) A red quark is attracted to an antired quark. This forms a meson whose quark structure is ($q\bar{q}$). (b) Three differently colored quarks attract each other to form a baryon.

Recall that the weak force is believed to be mediated by the W^+, W^-, and Z^0 bosons (spin 1 particles). These particles are said to have *weak charge* just as a quark has color charge. Thus, each elementary particle can have mass, electric charge, color charge, and weak charge. Of course, one or more of these could be zero. Scientists now believe that the truly elementary particles are leptons and quarks and the corresponding force mediators: gluon, photon, $W^\pm$, Z^0, and graviton. (Note that quarks and leptons are fermions with spin $\frac{1}{2}$, whereas the force mediators are bosons with spin 1 or higher.)

In 1979, Sheldon Glashow, Abdus Salam, and Steven Weinberg won a Nobel Prize for developing a theory that unified the electromagnetic and weak interactions. This so-called **electroweak theory** postulates that the weak and electromagnetic interactions have the same strength at very high particle energies. Thus, the two interactions are viewed as two different manifestations of a single unifying electroweak interaction. The photon and the three massive bosons ($W^\pm$ and Z^0) play a key role in the electroweak theory. The theory makes many concrete predictions, but perhaps the most spectacular is the prediction of the masses of the W and Z particles at about 82 GeV/c^2 and 93 GeV/c^2, respectively. The 1984 Nobel Prize in physics was awarded to Carlo Rubbia and Simon van der Meer for their work leading to the discovery of

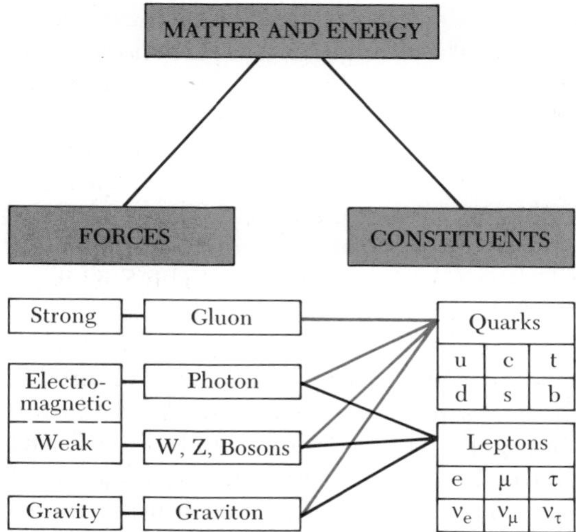

Figure 15.7 The Standard Model of particle physics.

these particles at just these masses at the CERN Laboratory in Geneva, Switzerland.

The combination of the electroweak theory and QCD for the strong interaction forms what is referred to in high-energy physics as the "Standard Model." Although the details of the Standard Model are complex, its essential ingredients can be summarized with the help of Figure 15.7. The strong force, mediated by gluons, holds quarks together to form composite particles such as protons, neutrons, and mesons. Leptons participate only in the electromagnetic and weak interactions. The electromagnetic force is mediated by photons, while the weak force is mediated by W and Z bosons. Note that all fundamental forces are mediated by spin 1 particles whose properties are given, to a large extent, by the symmetries involved in the theories.

However, the Standard Model does not answer all questions. A major problem is why the photon has no mass while the W and Z bosons are massive particles. Because of that mass difference, the electromagnetic and weak forces are quite distinct at low energies, but become similar in nature at very high energies. This behavior as one goes from low to high energies, called *symmetry breaking*, leaves open the question of the origin of particle masses. In order to resolve this problem, a hypothetical particle called the *Higgs boson* has been proposed, which provides a mechanism for breaking the electroweak symmetry. The Standard Model, including the Higgs mechanism, provides a logically consistent explanation of the massive nature of the W and Z bosons. Unfortunately, the Higgs boson has not yet been found, but physicists believe that its mass should be less than 1 TeV (10^{12} eV).

In order to determine whether the Higgs boson exists, two quarks of at least 1 TeV of energy must collide, but calculations show that this requires injecting 40 TeV of energy within the volume of a proton. The excess energy is needed because of the quarks and gluons contained in the proton. Although no existing accelerator can provide this energy, scientists in the United States are currently planning to construct the world's largest and most powerful particle accelerator (called the Superconducting Super Collider, or SSC), which will meet this need. The SSC will accelerate protons to 20 TeV (two protons traveling in opposite directions with this energy will give the required 40 TeV), will have a circumference of 82.944 km (about 52 miles), and will cost about 4 billion dollars! The proton energies in the SSC will be twenty times greater than is currently available at Fermilab, and physicists expect to be able to explore distances down to about 10^{-18} m (one thousandth the diameter of a proton). Thus, in effect it will be a microscope of unparalleled power. Figure 15.8 shows the evolution of the various stages of matter that scientists have been able to investigate with various types of microscopes. The Department of Energy endorsed detailed research on the SSC in 1983, and federal funds were recently allocated to support research and development on the superconducting magnets to be used in the accelerator. The SSC is expected to be completed in the mid-1990s if federal funding and Congressional authorization are received.

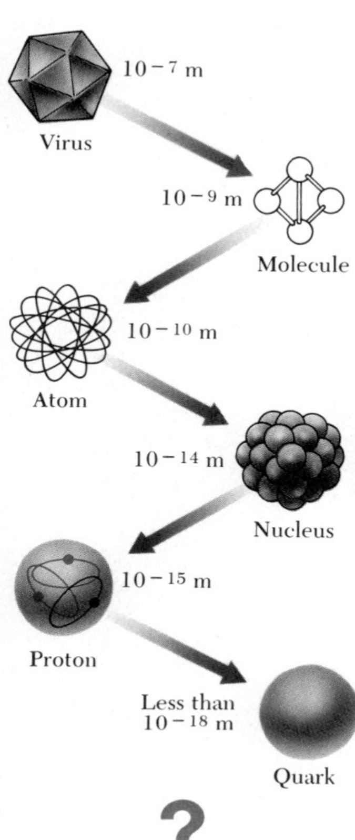

Figure 15.8 Looking at matter with various microscopes reveals structures ranging in size from the smallest living thing, a virus (at the top), down to a quark (at the bottom), which has not yet been observed as an isolated particle.

Following the success of the electroweak theory, scientists attempted to combine it with QCD in a *grand unification theory* known as GUT. In this model, the electroweak force was merged with the strong color force to form a grand unified force. One version of the theory considers leptons and quarks as members of the same family that are able to change into each other by exchanging an appropriate particle. Many GUT theories predict that protons are unstable, and will decay with a life time of about 10^{31} years. This is far greater than the age of the universe, and as yet proton decays have not been observed. The essay that follows this chapter describes ongoing experiments pertaining to attempts to observe nucleon decay.

15.10 THE COSMIC CONNECTION

Today cosmologists believe that the universe was born as the result of a single violent explosion known as the Big Bang. The evolution of the four fundamental forces from the Big Bang to the present is shown in Figure 15.9. During the first 10^{-43} s (the ultra-hot epoch where $T \approx 10^{32}$ K), it is presumed that the strong, electroweak, and gravity forces were joined to form a completely unified force. In the first 10^{-32} s following the Big Bang (the hot epoch where $T \approx 10^{29}$ K), gravity broke free of this unification while the strong and electroweak forces remained as one, described by a grand unification theory. This was a period when particle energies were so great ($> 10^{16}$ GeV) that very massive particles as well as light quarks, leptons, and their antiparticles existed. At the end of this fleeting instant of time, the very dense and hot universe could be squeezed into a sphere less than 10^{-4} m diameter, but was even smaller than this at an earlier time! Then the universe rapidly expanded and cooled during the warm epoch, which covers temperatures in the range from 10^{29} to 10^{15} K (corresponding to energies ranging from 10^{16} to 10^{2} GeV). During this epoch, the strong and electroweak forces parted company, and the grand unification scheme was broken. As the universe continued to cool, about 10^{-10} s after the Big Bang, the electroweak force split in two. This resulted in the four forces in the universe as we distinguish them today. During the first microsecond of cosmic history, there were only elementary particles interacting through fundamental forces. About one microsecond after the Big Bang, the quarks condensed into protons and neutrons, and a few minutes later the sea of protons and neutrons condensed to form atomic nuclei. You should read Appendix B, which provides a more detailed discussion of the Big Bang theory.

While particle physicists have been exploring the realm of the very small, cosmologists have been exploring cosmic history back to the first microsecond of the Big Bang. The observation of events that occur when two particles collide in an accelerator is essential in reconstructing the early moments in cosmic history. Perhaps the key to understanding the early universe is first to

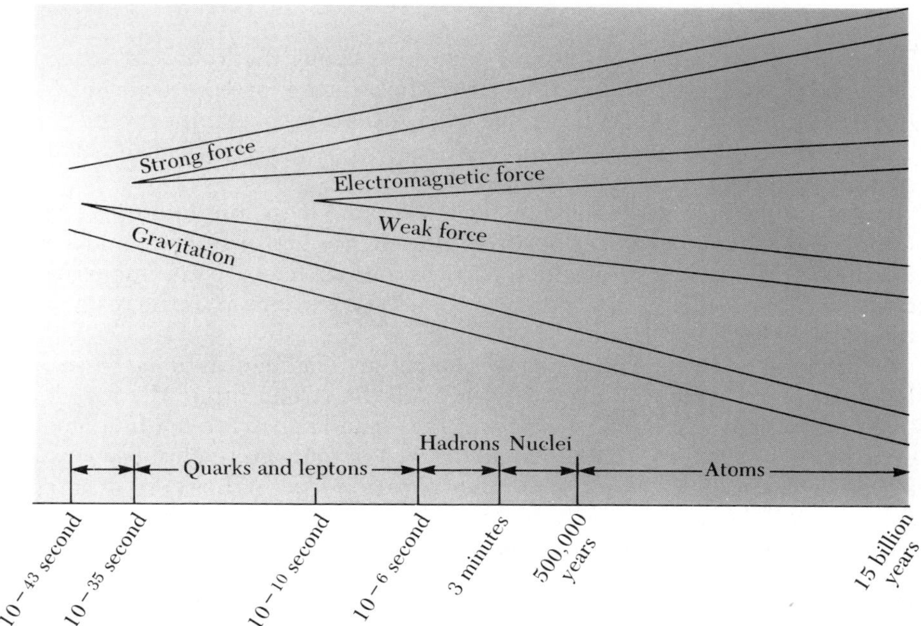

Figure 15.9 A brief history of the universe from the Big Bang to the present. The four forces became distinguishable during the first microsecond. Following this, all the quarks combined to form the strongly interacting particles. However, the leptons remained separate, and exist as individually observable particles to this day.

understand the world of elementary particles. Cosmologists and particle physicists now find they have many common goals and are joining hands to attempt to understand the physical world at its most fundamental level.

15.11 PROBLEMS AND PERSPECTIVES

Our understanding of physics at short distances is far from complete. Particle physics is faced with many questions. Why is there so little antimatter in the universe? Do neutrinos have a small rest mass, and if so, how do they contribute to the "dark matter" of the universe? Is it possible to unify the strong and electroweak theories in a logical and consistent manner? Why do quarks and leptons form three similar but distinct families? Are muons the same as electrons (apart from their difference in mass), or do they have other subtle differences that have not been detected? Why are some particles charged and others neutral? Why do quarks carry a fractional charge? What determines the masses of the fundamental constituents? Can isolated quarks exist? The questions go on and on. Because of the rapid advances and new discoveries in the field of particle physics, by the time you read this book some of these questions may be resolved.

An important and obvious question that remains is whether leptons and quarks have a substructure. If they have a substructure, one could envision an infinite number of deeper lying structure levels. However, if leptons and quarks are indeed the ultimate constituents of matter, as many physicists today believe, we should be able to construct a final theory of the structure of matter as Einstein himself dreamed. In the view of many physicists, the end of the road is in sight, but it is anyone's guess as to how long it will take to reach that goal.

15.12 SUMMARY

There are four fundamental forces in nature: strong (hadronic), electromagnetic, weak, and gravitational. The strong force is the force between nucleons that keeps the nucleus together. The weak force is responsible for beta decay. The electromagnetic and weak forces are now considered to be manifestations of a single force called the electroweak force.

An antiparticle and a particle have the same mass, but opposite charge. Other properties also have opposite values, such as lepton number and baryon number. It is possible to produce particle-antiparticle pairs in nuclear reactions if the available energy is greater than $2m_0c^2$, where m_0 is the rest mass of the particle (or antiparticle).

Particles other than photons, $W^{\pm}$, and Z^0 are classified as being *hadrons* or *leptons*. Hadrons interact primarily through the strong force. We now know that hadrons have size and internal structure, and hence are not fundamental constituents of matter. There are two types of hadrons, called *baryons* and *mesons*. Mesons have a baryon number zero and have either zero or integral spin. Baryons, which generally are the most massive particles, have a nonzero baryon number and a spin of $\frac{1}{2}$ or $\frac{3}{2}$. The neutron and proton are examples of baryons.

Leptons have no structure or size, and are considered to be truly elementary particles. Leptons interact only through the weak and electromagnetic forces. There are six leptons, the electron e^-, the muon μ^-, the tau τ^-, and their neutrinos ν_e, ν_μ, and ν_τ.

In all reactions and decays, quantities such as energy, linear momentum, angular momentum, electric charge, baryon number, and lepton number are strictly conserved. Certain particles have properties called *strangeness* and *charm*. These unusual properties are conserved only in those reactions and decays that occur via the strong force.

Recent theories in elementary particle physics have postulated that all hadrons are composed of smaller units known as *quarks*. Quarks have a fractional electric charge and a baryon number of $\frac{1}{3}$. There are six flavors of quarks, up (u), down (d), strange (s), charmed (c), top (t), and bottom (b). All baryons contain three quarks, while all mesons contain one quark and one antiquark.

According to the theory of *quantum chromodynamics*, quarks have a property called *color*, and the strong force between quarks is referred to as the *color force*.

Every fundamental interaction is said to be mediated by the exchange of field particles. The electromagnetic interaction is mediated by the photon; the weak interaction is mediated by the $W^{\pm}$ and Z^0 bosons; the gravitational interaction is mediated by gravitons; the strong interaction is mediated by gluons.

SUGGESTIONS FOR FURTHER READING

Frank Close, *The Cosmic Onion: Quarks and the Nature of the Universe*, The American Institute of Physics, 1986. A timely monograph on particle physics, including lively discussions of the Big Bang theory.

Harald Fritzsch, *Quarks, The Stuff of Matter*, London, Allen and Lane, 1983. An excellent introductory overview of elementary particle physics.

James S. Trefil, *From Atoms to Quarks*, New York, Scribner, 1980. This is an excellent introduction to the world of particle physics.

Steven Weinberg, *The Discovery of Elementary Particles*, New York, Scientific American Library, W. H. Freeman and Company, 1983. This book emphasizes the important discoveries, experiments, and intellectual exercises that reshaped physics in the 20th century.

Some *Scientific American* Articles

E. D. Bloom and G. J. Feldmann, "Quarkonium," *Sci. American*, May, 1982.

H. Harari, "The Structure of Quarks and Leptons," *Sci. American*, April, 1983.

J. David Jackson, Maury Tigner, and Stanley Wojcicki, "The Superconducting Supercollider," *Sci. American*, March, 1986.

Leon M. Lederman, "The Value of Fundamental Science," *Sci. American*, November, 1984.

N. B. Mistry, R. A. Poling, and E. H. Thorndike, "Particles with Naked Beauty," *Sci. American*, July, 1983.

Chris Quigg, "Elementary Particles and Forces," *Sci. American*, April, 1985.

Steven Weinberg, "The Decay of the Proton," *Sci. American*, June, 1981.

QUESTIONS

1. Name the four fundamental interactions and the particles that mediate each interaction.
2. Discuss the quark model of hadrons, and describe the properties of quarks.
3. Discuss the differences between hadrons and leptons.
4. Describe the properties of baryons and mesons and the important differences between them.
5. Particles known as resonances have very short lifetimes, of the order of 10^{-23} s. From this information, would you guess that they are hadrons or leptons? Explain.
6. The family of K mesons all decay into final states that contain no protons or neutrons. What is the baryon number of the K mesons?
7. The Ξ^0 particle decays by the weak interaction according to the decay mode $\Xi^0 \rightarrow \Lambda^0 + \pi^0$. Would you expect this decay to be fast or slow? Explain.
8. Identify the particle decays listed in Table 15.2 that occur by the weak interaction. Justify your answers.
9. Identify the particle decays listed in Table 15.2 that occur by the electromagnetic interaction. Justify your answers.
10. Two protons in a nucleus interact via the strong interaction. Are they also subject to the weak interaction?
11. Discuss the following conservation laws: energy, linear momentum, angular momentum, electric charge, baryon number, lepton number, and strangeness. Are all of these laws based on fundamental properties of nature? Explain.
12. An antibaryon interacts with a meson. Can a baryon be produced in such an interaction? Explain.
13. Discuss the essential features of the Standard Model of particle physics.
14. How many quarks are there in (a) a baryon, (b) an antibaryon, (c) a meson, and (c) an antimeson? How do you account for the fact that baryons have half-integral spins while mesons have spins of 0 or 1? (*Hint:* Quarks have spins of $\frac{1}{2}$.)
15. In the theory of quantum chromodynamics, quarks come in three colors. How would you justify the statement that "all baryons and mesons are colorless"?

PROBLEMS

1. Two photons are produced when a proton and antiproton annihilate each other. What are the maximum frequency and corresponding wavelength of each photon?
2. A photon produces a proton-antiproton pair according to the reaction $\gamma \rightarrow p + \bar{p}$. What is the frequency of the photon? What is its wavelength?
3. One of the mediators of the weak interaction is the Z^0 boson, whose mass is 96 GeV/c^2. Use this information to find an approximate value for the range of the weak interaction.
4. Occasionally, high-energy muons will collide with electrons and produce two neutrinos according to the reaction $\mu^+ + e \rightarrow 2\nu$. What kind of neutrinos are these?
5. When a high-energy proton or pion traveling near the speed of light collides with a nucleus, it travels an average distance of 3×10^{-15} m before interacting. From this information, estimate the time for the strong interaction to occur.
6. The neutral ρ meson decays by the strong interaction into two pions according to $\rho^0 \rightarrow \pi^+ + \pi^-$ with a half-life of about 10^{-23} s. The neutral K^0 meson also decays into two pions according to $K^0 \rightarrow \pi^+ + \pi^-$, but with a much longer half-life of about 10^{-10} s. How do you explain these observations?
7. Determine whether strangeness is conserved in the following decays and reactions.
   (a) $\Lambda^0 \rightarrow p + \pi^-$
   (b) $\pi^- + p \rightarrow \Lambda^0 + K^0$
   (c) $\bar{p} + p \rightarrow \bar{\Lambda}^0 + \Lambda^0$
   (d) $\pi^- + p \rightarrow \pi^- + \Sigma^+$
   (e) $\Xi^- \rightarrow \Lambda^0 + \pi^-$
   (f) $\Xi^0 \rightarrow p + \pi^-$
8. Each of the following decays is forbidden. For each process, determine a conservation law that is violated.
   (a) $\mu^- \rightarrow e + \gamma$
   (b) $n \rightarrow p + e + \nu_e$
   (c) $\Lambda^0 \rightarrow p + \pi^0$
   (d) $p \rightarrow e^+ + \pi^0$
   (e) $\Xi^0 \rightarrow n + \pi^0$

9. Each of the following reactions is forbidden. Determine a conservation law that is violated for each reaction.
(a) $p + \overline{p} \rightarrow \mu^+ + e$ (d) $p + p \rightarrow p + p + n$
(b) $\pi^- + p \rightarrow p + \pi^+$ (e) $\gamma + p \rightarrow n + \pi^0$
(c) $p + p \rightarrow p + \pi^+$

10. (a) Show that baryon number and charge are conserved in the following reactions of a pion with a proton.

$$\pi^- + p \longrightarrow K^- + \Sigma^+ \qquad (1)$$

$$\pi^- + p \longrightarrow \pi^- + \Sigma^+ \qquad (2)$$

(b) The first reaction is observed, but the second never occurs. Explain these observations.

11. The following reactions or decays involve one or more neutrinos. Supply the missing neutrinos (ν_e, ν_μ, or ν_τ).
(a) $\pi^- \rightarrow \mu^- + ?$ (d) $? + n \rightarrow p + e$
(b) $K^+ \rightarrow \mu^+ + ?$ (e) $? + n \rightarrow p + \mu^-$
(c) $? + p \rightarrow n + e^+$ (f) $\mu^- \rightarrow e + ? + ?$

12. Determine which of the reactions below can occur. For those that cannot occur, determine the conservation law (or laws) that each violates.
(a) $p \rightarrow \pi^+ + \pi^0$ (d) $\pi^+ \rightarrow \mu^+ + \nu_\mu$
(b) $p + p \rightarrow p + p + \pi^0$ (e) $n \rightarrow p + e + \bar{\nu}_e$
(c) $p + p \rightarrow p + \pi^+$ (f) $\pi^+ \rightarrow \mu^+ + n$

13. The quark composition of the proton is uud, while that of the neutron is udd. Show that the charge, baryon number, and strangeness of these particles equal the sums of these numbers for their quark constituents.

14. The quark compositions of the K^0 and Λ^0 particles are $d\bar{s}$ and uds, respectively. Show that the charge, baryon number, and strangeness of these particles equal the sums of these numbers for their quark constituents.

Essay

NUCLEON DECAY EXPERIMENTS

Hans Courant
University of Minnesota

Scientists in a number of laboratories around the world are testing the permanence of matter in the universe by searching for evidence of the spontaneous disintegration of protons and neutrons. The primary motivation for these investigations is to seek verification of one facet of the "Grand Unified Theory," which predicts such nucleon decay in nature.

According to the Big Bang theory, the universe exists in its present form some 15 billion years after its origin. Hence, one can conclude that a great deal of undisintegrated matter still exists and that the decay rate of matter (if it occurs at all) must be extremely slow. This implies that even a single disintegration of a nucleon in a large population of nucleons would be a very rare occurrence.

The "Grand Unified Theory" predicts that the probability for a nucleon to decay in a one-year time interval is approximately 10^{-31}. This means that a substantial portion of a collection of nucleons will have decayed in 10^{31} years. It is not easy to visualize this probability. Since the estimated life of the universe is 15 billion years ($= 1.5 \times 10^{10}$ years), then only the fraction $1.5 \times 10^{10} \times 10^{-31} = 1.5 \times 10^{-21}$ of the nucleons formed at the Big Bang would have decayed since the origin of the universe. Thus, practically *all* of the original matter of the universe still exists! In view of these numbers, it is hardly surprising that until this theory was propounded, nucleons were considered ultimately stable.

Standard instruments used in particle physics are also being used in an attempt to detect the occurrence of nucleon decay. Such events are difficult to identify because the detector is likely to be overwhelmed by vastly more numerous background decay events. Thus the apparatus needed to detect nucleon decay events must be designed to monitor a large number of nucleons while suppressing erroneous signals that mimic decay events. The detection of weak signals in the presence of extreme background noise is a particularly complex problem which must be overcome in the identification of proton decay.

There are many different nucleon decay detectors in operation in various laboratories around the world. All are located underground at depths of 2000 ft or more, and all consist of many tons of material. Experiments are conducted underground to reduce background noise rate by almost a million from the noise rate that would be experienced at the earth's surface. The likelihood of observing a nucleon decay is enhanced by monitoring many tons of material since each ton provides about 6×10^{29} nucleons for scrutiny.

Assuming a nucleon lifetime of 10^{31} years, a 1-kiloton detector (which contains 6×10^{32} nucleons) may be expected to detect perhaps 60 events in a period of one year. During that same period of time in an underground laboratory, 10 million or more other background events would be detected. Each of these would have to be identified as being distinctly different from a possible example of a nucleon decay. Thus, if the occurrence of nucleon decay is to be established with scientific certainty, the experiments must involve large masses of materials, shielded as well as possible from external sources of energetic particles. Furthermore, the detection apparatus must be able to discriminate between the millions of background events and the handful of genuine nucleon decay events which are expected.

Although the underground detector is shielded by thousands of feet of dirt and rock (and lake water in the case of one detector), it is not quiescently awaiting nucleon decay. Instead, the detector is constantly bombarded by cosmic radiation and by an occasional neutrino interaction, as well as by its own internal radioactivity. This background activity, which occurs at the rate of about 10^5 events each second, represents the proverbial "haystack" in which one hopes to find the "needle" of nucleon decay.

A nucleon decay event must have certain unique characteristics which give it a genuine "signature." These characteristics are then used to distinguish such an event from a similar, but spurious, background event. One such characteristic of a nucleon decay is the appearance of a spontaneous release of energy not related to any other activity in the detector. Therefore, an identified nucleon decay event must be isolated in place and in time; that is, the decay must occur entirely within the detector, having no connection whatsoever with anything on the exterior of the detector.

Another characteristic associated with the nucleon decay signature is the quantity of energy released. We know that the total energy released in a nucleon decay must equal the nucleon's rest energy, according to the Einstein relation $E = m_0c^2$. Hence an identified nucleon event must be one in which about 10^9 eV of energy is released in various forms. Furthermore, because momentum must also be conserved, the momentum of the decay product particles must total to near zero since the decaying nucleon is presumably nearly at rest.

As was mentioned earlier, the detection of nucleon decay consistent with a prediction from the "Grand Unified Theory" would be a very important and perhaps crucial endorsement of this theory. On the other hand, if it were established that nucleons are stable over time scales much longer than 10^{32} years, then some serious modifications of the theory would be required. Direct experimental tests of the theory are ruled out by the fact that the characteristic energy required by the theory is about 10^{15} GeV (or 10^{24} eV). Such energy is not available in the laboratory in existing high energy accelerators or those contemplated for construction. (The S. S. C. Superconducting Super Collider will produce particle energies of 10^4 GeV, far short of the required energy of 10^{15} GeV.) It should be noted that such large energy was prevalent up to the first 10^{-43} s of the "Big Bang" in the origin of the universe. Following that brief instant, the universe has cooled to much lower energies or temperatures. Since the extreme conditions of the early "Big Bang" can occur spontaneously in very tiny regions for an exceedingly short time, the probability of nucleon decay is predicted to be very small, perhaps 10^{-31} per year. The search for this rare event is what motivates physicists in their search for nucleon decay.

WAITING FOR NUCLEON DECAY

The original prototype nucleon decay detector, the Soudan I detector, is located 2000 ft underground in an abandoned iron mine at the centerpiece of Minnesota's Soudan State Park. This detector has been superseded by nucleon decay detectors built on a larger scale in several locations around the world. Although the Soudan I detector has not detected any nucleon decay events during its operation, accumulated cosmic ray data have revealed some interesting and fascinating results.

The Soudan I detector responds many times per hour to particles whose trajectories presumably begin above the ground, traversing the 2000 ft of earth and stone above the detector before passing through it and penetrating to greater depths. These tracks, which constitute a major part of the background events, allow researchers to verify the correct functioning of their instrument on an almost continuing basis. Since the observed tracks do not mimic those expected from nucleon decay, their source is identified as a part of cosmic radiation.

Because the Soudan I detector has excellent directional resolution, one can infer the original direction of the penetrating particles by mathematically extending the tracks about 3 m backward toward its source. After a large number of track recordings have been accumulated, it is then possible to chart the direction of origin on a map of the stars. In this manner, one can look for "hot spots" in the universe, corresponding to possible sources of energetic particles whose tracks have been observed. The analysis of the data following this procedure did not reveal any regions in space corresponding to a statistically significant excess of energetic particles.

It is possible to obtain additional and independent information in some cases to enhance the significance of the meager data taken with the Soudan I detector. One important example is the information obtained in the observation of a pulsating source of x-rays from the constellation Cygnus. This observation was first reported a few years ago by researchers using an x-ray telescope in a satellite orbiting the earth. A source of x-rays, named "Cygnus X-3," was observed to be intensity modulated, with successive maxima and minima occurring in intervals of 4.8 hours. The cause of this periodic pulsing was presumed to be a pair of stars rotating about each other with a period of 4.8 hours. Visual observation of this constellation is thought to be obscured by clouds of cosmic dust.

An analysis of subsequent data taken by the Soudan I detector of tracks coming from the direction of Cygnus strongly indicates bursts of tracks with a 4.8 hour period. This observation has resulted in considerable speculation on the part of high energy physicists, cosmic ray physicists, astrophysicists, and cosmologists. Although the particles detected are identified as being μ-mesons, they are certainly not the particles which have traversed the 10^4 light-years of space between the source and the detector. The manner in which these intense bursts of energy traverse space, and the form they take, is not understood at present. The detected μ-mesons are believed to be secondary particles generated by the primary energy carriers in the earth's atmosphere, or in the dirt or rock near the earth's surface above the detector.

WHAT DOES A LIFETIME OF 10^{31} YEARS MEAN?

Spontaneous decay of the type discussed in this essay can be described by a rate λ. For example, if the rate λ were one per hour, it would correspond to a decay probability of 1/3600 per second. Thus we would expect 1/3600 of a large sample of particles to decay in a time interval of one second. (Of course as time passes, the size of the residual sample of undecayed particles diminishes and the number of particles decaying per second diminishes correspondingly). After one hour has elapsed, the size of the sample will be reduced to $1/e = 1/2.7$ of its original size, and after a second hour has elapsed the sample size will be reduced to $1/e \times 1/e = 1/e^2 = 0.135$ of its original size. After each additional hour, the sample size is reduced by another factor of $1/e$. A lifetime of 10^{31} years means that it would take 10^{31} years to reduce the nucleon sample to $1/e$ of its original size.

EXAMPLE 1 Properties of the X particle

This example pertains to some calculations involving the X particle, the particle capable of transforming a constituent quark into an antiquark, thereby promoting the reconfiguration of the nucleon into its decay products. The Grand Unified Theory predicts that the rest energy of the X particle is $M_x c^2 = 10^{15}$ GeV $= 10^{24}$ eV. (a) What is the mass of the X-particle?

Solution:

$$M_x c^2 = (10^{24}\text{ eV})(1.6 \times 10^{-19}\text{ J/eV}) = 1.6 \times 10^5\text{ J}$$

$$M_x = \frac{1.6 \times 10^5\text{ J}}{(3 \times 10^8\text{ m/s}^2)^2} = 1.8 \times 10^{-12}\text{ kg}$$

(b) What is the characteristic time of the X-particle? That is, what is the time interval during which an energy of $M_x c^2$ may make a spontaneous and short appearance?

Solution: Using the uncertainty relation $\Delta E\ \Delta t \geq h$, with $\Delta E = M_x c^2$, and taking $\Delta t = \tau_x$, we find

$$\tau_x = \frac{h}{M_x c^2} = \frac{6.63 \times 10^{-34}\text{ J}\cdot\text{s}}{1.6 \times 10^5\text{ J}} = 4 \times 10^{-39}\text{ s}$$

(c) What is the characteristic time of a nucleon?

Solution: Taking $M_p c^2 \approxeq 10^9$ eV as the rest energy of a nucleon, we find

$$\tau_p = \frac{h}{M_p c^2} = 4 \times 10^{-24}\text{ s}$$

(d) What is the characteristic length of the X-particle? Note that the characteristic length is a measure of the "Compton wavelength" associated with the particle.

Solution: The characteristic length equals the distance the particle could travel in its characteristic time if moving with a velocity c. Thus, it is a measure of the range of action of the X particle. Using the deBroglie relation $\lambda = h/Mv$, we find

$$\lambda_x = h/M_x c = \tau_x c = (4 \times 10^{-39}\text{ s})(3 \times 10^8\text{ m/s}) = 1.2 \times 10^{-30}\text{ m}$$

(e) Estimate the time for a quark, confined to the "inside" of a nucleon, to collide with an X particle within the nucleon. Assume that the nucleon "size" is equal to its characteristic length.

Solution:

$$\tau = \underbrace{\left(\frac{\lambda_p}{\lambda_x}\right)^3}_{\text{Volume ratio of nuclear to X}} \times \underbrace{\left(\frac{\tau_p}{\tau_x}\right)}_{\text{Ratio of times}} \times \underbrace{\tau_p}_{\text{Time for quark to "traverse" nucleon}}$$

$$= \left(\frac{M_x}{M_p}\right)^3 \times \left(\frac{M_x}{M_p}\right) \times (4 \times 10^{-24}\ \text{s})$$

$$= 4 \times 10^{36}\ \text{s} = 10^{29}\ \text{years}$$

Note that these calculations represent a very simplistic picture. The result for τ is very approximate, and the value obtained is not thought to be a real measure of nucleon lifetime. Since the calculated value of τ depends on the fourth power of the mass of X, it is very sensitive to the value of M_x assumed for this calculation.

Suggestions for Further Reading

Frank Close, *The Cosmic Onion,* New York, American Institute of Physics, 1986.

James S. Trefil, *From Atoms to Quarks,* New York, Charles Scribner and James S. Trefil, *Moment of Creation,* New York, Macmillan, 1983.

Steven Weinberg, *The First Three Minutes,* New York, Bantam Books, 1979.

Essay Problems

1. Estimate the average number of nucleons (protons or neutrons) that will decay within the body of a 70-kg person in 60 years.
2. Calculate the size of a water sample containing a volume equivalent to that of 10^{31} drops. (This is the size of a water sample which will lose the equivalent of one drop in a time of one year). Take the nominal volume of one drop to be $1/20\ \text{cm}^3$.
3. Calculate the equivalent weight of one mole of X particles (see Example 1).

Appendix A

Calculation of the Number of Modes for Waves in a Cavity

We wish to solve for the number of standing waves that fall within the frequency range from f to $f + df$ for waves confined to a cubical cavity of side L. Since $N(f)$ is known experimentally to be the same for a cavity of any shape and for walls of any material, we can choose the simplest shape — a cube — and the simplest boundary conditions — waves that vanish at the boundaries. Starting with Maxwell's equations, it can be shown that the electric field obeys a wave equation that may be separated into time-dependent and time-independent parts. The time-dependent equation has a simple sinusoidal solution with frequency ω, and the time-independent components of the electric field satisfy equations of the type

$$\frac{\partial^2 E_x}{\partial x^2} + \frac{\partial^2 E_x}{\partial y^2} + \frac{\partial^2 E_x}{\partial z^2} + k^2 E_x = 0 \tag{A.1}$$

where

$$E_x = E_x(x, y, z)$$

and

$$k = \frac{2\pi}{\lambda} = \frac{2\pi f}{c}$$

Assuming that $E_x = u(x)v(y)w(z)$, we can separate Equation A.1 into three ordinary differential equations of the type

$$\frac{d^2 u}{dx^2} + k_x^2 u = 0 \tag{A.2}$$

where

$$k^2 = k_x^2 + k_y^2 + k_z^2$$

Equation A.2 is the simple harmonic oscillator equation and has the solution

$$u(x) = B \cos k_x x + C \sin k_x x$$

Imposing the boundary condition that E_x or u is 0 at $x = 0$ and at $x = L$ leads to $B = 0$ and $k_x L = n_x \pi$, where $n_x = 1, 2, 3, \ldots$. Similar solutions are obtained for $v(y)$ and $w(z)$, giving

$$E_x(x, y, z) = A \sin(k_x x) \sin(k_y y) \sin(k_z z)$$

where

$$k^2 = \frac{\pi^2}{L^2}(n_x{}^2 + n_y{}^2 + n_z{}^2) \tag{A.3}$$

and n_x, n_y, n_z are positive integers.

To determine the density of modes, we interpret Equation A.3 as giving the square of the distance from the origin to a point in k-space or "reciprocal" space. It is called reciprocal space because k has dimensions of $(\text{length})^{-1}$. As shown in Figure A.1, the axes in k-space are k_x, k_y, and k_z. Since $k_x = n_x\pi/L$, $k_y = n_y\,\pi/L$, and $k_z = n_z\pi/L$, the points in k-space (or modes) are separated by π/L along each axis, and there is one standing wave in k-space per $(\pi/L)^3$ of volume. The number of standing waves $N(k)$ having wave numbers between k and $k + dk$ is then simply the volume of k-space between k and $k + dk$ divided by $(\pi/L)^3$. The volume between k and $k + dk$ is $\frac{1}{8}$ of a spherical shell, so that

$$N(k)\,dk = \frac{\frac{1}{2}\pi k^2\,dk}{(\pi/L)^3} = \frac{Vk^2\,dk}{2\pi^2} \tag{A.4}$$

where $V = L^3$ is the cavity volume.

For electromagnetic waves there are two perpendicular polarizations for each mode, so that $N(k)$ in Equation A.4 must be increased by a factor of 2. Thus, we have for the number of standing waves per unit volume

$$\frac{N(k)\,dk}{V} = \frac{k^2\,dk}{\pi^2} \tag{A.5}$$

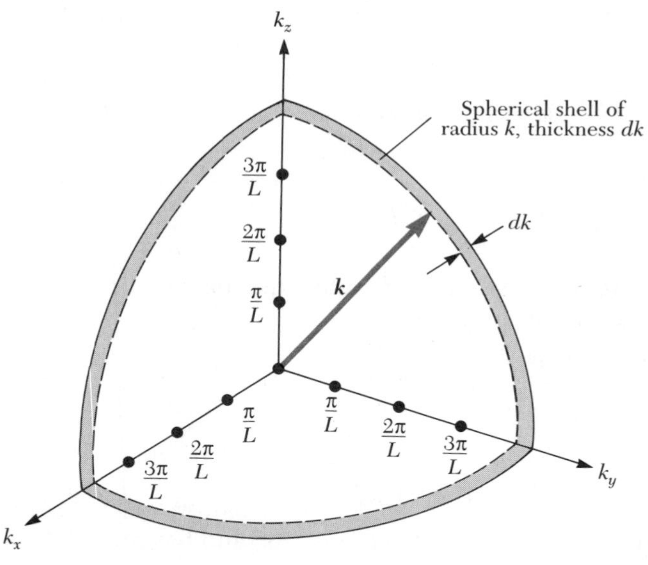

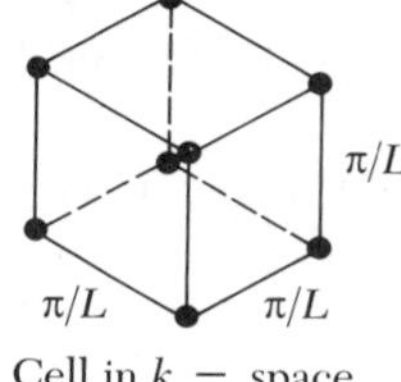

Figure A.1 A geometrical interpretation of $k^2 = \frac{\pi^2}{L^2}(n_x{}^2 + n_y{}^2 + n_z{}^2)$.

To find $N(f)$ we use $k = 2\pi f/c$ in Equation A.5 to obtain

$$N(f)\,df = \frac{8\pi f^2}{c^3}\,df \tag{A.6}$$

$N(\lambda)$, the number of modes per unit volume between λ and $\lambda + d\lambda$, may be obtained from Equation A.6 by using $f = c/\lambda$ to give

$$N(\lambda)\,d\lambda = -\frac{8\pi}{\lambda^4}\,d\lambda \tag{A.7}$$

Appendix B

Blackbody Radiation from the Big Bang

The theory of blackbody radiation developed in Chapter 2 has a straightforward application to one of the most fascinating theories in all of science — the big bang theory of the creation of the universe. This astonishing theory of cosmology states that the universe had a beginning, and further that the beginning was so cataclysmic that it is impossible to look back beyond the creation. The astronomer Robert Jastrow has put it nicely:

> No scientist can answer this Genesis question. We cannot tell whether the Prime Mover willed the world into being out of nothing or whether the creative agent was one of the more familiar forces of physics; in either case the astronomical evidence proves that the universe was created about 15 billion years ago in a fiery explosion, and in the searing heat of that first moment, all the evidence needed for a scientific study of the cause of the great explosion was melted down and destroyed.

Our intent in this appendix is to examine one theory of the creation of the universe, the big bang theory, and to look at the experimental evidence that supports it. Along the way we shall attempt a thumbnail sketch of the history of seminal cosmological discoveries and pause for an ancedote or two about the major figures involved.

Figure B.1 George Gamow (1904–1968), one of the first physicists to take the first half hour of the universe seriously.

As we have mentioned, the big bang theory of creation holds that the universe erupted from a pointlike singularity on the order of 15 to 20 billion years ago. The first few minutes after the big bang saw such extremes of energy that it is believed that all four interactions of physics (gravitational, electromagnetic, strong nuclear, and weak) were unified and that all matter melted down into an undifferentiated quark soup. In fact, considerations of the first few minutes are properly in the realm of GUTs (Grand Unified Theories) and have partly motivated the development of these theories, which attempt to unify the four different interactions of physics. During the first half hour of creation, the temperature was probably on the order of 10^{10} K and the universe could be viewed as a cosmic thermonuclear bomb fusing protons and neutrons into deuterium and then helium.[1]

[1] George Gamow (see Fig. B.1) and his students, Ralph Alpher and Robert Herman, were the first to take the first half hour of the universe seriously. In a mostly overlooked paper published in 1948, they made truly remarkable cosmological predictions. They correctly calculated the abundances of hydrogen and helium after the first half hour (75% H and 25% He), and predicted that radiation from the big bang should still be present with an apparent temperature of about 5 K. In Gamow's own words, the universe's supply of hydrogen and helium were created very quickly, "in less time than it takes to cook a dish of duck and roast potatoes." The comment is characteristic of this interesting physicist who is known as much for his explanation of alpha decay and theories of cosmology as for his delightful popular books, his cartoons, and his wonderful sense of humor. A classic Gamow story holds that having coauthored a paper with Alpher, he made Hans Bethe an honorary author so that the credits would read "Alpher, Bethe, and Gamow" (to be compared with the Greek letters α, β, and γ).

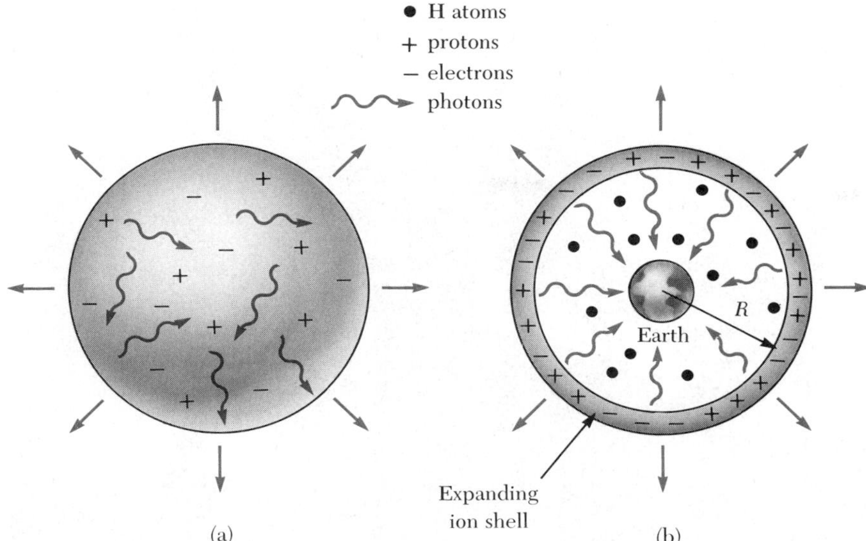

Figure B.2 Radiation- and matter-dominated stages in the evolution of the universe. (a) Radiation-dominated stage ($T > 3000$ K, age $< 700\,000$ years) is an expanding universe filled with protons, electrons, neutrons, photons, and neutrinos. Ions continually scatter photons, ensuring thermal equilibrium of radiation and matter. (b) Matter-dominated stage ($T < 3000$ K, age $>$ 700 000 years). Hydrogen atoms form, making the universe transparent to photons. Photons emitted from an expanding ion shell at $R \approx 14$ billion light years are currently seen on earth enormously Doppler shifted to the red.

Until about 700 000 years after the big bang, the universe was "radiation-dominated"; ions absorbed and reemitted photons, thereby ensuring thermal equilibrium of radiation and matter. Energetic radiation also prevented matter from forming clumps or even neutral hydrogen atoms. When the universe was about 700 000 years old, it had expanded and cooled to a temperature of 3000 K, and protons could bind to electrons and form neutral hydrogen. Since neutral atoms do not appreciably scatter photons, the universe suddenly became transparent to photons. Thus, the photons were frozen in time with a distribution of energies characteristic of a 3000 K black body. Radiation no longer dominated the universe and clumps of neutral matter steadily grew — first atoms, then molecules, gas clouds, stars, and finally galaxies. The state of affairs for radiation- and matter-dominated epochs is shown in Figure B.2.

First Observation of Fossil Radiation from the Primordial Fireball

When Arno A. Penzias and Robert W. Wilson of Bell Labs first detected the left-over glow from the big bang in 1965 at a wavelength of 7.35 cm (in the microwave region), they tried every trick in the book to get rid of this pesky signal, which was interfering with their satellite communications experiments. The microwave horn that served as their receiving antenna is shown in Figure B.3. Pointing the antenna in different directions made no difference in the intensity of the detected signal. The fact that the radiation was found to have equal strengths in all directions suggested that the entire universe was the source of this radiation. Booting a flock of pigeons from the 20-foot horn (bird droppings in the horn were a problem) and cooling the microwave detector both failed to remove the "spurious" signal and confirmed the resolve of the experimenters to explain their mysterious results. A casual conversation led Penzias and Wilson to discover that a group at Princeton under P. J. E. Peebles had predicted residual radiation from the big bang and that an experiment was being planned by another Princeton physicist, Robert Dicke, to seek confirmation of these theories. The excitement in the scientific community was high

Figure B.3 Robert W. Wilson (left) and Arno A. Penzias with Bell Telephone Laboratories horn-reflector antenna. (AT & T Bell Laboratories)

when Penzias and Wilson announced that they had already observed an excess microwave background compatible with a 3 K blackbody source.

The big bang photons reaching us today, as shown in Figure B.2, originated from a plasma shell that was expanding outward from the current position of the earth with a velocity very close to the velocity of light. Because of the relative motion of the radiation source away from the earth (relative to an earthbound observer), the radiation appears to be tremendously red-shifted by the Doppler effect, by a factor of roughly 1000. In fact, when viewed in a frame at rest with respect to the ions that emitted the radiation, the wavelength has not changed. In this frame the radiation has an intensity distribution typical of a black body with a 3000 K temperature and a maximum intensity wavelength of about 1 micron. In the next section we shall prove that a remarkable feature of a blackbody emitter receding from an observer is that it retains the characteristic blackbody intensity distribution, with an alteration in only the apparent temperature. Thus, from the point of view of an earthbound observer the maximum intensity wavelength is 1000 times longer or red-shifted to 1 millimeter, and hence the radiation appears to come from a black body 1000 times cooler at 3 K.

Because the measurements of Penzias and Wilson were taken at a single wavelength, they did not completely confirm the observed radiation as 3 K blackbody radiation. One can imagine the excitement in the late 1960s and 1970s as experimentalists fought obscuring atmospheric absorption and, point by point, added intensity data at different wavelengths. The results shown in Figure B.4 confirm that the intensity versus wavelength distribution of the background radiation is that of a black body at 2.7 K. This figure is, perhaps, the most clearcut evidence for the hot big bang theory of cosmology. Interestingly enough, it also presents the earliest view of the universe, a much earlier view than that available with the largest optical and radio telescopes.

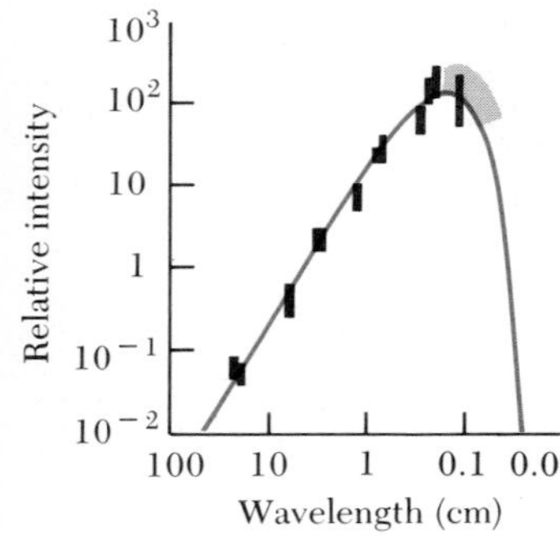

Figure B.4 Radiation spectrum of the big bang. The shaded areas are experimental results. The solid line is the spectrum calculated for a black body at 2.9 K.

Since the observed 3 K blackbody spectrum is the single strongest piece of evidence for the big bang universe, it is important to look at this topic in more detail. In particular, we wish to show theoretically what happens to a blackbody spectrum in an expanding universe.

Consider first a unit volume, V_0, of linear dimension ℓ_0, containing blackbody photons at a time when the universe first became transparent to photons (700 000 years after the big bang—A.B.B.!). The *energy per unit volume* of photons with wavelengths between λ and $\lambda + d\lambda$ is given by

$$du = \frac{8\pi hc}{\lambda^5}\left[\frac{1}{e^{hc/\lambda kT} - 1}\right] d\lambda \tag{B.1}$$

Suppose that between then (700 000 years A.B.B.) and now (15 000 000 000 years A.B.B.) the universe expands by a factor F in a linear dimension. What happens to the energy per unit volume? As shown in Figure B.5, the previous volume of V_0 now expands to F^3V_0. Also, from the point of view of an earthbound observer, the photon of original wavelength λ is Doppler-shifted to λ', where

$$\lambda' = \left(1 + \frac{v}{c}\right)\lambda \tag{B.2}$$

But $(1 + v/c) = F$, the expansion factor, so that Equation B.2 may be written in terms of F as

$$\lambda' = F\lambda \tag{B.3}$$

We can show that $(1 + v/c) = F$ with the following argument. To say that the universe expands by F in a linear dimension means that if you look at faraway objects (like other galaxies), you will find them to be receding. Consider the case of our galaxy taken to be at rest and a second galaxy receding at

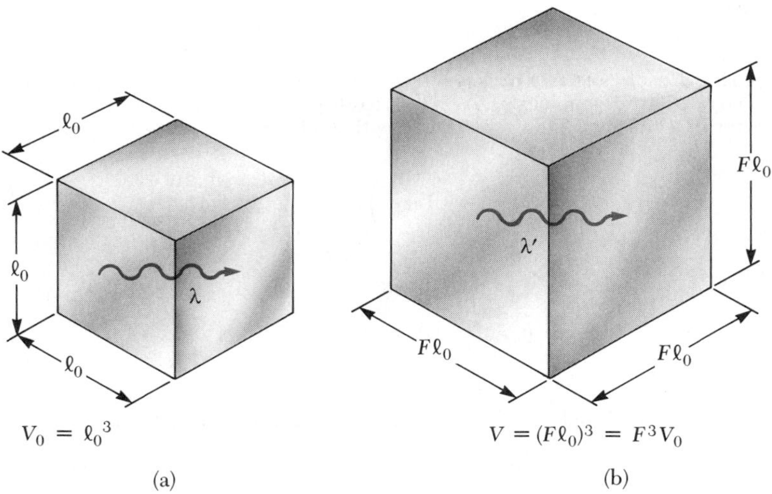

Figure B.5 Volume and wavelength change in an expanding universe. (a) Initially (700 000 A.B.B.), the universe has a volume V_0 and contains photons of wavelength λ. (b) At the present, the volume has expanded to F^3V_0. As the radiation source recedes, the photon is Doppler shifted to a wavelength $\lambda' = F\lambda$.

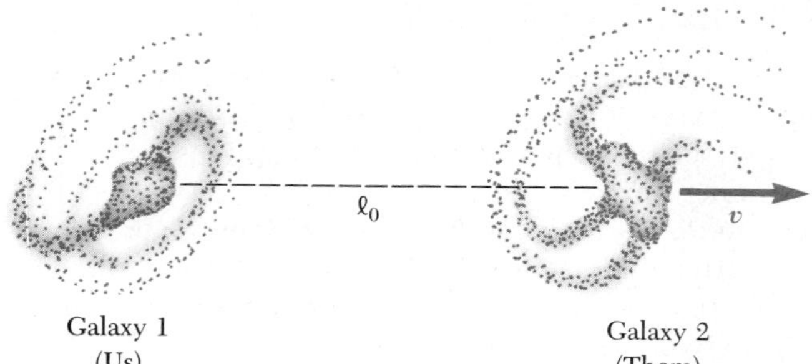

Figure B.6 Two galaxies in an expanding universe.

velocity v, as shown in Figure B.6. If ℓ_0 is the original separation, then a light signal may be sent from galaxy 2 to galaxy 1 in a time t, where $\ell_0 = ct$. In time t, however, the galaxies separate by an additional amount $\Delta\ell = vt$. Thus, we find for F

$$F = \frac{\text{current galaxy separation}}{\text{original separation}} = \frac{\ell_0 + \Delta\ell}{\ell_0} = \frac{ct + vt}{ct} = 1 + \frac{v}{c} \tag{B.4}$$

This proves our contention that the Doppler shift reduces to $\lambda' = F\lambda$ or that the photon wavelength increases by the universe expansion factor.[2]

Let du' be the current energy per unit volume of photons with wavelengths between λ' and $\lambda' + d\lambda'$; du' will be less than du because the volume has increased by a factor of F^3 and the energy per photon has decreased by a factor of F (since $E_{\text{photon}} = hc/\lambda$). Thus we have

$$du' = \frac{du}{F^4} = \frac{8\pi hc}{\lambda^5 F^4} \frac{d\lambda}{(e^{hc/\lambda kT} - 1)} \tag{B.5}$$

Since we want du' in terms of λ' to see what form the present spectral distribution takes, we may rewrite Equation B.5 as

$$du' = \frac{8\pi hc}{\lambda'^5} \frac{d\lambda'}{(e^{hc/\lambda' kT'} - 1)} \tag{B.6}$$

where $T' = T/F$ and $\lambda' = F\lambda$.

[2] $\lambda' = (1 + v/c)\lambda$ and $F = 1 + v/c$ are based on Newtonian mechanics. As $v \approx c$ we really should use relativity theory. The relativistic Doppler shift formula is easily obtained by noting that λ' must be decreased by a factor of $\sqrt{1 - v^2/c^2}$. Thus replacing λ' with $\lambda'\sqrt{1 - v^2/c^2}$ in the expression $\lambda' = (1 + v/c)\lambda$ gives

$$\lambda' \sqrt{1 - \frac{v^2}{c^2}} = \left(1 + \frac{v}{c}\right)\lambda$$

or

$$\lambda' = \frac{\sqrt{1 + \frac{v}{c}}}{\sqrt{1 - \frac{v}{c}}}\,\lambda$$

This equation is the relativistic Doppler shift formula. In a similar fashion you can show that

$$F = \frac{\sqrt{1 + \frac{v}{c}}}{\sqrt{1 - \frac{v}{c}}} \qquad \text{for } v \approx c$$

Thus the expression $\lambda' = F\lambda$ still holds when v is close to c.

Thus, we see that the radiation from a receding black body still has the same spectral distribution, but its temperature drops in proportion to the expansion factor, F. In a frame of reference moving with the ions that last scattered the radiation, the characteristic temperature of the radiation is about 3000 K. Equation B.6 shows that in our frame of reference the radiation appears to have a characteristic temperature of (3000 K)/F or 3 K, since we know that $F \approx 1000$.

Other Evidence For the Expanding Universe — American Experimentalists and European Theorists

Most of the key discoveries supporting the theory of an expanding universe, and indirectly the big bang theory of cosmology, were made in the 20th century. In 1913, Vesto Melvin Slipher, an American astronomer, reported that a survey of a dozen nebulae (distant luminous clouds that he believed to be solar systems in formation) showed that most were receding from the earth at speeds of up to several million miles per hour. Slipher was one of the first to use the methods of Doppler shifts systematically to measure velocities. The key to this method is to locate in nebular spectra the characteristic lines of some element whose wavelengths have been measured on earth. The velocity of the receding object is then directly proportional to the wavelength shift to the red, as shown in Figure B.7. Slipher's measurements of nebular velocities

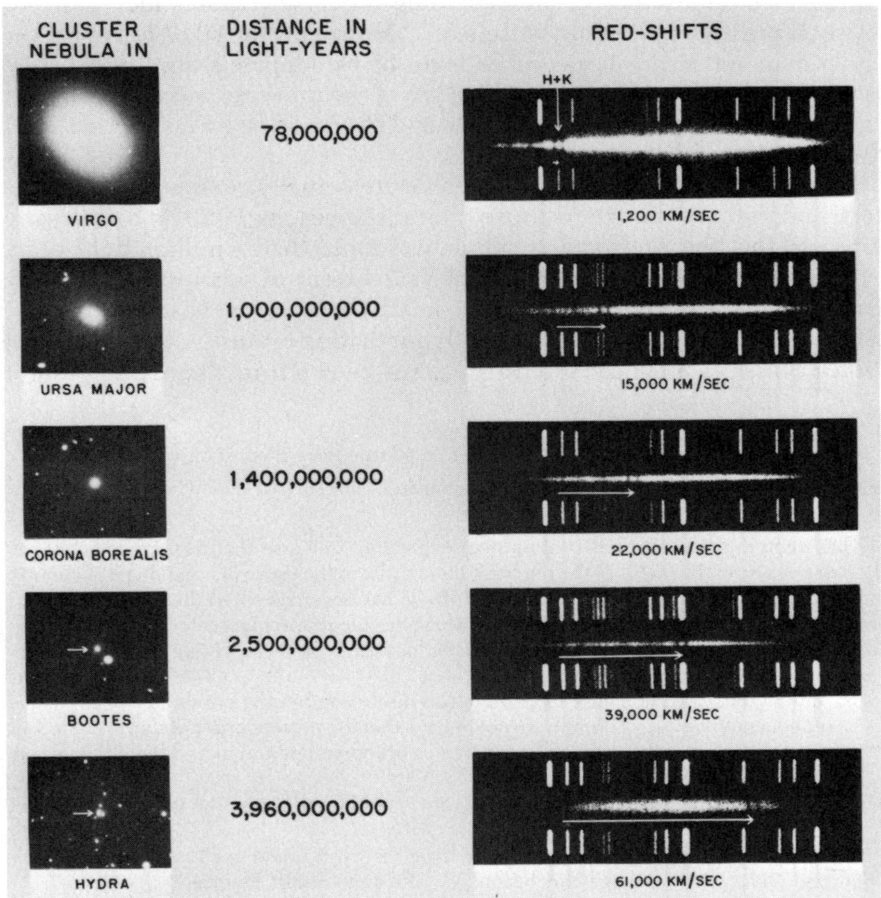

Figure B.7 Red shift of the K and H lines of calcium for four galaxies. Red shifts are expressed as velocities $c\,d\lambda/\lambda$. Calibration spectra taken in the laboratory are shown above and below each galaxy's spectrum.

were quite accurate but the distances to the nebulae were virtually unknown, although there was some indication that more distant objects had higher recessional velocities.

Shortly after Slipher's discovery, a flowering of ideas occurred in theoretical cosmology. In fact, one may say that Einstein, the founding father of relativity and quantum theory, was also the founder of theoretical cosmology. In 1916 Einstein published his general theory of relativity, in which he formulated the differential equations connecting the distribution of matter in the universe with the curvature of space-time. Einstein erroneously solved his general equations and found that in order to have a *universe static in time*[3] he had to add a "cosmic repulsion" force between galaxies. This was a rather unusual force, since it was independent of mass and grew stronger with *increasing* galaxy separation. In the same year, the Dutch mathematician and astronomer Willem de Sitter found another static solution to Einstein's equations. This solution predicted an increasing red shift for lines emitted by more distant objects. de Sitter's solution stirred much interest because it seemed capable of being confirmed experimentally.

Figure B.8 Milton L. Humason. The two American astronomers Hubble and Humason, who did the most to experimentally demonstrate the expanding universe, present an interesting contrast. Humason was originally a mule team driver and janitor at Mount Wilson. He taught himself astronomy, becoming a consummate experimentalist.

A major breakthrough in the cosmological problem came in 1922 when Alexander Friedmann,[4] a Russian mathematician, found an error in Einstein's work and was able to show that two *nonstatic* models were possible — an expanding universe model and a contracting universe. At first Einstein found it difficult to exchange the eternal, steady, static universe for an expanding, dynamic universe that presumably had a beginning in time. He wrote to de Sitter, "This circumstance of an expanding universe irritates me" and "to admit such possibilities seems senseless."[5] However, by 1923 Einstein was finally convinced of the impeccable logic of Friedmann's arguments. Any remaining doubt about the nonstatic nature of the universe was erased by the powerful experimental results of Hubble and Humason (see Figs. B.8 and B.9) on the red shifts of distant nebulae.

Hubble first established that nebulae were actually composed of billions of stars, and then went on to measure their distances. By 1923 he had already determined that the Andromeda nebula was more than 1 million light years distant, well beyond the 100 000 light year extent of our own Milky Way galaxy. Hubble measured the distance to nearby galaxies or "island universes" by observing the brightness of stars with fluctuating intensity, called *cepheid variables*. Since the intrinsic intensity I_{int} was known from the period of light

[3] The average distance between clumps of matter on a large scale is constant with time.

[4] Friedmann was an interesting fellow who was an amateur meteorologist and balloonist. He also taught the young Gamow relativity theory.

[5] There has been much scientific irritation over expanding universe theories. Presumably the irritation derives from the fact that the big bang theory places the cause of creation permanently beyond the reach of physical investigation. Fred Hoyle has never accepted the big bang theory, and prefers instead a steady state model of the universe in which matter is created out of nothing at the rate necessary to compensate for the increasing distances between visible galaxies. Others have expressed their irritation as follows:

Sir Arthur Eddington: "I have no axe to grind in this discussion but the notion of a beginning is repugnant to me. . . . I simply do not believe that the present order of things started off with a bang . . . the expanding universe is preposterous and incredible; it leaves me cold."

Walter Nernst: "To deny the infinite duration of time would be to betray the very foundation of science."

Philip Morrison: "I find it hard to accept the big bang theory; I would like to reject it."

Allan Sandage: "It is such a strange conclusion. . . . It cannot really be true."

Figure B.9 Edwin P. Hubble. Hubble was born to wealth, graduated from the University of Chicago, was a Rhodes Scholar, lawyer, athlete, soldier, and astronomer with an unerring eye for the important problems in astronomy.

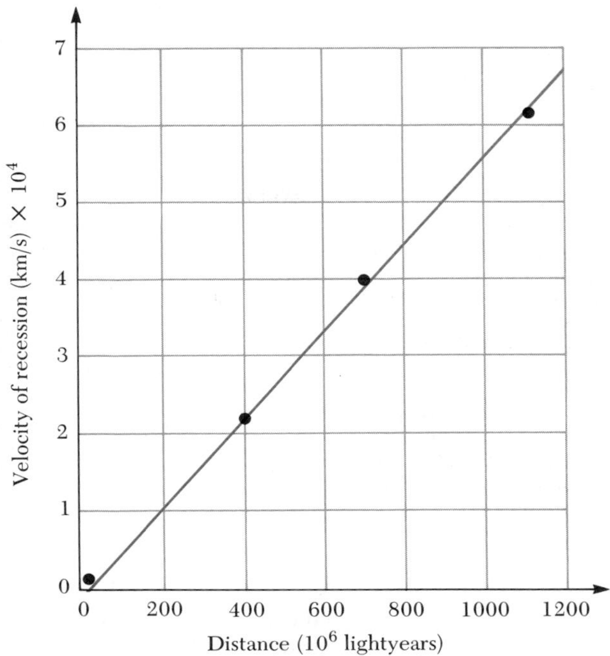

Figure B.10 Hubble's law: A plot of velocity of recession versus distance. The four galaxies shown in Figure B.7 are plotted here.

fluctuations, Hubble simply used the observed intensity I_{obs} and the inverse square law to find the distance, R:

$$I_{obs} = I_{int}/R^2 \quad (B.7)$$

In the late 1920s, Hubble made the bold assertion that the universe as a whole was expanding. From 1928 to 1936 he and Milton Humason toiled at Mount Wilson to prove this assertion until they reached the limits of the 100-inch telescope. Humason, with extreme skill and patience, laboriously photographed distant galaxies and determined red shifts. These galaxies were often invisible to the eye through the telescope and sometimes required week-long exposures in freezing weather. Hubble measured apparent brightness and by a variety of techniques determined distances out to half a billion light years. The results of this work and its continuation on the 200-inch Palomar telescope in the 1940s showed that the velocities of galaxies increase in direct proportion to the distance. (See Fig. B.10.) This linear relation is known as Hubble's law and may be written

$$v = HR \quad (B.8)$$

where H, Hubble's constant, is approximately (15 km/s)/million light years.

Will the Universe Expand Forever?

In the 1950s and 1960s Allan R. Sandage, using the 200-inch telescope at Mount Palomar, pushed the red shift distance measurements out from 1 billion to 6 billion light years. These measurements showed that galaxies at those

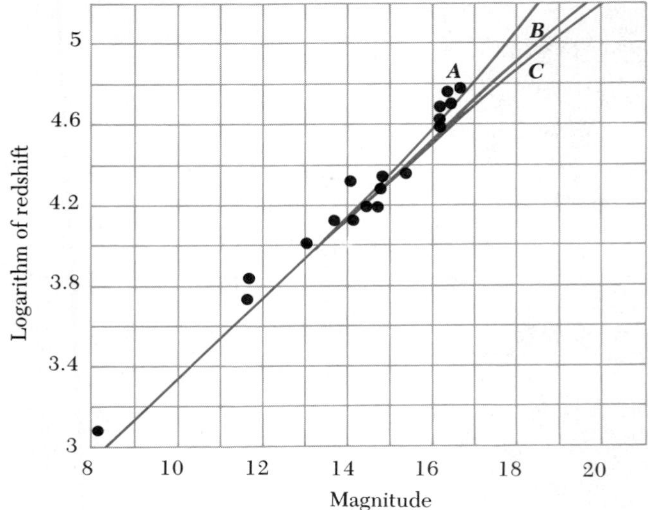

Figure B.11 Red shift or speed of recession versus apparent magnitude or distance of 18 faint clusters. Curve A is the trend suggested by the six faintest clusters of galaxies. Line C corresponds to a universe with a constant rate of expansion. If the data fall between B and C, the expansion slows but never stops. If the data fall to the left of B, expansion stops and contraction occurs.

distances were moving about 10 000 km/s faster than the Hubble law predicted. According to this result, the universe must have been expanding more rapidly 1 billion years ago and, consequently, the expansion is slowing.[6] (See Fig. B.11.) The question that remains concerns the rate at which the expansion is slowing. If the average mass density of the universe is less than the critical density (3 hydrogen atoms/1000 liters), the galaxies will slow in their outward rush but still escape to infinity. If the average density exceeds the critical value, the expansion will eventually stop and contraction will begin, possibly leading to a superdense state and another expansion or an oscillating universe. The critical mass density of the universe, ρ_c, can be estimated from energy considerations. Figure B.12 shows a large section of the universe with radius R, containing galaxies with a total mass M. A galaxy of mass m and velocity v at R will just escape to infinity with zero velocity if the sum of its kinetic energy and gravitational potential energy is zero. Thus,

$$E_{\text{total}} = 0 = K + U = \tfrac{1}{2}mv^2 - \frac{GmM}{R}$$

or

$$\tfrac{1}{2}mv^2 = \frac{Gm\frac{4}{3}\pi R^3\rho_c}{R}$$

or

$$v^2 = \frac{8\pi G}{3}R^2\rho_c \tag{B.9}$$

Since the galaxy of mass m obeys the Hubble law, $v = HR$, Equation B.9 becomes

$$H^2 = \frac{8\pi G}{3}\rho_c$$

[6] The data at large distances have large observational uncertainties and may be systematically in error from selection effects such as abnormal brightness in the most distant visible clusters.

or

$$\rho_c = \frac{3H^2}{8\pi G} \tag{B.10}$$

Using $H = 15$ km/s/million light years, where 1 light year $= 9.46 \times 10^{12}$ km, and $G = 6.67 \times 10^{-8}$ cm^3/g·s^2 yields $\rho_c = 4.5 \times 10^{-30}$ g/cm^3. As the mass of a hydrogen atom is 1.66×10^{-24} g, ρ_c corresponds to 2.7×10^{-6} hydrogen atoms per cm^3 or roughly 3 atoms per m^3.

The visible matter in galaxies averages out to 5×10^{-33} g/cm^3. The radiation in the universe has a mass equivalent of about 2% of the visible matter. Nonluminous matter (such as interstellar gas or black holes) may be estimated from the velocities of galaxies orbiting each other in a cluster. The higher the galaxy velocities, the more mass in the cluster. Results from measurements on the Coma cluster of galaxies, surprisingly, indicate that the amount of invisible matter is *20 to 30 times the amount* present in stars and luminous gas clouds. Yet even this large invisible component, if applied to the universe as a whole, leaves the observed mass density a factor of 10 less than ρ_c. This so-called missing mass has been the subject of intense theoretical and experimental work, with exotic particles such as axions, photinos, and superstring particles suggested as candidates for the missing mass. More mundane proposals have been that the missing mass is present in brown dwarfs, red dwarfs, or neutrinos. In fact, neutrinos are so abundant that a tiny neutrino rest mass on the order of 20 eV would furnish the missing mass and "close" the universe.

Although we are a bit more sure about the beginning of the universe, we are uncertain about its end. Will the universe expand forever? Will it collapse and repeat its expansion in an endless series of oscillations? Results and answers to these questions remain inconclusive and the exciting controversy continues.

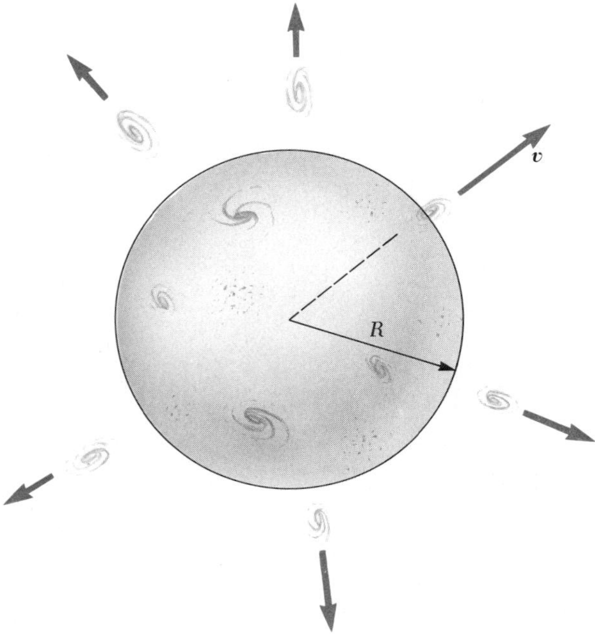

Figure B.12 A galaxy escaping from a large cluster contained within radius *R*. Only the mass within *R* slows the mass *m*.

SUGGESTIONS FOR FURTHER READING

1. David Layzer, *Constructing the Universe,* Scientific American Library, New York, W. H. Freeman and Co., 1984, Chapters 7 and 8.
2. George Gamow, "The Evolutionary Universe," *Scientific American,* September 1956.
3. Allan R. Sandage, "The Red-Shift," *Scientific American,* September 1956.
4. Carl Sagan and Frank Drake, "The Search for Extraterrestrial Intelligence," *Scientific American,* May 1975.
5. Richard A. Muller, "The Cosmic Background Radiation and the New Aether Drift," *Scientific American,* May 1978.
6. David L. Meier and Rashid A. Sunyaev, "Primeval Galaxies," *Scientific American,* November 1979.
7. John D. Barrow and Joseph Silk, "The Structure of the Early Universe," *Scientific American,* April 1980.

Appendix C

Quantization of Angular Momentum and Energy

We saw in Chapter 7 that angular momentum plays an important role in the quantization of many systems. Here we develop further the properties of angular momentum from the viewpoint of quantum mechanics, and show in more detail how angular momentum considerations facilitate the solution to Schrödinger's equation for central forces.

Our treatment is based on the operator methods introduced in Section 5.8, which the reader should review at this time. In particular, the eigenvalue condition

$$[Q]\Psi = q\Psi \tag{C.1}$$

used there as a *test* for sharp observables, becomes here a tool for discovering the form of the *unknown* wavefunction Ψ. We look upon Equation C.1 now as a *constraint* imposed on the wavefunction Ψ which guarantees that the observable Q will take the sharp value q in that state. The more sharp observables we can identify for some system, the more we can learn in advance about the waveform describing that system. But with few exceptions, the sharp quantities are just those which are conserved in the classical motion.

Consider the particle energy. If total energy is conserved, then E should be a sharp observable for Ψ. Otherwise, repeated measurements made on identical systems would reveal a statistical distribution of values for the particle energy, inconsistent with the idea of a conserved quantity. Thus, energy conservation suggests $\Delta E = 0$ which, in turn, requires Ψ to be an eigenfunction of the energy operator

$$[E]\Psi = E\Psi$$

Since $[E] = i\hbar\partial/\partial t$, this eigenvalue condition is met by the stationary waves $\psi(r)e^{-i\omega t}$, with $E = \hbar\omega$ the sharp value of particle energy.

The argument for energy applies equally well to other classically conserved quantities.[1] In the case of central forces, *angular momentum* about the force center also is conserved. This is a vector quantity $L = r \times p$, with rectangular components

$$L_z = (r \times p)_z = xp_y - yp_x,$$

[1] Exceptions to this rule do exist. For instance, an atomic electron cannot have all three angular momentum components sharp at once, even though in this case all components are conserved classically. For such *incompatible* observables, quantum mechanics adopts the broader interpretation of a conserved quantity as one whose *average value does not change over time,* no matter what may be the initial state of the system. With this definition, all components of angular momentum for an atomic electron remain conserved, but no more than one can be sharp in a given state.

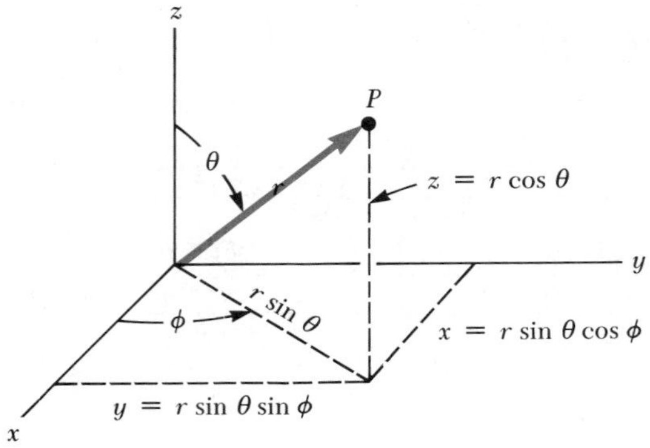

Figure C.1 The location of any point P may be described using the spherical coordinates r,θ,ϕ. The relation between the spherical and cartesian coordinates is obtained from the geometry of the figure.

and so on. In the same way, the *operators* for angular momentum are found from the coordinate and momentum *operators*. From Chapter 5, the operator for x is just the coordinate itself, while the operator for momentum in this direction is $(\hbar/i)\partial/\partial x$. Similar relations should apply to the directions labeled y and z,[2] so that the operator for L_z becomes

$$[L_z] = [x][p_y] - [y][p_x] = \frac{\hbar}{i}\left(x\frac{\partial}{\partial y} - y\frac{\partial}{\partial x}\right). \tag{C.2}$$

Angular momentum finds its natural expression in the spherical coordinates of Figure C.1. A little geometry applied to Figure C.1 shows that the spherical-to-cartesian coordinate transformation equations are

$$\begin{aligned} z &= r\cos\theta \\ x &= r\sin\theta\cos\phi \\ y &= r\sin\theta\sin\phi \end{aligned} \tag{C.3}$$

The inverse transformation is

$$\begin{aligned} r &= \{x^2 + y^2 + z^2\}^{1/2} \\ \cos\theta &= \frac{z}{r} = z\{x^2 + y^2 + z^2\}^{-1/2} \\ \tan\phi &= \frac{y}{x} \end{aligned} \tag{C.4}$$

The procedure for transcribing operators such as $[L_z]$ from cartesian to spherical form is straightforward, but tedious. For instance, an application of the chain rule gives

$$\frac{\partial f}{\partial z} = \frac{\partial r}{\partial z}\frac{\partial f}{\partial r} + \frac{\partial \theta}{\partial z}\frac{\partial f}{\partial \theta} + \frac{\partial \phi}{\partial z}\frac{\partial f}{\partial \phi}$$

for any function f. On the left we think of f expressed as a function of x, y, z, while on the right the same function is expressed in terms of r, θ, ϕ. The partial derivatives $\partial r/\partial z$, etc. are to be taken with the aid of the transformation equations, holding x and y fixed. The simplest of these is $\partial\phi/\partial z$. From the inverse

[2] For the momentum operator along each axis we use the one-dimensional form from Chapter 5, ($[p_x] = (\hbar/i)\partial/\partial x$, etc.), consistent with our belief that the cartesian axes label independent but logically equivalent directions in space.

transformations in Equations C.4, we see that ϕ is independent of z, so that $\partial\phi/\partial z = 0$. From the same equations we calculate

$$\frac{\partial r}{\partial z} = \frac{1}{2}\{x^2 + y^2 + z^2\}^{-1/2}(2z) = \frac{z}{r} = \cos\theta$$

Now, $\partial\theta/\partial z$ may be obtained from the second of Equations C.4 by implicit differentiation:

$$-\sin\theta\,\frac{\partial\theta}{\partial z} = \frac{\partial\{\cos\theta\}}{\partial z} = z(-\tfrac{1}{2})\{x^2 + y^2 + z^2\}^{-3/2}(2z) + \{x^2 + y^2 + z^2\}^{-1/2}$$

Converting the right-hand side back to spherical coordinates gives

$$-\sin\theta\,\frac{\partial\theta}{\partial z} = -\frac{(r\cos\theta)^2}{r^3} + \frac{1}{r} = \frac{1-\cos^2\theta}{r} = \frac{\sin^2\theta}{r}$$

or $\partial\theta/\partial z = -\sin\theta/r$. Collecting the above results, we have

$$\frac{\partial}{\partial z} = \frac{\partial r}{\partial z}\frac{\partial}{\partial r} + \frac{\partial\theta}{\partial z}\frac{\partial}{\partial\theta} = \cos\theta\,\frac{\partial}{\partial r} - \frac{\sin\theta}{r}\frac{\partial}{\partial\theta}$$

In like fashion we can obtain spherical representations for $\partial/\partial x$, $\partial/\partial y$, and ultimately for the angular momentum operators themselves:

$$\begin{aligned}
[L_x] &= i\hbar\left\{\sin\phi\,\frac{\partial}{\partial\theta} + \cot\theta\cos\phi\,\frac{\partial}{\partial\phi}\right\} \\
[L_y] &= -i\hbar\left\{\cos\phi\,\frac{\partial}{\partial\theta} - \cot\theta\sin\phi\,\frac{\partial}{\partial\phi}\right\} \qquad \text{(C.5)} \\
[L_z] &= -i\hbar\,\frac{\partial}{\partial\phi}
\end{aligned}$$

Notice that the operators for angular momentum in spherical form make no reference to the radial coordinate r—our reward for selecting the right system of coordinates! Thus, by insisting that angular momentum be sharp in the quantum state Ψ, we can discover how the wavefunction depends on the spherical angles θ and ϕ, *without knowing the details of the central force.*

L_z is Sharp—The Magnetic Quantum Number

For L_z to be sharp, our wavefunction Ψ must be an eigenfunction of $[L_z]$, or

$$-i\hbar\,\frac{\partial\Psi}{\partial\phi} = L_z\Psi \qquad \text{(C.6)}$$

Equation C.6 prescribes the functional dependence of ψ on the azimuthal angle ϕ to be

$$\psi(r) = Ce^{iL_z\phi/\hbar} \qquad \text{(C.7)}$$

with C still any function of r or θ. However, the values taken by L_z must be restricted. Since increasing ϕ by 2π radians returns us to the same place in space (see Fig. C.1), the wavefunction also must return to its initial value, that is, the solutions represented by Equation C.7 must be periodic in ϕ with period 2π. This will be true if $L_z/\hbar$ is any integer, say m_ℓ, or $L_z = m_\ell\hbar$. The *magnetic quantum number* m_ℓ indicates the (sharp) value for the z-component of angular momentum in the state described by ψ.

Since all components of L are conserved classically for central forces, we should continue by requiring the wave of Equation C.7 also to be an eigenfunction of the operator for a second angular momentum component, say $[L_x]$, so that L_z and L_x might both be sharp in the state ψ. For ψ to be an eigenfunction of $[L_x]$ with eigenvalue L_x requires

$$i\hbar\left\{\sin\phi\,\frac{\partial\psi}{\partial\theta} + \cot\theta\cos\phi\,\frac{\partial\psi}{\partial\phi}\right\} = L_x\psi$$

This relation must hold for all values of θ and ϕ. In particular, at $\phi = 0$ the requirement is

$$i\hbar\cot\theta\,\frac{\partial\psi}{\partial\phi} = L_x\psi$$

Since ψ varies exponentially with ϕ according to Equation C.7, the indicated derivative may be taken and evaluated at $\phi = 0$ to get

$$-L_z\cot\theta\,\psi = L_x\psi$$

which can be satisfied for *all* θ only if $L_x = L_z = 0$, or if ψ vanishes identically. A similar difficulty arises if we attempt to make L_y sharp together with L_z. Thus, *unless the angular momentum is exactly zero (all components), there is no waveform for which two or more components are simultaneously sharp!*

$|L|$ is Sharp—The Orbital Quantum Number

To make further progress, we must look to other classically conserved quantities. In addition to each component of angular momentum, the magnitude $|L|$ of the vector also is conserved, and becomes a candidate for an observable which can be made sharp together with L_z. Consider simply the squared-magnitude $L^2 = L\cdot L$. The operator for L^2 is

$$[L^2] = [L_x]^2 + [L_y]^2 + [L_z]^2$$

Using Equations C.5, we find the spherical coordinate form for $[L^2]$:

$$[L^2] = -\hbar^2\left\{\frac{\partial^2}{\partial\theta^2} + \cot\theta\,\frac{\partial}{\partial\theta} + \csc^2\theta\,\frac{\partial^2}{\partial\phi^2}\right\} \tag{C.8}$$

For L^2 to be sharp requires $[L^2]\psi = |L|^2\psi$ or

$$-\hbar^2\left\{\frac{\partial^2\psi}{\partial\theta^2} + \cot\theta\,\frac{\partial\psi}{\partial\theta} + \csc^2\theta\,\frac{\partial^2\psi}{\partial\phi^2}\right\} = |L|^2\psi$$

But from Equation C.6, we see that differentiating ψ with respect to ϕ is equivalent to multiplication by $iL_z/\hbar = im_\ell$, so that the last equation reduces to

$$-\hbar^2\left\{\frac{\partial^2\psi}{\partial\theta^2} + \cot\theta\,\frac{\partial\psi}{\partial\theta} - m_\ell^{\,2}\csc^2\theta\,\psi\right\} = |L|^2\psi \tag{C.9}$$

For waveforms satisfying Equation C.9, the magnitude of angular momentum will be sharp at the value $|L|$. The equation prescribes the polar angle or θ dependence of our wave. The solutions are not elementary functions, but can be investigated with the help of more advanced techniques. The results of such an investigation are reported in Chapter 7 and repeated here for completeness: physically acceptable solutions to Equation C.9 can be found provided $|L| = \sqrt{\ell(\ell+1)}\,\hbar$ where ℓ, the *orbital quantum number,* must be a nonnegative integer. Further, the magnetic quantum number, also appearing in Equation C.9, must be limited as $|m_\ell| \leq \ell$. With these restrictions, the solutions to Equation C.9 are the *associated Legendre polynomials* in $\cos\theta$, denoted $P_\ell^{m_\ell}(\cos\theta)$. Several of these polynomials are listed in Table C.1 for easy reference; you may verify by direct substitution that they satisfy Equation C.9 for the appropriate values of ℓ and m_ℓ.

The associated Legendre polynomials may be multiplied by $\exp(im_\ell\phi)$ and still satisfy the orbital equation Equation C.9. Indeed, the products $P_\ell^{m_\ell}(\cos\theta)\exp(im_\ell\phi)$ satisfy both Equations C.6 and C.9; thus, they represent waves for which $|L|$ and L_z are simultaneously sharp. Except for a multiplicative constant, these are just the spherical harmonics $Y_\ell^{m_\ell}(\theta, \phi)$ introduced in Section 7.2.

E is Sharp—The Radial Wave Equation

For energy to be conserved, E also must be a sharp observable. The stationary state form $\Psi(r, t) = \psi(r)e^{-i\omega t}$ results from imposing the eigenvalue condition for the energy operator $[E]$, but what of the other energy operator $[H] = [K] + [U]$ (the hamiltonian)? In fact, requiring $[H]\Psi = E\Psi$ is equivalent to writing the time-independent Schrödinger equation for $\psi(r)$

$$-\frac{\hbar^2}{2m}\nabla^2\psi + U(r)\psi = E\psi \tag{C.10}$$

since $[U] = U(r)$ and the kinetic energy operator $[K]$ in three dimensions is none other than

$$[K] = \frac{[p_x]^2 + [p_y]^2 + [p_z]^2}{2m} = \frac{(\hbar/i)^2\{(\partial/\partial x)^2 + (\partial/\partial y)^2 + (\partial/\partial z)^2\}}{2m} = -\frac{\hbar^2}{2m}\nabla^2$$

The spherical form of the Laplacian given in Chapter 7 may be used to write $[K]$ in spherical coordinates:

$$[K] = -\frac{\hbar^2}{2m}\left\{\frac{\partial^2}{\partial r^2} + \frac{2}{r}\frac{\partial}{\partial r} + \frac{1}{r^2}\left[\frac{\partial^2}{\partial\theta^2} + \cot\theta\frac{\partial}{\partial\theta} + \csc^2\theta\frac{\partial^2}{\partial\phi^2}\right]\right.$$

Comparing this with the spherical representation for $[L^2]$ from Equation C.8 shows that $[K]$ is the sum of two terms

$$[K] = [K^{\text{rad}}] + [K^{\text{orb}}] = [K^{\text{rad}}] + \frac{1}{2mr^2}[L^2], \tag{C.11}$$

representing the separate contributions to the energy from the orbital and radial components of motion. The comparison also furnishes the spherical form for $[K^{\text{rad}}]$:

$$[K^{\text{rad}}] = -\frac{\hbar^2}{2m}\left\{\frac{\partial^2}{\partial r^2} + \frac{2}{r}\frac{\partial}{\partial r}\right\} \tag{C.12}$$

These expressions for $[K]$ are completely general. When applied to waves for which $|L|$ is sharp, however, the *operator* $[L^2]$ may be replaced with the *number* $|L|^2 = \ell(\ell + 1)\hbar^2$. Thus, the Schrödinger equation (Eq. C.10) becomes

$$[K^{\text{rad}}]\psi(r) + \ell(\ell+1)\frac{\hbar^2}{2mr^2}\psi(r) + U(r)\psi(r) = E\psi(r) \tag{C.13}$$

For central forces all terms in this equation, *including* $U(r)$, involve only the spherical coordinate r: the angle variables θ and ϕ have been eliminated from

TABLE C.1 Some Associated Legendre Polynomials $P_\ell^{m_\ell}(\cos\theta)$

$P_0^0 = 1$
$P_1^0 = 2\cos\theta$
$P_1^1 = \sin\theta$
$P_2^0 = 4(3\cos^2\theta - 1)$
$P_2^1 = 4\sin\theta\cos\theta$
$P_2^2 = \sin^2\theta$
$P_3^0 = 24(5\cos^3\theta - 3\cos\theta)$
$P_3^1 = 6\sin\theta(5\cos^2\theta - 1)$
$P_3^2 = 6\sin^2\theta\cos\theta$
$P_3^3 = \sin^3\theta$

further consideration by the requirement that ψ be an eigenfunction of $[L^2]$! It follows that the solutions to Equation C.13 take the form of a radial wave $R(r)$ multiplied by a spherical harmonic:

$$\psi(\mathbf{r}) = R(r)Y_{\ell}^{m_{\ell}}(\theta, \phi), \tag{C.14}$$

with the radial wavefunction $R(r)$ also satisfying Equation C.13.

Equation C.13 for $R(r)$ is just the *radial wave equation* of Chapter 7; it determines the radial wavefunction $R(r)$ and the allowed particle energies E, once the potential energy function $U(r)$ is specifed. As the equation suggests, each angular momentum state will have a different radial wave and a different energy. Indeed, the "effective potential"

$$U_{\text{eff}}(r) = U(r) + \ell(\ell + 1)\frac{\hbar^2}{2mr^2} \tag{C.15}$$

is different for each orbital quantum number ℓ. By contrast, the magnetic quantum number m_{ℓ} appears nowhere in these equations. Thus, the radial waveform and particle energy remain the same for different m_{ℓ} values consistent with a given value of ℓ. In particular, for a fixed ℓ *the particle energy E is independent of* m_{ℓ}*, and so is at least* $2\ell + 1$*-fold degenerate.* Such degeneracy —a property of all central forces—stems from the spherical symmetry of the potential, and illustrates once again the deep-seated connection between symmetry and the degeneracy of quantum levels.

Appendix D

Overlap Integrals of Atomic Wavefunctions

As shown in Chapter 11, the normalization of molecular wavefunctions that are formed as combinations of atomic orbitals involves evaluation of so-called *overlap integrals* having the general form

$$\int \psi_a^*(r)\psi_b(r - \boldsymbol{R})\, dV \tag{D.1}$$

Such integrals are evaluated here for the case where ψ_a and ψ_b are wavefunctions for the hydrogen ground state:

$$\psi_a(r) = \psi_b(r) = \frac{1}{\sqrt{\pi}}\, a_0^{-3/2} e^{-r/a_0} \tag{D.2}$$

The analysis is most transparent when carried out in *atomic units,* where distances are measured in bohrs (a_0) and energies in rydbergs ($ke^2/2a_0$). In such units r/a_0 becomes simply r, and ke^2/r becomes $2/r$, etc.

Now the atomic wave of Equation D.2 is an s-state and, therefore, spherically symmetric; that is, ψ_a is a function only of $|\boldsymbol{r}| = r$. Likewise, $\psi_b(\boldsymbol{r} - \boldsymbol{R})$ will depend only on

$$|\boldsymbol{r} - \boldsymbol{R}| = \sqrt{r^2 + R^2 - 2\boldsymbol{r}\cdot\boldsymbol{R}}$$

If we orient the coordinate axes so that the z axis lies along the direction of $\boldsymbol{R}$, then $\boldsymbol{r}\cdot\boldsymbol{R} = rR\cos\theta$, where θ is the polar angle in spherical coordinates. Since the volume element in spherical coordinates is $dV = r^2\, dr\, d\Omega$, where the element of solid angle $d\Omega = \sin\theta\, d\theta\, d\phi$, the integration over angle can be carried out using the substitution $u^2 = r^2 + R^2 - 2rR\cos\theta$ (so that $udu = rR \sin\theta\, d\theta$). For $\theta = 0$, $u = |\boldsymbol{r} - \boldsymbol{R}|$ ($= u_1$), while $\theta = \pi$ implies $u = |\boldsymbol{r} + \boldsymbol{R}|$ ($= u_2$). Then

$$\begin{aligned}\int \psi_b(\boldsymbol{r} - \boldsymbol{R})\, d\Omega &= \frac{1}{\sqrt{\pi}}\, a_0^{-3/2}\left(\frac{2\pi}{rR}\right)\int_{u_1}^{u_2} ue^{-u}\, du = -\frac{2\sqrt{\pi}}{rR}\, a_0^{-3/2}(1+u)e^{-u}\Big|_{u_1}^{u_2} \\ &= \frac{2\sqrt{\pi}}{rR}\, a_0^{-3/2}\{(1 + |r - R|)e^{-|r-R|} - (1 + |r + R|)e^{-|r+R|}\}\end{aligned}$$

and the full overlap integral becomes

$$\begin{aligned}\int \psi_a^*(r)\psi_b(r - \boldsymbol{R})\, dV = \frac{2}{R}\bigg\{&\int_0^\infty re^{-r}(1 + |r - R|)e^{-|r-R|}\, dr \\ &- \int_0^\infty re^{-r}(1 + |r + R|)e^{-|r+R|}\, dr\bigg\}\end{aligned}$$

Since r and R are both nonnegative, $|r + R| = r + R$, and the second of the two integrals on the right can be integrated immediately using

$$\int_0^\infty x^n e^{-x}\,dx = n!$$

to get

$$\frac{2}{R}e^{-R}\int_0^\infty r(1+r+R)e^{-2r}\,dr = \frac{1}{R}e^{-R}\int_0^\infty (x/2)(1+\frac{x}{2}+R)e^{-x}\,dx$$
$$= \left(\frac{1}{R}+\frac{1}{2}\right)e^{-R}$$

The first integral must be evaluated separately in the two intervals $r < R$ and $r > R$. For $r < R$, we have $|r - R| = R - r$ and we find

$$\frac{2}{R}e^{-R}\int_0^R r(1+R-r)\,dr = [R(1+R) - 2R^2/3]e^{-R}$$

while for $r > R$ we obtain (with $x = r - R$)

$$\frac{2}{R}\int_R^\infty r(1+r-R)e^{(-2r+R)}\,dr = \frac{2}{R}e^{-R}\int_0^\infty (x+R)(1+x)e^{-2x}\,dx$$
$$= \left(\frac{1}{R}+\frac{3}{2}\right)e^{-R}$$

Combining all the above results gives the overlap integral as

$$\int \psi_a^*(r)\psi_b(\mathbf{r}-\mathbf{R})\,dV = (1+R+R^2/3)e^{-R} \tag{D.3}$$

With only minor alterations, the same method may be used to obtain the expressions given in the text for the energies E_+ and E_- of the lowest bonding and antibonding states, respectively, of $H_2{}^+$.

Appendix E

Conversion Factors

Length

	m	cm	km	in.	ft	mi
1 meter	1	10^2	10^{-3}	39.37	3.281	6.214×10^{-4}
1 centimeter	10^{-2}	1	10^{-5}	0.3937	3.281×10^{-2}	6.214×10^{-6}
1 kilometer	10^3	10^5	1	3.937×10^4	3.281×10^3	0.6214
1 inch	2.540×10^{-2}	2.540	2.540×10^{-5}	1	8.333×10^{-2}	1.578×10^{-5}
1 foot	0.3048	30.48	3.048×10^{-4}	12	1	1.894×10^{-4}
1 mile	1609	1.609×10^5	1.609	6.336×10^4	5280	1

Mass

	kg	g	slug	amu
1 kilogram	1	10^3	6.852×10^{-2}	6.024×10^{26}
1 gram	10^{-3}	1	6.852×10^{-5}	6.024×10^{23}
1 slug (lb/g)	14.59	1.459×10^4	1	8.789×10^{27}
1 atomic mass unit	1.660×10^{-27}	1.660×10^{-24}	1.137×10^{-28}	1

Time

	s	min	h	day	year
1 second	1	1.667×10^{-2}	2.778×10^{-4}	1.157×10^{-5}	3.169×10^{-8}
1 minute	60	1	1.667×10^{-2}	6.994×10^{-4}	1.901×10^{-6}
1 hour	3600	60	1	4.167×10^{-2}	1.141×10^{-4}
1 day	8.640×10^4	1440	24	1	2.738×10^{-3}
1 year	3.156×10^7	5.259×10^5	8.766×10^3	365.2	1

Speed

	m/s	cm/s	ft/s	mi/h
1 meter/second	1	10^2	3.281	2.237
1 centimeter/second	10^{-2}	1	3.281×10^{-2}	2.237×10^{-2}
1 foot/second	0.3048	30.48	1	0.6818
1 mile/hour	0.4470	44.70	1.467	1

Note: 1 mi/min = 60 mi/h = 88 ft/s.

Force

	N	dyn	lb
1 newton	1	10^5	0.2248
1 dyne	10^{-5}	1	2.248×10^{-6}
1 pound	4.448	4.448×10^5	1

Work, Energy, Heat

	J	erg	ft·lb
1 joule	1	10^{7}	0.7376
1 erg	10^{-7}	1	7.376×10^{-8}
1 ft·lb	1.356	1.356×10^{7}	1
1 eV	1.602×10^{-19}	1.602×10^{-12}	1.182×10^{-19}
1 cal	4.186	4.186×10^{7}	3.087
1 Btu	1.055×10^{3}	1.055×10^{10}	7.779×10^{2}
1 kWh	3.600×10^{6}	3.600×10^{13}	2.655×10^{6}

	eV	cal	Btu	kWh
1 joule	6.242×10^{18}	0.2389	9.481×10^{-4}	2.778×10^{-7}
1 erg	6.242×10^{11}	2.389×10^{-8}	9.481×10^{-11}	2.778×10^{-14}
1 ft·lb	8.464×10^{18}	0.3239	1.285×10^{-3}	3.766×10^{-7}
1 eV	1	3.827×10^{-20}	1.519×10^{-22}	4.450×10^{-26}
1 cal	2.613×10^{19}	1	3.968×10^{-3}	1.163×10^{-6}
1 Btu	6.585×10^{21}	2.520×10^{2}	1	2.930×10^{-4}
1 kWh	2.247×10^{25}	8.601×10^{5}	3.413×10^{2}	1

Pressure

	N/m²	dyn/cm²	atm
1 newton/meter²	1	10	9.869×10^{-6}
1 dyne/centimeter²	10^{-1}	1	9.869×10^{-7}
1 atmosphere	1.013×10^{5}	1.013×10^{6}	1
1 centimeter mercury*	1.333×10^{3}	1.333×10^{4}	1.316×10^{-2}
1 pound/inch²	6.895×10^{3}	6.895×10^{4}	6.805×10^{-2}
1 pound/foot²	47.88	4.788×10^{2}	4.725×10^{-4}

	cm Hg	lb/in.²	lb/ft²
1 newton/meter²	7.501×10^{-4}	1.450×10^{-4}	2.089×10^{-2}
1 dyne/centimeter²	7.501×10^{-5}	1.450×10^{-5}	2.089×10^{-3}
1 atmosphere	76	14.70	2.116×10^{3}
1 centimeter mercury*	1	0.1943	27.85
1 pound/inch²	5.171	1	144
1 pound/foot²	3.591×10^{-2}	6.944×10^{-3}	1

* At 0°C and at a location where the acceleration due to gravity has its "standard" value, 9.80665 m/s².

Appendix F

Current Values for Physical Constants

Parentheses indicate uncertainties in the last decimal places given.

speed of light, c	299 792 458 m/s (exact)
gravitational constant, G	$6.672\ 59(85) \times 10^{-11}\ \mathrm{N \cdot m^2 \cdot kg^{-2}}$
Planck's constant, h	$6.626\ 075\ 5(40) \times 10^{-34}\ \mathrm{J \cdot s}$
Boltzmann constant, k	$1.380\ 658(12) \times 10^{-23}\ \mathrm{J \cdot K^{-1}}$
Stefan-Boltzmann constant, σ	$5.670\ 51(19) \times 10^{-8}\ \mathrm{W \cdot m^{-2} \cdot K^{-4}}$
Rydberg constant, R_∞	$1.097\ 373\ 153\ 4(13) \times 10^{7}\ \mathrm{m^{-1}}$
fine-structure constant, α^{-1}	137.035 989 5(61)
Bohr radius, a_0	$5.291\ 772\ 49(24) \times 10^{-11}$ m
Avogadro's number, N_A	$6.022\ 136\ 7(36) \times 10^{23}\ \mathrm{mole^{-1}}$
mass of hydrogen atom, m_H	$1.673\ 534\ 0(10) \times 10^{-27}$ kg
mass of neutron, m_n	$1.674\ 928\ 6(10) \times 10^{-27}$ kg 939.565 63(28) MeV
mass of proton, m_p	$1.672\ 623\ 1(10) \times 10^{-27}$ kg 938.272 31(28) MeV
mass of electron, m_e	$9.109\ 389\ 7(54) \times 10^{-31}$ kg 0.510 999 06(15) MeV
proton-electron mass ratio, m_p/m_e	1836.152 701(37)
elementary charge, e	$1.602\ 177\ 33(49) \times 10^{-19}$ C
charge to mass, e/m_e	$1.758\ 819\ 62(53) \times 10^{11}\ \mathrm{C \cdot kg^{-1}}$
permeability of vacuum, $\mu_0 = 4\pi \times 10^{-7}$	$12.566\ 370\ 614 \ldots \times 10^{-7}$ H/m
permittivity of vacuum, $\epsilon_0 = 1/\mu_0 c^2$	$8.854\ 187\ 817 \ldots \times 10^{-12}$ F/m

Values from *The 1986 Adjustment of the Fundamental Physical Constants*, Report of the CODATA Task Group on Fundamental Constants, prepared by E. Richard Cohen and Barry N. Taylor. Committee on Data for Science and Technology CODATA Bulletin No. 63, Pergamon Press.

Appendix G

Table of Selected Atomic Masses*

Atomic Number Z	Element	Symbol	Mass Number A	Atomic Mass†	Percent Abundance, or Decay Mode (if radioactive)‡	Half-Life (if radioactive)
0	(Neutron)	n	1	1.008665	β^-	10.6 min
1	Hydrogen	H	1	1.007825	99.985	
	Deuterium	D	2	2.014102	0.015	
	Tritium	T	3	3.016049	β^-	12.33 yr
2	Helium	He	3	3.016029	0.00014	
			4	4.002603	≈ 100	
3	Lithium	Li	6	6.015123	7.5	
			7	7.016005	92.5	
4	Beryllium	Be	7	7.016930	EC, γ	53.3 days
			8	8.005305	2α	6.7×10^{-17} s
			9	9.012183	100	
5	Boron	B	10	10.012938	19.8	
			11	11.009305	80.2	
6	Carbon	C	11	11.011433	β^+, EC	20.4 min
			12	12.000000	98.89	
			13	13.003355	1.11	
			14	14.003242	β^-	5730 yr
7	Nitrogen	N	13	13.005739	β^+	9.96 min
			14	14.003074	99.63	
			15	15.000109	0.37	
8	Oxygen	O	15	15.003065	β^+, EC	122 s
			16	15.994915	99.76	
			18	17.999159	0.204	
9	Fluorine	F	19	18.998403	100	
10	Neon	Ne	20	19.992439	90.51	
			22	21.991384	9.22	
11	Sodium	Na	22	21.994435	β^+, EC, γ	2.602 yr
			23	22.989770	100	
			24	23.990964	β^-, γ	15.0 h
12	Magnesium	Mg	24	23.985045	78.99	
13	Aluminum	Al	27	26.981541	100	
14	Silicon	Si	28	27.976928	92.23	
			31	30.975364	β^-, γ	2.62 h
15	Phosphorus	P	31	30.973763	100	
			32	31.973908	β^-	14.28 days
16	Sulfur	S	32	31.972072	95.0	
			35	34.969033	β^-	87.4 days
17	Chlorine	Cl	35	34.968853	75.77	
			37	36.965903	24.23	

Atomic Number Z	Element	Symbol	Mass Number A	Atomic Mass†	Percent Abundance, or Decay Mode (if radioactive)‡	Half-Life (if radioactive)
18	Argon	Ar	40	39.962383	99.60	
19	Potassium	K	39	38.963708	93.26	
			40	39.964000	β^-, EC, γ, β^+	1.28×10^9 yr
20	Calcium	Ca	40	39.962591	96.94	
21	Scandium	Sc	45	44.955914	100	
22	Titanium	Ti	48	47.947947	73.7	
23	Vanadium	V	51	50.943963	99.75	
24	Chromium	Cr	52	51.940510	83.79	
25	Manganese	Mn	55	54.938046	100	
26	Iron	Fe	56	55.934939	91.8	
27	Cobalt	Co	59	58.933198	100	
			60	59.933820	β^-, γ	5.271 yr
28	Nickel	Ni	58	57.935347	68.3	
			60	59.930789	26.1	
			64	63.927968	0.91	
29	Copper	Cu	63	62.929599	69.2	
			64	63.929766	β^-, β^+	12.7 h
			65	64.927792	30.8	
30	Zinc	Zn	64	63.929145	48.6	
			66	65.926035	27.9	
31	Gallium	Ga	69	68.925581	60.1	
32	Germanium	Ge	72	71.922080	27.4	
			74	73.921179	36.5	
33	Arsenic	As	75	74.921596	100	
34	Selenium	Se	80	79.916521	49.8	
35	Bromine	Br	79	78.918336	50.69	
36	Krypton	Kr	84	83.911506	57.0	
			89	88.917563	β^-	3.2 min
37	Rubidium	Rb	85	84.911800	72.17	
38	Strontium	Sr	86	85.909273	9.8	
			88	87.905625	82.6	
			90	89.907746	β^-	28.8 yr
39	Yttrium	Y	89	88.905856	100	
40	Zirconium	Zr	90	89.904708	51.5	
41	Niobium	Nb	93	92.906378	100	
42	Molybdenum	Mo	98	97.905405	24.1	
43	Technetium	Tc	98	97.907210	β^-, γ	4.2×10^6 yr
44	Ruthenium	Ru	102	101.904348	31.6	
45	Rhodium	Rh	103	102.90550	100	
46	Palladium	Pd	106	105.90348	27.3	
47	Silver	Ag	107	106.905095	51.83	
			109	108.904754	48.17	
48	Cadmium	Cd	114	113.903361	28.7	
49	Indium	In	115	114.90388	95.7; β^-	5.1×10^{14} yr
50	Tin	Sn	120	119.902199	32.4	
51	Antimony	Sb	121	120.903824	57.3	
52	Tellurium	Te	130	129.90623	34.5; β^-	2×10^{21} yr
53	Iodine	I	127	126.904477	100	
			131	130.906118	β^-, γ	8.04 days
54	Xenon	Xe	132	131.90415	26.9	
			136	135.90722	8.9	

Atomic Number Z	Element	Symbol	Mass Number A	Atomic Mass†	Percent Abundance, or Decay Mode (if radioactive)‡	Half-Life (if radioactive)
55	Cesium	Cs	133	132.90543	100	
56	Barium	Ba	137	136.90582	11.2	
			138	137.90524	71.7	
			144	143.922673	β^-	11.9 s
57	Lanthanum	La	139	138.90636	99.911	
58	Cerium	Ce	140	139.90544	88.5	
59	Praseodymium	Pr	141	140.90766	100	
60	Neodymium	Nd	142	141.90773	27.2	
61	Promethium	Pm	145	144.91275	EC, α, γ	17.7 yr
62	Samarium	Sm	152	151.91974	26.6	
63	Europium	Eu	153	152.92124	52.1	
64	Gadolinium	Gd	158	157.92411	24.8	
65	Terbium	Tb	159	158.92535	100	
66	Dysprosium	Dy	164	163.92918	28.1	
67	Holmium	Ho	165	164.93033	100	
68	Erbium	Er	166	165.93031	33.4	
69	Thulium	Tm	169	168.93423	100	
70	Ytterbium	Yb	174	173.93887	31.6	
71	Lutecium	Lu	175	174.94079	97.39	
72	Hafnium	Hf	180	179.94656	35.2	
73	Tantalum	Ta	181	180.94801	99.988	
74	Tungsten (wolfram)	W	184	183.95095	30.7	
75	Rhenium	Re	187	186.95577	62.60, β^-	4×10^{10} yr
76	Osmium	Os	191	190.96094	β^-, γ	15.4 days
			192	191.96149	41.0	
77	Iridium	Ir	191	190.96060	37.3	
			193	192.96294	62.7	
78	Platinum	Pt	195	194.96479	33.8	
79	Gold	Au	197	196.96656	100	
80	Mercury	Hg	202	201.97063	29.8	
81	Thallium	Tl	205	204.97441	70.5	
82	Lead	Pb	210	209.990069	β^-	1.3 min
			204	203.973044	β^-, 1.48	1.4×10^{17} yr
			206	205.97446	24.1	
			207	206.97589	22.1	
			208	207.97664	52.3	
			210	209.98418	α, β^-, γ	22.3 yr
			211	210.98874	β^-, γ	36.1 min
			212	211.99188	β^-, γ	10.64 h
			214	213.99980	β^-, γ	26.8 min
83	Bismuth	Bi	209	208.98039	100	
			211	210.98726	α, β^-, γ	2.15 min
84	Polonium	Po	210	209.98286	α, γ	138.38 days
			214	213.99519	α, γ	164 μs
85	Astatine	At	218	218.00870	α, β^-	≈ 2 s
86	Radon	Rn	222	222.017574	α, γ	3.8235 days
87	Francium	Fr	223	223.019734	α, β^-, γ	21.8 min
88	Radium	Ra	226	226.025406	α, γ	1.60×10^3 yr
			228	228.031069	β^-	5.76 yr

Atomic Number Z	Element	Symbol	Mass Number A	Atomic Mass†	Percent Abundance, or Decay Mode (if radioactive)‡	Half-Life (if radioactive)
89	Actinium	Ac	227	227.027751	α, β^-, γ	21.773 yr
90	Thorium	Th	228	228.02873	α, γ	1.9131 yr
			232	232.038054	100, α, γ	1.41×10^{10} yr
91	Protactinium	Pa	231	231.035881	α, γ	3.28×10^4 yr
92	Uranium	U	232	232.03714	α, γ	72 yr
			233	233.039629	α, γ	1.592×10^5 yr
			235	235.043925	0.72; α, γ	7.038×10^8 yr
			236	236.045563	α, γ	2.342×10^7 yr
			238	238.050786	99.275; α, γ	4.468×10^9 yr
			239	239.054291	β^-, γ	23.5 min
93	Neptunium	Np	239	239.052932	β^-, γ	2.35 days
94	Plutonium	Pu	239	239.052158	α, γ	2.41×10^4 yr
95	Americium	Am	243	243.061374	α, γ	7.37×10^3 yr
96	Curium	Cm	245	245.065487	α, γ	8.5×10^3 yr
97	Berkelium	Bk	247	247.07003	α, γ	1.4×10^3 yr
98	Californium	Cf	249	249.074849	α, γ	351 yr
99	Einsteinium	Es	254	254.08802	α, γ, β^-	276 days
100	Fermium	Fm	253	253.08518	EC, α, γ	3.0 days
101	Mendelevium	Md	255	255.0911	EC, α	27 min
102	Nobelium	No	255	255.0933	EC, α	3.1 min
103	Lawrencium	Lr	257	257.0998	α	$\approx$35 s
104	Rutherfordium (?)	Rf	261	261.1087	α	1.1 min
105	Hahnium (?)	Ha	262	262.1138	α	0.7 min
106			263	263.1184	α	0.9 s
107			261	261	α	1 − 2 ms

° Data are taken from *Chart of the Nuclides*, 12th ed., General Electric, 1977, and from C. M. Lederer and V. S. Shirley, eds., *Table of Isotopes*, 7th ed., John Wiley & Sons, Inc., New York, 1978.

† The masses given in column 5 are those for the neutral atom, including the Z electrons.

‡ The abbreviation EC stands for "electron capture."

Appendix H

Nobel Prizes

All Nobel Prizes in physics are listed (and marked with a P), as well as relevant Nobel Prizes in Chemistry (C). The key dates for some of the scientific work are supplied; they often antedate the prize considerably.

1901 (P) *Wilhelm Roentgen* for discovering x-rays (1895).

1902 (P) *Hendrik A. Lorentz* for predicting the Zeeman effect and *Pieter Zeeman* for discovering the Zeeman effect, the splitting of spectral lines in magnetic fields.

1903 (P) *Antoine-Henri Becquerel* for discovering radioactivity (1896) and *Pierre* and *Marie Curie* for studying radioactivity.

1904 (P) *Lord Rayleigh* for studying the density of gases and discovering argon.
(C) *William Ramsay* for discovering the inert gas elements helium, neon, xenon, and krypton, and placing them in the periodic table.

1905 (P) *Philipp Lenard* for studying cathode rays, electrons (1898–1899).

1906 (P) *J. J. Thomson* for studying electrical discharge through gases and discovering the electron (1897).

1907 (P) *Albert A. Michelson* for inventing optical instruments and measuring the speed of light (1880s).

1908 (P) *Gabriel Lippmann* for making the first color photographic plate, using interference methods (1891).
(C) *Ernest Rutherford* for discovering that atoms can be broken apart by alpha rays and for studying radioactivity.

1909 (P) *Guglielmo Marconi* and *Carl Ferdinand Braun* for developing wireless telegraphy.

1910 (P) *Johannes D. van der Waals* for studying the equation of state for gases and liquids (1881).

1911 (P) *Wilhelm Wien* for discovering Wien's law giving the peak of a blackbody spectrum (1893).
(C) *Marie Curie* for discovering radium and polonium (1898) and isolating radium.

1912 (P) *Nils Dalén* for inventing automatic gas regulators for lighthouses.

1913 (P) *Heike Kamerlingh Onnes* for the discovery of superconductivity and liquefying helium (1908).

1914 (P) *Max T. F. von Laue* for studying x-rays from their diffraction by crystals, showing that x-rays are electromagnetic waves (1912).
(C) *Theodore W. Richards* for determining the atomic weights of sixty elements, indicating the existence of isotopes.

1915 (P) *William Henry Bragg* and *William Lawrence Bragg*, his son, for studying the diffraction of x-rays in crystals.

1917 (P) *Charles Barkla* for studying atoms by x-ray scattering (1906).

1918 (P) *Max Planck* for discovering energy quanta (1900).

1919 (P) *Johannes Stark*, for discovering the Stark effect, the splitting of spectral lines in electric fields (1913).

1920 (P) *Charles-Édouard Guillaume* for discovering invar, a nickel-steel alloy with low coefficient of expansion.
(C) *Walther Nernst* for studying heat changes in chemical reactions and formulating the third law of thermodynamics (1918).

1921 (P) *Albert Einstein* for explaining the photoelectric effect and for his services to theoretical physics (1905).
(C) *Frederick Soddy* for studying the chemistry of radioactive substances and discovering isotopes (1912).

1922 (P) *Niels Bohr* for his model of the atom and its radiation (1913).
(C) *Francis W. Aston* for using the mass spectrograph to study atomic weights, thus discovering 212 of the 287 naturally occurring isotopes.

1923 (P) *Robert A. Millikan* for measuring the charge on an electron (1911) and for studying the photoelectric effect experimentally (1914).

1924 (P) *Karl M. G. Siegbahn* for his work in x-ray spectroscopy.

1925 (P) *James Franck* and *Gustav Hertz* for discovering the Franck-Hertz effect in electron-atom collisions.

1926 (P) *Jean-Baptiste Perrin* for studying Brownian motion to validate the discontinuous structure of matter and measure the size of atoms.

1927 (P) *Arthur Holly Compton* for discovering the Compton effect on x-rays, their change in wavelength when they collide with matter (1922), and *Charles T. R. Wilson* for inventing the cloud chamber, used to study charged particles (1906).

1928 (P) *Owen W. Richardson* for studying the thermionic effect and electrons emitted by hot metals (1911).

1929 (P) *Louis Victor de Broglie* for discovering the wave nature of electrons (1923).

1930 (P) *Chandrasekhara Venkata Raman* for studying Raman scattering, the scattering of light by atoms and molecules with a change in wavelength (1928).

1932 (P) *Werner Heisenberg* for creating quantum mechanics (1925).

1933 (P) *Erwin Schrödinger* and *Paul A. M. Dirac* for developing wave mechanics (1925) and relativistic quantum mechanics (1927).
(C) *Harold Urey* for discovering heavy hydrogen, deuterium (1931).

1935 (P) *James Chadwick* for discovering the neutron (1932).
(C) *Irène* and *Frédéric Joliot-Curie* for synthesizing new radioactive elements.

1936 (P) *Carl D. Anderson* for discovering the positron in particular and antimatter in general (1932) and *Victor F. Hess* for discovering cosmic rays.
(C) *Peter J. W. Debye* for studying dipole moments and diffraction of x-rays and electrons in gases.

1937 (P) *Clinton Davisson* and *George Thomson* for discovering the diffraction of electrons by crystals, confirming de Broglie's hypothesis (1927).

1938 (P) *Enrico Fermi* for producing the transuranic radioactive elements by neutron irradiation (1934–1937).

1939 (P) *Ernest O. Lawrence* for inventing the cyclotron.

1943 (P) *Otto Stern* for developing molecular-beam studies (1923), and using them to discover the magnetic moment of the proton (1933).

1944 (P) *Isidor I. Rabi* for discovering nuclear magnetic resonance in atomic and molecular beams.
(C) *Otto Hahn* for discovering nuclear fission (1938).

1945 (P) *Wolfgang Pauli* for discovering the exclusion principle (1924).

1946 (P) *Percy W. Bridgman* for studying physics at high pressures.

1947 (P) *Edward V. Appleton* for studying the ionosphere.

1948 (P) *Patrick M. S. Blackett* for studying nuclear physics with cloud-chamber photographs of cosmic-ray interactions.

1949 (P) *Hideki Yukawa* for predicting the existence of mesons (1935).

1950 (P) *Cecil F. Powell* for developing the method of studying cosmic rays with photographic emulsions and discovering new mesons.

1951 (P) *John D. Cockcroft* and *Ernest T. S. Walton* for transmuting nuclei in an accelerator (1932).
(C) *Edwin M. McMillan* for producing neptunium (1940) and *Glenn T. Seaborg* for producing plutonium (1941) and further transuranic elements.

1952 (P) *Felix Bloch* and *Edward Mills Purcell* for discovering nuclear magnetic resonance in liquids and gases (1946).

1953 (P) *Frits Zernike* for inventing the phase-contrast microscope, which uses interference to provide high contrast.

1954 (P) *Max Born* for interpreting the wave function as a probability (1926) and other quantum-mechanical discoveries and *Walther Bothe* for developing the coincidence method to study subatomic particles (1930–1931), producing, in particular, the particle interpreted by Chadwick as the neutron.

1955 (P) *Willis E. Lamb, Jr.* for discovering the Lamb shift in the hydrogen spectrum (1947) and *Polykarp Kusch* for determining the magnetic moment of the electron (1947).

1956 (P) *John Bardeen, Walter H. Brattain,* and *William Shockley* for inventing the transistor (1956).

1957 (P) *T.-D. Lee* and *C.-N. Yang* for predicting that parity is not conserved in beta decay (1956).

1958 (P) *Pavel A. Čerenkov* for discovering Čerenkov radiation (1935) and *Ilya M. Frank* and *Igor Tamm* for interpreting it (1937).

1959 (P) *Emilio G. Segrè* and *Owen Chamberlain* for discovering the antiproton (1955).

1960 (P) *Donald A. Glaser* for inventing the bubble chamber to study elementary particles (1952).
(C) *Willard Libby* for developing radiocarbon dating (1947).

1961 (P) *Robert Hofstadter* for discovering internal structure in protons and neutrons and *Rudolf L. Mössbauer* for discovering the Mössbauer effect of recoilless gamma-ray emission (1957).

1962 (P) *Lev Davidovich Landau* for studying liquid helium and other condensed matter theoretically.

1963 (P) *Eugene P. Wigner* for applying symmetry principles to elementary-particle theory and *Maria Goeppert Mayer* and *J. Hans D. Jensen* for studying the shell model of nuclei (1947).

1964 (P) *Charles H. Townes, Nikolai G. Basov,* and *Alexandr M. Prokhorov* for developing masers (1951–1952) and lasers.

1965 (P) *Sin-itiro Tomonaga, Julian S. Schwinger,* and *Richard P. Feynman* for developing quantum electrodynamics (1948).

1966 (P) *Alfred Kastler* for his optical methods of studying atomic energy levels.

1967 (P) *Hans Albrecht Bethe* for discovering the routes of energy production in stars (1939).

1968 (P) *Luis W. Alvarez* for discovering resonance states of elementary particles.

1969 (P) *Murray Gell-Mann* for classifying elementary particles (1963).

1970 (P) *Hannes Alfvén* for developing magnetohydrodynamic theory and *Louis Eugène Félix Néel* for discovering antiferromagnetism and ferrimagnetism (1930s).

1971 (P) *Dennis Gabor* for developing holography (1947).
(C) *Gerhard Herzberg* for studying the structure of molecules spectroscopically.

1972 (P) *John Bardeen, Leon N. Cooper,* and *John Robert Schrieffer* for explaining superconductivity (1957).

1973 (P) *Leo Esaki* for discovering tunneling in semiconductors, *Ivar Giaever* for discovering tunneling in superconductors, and *Brian D. Josephson* for predicting the Josephson effect, which involves tunneling of paired electrons (1958–1962).

1974 (P) *Anthony Hewish* for discovering pulsars and *Martin Ryle* for developing radio interferometry.

1975 (P) *Aage N. Bohr, Ben R. Mottelson,* and *James Rainwater* for discovering why some nuclei take asymmetric shapes.

1976 (P) *Burton Richter* and *Samuel C. C. Ting* for discovering the J/psi particle, the first charmed particle (1974).

1977 (P) *John H. Van Vleck, Nevill F. Mott,* and *Philip W. Anderson* for studying solids quantum-mechanically.
(C) *Ilya Prigogine* for extending thermodynamics to show how life could arise in the face of the second law.

1978 (P) *Arno A. Penzias* and *Robert W. Wilson* for discovering the cosmic background radiation (1965) and *Pyotr Kapitsa* for his studies of liquid helium.

1979 (P) *Sheldon L. Glashow, Abdus Salam,* and *Steven Weinberg* for developing the theory that unified the weak and electromagnetic forces (1958–1971).

1980 (P) *Val Fitch* and *James W. Cronin* for discovering CP *(charge-parity) violation* (1964), which possibly explains the cosmological dominance of matter over antimatter.

1981 (P) *Nicolaas Bloembergen* and *Arthur L. Schawlow* for developing laser spectroscopy and *Kai M. Siegbahn* for developing high-resolution electron spectroscopy (1958).

1982 (P) *Kenneth G. Wilson* for developing a method of constructing theories of phase transitions to analyze critical phenomena.

1983 (P) *William A. Fowler* for theoretical studies of astrophysical nucleosynthesis and *Subramanyan Chandrasekhar* for studying physical processes of importance to stellar structure and evolution, including the prediction of white dwarf stars (1930).

1984 (P) *Carlo Rubbia* for discovering the W and Z particles, verifying the electroweak unification, and *Simon van der Meer,* for developing the method of stochastic cooling of the CERN beam that allowed the discovery (1982–1983).

1985 (P) *Klaus von Klitzing* for the quantized Hall effect, relating to conductivity in the presence of a magnetic field (1980).

1986 (P) *Ernst Ruska* for inventing the electron microscope (1931), and *Gerd Binnig* and *Heinrich Rohrer* for inventing the scanning-tunneling electron microscope (1981).

1987 (P) *J. Georg Bednorz* and *Karl Alex Müller* for the discovery of high temperature superconductivity (1986).

Answers to Odd-Numbered Problems

CHAPTER 1

5. $0.866c$
7. $0.0048c = 1.44 \times 10^6$ m/s
9. $0.436L_o$
11. (a) 8.33×10^{-8} s (b) 23.7 m
15. $-c/2$
17. $0.85c$
19. (a) 939 MeV (b) 3.01 GeV (c) 2.07 GeV
21. $0.745c$
23. (a) 0.183 MeV (b) 2.45 MeV
25. (a) 3.29 MeV (b) 2.77 MeV
27. (a) 70 beats/min (b) 30.5 beats/min
29. (a) $0.0236c$ (b) $6.18 \times 10^{-4}c = 1.85 \times 10^5$ m/s
37. 939.573 MeV (n) − 938.280 MeV (p) − 0.511 MeV (e^-) = 0.782 MeV

CHAPTER 2

1. (a) 2.07 eV (b) 4.14×10^{-5} eV (c) 1.24×10^{-7} eV
3. 1.61×10^{30} photons/s
5. 9410 nm
7. 4.05 eV
9. (a) 296 nm, 1.01×10^{15} Hz (b) 2.01 V
11. (a) For $\phi > 4.14$ eV (mercury) there is no photoelectric effect.
(b) K_{max} (Li) = 1.84 eV, K_{max} (Be) = 0.241 eV
13. 1.55 eV
15. 2.48 eV; 1.32×10^{-27} kg·m/s
17. (a) 3.25×10^{-4} nm (b) 2.78×10^5 eV (c) 22 keV
19. (a) 1.7 eV (b) 4.0×10^{-15} V·s (c) 730 nm
21. 1.35 eV
23. 0.0142 nm
29. (b) 2.90×10^{-3} m·K
31. 1.95×10^{-3} eV; 0.0861 nm
33. (a) 17.44 keV (b) 0.07596 nm (c) 16.32 keV (d) 1.12 keV
35. $\Delta E_{Compton} = 3.77 \times 10^{-5}$ eV; $\Delta E_{photo} = 3.10$ eV. The Compton process provides insufficient energy to overcome the work function.
37. (c) $\Delta\omega = 8.8 \times 10^{10}$ Hz; $\Delta\omega/\omega = 2.33 \times 10^{-5}$
(d) Electrons moving in the opposite sense are slowed by the induced field, while those in planes of rotation parallel to $\boldsymbol{B}$ are unaffected.
39. (a) $d_1 = d_0/\sqrt{2}$; $d_2 = d_0/\sqrt{5}$ (b) 51.4°, 57.8°, 64.3°
43. (a) 1470 m (b) 1.1×10^{17}

CHAPTER 3

1. 656.112 nm, 486.009 nm, 433.937 nm
3. (a) $n = 3$ (b) No, it cannot belong to either of these series.
5. (a) $E_n = -(54.4/n^2)$ eV (n = 1, 2, 3, . . .)
(b) ionization energy = 54.4 eV
7. (a) 0.0265 nm (b) 0.0176 nm (c) 0.0132 nm
9. (a) 1.89 eV (b) 658 nm (c) 4.56×10^{14} Hz
11. (a) $\lambda_{max} = 1874.606$ nm, $\lambda_{min} = 820.140$ nm
(b) 0.6627 eV, 1.515 eV
13. (a) 0.0529 nm (b) 1.99×10^{-24} kg·m/s
(c) 1.05×10^{-34} kg·m²/s (d) 13.6 eV
(e) −27.2 eV (f) −13.6 eV
17. (a) 2.56×10^{-3} Å; 2.815 keV
19. (a) 0.179 nm (b) 0.120 nm
21. (a) 1.62×10^{-6} m; 1.42×10^{-14} kg (b) Charges (in units of 10^{-19} C) are $q = 10.0, 13.4, 11.6, 16.7, 21.6$ (c) $e = 1.67 \times 10^{-19}$ C
23. 1.603×10^{-19} C, using Equation 3.1 with an amount of material (m) in grams numerically equal to its molar weight.
29. 6.00×10^{-15} m
31. 1.60×10^{14} Hz; $f_3 = 2.44 \times 10^{14}$ Hz and $f_4 = 1.03 \times 10^{14}$ Hz

CHAPTER 4

1. 3.97×10^{-13} m
3. 1.77×10^{-36} m
5. 0.0248 nm for the photon; 0.00548 nm for the electron
9. 134 eV, which is about ten times greater than the ground state energy of hydrogen (13.6 eV).
11. 151 V
13. (a) 9.93×10^{-7} m (b) 4.97 mm (c) We cannot say that the neutron passed through one slit or the other, but only that it passed through the slits.
15. 4.56×10^{-14} m
17. 3×10^{27}
19. (a) 0.0961 nm (b) 6.94 pm (c) 9.80×10^{-14} m
21. (a) $\Delta p \simeq \hbar/r$
(b) $K \simeq \hbar^2/2mr^2$; $U = -ke^2/r$; $E = K + U$
(c) $r = a_0 = 0.529$ Å minimizes E at $E_{min} = -ke^2/2a_0 = -13.6$ eV.
23. (b) 3.00×10^{-10} m = 0.300 nm
25. (a) $g(\omega) = \sqrt{2/\pi}\, V_0\{\sin(\omega\tau)\}/\omega$
(c) 2×10^6 Hz; 2×10^9 Hz

27. (a) $f(x) = \text{const} \exp\{ik_0x - x^2(\Delta k)^2/2\}$
(c) $\Delta k = \Delta p/\hbar$

29. $v_p = c\{1 + (mc/\hbar k)^2\}^{1/2}$, $v_g = c\{1 + (mc/\hbar k)^2\}^{-1/2}$ and $v_pv_g = c^2$

CHAPTER 5

1. (a) 0.126 nm (b) 5.26×10^{-24} kg·m/s (c) 95 eV

3. $U(x) = \dfrac{\hbar^2}{2mL^2}\left\{\dfrac{4x^2}{L^2} - 6\right\}$ is a parabola opening upward with its nose at $\left(0, -\dfrac{3\hbar^2}{mL^2}\right)$

7.

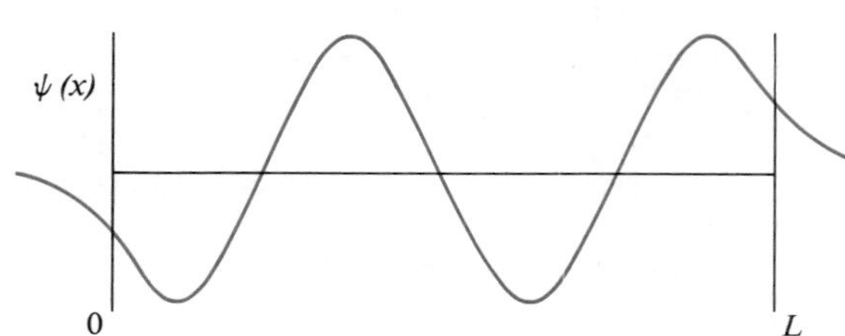

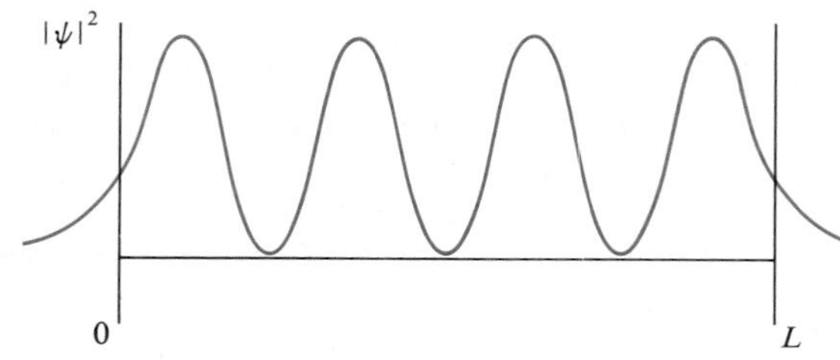

9. $\psi_1(x) = \sqrt{2/L}\cos(\pi x/L)$ $P_1(x) = (2/L)\cos^2(\pi x/L)$
$\psi_2(x) = \sqrt{2/L}\sin(2\pi x/L)$ $P_2(x) = (2/L)\sin^2(2\pi x/L)$
$\psi_3(x) = \sqrt{2/L}\cos(3\pi x/L)$ $P_3(x) = (2/L)\cos^2(3\pi x/L)$

11. (a) 5.13×10^{-3} eV (b) 9.40 eV (c) Much smaller electron mass

13. (a) $U = -\dfrac{7}{3}\dfrac{ke^2}{d}$ (b) $\dfrac{\hbar^2}{36md^2}$
(c) $d = \dfrac{\hbar^2}{42mke^2} = 0.050$ nm
(d) Lithium spacing $= a = 0.28$ nm

15. (b) $P_1 = 0.200$ (c) $P_2 = 0.351$
(d) $E_1 = 0.377$ eV; $E_2 = 1.51$ eV

17. The maximum probability density points are at

$$x = \frac{L}{n}\left(m + \frac{1}{2}\right) \qquad m = 0, 1, 2, \ldots, n-1$$

The minimum probability density points are at

$$x = \frac{mL}{n} \qquad m = 0, 1, 2, \ldots, n$$

19. $n = 4.27 \times 10^{28}$; excitation energy $= 4.69 \times 10^{-32}$ J

21. (b) $\langle x \rangle = L/2$, in agreement with the quantum result. $\langle x^2 \rangle = L^2/3$, larger than the quantum result by an amount $L^2/2(n\pi)^2$ which goes to zero for n large (correspondence principle).

23. (a) Not acceptable—diverges as $x \to \infty$
(b) Acceptable (c) Acceptable (d) Not acceptable—not single-valued (e) Not acceptable—wave is discontinuous (as is its slope).

25. $\psi(x) = A \sin kx$ for $0 \le x \le L$ and $\psi(x) = Ce^{-\alpha x}$ for $x > L$, where $k^2 = \dfrac{2mE}{\hbar^2}$ and $\alpha^2 = \dfrac{2m(U-E)}{\hbar^2}$.
Allowed energies satisfy $\dfrac{kL}{\sin(kL)} = \left[\dfrac{2mUL^2}{\hbar^2}\right]^{1/2}$,
which has solutions only if $\dfrac{2mUL^2}{\hbar^2} > 1$.

27. (c) and (d) are eigenfunctions of $[p]$ with eigenvalue $\hbar k$

29. (a) $\langle p \rangle = 0$ (b) $\langle p^2 \rangle = \dfrac{m\hbar\omega}{2}$ (c) $\Delta p = \sqrt{\dfrac{m\hbar\omega}{2}}$

31. (a) $\alpha = \dfrac{m\omega}{2\hbar}$; $E = \dfrac{3\alpha\hbar^2}{m} = \dfrac{3\hbar\omega}{2}$ (b) $C = \left(\dfrac{32\alpha^3}{\pi}\right)^{1/4}$

33. $P_c(x) = (1/\pi)(A^2 - x^2)^{-1/2}$

35. $x_0 = 0.5$ nm; $A = -0.18$ nm
$T_{\text{quant}} = 3.66 \times 10^{-15}$ s; $T_{\text{class}} = 3.47 \times 10^{-15}$ s.

CHAPTER 6

1. (b) $k = \dfrac{2mE}{\hbar^2}$ to the left and $k^2 = \dfrac{2m(U-E)}{\hbar^2}$ to the right of the step. $E/U = \frac{1}{2}$ (c) 1.44 fm

3. $R = 0.146$, $T = 0.854$ for both, provided their energies are the same.

5. (b) (1) 0.90 (2) 0.36 (3) 0.41 (4) $\simeq 10^{-(10^{32})}$

7. $\dfrac{2mUL^2}{\hbar^2} = 262.5$, so for energies not too near the top of the barrier the wide barrier result should apply.

9. $R = -\dfrac{E_0}{E}\left(1 - \dfrac{E_0}{E}\right)^{-1}$; $T = \left(1 - \dfrac{E_0}{E}\right)^{-1}$; with $E_0 = -\dfrac{mS^2}{2\hbar^2}$

11. 9.35×10^{20} Hz

13. 3.85×10^{30} Hz

CHAPTER 7

1. $E/E_0 = 6, 9, 9, 12, 14, 14$ where $E_0 = \dfrac{\pi^2\hbar^2}{8mL^2}$

3. (a) $E = \dfrac{11\pi^2\hbar^2}{2mL^2}$ (b) n_1, n_2, n_3 can have values (2, 2, 1); (2, 1, 2); (1, 2, 2)
(c) $\psi_{221} = A\sin(2\pi x/L)\sin(2\pi y/L)\sin(\pi z/L)$
$\psi_{212} = A\sin(2\pi x/L)\sin(\pi y/L)\sin(2\pi z/L)$
$\psi_{122} = A\sin(\pi x/L)\sin(2\pi y/L)\sin(2\pi z/L)$

5. (a) $E_{111} \simeq 1.5$ MeV (b) The states ψ_{221}, ψ_{122}, and ψ_{212} have the same energy $E = 3E_{111} \simeq 4.5$ MeV. The states ψ_{211}, ψ_{121}, ψ_{112} have the same energy $= 2E_{111} \simeq 3$ MeV. (c) Both states are threefold degenerate.

7. (a) For $n = 3$, ℓ can have the values of 0, 1, or 2. For $\ell = 0$, $m_\ell = 0$; for $\ell = 1$, $m_\ell = 1, 0, -1$; for $\ell = 2$, $m_\ell = 2, 1, 0, -1, -2$. (b) $E = -3.02$ eV

9. (a) 1312 nm (infrared) (b) 164 nm (ultraviolet)

11. $L_z = \pm 2\hbar, \pm\hbar$, or zero

13. (a) $\ell \simeq 4.58 \times 10^{65}$ (b) $\Delta L/L \simeq 1/\ell$
15. (a) $n = 6$ (b) -0.378 eV (c) $\ell = 4$; $L = \sqrt{20}\,\hbar$ (d) m_ℓ can be $-4, -3, -2, -1, 0, +1, +2, +3$, or $+4$ and $L_z = m_\ell\hbar$, giving $\theta = 153.4°, 132.1°, 116.6°, 102.9°, 90°, 77.1°, 63.4°, 47.9°, 26.6°$, respectively.
19. $\langle r \rangle = 4a_0 = 2.116$ Å
21. $\psi_1(x) = A \sin(k_1x) + B \cos(k_1x)$, and similarly for $\psi_2(y)$ and $\psi_3(z)$. (k_1, k_2, k_3) are components of a wavevector $\mathbf{k}$ with $E = \hbar^2|\mathbf{k}|^2/2m$, and $\phi(t) = e^{-iEt/\hbar}$. Sharp quantities are the energy E and the magnitude of particle momentum along each axis $|p_i| = \hbar|k_i|$.
23. $\langle U \rangle = -27.2$ eV and $\langle K \rangle = 13.6$ eV
25. $\Delta r = 0.866a_0$, more than half the average itself.
27. $P = 0.176$ for the 2s state and $P = 0.371$ for the 2p state.
29. $\psi(0) = 0$; $E_n = -\dfrac{mA^2}{2\hbar^2}\left[\dfrac{1}{n^2}\right]$ $(n = 1, 2, \ldots)$

CHAPTER 8

1. For $j = \frac{3}{2}$, $m_j = \frac{3}{2}, \frac{1}{2}, -\frac{1}{2}, -\frac{3}{2}$; for $j = \frac{1}{2}$, $m_j = \frac{1}{2}, -\frac{1}{2}$
3. (a) $7G_{9/2}$ (b) $6H_{9/2}$ and $6H_{11/2}$
5. (a) 2 (b) 8 (c) 18 (d) 32 (e) 50
7. [Ar]$3d^54s$ has the lower energy. The element is chromium. The lower energy configuration has six unpaired spins compared to four unpaired spins for [Ar]$3d^44s^2$.
11. (a) $-5\hbar/2, -3\hbar/2, -\hbar/2, +\hbar/2, +3\hbar/2, +5\hbar/2$ (b) $0, \hbar, 2\hbar, 3\hbar, 4\hbar$ (c) $\sqrt{20}\,\hbar = 4.72 \times 10^{-34}$ J·s
13. $\dfrac{dB_z}{dz} = 389$ T/m
15. (a) 394.8 eV

    $(n_1, n_2, n_3) = (1, 1, 1)$

    $(n_1, n_2, n_3) = (1, 1, 2), (1, 2, 1)$, or $(2, 1, 1)$

    (b) All eight particles go into the $(n_1, n_2, n_3) = (1, 1, 1)$ state, and so the energy is $8 \times E_{111} = 225.6$ eV.
17. 5.80×10^{-5} eV
19. (a) $L = \dfrac{1}{2}MR^2\omega$ (b) $\mu = \dfrac{QR^2}{2}\omega$; g = 2
21. $\Delta E_{calc} = 3.18 \times 10^{-5}$ eV; $\Delta E_{meas} = 5 \times 10^{-5}$ eV

CHAPTER 9

3. (a) 4.23 eV (b) 3.27×10^4
7. With Δ the level spacing, strengths of absorption are:

$T = 4$ K	$T = 1$ K
1.00 at $E = 3\Delta$	1.00 at $E = 3\Delta$
1.40 at $E = 2\Delta$	0.47 at $E = 2\Delta$
0.97 at $E = 1\Delta$	0.11 at $E = 1\Delta$

9. MB is valid for He gas at STP, but BE must be used for the liquid at 4.2 K where the density is greater by a factor of about 10^7.
11. (a) $kT(\pi^4/15)/2.40$ (b) 1.39 eV
13. $E_F = 33.4$ MeV; $E = 20$ MeV
15. 0.938
17. (a) 2.54×10^{22} electrons/cm^3 (b) 3.15 eV (c) $3.51 \times 10^{-3}c = 1.05 \times 10^6$ m/s
19. 2.89×10^{17} electrons

CHAPTER 10

3. (a) 1.78 eV (b) 5.35×10^{19} eV = 8.57 J (c) 171.4 MW
5. (a) No (b) 7.1×10^{15} ions/cm^3 (c) 13 W/cm^3 Random distribution of atoms in glasses leads to broader absorption lines and greater light loss than in single crystals.
7. 350
9. (b) 0.358 nm (c) 6.67×10^{-4} nm; the controlling factor is L

CHAPTER 11

1. $\mu = 1.22 \times 10^{-26}$ kg, $\mu_{calc} = 1.24 \times 10^{-26}$ kg
3. (a) 0.0120 nm (b) 0.00923 nm; HI has the weaker bond.
5. 0.95 u
7. (a) K(HI) = 292 N/m; K(NO) = 1550 N/m (b) A(HI) = 0.0123 nm; A(NO) = 0.00492 nm
9. $v = 7$ is the highest vibrational level that can be excited without the molecule coming apart.
11. $\Delta E_{rot} = \dfrac{5\hbar^2}{4mr^2} = 0.519$ MeV for H_2
13. (c) $E_{diss} = -E_{vib}$ (d) $U_0 = 4.79$ eV; $\alpha = 19.3$ nm^{-1}
15. Ground state: $E = \dfrac{2x^2\hbar^2}{mL^2}$ where x is the smallest root of the equation $\tan x = -\left(\dfrac{2\hbar^2}{mSL}\right)x$. First excited state: $E = \dfrac{2\pi^2\hbar^2}{mL^2}$. As $S \to \infty$, $x \to \pi$ and the two energies coincide; as $S \to 0$, $x \to \pi/2$, and the energies reduce to the ground and first excited states of an infinite well with no barrier.
17. (a) $R_0 = 1.44$ bohrs (b) $K = 1.03$ Ryd/bohr^2 = 801 N/m

CHAPTER 12

3. -7.84 eV
7. 6.70×10^{-3} eV
9. (a) -7.12 eV (b) -6.39 eV
11. (a) 3.80×10^{-14} s (b) 4.6 nm, or 10–20 lattice spacings; 1.2×10^5 m/s
13. (c) 3.9×10^3 m/s (d) 36 Ω·cm
15. (a) 0.094 eV for Si and 0.053 eV for Ge (b) 6.4 Å for Si and 8.5 Å for Ge
17. (a) 2.54×10^{22} electrons/cm^3 (b) 3.15 eV (c) 1.05×10^6 m/s (d) 3.33×10^{-14} s (e) 35.0 nm (f) 175 W/m·K

CHAPTER 13

1. (a) 1.90 fm (b) 7.44 fm (c) 3.92
3. 8.57×10^{13}
5. (a) 4.55×10^{-13} m = 455 fm (b) 6.03×10^{6} m/s
7. $E_b = 492$ MeV; $E_b/A = 8.79$ MeV/nucleon, which agrees with Figure 13.7
9. 3.53 MeV
11. 7.93 MeV
13. (a) ${}^{139}_{55}$Cs, for which $(A - Z)/Z = 1.53$
    (b) ^{139}La, which has $E_b/A = 8.379$ MeV/nucleon
    (c) ^{139}Cs, with a mass of 138.913 u
15. 160 MeV
17. (a) 0.805 h^{-1} (b) 0.861 h
19. (a) 1.55×10^{-5} s^{-1} (b) 2.39×10^{13} atoms
    (c) 1.87 mCi
21. 36.3 μg
23. 0.055 mCi
25. 9.46×10^{9} nuclei
27. (a) 8.07 MeV (b) -65.58 MeV
29. (a) $Q = -2.37$ MeV; cannot occur
    (b) $Q = -1.68$ MeV; cannot occur
    (c) $Q = 1.86$ MeV; can occur
31. 5.70 MeV
33. 17.35 MeV
35. (a) ${}^{1}_{1}\text{H} + {}^{18}_{8}\text{O} \rightarrow {}^{1}_{0}\text{n} + {}^{18}_{9}\text{F}$
    (b) (18.000934 ± 0.000002) u
37. ^{63}Cu 30%; ^{65}Cu 70%
39. (a) 5.70 MeV (b) 3.27 MeV, exothermic
41. (a) 6.40×10^{24} nuclei (b) 5.86×10^{12} decays/s
    (c) 1.10×10^{6} years
43. 221 years
45. 1.02×10^{3} Bq/g
47. (a) 2.75 fm (b) 152 N (c) 2.62 MeV
    (d) 7.44 fm; 379 N; 17.6 MeV
49. 2.2×10^{-6} eV
51. (a) 7.76×10^{4} eV (b) 4.62 MeV; 13.9 MeV
    (c) 1.03×10^{7} kWh
53. (a) 12.2 mg (b) 166 mW
55. 0.35%
57. (b) 4.79 MeV
59. (a) 9.7×10^{29} kg (b) 6.5×10^{11} m/s^2
    (c) 6.9×10^{41} J
61. 1.49×10^{3} Bq
63. 4.45×10^{-8} kg/h

CHAPTER 14

1. $N = N_0 e^{-\sigma(n_1 x_1 + n_2 x_2)}$
   For many layers, $N = N_0 e^{-\sigma \Sigma n_i x_i}$
3. (a) 7.37 cm (b) 38 bn
5. 0.01 cm
7. 200.6 MeV
9. (a) 996 g (b) 4.87 cm
11. 2664 years
13. (a) 1.11×10^{6} m/s (b) 9×10^{-8} s
15. (a) 2.4×10^{7} J (b) 10.6 kg (or 10.6 liters)
17. $E(X)/E(\gamma) = 0.50$
19. (a) 1.21 m (b) 1.83 m
21. 5 rad of heavy ions
23. (a) 9×10^{18} rad (b) 1.38×10^{7} rad
25. (a) $E/E_\beta = 3.1 \times 10^{7}$ (b) 3.1×10^{10} electrons
27. 371 Ci, which is much less than the fission inventory
29. 25 collisions
31. (b) 26 days
33. (b) 2.45 MeV
35. (a) 4.55×10^{-24} kg·m/s (b) 0.146 nm
37. (a) 2.24×10^{7} kWh (b) 17.6 MeV per D-T fusion event (c) 2.34×10^{8} kWh (d) 9.14 kWh
39. (a) 3.89×10^{38} protons (b) 4.44×10^{9} kg/s
41. (a) $v_\parallel = 1.90 \times 10^{5}$ m/s, $v_\perp = 1.10 \times 10^{5}$ m/s
    (b) 2.28×10^{-3} m (c) 2.49×10^{-2} m
43. $Q_1 = 0.931$ MeV, $Q_2 = 5.493$ MeV, $Q_3 = 19.285$ MeV, and $Q_4 = 12.861$ MeV. The total reaction is equivalent to either (1) + (2) + (3) or $2 \times (3) - (4)$, so that $Q_{\text{total}} = Q_1 + Q_2 + Q_3 = 25.709$ MeV or $Q_{\text{total}} = 2Q_3 - Q_4 = 25.709$ MeV
45. 95.2 s

CHAPTER 15

1. 2.26×10^{23} Hz, 1.32×10^{-15} m
3. 2×10^{-18} m
5. 1×10^{-23} s
7. (a) $-1 \neq 0$ (not conserved) (b) $0 = 0$ (conserved)
   (c) $0 = 0$ (conserved) (d) $0 \neq -1$ (not conserved)
   (e) $-2 \neq -1$ (not conserved) (f) $-2 \neq 0$ (not conserved)
9. (a) violates conservation of L_e and L_μ (b) violates conservation of charge (c) violates conservation of baryon number (d) violates conservation of baryon number (e) violates conservation of charge
11. (a) $\pi^- \rightarrow \mu^- + \bar{\nu}_\mu$
    L_μ: $0 \rightarrow 1 - 1$
    (b) $K^+ \rightarrow \mu^+ + \nu_\mu$
    L_μ: $0 \rightarrow -1 + 1$
    (c) $\bar{\nu}_e + p \rightarrow n + e^+$
    L_e: $-1 + 0 \rightarrow 0 - 1$
    (d) $\nu_e + p \rightarrow n + e$
    L_e: $1 + 0 \rightarrow 0 + 1$
    (e) $\nu_\mu + n \rightarrow p + \mu^-$
    L_μ: $1 + 0 \rightarrow 0 + 1$
    (f) $\mu^- \rightarrow e + \bar{\nu}_e + \nu_\mu$
    L_μ: $1 \rightarrow 0 + 0 + 1$
    L_e: $0 \rightarrow 1 - 1 + 0$
13. (a)

	Proton	Quarks u	u	d	Total (u + u + d)
Strangeness	0	0	0	0	0
Baryon number	1	$+\frac{1}{3}$	$+\frac{1}{3}$	$+\frac{1}{3}$	1
Charge	e	$+\frac{2}{3}$e	$+\frac{2}{3}$e	$-\frac{1}{3}$e	e

(b)

	Neutron	Quarks u	d	d	Total (u + d + d)
Strangeness	0	0	0	0	0
Baryon number	1	$+\frac{1}{3}$	$+\frac{1}{3}$	$+\frac{1}{3}$	1
Charge	0	$+\frac{2}{3}$e	$-\frac{1}{3}$e	$-\frac{1}{3}$e	0

Index

The Greek Alphabet

Alpha	A	α	Iota	I	ι	Rho	P	ρ
Beta	B	β	Kappa	K	κ	Sigma	Σ	σ
Gamma	Γ	γ	Lambda	Λ	λ	Tau	T	τ
Delta	Δ	δ	Mu	M	μ	Upsilon	Υ	υ
Epsilon	E	ϵ	Nu	N	ν	Phi	Φ	ϕ
Zeta	Z	ζ	Xi	Ξ	ξ	Chi	X	χ
Eta	H	η	Omicron	O	o	Psi	Ψ	ψ
Theta	Θ	θ	Pi	Π	π	Omega	Ω	ω

Mathematical Symbols Used in the Text and Their Meaning

Symbol	Meaning
$=$	is equal to
$\equiv$	is defined as
$\neq$	is not equal to
$\propto$	is proportional to
$>$	is greater than
$<$	is less than
$\gg$ ($\ll$)	is much greater (less) than
$\approx$	is approximately equal to
Δx	the change in x
$\sum_{i=1}^{N} x_i$	the sum of all quantities x_i from $i = 1$ to $i = N$
$\|x\|$	the magnitude of x (always a positive quantity)
$\Delta x \rightarrow 0$	Δx approaches zero
$\frac{dx}{dt}$	the derivative of x with respect to t
$\frac{\partial x}{\partial t}$	the partial derivative of x with respect to t
$\int$	integral